Encyclopedia of Industrial Automation

Encyclopedia of Industrial Automation

Glenn A. Graham CMfgE
Coopers & Lybrand

Edited by
Robert E. King
Society of Manufacturing Engineers

Longman Scientific & Technical,
Longman Group UK Limited,
Longman House, Burnt Mill, Harlow,
Essex CM20 2JE, England
and Associated Companies throughout the world.

Published in association with the Society of Manufacturing Engineers, and published in the USA under the title 'Automation Encyclopedia'

First published in 1988

British Library Cataloging in Publication Data

Graham, Glenn A.
Encyclopedia of industrial automation.
1. Manufacturing industries. Automation
I. Title II. King, Robert E. III. Society of Manufacturing Engineers
670.42'7

ISBN 0-582-03566-X

Printed in Great Britain at
The Bath Press, Avon

This book is dedicated to my wife Rochelle, whose support made this endeavor successful.

Glenn A. Graham

Introduction

The Society of Manufacturing Engineers

The Society of Manufacturing Engineers is an international technical society dedicated to advancing scientific knowledge in the field of manufacturing. SME has more than 77,000 members in 70 countries and serves as a forum for engineers and managers to share ideas, information and accomplishments.

Technology is constantly evolving. To be successful, today's engineers must keep pace with the torrent of information that appears each day. To meet this need, SME provides many opportunities in continuing education for its members.

This continuing education is provided through:

- Educational programs including seminars, clinics, programmed learning courses, as well as videotapes.

- Conferences and expositions which enable engineers and managers to examine the latest manufacturing concepts and technology.

- SME publications which include *Manufacturing Engineering* magazine, the *Journal of Manufacturing Systems*, the *Technical Digest*, and a wide range of books including the *Tool and Manufacturing Engineers Handbook*.

- Monthly meetings through five associations and their more than 300 chapters and 165 student chapters worldwide to provide a forum for membership participation and involvement.

- The SME Manufacturing Engineering Certification Institute formally recognizes manufacturing engineers and technologists for their technical expertise and knowledge acquired through experience and education.

As a leader among professional societies, SME assesses industry trends, then interprets and disseminates the information. SME members have discovered that their membership broadens their knowledge and experience throughout their careers. The Society of Manufacturing Engineers is truly industry's partner in productivity.

This encyclopedia has been compiled to serve as an invaluable, functional reference guide to the rapidly developing fields of industrial automation.

Acknowledgements

The following resources were used in the preparation of some of the material included in this book. Portions of these publications appear in this volume.

Automated Guided Vehicles and Automated Manufacturing. Author: Richard K. Miller. First edition, copyright 1987, Society of Manufacturing Engineers.

Commonly Used Terms in Robotics. Author: Jay Lee. First edition, Copyright, Jay Lee.

Computer Integrated Manufacturing: Glossary of Terms. Editor: Thomas V. Sobczak. First edition, copyright 1984, Society of Manufacturing Engineers.

Designing for Economical Production. Author: H. E. Trucks. Second edition, copyright 1987, Society of Manufacturing Engineers.

Low Cost Jigs, Fixtures, and Gages for Limited Production. Editor: W. E. Boyes. First edition, copyright 1986, Society of Manufacturing Engineers.

New Directions Through CAD/CAM. Authors: William Beeby and Phyllis Collier. First edition, copyright 1986, Society of Manufacturing Engineers.

Nontraditional Machining Process. Editor: E. J. Weller. Second edition, copyright 1984, Society of Manufacturing Engineers.

Tool and Manufacturing Engineers Handbook Volume 1: *Machining*. Editors: Thomas J. Drozda and Charles Wick. Fourth edition, copyright 1983, Society of Manufacturing Engineers.

Tool and Manufacturing Engineers Handbook Volume 2: *Forming*. Editors: Charles Wicks, John T. Benedict, Raymond F. Veilleux. Fourth edition, copyright 1984, Society of Manufacturing Engineers.

Tool and Manufacturing Engineers Handbook Volume 3: *Materials, Finishing and Coating*. Editors: Charles Wick and Raymond F. Veilleux. Fourth edition, copyright 1985, Society of Manufacturing Engineers.

Tool and Manufacturing Engineers Handbook Volume 4: *Quality Control and Assembly*. Editors: Charles Wick and Raymond F. Veilleux. Fourth edition, copyright 1987, Society of Manufacturing Engineers.

Tool and Manufacturing Engineers Handbook Volume 5: *Manufacturing Management*. Editors: Raymond F. Veilleux and Louis Petro. Fourth edition, copyright 1988, Society of Manufacturing Engineers.

Material in this volume is also reprinted from the **Dictionary of Manufacturing Terms**. The dictionary contains many definitions which were abstracted with permission from a variety of standards as well as association and company publications. Grateful acknowledgement is extended to the following publishers and publications.

Acoustical Society of America
Standards Secretariat
New York, NY

American National Standard Balancing Terminology, ANSI S2.7-1982

American National Standards Institute
New York, NY

American National Standard for Machine Tools, Power Press Brakes—Safety Requirements for Construction, Care, and Use B11.3-1982

Safety Requirements of the Construction, Care, and Use of Mechanical Presses B11.1-1982

American Society of Mechanical Engineers
New York, NY

Carbide Blanks For Twist Drills, Reamers, End Mills, Random Rod, Specification for ANSI B94.2-1983

Glossary of Mechanical Press Terms ANSI/ASME B5.49M-1984

Measurement of Out-of-Roundness ANSI B89-3.1-1972 (R1979)

Milling Cutters and End Mills ANSI/ASME B94.19-1985

Nomenclature, Definitions, and Letter Symbols for Screw Threads ANSI/ASME B1.7M-1984

Surface Texture ANSI/ASME B46.1-1985

Twist Drills—Straight Shank and Taper Shank, Combined Drills and Countersinks ANSI B94.11M-1979

American Welding Society
Miami, FL

Welding Terms and Definitions A3.0

American Society for Testing and Materials
Philadelphia, PA

Annual Book of ASTM Standards

Strippit-Houdaille, Inc.
Akron, NY

The Art of Forming, Patrick Oldenburg

E.W. Bliss Co.
Hastings, MI

Bliss Power Press Handbook, 1950

Battelle Memorial Institute
Columbus, OH

Forging Equipment, Materials, and Practices, T. Altan et al., 1973.

Forging Industries Association
Cleveland, OH

Forging Handbook, Thomas G. Byrer, 1985

American Gear Manufacturers Associations
Arlington, VA

Gear Handbook, Volume 1, AGMA 390.03*, 1980

*(This publication is being updated and will be known as AGMA 2000.)

ASM International
Metals Park, OH

Heat Treater's Guide, Paul M. Unterweiser, Howard E. Boyer, and James J. Kubbs, 1982

Metal Casting Society
Des Plaines, IL

Iron Castings Handbook, Charles F. Walton, 1981

National Tooling and Machining Association
Ft. Washington, MD

Modern Geometric Dimensioning and Tolerancing, 2nd ed., Lowell W. Foster, 1982

Bethlehem Steel Corporation
Bethlehem, PA

Modern Steels and Their Properties, 1980

Forging Industry Association
Cleveland, OH

Open Die Forging Manual, 3rd ed., 1982

Steel Founders' Society of America
Des Plaines, IL

Steel Castings Handbook, 5th ed., Peter F. Wieser, 1980

Hitchcock Publishing Co.
Wheaton, IL

Understanding Paint and Painting Processes, 2nd ed., Gerald L. Schneberger

The following served as reviewers for portions of this publication.

Ramon Bakerjian
Society of Manufacturing Engineers

David Dornfeld
University of California-Berkeley

Thomas J. Drozda
Society of Manufacturing Engineers

Margaret A. Eastwood
CIMCORP Inc.

James F. Fales
Ohio University

LaRoux K. Gillespie
Allied Corporation

James E. Heaton
ORACLE, Inc.

George J. Hess
Ingersoll Milling Machine Company

F.W. Houtz
AT&T

Nancy Lea Hyer
The University of North Carolina at Chapel Hill

Jake Krakauer
SiliconGraphics Computer Systems

Jack D. Lane
Robotic Integrated Systems Engineering Inc.

REVIEWERS

Jay Lee
U.S. Postal Service

Donald I. Manor
Deere & Company

H. Lee Martin
TeleRobotics International, Inc.

Elanor M. McLester
Boeing Computer Services

Roger N. Nagel
Lehigh University

J.E. Nicks
Ferris State College

John A. Piotrowski III
K.N. Aronson, Inc.

T.R. Pryor
Diffracto

Charles Savage
Digital Equipment Corp.

Mark Shaw
Society of Manufacturing Engineers

Warren L. Shrensker
General Electric

Walter W. Tucker
Eastern Michigan University

William O. Winchell
Alfred University

Dennis E. Wisnosky
Wizdom Systems, Inc.

Eugene J. Wittry
Caterpillar Inc.

Madden T. Works
Aerojet Electro Systems

Glenn Yeager
Integrated Automation Corp.

Dick Zakrzewski
The Timken Company

Nello Zuech
Vision Systems International

Coopers & Lybrand contributors include:
Paul Conroy
Donald DeWolfe
Peter Flentov
Dennis Harrison
William Hester
Per Johansen
Steve Keneally
Irvin Krause
Brian Lawn
Mary Ellen Manning
Annette Martens
Glenn Matto
Steve Meli
Nancy Merz
Gerald Michael
Andrew Nemtzow
Helen Ojha
Leonard Olin
Michael Schoonover
David Smith
Susan Stafford
Robert Stasey
Charles Waite
Donald Webb

SME staff who participated in the editorial development and production of this volume include:

EDITORIAL

Robert E. King
Editor

Rachel Subrin
Senior Publications Administrator

Toni Maffesoli
Editorial Secretary

TYPESETTING

Shari Smith
Administrative Coordinator

Cynthia Dagger
Typesetter

GRAPHICS

Cheryl Nizyborski
Graphic Designer

Kevin Rinna
Graphic Designer

Dan Tappen
Graphic Designer

Sandy Wallace
Keyliner

Trademarks

The following trademarks are registered or pending registration and are used in this book.

1-2-3, Lotus Development Corp.
Ada, Department of Defense.
AFP/SME, Society of Manufacturing Engineers.
AML, International Business Machines Corp.
ANSI, American National Standards Institute.
ANSYS, E. I. DuPont de Nemours & Co., Inc.
APAS, Westinghouse Electric Corp.
APL, International Business Machines Corp.
Apple, Apple Computers, Inc.
Applesoft BASIC, Apple Computers, Inc.
ARPANET, Department of Defense.
Ashton-Tate, Ashton-Tate.
ASQC, American Society for Quality Control.
Association for Finishing Processes of SME, Society of Manufacturing Engineers.
AT, International Business Machines Corp.
ATICTS, Data Enterprises.
AutoCAD, Autodesk, Inc.
AUTOFACT, Society of Manufacturing Engineers.
BASICA, International Business Machines Corp.
Bernoulli Drive, Iomega Corp.
BIOS, Tesco, Inc.
BISYNC, International Business Machines Corp.
Boeing Calc, Boeing Computer Services.
CADAM, CADAM, Inc.
CADDS4, Computervison.
CAMSCO, CAMSCO, Inc.
CASA/SME, Society of Manufacturing Engineers.
CATIA, Dassault Systemes.
CGA, International Business Machines Corp.
Compaq DeskPro, Compaq Computer Corp.
Compaq Plus, Compaq Computer Corp.
Compaq, Compaq Computer Corp.
Compuserve, Compuserve.
Computer and Automated Systems Association of SME, Society of Manufacturing Engineers.
Concurrent CP/M, Digital Research.
CP/M, Digital Research.
CRAY, Cray Computers.
dBase II, Ashton-Tate.
dBase III Plus, Ashton-Tate.
DEC, Digital Equipment Corp.
DECNET, Digital Equipment Corp.
Delrin, E. I. DuPont de Nemours & Co., Inc.
DESQview, Quarterdeck Office Systems
DNA, Digital Equipment Corp.
DOMAIN, Apollo Computer.

DOS, International Business Machines Corp.
EBCDIC, International Business Machines Corp.
EGA, International Business Machines Corp.
EIA, Electronics Industries Association.
Ethernet, Xerox Corporation.
Framework II, Ashton-Tate.
GEM, Digital Research, Inc.
GPSS, International Business Machines Corp.
GW BASIC, Microsoft Corp.
HDLC, International Business Machines Corp.
Helvetica, Eltra Corp.
Hercules Graphics Card, Hercules Computer Technology.
Hero, Heathkit.
Hyperbus, Network Systems.
Hyperchannel, Network Systems.
IBM, International Business Machines Corp.
IEEE, Institute of Electrical and Electronic Engineers.
Interactive EasyFlow, Haventree Software Limited.
Journal of Manufacturing Systems, Society of Manufacturing Engineers.
LaserJet, Hewlett-Packard, Inc.
Lotus, Lotus Development Corp.
LSI 11, Digital Equipment Corp.
M, International Business Machines Corp.
Machine Vision Association of SME, Society of Manufacturing Engineers.
Macintosh, Apple Computers, Inc.
Mac Plus, Apple Computers, Inc.
Manufacturing Engineering, Society of Manufacturing Engineers.
Manufacturing Insights, Society of Manufacturing Engineers.
MAP/1, Pritsker & Associates.
MARC, E. I. DuPont de Nemours & Co., Inc.
MAST, CMS Research.
MICLASS, Netherlands Central Organization for Applied Scientific Research.
Micro Channel, International Business Machines Corp.
Micro PDP-11, Digital Equipment Corp.
Microsoft BASIC, Microsoft Corp.
Microsoft Windows, Microsoft Corp.
MicroVax, Digital Equipment Corp.
MODWAY, Gould Electronics.
MS-DOS, Microsoft Corp.
Multiplan, Microsoft Corp.
Multiscan, Sony Corporation of America.
Multisynch, NEC Home Electronics (USA), Inc.
MVA/SME, Society of Manufacturing Engineers.
Mylar, E. I. DuPont de Nemours & Co., Inc.
NET/ONE, Ungerman-Bass.
OMNINET, Corvus Systems.
OS/2, International Business Machines Corp.
Palatino, Eltra Corporation.
PC, International Business Machines Corp.
PC DOS, International Business Machines Corp.
PCjr, International Business Machines Corp.
PCL, Hewlett-Packard Corp.

PDP, Digital Equipment Corp.
Perfect Circle, The Dana Corp.
PGA, International Business Machines Corp.
PL/1, International Business Machines Corp.
PLC, Allen-Bradley Corporation.
Polaroid Palette, Polaroid Corp.
post-it notes, 3M.
PRIMENET, Prime Computer, Inc.
PS/2, International Business Machines Corp.
Puma, Westinghouse Corp.
Q-bus, Digital Equipment Corp.
QuickBASIC, Microsoft Corp.
ReGis, Digital Equipment Corp.
Rhino, Scovill Manufacturing Co.
RIA, Robotics Industries Association.
RI/SME, Society of Manufacturing Engineers.
Robotic International of SME, Society of Manufacturing Engineers.
Robotics Today, Society of Manufacturing Engineers.
Rocketdyne, Rockwell International Corp.
RSX-11M, Digital Equipment Corp.
See Why, Istel, Inc.
Series 1/Ring, International Business Machines Corp.
Sidekick, Borland International, Inc.
SIMAN, Systems Modeling Corp.
SLAM, Pritsker & Associates.
SLAM II, Pritsker & Associates.
SNA, International Business Machines Corp.
SPEED, Horizon Software.
ST506/412, Seagate Technology.
Supercalc, Computer Associates International, Inc.
Teletype, The Teletype Corp.
TELEX, International Teleprinter Network.
The Source, The Source Information Network.
TI Basic, Texas Instruments, Inc.
TIM, Concord Data Systems.
Timeline, Breakthrough Software Corp.
TMEH, Society of Manufacturing Engineers.
TMS RMN, Hewlett-Packard Corp.
Tool and Manufacturing Engineers Handbook, Society of Manufacturing Engineers.
TopView, International Business Machines Corp.
Tron, Disney Studios.
True BASIC, Microsoft Corp.
Trumpf, Trumpf GMBH and Co.
Tuff Wheel II, Schwinn.
TWX, Teletypewriter Exchange Service.
UL, Underwriters' Laboratories, Inc.
Ultrix, Digital Equipment Corp.
UNIAPT, United Computing Corporation.
Unigraphics, McDonnell-Douglas Corp.
Unimate, Westinghouse Corp.
UNIVAC, Information Systems Group.

UNIX, AT&T.
USART, U.S. Art, Inc.
VAX, Digital Equipment Corp.
VAXBI, Digital Equipment Corp.
VAXbus, Digital Equipment Corp.
VGA, International Business Machines Corp.
Victor, The Dana Corp.
Visicalc, Personal Software, Inc.
VS, International Business Machines Corp.
WANGNET, Wang, Inc.
Witness, Istel, Inc.
Wordstar 2000, MicroPro International Corp.
XENIX, Microsoft Corp.
XT, International Business Machines Corp.

A

A Axis. A axis is an angle defining rotary motion of a machine tool member or slide around the X axis, such that a right-handed screw advanced in the positive A direction would be advanced in the positive X direction.

Also see: Axis, X Axis.

ABC Inventory Control. ABC inventory control analysis was derived from Vilfredo Pareto a nineteenth century engineer who was the first to document the Management Principle of Materiality which serve as a basis of ABC inventory control. The Management Principle of Materiality notes that controlling the relatively vital few will result in controlling the whole.

The ABC classification of the inventory items are in a decreasing order of annual dollar volume. Additional criteria may be used such as unit cost, scarcity of material, lead time, storage requirements, cost of stock out, and design volatility.

Class A inventory items are those items which have the highest annual dollar volume and receive the most attention for planning, inventory control including cycle counting, forecast evaluation and lead time reduction. Class B inventory items receive the same control activities with, but less frequency. Class C inventory items include the low-value items where the controls may include larger order quantities, floor stock and a planning rule of safety stock.

Also see: Inventory Control.

Abort. Abort is the stopping of a computer at an irregular point in its program, usually before the normal completion of the executing sequence. This occurrence may be due to human or machine initiation. Usually aborting a program requires restarting from a beginning entry point.

In data transmission, a function invoked by a primary or secondary sending station causing the recipient to discard (or ignore) all bit sequences transmitted by the sender following the preceding flag sequence.

When operating a computer from a keyboard terminal, the typical method of aborting a program is to press the Escape (Esc) key or to press Control-C (Ctrl-C) or to press the Break key (Break), or sometimes the Control-Break keys (Ctrl-Break). With some of these types of aborting, such as pressing the Break key, the terminal session may have to be reinstituted or the machine may have to be rebooted, instead of just restarting the program.

Also see: Application Program, Computers, Programming.

Abrasive. Abrasive is the material from which the grains ina grinding wheel are made—usually crystalline aluminum oxide, silicon carbide, or diamond.

Abrasive Flow Machining. Abrasive flow machining is a process for finishing holes, inaccessible areas or restricted passages by clamping the part in a fixture, then extruding semisolid abrasive media through the passage. Often, multiple parts are loaded into a single fixture and finished simultaneously.

Abrasive-wire Bandsawing. Abrasive-wire bandsawing is a variation of bandsawing that uses a small-diameter wire with diamond, cubic boron nitride or aluminum oxide abrasives bonded to the surface as the cutting blade. Abrasive-wire bandsawing is an alternative to electrical discharge machining for producing dies, stripper plates, electrodes, and cams from difficult-to-machine conductive and nonconductive materials.

Absolute Accuracy. Absolute accuracy is exactness as measured from a specified reference point.

Absolute Address. Absolute address is an address in a computer language that identifies a storage location or a device without the use of any intermediate reference. An address that is permanently assigned by the machine designer to a storage location.

The opposite of an absolute address is a relative address. Corresponding to any relative address is an absolute address. Typically the reference to the relationship between absolute and relative addresses is to speak of the relative address plus an offset value.

A popular type of computer program used today are TSR (Terminate and Stay Resident) programs, which typically desire to sit at the top of the program memory space (the programs desire to be loaded last). A term frequently used to describe these programs is ''well behaved'' or ''not well behaved'' which indicates whether these programs function and coexist well with other programs when loaded at a relative address location instead of an absolute address location.

Also see: Address, Computers.

Absolute Coding. Absolute coding is used to write machine language for subsequent use when instructions are executed by control circuits.

Also see: Computers.

Absolute Coordinates. Absolute coordinates define the location of a point in terms of x, y, and z distance from an established point of origin. The point of origin is usually a fixed reference relative to the earth, or the environment which is being considered. The absolute coordinates are different from relative coordinates which are x, y, and z distances from another point other than the established point of origin. One system's absolute coordinates may be another system's relative coordinates.

In a numerically controlled machining coordinate system, the absolute coordinates are the coordinates relative to the base of the machine. In an assembly system, the absolute coordinates may be relative to the building or shop floor as opposed to the base of any particular machine. In an electrical component insertion machine, the absolute coordinates are usually with respect to the machine base, with relative coordinates for each board in a multiple board fixture.

Also see: Industrial Robots.

Absolute Readout. An absolute readout is a presentation, by means of lights, cards, or other methods, of the true slide position as derived from position commands with a control system.

Also see: Acceptable Quality Level, Acceptance Test, Accuracy.

Acceptability. Acceptability is a term which refers to whether or not something is accepted or rejected. A certain level or quantity is defined for acceptability. If the subject is quality, greater than a certain number of rejects means an unacceptable yield for a certain process. Acceptability may vary in different circumstances. There is no such thing as absolute acceptability. Even a situation such as perfect (all correct, etc.) may be unacceptable if the purpose of the exercise or test is to determine the limit of a device, process, or person. In such a case, since there were no failures, the limit of capability is unknown. For planned obsolescence, consumable, or discarded products, it is important to know the extent of the reserve margin of functionality. Acceptability may even be defined by an equation or relationship to a variable. For example, an acceptable reserve of some quantity may be a function of the consumption and replenishment factors.

Acceptable Quality Level (AQL). An acceptable quality level is the stated percentage of minimum number of acceptable verses rejected items, from a quality perspective, which is allowable. This threshold percentage is a purely subjective value which differs among companies and industries. The concept of acceptable quality level is predicated upon the belief that 100% acceptable quality is not achievable.

Contemporary view of quality control are moving toward a belief in 100% quality. With this type of mentality, AQL becomes a moot factor (because it goes to 100%).

Many Far Eastern countries have adopted principles of 100% quality in relation to cost. 100% quality doesn't mean that every operation always happens perfectly on each initial attempt. What it does mean is no step of a manufacturing or assembly process is considered finished until that step has been performed 100% satisfactorily. In this style of operation, rework and repair are not independent processes, but are included and performed within each operation.

Also see: Quality Assurance, Quality Control.

Acceptance Sampling. Acceptance sampling is the practice of inspecting or testing a few items of a larger quantity lot to determine if the entire lot is acceptable. Typically, if an unacceptable item is found in the sample batch, the entire lot is rejected until it is 100% inspected. This is an application of statistical quality control methods.

The selection of samples may be random or it may be calculated and predetermined. Usually the method for the selection of the samples is based on historical learning which assists in using the most meaningful sampling methods.

Usually acceptance testing is not totally random and experimental, but utilizes some mathematical and statistical theory which improves the accuracy as well as calculates the probable error of the sample. Understanding the magnitude of the error is important because since by its definition, sampling is not 100% testing, the probable error must be included when determining if a sufficient percentage of sampling has been done to achieve a confident result.

Also see: Acceptability, Acceptable Quality Level, Acceptance Test, Statistical Process Control.

Acceptance Test. When referencing purchasing equipment or systems, the acceptance test is a predefined and agreed upon procedure for evaluating the system's performance, capabilities, and conformity to predefined specifications prior to acceptance of, and payment for, the system. When the product or system is large or of great cost there may be an acceptance test at the manufacturer's site and another acceptance test at the customer's site, to ensure the integrity of the system or product.

When referencing acceptance sampling, the acceptance test is the procedure used to determine whether an item is acceptable or not.

With contemporary material acquisition practices of ''ship to stock,'' incoming inspection or acceptance testing is not done. The idea is that activities which don't directly add value to the product are unnecessary and should not be done. Instead of acceptance testing, a procedure is implemented which ensures 100% acceptable quality of the delivered product.

Also see: Acceptability, Acceptance Sampling, Functional Specifications, Purchase Specifications.

Access Time. Access time is the time interval between the instant at which data are called for from a storage device and the instant delivery is completed, that is, the read time. Also, the time interval between the instant at which data are requested to be stored and the instant at which storage is completed, that is the write time.

If the storage device is a disk drive, the access time includes the seek time, which is the time for the read/write head to locate the proper track(s) and sector(s), as well as the actual read/write time. Since it stops turning a few seconds after an access is completed, a floppy disk drive additionally requires time to reach proper rotational speed before it can begin locating the appropriate track.

Since physical location on disk media isn't necessarily contiguous relative to a contiguous logical file, a "fragmented file" may require the device to access multiple tracks and sectors.

Furthermore, some devices verify each track that is written, as it is written, which doubles the write time as compared to the read time.

Also see: Disk, Diskette, Winchester Disk.

Accumulator. An accumulator is a location (register) in the ALU (arithmetic logic unit) where arithmetic operations are carried out. An accumulator is usually a register in which the result of a summing operation is kept. The word accumulator insinuates that some repetitious operation is being performed. Some computers may have multiple accumulators. The results of timers and clocks is sometimes kept in an accumulator. Error checking results are another type of data that is sometimes put in an accumulator. Identifying and using an accumulator allows the data to be in a known location if multiple entities desire access to the data.

Accumulators exist in various sizes, usually dependent upon the architecture of the computer in which they reside. Accumulators may be 8, 16, or 32 bit, for example. Some central processing unit chips have accumulators of different sizes. For example, a 32 bit CPU may have some multiple of two 16 bit accumulators. This may be done because the information which would be stored in the accumulators would never exceed 16 bits.

Also see: Computers.

Accuracy. Accuracy is a measure of lack of errors. When referencing computer programming and translating software and compilers, accuracy represents how closely the translated or compiled code performs the same functions as the source code.

This can mean conformance to a recognized standard.

When referencing a real number, accuracy denotes the number of digits to the right of the decimal point that can be considered significant in a particular value, or that can be supported by a particular algorithm, program, or system. For example, a controller may have 16 bit accuracy.

When applied to the field of robotics, this refers to the measurable difference between the actual endpoint position and the desired target endpoint position. Normally, the desired target position must be represented numerically within the controller. Errors in accuracy can be from improper calibration of the physical arm relative to the encoder and resolver signals, but also can be from nonlinearities in the resolution of the control system. For example, the accuracy of a particular robot is plus or minus .005 inches.

Also see: Calibration, Industrial Robots, Precision, Repeatability.

AC Motor. An alternating current (AC) motor is a motor that uses alternating current to drive the rotation of the motor. An AC motor uses the alternating current directly, without converting the AC to direct current (DC). Some DC motors can be connected to an AC supply since the motor control contains the circuitry to convert the AC to DC. AC motors exist in many types and configurations including ones with variable speed control. While speed control and direction in a DC motor can be accomplished easily by varying the voltage and current of the supply, an AC motor must continue to receive full voltage to avoid burning out the windings. The only reliable method of varying the speed in an AC motor is by varying the frequency of the alternating current source. While maximum speed may be obtained when the input is 60Hz (50Hz in Europe), varying the frequency down to 0Hz causes the motor to correspondingly slow to a complete stop, while the 0Hz (essentially DC) current holds the motor in a braked position. AC motors have a few significant differences from DC motors. AC motors usually are much less weight for the same torque size since the AC motor uses an electrical magnet instead of a permanent magnet as some DC motors use.

Also see: AC Servomotor, DC Motor.

Acoustic Coupler. An acoustic couple is an electronic modem device. It transmits and receives digital data through a standard telephone handset.

Also see: Modem.

Acronym. An acronym is a word formed from the initial letter or letters of each of the successive parts or major parts of a compound term. The government and the military profusely use acronyms, as well as do commercial businesses. Even some proper names began as acronyms.

Two examples: ACK =ACKnowledge;
CIM = Computer Integrated Manufacturing.

AC Servomotor. An AC servomotor is a two-phase induction motor that has two stator filed coils placed 90 electrical degrees apart. The two AC voltages are equal in magnitude and separated by a phase angle of 90 degrees. A two-phase induction motor runs at a speed slightly below the synchronous speed and is essentially a constant speed motor. When the unit is used as servomotor the speed must be proportioned to an input voltage. The two-phased motor is used as a servomotor by applying a AC voltage of fixed amplitude to one of the motor windings. When the other voltage is varied the torque and speed are a function of this voltage. The curve for zero controlled-field voltage goes through the origin of the torque-speed map with a negative slope. Therefore, when the control-field voltage becomes zero the motor develops a decelerating torque until it stops. The torque-speed curves must exhibit a large torque at zero speed in order for the servomotor to provide rapid acceleration. This is accomplished by building the rotor with a high resistance. The torque generated is a function of both the speed and the control-field voltage.

Active Storage. Active storage refers to data storage locations which hold data being transformed into motion.

Also see: Computers.

Actuator. An actuator is a mechanism that transforms energy into dynamic motion and force. Actuators may output rotary or linear motion, which can be either unidirectional or reciprocating.

Some typical actuators are electric solenoids, electric motors, hydraulic motors, pneumatic motors, hydraulic cylinders, and pneumatic cylinders. Some special types of actuators are sprague clutches which can be turned easily by the input shaft, however, the output shaft cannot easily drive the input shaft. Harmonic drives are other reduction drives which occupy small spaces, but accomplish great velocity or force reduction.

In Winchester disk drives, the technological drive for the smallest actuators with precision and speed has helped create such items as voice coil actuators, which are so named because they use a voice coil to sense the location of the tracks and provide feedback to the servomechanism.

Usually, any type of automated or self functioning device must use some sort of actuator to function.

A/D. See: Analog-to-Digital.

Ada. Ada is a programming language which was named for Augusta Ada Byron, Countess of Lovelace, and the inventor of the stored program. Ada is a high-level language which is usually compiled. It handles numerical data and is an imperative/algorithmic, procedural language. Ada was specifically designed for military embedded systems, but its use may broaden. Ada was authored by a design team at Cii Honeywell-Bull in France. The U. S. Department of Defense drove the development of Ada as it needed a single language which was tailored for programming embedded systems in real time and which could be easily maintained. The standard is ANSI Ada and the DoD (Department of Defense) holds a trademark on Ada, which prevents anyone else from legally producing a subset, superset, or variant of any kind. Validated compilers for Ada are slowly becoming available and unauthorized subsets are currently being sold for microcomputers.

Also see: Computer Languages.

Adapt. Adapt refers to the ability of something to successfully change as conditions in its environment dictate.

One example of an adaptable system is the adaptive control systems as it relates to robots, assembly machines, and arc welding systems. A robot which uses vision or some other type of sensor to grasp moving objects or pick random parts from a bin is exhibiting adaptability. The electronic component insertion machines which search for holes are utilizing adaptability.

Another example of a system which adapts is the multiscanning type RGB color monitors such as the NEC Multisynch and the SONY Multiscan. These monitors are capable of sensing the horizontal and vertical scanning frequency of the video driver circuitry and adapting to the frequency by locking their display onto that frequency. A side benefit of this type of monitor is that if there is a wider range of variance from product to product of the video display cards, the quality of the monitor display is unaffected as long as the actual frequency is within the upper and lower limits.

Adapt is also a computer-aided NC parts programming language similar to APT, but with fewer capabilities. Developed for small to medium-scale computers and used basically for two-axis contouring.

Also see: Numerical Control, Remote Center Compliance.

Adaptable Programmable Assembly System (APAS). Adaptable programmable assembly system is a term used by Westinghouse to describe an experimental research and development effort which they conducted to determine the feasibility of using a combination of robots and fixed automation to assemble a variety of electric motors. The system was developed and demonstrated in the Pittsburgh, PA R&D laboratory of the Industrial Automation Group of Westinghouse.

The system used robots with various end effectors, as the adaptable assembly devices, along with microcomputers and minicomputers to provide cell control and supervisory control. The intent of the experiment was to gather actual data regarding the issues and difficulties of designing and implementing a flexible manufacturing system using programmable automation (robots) instead of fixed automation. While there was considerable learning as a result of the APAS development effort, no such complete system has been sold to a customer. A similar concept has been developed by Allen-Bradley in Milwaukee, WS, for the assembly of electric motors.

Adapter. An adapter is a device which converts bits of information received serially into a bit format that can be used by the buffer.

Also see: Computers.

Adaptive Control. Adaptive control concepts represent one of the most sophisticated forms of automatic control. Such concepts imply an ability of the control system to ''learn'' or "adapt to'' an unknown environment such that adequate system control is maintained. Adaptive control systems are generally characterized by two important features.

First, the analytical tools used to synthesize adaptive control laws tend to be highly mathematical in nature. Nonlinear time-varying systems theory must often develop the adaptive control algorithms. Closed-form analytical solutions to the control analysis and synthesis problem tend to be unavailable. Consequently, computer simulation is often the only practical means of evaluating adaptive control stability and performance.

Second, due to their complexity, adaptive controls typically require sophisticated implementation technology. Until the 1960s and the 1970s, the prevailing mechanical implementation technologies proved a major barrier to implementation of adaptive controls. With the advent of digital electronic control systems in the 1960s and the microprocessor in the 1970s, implementation technology has ceased to be a major barrier to adaptive control concepts.

History. Adaptive controls have evolved from more basic feedback control theory. Consequently, to place adaptive control concepts in a proper historical perspective, it is necessary to provide a brief history of automatic feedback control. A feedback control system maintains control by comparing one variable to another and using the difference as a means of control. Typically, one variable is an actual or a sensed output variable. The other variable is a reference or desired value of the output variable; the difference between actual and desired values is used to control the system and drive the value of the actual variable to that of the desired variable.

James Watts' flyball governor for controlling speed, developed in 1788, was one of the very first widely used and reported automatic feedback control systems. Almost 100 years later, in 1876, the Russian engineer Wischnegradsky developed a detailed solution of the stability of a third-order flyball governor. During the latter part of that same century, steam power made possible larger and faster sailing vessels. Direct mechanical control of rudders became increasingly difficult and gave way to servomechanisms. Minorsky made one of the more significant contributions to feedback control systems in his study of automatic ship steering in 1922. Just a few years later in 1934, Hazen's classic paper *Theory of Servomechanisms* was published in the *Journal of the Franklin Institute*, marking the start of intense modern interest in automatic feedback control systems.

The early works of Watts, Minorsky, Hazen, and later pioneers benefited by important advances made in the science of mathematics during the 1700s and 1800s by Laplace, Fourier, and Cauchy. The Laplace and Fourier transforms, together with Cauchy's theory of functions of a complex variable, are the foundations upon which modern analysis of feedback control systems is based. The analytical tools with which post World War II engineers designed feedback control systems—the Routh-Hurwitz stability criterion, block diagram manipulation, root loci, Nyquist's stability criterion, and phase/margin analysis—all are derived from the fundamental works of Laplace, Fourier, and Cauchy. The major advantage of these mathematical methodologies is that for an important class of dynamic systems—namely, physical systems which can be adequately modeled by linear ordinary differential equations with constant coefficients—the methodologies transform the model from the time domain into the so-called frequency domain, and permit the designer to work with algebraic rather than differential equations.

Until the 1950s and early 1960s control system engineers relied exclusively upon the classical frequency domain techniques in the analysis, synthesis, and design of feedback control systems. However, control system researchers recognized that frequency domain analysis is applicable primarily to single input-single output linear time invariant systems. Spurred in part by the extraordinary guidance and control problems associated with the U.S. space program and the intense interest in advanced controls within the aerospace industry, the inherent limitations of frequency domain analysis led researchers in the late 1950s and early 1960s to search for an improved theory of control capable of handling in a unified fashion problems associated with the control of multi-input multi-output nonlinear time varying systems. The desire to incorporate this generality into the theory necessitated a return to the time domain and the development of the state-space description of a dynamic system. Thus, the efforts of Pontriagin, Bellman, Merriam, and Kalman resulted in what is now termed "modern control theory." Modern control theory is characterized in general by its ability to incorporate a broad class of control problems within a unified mathematical framework and, in particular, by a state-space description of the dynamic system, a time domain formulation and solution of the problem, and use of mathematical optimization methods to determine appropriate control algorithms.

The modern control concepts have provided, for the first time, analytical methodologies sufficiently powerful to address adaptive control design issues. Simultaneously, the advent of digital computer technologies has provided an equally powerful platform upon which to implement the complex algorithms typically associated with adaptive control laws. The analytical design methodologies and the implementation technology converged in the 1970s to make adaptive control analysis and synthesis technically feasible and economically practical for very many control problems.

Discussion. The design of adaptive controls starts with a mathematical model of the physical system which is to be controlled. The physical system itself may be mechanical, electrical, hydraulic, etc. Once the mathematical model has been derived, that is, once the physical system has been described by a set of mathematical equations, then the subsequent control analysis and synthesis is independent of whether the physical system is electrical, mechanical, etc. The mathematical description of the system encompasses the dynamic relationships between the system state vector x, the input vector u, and the output vector y. The state vector x consists of the state variables x, $x2$, ...xn of the system; the control vector u consists of the control variables u, $u2$... up of the system, and the output vector y consists of the output variables $y1$, $y2$, ... ym of the system. The system as described here is an nth order system with p inputs and m outputs. The inputs, outputs, and state variables are related by a set of ordinary differential and algebraic vector equations:

$$dx/dt = f(x,u,t) \text{ and } y = g(x,u,t).$$

Where t denotes time, d/dt denotes a derivative with respect to time, and the functions f and g are usually assumed continuous in all of their arguments with continuous first partial derivatives as well. Note that f and g may be nonlinear in any or all of their arguments.

When f and g are assumed perfectly known, then no adaptive control problem exists. For instance, consider the situation where the physical system is adequately modeled by a linear time invariant representation. Here

$$f(x,u,t) = Ax + Bu \text{ and } g(x,u,t) = Cx + Du$$

where A,B,C,D are matrices with known, constant elements. (This situation often occurs from linearization about a system operating point. In that case elements of the matrices A,B,C,D represent the partial derivatives of f and g with respect to x and u evaluated at the operating point.) With the algebraic forms of f and g as well as the elements of A,B,C,D known, the control designer need not worry about designing a control law for u that adapts to f,g,A,B,C, or D. For the given A,B,C,D, the designer need only develop a control law that (1) feeds back available state and/or output variables, and (2) optimizes some measure of system performance.

Consider, however, the more difficult situation that gives rise to a desire for an adaptive control system. Here the functional forms of f and/or g may be only partially known (e.g. linear vs. nonlinear); or the functional forms may be known but particular parameters (e.g. some elements in the A,B,C,D matrices) within the functions f,g, may be unknown. The control problem now becomes extraordinarily more difficult than that posed earlier. Unless a means is devised to accommodate (perhaps by the control system learning and adapting to) unknown parameters, serious closed-loop stability and control problems can arise.

To capture the essence of adaptive controls without becoming overly mired in mathematical complexities, consider a dynamic system description given by

$$dx/dt = f(x,u,p)$$

where p is a vector of potentially unknown parameters. It is desired to design a control u that feeds back the state vector x to achieve a specified closed-loop system performance and stability margin. Furthermore, the parameter vector p has a strong influence on system performance. Controls designed for a particular value of p prove inadequate for other values of p. Hence, design for a nominal p is not a viable approach.

It is clear that to obtain satisfactory closed loop system performance throughout the range of potential parameter values, the control u must explicitly or implicitly depend upon the parameter values. However, even for this conceptually straightforward problem—where the adaptation required is to a set of constant but unknown parameters—no simple methodology exists by which to design an adaptive controller. Moreover, no particular approach has achieved widespread acceptance within the community of control practitioners and researchers. "Model reference adaptive control" methods have been successfully developed for those situations where system dynamics are linear. In these methods a system model, based upon nominal parameter values, is incorporated into the control law. The control input is used to excite the model and the resulting model output is compared with the actual system model. If the actual parameter values differ from the nominal parameter values, then the actual system output will generally differ from the model output. This difference is fed back and used as the basis for control. In this manner, the actual output is driven to the model reference output. Under the assumptions of linear dynamics and constant coefficients, elaborate analyses have been carried out over the past 10 years to address issues of controllability, stability, and performance of model reference adaptive control systems.

Other approaches recognize more explicitly the unknown nature of the parameter vector. These approaches attempt to estimate the unknown parameters during system operation (i.e."learn" what the parameter values really are). The estimated parameter values are then used explicitly within the feedback control law. These approaches often recognize the true stochastic nature of most physical systems and make appropriate assumptions regarding the various noise processes within the physical systems. Moreover, in taking this type of an approach the control designer is often forced to confront nonlinear stochastic system models. Much of today's stochastic estimation methodologies are based upon the pioneering work of Kalman who addressed the stochastic estimation and control problem in the 1960s.

Before concluding this discussion of adaptive controls, it is important to note the difference between adaptive and feedback control systems. The difference is often not appreciated or clearly stated in texts and glossaries. A control law $u = u(x)$ that feeds back the system state variables, independent of changes in the dynamic system, is not an adaptive control—it is a feedback control. If sufficient changes in the system occur then that control may result in poor closed-loop performance. A control law that changes in response to or in anticipation of system changes (for instance feedback gains, rather than being constant, may be updated based upon estimated parameter changes) is an adaptive control.

Applications. Because of the mathematical sophistication of adaptive control, algorithms and the resulting required advanced implementation technologies, adaptive controls tend to be developed and applied primarily for those physical systems where stability and control are critical issues. In those cases, the control problem warrants high-volume engineering design time and potentially expensive implementation technology. Control problems associated with the precision machining of mechanical components, the process industries, and the aerospace industries often give rise to the need for adaptive controls. In the precision machining of high-value metal components to extremely close tolerances using computer numerical control (CNC) machine tools, proper control of the tool cutting path is a critical issue. An algorithm can be made that accommodates machine dynamics and directs the tool to follow a prescribed cutting path and can be designed in a relatively straightforward manner. However, tool performance and resulting part geometry and tolerances may depend upon the component material, part feeds, tool speeds, tool wear, tool cooling effectiveness, and other parameters and easily described within the usual context of control system analysis and synthesis methodologies. A cutting path controller unable to adapt to these conditions—either automatically or through operator adjustments—will likely deliver sub-par control and inadequate product. Within the process industries (oil, pulp and paper, steel, chemicals, etc.), generally characterized by conversion of raw material to finished product via a continuous flow process, flow control may typically involve appropriate mixing, heating, cooling, extruding, plating, shaping, rolling, etc. Parameters of the finished product may often depend strongly upon characteristics of the raw material including potential preprocessing and/or in-plant storage conditions.

Process controls must typically adjust to these and other conditions in order to insure a consistent finished product. Finally, the aerospace industry provides perhaps the greatest application of adaptive control concepts. High-performance military aircraft, rotorcraft, and propulsion systems are generally driven to the furthest extreme of potential operating envelopes through adaptive control systems. Flight control surfaces, engine/inlet variable geometric, and engine/afterburner fuel flows are often controlled by high-capability digital electronic control systems (e.g. so-called "fly-by- wire" flight controllers and full-authority digital electronic engine controllers). Performance seeking control systems adjust control schedules based upon position within the flight operating envelope (e.g. altitude, speed,) or critical vehicle/propulsion system parameters (e.g. control surface configurations, turbine temperatures, stability margins). These control schedule adjustments are programmed into the digital controller and are typically based upon in-flight senses or inferred parameters. The resulting adaptive control system enables near-optimum system performance under a wide variety of operating conditions.

Motion control for robots is a demanding field, as the system parameters vary with the load and the position. The most effective method for designing motion control systems for industrial robots is with adaptive control. Adaptive controllers continuously estimate the system parameters and adjust the control law accordingly. The parameter estimation provides the information required to detect fault conditions if the robot operation; thus fault detection is a natural byproduct of adaptive control.

Forrest D. Brummett of General Motors Corporation described applications of adaptive control systems in his SME Technical Paper: *Adaptive Controls Systems*. In it he wrote: "The conventional method of programming NC-CNC machines involves a programmer using a 'fixed' programmed feed rate concept. Through the use of this procedure, the programmer sets a machine tool feed rate that remains constant for the duration of the machining process. Under this procedure, factors contributing to this cutting torque variation include the depth of cut, type of tool used, tool wear, material composition, and material hardness. The NC programmer compensates for these variations in cutting torque factors by using sub-optimal feed rates contributing to significant productivity losses.

"The adaptive control feature is a programmable option which can cause the machine tool to automatically adapt its mode of operation to optimize piece-part production efficiency. On numerical control machines, adaptive control is a kind of 'think ahead' system that adapts the metal cutting operation to the actual conditions of the tool and material. It does this by measuring the force on the cutting tool in process and continually comparing this measured force to a reference force that was prescribed for the operation by a computer program. When measured forces are within 10-20% of the limiting reference force, the adaptive controller reduces the feed rate command to the NC unit in proportion to the error between measured and reference force signals. It acts to prevent the reference force from being exceeded. In addition, it prevents the spindle motor's rated horsepower from being exceeded by sensing motor current and comparing this signal to a reference value. By controlling force and horsepower in this way, it achieves feed rates in the cut that are from two to six times faster than those typically used in a conventionally programmed operation."

Brummett noted that: "The computer program that prescribes the reference force for particular cutter and cutting conditions also computes the maximum feed rate for the tape and the cutting speed and spindle speed to be used. It uses information about the type of workpiece material, size of cutter, and the maximum stock cross section to be experienced. This information is supplied to the computer program by a planner or programmer. Where APT, or other automatic programming languages are used, the programmer can input the information directly into his cutter path program with a special 'call' statement. In this case, feed rate and spindle speed (where possible) are also automatically put on tape. Reference force values and other commands to the adaptive controller are also automatically put on tape via insert statements. A printout of this information is also provided.

"The adaptive control system will improve the machine tool performance in the following manner:

- Reduces the machine cycle time by substantially increasing the feed rate and spindle speed whenever an air gap is encountered.
- Protects the machine and tool from damage due to heavy cuts which highly exceeds the programmed target horsepower.
- Displays constantly, valuable cutting data related to the machine tool which can be accumulated and stored away for future use. (Edit Memory feature.)
- Detects a dull tool by measuring the amount of cutting torque applied to that tool. This cutting torque, along with the spindle speed, will determine the amount of horsepower consumed. If the cutting horsepower is too high for a period of time, that tool will be considered dull and the operator will be alerted by the control unit indicator. The operator then changes the tool at the end of the machine cycle, thus preventing broken tools and machine wrecks causing excessive downtime.
- Reduced rework and improves quality significantly.
- Eliminates operator control of feed rates and simplifies program development on the production floor.
- Increases productivity by optimizing speeds and feed rates.
- Increases capacity—cost avoiding overtime and potential capital expenditures.
- In many instances, allows NC machining centers to be run on a percent man bases for significant labor savings."

Adaptive control is not without its drawbacks however. Added Brummett: "Torque control is not effective where the horsepower required is small relative to the available horsepower or the energy required to overcome the inertia of the spindle drive.

"A deflection control system cannot compute safe cutting forces or reliable limit feed rates for cutters having resultant forces parallel to the cutter axis, or for cutters on supported arbors. Since a deflection control system measures spindle displacement and, thus, senses cutting forces, perpendicular to the cutter axis only, it will not afford protection to cutter and machine when drilling, plunge end milling, or rampling at a steep angle."

The paper offered the following application selection guidelines. First, determine machining operations which offer the highest potential savings due to reduced cutting time par part. Concentrate on operations with the following characteristics: excessive cycle time; tools which require a high percentage of "available" horsepower; NC or CNC controls which provide a high degree of reliability, a high degree of variability in material composition or depth of cut.

Second, determine machining operations which are production bottlenecks due to excessive cutter wear, scrap or rework caused by cutting forces, or excessive machine downtime caused by machine wrecks.

Third, determine operations where machine and operator utilization could be improved by the ACS. Adaptive control eliminates the need for constant operator override of feed rate. In some cases, a two-man, two-machine complex can be upgraded to a percent man operation with the addition of an ACS to both machines.

It has been recognized that the adaptive control of EDM can improve EDM performance and reduce the production costs. Various systems are being attempted in university laboratories in cooperation with EDM manufacturers across Europe, Japan, and the U.S. Although the general approach seems to be the same, these systems widely differ in the areas of identification and monitoring the change of gap condition from sparking to arcing. These methods include measuring the voltage at two points during the pulse, measuring the radio frequency emmission and the stochastic analysis of the gap condition signals.

A sample and adaptive control EDM system was presented by K.P. Rajurkar (University of Nebraska) and S.M. Pandit (Michigan Technological University) in their paper presented at the North American Metalworking and Research Conference in 1985. Their paper titled *Adaptive Control of EDM–Technology and Research* noted: ''Although most of the Electrical Discharge Machines (EDM) available nowadays in the market have some kind of process control, still highly skilled operators are required to select and adjust optimal machine settings. In EDM, the objective of adaptive control is to optimize erosion rate while keeping the surface roughness, surface damage and tool wear within a reasonable limit.''

W. Worger and C. Y. Kuo, researchers at Arizona State University, reported in their SME Technical Paper *Application of Discrete Time Model Reference Adaptive Control to PUMA 560 Robot Manipulator* the successful use of adaptive control in manipulation a robot's arm.

Their findings show that the adaptive controller allows the arm to achieve faster response time than the PUMA controller allows while still maintaining consistent response and no overshoot, regardless of the arm configuration and payload.

Glossary. *frequency domain:* Refers to the analysis of control problems using algebraic equations that arise after applying Laplace and/or Fourier transformation. Theories to system differential equations.
computer numerical control: Computer numerical control (CNC) systems control a machine tool via commands from an electronic computer generally embedded within the machine tool. The commands or instructions may be generated by a CAD/CAM system or by some other media.
servomechanism: An automatic feedback system, typically mechanical in nature, used to control motion. The driving signal is a function of the difference between commanded position and/or rate and actual position and/or rate.
stochastic: A stochastic signal is one whose value at any given time is a random variable. It is often described in terms of statistical properties.

ADCCP. See: Advanced Data Communications Control Procedure.

Address. An address is a character or group of characters that identifies a register, a particular part of storage, or some other data source or destination. An address may be a label or number specifying where a unit of information is stored in a computer's memory. The address of memory is used more when programming in low-level languages such as Assembly Language and specific data must be placed in specific locations. When doing low-level or machine-level diagnostics, the contents of specific memory locations must be viewed, especially when the programmer is concerned with hardware interrupts and similar control aspects.

The memory of a computer is labelled with addresses. All computers start at memory location 0000 and are usually counted upwards in HEX, instead of decimal or octal notation. Some operating systems such as PC DOS (or MS-DOS) up to version 3.3 can only address 640K bytes of main memory.

Also see: Absolute Address, Accumulator, Computers, Register.

Address Index Pin. An address index pin is a screw-lock slide pin used to establish proper identification of a PLC I/O module.

Address Selector. An address selector is a switch, located at the top of each PLC I/O housing that establishes the address of the housing.

Address Word. An address word is a computer word containing only the memory location address.

Adhesion. Adhesion is the attractive force that exists between an electrodeposit and its substrate that can be measured as the force required to separate an electrodeposit and its substrate.

Also see: Automated Assembly, Industrial Robots.

Adhesive. See: Adhesion, Adhesive Joining, Automated Assembly, Industrial Robots.

Adhesive Joining. Adhesive joining is the family of processes for bonding surfaces using a substance capable of holding them together by surface attachment.

Also see: Adhesion.

Advanced Data Communication Control Procedures (ADCCP). The Advanced Data Communications Control Procedures are designed for nonpacket synchronous data communications over switched or leased facilities in a point-to-point or multipoint location, in either half-duplex or full-duplex operation.

ADCCP was approved by the American National Standards Institute (ANSI) in January 1979 after close to 10 years of review and revision. ADCCP is a bit-oriented protocol that handles synchronous data transmission between two points, or stations, connected by a transmission system. The protocol is fully transparent and independent of the code used, for both interactive and batch operation in full-duplex and half-duplex modes.

Before adoption as a standard, the ADCCP proposal heavily influenced the design of other bit-oriented protocols by many of the major vendors (such as the UDLC of Sperry Univac, BDLC or Burroughs, and SDLC of IBM). These protocols can be considered as compatible subsets or ADCCP, although they are not generally compatible with each other.

ADCCP is documented in ANSI standard X3.66-1979.

AGV. See: Automated Guided Vehicle.

AI. See: Artificial Intelligence.

Air Clamp. An air clamp is an air-operated device for holding workparts against a nonmagnetic driver.

Air Gaging. Air gaging is a method of measuring physical dimensions by precisely measuring pressure or flow and relating the measurement to the distance between the workpiece surface and a nozzle that directs airflow against the surface.

Also see: Gaging.

Air Motor. An air motor is a motor that uses compressed air as its source of energy to rotate the rotor. The motor is typically designed similar to a turbine with an inlet for the air which presents the air at a driving angle to the vanes of the rotor. The speed is altered by metering the volume of air that is allowed to enter the motor as a function of time. Small torque air motors have reasonable usefulness when constant speed under a varying load is not extremely critical, but large torque air motors require such a substantial amount of air that they are sometimes impractical. The compressibility of air makes air motors incapable of precise positioning such as electric motors are capable of performing. In applications where electricity is undesirable or when large torque is needed in a small space such as in air wrenches, air motors are very applicable, especially when the compressibility of air can be used to absorb shock loads, protect the equipment and lessen vibration to the operator.

ALGOL (ALGOrithmic Language). ALGOL, which stands for ALGOrithmic Language, is a high-level, procedural compiled language which is not extensible. The driving forces for the creation of ALGOL were the European need, for a general algebraic language and the American desire to produce a standard algebraic language. ALGOL is widely adopted in Europe, but has never really penetrated the wide acceptance of FORTRAN in the United States. ALGOL is most influential as a source for subsequent languages like Pascal and Ada. Current ALGOL evolved from earlier versions, which were initially called IAL, or International Algebraic Language. Many of the distinctive characteristics we associate with modern computer languages (including structured design and the rich use of data and control structures) are derived from ALGOL. Although there have been three published standards of ALGOL which are known as ALGOL 58, ALGOL 60, AND ALGOL 68, ALGOL 60 is the version which has been the most influential.

Also see: Computer Languages.

Algorithm. An algorithm is a set of rules expressed in symbolic form which perform a task as defined by the rules. An algorithm could be a mathematical equation or a procedure in a computer program. When particular quantities are assigned to the basic elements (called variables or parameters), the rules of the algorithm can be followed to produce the required result.

Sometimes in programming, an algorithm is a collection of statements which are used together as a subroutine or with an alias. In this scenario the algorithm might also be referred to as a macro.

Algorithms usually perform some unique function, such as the PID algorithms which provide proportional, integral, and derivative control, which also may include second or third order expressions, anti-windup coefficients, etc. These algorithms have been developed over some period of time and become the proverbial ''wheels'' which sometimes are reinvented.

Also see: Macro Instruction, Programming.

Alphanumeric. Alphanumeric refers to a character set that contains both letters and digits and that may contain control characters, special characters, and the space characters.

Usually when alphanumeric characters are mentioned relative to a CRT display or a printer, they are distinguished between alphanumeric characters and bit-mapped graphics.

When discussing typesetting and publishing, alphanumeric does not simply mean letters and numbers, but all the different fonts in the different typestyles with all the different attributes.

Also see: Characters, Graphics.

Alternating Current (AC). Alternating current is electrical current that periodically reverses direction, usually many times per second.

An *AC input module* is an PLC input/output rack module which converts the AC signals originating in user devices to the appropriate logic level for use with a processor.

ALU. See: Arithmetic and Logic Unit.

American National Standards Institute (ANSI). The American National Standards Institute is a voluntary body in the United States and is a member of the International Organization for Standardization (ISO). It develops standards and accepts standards proposals from other organizations in the United States. ANSI activities are extensive. In addition to standards on programming languages, such as COBOL and FORTRAN, ANSI has several committees and groups working with communications standards as well. Its body, ANSC X3 (American National Standards Committee 3), and subcommittees provide the organization and mechanism for members to develop communication standards in conjunction with the ISO groups. ANSI cooperates with CCITT through the State Department's Industry Advisory Group (IAG), which was created to allow ANSI and other standards groups more participation in CCITT issues.

One of the better-known standards endorsed by ANSI is the ASCII data representation standard. ANSI has also endorsed a standard for the functional attributes of a data display terminal.

American Standard Code for Information Interchange (ASCII). The American National Standard Code for Information Interchange is a seven-bit code with one additional bit added for error detection purposes, known as a parity bit. The code was first developed in 1963 and has become a standard. Since the ASCII set is made up of seven bit characters and one parity bit, this allows a maximum of 128 characters to be represented.

The ASCII set is used for information interchange among data processing systems, data communication systems, and associated equipment. The ASCII set consists of control characters and graphic characters and is properly an alphabet rather than a code.

The ASCII set is the U.S. implementation of the International Telegraph Alphabet Number 5.

Devices, such as displays and keyboards which are referred to as ASCII devices are capable of displaying the complete ASCII character set. Storage devices in personal computers utilize ASCII data representation, as opposed to IBM mainframes and peripherals which utilize EBCDIC data representation.

AML. AML, A Manufacturing Language, developed at IBM, was intended to be a general purpose plant floor automation language.

AML went into production use on an IBM manufacturing line in 1978. External test marketing began in 1980, and regular marketing started in 1982.

Also see: Industrial Robots.

Amphere. An amphere is the unit of electric current flowing through one ohm of resistance at one volt potential. An amphere also is a unit of electric current flow equal to a charge of one coulomb (6.24 x 1018 electrons) past a given point in a circuit in one second.

Amplifier. An amplifier is a signal gain device that is capable of enlarging the wave form of electric current, voltage, or power supplied as input.

Analog. Analog pertains to the representation of data or physical quantities in the form of a continuous signal. Usually, the instantaneous amplitude is a function of the value of the data or physical quantity being represented. An analog signal is a signal, which, when it varies, varies continuously.

Some of the first types of computers were analog machines, but now digital computers constitute the majority. In circuit design, devices such as power supplies typically have analog logic portions. Certain activities in circuit design, such as simulation and modeling, must be done separately for the analog portion and the digital portion of circuits. Servocontrollers are another example of analog circuitry and computing. Even integrated circuit chips (ICs) can have both digital and analog circuitry included on the wafer (Hybrids). While digital circuitry is usually TTL (Transistor-to-Transistor Logic) voltage levels of 0 to 5 volts or CMOS (Complementary Metal Oxide Semiconductor) voltage levels of 0 plus or minus 12 volts, analog circuits may utilize relatively higher voltages and currents.

Also see: Analog Communications, Analog Computer, Analog Signal, Analog-to-Digital.

Analog Communications. Analog communications is the transfer of information by means of a continuously variable quantity, such as electrical voltage or the intensity of a light beam in an optical fiber. Analog data may be considered as the information content of an analog signal, or the signal itself. The information content of the analog signal is conveyed by the value of the magnitude of some characteristics of the signal such as the amplitude, phase, or frequency of a voltage; the amplitude or duration of a pulse; the angular position of a shaft; or the pressure of a fluid.

Analog communications is affected greatly by transmission noise and electromagnetic interference. The hissing heard while playing an audio tape occurs due to the use of analog communications between the tape and the playback head.

Analog communications offers a continuous output signal as long as an input signal is present, as opposed to digital communication which, must be digitized using an analog-to-digital converter, and therefore represents many closely spaced instantaneous values of the input signal.

Analog Computer. An analog computer is a computer in which analog representation of data is mainly used.

Also see: Computers.

Analog Data. Analog data is the information content of an analog signal as conveyed by the value of magnitude of some characteristics of the signal, such as the amplitude, phase, or frequency of a voltage, the amplitude or duration of a pulse, the angular position of a shaft, or the pressure of a fluid. Analog data also is data representing, in continuous form, original information. For example, electrical signals on a telephone channel representing a voice.

Analog Input Module. An analog input module is a programmable controller I/O rack module which converts an analog signal to a three-digit binary coded decimal (BCD) number that can be manipulated by the processor.

Analog Output Module. An analog output module is a PLC I/O rack module which supplies output, based on and proportional to a three-digit BCD number manipulated by the processor and provided to the module.

Analog Signal. An analog signal is a continuous signal that varies in some direct correlation with an impressed phenomenon, stimulus, or event that bears intelligence. The signal may assume the form of an electric current variation, light intensity or wavelength variation, electromagnetic wave amplitude, phase, or frequency variation, mechanical displacement variation, or other physical phenomenon that varies continuously. Analog signals are susceptable to noise and other forms of interference, though digital signals are immune to some of the same types of interference.

A quasianalog signal is a digital signal that has been converted to a form that is suitable for transmission over a specified analog channel. The specification of the analog channel should include frequency range, frequency bandwidth, signal-to-noise (S/N), and envelop delay distortion requirements. When this form of signaling is used to convey message traffic over dial telephone systems, it is often referred to as voice data. A modem may be used for the conversion process.

Also see: Analog, Digital Signal, Digital-to-Analog.

Analog-to-Digital. The term analog-to-digital is used to describe both devices and processes which convert an input analog signal to an output digital signal with the same information content. An analog-to-digital, or A/D, converter is any device which is capable of converting signals from an analog form to a digital form.

There are two main type of techniques used for conversion. One method is called ''flash conversion'' and the other method is called ''successive approximation.'' Flash conversion consists of taking an instantaneous snapshot of the analog input signal and converting it to a corresponding digital value. The potential problem with flash conversion is if some noise or interference coincidentally altered the analog signal at the instant of sampling, an erroneous value is obtained. The desired capability of flash conversion is that it represents the fastest method to effect continuous sampling and conversion. Successive approximation or continuous slope method consists of taking continuous successive samples of smaller and smaller segments of the slope of the equation representing the analog input value and averaging this result to determine a representative digital value. The undesirable aspect of this method is the speed of conversion may not be sufficiently fast for the circumstances, while the positive aspect is this method is not very susceptible to extraneous noise and other forms of error.

Conversion of an analog signal to a digital signal is always referred to in terms of bits of accuracy where the bits correspond to the extent of decimal places used in the calculation.

Analog-to-Digital (A/D) Converter. An analog-to-digital (A/D) converter is a system hardware device sensing continuous analog signals and converting them to discrete digital signals.

Annealing. Annealing is a heat treatment process to reduce hardness or brittleness, relieve stresses, improve machinability, facilitate cold working, or produce a desired microstructure or properties. The process consists of heating to a suitable temperature, which is dependent upon the type of annealing, followed by slow cooling. Common types of annealing include:

Black annealing is box, close, or pot annealing of ferrous alloy sheet, strip, or wire.

Blue annealing is a process for softening hot rolled ferrous sheet by heating in an open furnace and then cooling in air. The formation of a bluish oxide on the surface is incidental.

Box, close, or pot annealing is a process of slow heating in a sealed container to minimize oxidation, followed by slow cooling.

Bright annealing is annealing in a furnace containing protective atmosphere to minimize surface oxidation and prevent discoloration of the bright surface.

Cycle annealing is annealing with a controlled time-temperature cycle to provide a specific microstructure or properties.

Flame annealing is annealing (softening) by heat from a high-temperature flame.

Full annealing is a term denoting annealing to produce minimum strength and hardness. For precise meaning, the condition of the starting material and the time-temperature cycle should be specified. The time-temperature cycle must produce complete transformation to austenite.

Graphitizing is annealing of ferrous alloys to precipitate some or all of the combined carbon as graphite; generally applicable only to cast irons.

Isothermal annealing is austenitizing a ferrous alloy and then cooling to and holding at a temperature at which the austenite transforms into a softer ferrite and carbide microstructure or a pearlite microstructure.

Malleablizing is annealing white cast iron to transform some or all of the combined carbon to graphite.

Partial annealing is a process for stress relieving cold worked material or for reducing its strength.

Process annealing is processes for softening metals to facilitate further cold working. The subcritical temperatures usually employed do not involve austenitization.

Quench annealing is annealing of austenitic ferrous alloys by solution heat treatment (heating followed by rapid quenching).

Recrystallization annealing is annealing cold worked metals to provide refined grain structures, primarily new, strain free grains.

Spheroidize annealing (spheroidizing) is a process to produce a rounded or globular form of carbide in the structure. This process generally requires austenitization, followed by heating to a subcritical temperature for an extended period of time.

Subcritical annealing is annealing of ferrous alloys at temperatures below those at which austenite begins to form. (See also partial and process annealing.)

Annotation. Annotation is the process of placing comments or such as text or special notes or identification flags on a drawing, image, diagram, or map. Usually annotation is done using a CAD/CAM system where the annotations may be identifying legends, titles, part numbers, or location codes.

When annotation is used on electrical CAD drawings, there are two times during the design when the annotations can be added; both during the original creation of the design as well as after the schematic has been captured. Adding annotations after the logic drawing and schematic have been created is referred to as back annotation.

Other methods of adding annotations to drawings are methods such as a software utility which adds a "soft" version of the yellow post-it notes by 3M. The notes can be added to screens such as Lotus 1-2-3 worksheets and word processing text. These annotations are visible when viewing the document on the monitor, but can be turned off if desired. However, these annotations are not visible on a print out.

Also see: Back Annotation.

ANSI. See: American National Standards Institute.

Anthropomorphic. Because robots are described or thought of as having human form or attributes, they are anthropomorphic. An anthropomorphic robot is generally a machine which is in some way physically or functionally similar to a human. Anthropomorphic robots perform tasks in a human like, more efficient manner.

Anthropomorphic robots are devices that perform tasks too dangerous, physically demanding, menial, or repetitive for a human to do effectively or efficiently. Industrial anthropomorphic robots usually consist of an arm, to which an end effector (gripper, spot welder, or drill) is attached, a power source supplying electrical or hydraulic power, and a control unit providing logical direction for the machine. The fact that this apparatus resembles and functions like a human arm reinforces the definition of anthropomorphic robot.

APAS. See: Adaptable Programmable Assembly System.

APL. APL, which stands for A Programming Language, is a high-level language which is interpreted, not compiled, and whose data is both numerical and character strings. APL is a functional/applicative, procedural language which is extensible. APL was developed by Kenneth Iverson of Harvard and Adin Falkoff of IBM. The driving forces for developing APL were to develop a language suitable for data processing that incorporates mathematical elegance. APL has a reputation of being a difficult language. APL has large memory requirements. It is a very logical language, but difficult to read, especially if the code is not well-documented. APL is generally available on mainframe computers and microcomputer version have recently become available. APL has a unique character set which can activate complex functions with a single keystroke. Using APL requires a special keyboard legend, and many programs can be written with one line of code.

Also see: Computer Languages.

Application Program. An application program is a computer program. It is written in some language which provides a focused, specific, and useful function for the user when executed, as opposed to a programming language or general purpose program which may have broad capabilities. Applications programs usually are written in some programming language (computer language) and run under an operating system. Applications programs may vary greatly as to the actual construction, while providing essentially the same results. The efficiency and speed of execution are sometimes desirable attributes of application programs.

Examples of some widely used application program are Lotus 1-2-3, Framework II, dBase III Plus, Wordstar 2000, and Timeline. While some people refer to these as programming languages, they are actually application languages. These application languages are themselves written in some programming language such as C or Modula 2. Since the value of computer hardware and software is not truly realized until useful work is accomplished, application programs are sometimes the crucial link in a computer systems solution.

Also see: Computer Languages.

Application Specific Integrated Circuit (ASIC). An application specific integrated circuit is one of the more recent categories of integrated circuit (IC). ASICs are specialized evolutions from custom and semicustom chips such as gate arrays. An ASIC is a chip which contains all the necessary circuitry for a particular application in silicon. The advantages are the greatly reduced size and the increased functionality of a single integrated circuit chip instead of a circuit on a printed circuit board, which may only partially use some of the functionality of the chips on the circuit. Another big advantage of creating a custom ASIC is the ability to conceal the exact circuitry which provides the functionality. The continuing decrease of the price vs. performance curve makes ASICs much more cost-effective than in the past. In some respects, ASICs can be thought of as nothing more than miniaturizing a circuit onto an integrated circuit chip. The same logic must be built regardless of which solution is selected.

APT. See: Automatically Programmed Tools.

AQL. See: Acceptable Quality Level.

Arc Cutting. Arc cutting is the process of severing and material shaping as well as sawing, drilling, and planning. There are seven cutting processes which utilize an arc. Some have been substituted by better and more efficient methods. These include oxygen arc cutting, air carbon arc cutting, gas metal and gas tungsten arc cutting, and plasma arc cutting.

Arc cutting can be done manually or automatically, and in both cases they can be brought to the work. The cut can be straight line, curved or bevelled. The cutting torch tip has a central orifice for the gas to be used and peripheral orifices for the preheat flame. The flame preheats the cutting surface to the ignition temperature, whereupon the gas converts the surface which issues from the bottom of the cut as a molten spray.

For highly alloyed and stainless steel cutting, it is often necessary to use a special process such as flux injection or powder cutting. The selection of the cutting process and the type of operation to be used depends on the material that is being cut and the ultimate end use of product.

Architecture. The architecture of a system is the basic arrangement of the major functional items and the intent of how the device will function. The operating boundaries, the basic size, and the capabilities to function and perform activities are defined by the architecture of a computer system. The architecture cannot be ignored or arbitrarily changed without physically changing the computer. The architecture of an automated manufacturing process or a local area network are the physical attributes and definitions of the respective objects.

Also see: Computers.

Archive. To archive is to place infrequently accessed data into permanent storage. Usually done on magnetic tape, at relatively low storage cost and slow access time.

Also see: Computers.

Arc Welding. Arc welding is a welding process in which metals are heated and joined by heat produced from an electric arc between the work and an electrode/filler metal. Contact is first made between the electrode/filler metal and the work to create an electric circuit, and then, by separating the conductors, an arc is formed. The electric energy is converted into intense heat in the arc, which attains a temperature of approximately 10,000 degrees F (5500 degrees C).

An *arc welding electrode* is a component of the welding circuit through which current is conducted between the electrode holder and the arc.

An *arc welding gun* is a device used in semiautomatic, machine, and automatic arc welding to transfer current, guide the consumable electrode, and direct shielding gas when used.

Either direct or alternating current can be used for arc welding, direct current being preferred for most purposes. A DC welder is a motor-generator set of constant energy type or a transformer rectifier type (constant potential machine), having the necessary characteristics to produce a stable arc. There should not be too great a current surge when the short circuit is made, and the machine should compensate to some extent for varying lengths of the arc. Direct current machines are built in capacities up to 1000 amperes, having an open circuit voltage of 40 to 95 volts. A 200 ampere machine has a rated current range of 40 to 250 amperes, according to the standard established by the National Electrical Manufacturers Association. While welding is going on, the arc voltage is 18 to 40 volts. In straight polarity, the electrode is connected to the negative terminal, whereas in reverse polarity the electrode is positive.

In automatic welding, the equipment performs the entire operation without the need for constant observation and adjustments by an operator. Such welding is generally warranted for high production volumes, when component fits and shapes allow, and the high cost of dedicated machines can be justified. Either the work or the gun, or both, are moved in automatic welding.

Automated arc welding is welding with equipment that performs the entire operation without human observation or adjustment of controls. The process was developed to meet increasing demands for improved productivity and higher quality welds. It makes use of microprocessor memories, sensing devices, and programmable and adaptive controls. Sensing devices may continuously monitor the operation and feed back information to the control system, which modifies welding parameters or motion patterns to provide necessary changes. Quality welds are produced even when the weldment conditions are less than optimum, such as poorly fitted joints. Programs for automated welding can be recorded, stored, and reused when required.

The automated arc welding process produces more consistent, higher quality welds than automatic welding. While automatic welding can be programmed to produce quality welds, the results depend on preparation of the components for welding. This preparation is usually not consistent. For example, springback can vary with different heats of metals to be joined, bend lines can differ and change joint fitup, flame-cut parts can vary because of torch tips, cutting dies can also wear, and workpiece warpage during welding can affect results. All of these variations are accommodated with automated welding.

Industrial robots are applied to arc welding applications, primarily because of their versatility. High-volume production is not required for economical robotic arc welding. The versatility of robots makes them economically justifiable for low to medium production requirements. They are well-suited for handling a variety of welding operations, small lot sizes, and parts having a variety of welds requiring different approach angles. Design changes are easily accommodated, and different parts are handled by changing the program and end-of-arm tooling.

Robots, however, are not applicable to all arc welding applications. Assemblies requiring welds that are difficult or awkward to reach and that necessitate long reaches are not generally suitable for robotic welding. Also, unlike a skilled operator, robots made with the present state of the art do not have the ability to sense what is happening during welding nor do they have the dexterity of an operator. Rapid advances in improved sensors and software capabilities, however, are narrowing the differences between robots and skilled welders.

Work cells or systems being used for robotic arc welding consist of five principal components: robot manipulator, controller, positioner, processing package, and tooling.

Through-the-arc sensing is a common method of weld joint tracking available on arc welding robots that use the gas metal arc welding proccess. This contacting (tactile) seam-tracking method oscillates the torch across the joint and measures the changes arc current as a result of changes in electrode stickout. The method is limited to joints with sidewalls that can be tracked. It cannot put a bead on a plate or a cover pass on a joint because the joints have no edges to track. Fillet welds are ideal for through-the-arc tracking.

Gas metal arc welding (GMAW) is the predominant welding process used in robotic arc welding. Occasionally, the *gas tungsten arc welding* (GTAW) or *plasma arc welding* (PAW) processes are also used with robots. Other welding processes are rarely used with welding robots. The high-frequency arc starter used with the GTAW and PAW processes can cause interference with the robot controller, which can make the controller unreliable. Robot controllers and high-frequency arc starters must be specially designed to overcome this problem.

Hazards exist with robotic arc welding because of possible malfunctions or operator carelessness. It is essential to ensure that nothing or no one is in the path of the robot movements. Protection can be provided by the use of proper warning and feedback devices. Interlocked barriers and presence-sensing devices are commonly used to surround the robot cell and prevent intrusion.

For either manual or automatic arc welding techniques, some tooling is usually necessary in the form of jigs, fixtures, clamps, dies, or gages. These tools frequently are inexpensive, and small-lot production lends itself to temporary tooling. The tooling itself is almost invariably of welded construction, which helps minimize cost and makes for versatility and flexibility.

The primary function of welding jigs and fixtures is to bring parts into accurate alignment and to present the assembled components in the best position for welding, which is downhand wherever possible, requiring joints to be in flat or near-flat position. In welding parlance, jigs are considered stationary while fixtures rotate, usually on trunnions, about a vertical or horizontal axis, either by hand operation or through motor and reduction-gear drive. Many standard types of welding positioners and tables can be purchased or built. They are readily adaptable to wide variety or work, at lower cost than with more intricate and special single-purpose devices.

Unlike a skilled welder, a robot does not recognize and does not correct for deficiencies in the tooling. If the tool does not position the part correctly, the robot may weld the part in the wrong location. If the tool controls the part position by rigidly fixturing the part prior to welding, the distortion from welding can cause the part to bind in the fixture and be difficult to remove. For a tool to be successful, the stackup of tolerances must be controlled, the part must be accurately positioned, and the part must not bind in the fixture after welding. After meeting all these requirements, the fixture must be capable of being quickly loaded and unloaded so that operation of the fixture does not cause the robot to the idle an excessive amount of time.

Tooling may restrict accessibility to the part. The motion of the torch and the robot around the part must be investigated to ensure adequate accessibility. In some cases, part of the fixture can be removed by an air cylinder midway through the weld cycle to improve access. The operation of the air cylinder could be part of the controller program. In other cases, manually tacking the part in a fixture and then robotically welding the tacked part in a simple fixture is the best solution to the accessibility problem. Fixtures for robotic arc welding are generally custom designed for specific parts, the major requirement being to accurately locate the parts to be joined in their best positions.

Tooling problems in automatic arc welding are so dependent on the nature of the job to be done that few basic principles for guidance can be detailed. Each job must be studied and the solution worked out for it. Engineers employed by the manufacturers of automatic welding equipment are available for consultation and help in designing tooling.

John M. Menhart of Barckhoff and Associates wrote in a *Manufacturing Engineering* (August 1987) article titled *Approaching Arc Welding Automation*: "Although automated arc welding applications are undeniably part-specific, there are recurring opportunities in any mechanization of arc welding mechanization and automation projects, robotic and hard-tooled, the following five general guidelines have been identified:

- Set clear objectives
- Monitor welding variables as they pass from human to machine control
- Stay alert to incremental opportunities to increase mechanization
- Compare results with earlier methods
- Review manufacturability of the part continuously.

"Consider the development of a system as three distinct phases. First is the welding process development, second is the tooling and handling approach, and third is the manipulator and controller. Each element is linked to the others, but can be developed independently.

"The welding process currently used on the weldment may be good for the proposed system. If changes may be beneficial, however, try them on the component or assemblies to see which process works best. Most welding processes can be used on any system within broad limits—processes suitable for mechanization will normally adapt well to flexible automation. When comparing processes or electrodes, favor those that can tolerate wider changes in such variables as electrode location or angle without causing weld quality problems.

"Try each process at both the lower and upper limits of its operating ranges before settling on preferred conditions. This not only helps to find the optimum conditions, but also provides data on the flexibility of the process for evaluating the manufacturability of the part. Determination of the correct process provides a good foundation for further workcenter design. In particular, the tooling must present the part to the welding system consistently. When the operator cannot control the electrode locations and angles, the accuracy of the part and tool is much more critical. When developing tooling, consider process accessibility, ease of loading, and accuracy of both components and final weldment. The workcenter loading and unloading methods are also a major part of the total system. Parts must be where the operator can reach them easily, and they must be oriented so that loading is as simple as possible. Unloaded parts should clear the area quickly.

''Finally, the manipulator must be capable of moving the welding process to the part accurately and consistently. Primary considerations are the work envelope, range of manipulation, and repeatability of motion.

''That a part, welding process, tooling, and manipulator must function together emphasizes the need for close cooperation between all functions. Refinements in the part or tooling may be needed to improve the operation. These refinements, or course, are part-specific, but in any job it may be possible to remove or redesign gussets or brackets, to change some intermittent welds to short continuous welds, or to move a fixture stop or clamp out of the way. Subassemblies, self-locating parts, or new configurations often provide relief while still meeting design requirements. Quick review and application of ideas from all sources will result in better performance of the system.

''The supplier of the system, or course, plays a vital role in the early application of a flexible system. Communication between the robot or automation producer and the user is essential, because each has the detailed knowledge of important details in the application and can provide guidance to the other. Define the role of each in the total application to avoid confusion and problems, especially where responsibilities overlap.

''Application engineers from robot suppliers will assist with such details as work envelope and positioning for their machines. They will also know the details of their systems' software and operating features. Combining the supplier's knowledge of the system with the user's knowledge of the design and manufacturing requirements results in a more effective application. Depending on the capability of the user, the supplier may participate in development of the welding process, tooling, and manipulator.

''By considering the arc welding process variables in the automated system, the requirements for the part, tooling, and welding system are easier to define. The system must produce sound welds to be successful, and all the welding variables must be controlled effectively to achieve that result. The tests with mechanized welding will show the requirements in part fitup and tooling in order to control the variables. These tests are comparatively low risk and allow changes without substantial cost. It also starts a process of evaluation and change at each step in the process. There is usually a need to re-evaluate the costs and benefits of flexible welding systems as development work proceeds. By having experience with many different approaches, managers will find options easier to evaluate, and more accurate judgments are usually possible.

''Welding applications are among the more difficult to automate because of the range of variables. Competition is requiring better welding methods. By combining well-developed welding processes with advantages of flexible automation, it is possible to reduce the cost of arc welding.''

Arithmetic and Logic Unit. The arithmetic and logic unit is the unit of a computing system that performs logic and mathematical functions.

Also see: Computers.

Arm. An arm is a polyarticular mechanism made up of segments of solid, rigid bodies that can move relative to the base of the arm and articulations or bilateral connection systems that limit the relative movement between segments. The role of the arm is to move the end effector.

An arm is made to move by an actuator, converting primary energy into mechanical energy and thus producing a rotational or translational movement that can move an articulator in either a rotational or linear way.

An arm is made up of various segments; they may include base, body, shoulder joint, base joint, upper arm, elbow joint, forearm, wrist and gripper.

For each arm, articulation mobility depends upon a primary source of energy (electrical, pneumatic, or hydraulic), conversion of the primary source of energy into mechanical energy, transmission of the mechanical energy to the appropriate articulation, and control of the characteristic values associated with movement.

Articulated arm robots are among the most versatile available, but usually with some sacrifice in repeatability. They are available with up to six axes of motion and payload capacities from 4 to 176 lb (2 to 80 kg). Such robots are generally more expensive than other types, and their complexity can be a disadvantage. However, they are well suited for applications requiring simultaneous movement of all axes. Also, unlike the large cylindrical coordinate manipulators, they are more efficient with respect to space requirements.

ARPANET. ARPANET (Advanced Research Projects Agency Network) is a communications network and computational resources owned and utilized by the Defense Advanced Research Products Agency (DARPA).

Also see: Networks.

Array. An array is an arrangement of values in a multi dimensional pattern with an equal number of items in each row along with an equal number of items in each column. Arrays are used in programming computers as an efficient method of entering data. Programmable controllers use arrays to provide table values which must be used in a certain relationship to each other. Charge-coupled devices are used in either one dimension or two dimension arrays. A camera which can image an area may be referred to as an *array camera*.

Array processors are special purpose computers. They are designed to optimize calculations done with arrays that require considerable number crunching. Vision systems, which have pixel information to be processed, may require array processors. When the vision system must perform its functions in real time, an array processor may be required if a general purpose computer cannot perform the necessary activities in a short enough time.

Artificial Intelligence (AI). Artificial intelligence is a field within computer science whose goals are to understand human intelligence by modeling it on a computer and to build computer systems that perform in an intelligent manner either alone or in consort with a human being. Artificial intelligence includes the development of specialized hardware, software, and techniques which support those goals. An expert system is the application of artificial intelligence methods in an attempt to perform tasks within a bounded domain. Artificial intelligence is the pure science of trying to create an intelligent capability within computers, while expert systems are the application of artificial intelligence to performing some activity.

James M. Rushing, Aluminum Company of America, wrote in his SME Technical Paper, *Implementation of AI in a Basic Metals Industrial Environment*: "The AI technology which is believed to hold the greatest potential for immediate gain is that of expert systems, also known as knowledge-based systems."

Artificial intelligence is often implemented by trying to provide as much relevant information about a subject as possible to the computer and providing routines which allow the computer to understand the relationships between the pieces of information. It is hoped that the computer will be able to determine the proper solutions in a logical manner to problems presented to it. The greatest problem in trying to mimic the way the human mind works is that there is usually far more past knowledge and experiences which contribute to the decisions that people make than is readily apparent. This means that most computer based artificial intelligence systems in reality do not have as much information and knowledge as the person who performs the same reasoning.

History. In the 1950s computer scientists generally viewed computers as number manipulators, and the standard measure of the complexity of a computer program was the number of computational operations it performed within a certain timeframe. With the development of LISP, a programming language specifically designed for symbol manipulation, AI scientists began to view computers as machines capable of symbolic processing. The first AI programs proved theorems and played games, yet these programs performed no mathematical functions. These programs legitimized the symbolic processing perspective. Adopting this new perspective was a critical step in accepting the potential for machine intelligence.

In the 1950s and 1960s, a good program was viewed as a computer embodiment of an algorithm. These programs could be guaranteed to solve a problem within a given time and space bound. Nonalgorithmic programs, on the other hand, were viewed as bad programs and the result of undisciplined programming. At this time, the prevailing view was that a good programmer would analyze the problem sufficiently to know precisely the behavior of his or her programs. Unfortunately, not all problems would lend themselves to the analytical scrutiny necessary to provide algorithmic solutions. For those problems which resisted an algorithmic approach, heuristic programs were constructed that were based upon approximate, partial knowledge that *might* lead to a solution. The approach of finding an approximate solution violated a standard engineering principle—analyze the problem to find the optimal solution. Moreover, it was often difficult to define precisely space and time bounds for these programs. Thus, providing heuristic programs, which were based upon rules of thumb, helped establish AI as a separate field within computer science.

In the mid-1960s most AI programs used one fundamental technique, heuristic search. A problem is formulated as being one of combinatorial search. For each problem this requires defining a search space, a set of operators, a goal state, and heuristics for selecting which operator to use next. In chess, for example, the search space is all possible board configurations, the set of operators is all legal chess moves, and the goal state consists of the checkmate board positions. The large size of the search space for chess precludes exhaustive search; consequently, heuristics are applied to help focus the search. These heuristics are rules describing strategy and tactics of chess. By limiting the search, however, the optimal solution may not be discovered. Good heuristics sacrifice a little in solution optimality for large reductions in the size of the search space.

During the 1960s, AI scientists were primarily preoccupied with developing methods for efficiently searching large, complex spaces. Programs based upon heuristic search generally had little knowledge of the task domain. During the 1970s many AI scientists began to believe that intelligence could not be achieved by simply searching through a large space, but rather by applying task specific knowledge, or problem solving expertise. That is, the emphasis shifted from control or search to knowledge. This shift represented the birth of expert systems—programs that provided guidance in solving narrowly defined problems by applying large amounts of highly specific knowledge. Expert systems consisted of an inference engine and a knowledge base. The inference engine, which was task independent, controlled the application of task specific knowledge contained in the knowledge base. For example, MYCIN, an expert system for diagnosing a small set of bacterial infections, contained a large but highly specific knowledge base of medical expertise which was reasoned over to provide diagnostic advice in a consultative mode.

During the mid-1970s, issues of knowledge representation began to emerge. Before then knowledge was generally represented in small chunks or fragments. In 1975, Marvin Minsky argued that knowledge should be represented in frames, which are substantial collections of integrated knowledge. By the late 1970s most AI research centered around knowledge representation. In general, good representations made important things explicit, were concise and facilitated computation. During this era, finding the appropriate representation for a problem was established as one of the most important themes of AI.

Prior to the mid 1970s most AI programs were built to solve "toy" problems—small illustrative tasks within simple, completely specified environments. The most famous "toy" problem involved stacking blocks. By working on toy problems, researchers believed they could develop approaches that could be applied toward solving larger "real-world" problems. Unfortunately, this was often not the case because the underlying assumption of toy problems, that the environment could be completely and explicitly represented, was generally violated in real-world problems. Currently, most AI programs are designed to solve complex real-world problems where complete knowledge of the environment is unavailable; heuristics or "rules of thumb" specific to the domain play a large role in dealing with the otherwise perplexing incompleteness.

Discussion. Researchers in vision and robotics are attempting to devise methods to enable a computer to sense and manipulate its environment intelligently. Vision systems recognize and reason about objects surrounding the viewer. The data available to a vision system consists of two-dimensional matrices of light intensities. From this two-dimensional representation, three-dimensional properties such as distances between surfaces, orientation, structure, and texture are recovered. This process is called *early vision*. These three-dimensional properties and others construct a model of the viewer's environment that will enable the viewer to navigate and manipulate objects. This process is called *high-level vision*.

Many questions concerning the mechanisms of early vision have been rigorously formulated. These questions include how to derive structure from stereo, structure from motion, structure from texture, and shape from shading, how to accomplish edge detection, and how to compute optical flow. Answering them provides spatial and geometrical information about the three-dimensional world. This information is sometimes referred to as the *two and one half-dimensional representation*. Early vision is generally considered to be a bottom-up process; that is, only low-level information about the scene is needed to compute a two and one half-dimensional representation.

High-level vision processes produce a world model from the two and one half-dimensional representation. Unfortunately, questions concerning the mechanisms constituting high-level vision have not been rigorously formulated. The world model provides the information needed for navigation in the environment, manipulation of objects, coarse object recognition, and reasoning about objects. Unlike early vision, high-level vision is generally considered to be a top-down process. That is, high-level vision depends heavily on firm expectations about what will be seen.

A robot with a world model will be able to manipulate its environment intelligently. The problem of moving a manipulator along a prescribed path requires the application of kinematics—relating joint angles to manipulator position and orientation. The manipulator motors apply torque. The computation of torques from joint angles is called dynamics. The formalization of dynamics is complicated and related to Newton's second law: force = mass x acceleration.

Besides kinematics and dynamics, robots also need spatial reasoning. Spatial reasoning is needed if the manipulator must move through a cluttered environment. Spatial reasoning has been simplified with the idea of configuration space objects, an alternate representation of the environment. The essence of the idea is that it is much easier to move a point, rather than a three-dimensional object, through an environment. The alternate representation is the result of enlarging the objects to be avoided and shrinking the manipulator until it becomes a point. The enlarged objects are referred to as *configuration space objects*. The spatial reasoning problem has now been transformed into a much easier problem of moving the point representing the manipulator through a space of enlarged objects.

The ability to learn is generally considered to be at the core of human intelligence and has provided the greatest challenge to AI researchers. There are three basic approaches to machine learning: parameter adjustment, concept formation, and evolution of structure.

The simplest and most popular approach adjusts values of parameters of a predetermined representation. Parameter adjustment is useful for some tasks, but it is clearly limited as a learning technique. The major limitation is that it can only affect the parameter values, not the structure of the predetermined representation.

The second approach, called concept formation, is motivated by people's ability to group related objects into categories. People learn concepts through experiences with members and nonmembers of the class denoted by the concept. A shortcoming of this approach is that the learnable concepts are limited to those that can be represented as combinations of the predetermined atomic objects and properties.

The third approach to learning involves neurocomputing, a style of mathematical computation modeled after the human nervous system. The defining feature of this style is parallel activity of large numbers of elements that communicate results among themselves. Systems based on this model, consist of many connected elements whose initial structure asks a representation. Systems created with this approach also employ parametric adjustment, but at the level of individual elements.

Applications. Richard J. Mayer, Don T. Phillips, and Robert E. Young of Texas A&M University report in their SME Technical Paper titled *Artificial Intelligence—Applications in Manufacturing*: "The techiques and theoretical results from the field of artificial intelligence (AI) offer a new and exciting technology for solving problems in manufacturing today. The technology however, is still in its infancy. Careful analysis of the characteristics of a problem area must be made to determine if the technology can actually be applied. For maximum benefits to be achieved, the AI-based applications must be integrated with existing manufacturing systems and practices.

"From the point of view of the solution to the manufacturing productivity problem, AI technology can provide two major contributions. The first contribution comes from the insight and understanding which AI theories provide us as to how people do planning, resource allocation, and general problem solving. By evaluating the AI results in these areas in the context of the manufacturing environment, we can develop a better understanding of the type and form of information which must be provided to improve the accuracy, timeliness, and effectiveness of manufacturing planning and control activities.

The second (and probably the more visible) contribution of the AI research to date comes from the potential for direct application of the computer algorithm and information representation schemes directly to manufacturing problems. These applications fall into three major classes:

1. Those applications which attempt to duplicate natural human abilities (vision, language processing, etc.).
2. Those applications which attempt to duplicate learned skills or expertise.
3. Those applications which simply improve the flexibility or effectiveness of traditional information management systems.

Research reported in the *Journal of Manufacturing Systems* has led to a working program based on AI techniques for automatically writing a part program for milling.

K. Preiss of the Ben Gurion University of the Negev (Beer Sheva, Israel) and E. Kaplansky of the Man-Machine Cooperation Limited (Beer Sheva, Israel) reported on this development in their paper titled *Automated Part Programming for CNC Milling by Artificial Intelligence Techniques*. The authors explained: "Research has led to a working program based on artificial intelligence techniques for automatically writing a part program for milling. The input to the program is the graphic representation of the part (a drawing), and user-defined items such as tool details, material type, and so on. The program has an initial state, the shape of the part. The program solves the problem of achieving the goal state from the initial state by using machining moves, and hence writes the part program. The current implementation produces a part program for a 2 1/2-dimensional part on a 3-axis CNC milling machine.

In his SME Techical Paper *Computer Vision Inspection from a CAD Database*, author Steven B. Curtis (Digital Equipment Corporation) reported: "The AI-1 module uses artificial intelligence techniques to derive camera locations and views from the intermediate database. It also ensures that the inspection system does not physically hit the part under inspection. A person can quite easily position a vision system and analyze its output. The exact algorithms that people use are not well understood, though. Therefore, a set of heuristic rules were derived by watching people position a vision system, and measure dimensions from the video output. The rules simulate the way people moved the vision system to inspect each feature."

At this time, while robots have some potential using expert systems, the robot/AI overlap is very little.

AI tends to work on high-level problems (such as planning) while robots deal with lower-level problems (such as compliance or friction).

Senior Editor B. Lee Martin discussed the future of AI as it related to CAD/CAM in an article in the December 1987 *Manufacturing Engineering*. In this feature titled *CAD/CAM-An Even Fuller Menu Ahead*, he writes: "How will artificial intelligence and expert systems impact CAD/CAM in the year 2000? It is obvious that future CAD/CAM systems will be smarter. They will have the ability to manage data from design through manufacture. Through the determined implementation and execution of rule-based artificial intelligence systems, the goal to establish a viable and practical connection between CAD and CAM will be significantly closer.

"Artificial intelligence and expert systems will build rules into software that will make it possible to let the software have more control and expand its capabilities. These new software systems will help to further automate the factory floor by making it possible for the design engineer to ned in mind the needs of the shop floor when proposing designs. This in an area where feature-based modeling will help."

He also noted: "Expert systems, according to some, have been overpromised and underdelivered. The academic definition of an expert system is a ruled-based or knowledge-based system where one utilizes a shell like Prolog or one of the other programs into which one inputs design rules and gets out guidance or a solution for an engineering problem. Another definition, a broader definition, says don't limit us by the tools, but instead think of capturing engineering expertise into a computer program. And forget about whether it was done in Prolog, Basic, or whatever. The academic definition is so restrictive in the type of engineering or manufacturing problem that can be solved by expert systems that it seems to narrow the applicability and benefits. If the broader definition is used, then special projects or special programs written on top of a vendor's software in a programming language allow an engineer or designer to take that software and top it off with a little programming designed in-house for each special application."

Robin P. Bergstrom of *Manufacturing Engineering* magazine discussed the future of AI in the January 1987 issue. The article titled: *AI: Shifting into High Gear* noted: "Projections for the AI Industry certainly seem bright on the surface. For example, DM Data, Inc. (Scottsdale, AZ) forecasts a market for AI products of more than $4 billion by 1990. It indicates that real growth will take place soon after 1986, as the cost performance of AI products improves. Since 1980, the industry has already achieved a startling growth rate of approximately 50% annually.

"International Resource Development, Inc. (Norwalk, CT) places the current market for AI services, software, and hardware at $725 million. By 1991, the market should grow to $4.2 billion.

"And a report issued by Arthur D. Little, Inc. (Cambridge, MA) suggest market by 1990. Further, in 1990 AI could account predicted for total computer revenues and employ 100,000 workers providing AI services and equipment.

"If the long-range market for AI looks promising, the short-range market by comparison is not so great. This largely because AI is still relatively new (conflicting definitions of AI, for example, are common) and much of the research on the subject is yet being done in university laboratories. Then, too, because AI is still in the research, it is so expensive, requiring substantial investment in equipment and personnel, the amount of interest being devoted to the technology by profit-oriented companies is, for the present, just beginning to grow.

"To be sure, a number of large corporations are funding their own AI projects—General Electric, IBM, Digital Equipment Corp., Xerox, Hewlett-Packard, Bell Laboratories, to name a few. In fact, as of 1986, two-thirds of the Fortune 500 companies have undertaken some level of AI research. And government agencies, such as DARPA (Defense Advanced Research Projects Agency), DoD (Department of Defense), and SCI (Strategic Computing Initiative), fuel the fire to one degree or another. University funding at institutions like Stanford, Carnegie-Mellon, and the Massachusetts Institute of Technology continued at whatever levels can be maintained, given the nature of academic budget constraints.

"Who, then, will fund the movement of AI from the research lab to the marketplace in the short term? Perhaps, as in other advanced technologies like robotics, vision, software, and sensors, the entrepreneur and the venture capitalist. Is there a tide of investors out there waiting to swell? No doubt. The point, however, is that by and large, they're waiting. Artificial intelligence, like another technology, old or new, will have to meet the strict, cold criteria of the real-world venture capitalist.

"The verdict on the future of AI may yet be out, but the projections make this much clear: Artificial intelligence, any way you look at is, is anything but a passing fad or fancy."

Technical paper author James Rushing wrote: "As large complex systems are approached in the industrial world, expert systems will tend to become very large with large number of rules and may start to slow down. Many processes will not be able to accept slower operation, particularly if they are operating in a real time regime. It may be advantageous to approach large systems from a distributed processing standpoint. This would allow the problem to be broken into a large number of smaller segments where each might run on its own dedicated microprocessor and communicate with a higher level controlling expert system through a blackboard architecture. This could continue to provide for high-speed operation of the various segments, yet attain the advantages of the level of coordination which only comes with the larger system. This type of structure would fit very nicely into a computer integrated manufacturing (CIM) environment. It is conceivable to have a very high level diagnostic expert system at the top level of the CIM hierarchy which would, upon completion of a diagnosis, direct the user to the next lower level in the hierarchy, continuing down through the various levels until it finally diagnoses a problem down at the machine or cell controller level.

"As the user community looks for potential applications for artificial intelligence, it is a good idea to develop a prototype system that can be used as a demonstration to sell the use of the technology to management and to colleagues. The first system need not be complex, but should be in an area that can have an impact on the cost of operations. If a system can be demonstrated that has the potential of paying the entire investment back from savings generated from one diagnostic consultation, considerable support will be gained for implementation of the technology. In a capital-intensive industry, the elimination of one instance of several hours of downtime on a major piece of equipment can easily recover the investment in a PC-based diagnostic expert system,'' he said.

Glossary. *algorithm:* A program that is guaranteed to solve a problem within given time and space bounds.
backward chaining: A goal-driven form of inference in which a query concerning the consequent or ''then'' portion of a rule establishes a search to find a literal match to the antecedent or ''if'' portion of the rule. The antecedent is established by matching it against facts or by further backward chaining. The process is repeated until no more matches can be found.
early vision: The computations that provide spatial and geometrical information.
expert system: An AI program whose goal is to simulate, to some extent, the intelligence of a human being as it is applied in the performance of a task, or as it is applied in supporting the performance of a task.
forward chaining: A form of inference in which the antecedent or ''if'' portion of a rule is matched against facts asserted in the global database to establish the truth of the consequent or ''then'' portion of a rule.
frames: Substantial collections of integrated knowledge about the world.
heuristic: Rule of thumb, strategy, trick, simplification, or any other kind of device which drastically limits the search for solutions in large problem spaces. Heuristics do not guarantee optimal solution.
heuristic search: A search problem which is focused with knowledge of heuristics. Often heuristic search techniques prevent combinatorial explosion of possible solutions.
higher-level vision: Tasks such as navigation in the environment, manipulation of objects, object recognition, and reasoning about objects.
inference: Reasoning from premises to conclusions.
knowledge engineering: The process of eliciting knowledge from individuals who are deemed to be experts in their field, and structuring that knowledge into a computable form.
neurocomputing: A style of mathematical computation modeled after the human nervous system.
production system: The computational model consisting of three major components: a global database of assertions, a rule-base of knowledge about the problem domain, and a rule interpreter that controls the problem solving process.
semantic net: A knowledge representation scheme which denotes objects and describes the relations which hold among them. The syntax consists of objects (nodes) and relations between pairs of objects (links). The semantics tends to be informal and intuition-based.
semantics: The meaning of a word as it is used in a phrase.
syntax: The pattern of sequences of tokens (words, punctuation marks, special symbols, characters) which are accepted by the grammar of a language.
symbolic processing: The view of computing as symbol manipulation, rather than number manipulation.

Also see: Computers, Expert Systems, Heuristic, Machine Vision, Industrial Robots, Software.

Artwork. Artwork can be graphical pictures of an electrical circuit which is to be photo-etched onto a printed circuit board or onto a silicon chip integrated circuit. The artwork is used for the photographic process and the visual quality of the artwork can affect the quality of the finished product. Artwork is produced by the computer aided design (CAD) system after the schematic design has been transformed into a physical layout and the routing of the circuit has been completed. The artwork is the final product from the design process which is passed along to the manufacturing process.

Also see: CAD, Postprocessing.

ASCII. See: American Standard Code for Information Interchange.

ASIC. See: Application Specific Integrated Circuit.

Aspect Ratio. The aspect ratio is the mathematical percentage which results from comparing an object's height to its width. An aspect ratio of 75 would indicate an object with width of one quarter of its height. Aspect ratio can apply to rectangular objects or round items such as automobile tires. By using the convention of defining the aspect ratio of tires, the width can be interpolated if the diameter of the tire is known. Certain thresholds for aspect ratios are used as guides and "rules of thumb" for such things as ensuring stability of an object by defining the maximum aspect ration (which determines the center of gravity) or determining the optimum width of a tire to have a desired pressure-per-square inch combined with desired flotation.

AS/RS. See: Automated Storage and Retrieval System.

Assembler. Assembler usually refers to the translator program which converts assembly language programming into executable machine language. Assemblers exist for virtually every microprocessor available. Cross assemblers are assembler program which can be run on one machine to assemble the code for a different target processor. The purpose of using cross assemblers is to allow the program development to occur on a larger, faster computer more suited for program development. The code may be then developed for a smaller microprocessor which can adequately execute it, but would be impractical to use as a development system. When the target hardware is satisfactory to use as a development system, possible problems with execution can sometimes be eliminated. A programmer using an assembler first writes the source code with an editor and then uses the assembler to assemble the source code into executable machine code.

Also see: Assembly Language, Compiler, Computer Languages, Computers.

Assembly Language. Assembly language is a programming language which is a highly procedural low-level language. Assembly language is translated by an assembler into machine language. As more high-level languages become available with low-level features, assembly language loses its attractiveness. Assembly language is different for each microprocessor. Assembly language is often confusing to learn due to its uniqueness. The language is much easier to understand if a person is knowledgeable about the computer hardware. Assembly language is available for every existing computer, although this software must be purchased separately for most microcomputers.

Assembly language was the first attempt to develop a method of programming without having to communicate with the computer directly in the machine language of ones (1) and zeros (0). Since assembly language is closer to machine language than higher level languages, programs written in assembler usually run faster since there are fewer stages of translation. Assembly language also gives the programmer greater control over what is happening in a program than do high-level languages. Another benefit, which is not of much recent concern is that assembly languages use less memory space than high level languages.

Also see: Assembler, Computer Languages, Computers.

Assembly Line. An assembly line is a sequential line of machines and/or manual operators that perform the necessary steps to assemble a part or product. The assembly line was first utilized by Henry Ford to produce Model A cars. The assembly line uses the philosophy that to have each operator perform a simple operation repetitively is a more efficient method of production than the individual craftsman approach where a single operator performs an extensive assembly operation of an entire product. The thought is that the operations are partitioned to created the greatest efficiencies for each operation.

In today's society of desiring meaningful work, job satisfaction, and greater job skills, the assembly line has begun to be reshaped into islands of skill and total product responsibility. The team approach to assembling a product with flexible systems and reconfigurable lines is beginning to prevail in many industries. Some industries and manufacturing quantities lend themselves very well to traditional assembly lines while others may be better served by the work cell approach.

Also see: Automated Assembly, Assembly Machine, Cell.

Assembly Machine. An assembly machine is a machine which performs an assembly operation. Examples of assembly machines are automatic component insertion equipment for electrical circuit boards or special purpose machines designed to assemble products such as printers, motors, or automobile subassemblies. Any machine that performs assembly operations can be categorized as an assembly machine.

Assembly machines require the presentation of all necessary components to assemble the desired part or product. Conveyors sometimes provide a convenient means of doing this and even transporting the finished product from the assembly machine. Some assembly operations such as assembling an integrated circuit chip, require the operation, along with all required equipment, to be located in a clean room to prevent contamination of the parts or products by dust particles during assembly.

As automation continues to be refined, assembly machines are becoming more complex and more capable. Some of the automatic assembly machines being used today can operate for more than a full shift without being tended by a human operator.

Also see: Assembly Line, Automated Assembly, Automation.

Assembly Robot. An assembly robot is a robot used primarily to assemble a device, subassembly or product as opposed to robots that are used for welding, machine loading, material movement, palletizing, spray painting and other activities. Assembling usually requires greater dexterity than many other tasks a robot may perform. The accuracy requirements and repeatability of an assembly robot are substantially greater than a palletizing or welding robot. A result of lapse in repeatability in a spotwelding robot might simply place a spot-weld in a different, but within tolerance, location while the same amount of inaccuracy in an assembly robot could either damage the parts, or, at the very least, simply not be capable of completing the assembly task. Assembly robots sometimes have other systems and aids such as vision, testing, or marking systems integrated in them. Assembly robots may be any size but usually are the smaller to midsized electric drive robots with servocontrolled joints.

Also see: Adaptable Programmable Assembly System, Assembly Machine, Automated Assembly, Industrial Robots.

Association for Finishing Processes of SME. The Association for Finishing Processes of the Society of Manufacturing Engineers (AFP/SME) was founded in 1975 to provide a resource to engineers working in the area of finishing process design and operation. It is an organization whose members are working in both contract finishing shops and captive finishing departments in more than a dozen finishing fields.

AFP/SME offers finishing professional educational resources and a forum for technical interchange. The association is designed to serve the educational needs of the individual member by following and assessing the trends and developments of the individual finishes and interpreting, publishing, and distribution any new information that develops in this rapidly changing field. AFP/SME conferences, publications workshops, and expositions serve to keep end-users of finishing equipment current of innovative applications for improved product quality and company productivity.

Associative Memory. Associative memory is a storage device in which a storage location is identified by its data content rather than its address.

Also see: Data.

Asynchronous Communication. Asynchronous communication is a type of data transmission which has an absence of continuous synchronization between the sender and receiver.

This allows a data character to be sent at any time without regards to any previous timing signal. Asynchronous communication is commonly found in unbuffered machines or terminals such as teletypes or teleprinters and low-speed computer terminals. Most of the modem communication used by personal computers is asynchronous. Its value lies in its simplicity. A disadvantage of asynchronous is that it is cannot achieve reliable transmission speeds as great as synchronous, over the same physical conditions.

An important part of any communications system is the clocking device. It continuously examines or samples the line for the presence or absence of the predefined signal levels. It also synchronizes all internal components. Asynchronous communication uses one clock source which is located at the receiving node, for the transmission circuit.

Also see: Start Bit, Stop Bit.

Asynchronous Operation. Asynchronous operation denotes that when multiple events occur, each event occurring with no regard for when other events occur. Asynchronous operation can refer to data communications messages, assembly processes or materials movement. The term asynchronous simply means not synchronous, or not occurring or happening at precisely the same time or period.

A push-along assembly line is an example of an asynchronous assembly process. Any process or procedure with buffers is usually capable of asynchronous operation, since an operation is not necessarily required to occur at precisely the same time as another. A transfer line or the final assembly line of an automobile plant are not examples of asynchronous operation, since the time window in which events must occur is not determined by each event.

A standard for serial asynchronous communication is RS232-C. Asynchronous communication can be controlled by a USART (Universal Synchronous-Asynchronous Receive-Transmit) or a UART (Universal Asynchronous Receive-Transmit) controller chip.

Also see: Universal Synchronous-Asynchronous Receive-Transmit.

ATC. See: Automatic Tool Change.

Attenuation. Attenuation is the decrease in power of a signal, its light beam, or lightwave, either absolutely or as a fraction of reference value. The decrease usually occurs as a result of absorption, reflection, diffusion, scattering, deflection, or dispersion from an original level, and usually not as a result of geometric spreading, i.e., the inverse square of the distance effect. In an optical fiber, attenuation is undesirable for transmission purposes but desirable for prevention of leakage or clandestine detection. Optical fibers have been classified as high-loss (over 100 dB/km), medium loss (20 to 100 dB/km), and low-loss (less than 20 dB/km).

Some causes of attenuation in radio signals can be clouds, echoes, and precipitation. An *attenuator* is the name given to a device which has an output signal that is a function of, and less than, its input signal. The attenuation is expressed as a transfer function and often in negative decibels. If the output is greater than the input, that is, if there is a gain in signal strength, the device is not an attenuator but is an amplifier.

Attribute. An attribute is a characteristic. For example, attributes of data include record length, record format, data set name, associated device type and volume identification, use and creation data.

Also see: Data.

Audit Trail. An audit trail is an historical record of activities or events which are recorded in a retrievable manner. This trail of information can be in a manual (source document) or data in a computer file.

The review of data in the audit trail takes the form of an official or a formal examination of accounts/transactions resulting in either a confirmation or reconciliation of that data. This activity entails a systematic investigation and appraisal of the transactions throughout the process. These steps will lead to the originating source documents, and result in substantiating individual transactions a well as lead to potential impact on related information in the process. Transactions identified for an audit can be made through a sampling selection process or by exception reporting from specified limits to be satisfied. The existence of backup documentation for these items becomes the audit trail itself.

Transactions with security sensitivity in a computer system have a step by step history with source documents, such as electronic logs and access records to restricted files. Specific steps are recommended and changes made to the documentation to prevent any loss of an audit trail. The loss or destruction of stored data in a computer system creates a significant loss in an audit trail. Key information, such as quantities or dates critical in maintaining a proper audit trail.

Also see: Computers, Data.

Auto-interactive Design. Auto-interactive design is when a manual operator interacts with an automated design process. The majority of auto-interactive design processes use templates and logical flow to guide the user through a standard design process. The user may have several options to choose from at various points in the design sequence. But the basic process is standardized such that a computer can be taught the basics to lead the user through the process.

Some types of auto-interactive design processes are the auto-routing of connections in the a design of a printed circuit board, where the user has the option of manually intervening at any point and can take over when the automated capabilities cannot continue. Some mechanical CAD design systems have the capability to automatically calculate the optimum profiles and paths for the machining tool to follow when creating the part. In the design of parts with several restrictions and requirements for attribute size and shape, the design system will automatically modify the designated sizes and shapes when the user tries to use an improper size or shape.

Also see: CAD, CAE.

Automated Assembly. Assembly in the manufacturing process consists of putting together all the component parts and subassemblies of a give product. Assembly includes fastening, performing inspections and functional tests, labeling, separating good assemblies from bad, and packing and/or preparing them for final use.

Assembly is unique compared to the methods of manufacturing such as machining, grinding, and welding in that most of these processes involve only a few disciplines and possibly only one. Most of these nonassembly operations cannot be performed without the aid of equipment, thus the development of automated methods has been necessary rather than optional. Assembly, on the other hand, may involve in one machine, many of fastening methods such as riveting, welding, screwdriving, and adhesive application. Automatic parts selection, probing, gaging, functional test, labeling, and packaging are also involved. The state of the art in assembly operations has not reached the level of standardization. Much manual work is still being performed.

Jack Lane, President of Robotic Integrated Systems Engineering, Incorporated, and the editor of the SME book *Automated Assembly*, (Second Edition) writes in the book's Preface: ''Assembly, though used extensively by all industries producing a finished product, is probably the least understood and most poorly documented technology applied in the manufacturing community. Very few colleges and/or universities offer courses or curriculums on this subject, due mainly to a lack of documented information available in the literature. Assembly is still thought of as being an ''art'' with very little documented scientific evidence to support any conclusions to the contrary.

''Prior to the last five years (1981-1986), very little new information evolved in the field of assembly and/or assembly processes. Manual assembly is predominant over hard automatic assembly techniques due to inherent assembly problems and the fact that these assembly machines lack the innate intelligence of a human operator and lack sufficient flexibility to changeover when product designs and market demands change. However, in the past five years, more and more emphasis and attention has been focused on assembly. Although many reasons are probably responsible for providing this impetus, new concepts, industrial robots, and the vision technology have predominated in terms of offering new approaches and capabilities for improvements in assembly operations.''

Authors William Beeby and Phyllis Collier in their book *New Directions Through CAD/CAM* wrote: ''Some of the highest direct labor costs for manufacturing in American industry accrue in assembly operations. For this reason, companies are looking to automated assembly processes in increasing numbers. Productivity figures associated with automatic assembly, as well as greater capabilities and reliability of the systems, encourage companies to use automated processes.''

In the SME *Manufacturing Insights Video Series*: *Automated Assembly*, Frank Riley, Senior Vice President of The Bodine Corporation, said: ''When you're talking about parts assembly, each individual component part has certain characteristics and it is the ability to handle the variations in part pieces, often coming off the same tool, that makes automatic assembly so difficult to achieve.

''Usually in assembly of products, people who enter (the field for) the first time find it difficult to realize that there is often failure of the actual insertion process which makes the product incomplete or defective. And very few assembly systems have the ability to correct an insertion defect. Usually, they have to segregate that assembly for possible hand repair, ejection to repair loops, or segregation for salvage or scrapping.

''It is this inability to self-correct, that makes automatic assembly somewhat unique and (the) unwillingness to accept this, leads to alot of problems for customers who look for absolute product guarantees on systems.''

Added Riley: ''It is very, very difficult when one buys a so-called automatic assembly system to guarantee specific net production.''

In the tape, Riley goes on to explain that automatic assembly systems can be expensive to set up, especially when a company can move off-shore to areas where manual assembly labor is cheap.

''On the other hand,'' he continued, ''there are several other factors leading to automatic assembly. At least on a corporate level, there are three major thrusts. One thrust is consistent and uniform quality levels. People have an expectation that if they buy a product, what they buy is going to work. Automatic assembly is a step in the consistency in the assembly process.''

He added that ''a second thrust is the reduction of in-process inventory. And the last item is, of course, more rapid response to customer order. Manufacturers want reduced inventory but they also want the ability (that) if a customer calls to be able to shoot large quantities of a product out very rapidly.

''Automatic assembly has a unique capability of addressing these three economic considerations. So we're seeing a trend away from direct labor reduction to the areas of quality, inventory control, and rapid response to customer orders,'' he said.

History. Assembly has traditionally been one of the highest areas of direct labor costs. In some cases, assembly accounts for 50% or more of manufacturing costs and typically 20-50%. However, closer cooperation between design and manufacturing engineers has resulted in reducing, and in some cases, eliminating altogether then need for assembly. When assembly is required, improved design or redesign of products has simplified automated (semiautomatic or automatic) assembly.

Discussion. Before automated assembly is adopted, several factors should be considered. These include practicality of the process for automation, simulation for economic considerations and justification, management involvement, and labor relations.

Determining the practicality of automated assembly requires careful evaluation of a number of factors. Chief among these are: the number of parts in the assembly; design of the parts with respect to producibility, assemblability, automatic handling, and testability; quality of parts to be assembled; availability of qualified personnel to be responsible for equipment operation, and total production and production-rate requirements.

Other factors include: product variations and frequency of design changes; joining methods required; assembly times and costs, and assembly line or system configuration, using simulation, including material handling.

The best candidate for successful and economical automated assembly are generally simple, small products that have a fairly stable design life. Such products are usually required in relatively large volumes and have a high labor content and/or a high reject rate because of their manual assembly. However, the development of flexible programmable, and robotic assembly systems can decrease production and product-life requirements.

The *Tool and Manufacturing Engineers Handbook*, Fourth Edition, Volume 1 lists potential advantages resulting from assembly automation. They are included in the following figure.

The high initial cost of automated assembly machines, systems, and equipment has been a deterrent to increased use, especially for smaller factories producing limited quantities of various products. However, the development of flexible manufacturing systems was expected to make automated assembly more economically justifiable.

Authors Peter B. Scott and Tom M. Husband of Imperial College, addressed the issue of economic evaluation in their SME Technical Paper titled *Robotic Assembly: Design, Analysis and Economic Evaluation*. The authors wrote: "Cost justification is further complicated by what a given company accountant will accept. Those used to dedicated machinery will expect early obsolescence and be unfamiliar with how best to satisfactorily take into account the robot's far greater flexibility. Likewise, straight line depreciation may sometimes be used, but the tax schemes of different countries may involve depreciation artificially weighted to early years, and special tax credits to encourage capital investment may also influence cost justification and the decision whether or not to buy."

Automatic assembly cannot be divorced from preceding operations because clean, consistently uniform, and high-quality parts are required for its success. The parts do not necessarily have to be held to closer tolerances, but uniformity from part to part is essential. Complete inspection of components on assembly machines is generally not practical or economical. It is usually better to

POTENTIAL ADVANTAGES RESULTING FROM ASSEMBLY AUTOMATION

1. Improved product quality, consistent product repeatability with fewer rejects, and a high degree of production reliability as a result of reducing or eliminating human errors. Component inspection and part testing during assembly prevents defective parts from being used. The consistently highquality products obtained reduce liability and warranty costs.
2. Reduced manufacturing costs resulting from decreased labor requirements and increased productivity. Savings result from reduction in both direct and indirect labor.
3. Improved safety and better working conditions by removing operators from hazardous operations.
4. More efficient production scheduling (such as Just In Time techniques) and reduced inventory requirements because of the ability of automated assembly systems to respond immediately to production demands.
5. Reduced floor space requirements.

Some possible advantages of automating assembly operations.

inspect the parts prior to assembly; only their critical dimensions, locations, and presence on assembly machines need be checked.

In addition to problems caused by excessive variations in parts, random machine failures (occasional misfeeds or incomplete assemblies) may be due to the presence of chips or foreign material, poor hopper construction, misadjusted or worn transfer units, or erratic operation of tooling. When assembly is combined with machining operations, care must be taken to prevent chips from entering the assemblies. Burrs, flash, or distorted parts, as well as thin or fragile parts, can be troublesome because they might interfere with tracking or escapement. In some cases, trouble may be avoided by prestacking the parts on mandrels, inserting them in magazines on the assembly machine, and using a shuttle mechanism to transfer parts (one at a time) from the top or bottom of the stack.

A major problem inherent in automating assembly is the incompatible design of the product and the resulting producibility of a given assembly. The design of products and their components will have more effect on their manufacturing costs than all the equipment and processing combines. Unfortunately, product design is usually concerned primarily with functional performance; the producibility of the product is secondary or neglected altogether.

Product designers working in close cooperation with manufacturing engineers in the initial design or redesign of a product and its components will play a major role in the success of automatic assembly. Parts should be designed for easy handling and orienting, with the provision of locating points, sufficient clearances for the assembly tooling, and realistic tolerances (including torque and press-fit tolerance ranges). In some cases, fasteners can be redesigned or even eliminated to facilitate assembly.

J. Ronald Bailey, IBM Corporation, wrote in his SME Technical Paper *Product Design for Robotic Assembly*: ''Increasing international competition is a primary driving force in today's manufacturing environment. As industry strives to meet this competition, tremendous pressure is now placed on the designer to deliver innovative products which can be manufactured in a most efficient manner.

''In addition, the design/development cycle has been compressed in many industries because of the accelerated rate at which new products are introduced. This accelerated pace puts great emphasis on 'doing things right the first time' since product life tends to be short and production decisions tend to be based on delivery schedules. There is a temptation in this environment to relax design requirements for producibility, particularly when sourcing decisions are made prematurely. For example, design for assembly may be given a low priority if it has been decided that a product will be manufactured off shore. However, this relaxation may prove to be strategically foolish because of the exposure to value engineered copies by the competition with the accompanying pressure on profit margins and market share.''

Henry W. Stoll of the Industrial Technology Institute wrote in his SME Technical Paper *Automation: Teach and Test for Good Product Design*: ''In the late 1970s, American manufacturers suddenly realized that their preeminent position in the global marketplace was being challenged by declining U.S. productivity, a threat from world-class manufacturers, and the emergence of new automation processes. This precipitated a quest for short-term solutions that quickly focused on fledgling technologies such as robotics, vision and optical processing, and flexible manufacturing systems. Failure to achieve promised productivity improvements using these and other advanced automation technology in a variety of industrial settings has taught much about the product design/automation interface.

''Perhaps the most important lesson learned is that the design, function and implementation of advanced automation technology is directly related to the product being manufactured. Hence, when implementing these technologies, product design and automation design can no longer be treated as separate entities, but rather, must be integrated into one common activity. It is especially essential that interactions between product and process enter into consideration very early in the product design process. Product concept and automation decisions must be made in parallel in order to obtain an integrated manufacturing system optimally configured to satisfy both product and process needs.

"Design for automation, or 'automation friendly' design, is concerned with defining product design alternatives which facilitate optimization of the product/process as a whole," he wrote. Stoll went on to describe an approach to "automation friendly" product design which involves the application of concisely stated design principles and guidelines to each decision-making step. The approach has three objectives.

First, identify the product concepts which are inherently easy to automate.

Second, focus on component design for ease of automated processing.

Third, integrate automation system design with productivity design to achieve optimal matching of needs and requirements.

"In developing this approach," wrote Stoll, "the problem of designing for automation is first briefly reviewed. A set of codified design principles and guidelines, distilled from automation experience, are then concisely stated and discussed to show how they can be used to both teach and test for automation friendly design.

"Using these principles and guidelines as a basis, a design for automation methodology is then proposed and its use discussed," he said.

Another important consideration in assembly automation is the system concept and selection. Many factors influence this consideration, such as volume of parts per unit of time, product life, frequency of design changes, available labor and costs, management attitudes, and competitive pressures. The suitability of automatic assembly for a specific product also depends on total production requirements, the complexity of the assembly, the number of parts in the assembly, and the cost of handling, feeding, orienting, inserting, and joining the parts. Assembly machine costs generally increase in proportion to the number of parts in the product.

In the area of robotic assembly, Baily wrote: "The most important general design guideline for robotic assembly is to minimize the number of parts. Fewer parts to procure, fabricate, transport, orient feed and assemble will almost always lead to lower product costs."

Value analysis programs are being used to determine simpler product designs for manufacturing methods that can reduce costs. The most effective technique for minimizing the cost of assembly is to reduce the number of parts to be assembled. The function of all parts in an assembly should be determined and consideration given to the possibility of combining the functions of several parts into a single component. When considering the economic aspects of parts count reduction and integration, also consider the total cost of each part (direct and indirect costs), not just a purchased or fabricated cost.

The optimum product design is one that eliminates the need for assembly or reduces the number of parts to be assembled to a minimum. One simple example is illustrated in the figure below which shows a single stamping that replaces a two-part assembly. Such designs usually reduce total product and assembly costs.

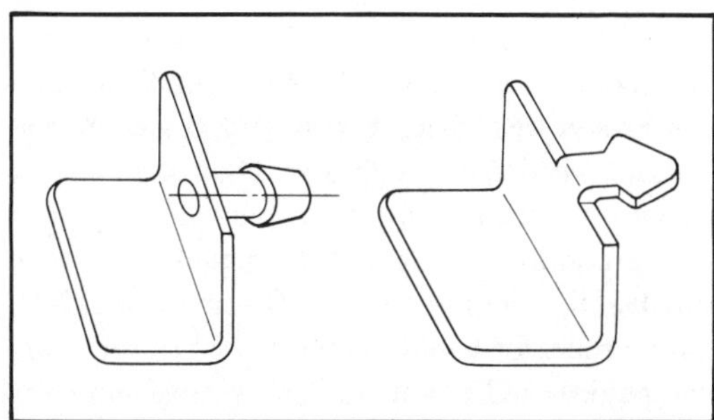

Two-part assembly is replaced by a single stamping.

When single-component products are impossible or uneconomical, the number of parts required should generally be kept as low as possible, and their complexity would be minimized. There are rare occasions, however, where it may be more economical to manufacture two or more pieces to replace one. The reason for minimizing the number of parts is to improve the remaining, more functional parts and to eliminate nonfunctional ones.

In the area of industrial robotics, the robot is available with electric, pneumatic, or hydraulic actuators to move the robot arms through desired paths. Servoelectric drive mechanisms are the most popular drive for assembly applications. However, pneumatic or hydraulic drives may be economically desirable for some applications if clean air or oil pressure sources are available.

Robots may or may not be the wave of the future in automated assembly. Gerald Michael, Senior Consultant, Arthur D. Little, noted in the *Automated Assembly* videotape: "When you compare the cost of robotic assembly against traditional manual methods or high-volume dedicated machinery, robots usually lose out."

A variety of control products are being used for the automation of assembly operations. Primary objectives of the controls include maintaining or improving product quality, increasing output, reducing inventories and scrap, and creating and processing of the production information (database).

Depending on the degree of sophistication needed or desired, automated control systems can be provided in a hierarchy of interactive levels. Cam-actuated assembly machines generally require a minimum of control sophisticated. Power-and-free machines having independent stations need more extensive control systems. Flexible assembly systems require programmable controls. Integrated automation systems involve the coordinated operation of multiple microprocessor-based controls and rely on computer-based production scheduling and data management linked by interactive communication networks.

Modern hierarchical control systems, exclusive of the actual mechanical elements, can include the following: programmable logic controllers; motion or robot controls; computers, and communication systems.

Essential to any automated assembly machine and forming the base for hierarchical control systems are real-time control devices that interface to programmable controllers and other control devices. These devices are generally connected via I/O modules that interface limit switches, pushbuttons, proximity switches, solenoids, motor starters, pilot lights, and other sensing and drive devices. The devices typically involve digital I/O points or on/off controls.

Also essential to any automated assembly machine and forming a next higher level for hierarchical control systems are devices for real-time control at the station level. This in a more complex level and generally involves the use of intelligent I/O modules, real-time, logic-solving programmable controllers, and various motion control systems.

At the cell level, control emphasis shifts to the coordination of multiple stations, with the level of control moving upward from real-time operations to supervisory tasks. High-capability supervisory programmable controllers or computers use either data highway or peer-to-peer communications to direct the operations of station-level programmable controllers or computers. This is the basis for the concept of distributed control and computer-integrated manufacturing (CIM).

Above the cell level, computers at the assembly center and plant levels coordinate larger areas of intraplant activity. Planning and management functions are usually implemented at the assembly center level, while strategies and direction of the complete overall functions are generally done at the plant level.

At these upper levels of control, a prominent issue centers around high-level broadband communications between diverse automated devices from various manufacturers. The issue is protocol standardization of the variety of schemes available for intraplant (multilevel) communications.

The Manufacturing Automation Protocol (MAP) standardization program is an international effort geared toward dealing with this issue. The MAP program is a multivendor communications standard that can reduce communication costs and increase the uptime of the application by effectively utilizing VLSI (very large scale integration) technology and communication standards.

While the industrial robot is an important piece of many automated assembly puzzles, other forms of automation are used in assembly. An important adjunct to the assembly operation is the parts feeders. These relatively inexpensive feeders often perform the same function as the expensive vision systems and robots that were developed in the early 1980s.

Rotary, orbital and vibratory bowl feeders have been available for many years. Other part feeders such as straightline vibratory and belt conveyor orientors are more recent developments.

At the SME Assemblex VIII Conference in 1982, Richard D. Zimmerman, Spectrum Automation Company reported on the types of feeders in his technical paper titled *Updated Art—Parts Feeding*.

Zimmerman wrote: ''Rotary bowls use agitator slots to lift and tumble parts onto a gravity orienting track. Simple parts, at reasonable production rates, are most frequently fed with these units. Noise, parts abuse and tooling wear are good points for critical consideration. Several feeder firms offer rotary bowls for users that prefer them to vibratory units.

''Orbital bowls operate with action from a vertical axis, rotating disc within a fixed perimeter wall. Parts are moved to and along the wall by the disc's friction surface. Tooling, similar to that of a vibratory bowl, gets parts from the disc action. Certain part shapes have been fed at near 1000 pieces, per minute.

''Vibratory bowls have added tooling features to inspect, sort and even assemble certain parts. Tooling is usually built by artisans, to suit the function, with work hardening stainless steel. Complex helical or spiral tooling cannot be heat treated without shape warpage.

''Straightline vibratory orientors use the same principal of operation as bowls with some possible extra benefits. In-line orientor tooling permits simplified, true design records for preorent, orient, sort, inspect, distribute to multiple paths and live feed functions. The straightline concept permits quickchange, bolt-in tooling plates for families of parts. It also permits fully hardened part contact surfaces for long-term accuracy and wear life.''

Debugging or troubleshooting refers to the work required to obtain optimum production from a new assembly machine or system in the shortest possible time. Proper design, construction, and installation of the machine or system will minimize debugging needs; however, variations or changes in the parts to be assembled and the machine components and tooling generally necessitate varying degrees of debugging. Debugging should start with the design and construction of the machine and tooling and continue through the installation and operation. In many production systems, debugging is often a continuing process after initial operation. With custom-designed assembly machines, debugging may cost up to 25% of total machine cost.

It is desirable to eliminate or minimize joining and fastening operations in automatic assembly whenever possible to reduce costs and increase productivity. When joining or fastening is required, select the easiest, fastest, and most economical method possible. The best product designs for assembly are those that require simple, conventional tools and no special tooling for assembly. For example, the use of snap or interference fits is less costly and requires less time than fasteners. Staking, spinning, or welding are also often preferred to the costly feeding and driving of fasteners such as screws, bolts and nuts. the use of clamps to hold parts in place while an adhesive cures should be avoided. Consideration should be given to using a faster curing adhesive that will provide the bonding strength needed.

Requirements for disassembly, maintenance, or repairability of a product limit the joining methods that may be used. When required, joining operations should be performed at separate stations, preceded by devices that automatically inspect the presence and position of the components needed.

Although snap and press fits can often be made as part of the normal machine function, many assemblies require the use of joining and fastening processes. When required, joining and fastening operations should generally be performed at separate stations, preceded by devices that automatically check the presence and position of the components.

Practically every known method of joining and fastening is being performed on assembly machines.

Selection of a particular joining or fastening method for a specific application depends on several factors, including the following:

- The materials to be joined.
- Size, weight, and geometry of the components to be assembled.
- Joint designs and accessibility.
- Functional requirements of the assembled product, including strength, reliability, environment, appearance, and whether it has to be dismantled for maintenance or repair.
- Production requirements (rate and total).
- Edge and surface preparations necessary.
- Adaptability and compatibility of the joining method to automation, and effects on joint properties.
- Available equipment.
- Tooling requirements.
- Costs.
- Safety considerations.

Automated guided vehicles (AGVs) have grown in importance in automated assembly. Clifford T. Anglewicz of Volvo of America Corporation, wrote in his SME Technical Paper titled *Wire Guided Vehicles in Assembly*: ''Today, there is a new focus on efficient material handling methods in assembly processes. Just In Time delivery concepts are receiving more attention. Wire-guided carriers offer a high degree of flexibility in material inventory management within the plant. Flexible carrier control systems) programmed software and operator input terminals) can be designed to insure that material is delivered to workstations in the exact amounts required and at the exact time it will be required, thereby reducing required floor space for material storage.

''Or, flexible wire-guided carrier assembly lines can move through material storage areas, when desired. Such a design is particularly appropriate for use with materials with special handling requirements such as those that come in bulky, or odd-shaped packages, or those which require special care, such as acid. In addition, the carrier assembly line can move through subasembly buildup areas. Handling time and loss of material due to damage in transport to the main assembly line is virtually eliminated.''

Automatic storage and retrieval (AS/RS) is a key element in the computer-directed material handling system that is needed to make computer integrated manufacturing and the automated factory a reality.

Darrell B. Searls, Jervis B. Webb Company, wrote in his SME Technical Paper, *Automatic Storage and Retrieval in the Automated Factory*: ''The automated factory implies the continuous controlled movement of material to and through the factory processes. Many of these processes have been highly automated for some time, giving rise to the concept of 'islands of automation.' The automated factory integrates these islands of automation with a material handling system that provides for the scheduled flow of material through the factory with continuous feedback of information to regulate and adjust all of the factory elements to achieve maximum productivity with the expenditure of a minimum of resources.

''This results in reduced cycle time; reduced inventories; smaller physical plant, and better overall utilization of capital and labor.

''Large production units combine processes with different operating rates and reliability and, in the event of production problems, provision must be made to store material between processes until the total system can be brought back into equilibrium. Automated storage and retrieval, then, is required in the automated factory wherever buffering between processes is necessary and wherever there may be an unscheduled pause in the flow of material.''

A view of the role of machine vision in automated assembly was presented by Tim Pryor and Walt Pastorius of Diffracto Limited in their SME Technical Paper *Applications of Machine Vision to Parts Inspection and Machine Control in the Piece Part Manufacturing Industries*. In this paper, they wrote: ''The electro-optical dimensional gage or inspection system has a large number of advantages over both its dimensional inspection competitors (commonly the LVDT or ''electronic gage,'' and the air gage), and human visual inspection. ''The usual advantages are: noncontact sensing; high-speed operation; dimensional accuracy (short and long-term); accuracy relative to human inspection, large stand-off, and the ability to 'see' in small or restricted areas.

''Relative to human visual inspection, electro-optical inspection also provides consistency of inspection, both in providing a true, 100% check and in checking each part in a consistent manner. Neither is possible with subjective inspectors.''

Applications. The variety of automation concepts continues to increase with technological development. Currently, assembly systems can be divided into two major subdivisions: continuous assembly, such as with bottling and cigarette manufacturing, and intermittent assembly, such as indexing units. Each of these categories offers the choice of a single station, multiple rotary stations, multiple in-line machines, multiple carousel lines, or the power-and-free conveyor design that permits accumulation of the parts with a combination of intermittent and continuous operation. The degree of manual work in the process will strongly influence the concept selection because how the equipment paces the operator is important.

Several examples of automated or automatic assembly follow.

A robotic assembly systems at the Fisher Body Division of General Motors Corporation is assembling 12 different door trim panels at a rate of more than 400 assemblies per hour. The system consists of 37-station nonsynchronous conveyor with 13 robots. Three optical sensors at one station determine which car line and style of trim panels are to be assembled. Data are relayed to a master programmable controller that controls subsequent operations. A vision systems with two cameras at another station scans the trim panels to determine locations of the studs. If the studs are improperly located, the vision systems alters the robot program for proper location. The various robots install stamped nuts and plastic fasteners supplied from overhead hopper bowls and spray adhesive onto the panels. A communications network system provides diagnostic information and production data.

Westinghouse Electric Corporation, with funding from the National Science Foundation, developed a programmable assembly system for the low-volume or batch production of small, fractional-horsepower electric motors having a number of variations. The system consists of six computer-controlled workstations, robots at four of the six stations, part presentation equipment, fixtures and tools, transfer conveyors, and sensory devices in a complete assembly line (see figure on the next page).

A robotic cell for assembling electrical contactors is in use at the Allen-Bradley Co. A material handling robot in the center of the cell retrieves pallets of unassembled parts delivered from a warehouse. The pallets are individually presented to one of three assembly robots.

A computer-integrated flexible system is being used by the Industrial Control Division of Allen-Bradley Company to assemble, test, mark, and package electromechanical contactors and relays for electric motors at rates of about 600 units per hour. With a hierarchical control system, the line can produce 125 variations of the products in two sizes.

This Just In Time process permits the fast production of small lot sizes and eliminates work-in-progress and finished parts inventories as well as direct labor. Only a few attendants are necessary to oversee operations. Color graphics interfaces permit human intervention at any state in the process by means of on-line monitoring,

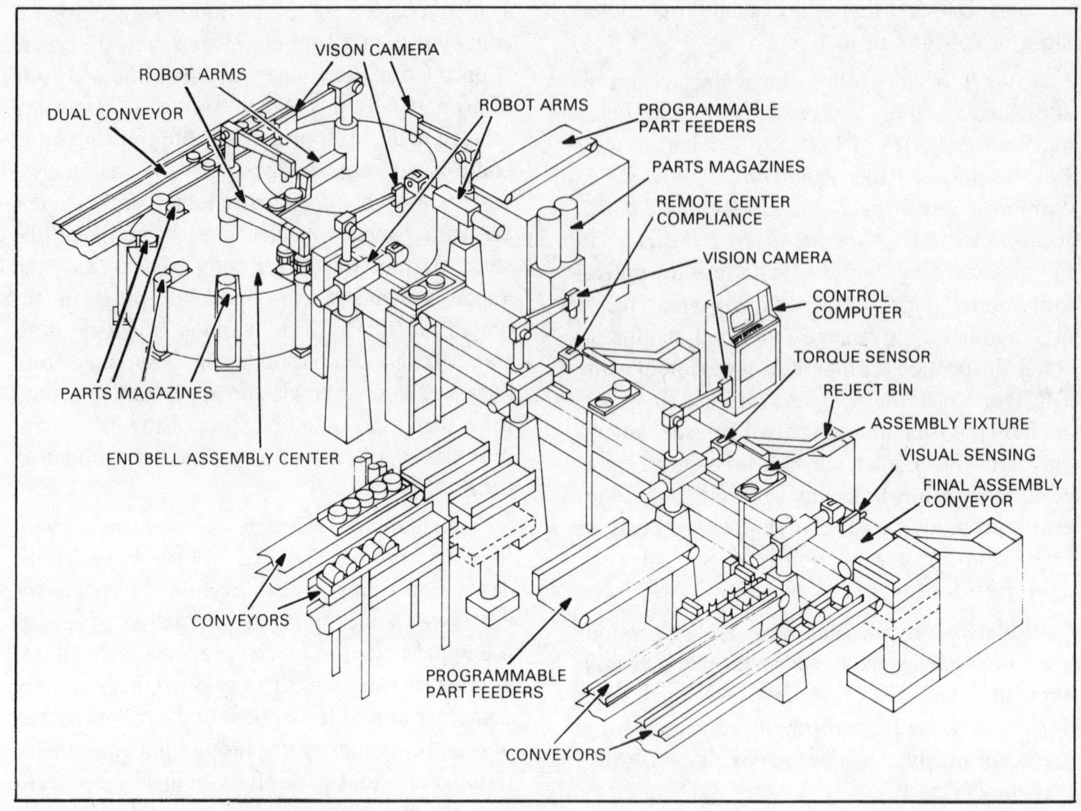

Robotic assembly system for small electric motors.

diagnostics, and control. Warranty expenses have been lowered through improved quality. A possible problem with Just In Time methods is the a shortage of any part will stop production.

The foundation of the assembly system is a center-level computer into which customer orders are received and stored. This computer does all the necessary recording, accounting, and downloading of customer orders to a supervisory programmable controller at the cell level. Production of all products does not begin until an actual customer order has been received and processed. The accounting system is entirely automated. In addition to the initial logging of customer orders, shipping records and billings are entirely computer generated.

Many labor-intensive applications, such as wire harness fabrication, have demonstrated the feasibility of using automated assembly techniques to integrate the myriad of processes into a single manufacturing system.

Because wire harness fabrication is one of the most labor-intensive activities in manufacturing, several companies in the U.S. including Hughes Aircraft, are working to develop a cost-effective, computer-controlled system for producing wire harnesses using a robot arm to feed, mark, strip, terminate, insert, route, tie, and inspect wires. Other goals include developing off-line programming techniques, system configurations that will accommodate large harnesses, and cost and time estimating methods.

As automatic harness assembly is increasingly capable of achieving cost reductions of more that 10:1 over manual mekthods, and providing greater accuracy; it appears likely thatdesigners and connector manufacturers will consider this method of assembly automation.

In another application, provided in an SME Technical Paper titled *The Architecture of a Small Automated Assembly Line*, Theodore W. Leverett of IBM Corporation described a four-station automated assembly line (using IBM 7547 Robotic Systems) that was put into use at IBM Charlotte. This line is assembling hammer blocks for a line printer. This subassembly is the size of a blackboard eraser and has 74 parts, including seven pieces that are bonded in place and 45 screws. This is a buffered serial line using a conventional transport system, and featuring identical robot workstations, minimum computer and communications overhead, each of expansion, and the ability to quickly changed from on product to another. The line schematic and a workstation are illustrated in the following two figures on pages 44 and 45 .

In April 1983, *Manufacturing Engineering* reported on how the use of probes and sensing systems was gaining in importance as the need increased to automatically verify the presence, position, and function of assembled components.

The article explained that welding was growing in interest as a means of joining parts in automatic assembly; the percentage of welded joints was continuing to increase while the demand for screwed and riveted joints was decreasing. In laser welding, two parts are melted and then fused together.

The article, written by Harry Waldman, reported: "Extending the precision and productivity of laser welding, the Ford Motor Company (Dearborn, MI) is installing a CNC laser system to automatically join a six-gear shaft to a planet-gear cage.

"In the testing stage since November 1981, the system, developed by Letnan Industries, Incorporated (Sterling Heights, MI), is a six-station dial-index machine controlled by an Allen-Bradley PLC 220) that is automatically loaded by a robot to weld 200 planet-cages per hour. Set to run for one shift per day, it will replace a TIG welder that, each day, saw 250 carriers rejected out of a production lot of 1100 to 1200."

Three nonrobotic automated assembly applications were described in the *Automated Assembly* videotape.

At the Tandy/Bell and Howell Home Video plant, the production of cassette shells for home video is completed. An assembly machine takes advantage of the fact that close to 90% of the assembly operations are generic including insertion, tapping, pick and place, and so forth. The assembly machine is built around modular units or bays.

At Hubbell Corporation, two assembly machines are assembling 110-volt electrical power receptacles. A vibrating circular bowl orients all the parts in the proper direction. A hole is tapped and the grounding screw is inserted and tightened. One of the ground contact assemblies is placed on a pin on the fixture, a rivet is inserted, and the part is inspected.

A pick and place unit removes the defective units and discards them.

Remmele Engineering Incorporated is assembling fuses for automobiles using a four-part process. Fusible link material attached to the fuse body with automated resistance spot welding, a punch and die station, and parts feeders. Sensors qualify the presence of material in workstations. In a vertical indexer station, an individual fuse is cut from the strip. At the same time, the fuse blades are end cut to proper form. Scrap is automatically disposed of. The entire process at Remmele is supervised with a programmable controller. Sensors send information to the controller at each step of the assembly process.

Glossary. *adhesion:* The attractive force that exists between an electrodeposit and its substrate that can be measured as the force required to separate an electrodeposit and its substrate.

air gaging: A method of measuring physical dimensions by precisely measuring pressure or flow and relating the measurement to the distance between the workpiece surface and a nozzle that directs airflow against the surface.

burr: An undesirable projection of material that results from a cutting, forming, blanking, or shearing operation.

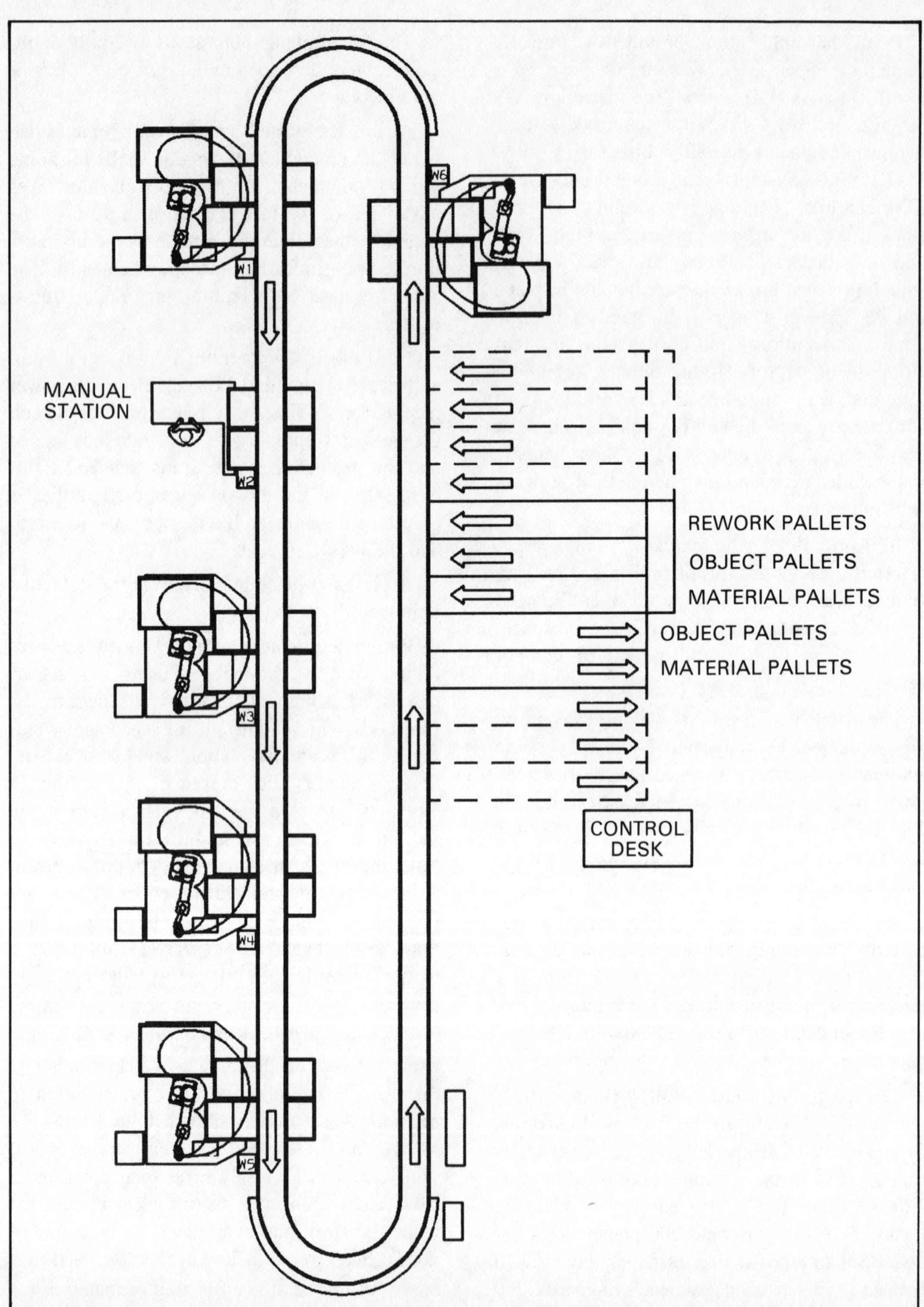

An automated assembly line schematic.

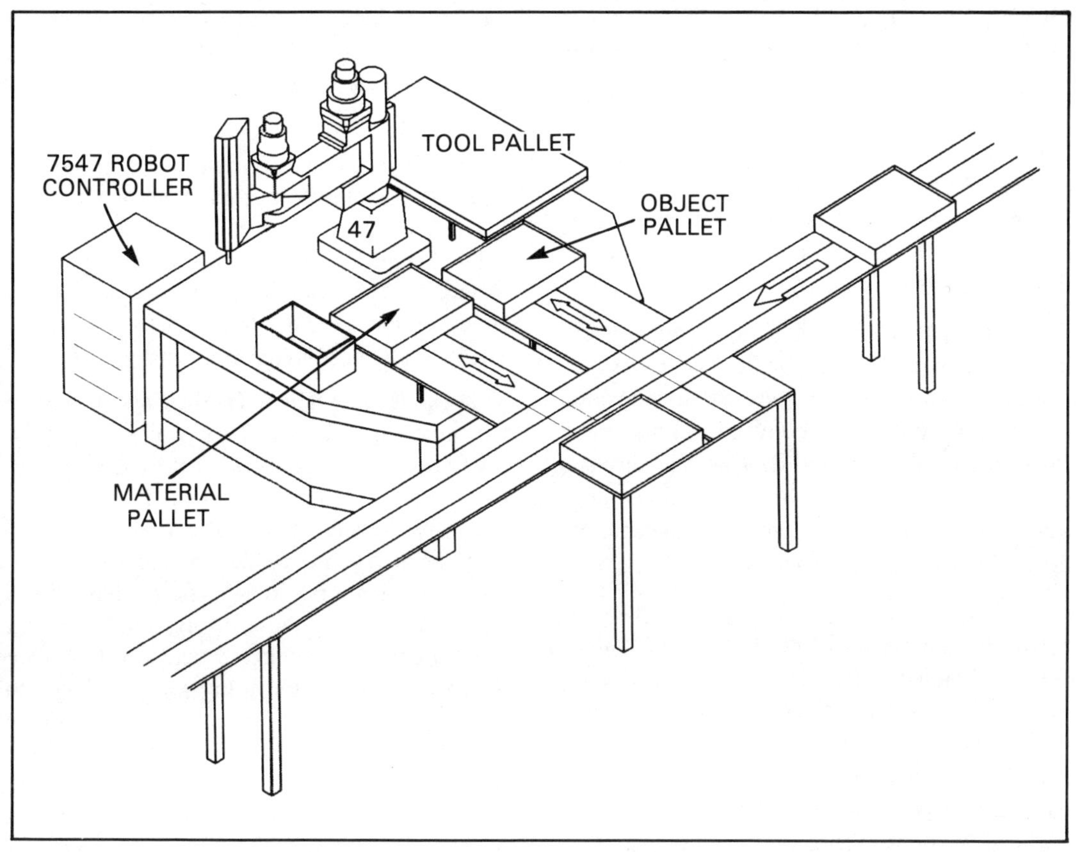

A sample automated assembly workstation.

database: A complete collection of information, such as that contained in libraries, stored in drums, disks, or other storage media for computer processing.

flash: The excess material squeezed out of a cavity as a compression mold closes or as pressure is applied to a transfer of injection mold.

gaging: Checking the workpiece for size with gages.

programmable controllers: A solid-state industrial control system with a memory which can be set to operate in a specific manner to store instruction that implement function such as I/O control logic, timing, counting, arithmetic, and data manipulation.

rivet: A one-piece, unthreaded, permanent fastener consisting of a head and a body. A rivet holds two parts together by passing the body through matching holes in the two parts, then forming a second head on the body end.

simulation: The representation of physical systems and phenomena by computers or other equipment.

very large-scale integration: High density integrated circuits characterized by relatively large size and high complexity.

welding: A localized coalescence (joining) of metals or nonmetals produced either by heating the materials to suitable temperatures, with or without application of pressure or by the application of pressure alone and with or without the use of filler material.

Automated Guided Vehicle (AGV). An automated guided vehicle is a programmed carrier which transports goods from point to point. Automated guided vehicles are a part of technology called automated materials handling (AMH). This technology is responsible for the physical movement of materials in a factory, from original receipt of raw materials and purchased components through to the shipment of the end product. The systems that comprise automated materials handling include high tech computer controlled systems to conventional materials handling systems.

Automated materials handling technologies include: automated storage and retrieval systems (AS/RS), mini-loaders, Automated guided vehicle systems (AGVS) and conventional materials/part storage and transport designs. The discussion that follows will be centered around automated guided vehicles which are the core of the automated guided vehicle system.

Michael C. Dempsey of Portec, Incorporated noted in his SME Technical Paper, *Automated Guided Vehicle Control Systems Design, Application, and Selection*: ''Automated guided vehicle systems have become an integral part of today' automated factory. AGVS often forms the backbone of the automated material handling system. It is the skeleton around which the individual production units are tied and linked to form a system. The increased utilization of AGVS as the material handling integrator has been due to the real-time monitoring and system management opportunities provided by the AGV control system. The advancement of computer technologies for application in material handling has provided the key to create flexible control systems tailored to the individual needs of the user. Moreover, the modular development of the control functions and associated hardware allow the AGVS to grow with the facility as changes in production processes or interface requirements occur.''

History. Automated guided vehicles were invented about three decades ago. As electronic technology progressed so too did the AGV. During their earliest stages these driverless vehicles were closely tied to vacuum tube technology. As technology progressed, the AGV moved through the transistor to the integrated circuit technology. Today, AGVs are controlled by optics or by a wire imbedded in the floor of the facility.

Discussion. Optically guided system rely on tape or a line of reflective paint put on the floor. Each AGV has a device to track the tape or paint. Therefore, dirt or trash covering the paint or tape can provide a drawback to the system. However the system has its relatively easy to establish and cost-effective air many cases.

Dr. Pius J. Egbelu of The Pennsylvania State University in his SME Technical Paper titled: *Unit Load Design for Minimum Cost Manufacture in an Automated Guided Vehicle-Based Transport System* wrote: ''The vital role played by automated guided vehicles in the operation of integrated manufacturing systems have been well recognized by those involved in the design of such systems. Automated guided vehicles are increasingly being used as the transport device to integrate machines, workstations, and storage systems in warehousing and manufacturing operations. By providing the linkage between production units, automated guided vehicles help to bring the concept of integrated manufacturing system into reality. Among the benefits ascribed to the use of automated guided vehicle system in manufacturing are increased system productivity and flexibility. An AGV system can be interfaced manually or automatically with other machinery and handling equipment. Movement of parts is done accurately, safely, and with minimum material damage. Once programmed, the vehicles can perform the handling operations without any human intervention. In computer controlled systems, on-line tracking of load movements by the vehicles provide real time inventory control.''

AGVs have a variety of uses in industrial applications. The importance of AGVs in the automated factory rests in their ability to operate under computer control, interface automatically with production machines and storage equipment, deliver materials over flexible routes, and the ability to operate in hazardous and special environments.

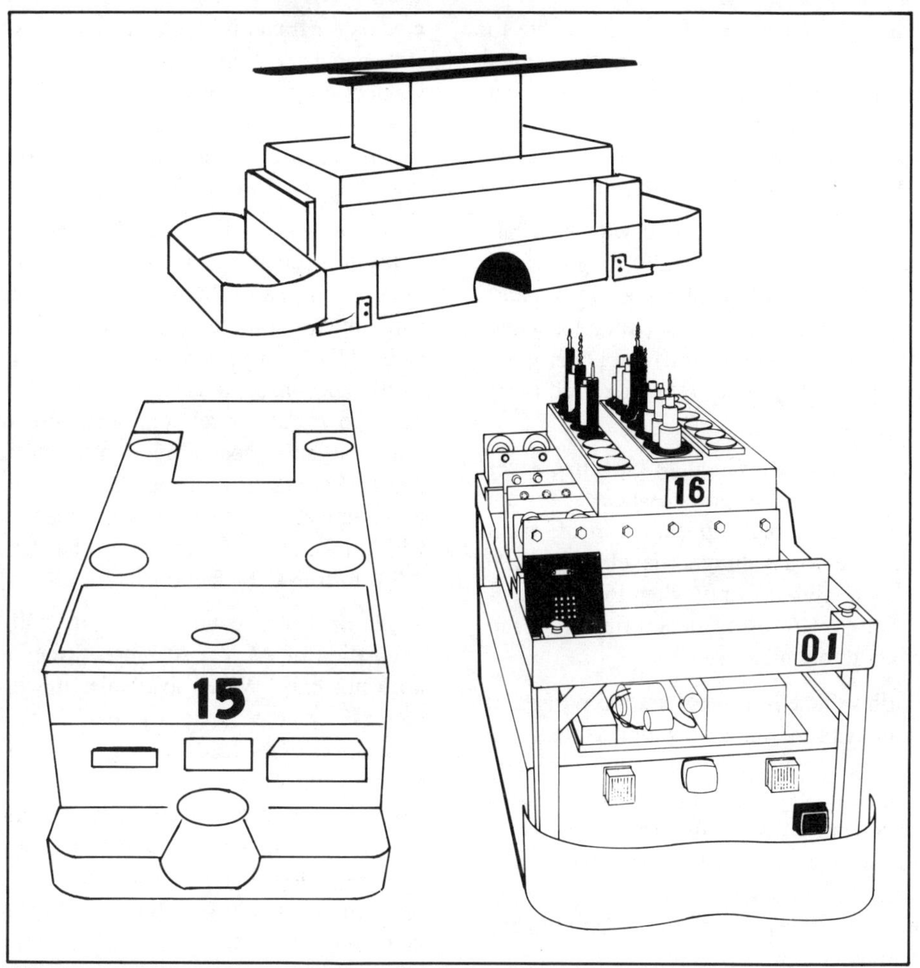

Sample automated guided vehicles.

There are six basic vehicle types presently in use. They are towing vehicles, unit load vehicles, pallet trucks, fork trucks, light load vehicles, and assembly line vehicles.

Towing vehicles the most numerous type. These units can pull, depending on make and model, 8,000 to 50,000 lbs. and are generally used with the bulk movement of product into and out of warehouses.

Chain movements of product with AGVS trains is also popular. In this case, the AGVS trains are loaded with product destined for specific destinations along the guide path route. The train will make several stops in order for the product to be unloaded at the correct locations.

Trains systems are generally used where movement of product is over very long distances, sometimes between buildings, outdoors, or in very large distributed systems where the runs are quite long. Since each train can move as much as ten pallet loads at a given time, this becomes a very efficient method and can usually be justified easily based on the elimination of fork trucks or manual trains and operators.

The unit load AGV works in the area of individual pallet movements.

Normally, the unit load carriers work in warehousing and distribution systems. It is here that the guide path lengths are short and the volume is high. This type of vehicle is able to work in tight quarters (where fork trucks cannot easily maneuver). System versatility for product movement is increased since unit load vehicles can work independently of each other and even pass each other during their operation.

The pallet truck is associated with distribution systems. This vehicle delivers goods by moving to its destination along a guide path, onto a spur, setting down its pallet load and driving to the next assignment. The pallet truck increases system flexibility by being used manually.

The fork truck uses a guide path like other AGVs. However, it is used when a system requires automatic pick up and drop off from shop floor level to some higher level. This AGV is generally considered only when total automation is required. However, greater flexibility can be obtained through its use.

The fifth vehicle is a light load AGV (used in light manufacturing areas). Smaller than the other vehicles, this AGV has only a several hundred pound capacity. It can move small component parts in a container, tray or basket, to an assembly area. Similar to a unit load AGV, the system will operate in tight areas. Proper applications includes: electronic assembly, and other light assembly.

The final AGV to be discussed is the assembly line AGV—a truck adapted from the light load vehicles. The vehicles are used to transfer major subassemblies (usually motors).

This type of vehicle allows for parallel operations thus giving better flexibility to a manufacturing process. This vehicle is generally integrated into a larger system of automated assembly. Computer control and heavy planning are required.

Future trends in the industry should show little variation from the present day vehicles. As mechanical capability of the units are improved, the vehicles will cease to be passive horizontal transporters of material. Some elemental capabilities emerging in today's market are lifting loads in elevated storage bins, side loading models for travel through narrow storage aisles, and articulated frames. Since it is believed that vendors may encounter difficulty differentiating themselves purely on mechanical capabilities, the mechanical differentiation will be targeted toward specialized market requirements.

Major vendor differentiation has the opportunity to occur through AGV control and communication capabilities. Some current and emerging capabilities are limited off-the-guide path travel, increased on board control and intelligence through use of microprocessors, and more powerful central control stations with real time instruction teams mission to and from the AGV. An interesting capability is the computer to computer interface between the AGV and CNC tools, MRP systems, etc., thus lending the AGV to true CIM facilities.

An important control capability will be self-navigation of AGVs. At present, the capability does not exist. When available, it will greatly increase system flexibility and elimination of optical or embedded wire guidance systems. Along with an increased flexibility cost benefits will be gained. Ongoing controversy exists regarding both the accuracy and safety of the off-the-guidepath travel using today's technology. Improvements over time can, perhaps overcome these obstacles.

Garry A. Koff of Mannesmann Demag discussed present day AGV guidance in his SME Technical Paper *Basics of AGVS*. He explained: "AGVS steering control allows AGVS vehicles to physically maneuver in different ways. There are two basic types of AGVS steering control, *differential speed steer control* and *steered wheel steer control*.

"Differential speed control uses two fixed wheel drives and varies the speeds between the two drives on either side of the guide part to permit the vehicle to negotiate a turn; much in the way a tank or a tracked vehicle turns. An amplitude detection type of guidance sensor is used to give guidance information to the steer system. Amplitude detection guidance is based on balancing the signal received from a left and right sensor in front of the vehicle. When one or the other signal is greater, the steering system compensates by correcting the steering until the amplitude of the left or right signals are equal.

"The steered wheel control uses automotive type steering control in which a front steered wheel turns to follow the guide path. Phase detection guidance is usually used for this type of steering. The phase detection guidance sensor can differentiate whether the vehicle is to the left or right of the path by detecting the positive or negative phase of the signal received from the guide path wire. This information causes the vehicle to correct its steering so that there is no phase difference.

"In either case, the guide parts look the same for most applications. Steered wheel control is used in all type of AGVS vehicles, however, differential steer control is not used in towing vehicles applications or on vehicles which have man onboard controls."

Koff reported that there are three types of traffic management in general use. They are: zone control, forward sensing, and combination control. Zone control is the most popular and widely used type of traffic management. The rules of zone control are that only one vehicle is permitted in a given zone at a time.

When a vehicle occupies a zone the closest a trailing vehicle can get is into the next completely unoccupied zone behind the lead vehicle. The lead vehicle must then proceed into the next zone before the trailing vehicle can move ahead into its next zone. A zone may have multiple stop stations in it. If a vehicle is allowed to occupy a zone, it can proceed to any stop in that zone.

Forward sensing control is useful where there is alot of straight path in the system and where that straight path is not interrupted by intersecting curves. In these methods, whenever the sensing system sees a vehicle in front of it the vehicle goes into a hold. When the blocking vehicle moves out of the range of the sensing device, the holding vehicle automatically restarts.

One advantage of forward sensing for traffic management is that vehicles can come quite close to one another for there are no fixed holding points as in zone control methods. This permits a greater density of vehicles in a given area. A drawback is that it can be used only on simple paths.

Most systems are done entirely with one method only. However, a system layout can, occasionally, benefit from both types of traffic control or a combination control traffic management system. There may be long runs in which forward sensing control can be used, reducing the expense of the traffic management system significantly. In the remaining areas of the path where there are divergence path and convergence path areas, a form of zone control would be used. The vehicles would response to both types of traffic control depending upon the area in which they are operating.

AGVs system management can be divided into two areas: vehicle dispatch methods and system monitoring methods.

Vehicle dispatch methods are as follows:

- Onboard Dispatch.
- Offboard Call Systems.
- Remote Terminal.
- Central Computer.
- Combination.

Onboard dispatch selector involves a control panel on each vehicle used by an operator to dispatch the vehicle to a single or series of stop stations. He/she may also select the function they wish the vehicle to perform at the stop station. This is the most common form of system management and is usually the most flexible type.

An *offboard call system* can also be used for vehicle management in AGVs systems. The call systems vary in complexity from simple types, which involve only a push button at a call station to stop the passing guided vehicle, all the way to call box controls which not only can call a specific vehicle, but can also remotely dispatch that vehicle to other destinations into the call box panel at their station which will communicate those destinations to the guided vehicle after it leaves their stop location. This is useful in systems where load transfer is automatic.

A *remote dispatch terminal* can give a degree of centralized dispatching to the control of the guided vehicles in the system. The keyboard remote terminal approach allows an operator to control the individual guided vehicles from a central location. In order to do this the operator must have some visual status of the vehicle's location and condition so that they can be effectively dispatched to specific areas in the system where they are needed. This is usually to a CRT graphic display or locator panel. Many times this approach is used in systems which can not totally justify computer controlled solutions, yet where a high degree of selective movement with automatic transfer is required. For example, if loads are brought to a receiving area and then taken to selective locations in a storage area with an automatic load transfer unit used by the AGVS vehicles, then a centralized control may be an effective solution. A central operator would use the keyboard to send the vehicles to the receiving area to automatically pickup loads and then dispatch those vehicles to selected storage locations where the loads would be automatically dropped off.

Central computer control for system management is the highest level of control possible. The AGVs would respond to a central computer's commands for where they should go and what they should do. This would be automating the previous case and eliminating the central dispatcher completely. Many integrated material handling systems incorporate central computer control for AGV system management. Most of these systems interface with an automated storage and retrieval system taking loads in and out of storage. A computer directs the vehicles in coordination with the loads moving in and out of the AS/RS system. This level of control permits automatic tracking of every load in the system. Where the volume is high and automatic load transfer occurs, the computer form of system management is very efficient.

It is also possible to mix these various forms of system management together at the same time in one system. This depends on the type of control offered by a particular vendor, but in many cases systems can be operated in several different modes at the same time. This is particularly useful is you have a sophisticated level of operation in your system and wish to have a backup to it in case the main level of operation control fails. For example, if your system is computer controlled and the central computer fails, then resorting to remote terminal control or onboard vehicle control would allow the AGV system to continue to operate in the integrated mode. Not all central computer control systems permit this type of operation but is is very important in certain situations.

AGV systems monitoring is an important consideration in many systems. Simple systems do not require extensive monitoring or control. Sophisticated systems benefit greatly from monitoring. When a sophisticated system is installed, there is usually a high degree of automation and throughput required. A breakdown or slowdown in the system could cause serious problems if not detected immediately. There are three approaches to monitoring:

- Locator Panel.
- CRT Color Graphics Display.
- Central Logging and Report.

A simple monitoring system for AGV systems can be a *locator panel* which merely indicates if a vehicle is in a given area of the guide path. It does not identify the vehicle specifically or its condition. A light next to the zone that a vehicle occupies illuminates to indicate that there is a vehicle in that zone. Sometimes a timer is used for each zone to indicate if that zone has been occupied for too long, indicating that a vehicle problem may exist in the zone.

A *CRT color graphics display* is a newer development which specifically shows where each vehicle is and its status. CRT color graphics is a real time monitoring capability which can instantly detect a problem, identify specific vehicles, and show the location of the failure on the graphics display. Other useful information includes whether the vehicle is moving or is blocked by other traffic, whether the vehicle is loaded or empty, if the battery is O.K., and where the vehicle is going. Operators can spot blockages and slowdowns quickly and take corrective action as required.

The CRT color Monitor can show system condition either in graphic or table form. The table form lists each vehicle ID, its location, destination, condition, mode of control, load status, and alarm condition in a column format listing which is updated once per second for absolute current system monitoring.

A *central logging and report* capability for monitoring an AGV system is helpful when attempting to develop historical data on the system's performance. Performance reports can be printed out periodically indicating such things as how often the vehicles were moving, how many loads a given vehicle transported in a given period, how many loads were received or issued from a given stop station, where were they taken, when a vehicle's battery was low, and when did a specific load get picked up and what was its destination? Performance data can be quite useful for keeping system efficiency at the highest possible levels.

AGV control was also discussed William J. Higgins (Mannesmann Derag, Grand Rapids, MI) in the article *Integration and Control of AGV Systems* which appeared in the April 1987 *Robotics Today*. Higgins wrote: "Vehicle guidance control is always a machine-level function. In wire-guided systems, wires carrying low-power radio frequency signals are embedded in the floor along the desired paths. The vehicle is equipped with an antenna that detects either a phase differential or an amplitude differential when is it moved from side to side over the wire. This signal drives the servo circuit attached to the steering mechanism, maintaining the vehicle directly above the work. Photo-optical guidance is similar, except that paint or reflective tape is used to describe the desired paths, and optical; sensors detect a brightness differential between a path and the floor to generated the guidance signal.

"To determine position within the system, vehicles are typically equipped with controls that read floor-based codes. These codes may be magnets or responders for wire-guided vehicles, or special bar codes for optically guided units.

"In special cases, vehicles leave the guidepath temporarily, using locked steering or dead reckoning. Guidance technology is also being developed that would eliminate the requirement for a floor-based guidepath. It incorporates elevated targets positioned along the desired path and a vehicle-based scanner to read the relative position of the targets. An onboard computer calculates positioning information using triangulation and generates the guidance signal.

"Safety is handled at machine level with bumpers and other contact sensors, or range sensors such as photo-optic and sonar systems. Machine-level controls are also used for monitoring battery charge and the electronic status of the vehicle and for providing diagnostic data for maintenance.

"Vehicle routing may be controlled either at the machine level or system control level. In wire-guided systems, routing is accomplished by one of two methods—path switching or frequency selection. Path switching is associated with system-level controls; frequency selection is associated with system controls at the machine level.

"In path switching, the central control monitors vehicle location and destination, determines routing, and provides only one path for the vehicle to follow by switching segments of the guidepath wire on or off in front of the vehicle. In frequency selection, the vehicle knows its location and destination relative to a system map in memory, determines routing, and selects the frequency that is associated with the desired route. Since these various frequencies are always on, machine controls must switch internal circuits to monitor only on frequency at a time.

"In optically guided systems, routing can also be controlled either at the machine level or the system level. In either case, when a vehicle reaches a point in the system with alternate routes, preset lock steering is momentarily employed to route the vehicle onto the proper path.

reaches a point in the system with alternate routes, preset lock steering is momentarily employed to route the vehicle onto the proper path.

"When AGV systems support more than one vehicle, traffic control is required to ensure that vehicles do not collide with each other. The traffic control method varies with the location of the system control intelligence. In centralized intelligence systems, traffic control decisions are made at the system control level. This controller then either directs a distributed network of stop/proceed signal devices or communicates instructions directly to semi-intelligent vehicles. In smart vehiclesystems, each vehicle can determine the location of all other vehicles in the system. Using this data, vehicle-level controls can make decisions to stop or proceed.

"Limited traffic control can also be accomplished by utilizing forward-sensing devices associated with safety. For example, safety bumpers are sometimes used for traffic control in systems incorporating slow, lightweight vehicles.

"The traffic control function requires that vehicles have some means of communicating, either to the system-level control or to one another. Communication serves other system control functions as well and is accomplished through the guide wire, through other floor-installed devices, or by broadcasting light or radio signals. By communicating through the guide wire, constant communication can be maintained throughout thesystem. Communication with light or floor-installed devices other than the guide wire results in communication only at specific locations in the system and may require the vehicl to stop during communication. Constant communication offers many benefits relating to system efficiency and is an essential element of smart vehicl control technology.

"Many smart vehicle systems operate with no integration to higher-level controls. This is particularly true of tugger systems, becausethese systems are generally loaded and unloaded manually and vehicles can be conveniently dispatched at the local level. Unit load systems can also be operated by smart vehicle controls by incorporating simple operation strategies. For example, individual vehicles in a system can be given specific from-to assignments that are simply repeated. Local interface signals are used to initiate the cycle and prevent double loading at the destination.

"Another method of operating smart vehicle systems without system-level controls is to have vehicles recirculate looking for a local interface that indicates the presence of a load. This is particularly well-suited to systems in which there are a large number of from-to points for which the from-to assignment is constant. If a means of communicating the destination at the local level is utilized, this method can also be used with from-to assignments.

"The first level of control above basic system functionality (guidance, routing, and traffic control) is what might be called system management. These controls monitor system function controls and initiate actions to improve system efficiency. System management controls are usually linked to manual or automatic vehicle call systems. Calls may also be generated directly from system management controls on associated equipment such as conveyor systems. When a call is received, the system manager reviews the location and status of all vehiclesand dispatches the most appropriate one. A constant communication method at the functional level is useful to system management controls both in terms of acquiring real-time information and to permit dispatching instructions in real time.

"An example of system management integration is a system in operation at the U.S. Postal Service mail distribution facility in Baltimore. The primary activity of this manager is to monitor the status of multiple manual call stations and optimize dispatching of vehicles in response to calls. This system incorporates smart tugger vehicles, and the system management hardware is a programmable logic controller.

"At the next level of integration, control systems are implemented to direct the activity of the AGV system, and usually other subsystems, relative to material flow. These controls receive data relative to the material in the transport directory from the AGV system and often from inputs external to the AGV system. The inputs may be generated automatically by load identification equipment such as bar code scanners or manually at computer terminals.

"System director controls serve as an operations management tool. Material flow requirements established by operations management are analyzed relative to material in storage and in transit, and appropriate commands to system management control are generated to establish the required material movement. These controls also provide real-time tracking of material and summary reports of material flow.

"At the top of the control hierarchy may be one or more additional levels of control dealing with scheduling and planning of material flow, inventory management, production management, and financial analysis. These controls play a key role as corporate management controls.

"An example of an AGV system that is fully integrated to all of these control levels is a system in operation at an automotive plastics plant. This highly automated facility includes a system of 20 smart unit load carrier vehicles and other equipment, such as an automatic storage and retrieval system, a sophisticated conveyor system, and an electrified monorail system. Programmable controllers are used at the system management level, microcomputers at the system director level, and a minicomputer at the corporate management control level. Both of the latter two control levels incorporate backup capabilities to take over in the event of a failure of the main system."

Applications. Robert R. Lasecki of The Austin Company presented some systems consideration for *AGV System Selection Methodology*. Lasecki wrote: "A major factor to be considered in selecting AGV types and system configurations is the guidepath layout and vehicle flow. The orderly flow of vehicles can provide prompt and efficient delivery of material. Intersections with merging vehicles tend to slow movement. Intersections with crossing vehicles cause severe restrictions in vehicle flow and can, in complex systems having many vehicles, result in gridlock situations. Those situations are those where vehicle traffic reaches a congestion point and flow stops. It is important to note that any time congestion in a system exists, adding more vehicles will only compound the problem. Congestion points can be alleviated by relocating either source or delivery points or by altering the route selected. When computing the number of vehicles based on speeds, distance of load movements, distances of empty moves, number of loads moved, and load transfer times allowances must be made for communications times and inter-vehicle blocking. Since algorithms and most blocking systems are customized, it is difficult to estimate the actual amount of time that vehicles will spend waiting for access to stations or yielding to other vehicles. For well designed systems having high flow, the blocking factor can be as little as 20% of the total travel time. For systems with high congestion and synchronous interface requirements with other equipment, the blocking time has been known to excess 300% of the travel time.

"System performance can be evaluated prior to commitment to physical hardware by using computer simulation. A simulation is a mathematical model of the system that will depict performance over time to validate a design approach. More than 30 simulation packages and languages are available. In order to properly represent a system, an accurate mathematical model of the system and proposed vehicle must be prepared. Poorly developed models or those not representative of the system to be simulated will only yield erroneous results. Simulation models must not only account for vehicle dynamics and load stand interactions, but must also accurately represent the control algorithms used for vehicle dispatch and management. Often multiple iterations of a simulation package are used to develop custom vehicle management algorithms for complex systems. Although generally expensive to perform properly, a good simulation can provide design validation prior to hardware commitment, and prevent a potential disaster in poor system performance following installation.

"In summary, if appropriate facility planning is executed and an objective and thorough methodology is followed during critical evaluation and design stages, a highly efficient and successful AGV system implementation can be achieved."

In another application, computer-controlled automated guided vehicle system (AGVS) is implemented in one of Intel's integrated circuit packaging facilities. The AGVS is designed so that materials are delivered in a Just In Time fashion. The automated guided vehicle (AGV) is equipped with two shuttle arms to transfer loads. Since the AGV delivers material to clean rooms, potential contamination must be avoided.

Shap-Ping T. Wang of Intel Corporation described the AGVs in his SME Technical Paper, *An AGVS in Integrated Circuit Packaging Process*. The author noted that the AGV "can follow a preassigned guidepath to carry out material handling assignments such as bringing a tote of material from one place to the other. Automated guided vehicle system (AGVS) provides a very flexible manufacturing environment since it can easily adapt to layout changes. Most important of all, it is generally cost effective and space saving compared to a conveyor."

The 35 unit AGV system delivers automotive seat components to and from 15 individual build stations at a General Motors plant in Windsor, Ontario. The system utilizes bar codes, interfaces with a robot, an on-line computer, and automatic conveyors. The AGVs allow the incorporation of the individual build stations which have produced a quantum leap in quality, productivity, and pride in workmanship. The flexibility of the AGV system has allowed rapid response to processing and engineering changes. This is particularly important in that the plant supplying built-up seats for luxury vehicles in line sequence, and on a Just In Time basis, to a nearby plant.

Jack Pikaart of General Motors of Canada described the system in his SME Technical Paper titled: *Automated Guided Vehicles in Seat Assembly*. In it he wrote: "When the vehicle leaves the customer's paint shop, an order for the seat is broadcast via communication lines. A build sheet containing the order and the bill of material, is printed at the beginning of the system. Pertinent information is entered into the computer system by scanning bar coded data.

"The operator at this first workstation loads two totes containing the proper seat adjusters, then inserts the build sheet into a pocket on top of the AGV and pushes a button to release the vehicle. A bar code scanner reads the code indicating the seat frames required to build these seats. This information is fed to a robot, which loads two totes to the AGV top. These totes contain the seat cushion frame and the seat back frame for both the passenger and the driver front seats. Two nested empty totes are extracted by vacuum cups from the AGV and deposited to a return conveyor.

"At the third stop, the AGV takes on a tote containing both front seat cushion covers and ejects two nested empty totes. A few feet further, the AGV stops again to take on a tote with the front seat back covers.

"At the fifth stop, the carrier stops for the proper side finish panels. At this point, the AGV contains all the necessary components to build a complete passenger and driver seat. At this time, the computer system selects which of the 15 build stations will receive the AGV. The algorithm will first look for empty stations; if no stations are empty, the computer will then select the station that will have the earliest completion time. Each station has capacity for two AGVs; one being worked, and a reserve vehicle.

"When the AGV comes into the 'being worked on' position a set of brushes makes contact with a floor mounted plate to charge the AGV's batteries. Employes in the build station remove the components from the AGV, completely build up the seats, nest the empty totes, place them to the lower openings, and place the built-up seats on top of the AGV. Twenty seconds before the AGV is scheduled to leave, a beeper sounds on the carrier. If the employes will finish their work in 20 seconds, they continue; if however, more time is required, they press a hold button. This action causes an alarm to be printed on the supervisor's screen, and activates a red light at the build station, so that the necessary assistance can be provided. Upon completion, the employees release the hold button, to send the vehicle on.''

Material Handling Magazine (April 1985) described an AGV Pilot System for Buick City, a General Motors assembly plant in Flint, Michigan. The line is an advanced concept for the use of AGVs for auto assembly specifically for the engine dress line. There were managerial breakthroughs necessary for the U.S. since assembly on an AGV meant a team concept needed to be utilized. Similar teams had been used in Europe where assembly on AGVs has been used for some time.

The dressing of engines (dressing is the term used for adding wiring, power steering, pumps, etc.) is done at the beginning of the assembly line as components are received from other GM divisions and suppliers. A printer terminal at the beginning of the AGV assembly line informs the assemblers of the sequence of that engines are to be built and what parts will be put on the engines.

The first step is to place an engine into an AGV and send it to an area where a transmission is added. When complete, the AGV is sent to the next progressive assembly step. The AGV are computer controlled and even though they can be lined up three deep at an assembly station the computer releases the engines in the proper sequence. No separate changing stations for the AGVs are necessary since brass shoes on the floor are mated with contacts on the AGV to continuously charge the battery.

The AGVs have power lift beds so the engine can be set at the proper work height. All of the hardware necessary for assembly are in carts along the guide path and pneumatic assembly tools are suspended overhead. The actual assembly of the engine may be done quickly but the AGV will not be released until it can take its proper place in queue. The assembly will not be released from that station if the assemblers are dissatisfied with its work.

The last stop for the AGV is at a point parallel to an overhead conveyor. Hangers are placed on the engine and it is carried to a point in the assembly where it can be assembled into a car. The AGV itself returns to the start of the dress line and begin their journey once again.

Some comments on this pilot line are that it is extremely quiet, and that the vehicles require no regular service since the AGV is continuously charged. The biggest problem with the vehicles was electronic in nature causing three or four of the 60 pilot line vehicles to be out of service at any one time. Onboard batteries are used for the computers to avoid loss of memory due to line power failure. In conclusion, the actual process of the AGV line was created by the assemblers designing their own jobs around the use of a new technology, the automated guided vehicle.

Author Dempsey also noted: ''As a greater variety of AGV systems have been installed and the user experience level increased, significantly noticeable trends have developed in the control marketplace. These trends have been primarily driven by the users themselves, but also vendor-generated to reduce costs through application of computer knowledge from outside of the AGV area.''

He listed future trends. These included:

- Increased diagnostics;
- Smaller packaging of hardware on the vehicle;
- Vehicle-to-system communication speeds;
- Off-wire guidance;
- The application of larger and more hardware redundant systems at the system management and host levels;
- The ability to download a new 'route map' and path optimization information to the vehicle automatically;
- Increased use of horizontal interfaces to PLC or other microprocessor-controlled devices;
- The use of complex vehicle task optimization and prioritization;
- Improved monitoring capabilities of the system, for the vehicles as well as individual components;

- The use of computer emulators to pretest the traffic and system management software prior to field installation.

Other applications of the AGVs follow. They are from the SME book *Automated Guided Vehicles and Automated Manufacturing* by Richard Miller.

Four U.S. manufacturers of AGVs specialize in the transport of loads in excess of 10,000 pounds (4,536 kg): Aero-Go, Elwell-Parker, Mentor Products, and Volvo Automated Systems. An Aero-Go vehicle uses four air bearings rather than wheels and transports 72,000 lb (32,658 kg) steel coils.

Volvo Automated Systems manufactures a heavy load AGV that can carry a 20,000-point payload. In one instance, the AGV carries a coil load of steel blanks. Komatsu Forklift Incorporated has set up a complete factory system utilizing Komatsu AGVs. The key items in the system include computer-controlled loading/unloading and storage functions with accuracy in high stacking.

Within the FMS, there are many possible pallet movement designs. The three principal categories are carts, roller conveyors, and robots. Guidance and control of carts can take many distinct forms. Carts can move along tracks, energized and externally controlled by the central computer. Sensors (optical, proximity, or limit switch) located at appropriate points along the track identify the precise location of the cart. The sensors can also be used to position the cart in the required tolerance. The tolerance is typically 0.06 inch (1.5 mm) to transfer pallets to a machine or unload station. Wheel encoders can be used to less precise feedback for the drive system and its programmed speeds. All carts should have dead-man bumpers at each end to help prevent accidents.

Battery-powered carts (AGVs) can be moved along a flat floor, guided by an antenna that detects a wire embedded below the surface. Position sensors still must be used to control pallet transfer. The Cincinnati Milacron (Cincinnati, OH) Variable Mission System uses this type of material handling system, specifically—the Eaton-Kenway (Salt Lake City, UT) Robocarrier.

Another cart design uses a tow chain in a trough under the floor. The chain moves continuously and cart movement is controlled by extending a drive pin from the cart down into the chain. At specific points along the guideway, computer-operated cam-type stop mechanisms raise the drive pins to halt cart movement. On advantage of this system is that it provides automatic buffering with stationary carts along the track. This system is used by Kearney and Trecker (Milwaukee, WI).

A variation on this method from SI Handling (Eaton, PA) uses a floor-mounted spinning cylinder that imparts motion to the roller drive of individual carts. A computer initiates drive disengagement, and the carts can be stopped at selectable locations.

The LTV Aircraft Products Group implemented its 40,000-square-foot flexible machining cell in July, 1984. The integrated system is totally computer controlled and consists of highly versatile machines which direct the automated machining inspection of more than 1,300 military and commercial aircraft parts. The automated cell has reduced 200,000 hours of conventional machining time to 70,000 hours for a 3:1 productivity improvement and cost-reduction of more than $20 million. The savings are being shared with the U.S. Air Force.

LTV builds the aft and the aft intermediate fuselage sections of the bomber under subcontract with Rockwell International Corporation. Cincinnati Milacron, Incorporated, supplied the machinery and installed the system under a contract with LTV.

The equipment in the cell includes with machining centers, each with automatic tool changing and a 90-tool magazine; and automated cleaning station; two automated inspection stations capable of checking parts to engineering design tolerances; two 10-pallet carousels for parts pickup and delivery; and a central ship-collecting system which segregates steel and aluminum chips for reclamation.

An automated transportation system and mobile robotics are another integral part of the flexible machining cell. In operation, four robot carriers, each capable of carrying 5,000 lbs., shuttle parts-laden pallets throughout the system, following a wire network embedded in the floor.

A guided vehicle is then instructed to take the finished castings to a wash module where high-pressure water will remove debris. Next, an automatic coordinate measuring machine probes each part for dimensional accuracy. The next manufacturing cell includes automatic storage and retrieval systems (AS/RS) for raw stock and finished castings.

The a guided vehicle is signalled to pick up a particular plate size staged outside the appropriate AS/RS and deliver it to a numerically controlled cutting machine for cutting and shaping into a finished blank. The blanks are then delivered to a central AS/RS which supplies a manufacturing cell.

Glossary. *guide path:* Optical or imbedded-wire path which the AGV will follow.
routing: The vehicle's ability to make decisions along the guidance path in order to select optimum routes to specific destinations.
traffic management: A system or vehicle ability to avoid collisions with other vehicles while, at the same time, maximizing vehicle flow and therefore load movement throughout the system.

Automated Process Planning. Automated process planning is a creation of process plans, with partial or total computer assistance, for items in a particular family.

Automated Storage and Retrieval System (AS/RS). Two categories of automated storage and retrieval systems exist, based on system size and the size of goods that they store. The larger high rise systems are referred to as a *unit loader*, while smaller versions, handling binnacle materials, are called *mini-loaders*. Essentially an AS/RS is a vast rack system into which parts are loaded and unloaded automatically.

System designs vary. An unmanned system is considered to be state of the art while other systems requiring human intervention are a variation on a state of the art units. Bar codes and optical character readers can identify loads as the loads are moved through the AS/RS system. Product loads can be stored in various manners including pallet, bins, pans or bundles hung from racks.

An AS/RS is intended to reduce costs of inventorying goods by improving storage and retrieval efficiencies, maximization of storage space, and improved inventory control. Work in process can be removed from aisles and shop floors and placed in an AS/RS.

The major draw back of an Automated Storage and Retrieval System is the capital expense. A misjudgment during a company's planning stage for the AS/RS can create an insufficiently sized system or an under-utilized one.

Automatic Acceleration and Deceleration. Automatic acceleration and deceleration is a control system that provides for smooth changes in the velocity of the machine toolslide.

Automatically Programmed Tools. In today's automated manufacturing environment, the initials APT are likely to have greater recognition than the term automatically programmed tools from which the initials are derived. APT is a programming language widely used in computer aided manufacturing to program numerically controlled (NC) and computer numerically controlled (CNC) machine tools. That is, once a part has been designed, analyzed, and approved for manufacturing, computer programs for the machine tools that are to process the part—cut, grind, mill, drill, bore, etc.—must be prepared. These computer programs will control tool motion of the NC machines according to the geometry of the part to be manufactured and the particular characteristics of the machine tool. An engineer, typically called a *part programmer* will utilize the APT language to create computer code that reflects the geometry of the part under consideration. That APT code will then be post-processed according to the specific control processor in the particular machine tool that is scheduled to manufacture the part. The post-processed code will provide motion control for the tool. APT has experienced wide spread success and industry acceptance because of its user-oriented command structure and its resulting ease of use by manufacturing and other noncomputer aided technical personnel.

Automatic Control. An automatic control system has one or more automatic controllers connected in closed loops with one or more processes. The individual controllers are instruments or groups of instruments which continuously measure the value of a variable quantity or condition and then automatically act on the controlled equipment to correct any deviation from a desired, preset value.

Automation implies the performance of a task without human assistance. Manufacturing processes are classified as manual, semiautomatic, or automatic depending on the extent of human involvement in an ongoing operation.

The primary reasons for automating a manufacturing process are to:

- Reduce the cost of the manufactured product through savings in both material and labor.
- Improve the quality of manufactured product by eliminating errors and reducing the variability in product quality.
- Reduce the lead time for manufactured product, thus providing better service for customers.
- Increase the rate of production.
- Make the workplace safer.

The economic reality of the marketplace has provided the incentive for industry to automate its manufacturing processes. The strategy of skilled labor sparked the drive toward automation in Japan and West Germany. Stern competition from Japanese and West German manufacturers in terms of both product cost and product quality has necessitated automation in the United States. Whatever the reasons, a strong movement toward automated manufacturing processes is being witnessed throughout the industrial nations of the world.

Automatic Dimensioning. Automatic dimensioning occurs when a CAD system automatically places the dimensions, guide lines and arrows on a drawing that has been created on the system by a user. The CAD system knows the grid dimensions of the workspace that the user has created the drawing on. The major portion of the automatic dimensioning capability is not readily visible to the outside world, such as the intelligence to not place dimensioning lines and numbers on top of drawing parts and to dimension objects at the most acceptable locations to aid in the viewing of the drawing. Automatic dimensioning makes dual-dimensioning very easy since the conversion between two different units of measure can be done quickly and easily by the system. Automatic dimensioning also has the ability to ensure that all of the necessary dimensions are placed on the drawing. This can eliminate rework required to correct the drawing at a later time.

Also see: CAD.

Automatic Identification. Automatic identification occurs when an unknown device or item is viewed or sensed by mechanical means and its identification is determined automatically by the system. Automatic identification can use sophisticated methods, such as video cameras, or simpler methods, such as bar codes or magnetic strips, to provide the necessary data. The UPC codes on products in supermarkets allow the laser bar-code scanner, in conjunction with the computer system of the point-of-sale data terminal, to identify the product and relate the description and price from a database. Railroad cars use either large colored bar codes or a semiconductor memory device which can be sensed, written to and read by a scanner. Automatic identification is a necessary part of developing total automation for some processes. As computers become faster and more capable, automatic identification in real time becomes feasible for use in automation systems in complex production process.

Also see: Automation.

Automatic Screw Machines. Turret lathes are often called screw machines when they are set up for bar stock. Turret lathes that are adaptable only to bar stock work are constructed for light work. They have spring collects for holding the bars during machining and friction fingers or rollers to feed the bar stock forward. Some bar-stock feeding devices are operated by hand and others are operated semiautomatically. Some screw machines are provided with power feed, but more often, both turret and cross slide are operated manually.

Automatic screw machines may be classified as single-spindle or multiple-spindle. Single-spindle machines rotate the bar stock. The tools are carried on a turret and/or cross slides, on a circular drum and on cross slides. Multiple spindle machines can have four, five, six, or eight spindles, each carrying a piece of bar stock. Capacities range from 1/8 to 6 inch (3.2 to 152.4 mm) diameter bar stock.

The feeds of the forming tools on automatic screw machines vary with the width of the cut and the width of the forming tool. Also the smaller the diameter of the stock, the smaller the feed should be. On multiple spindle machines where many machines are working simultaneously, the feeds should be such as to reduce the actual cutting time to a minimum. Often, only one or two tools in a set are working up to capacity as far as actual speed and feed are concerned.

A cutting fluid should always be used to increase tool life and to provide a better finish on the product. A rich emulsion is often satisfactory for general work provided that it does not interfere with the lubrication of the machinc. A light paraffin oil is satisfactory for brass. For aluminum, a paraffin oil plus 5 to 10 % lard oil is very satisfactory. On most steels, a sulfurous mineral oil or a sulfurous-chlorinated oil with some addition of fatty oil is best for general use.

The Swiss-type single-spindle automatic screw machine is not, strictly speaking, a screw machine, since it does not perform the ordinary screw jobs with the same facility as the conventional type of screw machine. Owing to the method by which the stock is fed, while rotating, by a sliding headstock through an adjustable carbide-lined guide bushing into the cutting tools, it should probably be known as a Swiss bushing-type or precision sliding-headstock automatic lathe. The Swiss-type automatic differs from the traditional automatic screw machine, which employs form tools, indexing turrets, and box tools, in that it uses single-point tools.

Swiss-type automatics with CNC eliminate the need for cams, minimize setup times and necessary operator skill, and are being used extensively for lower volume production requirements. Another important advantage of machines equipped with CNC is that the restriction to 360 degree layouts on cam-controlled machines is eliminated. This permits operations to be performed that are difficult or impossible to perform on cam-controlled machines. These operation include certain deep-hole drilling, single-point threading (both left- and right-hand), taper cutting, and contouring.

Swiss-type automatics with CNC made by one machine tool builder have simultaneous two-axis control for the X-axis toolholder and the Z-axis headstock. This permits straight, taper, and circular turning, as well as threading. Five tool-holders, arranged in a radial configuration on these machines, are controlled by a servodisc that comes with the machines and does not have to be designed and built. Programmed signals from the CNC unit command the disc to select the predetermined toolholders, which travel the required distances by rotation of the disc.

Attachments are available for the CNC machines. These include cross milling and drilling attachments, rotary guide bushings, and automatic bar feeders. Also available is a sliding attachment with four dead (nonrotating) spindles that moves laterally (front to back) and provides single-spindle back-drilling capability. Each spindle is positioned by hydraulic cylinder and locator screw.

Different designs of Swiss-type automatics are offered by other machine tool builders. With one design, movements of the toolslides are individually controlled by ballscrews. This permits repeatability of toolslide positioning within 0.00005 inch (0.0013 mm).

On another machine, there are no radial slides. Tooling is mounted on two five-tool turrets, one turret being indexed while tools on the other turret are cutting. On this machine, the headstock is fixed, but the guide bushing is mounted on a slanted carriage and moves with the tools.

Another builder offers a machine with a 12 station, 30 degree slanted, universal turret with servocontrolled movement in the X axis. The Z axis movement is derived from a servocontrolled sliding headstock. The headstock spindle provides for the C axis which permits (in conjunction with the turret tools) cross-machining operations. A four-position end working attachment with lateral tool positioning is also provided.

Some CNC machines permit programming by manual data input (MDI) or punched tape. The six radial toolslides on one machine are controlled by two axes, allowing simultaneous use of two tools. Another machine allows independent movement of six radial toolslides, the headstock, and a four-spindle endworking attachment. Any five of the eight programmable axes can be operated simultaneously at varying feed rates. a dual memory in the control permits production from one memory while the second memory is being programmed for the next workpiece.

Automatic Size Correction. Automatic size correction is an automatic device for making corrections in the machine or the cycle in response to gage signals in order to correct drift.

Automatic System for Position of Tools (AUTOSPOT). Automatic system for position of tools is a general-purpose computer program used in preparing instructions for NC positioning and straight-cut systems.

Also see: Numerical Control.

Automatic Tape Rewind. Automatic tape rewind is a system feature causing the input tape to be rewound to the initial starting block after reaching end of program.

Automatic Tool Changer. An automatic tool changer is an automatic mechanism typically included in a machining center which, upon the appropriate command, will remove one cutting tool from the spindle nose and replace it with another. The changer restores the used tool to the magazine, and selects and withdraws the next desired tool from the storage magazine. The tool changer is controlled by a set of prerecorded/predetermined instructions associated with the part(s) to be produced.

The use of automatic tool changers has a broad application, one which has expanded significantly from their original use as metal cutting tools. Automatic tool changers have been applied to productive equipment such as assembly equipment, measuring equipment, inspection equipment, drafting machines, typesetters, electrical wiring machines, tube-and-wire-bending machines, and many others.

When first introduced, automatic tool changers were typically controlled by numerical control (NC) tapes. Today, the recording and transmission of instructions to the changer may be done via a number of vehicles including NC tape, a direct link to a computer, or by a micro processor programmed with the instructions within the machine.

In all cases, the use of automatic tool changers can reduce the need for human intervention in the setup process. The results can be more precise alignment of tools, reduced setup/labor costs, and an overall increase in production throughput capability.

Also see: Numerical Control.

Automation. Automation is the process of making an activity, process, or machine function without the need for intervention of manual operators. Automated processes may interact with humans, either to disseminate information or to require certain activities. The majority of the operation is self-running and the process is dictating to the human when to do certain activities as opposed to the other way around. Some types of automation can function unattended for long periods of time such as pipelines and pumping stations. Due to the progressively higher cost of human labor, automation in industry continues to be a goal which is strived for by those attempting to decrease product costs, improve quality, and reduce time-to-market. In certain industries, automation has made it possible to achieve repeatability and quality which was nearly impossible when using manual operators. Automation also provides a guarantee of repeatability which eliminates the "human factor" of the possibility of a human intentionally or inadvertently altering the operation.

Auxiliary Function. An auxiliary function is a function of a machine other than the control of the coordinates of a workpiece or tool. Usually on/off type operations such as starting and stopping a spindle or coolant pump.

Axis. There are three definitions related to the term *axis*. First, it is the general direction of relative motion between a cutting tool and the workpiece.

Second, it is the reference line determining a coordinate, obtained by setting all coordinates to zero.

Third, it is the imaginary straight line which forms the longitudinal centerline of the tool or threaded part.

Axis interchange refers to the capability to put information related to one axis into the storage of another axis.

An axis is a traveled path in space; usually referred to as a linear direction of travel in any of three dimensions. Labels of "X," "Y," and "Z" are commonly used to depict directions relative to the Earth. "X" refers to a directional plane or line parallel to the earth; "Y" refers to a directional plane or line that is parallel to earth and perpendicular to "X"; "Z" refers to a directional plane or line that is vertical to and perpendicular to the Earth' surface. The sometimes used term *world coordinate* relates to the intersecting points of these three dimensions to the Earth surface.

An example of the X, Y, and Z axes is provided in the diagram which is seen below. The figure shows a robot performing plasma-arc cutting.

When defining robot axes, the phrase *degrees of freedom* is used. Degrees of freedom is meant

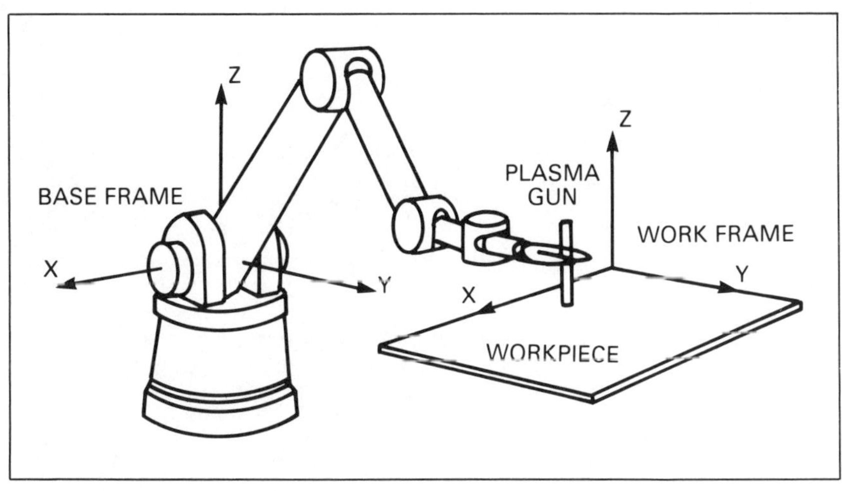

A robot performing plasma arc contouring with the X, Y, and Z axis illustrated.

to define the number of independent directions the end effector can be moved. The number of axes or degrees of freedom should be subdivided into:

- Major axes (motions): These axes are the number of independent directions the arm can move the attached wrist and end effector relative to a point of origin of the manipulator such as the base. The number of robot arm axes required to reach world coordinate points is dependent on the design of robot and configuration.
- Minor axes (motion): The axes may be described as the number of independent attitudes the wrist can orient the attached end effector. Relative to the mounting point of the wrist assembly on the arm.

Axis Inhibit. Axis inhibit is a feature of NC control which prevents movement of selected machine tool slides when power is on.

Also see: Numerical Control.

B

Back-annotation. Back-annotation is a CAD process by which data (text) is automatically extracted from a completed PC board design or wiring diagram stored on the system and used to update logic elements on the schematic created earlier in the design process. Text information can also be backannotated into piping drawings and 3-D models created on the system.

Also see: CAD, CAD-D.

Back-gage. A back-gage is a surface on two or more supports located behind the shear that can be positioned accurately either manually or automatically to control part size.

Also see: Gaging.

Background Processing. Background processing is the processing mode for executing lower priority programs which do not require user interaction.

Also see: Computers.

Backhand Welding. Backhand welding is a technique in which the welding torch or gun is directed opposite to the progress of welding. It is sometimes referred to as the *pull gun technique* in the gas metal arc welding (GMAW) and the flux cored arc welding (FCAW) processes.

Also see: Arc Welding, Welding.

Backlash. Backlash describes a reaction in dynamic motion systems where potential energy which has been created while the object was in motion is released when the object stops. The release of this potential energy or inertia causes the device to quickly snap backwards relative to the last direction of motion. Backlash can cause a system's final resting position to be different from intended and from where the control system intended to stop the device. In a servocontrol system, the servo constantly maintains the position using feedback from the encoder or resolver, so backlash error is quickly and automatically corrected. In a stepper control system, which is an open-loop control system, the control system is often unaware of the backlash condition. Not only does the system not correct for the problem, but when the next move is carried out, the stepper control system begins at an erroneous location. This causes the next stopping position to possibly also be in error unless the coincidence occurs of moving in the opposite direction of the previous move and experiencing precisely the same backlash. But now, it is in the opposite direction which cancels the original error.

Also see: Computers, Servocontrol, Stepper Control.

Backplane. A backplane is a printed circuit card located in the back of a rack or cardcage. A backplane has sockets into which specific boards fit for interconnection.

Also see: Computers.

Backup Copy. A backup copy is a copy of a file or data set that is kept for reference in case the original file or data set is destroyed.

Also see: Computers.

Bag Molding. Bag molding is a technique of molding reinforced plastics composites using a flexible cover (bag) over a rigid mold. The composites material is placed in the mold and covered with the bag; pressure is applied by vacuum, autoclave, press, or by inflating the bag.

Also see: Vacuum Bag Molding.

Band. Band is the range of frequencies between two defined lines. Band is also a group of circular recording tracks on a storage device such as magnetic drum, disk, or tape loop.

Also see: Computers.

Band Machining. See: Bandsawing.

Band Polishing. Band polishing is a variation of bandsawing that uses an abrasive band to smooth or polish parts previously sawed or filed.

Also see: Bandsawing.

Bandsawing. Power bandsawing, often called *band machining*, uses a long endless band with many small teeth traveling over two or more wheels (one is a driven wheel, and the others are idlers) in one direction. The band, with only a portion exposed, produces a continuous and uniform cutting action with evenly distributed low, individual tooth loads. Bandsawing machines are available in a wide variety of types to suit many different applications.

The cutting action of bandsawing differs from other sawing methods in that its continuous, single-direction cutting action, combined with blade guiding and tensioning, gives it the ability to follow a path that cannot be duplicated with power hacksawing and circular sawing. The bandsaw blade or band can follow the cutting teeth along any path over which it is guided, making radii or contour cuts possible. Band teeth cut with a shearing action and tend to take a full, uniform chip.

A variation of the vertical bandsawing machine uses computer numerical control (CNC). These CNC machines are used for contouring intricate shapes in materials that are difficult to machine. Such machines can often provide lower cost production than other methods for cutting various dies, stripper plates, electrodes, cams, and other complex-shaped parts.

Also see: Numerical Control.

Bandwidth. Bandwidth is a range of frequencies, usually specifying the number of hertz of the band or the upper and lower limiting frequencies. Bandwidth describes the range of frequencies that a device is capable of generating, handling, passing, or allowing. Usually with the range of frequencies in which the responsibility is not reduced greater than 3 dB from the maximum response. It is the difference between the limiting frequencies of a continuous frequency band in which performance with respect to some characteristic falls within specified limits.

In transmission circuits the term bandwidth-limited pertains to an ability to pass signals with frequency components that lie only within certain limits. An optical filter that passes only a narrow frequency band—for example, in the blue or red region only—is bandwidth-limited. Many light sources and photodetectors are bandwidth-limited because of the characteristics of the material used in their construction. Sometimes signals that are not themselves originally bandwidth-limited may be passed through filters or devices that introduce band limiting at either high or low frequencies. Hence, distortion is introduced.

Also see: Band, Hertz.

Bar Coding. Bar coding is an automatic identification technology. Bar coding is the predominant automatic ID technology in use today, even though several other methods of automatic identification exist and are in use also. Some of these other methods are machine vision, radio frequency identification, radio frequency data communications, magnetic stripe and smart cards, optical character recognition, and voice data entry.

Bar codes are a series of lines and spaces that represent information. The manner in which the lines and spaces are arranged is called a symbology, of which there are several defined symbologies. Some of the more well known are the Universal Product Code (UPC) which is used on most retail products. The European counterpart to the UPC is the European Article Numbering system (EAN) and the Japanese counterpart is the Japanese Article Numbering system (JAN).

Bar codes are read or "scanned" by passing a beam of light across the code and interpreting the fluctuations in reflection of the beam as it crosses the bars and spaces. The output of the scanner can be analog, digital, or ASCII information. If the output is either digital or analog, the data must be decoded and compared to the known representation of numbers and letters which the particular code represents to translate into ASCII information. In some scanners, the decoding function is done by the scanning head as a single integrated unit. The data is then entered into a personal computer, controller, or host computer system.

Bar code scanners use one of several light sources including incandescent, visible red LED, infrared LED, and laser such as helium-neon or solid state diode (a form of infrared) to read the symbol. Some scanners must contact the label to read, while others can read from a distance, sometimes up to several feet. Some readers are stationary, while others are hand-held. The beam of light from the reader can be fixed or moving. A fixed-beam reader must have either the reader (and its beam) moved across the bar code, or the bar code moved through the field of view of the reader. A moving beam reader can accommodate reading either a stationary or moving object.

While bar codes can be printed directly on many objects and surfaces, the more common method of applying a bar code is to use labels which are preprinted and then placed on the object to be identified. The use of preprinted labels allows the use of highly contrastive surfaces and nonsmearing label stock. Some objects would be difficult to print a bar code directly on due to their shape or size. Most pre-printed bar code labels are verified for satisfactory readability before being placed into service.

Discussion. Bar coding is essentially a digital representation of data used to represent more complex information. The use of dark and light bars is a binary representation. The relationship between narrow and wide bars and narrow and wide spaces is also a binary encoding scheme. However, the possible combinations of lines and spaces in turn represent ASCII data or in some cases, specially defined information.

In some applications where the security of the bar code label data must be maintained, the bar code is printed using black ink which is sensitive to infrared light while the entire area is then overprinted using ink or a translucent mylar strip. The strip appears to be opaque to the human eye, but allows the infrared light to pass through. These types of labels are sometimes referred to as "secure strip" type labels.

Several types of bar code readers are available to the consumer. Hand-held wands using solid-state LED light sources are the most common and the least expensive. However, with the continued reduction in cost of laser light source scanners, the quicker input and ability to read an entire label at the same time as opposed to having to scan a label from end to end, makes laser readers more desirable in most situations where the cost isn't an extremely critical factor. When trying to determine the labor cost per item of having a person scan a label with an LED wand verses a laser scanner, the cost of productivity is usually much less when the person uses a laser scanner. Applications such as retail point-of-sale customer check-out are much more effective when the clerk has both hands free to handle the merchandise.

Bar code scanners can be fixed-head type, or moving head. They can be hand-held or stationary-mounted. Either the scanner must move past the bar code label, or the label must move past the scanner. A laser scanner which scans an entire label at once must simply have its scanning beam coincide with the entire label at the same moment in time. Typically a laser reader scans many times per second, and decodes multiple times per second, although usually the number of scans per second is much greater than the number of decodes per second.

While some wands are contact readers, in that the tip of the wand must be placed in contact with the label surface to maintain the proper distance between the read head and the label, laser scanners are usually noncontact readers, plus the typical laser scanner has a considerable depth of field through which the scanner can perform an accurate read.

The performance of bar code scanners and readers varies among manufacturers. One of the parameters for judging the functional quality of a bar code reader is the first-pass readability. Factors such as the depth of field of the scanner and the method by which a wand decodes the scanning speed contribute to these functional specifications. Some wands use an averaging algorithm which recalculates the scanning velocity as each character is scanned. This practice allows a greater variation from a constant speed of the relative movement between the bar code and the reader.

There are several types of bar codes in use such as the LOGMARS, which is used more for government applications; and the UPC, which is used for retail consumer products in this country. One of the most widely used codes is Code 39, which can display the alphanumeric ASCII set. Code 93 is simply a smaller physical size (high-density code 39) version of Code 39. Some of the other codes which aren't as widely used, are ones such as Code 2 of 11 and Interleaved code 2 of 11.

Code 39 is so-named because there are a total of nine elements used to indicate a character. Of these nine elements (both spaces and lines), three are wide and the remaining six are narrow. This rule is adhered to for all characters in the Code 39. At the beginning of the code are two

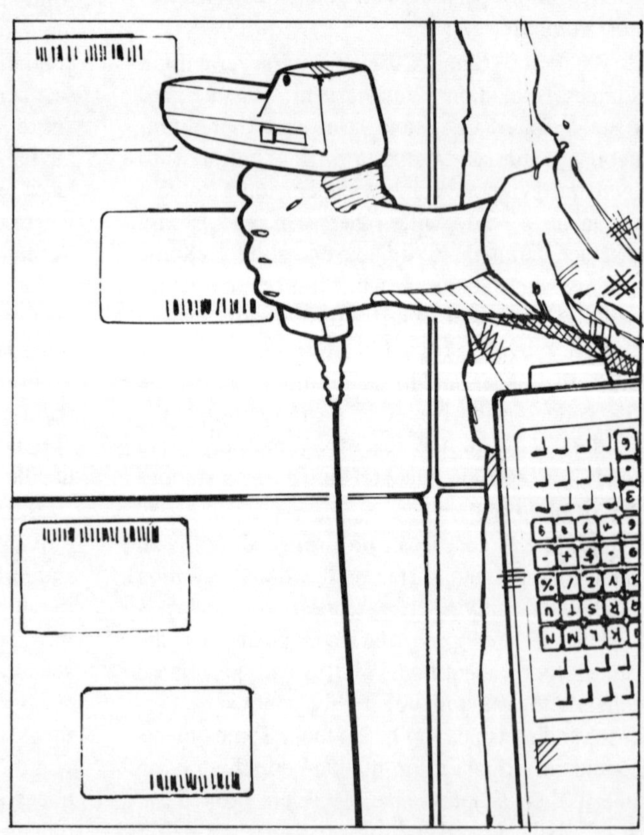

Bar coding at work in a warehouse.

bars and two spaces which are simply blocking elements to indicate the beginning of the code. This is useful since some wands read codes bidirectionally. For most wands and readers to function properly, the device must only see white before and after the code, at the time when the read is made.

Applications. Some of the applications for bar coding include the following.

Access control is the entrance to and egress from buildings or secured areas via automated identification technologies. Parking and revenue control such as parking structures that bill on a monthly basis or on an aggregate time usage basis can use bar coded cards to easily record entry and exit from parking structures. In personnel identification applications, bar codes are used on ID badges to allow automatic verification and recording of personnel activities. Time and attendance is a necessary daily activity which can be automated by use of bar code labels on time cards and ID badges.

Cataloging is the identification and retrieval of items and documents which are put into semipermanent storage. Library books are an example of the use of bar codes for identification, which is utilized to automate the checkout of the material, as well as inventory control. In some cases, codes assist the librarian in replacing the book into its proper place on the shelf. Some borrower's cards also have bar code identification labels which are read by the checkout system.

In data acquisition, bar codes are used as high-speed, low-error automated data entry instead of keyboard or other manual data entry. Sometimes test routines or extended command entries are preprinted as bar codes and used as menu selections. This practice is done when bar code wands are in use for data collection and therefore the operator does not have to put down the bar code input device to type on a keyboard.

Document tracking, such as sales orders and routing sheets for manufacturing operations are tracked using bar code identification labels on the sheet, which allows a central repository of data to be updated as to the whereabouts of the document at any point in time and to be aware of when a document changes defined locations. Item tracking is the specific task of tracking an item, whether it is an inventory item, a product, or a container such as a tote in a factory or a railroad car. The item's status, location, and relevant operator information as a function of date and time is maintained and recorded. Records keeping can be greatly assisted by the automated data entry capabilities and integrity of entered data using bar code labels.

Inspection is an application that may use bar code menus to report defects and conditions of a part during the inspection process, to avoid keyboard entry and ensure correctness of the data. The selection of the appropriate data is still the responsibility of the human operator. Machine utilization is a recording activity that can be easily accomplished by using bar code labels and scanning the appropriate information at the start and at the finish of the job. This information is summarized to determine the utilization of the machine, as well as the time required for each individual job. Production control such as scheduling, work-in-process tracking, inventory status and control, raw materials tracking and component parts management can be improved by using bar code labels on the parts or their containers. Quality assurance can be enhanced by using automated data collection techniques for predefined QA procedures and techniques. Work-in-process can be tracked the same as warehoused items by continuing to track the bar code labelled items throughout the process.

Inventory control is an area where bar codes have been used for some time with excellent improvements in the accuracy of the management of the data. Receiving and tracking inventory is done with bar code labels that are sometimes applied by the vendor before delivery to the customer. Most MRPII systems and shop floor control systems can accommodate the volume of data which is automatically generated by bar codes being read at each major point in the movement of the material. Package control systems, such as common carriers like Federal

Express and United Parcel Service use bar codes to track packages and the same principle applies in tracking material in a factory.

The use of bar code labels allows easy and accurate input of the data at each transfer point, which ensures the central tracking computer knows where each package is at any point in time. Shipping and Receiving functions can be accelerated by using bar code labels on all packages which enhances accuracy. Software distribution using bar coded labels eases data entry. Warehousing is an area which can be computerized by applying bar code labels to all items stored. A computer can keep track of the movement of warehoused materials.

Job costing is a function which is enhanced by the use of bar code labels on the component parts which allows an assembly or subassembly to be scanned to tally the cost of all the component parts. Order entry uses bar code labels to effect high-speed, low-error rate entry of data. Menu sheets of bar code labels are usually scanned to select the valid descriptions and allowable order quantities of individual items.

Consumer retail, uses product or SKU number identification, which allows the pricing and inventory control activities to be automated and integrated with the sale and check-out routines. The Universal Product Code (UPC) is an accepted standard listing of codes for types and sizes of products which is utilized by most U.S. merchants. Their individual point-of-sale computer systems apply the merchant's desired pricing to the product by using a table relating products to prices. Pricing changes can easily be instituted, with no need to correct existing price labels on the products. Only the price label on the shelf must be corrected. Federal law requires that the price be displayed somewhere, so if the product does not contain it, the display shelf must indicate the price.

Property identification can be effectively done using bar code labels such as for ''brass tag'' capital equipment identification as business assets. Vehicle identification such as automobiles or railroad cars can be done using bar coded identification.

The use of bar code labels provides greater accuracy of data transfer and allows some types of information to be exchanged without the knowledge of or any additional effort on the part of the human participants.

One example of the use of bar codes in industry was provided by the SME Technical Paper *Bar Code Applications in a Manufacturing Environment*. The paper was written by David J. Czaplicki of the INTERMEC Corporation.

Wrote Czaplicki: ''The use of bar code technology has exploded because it provides management with the timely and accurate information needed to survive in today's competitive marketplace. Nine different bar code applications have been utilized by an electronics manufacturer, starting with the visitors' security control system. Next is the dock-to-stock process along with receiving inspection, and inventory control. The toolroom, automatic testing equipment, and a paperless repair procedure also use bar coding. Shop floor data collection and in-process inspection also use bar coding.''

Czaplicki continued: ''INTERMEC began using bar code in its Lynnwood facility about five years ago. We did so for the following reasons:

- We believe that it really is helping us to manufacture better products today and to capture information more accurately.
- It is also giving us a vehicle to test new applications.
- It makes our employees more familiar with the products they're manufacturing.
- It gives us a showcase to exhibit to prospective customers.''

Czaplicki noted the following example: ''A truck makes a delivery to the receiving dock. When the product arrives it is usually shipped with a packing list and may have a bar code label on it. But, let's assume for the moment that it does not.

''The receiving clerk takes that packing list, goes to a terminal hooked to the System 38 and fills in the blanks on the screen. He is entering the purchase order number, the actual INTERMEC part number received, etc. This is not batch—this is real time. The next thing that happens is that a thermal printer, connected on-line to the system with the same application program, then prints up a label for every separate package that has been received. On this label are fields of human readable information that include a description of what is inside, its INTERMEC part number and a number also in bar code. This number is a 'license plate' that is sequentially assigned. It is an arbitrary number that is assigned by the computer. That number is linked to that data. Now by just scanning the bar code symbol, we know exactly what is in the package.

"The clerk then places one of the labels on every one of the cartons. He also tells the computer which cart he has placed the package on. One of the things that we are trying to do now is to send bar coded labels with our purchase orders to our vendors. That can save us one operation. Then, all we have to do is scan bar code labels when they arrive.

"Next, the cart moves to the next work center which is Receiving/Inspection.

"The tool room is another interesting application. This is somewhat similar to a library application. Basically you're trying to keep track of some valuable assets.

"Like most companies, we have tools, fixtures, calibration gages, etc., which need to be maintained and tracked and also need to be issued in an efficient manner to the production floor for their use. But we need to know who has what, and we need to know how often it is used, so that we can monitor calibration cycles. The way it works is that every tool has a bar code symbol on it. It has a flag character which begins with 'T,' so that we can uniquely identify it. If an employee comes up and wants to check a tool out, their bar code identification badge is scanned. The bar code symbol on the tool is scanned and a transaction code is scanned. So if the tool is being taken, that transaction code is scanned which tells the computer that this item is being checked-out. When it returns you scan the checked-in code. By going to a terminal it is possible to understand the status of any given tool."

The SME Technical Paper *Managing Tooling in an FMS* by Robert Olker of Kearney & Trecker Corporation provided an example of a bar coding application.

Noted Olker: "Electronic tool identification builds upon basic tool data parameters by providing a means of automatically identifying individual tools when they appear at the various stations, and then triggering the automatic transfer of the necessary parameters for that tool to the station."

He continued: "Machine-mounted sensors can be used at the transfer of tools from the delivery system to machine station storage to verify correct delivery and establish physical presence in the storage system." The figure on the next page provides an example of a bar code and laser reader.

Eric W. Roberts of ASK Computer Systems Incorporated discussed bar code applications in his SME Technical Paper, "Quality Assurance, Bar Coding, and MRPII: A Systems Approach to Integration." Roberts argued: "Computerized quality systems must be connected to MRPII systems in order to achieve true computer integrated manufacturing; and installation of a plant-wide bar code system can facilitate such connections by enabling fast and accurate data entry."

He continued: "Bar codes can play an important role in achieving the benefits of linking an MRPII system with a quality system. They can facilitate collection of the potentially voluminous amounts of data that a quality system can generate. And in doing so, bar codes can act as a 'catalyst' to spur reluctant manufacturers into action. To date, many manufacturers have balked at implementing computer-based quality systems due to concerns about the volume and accuracy of data. Bar coding systems may make them much more receptive to such implementation.

"The production and inventory control area would also benefit from such a link. Work-in-process quality control systems have one pri-

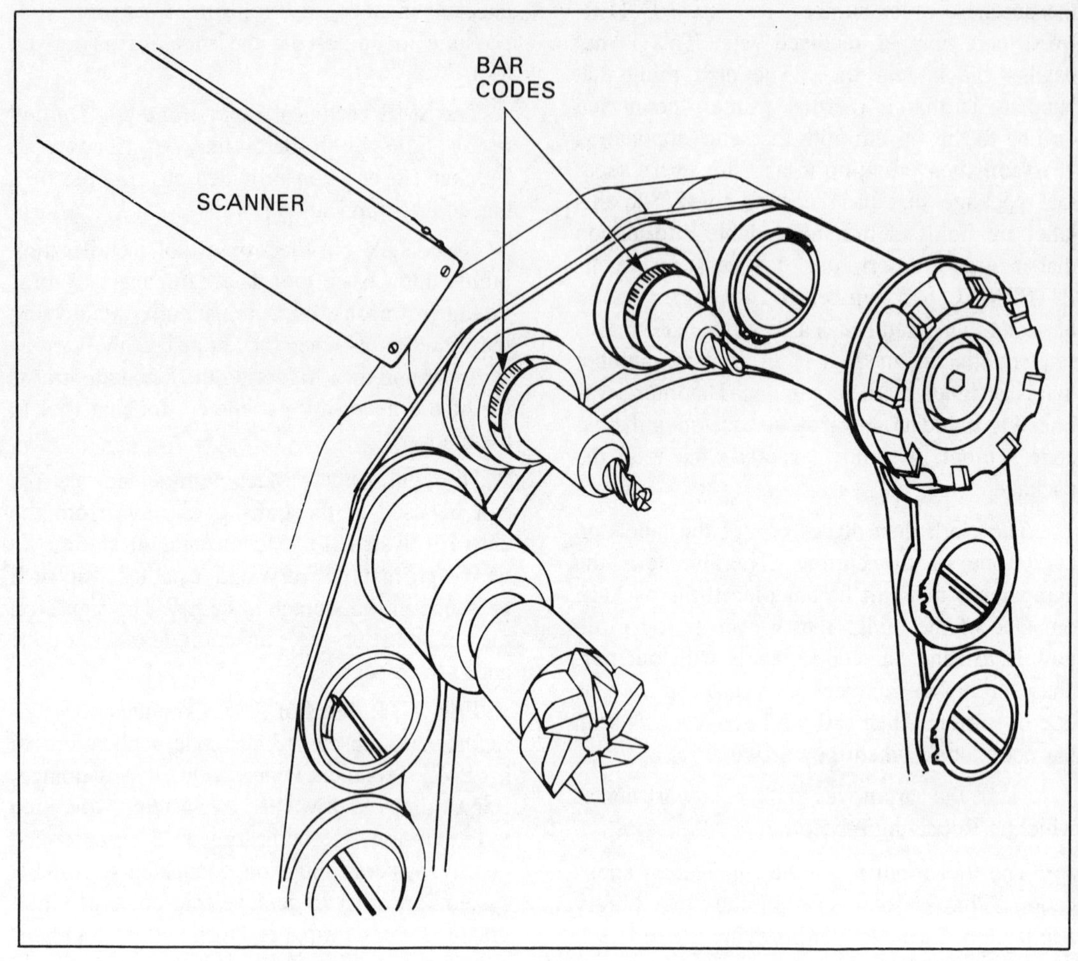

A bar code and scanner system is used at the transfer of tools from a delivery system to a machine station storage. The system verifies correct delivery and establishes physical presence in this storage system.

mary purpose, and that is the elimination of scrap and rework. The objective is to isolate and correct the root causes of product defects. Tying the quality system to the manufacturing system streamlines the collection of shop floor data. For example, a production operator could record data on quality characteristics while simultaneously recording operational data such as work order number, quantity completed, scrap dollars, and labor expended.

"One way to optimize the benefits of linking flexiblity to remain economical at widely ranging volumes will certainly increase the likelihood of installation.a quality system with an MRPII system is by installing a plant-wide bar code system. This will serve to facilitate the collection of quality data, as well as data required by other MRPII applications.

"There are three primary sources of bar coded information on the factory floor: menus, turnaround documents, and labels. A properly designed bar code system will integrate all of them. Following are definitions of each of these sources and suggestions as to how each can be used in a quality data collection system.

"A bar code menu is a page of frequently recurring data represented in bar code form. Examples in a quality application include fault codes, test area ID's, inspection center ID's, inspector numbers, operator numbers, machine numbers, work center numbers, and account numbers for posting dollars of scrap.

''Customized menus can be generated for each data entry station, or the same menu can be reproduced for use at multiple stations. Menus can be edited and regenerated instantly on-site. Since they're relatively permanent documents, menus can be inserted in plastic document protectors or permanently laminated. If a given station requires multiple menus, they can be placed in a three ring binder.

''Turnaround documents (TADs) are defined as any piece of paper printed locally containing bar codes from which data will be 'turned around' at some later time back to the host computer which originally printed the code. TADs typically have a much shorter life expectancy than menus, since they are associated with transitory orders such as sales orders, purchase orders, and work orders. Once the order is satisfied, the TAD is destroyed, unless it is needed as a paper archive to backup the computer record.

''The integration of quality and MRPII via bar codes is an exciting entry point into the world of CIM. However, eliminating manual methods of data collection with keyboards and bar code technologies is only the beginning. In the not-too-distant future, keyboard data entry will be obsoleted as plants convert entirely to bar code systems. Also in the near future, bar code systems will be complemented by voice data entry.

''Farther down the road, manual and semi-manual data collection methods will be entirely eliminated by automatic data collection. Data will be recorded at lightning speed with zero error by machines that talk to machines. And human operators will be completely relieved of data collection responsibilities.

''Of course, this scenario should not be interpreted as machines obsoleting humans. It will actually be a case of machines freeing humans from performing mundane tasks. Instead of collecting data, workers will be able to concentrate on higher-level job functions. They will be able to make better use of their most valuable resources—experience and reasoning.

''Bar codes are not an end unto themselves. They are not a panacea to cure all the ills of the shop floor. But they can help bring about a connection between quality systems and MRPII systems. A bar code system for quality purposes can be implemented as an integral part of a plant-wide bar code system. In contrast to a bar code system dedicated solely to quality control, a plant-wide bar code systems is far more cost effective and represents an important step on the road to CIM.''

Nancy E. Ryan reported on several industrial uses of bar codes in the article *Bar Coding Labeled Efficient and Flexible* (*Manufacturing Engineering*, June 1987).

The article explains: ''Various market reports indicate that the bar coding industry is in the midst of healthy growth. For example, a report recently published by Ross Associates (Needham, MA) estimates that total US bar code shipments (readers, printers, and labels) were $511 million in 1986 and will exceed $1 billion by 1990, with an average annual increase of 19%. *The Market for Bar Code Products: A Strategic Analysis*, a new study from Venture Development Corp. (Natick, MA), projects that US shipments of bar code products (input devices; printers; labels and other consumables; and software, system integration, and miscellaneous products) equaled $936 million in 1986 and will increase by 18.6% per year to $2.6 billion in 1992.

Several examples are also reviewed.

''General Dynamics Land Systems Division (Troy, MI) turned to bar coding to improve availability of materials needed in production, to provide better control of materials, to standardize tool crib inventory control operations, and to reduce related production costs. The Land Systems Division, which has an estimated $9.1 million tool inventory, has plants in Warren and Sterling Heights, MI; Lima, OH; and Eynon, PA. Bar coding is currently in use at the Detroit Army Tank Plant (DATP), which performs component machining, vehicle assembly, testing, finishing, and shipping of M-1 and M-60 tanks for the DoD. Divisional implementation of bar coding is slated for August 1987.

"The DATP tool crib is serviced by two terminals located at the crib's front window, each of which supports 20 satellite bar code wands. Production workers at DATP wear special bar coded badges that feature their six-digit employee numbers. To check out tools, a worker brings a completed requisition form to the tool crib window and displays his or her badge to an attendant stationed at one of the two terminals. The tool crib attendant wands the worker's badge; the worker then proceeds to the high-density tool cabinets that house the tools. Bar code labels representing tool identification numbers are adhered to the cabinets' interiors; attaching labels directly to tools is not effective. Labels representing quantities are located atop the cabinets, as are wands with extension cords. The worker scans the appropriate label with a wand, indicating the tools being checked out, and returns to the front window to complete the transaction. The worker can check in tools at the front window, too; the tools are designated as OK, lost, or broken, which is then recorded on ATICTS.

"Producto Machine Company (Bridgeport, CT), which manufactures special machine tools as well as tooling and precision components, is using bar coding to streamline shop floor control activities. Producto, a family-owned company with approximately 400 employees, has two main plants."

The article continues: "'Anybody can sit down at a terminal and find out exactly when a particular job was opened up,' Joe Brignola, system manager at Producto explains. They can determine exactly where that job is at any given moment. Bar coding helps us track all the jobs through all the operations.'"

Also see: Flexible Manufacturing System.

Barrel Finishing. Barrel finishing is a mass finishing process. It involves low-pressure abrasion resulting from tumbling workpieces in a barrel (usually of hexagonal or octagonal cross section) together with an abrasive slurry.

Also see: Finishing.

Base. A base is the quantity of characters for use in each of the digital positions of a numbering system. A base is also a number multiplied by itself as many times as indicated by an exponent.

Base Address. Base address is a given address in a computer instruction serving as a base, index, initial or staring point for subsequent addresses that are to be modified. Base address is also a number used in symbolic coding in conjunction with a relative address.

Also see: Computers.

Baseband. The bandwidth of a communication channel is the frequency range of a channel and is the difference between the highest frequencies and the lowest frequencies in the channel.

If pertaining to a radio signal, baseband, or narrowband, pertains to a range of frequencies that is narrower than 1% of the midband value of the range. Another characteristic of baseband is that it generates digital waveforms and uses time division multiplexing (TDM) schemes.

Baseband is also the band of frequencies associated with or comprising an original signal from a modulated source. In the process of modulation, the baseband is occupied by the aggregate of the transmitted signals used to modulate a carrier. In demodulation, it is the recovered aggregate of the transmitted signals. The term is commonly applied to cases where the ratio of the upper to the lower limit of the frequency band is large compared to unity.

Baseband bandwidth contains the narrow band (0 to 300 Hz) and voice band or grade (300 to 4000 Hz). Therefore, any bandwidth of less than 4 kHz would be considered to be a baseband. Typically video images are unable to be transmitted on a baseband communications network.

A radio baseband is the frequency band that is available for the transmission of all calls in the combined radio telephone or other radio signaling or data channels.

Also see: Manufacturing Automation Protocol, Technical and Office Protocol.

BASIC. BASIC or Beginner's All-Purpose Symbolic Instruction Code is the most widely known computer programming language. BASIC is a high-level language which is generally interpreted (originally BASIC was compiled) although compiled versions are becoming more prevalent as BASIC is becoming a more accepted language in which to write commercial applications programs. BASIC is a statement language, which was originally unstructured, but now is often structured. The data types are numerical and simple graphics. BASIC is an imperative/algorithmic, procedural language which is not extensible. BASIC evolved and was created from FORTRAN and ALGOL. BASIC has small memory requirements and is easy to learn and use. BASIC is readily available to run on most computers.

CBASIC and SBASIC are versions of the BASIC language. Proprietary hardware system implementations of BASIC are Applesoft BASIC, TI BASIC, and IBM BASICA. Other BASICs such as GW BASIC, True BASIC, and Microsoft's Microsoft BASIC and Quick BASIC are various offerings by software houses.

Also see: Computer Languages, Computers, Depreciation, Programming.

Batch Manufacturing. Batch manufacturing produces products through a process of controlled lot size (or batch size). This process is commonly used in the food industry (such as canning), paint industry, rubber, pharmaceuticals, glass, and steel.

The term *batch* is used to describe batches of parts in a discrete manufacturing operation. This definition can be inappropriate for the discrete grouping of items in an assembly application.

Batching in process manufacturing requires a predetermined mix of materials to produce the product. Specific factors such as heat, potential contamination, and time are critical variables taken into consideration. These factors may require a cyclical approach to the batch process and may effect equipment scheduling techniques. The term continuous batch is used in context with batch manufacturing in depicting the cycle as an ongoing process until specific operations are complete. The exact size of batches thus requires a planning process to match capacities with available equipment. The control systems used are dependent upon load control and take into consideration items such as shrinkage or potential waste. Batch manufacturing often includes discrete operations as part of the overall process.

Batch Processing. Batch processing is the technique of processing an entire group (batch) of similar or related jobs or input items on a system at one time without operator interaction. Batch processing was a typical method of running computer programs when large mainframes were used for centralized processing. The typical programs required extensive resources which had to be shared. Programs were queued into mainframes and at some later time the output was made available to the user. With today's proliferation of personal computers, distributed processing and personal computing have become the standard practice for many types of computing. However, for certain types of applications, batch processing is still an effective and efficient method. For example, a weekly payroll, a quarterly financial reconciliation or a monthly billing can best be done by running a program which handles the entire task in a batch mode at the appropriate time. Sometimes, the batch processing can occur during nonpeak user times such as overnight, which not only provides for better utilization of computing resources, but also allows the priority during peak times to be given to individual users who may desire interactive computing.

Also see: Computers, Programming.

Baud. A baud is a unit of signalling speed equal to the number of code elements (bits) per second. Normally, one data bit per cycle is transmitted but special equipment allows the transmittal of two to four bits per cycle in an octal train of signals.

Also see: Baud, Rate, Computers.

Baudot Code. A baudot code is used in transmitting data in which five equal length bits represent one character.

Baud Rate. A baud is the basic unit of modulation rate or signaling speed. One baud corresponds to a rate of one unit interval per second, where the modulation rate is expressed as the reciprocal of the duration in seconds of the smallest unit interval—i.e., of the shortest signal element. Thus, if the shortest signal-element unit interval is 20 msec, the modulation rate is 50 baud; in Morse code, if a dot is 10 msec, the speed is 100 baud. As a result, the speed in baud is the number of shortest discrete conditions or signal events per second. In bilevel signaling, the baud rate is always equal to or greater than the bit rate, since several signal elements may be required to represent a bit. It is an instantaneous rate, not an average rate. It can change from instant to instant depending on the manner of transmission and signal representation of pulses or bits.

Also see: ps (bits per second), Computers.

BCD. See: Binary Coded Decimal.

Behind The Tape Reader. Behind the tape reader is a means of putting data directly into a machine control unit from an external source other than a tape reader.

Benchmark. A benchmark is a reference that serves as a standard against which measurements or comparisons can be made. Benchmarks are usually thought of relative to dynamic performance of a device or system. When referring to computer systems, benchmarks represent a relevant, defined procedure whose result can be related to the typical activities that a user may undertake. A benchmark also may be designed to exercise certain portions of a device or system to illustrate the performance of that specific portion, such as access times for a disk drive or the time to repaint a graphics image on a monitor, or the time required to do a complicated mathematical calculation. A well-known benchmark includes floating point calculations or similar tests which are reproducible on different machines. Whether a system can perform a benchmark test is, at times, more important than the results of the test.

Also see: Computers.

Beta Site. This term denotes the start-up site for a newly developed product (software or hardware). It is most commonly associated with the phase of testing a newly released product at selected user site(s). This site becomes the proving grounds for the product and usually includes considerable debugging and further product enhancements to accomplish the results at that initial site(s). The initial testing is referred to as the alpha testing or in-house testing. This differs from the beta site in that the beta application is an actual implementation with real-life user requirements. Beta sites are noted for having problems and are often intentionally avoided by potential users of new software or hardware for those reasons. Added promotional incentives are commonly included to enhance the new users acceptance of beta testing the product. This commonly includes a concentrated support by the developer of the product during the installation and implementation stages.

Also see: Hardware, Software.

Bill Of Material. A bill of material is a list specifying the component parts (raw material and subassemblies) which are part of the assembly. The assembly is called the parent and the component parts are a list which will include at a minimum the part number and description of the material and the quantity required for one unit of the parent.

The bill of material file is a prerequisite for material requirements planning. The bill of material has many uses, including material requirements planning, production planning, order entry, product costing and disbursement of materials to the shop floor through pick or kit lists.

The bill of material should represent the actual components and subassemblies which are used and produced in the manufacturing process. The bill of material is essentially an engineering document. It must be defined to fit the needs of planning and manufacturing.

Bin Picking. Bin picking describes the process of obtaining goods/materials from predetermined locations (bins) within a stockroom or warehouse. The term bin depicts the actual type of container used to hold the material. Bin has been expanded to mean any type of physical location where material is stored. The term picking is the locating and physical movement of the material from its designated location.

A locator address system supports the picking activity in an effective manner. The address system identifies a designation for each location in a warehouse or stockroom. Specific considerations are made to determine proper lot sizing and space utilization of the stored material. The considerations take into account the type of bin picking process. Bin picking can take the form of a manual or automated process. The material handling equipment used must satisfy the space and structure and limitations of the storage area. The cost of labor can be a significant consideration in selecting and establishing a bin picking system. Automatic systems are designed with labor saving factors as a justification for expenditures for equipment. Automated forms of bin picking include two major areas—move the material to the picker or move the picker to the material. These may include moveable bays, bar coding, programmed robotics, and automated tracking movement systems. Bin picking processes use conveyors as an effective interface between manual and automated picking approaches.

The bin picking process records completed issues for each picking activity to document all transactions for proper accountability of costs. These transactions also drive material replenishment signals to both manual and computer manufacturing planning and control systems.

Also see: Bar Coding, Industrial Robots.

Binary. Binary is a numerical system pertaining to characteristics involving a selection or condition in which two possibilities exist.

Also see: Binary Circuit, Binary Code, Binary Coded Digit, Binary Digit, Computers.

Binary Circuit. A binary circuit operates as a switch indicating either of two modes: on or off. It also is known as *direct circuit* and *digital circuit*.

Binary Code. A binary code is composed by selection and configuration of an entity that can assume either one of two possible states. In binary, which is a base two numbering system, digits can be either zeros or ones. Since each place digit in binary communication can only have two states, large numbers in binary representation tend to have many digits. The words *binary digit* are sometimes abbreviated as "bit." Digital computers use binary code to perform all calculations since they are capable of manipulating only ones and zeros. Digital communication is a succession of basic pulses representing ones and zeros. In control systems such as industrial systems which use limit switches and operator pushbuttons, the two-state signal which is emanated by the device is a binary signal of either on or off.

Also see: Binary Coded Decimal, Binary Digit, Computers, Hexidecimal Numbering System.

Binary Coded Decimal. Binary coded decimal (BCD) is a decimal notation in which the individual decimal digits are represented by a pattern of ones and zeroes. Each group of four binary positions represents a single decimal integer. This code is one of the most common variations of the binary system. It employs only the first four binary positions with respective values of 1, 2, 4, and 8. Any decimal digit from zero to nine can be represented by a combination of these four values. In this system a separate binary equivalent is required for each digit of the decimal number being expressed.

For example, in binary coded decimal notation that uses the weights 8-4-2-1, the number 21 is represented by 00100001. This is compared with 10101 in the pure binary numbering system. Essentially in binary coded decimal, each digit of a decimal number is coded separately into a four digit binary number.

Coding in BCD is useful for situations where a translation must be made between a base two representation of a number and a base ten representation of a number.

Binary Digit. A binary digit is one of the members of a set of two in the binary numbering system; e.g. either of the digits zero or one. In lightwave communication systems, the presence or the absence of a pulse of light at a particular place and instant of time may be used to represent binary digits, patterns or sequences which can be used to code characters in the same manner as electrical pulses in wires or electromagnetic pulses in free space. The digit may be the individual light or electrical pulse itself. In the binary system of numbering, each digit can have only one of two values instead of one of ten values as in the decimal system. The binary digit is a member selected from a binary set, a set of two and only two items. The words *binary digit* are abbreviated as *bit*. The bit is a unit of information equal to one binary decision or the designation of one of two possible and perhaps equally likely states of anything used to store, convey, or simply represent information.

Binary Synchronous Communication Protocol (BISYNC). Binary Synchronous Communication Protocol is an early standard protocol for half-duplex communication, developed by IBM around 1965 and in wide use today.

Also see: Binary, Binary Synchronous Communications, Computers.

Binary Synchronous Communications (BSC). BSC is a data link protocol. Binary Synchronous Communications is one of the most widely used data link control methods in the U.S. Until the advent of SDLC, BSC was IBM's major data link control offering.

BSC is intended for half-duplex, point-to-point, or multipoint lines. BSC operates with EBCDIC, ASCII, or Transcode. All stations on a line must use the same code. If ASCII is used, error checking is accomplished by a vertical redundancy check (VRC) and a longitudinal redundancy check (LRC). If EBCDIC or Transcode are used, error checking is accomplished by cyclic redundancy checking (CRC).

BSC uses alternating sequence acknowledgment. The typical BSC message format is as follows: SYN (synchronizing), SOH (start of header), the header, STX (start of text), user data, ETB (end of block) or ETX (end of text).

Other BSC control functions are: EOT (end of transmission), ACK0 & ACK1 (positive acknowledgment), NAK (negative acknowledgment), ENQ (enquiry), and DLE (data line escape).

Also see: Binary.

Bipolar. Bipolar describes the construction of one of the two types of integrated circuits. The other type of integrated circuit is the MOS type integrated circuit. Bipolar integrated circuits (ICs) are characterized by the use of the principle element which is the bipolar junction transistor. Bipolar circuits are generally used where highest logic speed is desired, and MOS for largest-scale integration or lowest power dissipation. Linear circuits are mostly bipolar, but MOS devices are used extensively in switched-capacitor filters.

Large mainframe computers and supercomputers such as the Cray family of computers, use bipolar circuitry since the primary need is for computing speed. The large amount of heat generated by these machines is sometimes removed by liquid nitrogen or other refrigerants being applied directly to the chips. In most supercomputer circuits, excessive heat is the limiting factor for the throughput of the chip as opposed to the theoretical capability of the chip.

Also see: Integrated Circuit, Metal Oxide Semiconductor.

Biquinary System. A biquinary system is a method of representing decimal numbers, each digit of which is represented by one of two binary numbers plus one of five quinary numbers.

BISYNC. See: Binary Synchronous Communication Protocol.

Bit. See: Binary Digit.

Bit Map. A bit map is a grid pattern of bits (i.e. ON and/or OFF) stored in memory and used to generate the image on a raster scan display. Each bit corresponds to a dot in a raster display image. In addition, attributes of each picture element (pixel) can be represented by a group of bits.

The more sophisticated the attributes, such as color and hues, as opposed to shades of gray, or just black and white, the greater the number of bits required for the representation of the image. A large, detailed, color display requires a substantial amount of video memory, whose speed determines the performance of screen paints.

An example of a bit mapped display is the image created by an RGB (red, green, blue) monitor, where each location on the screen grid is represented by a pixel of each of the three colors; red, green, and blue. In addition, the intensity of each pixel color is respectively represented by additional bits.

Also see: Binary, Binary Digit (Bit), Computers.

Bit Plane. A bit plane refers to the virtual image planes within a graphic display monitor, which allow the conversion of a logical pattern of bits to be displayed as a corresponding array of pixels.

Also see: Bit Map, Computers.

Bit Stream. A bit stream is an uninterrupted sequence of pulses representing binary digits transmitted in a transmission medium, for example, a continuous sequence of bits in a wireline or optical fiber.

Serial communication is another example of a bit stream, as opposed to parallel communication, which is a byte stream, or word stream.

Also see: Bit, Bit String, Byte, Computers, Word.

Bit String. A bit string is a linear sequence of bits.

Also see: Bit, Bit Stream, Computers.

Block. In data communications, block is a group of bits or digits transmitted as a unit. An encoding procedure is generally applied to the group of bits of digits for error control purposes.

In computer programming languages, a sequence of one or more declarations or instructions that may be treated as a unit.

In computers, a block may refer to a group of information which is manipulated upon as one contiguous unit. The term block may also refer to segments of memory which are utilized as a continuous segment.

In computer applications there are preset rules for the formation of blocks for purposes of transmission, storage, checking or other functions regarding data.

In packet data transmission, the packets are blocks of data of a predefined size, which are constructed from the various messages requiring transmission. The packets are assembled at the head end and disassembled at the receiving end where the original messages are reconstructed.

Also see: Block Address Format, Block Delete, Block Diagram, Computers.

Block Address Format. Block address format is a tape programming method for numerical control (NC) systems in which only instructions that need changing are punched into the tape. Block address format is a means of identifying words by use of an address specifying the format and meaning of words in the block.

Also see: Block, Computers, Numerical Control, Programming.

Block Delete. Block delete is a feature permitting selected blocks of tape to be ignored by the control system when indicated by the operator.

Also see: Block.

Block Diagram. A block diagram is an orderly, structured geometrical representation of a process, procedure, or system. A flowchart is a specific type of block diagram which shows logical decision-making processes. Block diagrams may be used to assist in repairing products or in the assembly of a device. A block diagram is usually used to allow the creation of a representation without the extensive time and work of drawing detailed pictures or using actual photographs.

Structured design techniques employed today for such tasks as software design and systems design have software tools to aid in the creation of block diagrams. Some of these tools prompt the user in the creation of block diagrams as well as provide some functions automatically such as connection elements.

Operator panels sometimes display block diagrams to aid in the recognition of the function of controls. Automobiles and similar complex products utilize block diagrams to show information such as electrical wire routing and electrical wire connections.

Also see: Flowchart.

BOM. See: Bill Of Material.

Bomb Out. A bomb out is a complete failure of a computer routine resulting in the need to restart or reprogram the computer.

Also see: Computers, Programming.

Boolean Algebra. Boolean algebra is a process of deduction reasoning using a symbolic logic and dealing with classes, propositions, or on-off circuit elements such as AND, OR, NOT, EXCEPT, IF, THEN, etc., to permit mathematical calculations.

Boolean algebra was developed by George Boole, a mathematician, around 1854. Boole's purpose was to enable logical problems to be dealt with in algebraic notation. Boolean algebra established the basis of mathematical logic.

An example of applying Boolean logic is to consider the ''OR'' operator which when used with two inputs, the output is true when either one or the other input is true (or also when both inputs are true). Another example of Boolean logic is to consider the ''AND'' operator which when used with two inputs, the output is true only when both inputs are true. Normally truth tables are used to show the matrix of possibilities of the Boolean logic.

Boolean algebra allows the combination of multiple Boolean operators into complicated algebraic statements.

Also see: Computers.

Boot. A boot is a program used to start the computer, usually by clearing the memory, checking devices and loading the operating system into internal RAM from internal or external memory.

Also see: Bootstrap, Computers.

Bootstrap. A bootstrap is a technique for loading the first few instructions of a routine into storage, then using these instructions to bring the rest of the routine into the computer from an input device. This usually involves either the entering of a few instructions manually or the use of a special key on the console.

Also see: Boot, Computers.

Boring. Boring is the enlarging of a hole that has already been drilled or cored. Generally, it is an operation of truing a hole, that has been drilled previously, with a single point lathe-type tool. Boring is essentially internal turning, in that usually a single point cutting tool forms the internal shape.

Drilled holes are frequently bored to eliminate any possible eccentricity and to enlarge the hole to reaming size. Boring is also used to finish holes to correct size, as is frequently done on large holes or on odd-sized holes for which no reamer is available.

Most boring operations use a single-point cutter because they are simple to set up and maintain. The cutter is mounted in either a heavy bar or a boring head. The bar or head serves to transmit power from the machine spindle to the cutter as well as to hold it rigidly during the cutting operation. The workpiece is normally stationary, and the rotating cutter is fed through the hold. For precision boring work, it is necessary to use a tool with a micrometer adjustment. Such tools are held in a cutter head and rotate. Hence, any increase in hole size is obtained by micrometer adjustment of the tool radially from its center. In production work, boring cutters with multiple-cutting edges are widely used. These cutters resemble shell reamers in appearance but are usually provided with inserted tooth cutters which may be adjusted radially to compensate for wear and variations of diameter. Boring tools of this type have longer life than single-pointed tools and are more economical for production jobs.

Also see: Boring Machines, Numerical Control.

Boring Machines. Boring machines make round holes in parts. Several machines have been developed that are specially adapted to boring work. One of them, known as a jig borer, is constructed for precision work on jigs and fixtures. Similar in appearance to a drill press, it will do both drilling and end-milling work in addition to boring. The vertical boring mill and the horizontal boring machine are adapted to large work. Although the operations that these machines perform can be done on lathes and other machines, their construction is justified by the ease of economy obtained in holding and machining the work.

Jig boring machines resemble a vertical milling machine, but are constructed with greater precision and are equipped with accurate measuring devices for controlling table movements. These machines are designed for locating and boring holes in jigs, fixtures, dies, gages, and other precision parts.

The vertical boring mill is so named because the work rotates on a horizontal table similar to an old pottery wheel. The cutting tools are stationary, except for feed movements, and are mounted on a vertical tool head which is in turn mounted on an adjustable-height cross slide. The vertical boring mill is able to hold large heavy parts, since the work can be placed on the table with a crane and does not require much fixturing to hold it in place.

The horizontal boring machine, which differs from the vertical boring mill in that the work is stationary and the tool is revolved, is adapted to the boring of horizontal holes. The horizontal spindle for holding the tools is supported in an assembly at one end which can be adjusted vertically within the limits of the machine. A worktable having longitudinal and crosswise movements is supported on the bed of the machine. At the opposite end of the machine from the spindle column is an upright to support the outer end of a boring bar when boring through holes in large castings.

One type of numerically controlled precision-boring machine is a horizontal, ballscrew-actuated, CNC machine. Often, such machines have multiple spindles for high-volume production. These machines are also available with one, two, or four spindles and can operate at high metal removal rates with close tolerance capabilities. They also offer the flexibility necessary for high-volume production. Electric servo-axes and variable-speed spindle drives allow infinite programmable variations of spindle speeds, feed rates, and tooling-path control.

A versatile CNC boring machine for machining odd-shaped workpieces consists basically of a rotating head with a feed-out slide on which various tools can be mounted. The workpieces are held stationary. The machine has two axes of CNC motion. The first axis is the slide that moves radially on the rotating head to control bore sizes and facing feed rates. The second axis is the entire headstock moving relative to the workpiece or the workpiece moving relative to the tool. This provides axial motion for boring feed rates and controlling the depths on flanges, faces, etc. The combination of both motions can produce tapers or contours.

Modular design allows the machine to be built with a stationary head and rotary table on a slide; the head movable on a slide and the rotary table stationary; the head movable on a slide and the rotary table mounted on a cross slide; dual heads to machine both ends of a workpiece simultaneously; and other configurations.

Also see: Boring, Jig Boring, Numerical Control.

Bpi (bits per inch). Bits per inch is the number of bits that can be stored per inch of a magnetic tape. This is typically used to refer to nine-track reel-to-reel magnetic tape.

Also see: Computers, Magnetic Tape, Recording Density.

Bps (Bits per second). Bits per second is the number of bits occurring at or passing a point per second. Values of modulation rate in baud and in bits per second are numerically the same if, and only if, all of the three following conditions are met: (1.) all pulses (bits) are the same length, (2.) all pulses (bits) are equal to the unit interval (the time element between the same two significant instants of adjacent pulses). (3.) binary operation is used.

In n-ary operation, the bits per second (b/s) equals the modulation rate in baud multiplied by the logarithm to the base 2 of N, where N is the number of distinct states used in the digital modulation process. A distinct state might be the amplitude, phase, or frequency.

A common error by most people is to assume that baud and bits per second are interchangeable terms in any and all circumstances. For example, when communicating using a modem, the actual data transfer rate may not be as great as the baud rate, due to retransmission for errors, and the fact that some data may require multiple electrical pulses to represent a unit of information.

Also see: Baud Rate.

Brake Forming. Brake forming is the use of a pan brake to bend sheet metal. A brake may be manual or powered, especially for heavier stock. A brake does not cut the metal as a shear does, only bend the stock into any preferred angle. If the material is very stiff or thick, a progressive bending sequence is sometimes used which means multiple bends are made to develop the finished angle.

When bending metal, the amount of stock used by the bend must be calculated and considered when the size of the flat stock was determined. Another allowance which must be made when bending metal or other springy substance is to allow for spring-back. This means the initial bend must over-bend so the final spring-back position is at the desired angle.

Also see: Forming.

Branch. When referring to computers, a branch is a part of a computer program executed as the result of a program decision. The difference between a branch and a subroutine is that a branch diverts the flow and does not provide for any return to the point of diversion, whereas a subroutine executes and then upon completion, returns program flow to the statement immediately following the statement calling the subroutine. However, program flow can be directed to any point in the program at the end of a branch, including the statement following the branching decision statement by merely including a ''Goto'' statement as the last statement in the branch.

Programs can have an unlimited number of branches and branches may exist for various reasons. Structured programming techniques foster the use of subroutines instead of branches to eliminate the situation of branching to a part of a program and not being able to return to the continuing program flow, getting into an endless loop, or simply completely loosing track of the logic when trying to program a complicated program.

Also see: Computers, Programming.

Braze Welding. Braze welding is a process variation in which a filler metal, having liquids above 840° F (450° C) and below the solids of the base metal, is used. Unlike brazing, in braze welding the filler metal is not distributed in the joint by capillary action.

Also see: Brazing, Welding.

Brazing. In brazing, a nonferrous alloy is introduced in a liquid state between the pieces of metal to be joined and allowed to solidify. The filler metal having a melting temperature of over 800 degrees F (430 degrees C), but lower than the melting temperature of the parent metal, is distributed between the surfaces by capillary action. *Braze welding* is similar to ordinary brazing except that the filler metal is not distributed by capillary action. In this process, filler metal is melted and deposited at the point where the weld is to be made. In both cases special fluxes are required to remove surface oxide and to give the filler metal the fluidity to wet the joint surfaces completely.

In *torch brazing* the heat is applied to the base metal by an oxyacetylene or other gas flame supply. In *furnace brazing* the work is placed in a controlled atmosphere furnace with the filler metal properly positioned on the joint and brought up to a suitable temperature. *Dip brazing* consists of immersing pre-assembled parts in a bath of molten brazing metal. In *induction brazing*, the joint is rapidly heated by an induction current applied directly to the joint by an electric coil. Filler metal is applied to the joint prior to induction of current.

One advantage of the brazing process is the effecting of joints in materials difficult to weld, in dissimilar metals, and in extremely thin sections of metal. In addition, the process is rapid and results in a neat appearance that requires little or no finishing work.

Applications for brazing are widely varied, and the process is used in practically every industry, from jewelry to aerospace. High-vacuum, refrigeration, and air-conditioning equipment manufacturers use the process extensively in making leak-tight joints. Applications in the automotive industry include the production of accessories and steering wheels and in joining tubing. Aircraft and aerospace uses include making honeycomb structures, engine nozzles, tubing joints, and many other brazements.

Heating and plumbing manufacturers and contractors use brazing for joining pipe, tubing and headers. Seams and spud connections for water heaters and tanks are made by brazing. Pasteurizers, separators, and tanks for dairy equipment are also brazed. Electrical wires, cables, and bus bars are joined by brazing, and the chemical industry uses the process for producing tanks, vats, and piping.

The selection of brazing versus welding or other joining processes depends primarily on the sizes of the components and the materials to be joined, configurations of the joints, thicknesses of the sections, the number of joints to be made, and service requirements. Brazing is generally more suitable for joining smaller parts because of the difficulty in applying the required heat to larger components.

However, parts as large as 70 inches (1780 mm) in diameter and 15 feet (4.6 m) long are brazed. The process is also usually preferred for joining thin sections to thick sections because reduced heat requirements minimize distortion.

The use of lasers as a source of heat for brazing and soldering is a relatively recent development. A major advantage of using lasers is the capability of selectively applying heating energy to small areas without heating the entire component. Another advantage is close control of heat input and the ability to locate or position the heat accurately. The use of lasers also controls grain structure and intermetallic formations.

Because of their advantages, lasers are used for the production of miniature and thin precision parts, to make joints near thermally-sensitive connections, and for connections inside evacuated or pressurized vessels or containers.

Brazing is also being done automatically with various heating methods, including torches, furnaces (controlled atmosphere, including vacuum), resistance, induction, dip, and infrared methods. Accurate control of temperature and time at temperature is essential for any automated process. Filler metals are either preplaced at the joint during assembly of the components, or, in some processes, are automatically fed into the joints at the brazing temperature.

Automatic applicators for paste alloys (mixtures of powdered filler metal and flux in a paste binder) have expanded the use of automatic brazing and soldering systems. Such systems include rotary index, in-line, fixed station, and shuttle machines.

Various degrees of automation are being used. Completely automated systems can include automatic fluxing (if paste alloys are not used), in-line inspection and cleaning, simultaneous brazing of multiple joints, and continuous brazing operations. The cost of automating is generally justified by increased productivity and, in many cases, by the material and energy savings.

Also see: Braze Welding, Lasers, Welding.

Breadboard. A breadboard describes the rough model for constructing another device, usually referring to a circuit board that has a specific configuration on it. A breadboard is created to test or prove functionality. A model is fabricated to illustrate appearance. In some situations a physical model of an idea or product may be required to illustrate or demonstrate the product's functionality when the product is too difficult to visualize.

Before computer simulation was available, the standard method for creating an electrical circuit was to design the circuitry on paper, then build a breadboard to debug the logic and timing of the circuit. Since the advent of behavioral modeling, timing analysis, and other types of electrical circuit simulation, breadboarding is a practice which is now considered outdated since the overall development time and costs are greater if a physical breadboard must be built and manually tested and evaluated.

Also see: Computers, Computer Simulation, Model.

B Register. B register is a hardware element which holds a number that can be added to or subtracted from the operand address prior to or during execution of a computer instruction.

Also see: Computers, Hardware.

Brinell Hardness. In the Brinell Hardness Test, a known load is applied for a given length of time to the surface of a specimen through a hardened steel or carbide ball of known diameter. The diameter of the resulting permanent indentation is then converted to the Brinell hardness number by the use of standard tables provided in a handbook with the testing equipment.

There are two types of Brinell testing machines in general use: the hydraulic and the dead-lever loading types. Providing rings are generally used for calibrating the hardness-testing machines for load. A microscope specially designed for measuring Brinell indentations in generally used for diameter measurements. It contains a scale having 0.1 mm graduations and permits an estimation of 0.01 mm to be made. A separate graduated scale is usually supplied for calibrating the microscope.

The standard ball diameter is 10 +0.01 mm with a permissible variation in the diameter of any one ball of +0.0025 mm. The hardness of the ball should be such that, when it is pressed with a load of 3,000 KG against a piece of steel having a Brinell hardness number of 500, the permanent change in diameter should be no more than .00025 mm. A tungsten-carbide ball is generally used for measuring Brinell hardness numbers greater than 500. The type of ball used is generally reported for Brinell hardness numbers greater than 500.

Broaching. Broaching is the operation of removing material by an elongated tool having a number of successive teeth of increasing size which cut in a fixed path. It is a method of stock removal that has its most common applications in metal working but can also be used as a method to work wood, hard vulcanized rubber, graphite, and some plastics.

The machine used may be hand-operated (arbor press) or electromechanically or hydraulically operated. Both the cutting tool (broach) and the work piece are held in rigid fixtures. The principal function on the machine is to provide the speed necessary for cutting to take place. Cutting speed or travel of the broach across the work is usually 15 to 25 times faster than in milling. Speeds of 200 SFPM or more are common, although most broaching operations are done at lower speed. The feed (or material removal rate) is regulated by the broach, which consists of a hardened steel bar with a series of cutting teeth. Each tooth is made progressively higher, to remove successfully larger amounts of materials. Thus, the cut grows deeper as the operation progresses.

The process of surface finishing by broaching may be compared roughly to planing, except that in broaching a multitoothed tool is used. In both cases, the cutting tool may be either passed across a fixed workpiece or the tool may be held stationary and the work moved in a continuous stroke. In some machines, the tool is pushed along the surface to be broached while other machines pull the broaches through or over the work.

Machine control systems for broaching can range from single to sophisticated. Advanced control systems available on some machines can accommodate any required sequence of ram motions, leading, clamping, workpiece movement, unclamping, and unloading. Programmable controllers are used on some machines. While numerical control (NC) and computer numerical control (CNC) do not ordinarily lend themselves to the simple requirements of broaching, they have been used for some special applications.

In addition, many different material handling devices are available to automate the broaching process. Broaching machines have been incorporated into transfer lines that also perform other operations such as drilling, boring, and milling.

The April 1982, *Robotics Today* reported that robot loading and unloading was contributing to fully automatic operation of a blind spline broaching machine.

The article noted: "Representing the latest advancement in blind spline broaching, the machine can be used for a variety of workpiece configurations where the part cannot be passed completely over or through the broach. Typical applications include keyways, gears and cams, as well as splines. The basic broaching technique permits production of blind splines at speeds up to 10 times faster than operations such as milling and hobbing.

"Blank parts are delivered by gravity roller conveyor and rotated into a vertical position for the pick-and-place robot to position the part in the load station. The table indexes the workpiece beneath the broaching ram, which moves down to pick up the part with a hydraulically expanding collet. As the table indexes form station to station, the part is successively inserted and withdrawn from each tool to generate the depth of cut and the involute forms.

"When the splines are completed, the ram lowers the part into the final unload station, and the collet releases it. At the same time, the load robot places a new blank part into the load station. On the next index, the unload robot picks up the finished part and deposits it onto another gravity conveyor. Simultaneously, the blank part moves under the ram, and the cycle repeats.

"The load and unload systems, which are positive action cam-operated, are protected from overload. Although the robots are synchronous in operation, inlet and outlet conditions control the machine. For example, if there are no parts on the inlet conveyor, or if the outlet conveyor is blocked, the machine will not cycle."

Also see: Industrial Robots.

Broadband. In regard to a radio signal, broadband, or wideband, pertains to a range of frequencies that is broader than 1% of the midband value of the range.

If pertaining to a voice frequency signal, broadband is a group of voice frequency channels that use a bandwidth greater than 4kHz or greater than 20kHz. Both 4kHz and 20kHz have been proposed as the limiting width between broadband and narrow band for voice channels.

In telephone systems, broadband can also refer to the transmission of signals over a frequency range that is produced by a modulated carrier system, usually in the range from 60kHz to 20MHz.

Another distinguishing characteristic of broadband is the method of signal generation. Broadband uses a radio frequency (RF) modem to generate an AC signal. The signal lies within a preassigned bandwidth and data is used to modulate the RF carrier at that fixed frequency.

In most applications broadband communications channels are usually divided into several smaller, slower channels, as required. Broadband uses frequency division multiplexing (FDM) to provide multiple channels. This allows a single communications cable to be utilized for multiple applications and different devices.

Examples of broadband communications systems include fiber optic cable networks, multichannel telephone cable, and multichannel line-of-sight microwave systems.

Also see: Fiber Optics, Manufacturing Automation Protocol.

Brushing. Brushing was developed where it effectively and economically replaced the older sandpaper and fill methods of removing burrs, scratches and similar mechanical imperfections from precision and highly stressed components. The greatest appreciation has been made in the manufacture of gears and bearing races where the removal of sharp edges and stress raisers by power methods has increased the speed of the operation.

Wheels with suitable abrasive compound and wire-bristle wheels are used. The wheels can be operated on flexible shafts, bench grinders, polishing lathes or automatic polishing machines. For any particular operation there is usually an optimum wheel and bristle material size, operating speed, and technique. Some applications for brushing are deburring of engine pushrods, deburring the ends of airplane tubing, removing hard and tough insulation from the ends of instrument wire feed, removing insulating chemical finish, and the removing excess silver plating.

BSC. See: Binary Synchronous Communications.

BTR. See: Behind The Tape Reader.

Bubble Memory. Bubble memory is a type of memory device using discrete, microscopic magnetic cells in an aluminum garnet substrate. Bubble memory was developed by IBM as a type of memory which could be utilized in areas of high vibration, and other conditions typically adverse to computer memory storage devices. Bubble memory is a nonvolatile memory, which means the contents of the memory are not lost when power is removed from the bubble memory device. While the concept of bubble memory has been proven and products have been in use for some time, the price/performance relationship hasn't propelled this type of memory into the forefront as a preferred solution for most situations. However, for situations of severe shock and vibration and where a completely sealed, nonrotating memory device is needed such as in factories or on military equipment, bubble memory does continue to be the product of choice.

Also see: Computers.

Buffer Memory. Buffer memory is a section of memory used for temporary storage of data allowing transfer between the main CPU and storage devices, without slowing the CPU to the maximum transfer speed of the device. Buffer memory is sometimes referred to as cache memory, especially when it is used as a queue between the main RAM memory and a hard disk or as a stack queue for instructions waiting to be executed by the main CPU. Buffer memory often has faster speed memory than the remainder of the main memory to allow fast execution by the processor, without the expense of the entire RAM memory being of such high speed and higher cost. Buffer memory is also used as an interface for data between dissimilar computing entities or communication protocols or formats. Buffer memory is used as a means to accelerate the transfer of information between a RAM disk and main memory.

Also see: Computers.

Buffing. Buffing is the smoothing and brightening of a surface by an abrasive compound pressed against the work or material to be buffed by means of a soft wheel or belt. The abrasive is a fine powder or flour mixed with tallow or wax which forms a smooth composition or paste. This composition or paste is then applied as required to the buffing wheel or belt, which is made of a pliable material, such as soft leather, linen, felt, or muslin. Buffing is possible on the same types of machines as polishing and frequently both types of wheels are included on one machine.

Buffing lathes, jack stands, buffers and buffing heads are available in both bench and floor types with motor, shaft, or with V-belt drive.

There are many different types of buffing wheels. They include: fabrics, full-disk buffs, packed buffs, pieced buffs, bias buffs, finger or spoke buffs, and goblet buffs.

Also see: Finishing.

Bug. A bug is an anomaly within a program which causes results which are unexpected and undesired. A bug may not always be obvious and readily visible, however the effect of the bug may be extremely obvious. A bug can be caused by improper syntax in the program, improper program structure, or even faults in the programming language itself. With compiled languages, it is fairly common to have one or more bugs in the compiler itself. Sometimes the bugs do not manifest themselves right away, which can cause serious problems for the user of the program. Any hardware or software used for life-sustaining types of activities is subject to significant liability if the functioning isn't exactly as it should be.

A special type of bug is what is termed a software virus. A virus is an intentionally created bug which may reside dormant with no activity until certain conditions are met, such as after the software has been operating for a certain length of time or until a particular point in the program is reached, or until a certain point in calendar time has come to pass. Once the predetermined factor has occurred, the software virus comes to life to wreak havoc on the program, and furthermore, the virus spreads throughout the program, hence the name virus. Usually the creation of a software virus is done only with malicious intent.

Also see: Computer Languages, Computers, Programming, Software.

Building Block System. A building block system is a computer system that can be made larger and more complex by adding modules that permit additional operations to be performed.

Also see: Computers.

Bulk Memory. Bulk memory is a high-capacity memory device, such as magnetic tape, disk, or drum, used to store large quantities of data.

Also see: Computers.

Burn-in. Burn-in is operating a device at an elevated temperature to improve the probability that any device weakness will cause a failure. If this failure can be made to occur during the burn-in period (infant mortality), the defective component can be replaced before the unit is shipped to the final customer. There are many different opinions regarding how long a burn-in period should last and what the parameters of the burn-in should be. There is also varied opinion regarding whether a product should be powered-up, or even operational (exercised) during the burn-in period. Burn-ins are sometimes conducted at elevated temperatures, both to try to force any marginal failures, and also because of the premise that if the product does not fail at a temperature which is higher than any normal operating temperature, then the product should not fail during normal operation at normal operating temperatures.

Also see: Acceptance Test, Computers.

Burr. A burr is an undesirable projection of material that results from a cutting, forming, blanking, or shearing operation.

Also see: Deburring, Fettling, Finishing.

Burst. In a data communications system, a burst is a sequence of signals counted as a unit in accordance with some specific criterion or measure.

This term is sometimes used to represent the maximum speed of data transmission assuming ideal conditions with no retransmissions due to errors and no propagation delays.

A burst is usually unmodulated and uncontrolled and simply output at the slow rate.

Also see: Baud, Bps (Bits per second).

Bus. A bus is a circuit or group of circuits providing a communications path between two or more devices, such as between a CPU, peripherals, and memory, or between functions on a single PC board.

Some well-known bus structures are the STD bus, the S-100 bus, and the VME bus. The Q-bus, VAX bus, and the VAXBI (VAX Bus Interconnect) are well-known bus structures of Digital Equipment Corp. Personal computers have provided the PC bus and the AT bus, and more recently, the Microchannel bus of the PS/2 which is the new generation of personal systems from IBM.

Also see: Computers, Manufacturing Automation Protocol, Network, Technical and Office Protocol.

Byte. A byte is a sequence of adjacent bits, usually eight, representing a character that is operated on as a unit. The byte may be a group or string of pulses. Bytes are used to constitute alphanumeric characters, for example, to represent decimal digits or letters. The American National Standard Code for Information Interchange (ASCII) and International Telegraph Alphabet Number 5 use 7-bit bytes (8-bit bytes including the parity bit) to represent the characters and control signals.

An *n*-bit byte is a byte that is composed of *n* binary digits, where *n* can assume any positive integral value. Normally *n* is not greater than 16, while 4 and 8 are common. A byte is usually shorter than a word.

A byte is a measure of the memory capacity of a system, or of an individual storage unit.

Also see: Bit, Computers, Word.

C

C. C is a computer language used for systems programming such as writing compilers, operating systems, and editors. C is also used for application programs.

Also see: Computer Languages.

Cable. A cable is an assembly of electrical conductors that are insulated from each other, but usually twisted together around a central core and wrapped in a heavy insulation.

Also see: Computers.

Cache Memory. Cache memory is a high-speed memory used as a buffer between a fast central processing unit (CPU) and a slower main memory in a computer system.

CAD. See: Computer Aided Design.

CADAM. See: Computer Graphics Augmented Design and Manufacturing.

CAD/CAM. CAD/CAM (Computer Aided Design/Computer Aided Manufacturing) is the use of computer based techniques to aid design engineering activities and manufacturing engineering activities.

Though many CAE (Computer Aided Engineering) analytical techniques may be resident on a CAD/CAM system, the distinguishing feature of CAD/CAM is its display terminal which displays a graphic representation of a part or assembly. Using a variety of input devices, the designer creates a design for a part, an assembly, an electronic schematic or a layout, complete with dimensions and notes. The workstation or terminal displays a graphical representation of the design that interactively reflects the changes the designer makes, allowing one to monitor one's work.

The CAD/CAM system creates a geometric model of the design which is stored in a database. Once a design is created, it is never necessary to redraw it; it is simply recalled from the database and reused. From the information contained in this model, the system is capable of generating multiview representations of the part automatically. The model is also available for retrieval and modification or reuse in similar designs, or by other activities.

History. The enabling technologies for CAD/CAM—interactive processing, graphics display terminals, and means of representing product geometries or models—were not developed until the early 1960s. In this period, the expense of the hardware, the cost of software development, and the limited availability of the vector refresh technology for graphics terminals severely restricted the use of CAD/CAM.

In the early 1970s, the CAD/CAM industry was born with the formation of the first commercial suppliers of CAD/CAM systems. These suppliers took advantage of developing technologies—the minicomputer and the storage graphic screen—added CAD/CAM software, and offered turnkey systems. The cost of these systems was still high, and installations were generally limited to aerospace and automotive companies.

In the late 1970s, the use of CAD/CAM began to spread, fueled by its acceptance among early users and the increase in the performance price ratio of computers. The early 1980s saw the advent of two new hardware technologies which represented large leaps in cost-effectiveness to end-users: the microprocessor based intelligent workstations and raster display screens. These technological advances, coupled with continued market acceptance, management concerns regarding engineering productivity, and recognition of some of the strategic advantages of CAD/CAM, caused the breadth and depth of market penetration to increase. At the same time, the development of the solids modeler represents a next generation software for mechanical CAD/CAM which is being implemented by innovative users.

Discussion. Turnkey systems with the supplier providing hardware, software, support, and perhaps user-specific applications software are commercially available for virtually all design areas.

CAD/CAM systems are all computer based, with CPU, memory, input devices, output devices, and communications equipment. The distinguishing feature of all CAD/CAM systems is the graphical display hardware and software.

Initial systems used dumb graphic terminals connected to powerful mainframe computers. This configuration evolved to minicomputers with dumb terminals in the early 1970s. As microprocessors became more powerful, intelligence was added to the terminal, evolving first to smart terminals and then to workstations. Workstations may be fully stand-alone, or linked to a centralized processing unit. Today, depending upon the level of sophistication, CAD/CAM systems run on everything from mainframe computers to desktop PCs, with many in-between configurations available.

Unlike analytical and business computing technology, CAD/CAM requires the input of graphical information. CAD/CAM systems, in addition to a terminal keyboard, typically have a stylus, joystick, or mouse available as means of entering graphical data.

CAD/CAM software consists of six levels: the operating system, interactive graphics software, CAD/CAM systems software, the modeler, applications software, and user-developed software. The last four levels are the CAD/CAM specific software.

The interactive graphics software draws the images on the screen, manages the input/output devices associated with the screen, and communicates with the CPU. In some configurations, the graphics software resides in a smart terminal or workstation with its own CPU. The CAD/CAM systems software manages the user interface and database.

The management of the CAD/CAM database is a key issue in the implementation. Interfaces between CAD/CAM and other systems, MRP for example, are frequently desirable, if not required. Also, one of CAD/CAM's greatest strengths—the ability to retrieve an old design for modification—will be directly affected by the database management. Finally, CAD/CAM allows the creation of a database of standard parts. These standard parts can be incorporated into assemblies without being redrawn or redesigned, greatly increasing designer productivity.

Modelers are either two-dimensional, three-dimensional, or solids' modelers. All modelers have restrictions, in terms of points, edges, faces, and bodies per part. Two-dimensional modelers are used in applications where three or more views are not essential to product definition, e.g., design of printed circuit boards, electrical schematics, facilities layout, and map making. Traditional three-dimensional modelers are wireframe and surface modelers.

With wireframe modeling, lines and arcs are used to describe the edges of an object. Wire frame modelers, because they do not show surfaces, generally display hidden lines. Surface modeling provides a definition of the surface of a part. Sculptured surfaces such as rotor blades and aircraft wings can be modeled as complete geometrical information is available for each point of the surface.

Solid modeling describes not only a part's edges and surfaces, but also its internal space.

Solid modeling, being the most comprehensive, is also the most complicated and generally requires a powerful computer system.

Overall, CAD/CAM is a tool for improving the quality of engineering information and its dissemination. As such, it decreases design turnaround time, decreases the number of engineering design changes (and the production problems that caused the engineering design changes), decreases rework costs and costs of errors due to disinformation, and increases product quality.

In his SME Technical Paper *Features of Successful Use of CAD/CAM in Industry*, John Stark of Coopers & Lybrand Associates Europe notes:

"For CAD/CAM to work well within a company, it is necessary to understand not only what CAD/CAM is, but also what it is not. Without such understanding, it will almost certainly be badly organized, badly managed, and badly used. It is as different from business applications of computing (such as payroll) and other classical EDP department applications (such as production planning), as engineering as an activity is different from financial activities and manufacturing planning activities. This difference means that it may not be possible to manage it in exactly the same way as classical EDP applications are managed.

"It is also important to understand that there is not just one CAD/CAM solution, any more than there is just one CAD/CAM system available on the market. There are many systems available and they are all different. Some are more applicable to particular industries and applications than others.

"It is also necessary to understand that CAD-/CAM is not a turnkey solution to all design engineering and manufacturing engineering activities. It is just a technique that can be used in these areas and, like all techniques, will not lead to success without good management, trained users, and a lot of hard work.

"It must be understood that CAD/CAM is an information management technique, and has company-wide implications greatly exceeding those of computerized drafting. There are computer programs that can be used just to improve drafting productivity, but they do no more than that, do not offer all the benefits of CAD/CAM, and should not be thought of as CAD/CAM programs.

"It is useful to describe briefly some of the major factors leading to successful CAD/CAM. Although these are described one by one, it must be stressed that CAD/CAM is an ongoing activity, and that many of these factors relate to activities that are not on-off, but will be repeated several times or on a continuing basis.

"One extremely important factor is top management commitment to CAD/CAM and corresponding involvement in the CAD/CAM implementation and management process. This commitment is necessary from the earliest stages, ensuring that the right financial approach is taken to the justification of potential investment in CAD/CAM. Implementation of CAD/CAM, in particular successful implementation of CAD/CAM, will lead to change, and top management must manage such changes. It may be necessary to assign one top manager the responsibility for developing an organization in which the CAD/CAM technique can flourish. Middle management must be trained to manage projects using CAD/CAM. New procedures will affect not only engineering staff, but also managers and staff in other functions (such as marketing and purchasing) who use data derived from the CAD/CAM system. They too must be committed to CAD/CAM success.

"Without a thorough understanding of the company's CAD/CAM requirements, an investment in a system can easily turn out to be a waste of money. There are all sorts of CAD/CAM systems, some good at some applications, some good with particular products. Without taking the time to find out just what the company requires, it will not be possible to select the most suitable system.

"An important but often overlooked factor in successful use of CAD/CAM is good basic knowledge of the applications (i.e., of drafting, NC programming, etc.). Some managers believe that CAD/CAM will allow them to dispense with highly skilled workers, replacing them with lowly paid unskilled staff. Such a belief is both false and self-defeating. It is just those people with the best knowledge of the basic applications who can make the most of CAD/CAM, and in so doing save time and money. This is particularly true for innovative products.

"It is important that the users of CAD/CAM have a positive attitude towards it. This can only be brought about by making them aware, from the start, of the reasons for its introduction, and by showing that both the CAD/CAM system and its users are necessary components of the company's future. It is important during the initial training period that users learn to use the system efficiently and confidently. Experienced staff should be available to help them when they finish the training period and start to use the system on everyday project work. Special motivation should be provided to attract users to CAD/CAM, e.g., the possibility to work flexible hours, the possibility of career development in CAD/CAM. In addition to initial training for the users (and it should be noted that depending on the system this requires from a few days to several months), training and education in CAD/CAM need to carried out on an ongoing company-wide basis.

"All CAD/CAM systems are different, just as all companies are different. Selection of the right system to meet a particular company's requirements is a major factor of eventual success. Key issues in system selection include the basic hardware and software, the type of product modeller required, the way the data is managed, the level of competence required to use the system, the possibilities to interface the system to other applications in the company, and the possibilities of future development of the system.

"Support staff must be available to assist users to make good everyday use of CAD/CAM. Not only must they be responsible for short-term operational tasks such as workstation scheduling, but they must be able to decide with middle managers which projects would gain most from use of CAD/CAM. The support staff must be well organized, well managed, and should be firm believers in the usefulness of CAD/CAM. They must develop and document the procedures leading to efficient system use.

"CAD/CAM systems have become more and more powerful over the last few years, and there is every reason to believe that this trend will continue in the future. To ensure that a company benefits from such improvements, it is important to select a system with a clear potential for further development and a good track record of past development. Hopefully, most of the future developments required by the company will be carried out by the system vendor, but many companies will decide to carry out a few company-specific developments themselves. They will only be able to do this if means are provided by the vendor to include company-specific software in the system, and to access the system's database.

"Finally, it is worthwhile pointing out that selection and implementation of a CAD/CAM system are among the very earliest steps along the road to successful use of CAD/CAM. Many companies have behaved as if they were the only such steps. However, it is clear that it is only when the CAD/CAM technique is in everyday use, that it can start to produce productivity gains. Management must ensure on an ongoing basis that savings really are being made, and that CAD/CAM is meeting the original objectives and targets. Since the investment in CAD/CAM is so high, underutilization of CAD/CAM means that large sums of money which are needed elsewhere in furthering business opportunities will also be lost. One way to check whether CAD/CAM use is meeting objectives is to carry out a CAD/CAM Audit."

Applications. CAD/CAM applications include electronic design and manufacture, mechanical design and manufacture, and architectural design.

In its simplest form, CAD is merely a drafting tool—an automated way of generating a schematic or multiview drawing, complete with dimensions and notes. Tolerance checking and design rule checking add sophistication to CAD as a drafting tool. Additional checking applications include interference checking, tolerance build-up analysis, gap checking, and collision checking.

In addition to drafting and checking of drafting, CAD/CAM database management adds information management capabilities. Configuration control that details the changes between different revision levels of a design has been incorporated into some CAD/CAM systems. Group technology coding and classification, making retrieval of past designs easier, also is available.

In the engineering design function, CAD/CAM may provide kinematics analysis, simulation of how parts move relative to one another, and finite element analysis for computation of stress deflection, or temperature distribution throughout a part.

Unlike traditional engineering, where a model of a design is recreated by different people in different stages in the manufacturing process, with CAD/CAM the model is created on one database and is accessible by different people. Thus the model created by design engineers is used by manufacturing engineers for their own purposes. CAD/CAM applications software may include several manufacturing engineering applications programs. Flat pattern generation in which a three-dimensional sheet metal part is unfolded into its corresponding flat pattern aids in planning the manufacturing process, as do process planning software, aids for tool and fixture design, and NC machine programming.

Most CAD/CAM systems available today are capable of generating NC part programs to machine a component designed on the system. This achieves apparent integration between CAD and CAM. However, the user still must manually enter crucial manufacturing data such as cutting conditions, tooling, and machine information. An automated process planning system has the capacity to provide this information.

The major impact of CAD on tool design will be in the areas of tools and techniques used in accomplishing the design process. Design tools such as pencils, scales, sketch pads, and pocket calculators will eventually become obsolete as they are replaced by computer tools such as keyboards, system displays, digitizers, and plotters. In addition the techniques used in designing a tool and documenting the design will also undergo significant change.

CAD is a rapidly developing and changing technology. The future of CAD in tool design will be one of continuous change and improvement. Hardware development has produced equipment that is vastly superior to its predecessors, and hardware technology can be expected to continue improving.

Improvements to processors will provide more memory, allowing many more workstations to be interfaced with even smaller, more powerful central processors. Along with the additional memory will come faster processing time, decreasing even further the amount of time required to design closely toleranced tools. Color displays with larger, thinner screens, but high resolution, can also be expected. Finally, voice-activated function menus will become a standard feature of CAD systems.

Software developments will probably overshadow hardware developments, however. Improved three-dimensional modeling programs will become more readily available and less expensive, as will programs for motion and testing through simulation. However, the most significant trend with regard to the computer in tool design will be the continuously improving partnership between CAD and its manufacturing counterpart CAM.

Automation in the manufacturing of tools is a concept that is even older than automation in tool design. Numerical control (NC) machining processes have been used since the early 1950s. These processes were improved to include computer numerical control (CNC) and direct numerical control (DNC). The current phase is computer aided manufacturing or CAM.

CAM involves using the database created when designing a tool on the computer to prepare the instructions for operating automated machines which will make the tool. At present, this is a concept that is being used principally by the largest, most advanced companies. However, continuous technological developments, coupled with the need for constant improvement of productivity, will make broad-based adoption and use of CAD/CAM techniques commonplace.

What is the future in CAD/CAM? Senior Editor R. Lee Martin addressed this point in the December 1987 *Manufacturing Engineering* feature titled *CAD/CAM—An Even Fuller Menu Ahead*. In this article he wrote:

"Will the idea of integration seriously impact the future of CAD/CAM? Both vendors and users alike talk about integration—creating interfaces so that one vendor's system will communicate with another's. In the U.S., more so than in Europe or Japan, it appears that users are still buying the best niche. They continue to look for the pop, bang, and sizzle in a variety of products rather than consider how they will later get them to communicate, to work together. They figure they'll solve the problem in-house, or that the standards issue will solve the integration problem for them. But it's not as easy as that. The types of platforms coming out on the market today are becoming very large and complex—everything from the IBM/PC-XT to the PS/2 to the new 386 lines, from RISC architecture to the mainframe.

"Just imagine what they will be like in the year 2000. Says Dick Bilden, technology project manager at Dewberry and Davis (Fairfax, VA), "I've been involved with AutoCAD for years. Its ability to move CAD/CAM operations from the computer room to the engineer's desk has opened many doors. But a problem I'm beginning to see is that the technology—hardware and software—is coming on so fast that people might not be ready. The technology is advancing almost faster than our ability to absorb and implement it.'

"Further complicating this is the fact that there are two views on integration—one held by traditional electronic data processing (EDP) people and one by the engineering community. EDP looks at integration as the ability to have centralized file management with archiving and backup all in one system that doesn't require running around from machine to machine, building to building, or office to office. It sees centralized electronic mail and communications, the ability to mix and match hardware platforms of different speeds and different operating systems all on a single network. But that is computer integration, not CAD/CAM data integration. If you talk to the engineering community, you'll find it not only wants to be able to have those systems talking to one another, but if the design is on an IBM/PC in the design system and it needs to be uploaded to the manufacturing system on a Sun workstation or any other machine, the user doesn't want to have to go through intermediate steps to move the file from platform to platform. This problem gets even worse when you start to think about revisions and the inability to go immediately from a DEC format to an IEEE format. What will happen to those changes while the company waits for someone to write the conversion? Now couple this with the reality of having to deal with 15,000 -20,000 databases per customer, and you begin to see the challenge of CAD/CAM data integration.

"In many cases, integration means that the company will have to make a long-term, large-scale corporate commitment to choose a vendor who is going to provide that key integration, not just across platforms—PC to workstation to mainframe—but across all systems using a common database with a transparent conversion process with no manual intervention. That will mean choosing what Jacobsen (Carl Jacobsen, Computervision) refers to as 'strategic partners'—either a single vendor, a group of vendors, or in-house experts—who will provide that absolutely necessary, key integration.

"From a practical point of view, future CAD/CAM users will probably opt for a variety of components to configure their systems. This will force vendors to make systems more compatible and to integrate them and establish standards. According to (John West, Cimlinc, Incorporated, Elk Grove Village, IL) the day of the proprietary product is over: 'Users have become sophisticated enough to recognize that they don't have to be locked into just one vendor. And they don't want to be. This new attitude will force the issue of standards and help make network computing work.' In the April 1987 edition of *Computer Aided Design Report,* consultant Dave Albert offers these comments concerning single vendors: 'I think what people are going to do in the long run is look for the optimal solution for a portion of their design problems and then live with the problems of transferring data between various CAD/CAM systems as opposed to going for one grandiose solution that has a common database that satisfies all problems.' David Gossard, a professor of mechanical engineering at MIT and Director of its CAD laboratory, agrees: 'What we are engaged in here—call it CAD/CAM, CAE, or whatever the acronym of the next decade will be—is a continuous process. In the '70s we had drafting systems, in the '80s came solids modeling, and there will be other waves of new technologies and new ideas that come along as the user community becomes more sophisticated and more educated.' The wise buyer will identify what tools are available at a given moment,' he adds, 'and try to use themost cost-effective ones for his company's business.'

"Standards will be part of what helps to drive integration. Certainly, as users become more sophisticated and require more integration from vendors, standards will begin to develop. But in this situation, standards will be user-driven as opposed to vendor-driven. Part of this stems from a lack of industry-wide standard manufacturing methods. If such standards could be developed and adopted, vendors could more easily justify R&D funding knowing that their work will be valuable for a large portion of the market. In the absence of such standards, the best a vendor can do is to provide generic tools that can be customized to meet a specific company's requirements.

"By the year 2000, the IGES (Initial Graphics Exchange Standard) Committee will have advanced graphics standards as well as product definition exchange specifications (PDES). Plotting formats, standard spoolers for plotters, and standard plotter files that are either de facto standards or industry standards are all going to appear. According to Dr. Joel Orr of Orr Associates in the April 1987 issue of Datapro's *Manufacturing Automation News and Perspectives,* 'As for standardization, there are vendor representatives who are saying, "It's not in my company's best interest to cooperate with the standards effort, and, therefore, I'm not going to go along. I'm going to gamble that my additional features and efficiency and lower price will prove sufficiently attractive to keep a strong customer base in spite of the fact that everybody else is going some other way."' Orr indicates that that attitude presents opportunity for business, and the driving force is the business force. People, he indicates, are in business to make money. Although people can be idealistic about engineering and networking and databases, their only value is what they contribute to a company's ability to stay in business by using them. That, ultimately, is the driving force. Vendors who remember that it's the driving force for their customers as well will probably survive better than those who don't.

"In this environment, there will be a variety of software packages being utilized on a number of workstations in a common network. Because of this, a high degree of integration will be necessary between the various pieces of software and the various platforms that the software is run on. Tony Affuso, Electronic Data Systems (Troy, MI), offers this opinion: 'CAD/CAM will become very critical as a means of communications in the product delivery cycle. We envision the information that is created in product delivery will be delivered from CAD/CAM. Therefore, CAD will become an integral element in further automation. In order to accomplish this, however, there must exist an extremely high degree of system integration."

Karl H. Schultz, Schultz Associates in his SME Technical Paper, *Complementing Computer Aided Manufacturing (CAM)* explains that "the interlinking of CAD and CAM will create the need for change in an organization. It is generally understood, that the biggest barriers to using new manufacturing technology are managerial not technical. This is managerial at all levels—foreman to the top executive. There is also a need to break down the functional 'wall'/ 'turf' between engineering and manufacturing if there is any. Implementing CAM, as part of CAD, will be most difficult if there are strong functional walls. Some organizations are joining product and manufacturing engineering under one manager to breakdown these walls.

"People must have a win/win versus a win/ lose attitude because this program crosses many-functional barriers. Communicate with people at all levels and keep people informed along each step of the program."

In the book, *New Directions Through CAD/ CAM*, authors William Beeby and Phyllis Collier discuss other CAD/CAM applications.

The authors note: "In the mid-seventies, John Deere & Company engaged in a review of the company's computer developments that included more than 200 employees. Deere brought in consulting firm Arthur D. Little to help in the review. Participants classified their systems according to two categories: corporate-wide CAD/ CAM systems and stand-alone systems whose-data were not normally required by other groups in the company. Furthermore, they determined four separate building blocks necessary to a computer-integrated manufacturing system: a geometric modeling system to create surface definition data, group technology for easier parts identification, process planning to establish a process database, and a computerized inspection and reporting system. Deere began an educational effort to acquaint employees with the findings and plans resulting from the review.

"As an example of productivity gains, after a successful evolutionary process, and with the use of a personal computer, the tool-grinding department was able to reduce the overall tool inventory at Deere from 32,000 to 16,000, though the same number of tools are finished—about 13,000—per week. And using a Digital Equipment PDP 11/23 stand-alone system, the maintenance shop successfully combined three shops into one, more efficient shop. These are just two examples of Deere's realization of productivity gains due to a systematic and well thought out computer enhancement to the workplace.

"Another organization, GE's Computer Aided Engineering and Manufacturing Council, determined from a survey of 160 GE businesses that 56% of primary benefits due to CAD/CAM implementation was because of productivity increases.

"Using seven interactive graphics terminals, Grumman Aerospace Corporation was able to reduce total design time by 50% and drafting time by 7,000 hours, or 10% compared with manual methods. The manufacturing engineering group, using only two terminals, was able to pare tool design time by 12% and programming time by 50%.

"At Boeing, in the last six months of 1980, 350 distinct part numbers were programmed into an interactive graphics system, which eventually produced 15,000 parts. The graphics system has reduced the average parts programming time by more than two hours per part.

"Repetitions of numerical control tape tryouts have been reduced from 3.3 per tape to 2.3 per tape. In 1981, this tape tryout reduction translated into an increase of approximately 1,000 hours of machine capacity.

"Also at Boeing, an automated floor drilling procedure was established when engineering began transmitting geometric data to the large host computer for availability to the shop. Next, a drill unit was designed to take instructions from a microcomputer. Servomotors move the drill heads on the drill unit framework, also directed by the microcomputer. The main deck of the 747 contains 345 panels and 23,000 fasteners that require 28-1/2 workhours per aircraft to install. This automated system drills 30 holes per minute for a cost savings of 85% over the manual drilling operation. Drilling 23,000 holes in the floor is a highly repetitious process, making it a made-to-order computer application.

"An example of savings in the design of printed circuit boards was realized when Potter & Brumfield Division of AMF installed a 3D color-graphics system to aid in the creation of such electromechanical components as solenoids and relays, and such shop components as pallets, tools, and fixtures. The system is expected to save more than 11,000 labor hours on 25% of the company's annual production of new printed circuit boards. Potter & Brumfield justified the purchase of the system by calculating anticipated savings. The design of a 5- by 10-inch printed circuit board containing 50 integrated circuits and 35 discrete components would require 115 hours manually and 30 hours using CAD/CAM.

''Whiting Corporation, a subsidiary of Wheelabrator-Frye, chose a middle-size minicomputer to program NC tapes, FORTRAN routines, and finite-element structural analysis in the company's manufacture of overhead cranes, metallurgical equipment, and chemical processing systems. The system supports up to 32 online terminals and peripherals. Installed in January 1980, the system enabled Whiting to reduce tape preparation for its 22 NC machines from days to hours. The company has reduced the cost of outside finite-element analysis services from more than $2,000 to $100 to $200 per month.

''Sikorsky Aircraft cites greater productivity as the major reason for its increase in computer graphics terminals from 13 to 54. Manager of air vehicle design and development, Kenneth M. Rosen believes a designer is, on the average, three times more productive at a terminal than on a drawing board. From the expansion, Sikorsky realized a saving of 500,000 hours in the 1981-1985 period alone.

''Increases in productivity do not necessarily mean decreases in employment. According to Robert M. Ronningen, president of Ronningen Research & Development Co. (Vicksburg, MI), whose company makes plastic molds, parts, and prototypes for products varying from steering wheels to computer terminals, his employee count has tripled in size since installing a minicomputer-based CAD/CAM system.

''The foregoing examples reveal possibilities that exist in saving money and labor hours using CAD/CAM. But they by no means paint the entire picture—they are details from the broad canvas.''

There are however difficulties in measuring CAD/CAM benefits. Beeby and Collier address this issue.

''Suppliers of turnkey CAD/CAM systems liberally quote productivity improvements of 3:1 or 4:1. Users occasionally report benefits as high as 20:1. With the promise of increases in engineering and operations productivity as the primary justification for heavy capital investments in CAD/CAM equipment, it is no wonder that everyone involved wants to tout as high gains as they can muster up in as short a time as possible.

''The truth in reporting cost-effectiveness and performance figures depends on a large number of variables, including the way the system is used. Another truth is that measuring the productivity of a given system in any way meaningful for the user company is extremely difficult.

''For example, two-dimensional line drawings can often be drafted as fast manually as they can using an interactive graphics device. However, the graphics system can be used within the same group to good advantage for drafting complex three-dimensional designs and to create different versions of the same basic drawing.

''These differences make evaluating a CAD/CAM setup difficult. Most companies starting out with a system to perform these varying functions begin with low-cost, bare-bones software until they have evaluated their own needs.''

The authors continue: ''In 1979, when Rocketdyne Division of Rockwell International Corporation established a CAD/CAM implementation committee, the company laid down specific goals, some of which missed the mark, but were at least a beginning. One important criterion to the committee's success was that the policies were reviewed by all top management and signed by the president of the company.

''Originally, Rocketdyne set specific goals for the CAD/CAM productivity measurement, such as: cost reductions generated in user areas, specific amounts of CAD basic training, continual use for skills improvement, and specific productivity ratios within specific time periods. Rocketdyne soon discovered that too many people had been trained for too few interactive graphics terminals. Another discovery was the need for better screening for user attitude. Surveys were devised and initiated to solve the human factor problems of interactive graphics operation.

"System payback was based on 2:1 productivity ratio for creating drawings, (CAD/CAM versus board production). Factors that could not be measured—in addition to attitude—were design quality and drawing accuracy. Rocketdyne also discovered that the increased interactions between design and manufacturing reduced the number of engineering changes, another difficult-to-quantify aspect that affects tooling, machine loading, scheduling, and parts assembly. Other intangible and peripheral benefits include:

- Better proposal support, designs, and design and data retrieval
- Improved problem-solving capability
- Augmentation of critical skills
- Automation of bills of materials, indented parts lists, etc.
- Reduced floor space requirements.

"What Rocketdyne and the host of other CAD/CAM users have proven is that effective productivity tracking and measurement systems—and hence the data on which to base computer implementation decisions—have lagged behind the rapid growth of CAD/CAM technology. The first step toward improving productivity is developing some means to measure it.

"The General Electric Company Ordnance Systems Department in Massachusetts uses a CAD/CAM interactive graphics system to link an engineering database with manufacturing. The master database performs mechanical analysis, tool design, numerical control programming, and the manufacture and documentation of mechanical piece parts and assemblies.

"In the electrical and electromechanical applications, such as printed circuit boards, hybrid circuits, logic design, wire wrapped plates, and so forth, CAD/CAM equipment is used. But the experience of the G.E. Ordnance System reveals that the progress for use of CAD/CAM in mechanical design is lagging behind that of electrical/electronic design and production.

"The Ordnance System has been using interactive graphics since 1974 and has invested heavily in hardware, software, and personnel for the graphics system facility. Great progress in productivity has been achieved in the electrical area, but it has been only recently that cost-effective mechanical design has been realized. In activities where traditional manual methods had been used, the introduction of three-dimensional capabilities in graphics systems has made possible a cost-effective use of ICG equipment for creation of complex geometric data.

"The Ordnance System process for designing mechanical parts is, by the department's admission, a simple method and is an important key to its success. Designers build mechanical parts on an interactive graphics terminal. Detail parts are then extracted, dimensioned, and plotted. Tool designers next retrieve piece part geometries in the graphics database for jig and fixture design.

"NC programmers take parts from the database that are to be produced by NC machines so that they may program the parts using a graphics terminal. Coded data are converted to a program called UNIAPT for processing into machine tool language, the output of which is an NC machine control tape. One of the most important aspects is that the entire process takes place within one room and includes no manual coding of data from design to the production of NC tapes.

"Mechanical analysts who require design geometry to perform stress and heat-transfer analysis to verify design also may retrieve the data.

"Dedicated work and storage areas have been prescribed for each function that uses the database, such as product design, tool design, NC programming, and mechanical analysis. Standards have been set where various kinds of information may be stored in these work and storage areas. Such standards achieve several goals:

- ensuring that tool designers and NC programmers have access to the geometry they need without being burdened with data that are not pertinent to their tasks
- allowing users to manipulate data without jeopardizing or being burdened with other users' data
- enabling users to make hard copies as necessary
- allowing more than one operator to work on the same task, (one continuing where another left off)."

Goals and directions were also addressed by the authors.

"The two basic methods of approach for company operating policies are the technology-driven system, which emphasizes the role of CAD/CAM in a computer-integrated manufacturing setting; and the management-driven system, which focuses on the people—within the context of management policies, practices, and methods. Each philosophy differs in its primary objectives. Regardless of which of these two approaches they use, American industries have realized that survival means being able to respond to the world market with products that are low in cost and high in quality—and that can respond quickly to what the consumer demands. Of course, these objectives have always existed, but with today's increased capital and labor costs and declining productivity rates and great foreign competition, the objectives have turned into goals for survival. The survivors will be companies that are able to achieve such goals.

"Market demands will force management to develop new organizational and technological philosophies. Today's managers are asking such questions as:

- What is CAD/CAM in the context of the new objectives?
- What role should it play in meeting the objectives?
- How can the company be reorganized to meet these new demands?

"Two points are germane to the understanding of the role of CAD/CAM. CAD/CAM should be defined as a total system, complex in each of its details. But each operation is so intertwined that no single entity can operate outside the context of the others. CAD/CAM is a total system. In the case of a manufacturing company, it encompasses the product from initial development as an idea to field service and customer support. CAD/CAM is really nothing more than a series of information transfers—creating, transmitting and receiving, translating, storing, interpreting, and consolidating, or making tangible. And computer-integrated manufacturing is connecting the many points of information transfer.

"In response to the need for early involvement in the design stage of a product, manufacturers are often finding it advantageous to employ a task force or project team approach. Such a unit comprises members of functional groups other than engineering design to work directly with the designer. This approach permits cost and quality to be considered early in the design phase. But despite the benefits of shorter design cycles and fewer downstream errors, companies have shied away from a task force approach because of cost factors and time that manufacturing personnel must devote to it.

"Important goals are integrating organizations and staffs as well as the processes by senior management and positive participation of middle-level managers. To achieve these goals, industry must work with the university community to strongly encourage teaching of product design and process design integration. Corporate and continuing education need to assist in this regard.

"Appointing corporate executives to oversee the separate endeavors of technology and productivity fosters success in these efforts, as well as helping to control cost.

"Because a specific workable form of organization has not been defined for all manufacturers, companies should try new organizational patterns so long as they work for optimal integration of engineering and manufacturing.

"Working CAD/CAM into the system has made clearer the gap between product design and manufacturing engineering. A first-step goal should be improving communications between these functions before productivity improvements can be seen.

"The CAD/CAM environment of today must be seen from the perspective of the total system, one that is integrated and interactive. Those endeavors that are now carried out separately must begin to lose their old identities; the old organizational and informational barriers should now begin to erode. Companies must begin to achieve high quality in their products as well as low cost, reliability, and value. Therefore, the key goal for survival should be to effect this cohesiveness and to make sure that operating policies are consistent and are supported by the company's decision making community."

Beeby and Collier also note: "We have learned that CAD/CAM is not merely NC tooling in the factory, nor is it interactive graphics alone. It is the act of bridging all the various gaps or many disciplines that have been computerized in the most efficient way.

"In the last few years, industry has experienced an upsurge of new communications among functions and groups that have barely spoken to one another in the past, whose responsibilities rarely touched. In this history-making cooperation of functions lies the key to the future.

"Because of the advent of more specialization, the geographic separation of people and activities, the magnitude of production, and the diverse languages and processes our computers precipitate, companies have had to devise new ways to pass information along. In the past, communication among the varying disciplines of manufacturing was simply a matter of walking a short distance from one person to the other in the same building and oral communication sufficed. Now, at least in large companies, this simple world no longer exists. Communications among people, systems, and activities had to be reorganized."

Glossary. *B-spline:* A mathematical representation of a smooth curve.
data tablet: An input device with a direct correspondence between positions on it and the display surface of the CRT. Typically used in conjunction with a mouse or stylus for indicating positions on the CRT, for digitizing input of drawings, or for menu selection.
fillet: Rounded corner or arc that blends together two intersecting curves or lines. In three dimensions, a fillet surface is a transition surface which blends together two surfaces.
finite element analysis: Method for determining the structural integrity of a mechanical part by mathematical simulation.
interactive: Denotes two-way communications between a system and its operator or user.
joystick: A data entry device employing a hand-controlled lever used to manually enter X, Y, and Z coordinates.
kinematic analysis: A process for plotting or animating the motion of parts in a machine or structure under design. Kinematics simulation allows the motion of mechanisms to be studied for interference, acceleration, and force determinations while still in the design stage.
mirroring: A CAD aid which automatically creates a mirror image of a graphic entity.
mouse: A hand-held data entry device used to position a cursor on a data tablet.
raster display screen: A graphics display screen that uses a coordinate grid to divide the screen and a predetermined regular pattern of scanning.
storage graphic screen: A CRT that can retain a visual image for some length of time, not requiring continual refreshing. The image tends to be slow and rather dim, and no single element can be modified or deleted without refreshing the entire screen.
stylus: A hand-held object which provides coordinate input to the display device.
vector refresh: A CRT in which the image is produced by the CRT beam drawing straight-line vectors between given data points. The image on

the CRT is refreshed often (approximately 60 times per second).
workstation: A station which can perform certain data processing functions in a stand-alone mode, independent of another computer. Contains a built-in computer, usually a microprocessor or minicomputer, and dedicated memory.
zoom: A capability that proportionally enlarges or reduces a figure displayed on a CRT screen.

Also see: Computers, Numerical Control, Solid Modeling, Workstation.

CAD-D. See: Computer Aided Design and Drafting.

CAE. See: Computer Aided Engineering.

CALB. See: Computer Aided Line Balancing.

Calibration. Calibration is the act or process of calibrating. Calibration is also the state of being calibrated. When a device is calibrated, the device has been adjusted or referenced relative to a known standard. A calibrated device can be utilized for some purpose and the resulting output has known meaning and reference. Calibration is also used as a means to know the accuracy and the performance of a device.

The National Bureau of Standards (NBS) in Washington, D. C. is a governing body which establishes calibration standards and provides some reference standards for calibrating other calibration standards. For example, the highest standard for certain devices is a reference standard at the NBS, against which any calibration standard device that will be used to calibrate other devices in the field must first itself be calibrated. The Department of Weights and Measures certifies the accuracy of instruments used in commercial transactions, such as scales and petroleum pump flow meters.

Call. Call is a term which is used when discussing computer programs. Normally, ''call'' references the statement in a program which directs the flow of execution to a subroutine. Logical calls may occur within a machine or may be transmitted from one machine to another. Utilities and complicated procedures which have been captured by a subroutine are the typical circumstances which call a subroutine.

When referencing data and voice communication, the term ''call'' refers to the process by which a station, that desires to initiate conversation with another station, makes its intention known over the communication link. A call may not contain any data other than simply to indicate a desire to establish a communications link. A call may be to a specific end station, or it may be over a network to any listening station(s); this is referred to as a broadcast message. The response to a call is some form of answer, either from a responding communication station, or by the execution of a subroutine.

Also see: Calling Sequence, Programming, Subroutine.

Calligraphic. Calligraphic comes from the term calligraphy, meaning writing by style, and is used to describe the method by which graphics monitors draw images on the video screen. The other method used by graphics terminals to create images is raster scanning. Calligraphic creation of images has several advantages for high resolution graphics, and especially for animation or multilayer images. Calligraphic images are drawn as if using an imaginary stylus. The corners and curves of calligraphic images are cleaner and smoother than raster-generated images. The ultimate resolution of calligraphic images is greater than raster images, also. Since a calligraphic image is drawn on the screen as it is needed, and only the image is drawn, the time to create a new image is less than if the entire screen is being repainted with the modified image (as in raster graphics). This advantage becomes a negative if a majority of the screen must be changed; in this case, raster graphics can occur faster than calligraphic graphics. Depending upon the majority of the uses of the application, one type may be preferred or the other.

Also see: Graphics Terminal.

Calling Sequence. A calling sequence is a specific set of instructions used to begin, initialize, or transfer to a given subroutine and return form the subroutine after it is executed.

Also see: Programming.

CAM. See: Computer Aided Manufacturing.

Canned Cycle. In numerical control, canned cycle refers to a set of operations preset in hardware or software and initiated by a single command. Several operations are performed in a predetermined sequence; the function ends with a return to the beginning condition.

Also see: Numerical Control.

CAP. See: Computer Aided Planning.

Capacity Planning Operations Sequencing System. Order sequencing systems are the execution phase of capacity planning operations. Sequencing sets the priorities for a plant or work center to process a variety of different jobs. A sequencing system includes the communication of sequencing and order release information to the plant. Orders are released to manufacturing and to suppliers; order release principles and decision techniques are relevant to production and to purchasing.

There is a delicate balance between the order release planning or order sequencing that is the basis for the short-range schedule and the actual order release. Numerous factors such as scrap, late vendor delivery of material, machine failure and bottlenecks can change the order release decision, up to the scheduled release date. Work-in-process, queue lengths, work center utilization, and the completion status of related material should be in the consideration of the sequencing system. Input-output control, operation splitting, operation overlapping, order priority rules, mathematical programming, and simulation can be helpful in making the sequencing system more effective.

Capacity Requirements Planning. Capacity requirements planning establishes, measures, and adjusts limits or levels of human or machine capacity that are in line with the production plan. Capacity requirements planning is the measuring of the labor and/or machine resources needed to manufacture the Master Production Schedule. Capacity requirements planning determines the capacity required in each work center by time period to make a given output.

Capacity requirements planning takes the routing files, work center file and purchase file and backward schedules each item requirement from the material requirement plan (including planned, firm planned, and release manufacturing orders). This loads all capacity centers without capacity constraints (infinite loading). Capacity planning then compares with available capacity (finite loading) and if capacity is sufficient the master production schedule is a fit, if capacity is insufficient capacity is increased or the schedule revised.

Capacity requirements planning is a review of the required capacities by work center by period, in the short-to-medium range, to meet the master production schedule. Resource requirements planning is the tool to review long-range planning of capacity.

CAPOSS. See: Capacity Planning Operations Sequencing System.

CAPP. See: Computer Aided Process Planning.

Card. A card is an information-carrying medium that introduces instructions to computers, either directly or indirectly and often via punched codes. A *card punch* is a device used to punch holes in a card according to a standard code. A *card reader* is a device to sense and transmit information from code punched on cards.

Also see: Computers.

Card-to-tape Conversion. Card-to-tape conversion is an operation in which data is converted directly from punched cards to punched or magnetic tape by means of a utility program.

Also see: Card, Computers.

Carrier Frequency. A carrier frequency is the frequency of an unmodulated carrier wave. It is also a frequency capable of being modulated by an information-bearing signal of another frequency or frequencies. In frequency modulation, the carrier frequency is also referred to as the *center frequency*.

With regard to communication, a carrier is a wave, pulse train, or other signal suitable for modulation by an information-bearing signal to be transmitted over a communication system. A carrier can be an unmodulated emission. Usually a carrier is a sinusoidal wave, a recurring series of pulses, or a direct current signal.

When a carrier is modulated, there is a frequency band produced both above and below the modulated frequency which is referred to as the sideband.

In some applications, to increase the power, range, and effectiveness of the transmission, only a sideband is transmitted, while the other sideband and the carrier frequency are suppressed.

Another variation of carrier transmission is to use a pulsed carrier, which is used as a modulating wave to modulate another carrier wave.

Also see: Modulation.

Cartesian Coordinates. Cartesian coordinates are two or three axes that intersect each other at right angles forming rectangles. Any point within the rectangular space can be identified by the distance and direction from any other point. Cartesian coordinates are also known as *rectangular coordinates*.

Cascading. Cascading refers to an automatic controlling system in which control units are linked to regulate the operation of an adjacent unit. Cascading also refers to a programming technique that extends the timing and counting ranges beyond the maximum values that can be accumulated.

Also see: Programming.

Casting. Casting is the process of forming a liquid, plastic, or metal substance into a fixed shape by letting it cool in the mold. Two broad categories of casting processes exist: ingot casting (includes continuous casting) and casting to shape. Ingot casting is the process of pouring a substance into a permanent or reusable mold. Following solidification, this substance is further processed mechanically into many new shapes. Casting to shape involves pouring a substance into molds in which the cavity provides the final shape, followed by machining or welding for the specific application.

The design considerations of production for ingot castings depend upon several factors. In the case of the high technologically complicated semicontinuous or continuous casting process, the design of the cooling system, choice of the lubrication, and the rate of ingot movement all play a significant role in the final output.

In the production of castings which are not processed mechanically, the construction of a pattern is a most important step toward successful production of the casting.

Computer-based systems are gaining acceptance in virtually all aspects of foundry operations. Computer use extends into all functional areas, from sand control to final quality inspection; and the computer now has an integral role in the entire process, from design to the manufacture of castings.

The capability for producing intricate shapes, both internal and external, is an important advantage of casting processes. A number of different casting processes may be used to make castings with complex geometric configurations. Investment casting, die casting, gravity metal-mold casting, and sand casting processes present an array of possibilities for near-net-shape component manufacturing. While each of the casting processes has advantages and limitations, the extent to which the complexity of the geometry affects the capability for near-net-shape processing is a common factor. Generally, the more complex the shape the more likely the raw shape will be different from the final shape. For complex parts, the near-net-shape casting processes offer a significant potential for saving material and lessening the need for secondary operations.

Another common problem with near-net-shape processing of complex geometries stems from representing three-dimensional shapes on a two-dimensional medium, drawing or blueprint. The blending of external and internal joining sections is commonly done on the drawing by notation alone and must later be more fully defined by the craftsman, who is the pattern-maker in the case of castings. Subsequent lay-outs of the component frequently do not deal with questions of near-net-shape processing, but rather whether the casting reasonably represents the drawing and whether the important dimensions to be machined are as required. Process capabilities may be ignored or accepted at this stage rather than used to their limits. It is here that the possibilities that come from computer aided engineering can bring about substantial improvements in near-net-shape casting of components.

CAT. See: Computer Aided Testing.

Cathode Ray Tube. A cathode ray tube (CRT) is an electron tube consisting essentially of an electron gun that produces a concentrated stream of high-speed electrons (an electron beam or cathode ray) that impinges on a phosphorescent coating on the back of a viewing face (screen). The excitation of the phosphor produces light. The intensity of the light is controlled by regulating the flow rate of the electrons. Deflection of the beam is achieved either electromagnetically, by currents in coils outside the tube, or electrostatically, by electric fields produced by voltages on pairs of deflection plates (horizontal and vertical) between which the beam passes inside the tube on its way to the phosphor-coated screen.

The acronym CRT is often used to refer to a monitor or terminal, or any type of output display device which contains a vacuum tube display. CRT displays may be either calligraphic or raster devices. Different colors of monochrome CRT displays are achieved by utilizing different compositions of phosphor coating.

There are several types of tubes. The refresh tube (or vector or direct-beam tube) paints and repaints the image. Images can continually change in dynamic motion, an advantage in such analyses as kinematics, or clearances of moving parts. The disadvantage is flickering that occurs because of the length of time reqiured to repaint the lines in a continuous mode.

Raster tubes behave similarily to television screens. Rather than scanning as refresh tubes do, a raster screen uses a matrix of tiny dots called *pixels* to create an image. Animation is possible with a raster tube display. Images do not flicker, but line quality is poor because of the horizontal/vertical-only matrix of pixels.

The storage or direct-view tube displays an image in a one-time, line-for-line mode. Storage tubes have excellent resolution and require little computer memory and processing. The primary limitation of the storage tube is its inability to retrace or redraw any changes without repainting the entire image after each change. Also, low contrast levels of the display require dimming the room lights for comfortable viewing.

Existing monochromatic CRTs can be converted to color with a technology that uses special phosphors that change color depending on the current density of the beam from an electron gun. These phosphors allow the CRT to employ a single electron gun rather than the usual three guns for full-color CRTs.

Also see: Monitor.

CATIA. See: Computer Graphics Aided Three Dimensional Interactive Application.

CATV. CATV is the acronym for cable television and also implies any of the defined standards regarding cable television transmission, connectors, media, etc.

CCD (Charge Coupled Device). A charge coupled device is a solid-state device with a direct digital output, where the amount of charge is directly related to the signal applied. One application of the charge coupled device is with used as the imaging device in machine vision cameras. A charge coupled device or CCD camera has an integrated circuit chip with an array of CCDs upon which the image from the lens is focused. The CCD generates a matrix of digital signals which are in proportion to the intensity of light striking each pixel. This signal is then processed by the accompanying circuitry to generate a useful output signal.

Solid-state imagers offer many advantages over other imaging methods, including high reliability, high sensitivity, low power dissipation, and good spectral response over the entire visible spectrum. Charge coupled devices are produced in two formats, as linear arrays and as area arrays. Some of the products that use linear CCD imaging arrays are facsimile equipment, photocopiers, mail sorters, and bar code readers. Area CCD imaging arrays are used in closed circuit television cameras, video cameras, and for astronomical observations.

Also see: Machine Vision.

CCITT (The International Consultative Committefor Telegraphy and Telephony). The International Consultative Committee for Telegraphy and Telephony is a committee of the International Telecommunication Union (ITU) located in Geneva, Switzerland.

CD-ROM. A CD-ROM, (Compact Disk—Read-only Memory) is a read-only type of memory which utilizes a compact disk, similar to the audio compact disk (CD), as the storage medium. A CD-ROM is different from an erasable optical disk and a WORM disk in that a CD-ROM can only be written to (once) by specialized factory equipment. The master CD-ROM can then be economically reproduced in volume. An erasable optical disk and a WORM disk are designed to be written to by the end user. The one-way data flow in using CD-ROMs allows much less expensive drives to be utilized than with WORM drives. CD-ROMs offer a unique media ability to store and distribute many hundreds of megabytes of published text, numbers, and graphics images, combined perhaps with audio and visual material. By being an optical, not a magnetic data storage medium, a CD-ROM is impervious to hazards of magnetism and contaminated handling or similar damage.

Also see: Erasable Optical Disk, Optical Disk Drive, WORM.

Cell. A cell can be different things depending upon the discipline which uses the term and the intended reference. In all cases, a cell generically refers to a related natural grouping of devices which generate while interacting with each other.

In electronic circuit design, a cell refers to the portion of an integrated circuit chip which performs a certain basic function. Integrated circuits are designed using standard cells or templates which can be interconnected to achieve the desired overall functionality. The use of standard cells ensures that the chip will function properly, since the standard cells have been previously used and proven correct. The use of standard cell building blocks avoids having to create all the detailed circuitry from scratch.

In the manufacturing world, a cell usually refers to a grouping of machines which have been arranged in a configuration which supports performing the desired special functions. A machining cell performs machining, while inspection or plating cells perform their respective functions.

Also see: Flexible Manufacturing System.

Cell Controller. A cell controller is a computer device that is designed to be used as the interfacing controller between intercell and intracell control and data transmission needs. The cell controller is usually capable of being connected to a network to receive high-level communication and also of driving the machine controllers in the cell. Typical activities performed by a cell controller are downloading configurations and part manufacturing sequences to the controllers in the cell, monitoring the performance and material movement into and out of the cell, and being a single source for all necessary management data. Cell controllers typically must be capable of surviving on a harsh factory floor and performing in highly susceptible environments. Various products are used as cell controllers. The major programmable controller manufacturers have designed special-purpose cell controllers. The major computer companies and some systems houses have developed factory hardened microcomputers and minicomputers which can provide cell control. Cell controllers are an integral and basic component of a hierarchical control system.

Also see: Flexible Manufacturing System.

Centralized Control. Centralized control, in a computerized environment, denotes the processing of all data at one geographical location. This can relate to a corporate or home office computer center.

The approach used in the centralized control of data is a concentration of resources such as equipment and personnel in one area. Data entry is a common example of using a central control point in contrast to a decentralized entry from various locations. This depicts a centralized control of source documentation files and error reconciliation. The centralized control of that data reduces the training and avoids maintaining procedures at multiple locations. A centralized control for processing data can exist, however, with decentralized data entry.

The support of hardware and processing of data in one environment depicts centralized control. A perceived advantage is that of independence and objectivity from local management or for competing services. This would include the reduced risk of duplicating information and consistency in management reporting of data. The centralized control of managing volumes and documenting trends can be a result of centralized control.

Central Processing Unit (CPU). A central processing unit is the main control unit in a von Neumann architecture computer system; it includes at least the ALU and CU and also the main memory in some systems. .

Also see: Architecture, Computers, von Neumann Bottleneck.

Chad. A chad is a piece of material that is removed from a storage medium such as a punched card or tape, when a hole or notch is formed. Numerical control systems use *chad detectors* to check for pieces of material that remain in the hole of a medium, such as a card or tape after it was punched.

Also see: Card, Computers, Numerical Control.

Chaining. Chaining is a system of storing records in which each record belongs to a list or group of records and has a linking field for tracing the chain.

Also see: Computers.

Chamfering. Chamfers are angles formed or cut on cylindrical forms and holes to provide relief from sharp edges of perpendicular surfaces. The chamfer angle generally used in practice is 45 degrees. A primary example of the application of chamfering is in reaming. Practically all cutting done by a reamer is accomplished by the chamfer, which forms a truncated cone on the end of the reamer and carries the cutting edges. The chamfer is shaped so that the reamer will start properly in the hole, and the major portion of the cutting is done by the chamfer length.

A chamfer is also used to taper the end of a tap by cutting away the crest of the first few threads to distribute the cutting action over several teeth. It also acts as a guide to start the tap into a hole. Chamfering is also used to break the sharp angle formed by the edge of the hole and the surface in which it has been bored or drilled. This is done to provide for easy guidance of mating parts into the hole or for safety reasons, to eliminate sharp cutting edges.

Channel. A channel, in data transmission, describes a means of performing simplex transmission in one preassigned direction. In addition, a channel may represent a single unidirectional or bidirectional path for transmitting or receiving electrical signals. In one-way-at-a-time transmission, the single channel may be used for two-way transmission.

A connection between two nodes in a network can be referred to as a channel. When there is the capability for multiple data transmissions concurrently, each single path of transmission is referred to as a channel.

Any structured, defined path along which data can be sent is called a channel.

Standard NC tape has eight channels.

Also see: Baseband, Broadband.

Character. A character is one member of a set of elements used for the organization, control, or representation of data. Characters may be letters, digits, punctuation marks, control signals, or other signs and symbols often represented in a spatial arrangement of adjacent or connected strokes, or in the form of other physical conditions or events, such as time or space sequences or electrical pulses or patterns of magnetized spots, or punched holes in cards, tapes, or other data media.

Characters represent data, which, in turn, represent information when meaning is assigned to the data. Characters are used for the organization, control, representation, transmission, and storage of information. In most communication systems, characters are represented as strings or sequences of pulses or as graphic symbols on a display surface. Characters are represented in optical fiber systems as light pulses in single fibers or as a picture element (pixel) in optical cables.

A character can also be thought of as a symbol such as a letter, numeral, punctuation mark, and by extension, a nonprinting control character, such as a space, shift, carriage return, or line-feed character.

Characters Per Second. Characters per second is a measure of the speed at which ASCII data is transmitted. If multiple electrical impulses or bits are required to represent a character, the character per second (cps) speed will be less than the respective bits per second (bps).

Also see: ASCII, Band Bps (bits per second), Character.

Check Sum. A check sum is an error detection code that sums all one bits of a group of data storage location. Summing is done without carrying over from one column to another. The known result is stored; any variance from this result indicates data has been altered. Check sums can be prepared for any portion of logic memory, storage memory, or register content—in essence any block of memory.

Also see: Data Error Handling.

Chip. A chip is a small piece of silicon impregnated with impurities in a pattern to form transistors, diodes, and resistors. Electrical paths are formed on it by depositing thin layers of aluminum or gold. Chip is the commonly used name for an integrated circuit.

Also see: Computers.

CIM. See: Computer Integrated Manufacturing.

Circuit. A circuit is a particular interconnection of electronic component parts that form a particular piece of electronic equipment, showing where and how they are connected and what function they serve. A circuit is usually a logical connection of electrical components which includes a power source and a power sink. A circuit may be designed to perform various functions.

There are many basic "building blocks" of simple circuits, from which most complicated circuits are developed. Some of these basic circuits are amplifier, audio control, biasing, control, coupling, frequency conversion and multiplication, indicator, logic, modulation and demodulation, power supply, signal conditioning, and signal generation circuits. Some of the specific circuits are well known such as the Schmitt trigger, the Wein bridge oscillator, the Pierce crystal oscillator, and the Darlington circuit. Some of the more contemporary developments in circuits include the phase-locked loop (PLL) which most FM tuners now employ. Standard logic has given way somewhat to CMOS circuitry as technology continues to advance.

Circuit Board. See: Printed Circuit Board.

Circuit Card. A circuit card is usually a finished, functional product consisting of a printed wiring board assembled with components soldered into place. A circuit card is a single subassembly which may be plugged into a backplane with multiple circuit cards to provide a complete product.

Some companies such as Digital use the term *module* to describe the identical product which others may call a circuit card. The distinction between what is an unpopulated board, a populated soldered board, and a product made up of a collection of boards is sometimes simply corporate custom and semantics. Calling an item by a different name does not change the reality of what it is.

Also see: Backplane, Printed Circuit Board.

Circular Interpolation. Circular interpolation is a function automatically performed in the control of defining the continuum of points in a radius based on a minimum of three taught coordinate positions.

Cladding. Cladding is an optically conductive material, with a refractive index lower than that of the core, placed over or outside the core material of an optical waveguide that serves to reflect or refract lightwaves in order to confine them to the core. The cladding also serves to protect the core.

Also see: Fiber Optics.

Cleaning. Cleaning is the act of removing unwanted surface impurities or contaminates. Both before and after a circuit board is soldered, it must be cleaned. Some of the undesirable elements which adhere to the solder side of the board could cause corrosion or a lack of proper conductivity. Even components are sometimes cleaned and tinned before insertion. Materials which are to receive chemical treatments or surface finishes, such as plating and painting or surface hardening, must be cleaned both physically and chemically. When glass is to be glued or finished, it must sometimes be ''etched,'' which is washing with acid to remove the top oxidation layer. When fabricating integrated circuit wafers, not only does the product have to be kept clean, but so does the room and its environment.

Items such as the read/write heads on disk drives and optical input devices must be kept clean for proper working. While a floppy-disk drive head contacts the media, when the diskette is spinning, the working distance between a hard disk and the read/write head is a few millimeters, which is so close even a particle of dust would not pass between the media and head. The catastrophic result of a particle of dust wedging between the head and the spinning media is what is referred to as a *head crash*.

Also see: Clean Room.

Clean Room. There are numerous applications that exist in industry, research, and public services that require environments protected and isolated from contaminants that are present in the air and on surfaces within these environments. There are various degrees of cleanliness, and these are referred to as classes defined in the Federal Standard 209. This standard defines three classes and two divisions of clean room requirements. For a class 100 AF standard clean workstation the maximum number of contaminant particles of a diameter of 0.5 microns and larger must be less than 100 particles/cubic feet. For particles with a diameter greater than 5.0 microns the restriction is zero. Class 10,000 clean rooms require less than 10,000 particles/cubic foot for particles 0.5 micron and larger; and no more than 65 particles/cubic foot for those of 5.0 microns or greater. Class 100,000 requires less than 100,000 articles/cubic foot of 0.5 microns or greater. By way of comparison, one might think of the standard hospital operating room which would have requirements of no more than 1,000,000 particles/cubic feet of 0.5 micron or greater and no more than 10,000 particles/cubic foot of 5.0 microns or greater.

These requirements are generally met through the use of specially designed air filtration systems, positive air pressure in the room, synchronized double doors which do not allow the inner doors to open until the outer doors have closed, special clothing requirements for personnel, and unique cleaning materials and schedules.

Dr. Tom Papanek of Adept Technology described clean room conditions in his SME Technical Paper *Design and Test of a Robot for Class 10 Clean Rooms*. He wrote: "Over the past decade, there has been an increasing need for clean manufacturing processes. The economics of semiconductor wafer fabrication depends critically on the product yield, and as device line widths have shrunk to the micron and submicron range, the effect of particle contamination on yield has become increasingly important. The high cost of capital equipment in the semiconductor industry also demands high product throughput from fabrication equipment. These two needs, high throughput and low particle levels, can best be achieved with automated equipment. Until recently, most wafer handling devices were highly specialized, developed by capital equipment suppliers for each application. General-purpose robotic devices for low particulate wafer handling are a fairly recent development, and the design rules for Class 10 and Class 1 equipment are still evolving.

"Achieving these low particles levels is not an easy task. By way of comparison, a typical office environment is Class 100,000, that is, there are 100,000 particles of size 0.5 micron or larger in every cubic foot of air. To understand the particle sizes involved, consider that common house dust is .01 to 5 microns in size, tobacco smoke is .01 to 0.5 microns and human skin cells range from 1 to 50 microns. A typical Class 10 clean room will have ceiling to floor laminar air return, sealed walls and air lock entrances. Any personnel in the room will be wearing full overalls (commonly called "bunny suits"), boots, hoods, and gloves of low particle shedding materials.

"Notetaking is done not on paper, but on sheets of pressed platic fiber. Personnel are requested to not smoke for several hours before entering the room, nor use talcum powder in the morning, as the resulting particles from the breath or skin would degrade the room air. The even more stringent requirements of Class 1 can realistically only be achieved by eliminating all people from the room and depending solely on automated handling."

Clock Rate. Clock rate refers to the frequency of the reference timing signal or clock used by a machine or system. A clock rate is also required for data transmission systems and is used by a computer to control the timing of certain functions such as the control of the duration of signal elements, the sampling rate, or the execution of instructions.

A clock rate may be utilized for synchronization of periodic signals. If a system consists of multiple clock sources, sometimes a master clock or reference clock is used to synchronize other clocks. The stability of a clock source is the ability of a device to produce the same number of equally spaced timing pulses in equal and relatively large time intervals. Clock stability is usually expressed in parts per million, with a positive number indicating the number of excess timing pulses per million of a standard reference clock.

On some data recording, a clock track is recorded to provide a readily available, imbedded reference timing signal for use during reading and writing.

Also see: Computers.

Closed Loop Control. Closed loop control is a type of control logic in which there is feedback from the output to the input, which allows the controlled parameters to be varied based on the magnitude and the change of the output (the first and second derivatives). A closed control loop may have the output tied directly to the input, or the output may be scaled or conditioned before it is in a form which can be related to the input parameters. Closed loop control prevents such things as "run-away," which is the unchecked changing of the output, with no ability of influencing the input signal in a compensating direction. Closed loop control is the type of control which is implemented when using feedback control systems. Constant speed control of a varying load would be impossible without closed loop control. While eventually an input value could be derived which should provide constant speed for a constant load, the startup phase would be either impossible or prohibitively long without employing greater torque for startup.

Also see: Open-loop Control.

CMM. See: Coordinate Measuring Machine.

CMOS (Complementary MOS). See: Metal-oxide Semiconductor.

CNC. See: Computer Numerical Control.

Coating. A coating is a thin layer which is deposited on the surface of a device. The coating may be painted on or plated such as electroplating. Some coatings are to prevent rust and corrosion while other coatings are to provide a different property for the outer surface such as lubrication or reflectiveness. Products such as magnetic recording media receive a coating of iron oxide or similar material to allow the recording process to be performed on the substrate. Integrated circuit chips would be impossible without the ability to coat the base substrate with different layers of conductive, resistive, and nonconductive materials. Coatings can be applied by dipping, spraying, or even ion attraction. When coatings become significantly thick compared to the item which they are on, they are sometimes considered to be cladding, which usually means a rigid outer shell which may be applied in sections and usually is not tightly bonded to the entire surface of the underlying device.

Also see: Cladding, Finishing, Plating.

COAX. See: Coaxial Cable.

Coaxial Cable. A cable consisting of an insulated central conductor with additional insulation on the outside and covered with an outer sheath.

For example, an electrical or electromagnetic transmission line consisting of one conductor (usually a small copper tube or wire) within and insulated from another conductor, using a dielectric as an insulator; the outer conductor usually consists of copper tubing or copper braid and may be grounded. The inside diameter of the outer conductor is sufficiently larger than the outside diameter of the inner conductor so as to allow space for the insulating material between them.

Coaxial cable is the most widely used media for both broadband and baseband data transmission networks. The cable for baseband is 0.375 inch in diameter and is surrounded by a copper cover. The broadband cable is slightly larger, is covered with aluminum, and costs approximately 50% more than the baseband cable. Cable television (CATV) is a familiar use of coaxial cable.

COBOL. COBOL, an acronym for COmmon Business Oriented Language, is a computer language which was the first major business oriented programming language. COBOL was authored by CODASYL (COmmittee/COnference on DAta SYstems Languages) and Grace Hopper, who had authored a language known as FLOW-MATIC. COBOL was developed by three committees; a Short-Range Committee, an Intermediate-Range Committee, and a Long-Range Committee. The results of the Short-Range Committee, whose charge was to discover what was right or wrong with three existing business languages, FLOW-MATIC, AIMACO (standing for AIr MAterial COmmand), and COMTRAN (COMmercial TRANslater) and make a composite of these three languages which could be used in the short run ("good for at least the next year or so"). Although the intent was that, the result of the Short-Range Committee, known as COBOL 60 (circa 1960), was to be an interim language until there was time for the other two committees to design a new language, history shows that only modification to the original COBOL 60 has occurred since that time.

The DoD (Department of Defense) is currently the largest user of COBOL, but the DoD did not initiate the idea of creating COBOL. Academic, commercial, and manufacturing users originated the idea of designing a common business language and asked the Director of the

Data Systems Research Staff of the Office of the Assistant Secretary of Defense to ask the DoD to sponsor the meeting of all interested parties.

Also see: Computer Languages.

Code. A code is a system of organized symbols (bits) representing information in a language that can be understood and handled by a control system. A code is also a system of symbols that can be used by machines, such as computers. Special external meaning is dictated by the specific arrangement of the symbols. *Code conversion* is the means of changing a character bit grouping in one code into a corresponding character bit grouping in another code.

Also see: Absolute Coding, Bar Coding, Baudot Code, Binary Code, Binary Coded Decimal, Coding, Computers, Decimal Code, EBCDIC, End of Line Code, Excess-three Code, G Code, Magnetic Three Feed Rate Coding, Master Code Relay, Mnemonic, Object Code, Operation Code, Pseudocode, Source Code, Symbolic Coding.

Coding. Coding is the preparation of a set of program instructions, by means of a specialized language, into an accurate representation of the program thereby allowing a given action to be taken or problem to be solved.

Also see: Code, Computers.

Coining. Coining is a closed die squeezing operation in which all surfaces of the work are confined or restrained. Coining also refers to a press brake bending operation in which the punch bottoms against the workpiece and the die. It also refers to a process similar to bottoming although greater force is applied. Coining alters the radius, and bottoming sets the bend open but does not affect shape.

Coining is the process of applying necessary pressure to all or some portion of a forging's surface to obtain closer tolerances and smoother surfaces or to eliminate draft. Coining may be done while forgings are hot or cold and is usually performed on surfaces parallel with the parting line of the forging.

Coining is used for sheet metal workings as well as for bulk forming. During this operation, metal is intentionally thinned or thickened to achieve the required indentations or raised designs. Coining is widely used for lettering on sheet metal or components such as coins. Since it is done in a closed die in which all surfaces of the sheet metal are defined or restricted, a well-defined imprint of the die on the workpiece results.

In addition, coining is the process of pressing of a sintered compact to obtain a definite surface configuration or density.

Also see: Forging.

Cold Welding. Cold welding is a solid-state joining process that bonds metals by application of mechanical force alone. Welding is done at or near room temperature and there is extensive plastic deformation.

Also see: Welding.

Command. A command is a signal from a machine control unit initiating a movement or function.

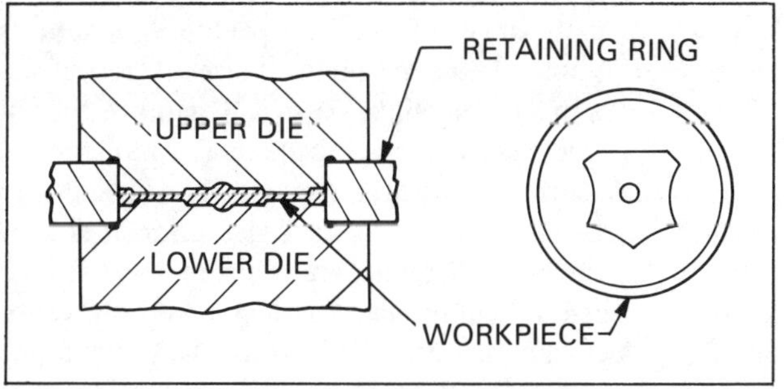

Coining is used for sheet metal working as well as for bulk forming.

Command Readout. Command readout refers to a visual display of input data read from the tape of a control system.

Also see: Command, Computers.

Communication Protocol. In communications systems, communication protocol is the rules for communication system operation that must be followed if communication is to be effected. Protocols may govern portions of a network, types of service, or administrative procedures.

For example, a data link protocol is the specification of methods whereby data communication over a data link is performed in terms of the particular transmission mode, control procedures, and recovery procedures. Protocols are designed to control the layers of a communication network or to control the exchange of data among computers in an application network.

In modern data networks, such as MAP/TOP, protocols are layered to allow a more structured method of communication interconnection. This layering reduces complexity and provides peer-to-peer layer interaction across nodes, as well as allowing changes to be made in one layer without affecting other layers.

Also see: Open Systems Interconnection.

Communications Link. A communications link is a formally connected line or channel between two points or devices which allows information or data to flow between them when desired. If related to the OSI model, a communications link is the lower-level layer(s) of the seven-layer model.

Also see: Channel, Open Systems Interconnection.

Comparator. A comparator is a device or network used to compare information from two sources.

Compare Function. A compare function is a user-programmed instruction which equates numerical values for "equal" or "less than" relationships for the purpose of varying an operation sequence or application.

Compatibility. Compatibility refers to the degree to which tapes, languages, programming, and various specified units can be interchanged between various machine tools and various numerical control systems with minimal reduction in capability.

Also see: Numerical Control.

Compensation. Compensation refers to electrical filter circuits used in servocontrollers to improve the performance of closed-loop systems.

When contouring, conpensation refers to the displacement always normal to the cutting path and workpiece programmed surface, that accounts for the variance between the actual and programmed radius and dimension of the cutting tool.

Compensator. A compensator is a device that advances (crossfeeds) the wheel to compensate for wear and dressing.

Also see: Grinding.

Compiler. A compiler is a computer program which converts a program written in a high level language into binary coded instructions or machine language to allow execution by a computer. Some high level languages such as FORTRAN, C, PASCAL etc. must be compiled before they can be executed. Other languages are interpretive languages. Some languages such as BASIC are available as either a compiled or an interpretive language. Compilers are not all the same, even when compiling the same language. Some compilers produce more dense, compact code than others. Compact code is usually more efficient in operation, as well as simply a more economical use of resources. Compilers sometimes parse and optimize code in a first pass, before doing the final compilation in a successive pass or passes. The difficulty in debugging compiled code is, error messages given by the compiler are sometimes sparse, and when executing the finished code, the error messages are often difficult to relate back to the original source code.

Also see: Interpretive Language.

Composites. Composites heterogeneous, solid structural material, consisting of two or more distinct components that are mechanically or metallurgically bonded together. Provided the individual pieces of the two phases are not easily seen with the naked eye, the resulting composite can itself be regarded as a homogeneous material. Such materials are familiar; most natural materials such as wood are composites, as are tires, glassfiber reinforced plastics, the cemented carbides, and paper are composites consisting of cellulose fibers.

It is sometimes difficult to distinguish between a composite material and an engineered structure, which contains more than one material and is designed to perform a particular function. The combination is usually spoken of as a composite material provided that it has its own distinctive properties, such as being tougher than any of the individual materials used alone, having a negative thermal expansion coefficient or having some other property not clearly shown by any of the component materials. The term is often used to mean specifically the combination of very strong and stiff fibers with a weak matrix designed to hold the fibers together. This type of composite combines the extreme strength of the fibers, and, due to the presence of the weaker matrix, shows much greater toughness than would otherwise be attainable.

Computer Aided Design (CAD). Computer aided design is a term which has been in use for some time. The term *CAD* is frequently linked with CAM (computer aided manufacturing), which indicated the typical use of CAD as a front-end for the design of manufactured products. Now, CAD and a more specialized area, CADD (computer aided design and drafting), are becoming stand-alone areas. Architectural design and illustrating arts are some of the areas that use CAD as the means to the end, and the end, are both direct output of the same tool, CAD (as opposed to CAD/CAM).

CAD used to be a very expensive tool which was only available on mainframe computers. Now, personal computer-based CAD systems provide the same capabilities, and more, than mainframe systems of only a few years ago. Networking and connectivity for data sharing is the major weak point in using PC-based CAD systems. AutoCAD by Autodesk of Sausalito, CA started the revolution in, and the acceptability of, PC-based CAD.

Suppliers of CAD equipment for project design work view the CAD/CAM market in terms of both industry use and applications. Evolving a set of application categories that is meaningful to both designers and manufacturers has been difficult. For example, an application using tubing packages may work in architectural, aeronautical, and petrochemical projects. Vendors and users have derived the following categories as best representing unique market segments:

- mechanical design, which includes design and manufacturer of discrete products;
- electrical, such as integrated circuit design and printed circuit board layout;
- architecture and engineering;
- mapping;
- petrochemical, including analysis of seismic data and other earth studies, and
- other, a broad but important category that includes the previously mentioned projects involving transportation, urban planning, and educational fields.

In 1979, the U.S.-installed design systems accounted for 75% of the total throughout the world. But that percentage is changing as other nations are catching on to the savings to be gained. At that time, the application ratio also shifted toward the disciplines of architecture and engineering, mapping, and petrochemical, away from the early pioneer electrical and manufacturing markets.

Other expanding areas of CAD use include marketing and presentation graphics. General Mills in Minneapolis, Minnesota, has made effective use of computer-generated business graphics, initially applying the technique to marketing graphics and has since expanded its efforts to include presentation, financial, and manufacturing applications.

We have just scratched the surface in the uses of CAD processes and equipment for design work in the many arenas of American industry. Many hopeful, instructive, and helpful stories lie behind the summaries told here and in those not yet publicized. The next CAD/CAM generation chroniclers will have an even greater anthology of stories to tell.

Applications in the area of CAD are many. William Beeby and Phyllis Collier discussed many of the applications in their book *New Directions Through CAD/CAM*: "Combining a CAD system with a simulation system allowed Deere to create the now famous computer animation film of a plowshare hitting a stone. The CAD system was used to create the geometric model of the plow and to determine the effectiveness of a plow reset mechanism.

"Deere & Company and other companies that have developed and implemented CAD modeling graphics systems have found that these systems can predict structural behavior by subjecting a mathematical model of a part to simulated loads. To simplify the mathematical analysis, the model is sectioned off into finite numbers of rectangles, hence the term finite-element models. An efficient process is one that marries a CAD graphics system to a structural analysis system so that the finite-element model can be created from the geometric model produced on the CAD system. Such output can be color-enhanced and shaded to indicate stress and contours, allowing a designer to spot abnormalities faster and more easily than with numeric printouts, as was previously done.

"The term 'project design' means so many things to so many industries that the best and only explanation is a description of how some of these various companies are using CAD equipment to perform their design jobs.

"Many companies enjoy the advantage of preliminary work done on previous programs that have led up to a workable technology for current systems. The Boeing 767 and 757 programs, for example, shared the advantage of experience of previous aircraft programs. Their history assisted 767 and 757 project engineering in determining the percentage of CAD penetration, or the ratio of computerization to manual techniques.

"In the airplane industry, part cards and drawing sheets are indications of the amount of design work and geometric entities included. Historical data from the YC-14 are available for analysis and comparisons. These data have been accumulated in four major areas of endeavor: structures, payloads, systems, and propulsion.

"Developmental work was done in the 1970s with various kinds of interactive systems, and before that, with batch methods. Boeing found that each of these systems is more successful in specific applications than in others. This conclusion was reached in part by in-depth studies and partly through trail and error. Companies are gaining from experiences in the real-time, production environment. Industries are learning the most effective combinations of interactive and batch, of divergent kinds of graphics systems and plotters, and of manual and computerized drafting. More generally, they are learning that computers are very powerful in terms of geometric manipulations, especially for self-checking capabilities. They've learned, as with any tool, that there is a right way and a wrong way to use computers. They try to find applications that are productive. In some instances, it takes longer to accomplish the task by computer than by hand as with, single-end-item parts.

"Using the batch-CAD process and a combination of its languages, designers program families of parts that are stored on the computer so that operations can use those data to create the actual part. Designers can ask the computer to give a cut of this loft, or a layout at every place where, for instance, there is going to be a rib. Then, they can program a rib in the computer so that it produces a layout of that particular rib. But it can also draw 59 other ones of different sizes from the one program. The engineer could select a rib at each of 60 stations. Then the computer selects the right line for that station, and builds that rib around it.

"The value is that if drafters were to do the job, it would be very difficult. Four drafters might work on it so that everyone does the same thing at 60 different locations. This way it is certain, from one program, that the rib is designed with exactly the same ground rules.

"There are 38 ribs in different sizes in the stabilizer of the 767. The computer drew all of them. Of course, the computer shows mistakes right away, and designers have caught a lot of the errors that a manual drafter will correct by making something fit, by just blending it in.

"In mockup, the manufacturing division actually uses full-sized drawings to lay down and use as a tool. For example, the computer drew the floor segments of the 130 body frames in the 767. The consistency between ribs is extremely high. Lines are within 0.003 inch (.0762 mm) of what the engineer describes. This allows the engineer to ask for a cut through the wing at every station where there is a rib. The computer will then draw it full size on one piece of paper. The engineer can use this for routing plumbing and checking for interferences that might occur. A drafter would not lay it out that accurately. On the 767, all of the body structure, including the floor beams, the skins, stringers and frames are in the computer database.

"These batch programs, which pioneered the use of batch CAD, were initiated during the development of the YC-14. However, due to geometry differences and software construction these programs were not readily usable on subsequent aircraft design programs. Following this experience, engineers began creating several large general programs that incorporated macros, (parts or components that are preprogrammed). These general programs could then be adapted to any Boeing commercial aircraft configuration. These first general batch programs were for wing box structures and body monocoque.

"Airplanes are just one of many examples of CAD project design. In the early 1960s, Litton Industries in Woodland Hills, California, began a CAD effort to design and produce printed circuit boards from string lists using computer aided interconnect design. As a result of this early work, many batch processes are still used for interconnect artwork generation. Interactive computer graphics facilitated interconnect artwork design and has become an essential factor in integrated circuit design. An industry that involves one million active elements per chip would not be possible without the technology of ICG.

"However, because of high startup costs of implementing ICG equipment, smaller firms are digitizing manually generated artwork and performing final work using computer-controlled photoplotters. Only firms with large artwork design requirements are fully utilizing ICG systems. Another reason for hesitancy on the part of companies to implement ICGs is the fact that fully automated computerized designs are not as efficient in terms of via-usage and track-path assignments as are manual methods.

"Just what does the printed circuit board schematic process include?

"Schematics for printed circuit boards are usually produced by a circuit design engineer. Many ICG programs offer logic verification and simulation capabilities. This software generates circuit assignment and a schematic. A typical PCB design takes two to three weeks. The design can then be drafted, checked, and revised if necessary. The designer can digitize a PCB schematic on an ICG system using a sketch of the schematic.

"Component lead assignments, which must be made before finishing the schematic, can be either assigned and placed manually or automatically during a computer aided circuit design process. An approach called 'skeletal equations' can be used, entering only signal names on the page. This method eliminates pin assignment errors. Many interactive graphics systems can then generate string lists based on the lead assignments.

"Before the PCB designer places components, the engineer must define the physical configuration, taking into consideration many restrictions, such as: impedance factors, layer spacing, center-to-center track distances and tolerances, number of layers, ground and power distribution, pad sizes, via sizes and limitations, dual-in-line package (DIP), placement patterns, and decoupling capacitor placement. These board parameters are deviations that decrease fully automated design capability.

"A variety of combinations in the method of component placement is possible. For example, the board may be digitized from a sketch, a photograph, or it may be entered by keying in the coordinate data, or by custom execute-type files. The job of component placement still requires the skill of an experienced circuit designer, who uses the computer as another tool to get the job done. The computer just helps to eliminate a lot of the guesswork.

"In addition to the foregoing processes using interactive systems, computerized batch processes have been used for many years in component placement. Designers learn how to manipulate and make revisions using batch processes so that they often make excellent placement devices.

"At Fairchild Camera & Instrument Corporation in Mountain View, California, technicians use a turnkey CAD system to autoroute, interactively route, and hookup check two-layer interconnection metal on Fairchild's 4000-gate integrated-circuit gate-array chip. The graphics database includes gate-array diffusion-mask layers and interconnection metal patterns, such as flip-flops and multiplexers, which have been entered into a macrocell library. The database was converted from an integrated circuit software database to a printed circuit electrical schematic software database. The macrocell library parts also include connect nodes and nodal names. The computer extracts a net list from the logic diagram, which it uses to direct the autoroute, and the interactive route, as well as to hookup check the software. Then a design-rules check verifies that minimum line widths and correct gaps and overlaps have remained during the final editing operation. When the interconnected metal is approved, the metal and via graphics data are converted back to the integrated-circuit software for standard postprocessing.

"Ford Motor Company in Dearborn, Michigan, provides another story of industry plunging into the CAD realm in its design work. At Ford, more than 250 Ford-developed workstations are in operation, and new stations are installed at the rate of two per week. The Ford system is based on distributed processing with centralized data control and shared-resource management.

"The Ford Graphics System is the major medium for CAD/CAM work, and has evolved from the user community and its requirements. The system that has resulted is a powerful, efficient, and user-friendly CAD tool. Development began in 1967 following an initial research phase during which all work was performed on the project basis. In other words, all development costs were recovered from resultant savings.

"In the beginning, all design stations comprised a CDC minicomputer, a disk drive, a tape drive, a teletype, and a vector refresh display tube. A more sophisticated system at that time would have cost two to four times more per design station than the system selected. Benefits of the stand-alone configuration graphics system included simplicity of operation, location flexibility, data security by virtue of isolation, invulnerability to computer failure, and quick response.

"In that early period, the systems had to be used 20 hours a day to be cost effective, and applications were carefully selected. The first application selected was body surface design for which a development team was formed within the drafting department of the body engineering group. The user worked closely with the development team so that the resulting system would be user compatible.

"Since the time of the initial effort, hardware costs have decreased as capabilities have been added so that mechanical design, detailing, body electrical-circuitry design, plant layout, numerical control, and finite-element analysis of structures tasks can be performed.

"Number of users and sophistication required has increased, along with requests for further enhancements. Therefore, in 1980 a replacement computer system was deemed necessary. This system includes virtual memory, a software-compatible network of computers, increased response time, networking capabilities, and a 32-bit capacity architecture.

"In October 1981, the first prime development computers were installed at Ford, forming the Ford Graphics System, to begin production work. System enhancements include automatic menu sequencing; database links among graphic and nongraphic systems; releasing and problem-tracking system; expansions in engineering analysis capabilities; structural, aerodynamic, and kinematic analysis capabilities; and the development of communication links between design stations and organizations in the network.

The authors continued: "One of the most noteworthy exploitations of CAD technology occurred when the crucial thermal-protection tiles for the Space Shuttle were programmed. NASA space travel, notably the Space Shuttle, has become nearly routine. The technologies now are beginning to focus more closely on the modules themselves and how they can better withstand extremes of velocities and temperatures during takeoff, orbit, and re-entry. One of the techniques that will figure in the keen competition for future Space Shuttle contracts is the development of CAD/CAM technologies for use in design and analysis of the 24,000 uniquely shaped silica tiles in the thermal-protection system of each orbiter, and other similar applications.

"Lockheed Missiles & Space Company, Incorporated has been producing the tiles that must withstand re-entry heat of up to 2,300 degrees F (1,260 degrees C).

The Space Shuttle tile program provides an excellent example of slightly varying parts of the same family. Each tile on the orbiter is unique and is produced by numerically controlled milling machines to tolerances of 0.016 inch (.406 mm). Nearly half the tiles are created in mirror image; flipping the figures on the X or Y axis can yield two tiles. However, 6,000 or so are highly complex. About 75% require five-axis machines to produce.

"Each tile had to be programmed for the NC machine by a programmer using an orbiter's loft data, master dimensions definitions that describe the structural and aerodynamic surfaces of the shuttle. These MD data were taken from Rockwell magnetic tapes, documents, and punch cards. Ninety programmers worked 10 months, logging in more than 160 hours per week of Univac 1108 computer time.

"Lockheed is developing an integrated process for tile design and fabrication that would permit significant reductions in both NC programming and inspection costs.

"Not all of the design work was performed using interactive graphics systems. As mentioned earlier, CAD does not mean graphics alone. For the *Columbia* orbiter, Rockwell, the prime contractor, gave Lockheed a group of drawings presenting general planform arrangements, together with magnetic tapes, punch cards, and documentation defining inner and outer mold lines. In some sections, equations were used to define surface contours. Lockheed then took these data with the side planes and other surfaces to produce the tile configurations.

"Though most of the tile programs were written from magnetic tapes, 5,800 were programmed using data from the documentation. Programmers experienced some failures before being able to make the programs work. Some 6,000 tool tries were made for developing the *Columbia* tiles, a number that was cut to about 1,400 on the *Challenger*, and approximately 800 on the *Discovery*. Programming and machining the 24,000 different tiles for an orbiter are complicated further by the fact that the 0.06 to 0.22-inch (1.52-5.59 mm) coating covering each tile must be glazed for about two hours at 2,000 degrees F, (1,204 degrees C) causing distortion due to shrinkage. To compensate for the distortion, tiles are machined with a circular arc to produce concave edges in planform and are undercut on the sides using a one-quarter-degree conical cutter.

"Performing analysis work in finite-element modes is fast becoming one of the most effective of CAD technology. With a CAD system, companies are finding that several potential designs can be evaluated quickly by building them on the computer. The best design can be selected without building a solid model, running extensive lab tests, or cutting metal.

"Typically, such analysis begins with generating a finite-element model of the component on the CAD system. The analysis work is done on a large host mainframe computer. Results of the computer output can then be displayed on the graphics scope for the designer to use. Finite-element analysis provides as realistic a model as the tangible one. Once component designs are set, a simulation program can be run to determine performance.

"Many such programs are available from vendors and several have been developed by industries in-house. Finite-element analysis has become the industry standard technique of sophisticated design analysis. It has become widely accepted as a tool in many kinds of design activity, including aerospace and aircraft, automobiles, and heavy industry products made from metals and nonmetal composites. Plastics have not yet proven strong enough to withstand the rigors of finite-element testing, nor has the process proven to be cost effective for plastics.

"However, since many industries are becoming more weight conscious, methods are being devised to adapt the time- and cost-saving aspects of finite-element analysis to plastic applications. Attention is centering on injection-molded parts which have complex configurations where finite-element analysis provides the only feasible means of analysis. Toward this goal, DuPont's Polymer Products Department's Technical Services Laboratory developed finite-element analysis methods for thermoplastic resins. Because of the vast disparities in plastics materials and metals, analyses must proceed in a different manner. Static loads, impact loads, long-term fatigue, and other elements may be examined by computer before working with metals. But the computer does not provide an easy solution for such analysis in plastics. Because prototype models often must be made for some design activities, caution must be exercised when adopting analysis systems to the design of plastics.

"After the hardware is selected, the appropriate engineering analysis programs must be chosen. Such choices are made more difficult with plastics than with metals because many of the unreinforced polymers have viscoelastic responses, indicated by nonlinear stress-strain curves. Those that are glass-reinforced usually are anisotropic, meaning their mechanical properties vary with glass orientation. The analysis software has not always been able to handle this combination. DuPont has developed two programs that accommodate these conditions: MARC and ANSYS.

"Despite the problems in marrying CAD analysis systems to plastics design, several successful applications have already emerged. One valuable use of finite-element analysis is in showing designers how to use less material, thus lowering weight. Skyway Recreation Products, for example, has produced a bicycle wheel for use on Schwinn and other bikes. An earlier version of the same wheel weighed 2.5 lb (1.13 kg); the new one, called Tuff Wheel II, weighs 1.8 lb (0.81 kg), 28% lighter. Finite-element analysis verified in the design stage that the lighter wheel would be as structurally sound.

"Other such analysis has shown the correct material for a design by substituting stress-strain curves for different materials. The correct material, as determined by minimizing deflection while retaining impact strength, was chosen without using tangible models.

"In the analysis stage of aircraft design at Boeing, technology staff engineers begin studying performance information for propulsion, acoustics (noise levels in decibels), weight, aerodynamics, speed, and range. The result of these studies is the theoretical shape of the airplane. Using a sophisticated three-dimensional refresh graphics analysis system, the following questions are considered. How fuel efficient can we make the airplane? What gage should stringers, skins, or spars, be? What is the fuselage diameter? The tech staff analyzes elements—sheer flow, tensions, aerolastic structure, and stiffness. If the engineer wants to investigate, for example, temperature versus altitude on a certain pressure probe in the engine, those two variables can be selected and within 10 seconds the system displays and plots the results.

"Aerodynamics produces a loft of the wing. Materials testing uses this information to determine the effect of stress and temperatures. The wind tunnel staff and project design group then step into their roles. A process called geometry control system, or GCS, which is a computer program system that numerically defines airplane external geometry, provides preliminary configuration data for wind tunnel testing. Numerous configurations are modeled in wind tunnel tests.

"Production drawings consist of detail and assembly drawings of model components. Nacelles, nacelle struts, and wing families are examples. The largest production payoff so far has been new airplane contours and the immediate availability of drawings of different scaled models. This technique has been used extensively on nacelle struts, aft body contours, and wing-to-body fairings. Through a link with the large host computer, files can be transferred to and from the interactive systems.

"Motion and clearance studies are an important task for which CAD graphics is used as tool. Landing gears, doors, overhead storage, elevators, trailing and leading edge actuators, and other mechanisms are designed and studied using an ICG scope and plotting them out on a plotter. The engine bleed manifold on the 767 program is a typical example. The engineer studied the engine nacelle and strut definition and then merged the engine model data, which included engine bleed parts. Working with this information, the designer was able to accurately create the manifold geometry in three dimensions and check for interferences.

"A number of control and drive systems have been defined using the interactive design system. Definition engineers studied control locations and dynamics of the alternate landing gear release system. The engineers located the three-dimensional centerlines of the body coordinates of all components in this assembly. Other projects and staff groups are able to use plots of these data. To complete the design, engineers took 120 hours using a computer. The job would have taken an estimated 360 hours manually.

"A good example of airplane propulsion analysis using interactive graphics is the eight-stage pneumatic duct. Designers made an installation drawing of the duct for the 757 mockup. Using the leveling capability and based on a 3-D model, they created a two-dimensional drawing for the installation. Ratio of computer to manual time on this project was 24:35.

"The Dana Corporation, a Fortune 500 company, has been using a Gerber Interactive Design System since December 1977 in its Spicer Axle Division to perform finite-element analysis. The corporation has many divisions and product lines, including speed controls for automobiles, Perfect Circle piston rings and pistons, Victor oil seals and gasket, Spicer transmission, Parish frames, and driving axles for the automotive and light truck industry. Some of Dana's customers include the Big Three automakers.

"The first application using interactive graphics for finite-element analysis occurred in the product engineering department. This first system comprised a central processing unit, a Control Data disk drive, magnetic tape drive, line printer, hard copy unit, and three workstations. A high-speed plotter produced finished drawings.

"The Dana Corporation workload was divided into the following jobs: two-dimensional drafting and design, three-dimensional design and layout, and finite-element model generation. The previous method was primarily based on design experience using similar parts. If the part needed more strength, mass was added to the former design. After design and drafting were completed, prints would be released for castings or forgings. Parts were made and assembled, then sent to the engineering test laboratory for functional testing. Parts that were not adequate would be made again.

''The new system incorporating the interactive graphics system and finite-element analysis capabilities follows the same process as before, but with computer. The new process begins with creation of a finite-element model and uses the computer to determine the results of the design. Using computer simulations to test conditions, revisions to the model can be made quickly.''

In another application, a CAD system utilized at the Automotive Group of Wickes Manufacturing Co. (Southfield, MI) is helping to produce automotive components and assemblies faster, more accurately, and to a higher quality than previously possible. Products such as parking brakes, transmission shifters, and seating suspension systems are designed and tested at the company's Product Development Center in Troy, MI. Computer magnetic tapes containing the new designs are then delivered to the customer for merging with overall system designs generated on the customer's CAD system or direct to a supplier for manufacturing special tooling or components.

One product developed on Wickes' four McDonnell Douglas Unigraphics workstations at the Product Development Center was a multi-component parking brake for a major automobile manufacturer. The ''black box'' design was performed on the CAD system. In this concept, the customer provides the space in the vehicle into which Wickes designs a black box—a component or assembly. To do this, the automaker provided necessary vehicle information on magnetic tape formatted in IGES, a neutral language. An IGES translator converted this information into the Unigraphics format.

Miles Doolittle of Wickes Manufacturing described the system in an article in the February 1987 *Manufacturing Engineering*. In the article titled: *CAD System Slashes Design Time*, Doolittle wrote: ''Wickes engineers and designers reviewed vehicle data, and the project team went to work on the allotted space. They reviewed information about the firewall, steering column, and parking brake and identified objectives, concepts, and end results. Many hypotheticals were also examined to find the optimum design. Using dynamic rotation, they viewed the parking brake installation area from different angles. Unnecessary file information was deleted. Components that had to be moved to accommodate the parking brake, such as electrical cables, were identified and moved. The team calculated maximum travels, clearances, torques, moments of inertia, and radii of gyration.''

Finite element modeling determines the stress levels in critical parts of the brake, as well as whether any portions of the brake exceed the yield strength of the material. Such analysis determines whether the parking brake will bend or break. Similarly, it determines points that are too strong. Correction of such problems typically calls for changing the metal gage of a bracket or modifying a flange.

Doolittle noted: ''Brake testing revealed opportunities to refine the design. By reinforcing the brake stop bumpers, for example, reliability improved. Special two-part bumpers made of high-tech plastic were added to attenuate noise. Test documentation was sent to the customer with parking brake prototypes for further examination.

'' 'After receiving customer approval, magnetic tapes of the brake design were sent to X-Cel Tooling, Incorporated (Iron Ridge, WI) to produce brake tooling. Their operation is completely paperless,' according to Terry Grimes, X-Cel's president.

''We move information from one department to another with a single database with two 5X Computervision CAD systems connected via a GNA link,'' Grimes explains. ''This cuts our design and production time in half. Furthermore, any last-minute engineering changes can be instantaneously made via the telephone, by transferring part files through modems.''

Information from Wickes was first run through a postprocessor and put into Computervision format. The progressive die strip layout was developed to determine where notching, cutting, and forming occur. Press tonnages were examined to ensure that the die would work on a particular press. A top die, section view, and layout were designed on the CAD system. The die was then broken down into complete details for MRP and CAM programming.

"What makes this unique is that we use Wickes' finished part—the exact graphic model the engineer designed," says Grimes. "We just postprocessed their data again, rather than input lines, points, circles, and dimensions like a normal programming operation."

In another example the complexities of automating a magnet factory were brought under control through the use of a CAD system for design, simulation, and off-line programming.

Magma (Gesher, Israel) is an enterprise specializing in the production of hard ferrite permanent magnets. The manufacturing processes at the factory were highly labor intensive, with workers carrying out all loading, unloading, material handling, and process control tasks.

Magma is now reorienting its operations toward fully automated processes. The entire plant is being redesigned, with implementation to occur in two stages. When the plan is fully realized, only three to four workers, mostly part-time, will be required, and they will be employed in supervision and quality control.

The company's CAD system performed work cell design, simulation, and off-line programming. The system has four critical capabilities:

- The design of each individual work cell component (conveyors, pallets, presses, and others)
- The layout of the work cell and definition of the tasks of the robot and the mechanisms of the presses and the conveyor
- The simulation of the complete process
- The downloading to the robot controller.

CAD systems applied to architectural applications proved the tools to model a design concept, produce the architectural production drawings, perform space and facilities planning, produce the contract drawings, and to produce schedules and bills of materials that are required for the electrical, plumbing, and structural work in a facility.

Technical publications is a growing area of importance for CAD systems. Almost every manufacturer is faced with the need to provide technical documentation on products, write proposals for sales or other business activities, and to publish product literature as well as annual reports.

Specialized software allows a CAD system to take a design (a product design or an illustrated parts breakdown) that resides in the CAD engineering database, merge that design with text on the CAD screen, and through this merger, develop the exact format of a document that can then be printed out on a laser printer or sent to a phototypesetter for subsequent publishing. CAD system software exists that also allows business graphics to be generated and halftone photographs to be merged with text.

Over the past 10 years, the use of CAD systems has increased dramatically. At the time many of these systems were installed, they were selected to solve specific problems; integration or communication between systems was not an issue. The consequence of these diverse acquisitions was that several varieties of heterogeneous CAD systems were employed, within each industrial enterprise and within the industry as a whole.

As a result, CAD systems used by the engineering department of a company to design and define products could not electronically communicate this information to CAM systems used in manufacturing or other downstream functions. Rather, engineering was required to produce transitional paper drawings so that design data could be transmitted to manufacturing. As organizations began to realize the value of integrating different business functions, the need to communicate between the computer-based systems that controlled those functions became obvious. Several efforts were initiated in the early 1980s to address this problem. One such effort resulted in today's IGES organization and standard. IGES, or Initial Graphics Exchange Standard, was intended as a standard for exchanging graphical and textual data between heterogeneous CAD systems. After diligent work by some 70 companies and hundreds of individuals, the first versions of the IGES standard were released under the auspices of the

National Bureau of Standards. IGES has been adopted as an ANSI standard.

At this point, Version 3.0 is being specified in a number of DOD, DOE, and NASA procurement contracts and is consequently the subject of widespread user and vendor attention. IGES Version 3.0 specifies a standard format for exchanging graphic entities such as lines, curves, circles, 3-D wire frames, and surfaces as well as textual data. It is used principally for transmitting engineering drawings, technical orders, schematics, and similar graphic documentation between CAD systems.

As useful as IGES is, it has some drawbacks. The more sophisticated manufacturers have seen a need for another standard that will allow exchange not just of graphic entities, but of complete product definitions. Consequently, a new effort has been launched, called PDES or Product Definition Exchange Specification. The PDES program was begun in 1985 as a research initiative within the IGES community. PDES concerns not just graphic and textual entities, but the entire array of product definition information relating to the complete lifecycle of the product. PDES information is to be based on solid geometric model representations of detail parts logically associated with a wide range of related information to directly support design, analysis, manufacturing, and logistics applications.

IGES and PDES are individually important but essentially different products. They are based on different technical approaches and are presently in widely separated stages of development.

Also see: CAD/CAM, Computer Aided Engineering, Electronic Design Interchange Format.

Computer Aided Design and Drafting (CADD). Computer aided design and drafting is a term usually applied more to the specific drafting usage of CAD systems, as opposed to the more general aspects of CAD, which may or may not entail actually drawing or drafting a picture. CADD is largely associated with mechanical drafting, but also can be applied to the drawing of an electrical schematic. While some design and CAE type activities may be required to generate a drawing of a device, the CADD portion of activities is the specific actions associated with the picture creation.

CADD systems may be two-dimensional or three-dimensional. Two-dimensional systems are usually the type used with CADD, since many of the items drafted only have two dimensions of relevant description. A CADD system typically can only generate drawings and store the drawing files, but is not capable of analysis such as finite element analyses or timing analyses, which are CAE functions.

Also see: Computer Aided Design, Computer Aided Engineering.

Computer Aided Engineering (CAE). Computer aided engineering is the practice of using the computer to assist in engineering design, development, calculations, testing, and experimentation. CAE is probably best known relative to the design of electrical integrated circuits and printed circuit boards. The primary reason is that the complexity and design parameters for these products cannot be managed without the assistance of computer based tools. Another reason for the continuing emergence of CAE is the need to develop products in shorter timeframes to match smaller opportunity windows and with decreased development costs. The demand for greater reliability and performance by the customers adds to the CAE revolution. Other aspects of the design, manufacturing, and support process are supported by the use of CAE tools, such as engineering changes, documentation, and the creation of user manuals, and manufacturing and test aids. The large field of CAE vendors will undoubtedly experience some shakeout as the strong survive and the weak are eliminated.

Also see: Computer Aided Design.

Computer Aided Line Balancing. Computer aided line balancing is the process of balancing the sequence of a manufacturing line with the aid of a computer and a computer database. Using a concept called total processing, the manufacturing engineer identifies each processing requirement that interfaces with the computer database corresponding to the original design and process time in each operation. Programs will include the logic which automatically commands an optimum machining sequence of processing operations, selects appropriate feeds, and regulates a number of other machining parameters. Correlation of the design phase, the part processing phase with the manufacturing process and time phase has significantly affected the production cycle. Computer aided line balancing permits the manufacturing of a product to progress at a more rapid rate from raw material to a finished product.

Computer Aided Manufacturing (CAM). Computer aided manufacturing is when computers are used to assist in any part of the manufacturing operation. Computer Aided Manufacturing is one of the integral components of Computer Integrated Manufacturing (CIM). CAM is often used in conjunction with Computer Aided Design as CAD/CAM. Usually the input to a CAM system comes from the CAD activities which occur before the manufacturing effort. CAM may be simply the translation of the design into information which an automated assembly machine or numerically controlled machine can utilize as input, or CAM may provide many sophisticated functions in addition, such as optimizing tool paths and material usage to minimize waste. As the desire for increased productivity continues, CAM must be a part of most manufacturing activities, whether in the form of an MRPII system, an automated cost accounting system, or CNC links to the design activity. CAM is not an entity unto itself, but is simply when computers are utilized to assist in any manufacturing activity.

Karl H. Schultz of Schultz Associates explained in his SME Technical Paper (*Implementing Computer-aided Manufacturing (CAM)*): "In starting a CAM project, a team (task force/steering committee) should be developed. The leader, a "champion", reports to the top manufacturing manager at the location. He/she must be of high rank and have a good technical background. It is important that he/she also be people oriented and have a good ability to communicate. The other members of the team should be:

Manufacturing/industrial engineering—Should be the manager. Can call upon his/her sources such as the CNC programmer(s), industrial engineering for time standards generation, tool design, and plant layout.

Factory supervision—The plant superintendent should be the representative. This person, and his/her peoples' acceptance, will help greatly in the success of the CAM program.

Engineering design (CAD manager)—This person will have the best knowledge of CAD and be in a position to make any engineering changes to facilitate manufacturing.

Data systems—This person is needed because of the many possible types of hardware and software available and to ensure that there is a CAD/CAM systems capability with the business system.

Materials management—This is necessary because of the possible interface with MRP and BOM.

Accounting/finance—will probably be the cost supervisor. This person's input will ensure that uniform product cost data can be developed.

Factory/union—A CAM system will obviously have a great effect on the manufacturing floor. Maintaining contact with the manufacturing labor force will assist in the installation by maintaining open communication. This is whether a union is present or not."

Team meetings should be held once a week. Every meeting should have the actions accomplished, and actions required with responsibilities and timing. The published minutes should not be more than two pages and distributed to all key people. A monthly meeting with top key management should be held.

Authors William Beeby and Phyllis Collier write in their book *New Directions Through CAD/CAM*: "The many building blocks that make up computer-aided manufacturing, CAM, represent functions that must fit together if they are to avoid collapse. CAM involves may disparate activities. These include: online planning, computer numerical control, automated assembly, computer-aided process planning, scheduling, automated template making, tool design and production, automated materials handling, and robotics, to name a few. American industry's responsibility to come to a meeting of the minds—and objectives—has never been greater.

''For any one of the foregoing disciplines to define CAM in terms of its own operation alone is to make a grave error. Fortunately, the current effort toward integrating CAD and CAM and the various disciplines within them, as well as discernible trends noted in the literature, reveal that manufacturers who might once have clung to their own ways of doing things are coming forward with cooperative methods that work well for all.

''Any discussion of CAM must distinguish between information automation, which refers to data transfer, and process automation, meaning robots or handling systems. One must keep well in mind that CAM is not limited to one or the other type of operation, but ideally combines both, utilizing the capabilities of computers for tool, part, and template making; assembly, inspection, and testing; storage and retrieval functions and other typical robotic tasks such as spot welding, spray painting, and deburring.

''Emphasis in manufacturing is on the development of computer graphics use in tooling areas. Often the systems start out small, but as groups begin finding ways to apply the graphics system to their work, they purchase more systems, terminals, and upgraded equipment.Applications include drafting two and three-dimensional parts and tools and creating them on the machines in the factory. One of the first jobs of a shop adding graphics systems is to develop the software needed to perform these operations. Now, for some of the parts, CAM shops can go from the engineering definition to a completed blue-streak (special order or custom) part in a matter of minutes. Bonded assemblies, sheet metal parts, skins and spars, gears, welded ducts, and fiberglass parts are examples of the processes that lend themselves to CAM systems.

''Engineering does the design work and turns the geometric, material, and mathematical information over to production engineering. Engineering and manufacturing work together to see that the information is correct, that the part gets completed both accurately and on time. They must determine which tools the job requires, what parts are needed for the assembly, and how long it must take to make the part.

''These and other questions are answered by representatives of the tooling shop, the factory itself, tool design, shop load (industrial engineering), numerical control, and quality assurance. All of these disciplines must make sure that the end product is checked throughout the routine, that all the pieces fit together.

''Because the journey of parts from start to destination is a complicated one, manufacturing uses product managers' flow charts to condense the pertinent information into one reference. Each end item has its own flow chart. These charts contain milestones which are strictly adhered to and they illustrate graphically the flow of the part, including its creation, necessary paperwork, and ultimate transportation. Another byproduct of the computer development effort, the calculations and drafting of the chart are done quickly on the interactive graphics system. The information is stored and available to all concerned.

''Many shop routines and processes are using computer graphics, but the one that holds the most promise so far in terms of time and effort saved is the template-making functions. Graphics automated template systems are CAD/CAM at its finest. The job that had once taken several days can now be done in a half hour. Engineering creates the flat pattern, and sends the information over a communications link through the data base to the fabrication area where the template maker creates the geometry on the system. Or it sends a photo contact master, or PCM, which is a full-size copy of the engineering geometry, for the shop to use.

"Then, with a number of computer aids, the template is 'computerized.' The aids are on the system to be used repeatedly.

Also see: CAD/CAM, Computer Aided Design, Computer Integrated Manufacturing.

Computer Aided Planning. Manufacturing planning is an important step in the transformation of a new functional design into a marketable product. Manufacturing planning involves the identification of the best ways to produce the product so that quality and reliability are ensured at a competitive profitable price. Good manufacturing planning will help ensure the services of the enterprise, while poor manufacturing planning may result in costs that are so prohibitive that the product may not sell.

Manufacturing planning must consider several alternatives in order to select the best procedure to produce a part. Furthermore, there is seldom much time to perform this function. The product usually is required by the customer shortly after design approval. The computer permits process planning rapidly while considering all possible alternatives.

Software can be developed that considers all alternatives. The most favorable process may be selected for new design during the initial design phases. In this way the best process can be selected well in advance of completion of the design, and the functional designer will be able to incorporate design requirements characteristics of the processes selected.

Also see: Computer Aided Process Planning.

Computer Aided Process Planning (CAPP). Computer aided process planning refers to those activities associated with planning how a part is to be manufactured. A part, once designed, will generally call for a variety of manufacturing operations and associated equipment for its production. Neither the operations nor the sequence of operations will be unique. That is, it may be possible to employ a number of competing operations to shape a particular part; once the particular operations are specified, alternative routing sequences may be possible. Computer aided process planning (CAPP) is the use of computers to develop the plan for manufacturing a part. The process planner, starting from the design geometry and specified characteristics of the part to be manufactured, will typically work interactively at a computer terminal with a screen display, and will develop the process plan based upon the rules inherent in the particular CAPP systems in use today employ a variant approach based upon group technology concepts. In these systems, process plans for new parts are created, primarily on the new parts' similarity to older parts whose process plans have already been developed and now reside in the CAPP system's database. More advanced CAPP approaches that do not start from a previous similar part's process plan are termed generative approaches. For the most part, generative approaches are still in the research stage and are not widely commercially available.

One of the bigger hurdles that computer integrated manufacturing (CIM) must overcome is closing the gap between CAD and CAM. Technologies exist to minimize this hurdle, two of the most important being computer-aided process planning (CAPP) and group technology (GT).

Computer-aided process planning systems are, in effect, expert systems that capture the knowledge of a specific manufacturing environment plus generic manufacturing engineering principles. This knowledge is then applied to a new part design to create the plan for the physical manufacture of the part. This plan specifies the actual machinery employed in the part production, the sequence of operations to be performed, the tooling, speed and feeds, and any other data that is required to transform the design to a finished product. To use CAPP most effectively in a CIM environment, the design should originate on a CAD system and be electronically transferred to the CAPP system from the database.

CAPP draws on the geometric model of the part to be produced, generated in the CAD system, and matches the characteristics and components of that part to the production machinery on the factory floor. This technology will develop the process sheets or routings needed to manufacture a part.

Because a computer-aided process planning system contains information on all parts being produced at any given time, it is able to contribute to more efficient machinery utilization. Because CAPP determines how a part will be made, it is a factor in determining the cost of manufacturing the product.

Process planning is not a new technology. Successful process planning has been performed in many manufacturing companies by individuals with a great deal of experience in manufacturing and in the manufacturing resources and practices of the individual company. However, such a manual system is time consuming and totally dependent on the knowledge and experience of individuals who will eventually retire and leave the organization. CAPP systems capture the knowledge and experience of these individuals. They also allow a planner to generate a process plan for a new part by modifying an existing plan for a similar part.

A CAPP system will provide a set of instructions on how to make a part, in what sequence the process steps shall be executed, and what machines, tools/fixtures, workcenters, and labor skills will be required. A process planning system, manual or automatic, must be constantly updated to reflect a company's total manufacturing capability. Process plans must change as, for example, new machinery is procured and new quality control procedures or robots are introduced.

Computer-aided process planning systems are of two basic types: variant and generative. The variant method of process planning is the most commonly used method today. This method develops a process plan by modifying an existing plan that is selected using group technology principles, namely coding and classification. Some form of parts classification and coding is essential to expedite the location of previously developed plans, which are to be adopted as is or modified and adopted, and to specify the processing plan for a new part.

Generative process planning systems incorporate into their database a body of manufacturing logic, the capacities of existing machinery, standards, specifications, and the like. Based on the part description (geometry and material) and finished specifications, the computer then selects from this stored knowledge the optimum method of producing the part and automatically generates the process plan.

Automated process planning is useful to small and medium-sized companies as well as those with large manufacturing facilities. The more complex the planning function and the more factors to be considered, the better suited a computer is to the job. Some aircraft manufacturers, for example, may generate many hundreds of plans in a week for parts that will have limited production runs. In this kind of situation, automated process planning speeds the planning process by as much as 75% and makes the plans consistent. It also saves labor, scrap, and rework. Therefore, CAPP substantially lowers the cost of getting the design to the manufacturing floor and allows user companies to be more flexible in their parts production.

CAPP systems are actually expert systems that capture the knowledge of process planning and allow that knowledge to be applied to new situations. CAPP systems are probably the major application of artificial intelligence principles in manufacturing.

Mukasa E. Ssemakul and Philippe O. Sivac of the University of Maryland discussed Computer-Aided Process Planning in their SME Technical Paper titled *Automatic Generation of NC Part Programs*. In this paper they wrote: "Process Planning is defined as the task of translating part design specifications into manufacturing instructions required to convert a part from a rough state to a finished state. This function takes up approximately 40% of the preparation time for a new part. Traditionally, it is performed manually by highly skilled workers who possess in-depth knowledge of the manufacturing processes involved and the capabilities of the machine tools. On the basis of their experience, they are expected to develop a detailed process plan to be used in manufacturing the part. This involves selecting the appropriate machine tools, jigs and fixtures, determining the appropriate machining operations and associated cutting conditions, as well as calculating the cutting times and the noncutting times. A detailed knowledge of the particular manufacturing environment is necessary because it imposes constraints on available alternatives. The result is a detailed process plan which includes machine tools to be used, sequence of operations, tool and fixture selection, and cutting conditions. This activity is very labor intensive and becomes often tedious when dealing with a large number of process plans and revisions to those plans. Rather than carry out an exhaustive analysis and arrive at optimal values, which would be too time consuming, process planners all too often tend to play safe by using conservative values. What is considered to be conservative varies with the particular manufacturing environment. This has the unfortunate side effect that if several process planners with different backgrounds are given the same part to plan, they will likely come up with different plans. Even more significantly, the same planner, if required to generate a plan for the same part on different occasions will probably come up with inconsistent results. Consequently, the speed and consistency brought about by the computer was sought in process planning. Another factor is the lack of qualified and experienced process planners. Another advantage of the use of computers in the process planning function is that it can reduce the required skill of a planner.

CAPP can be implemented at two levels:

- Stand-alone systems. As a stand-alone system, CAPP can generate process plans faster and with better consistency than human process planners. CAPP systems, for example, can calculate optimum cutting conditions based on varying criterias such as minimizing costs, minimum tool wear, or shortest cutting time. With the use of artificial intelligence (AI) techniques, they can incorporate manufacturing rules to select and sequence the appropriate processes to complete the process plan. Computers can thus free expert process planners to work on cases which fall outside the scope of the CAPP system.
- Within the Computer Integrated Manufacturing (CIM) architecture. CAPP's function is to transform the part design into manufacturing instructions. This requires the use of the part geometry as well as design information. Based on this the system calculates the detailed data required for manufacturing as outlined above. CAPP appears thus as the ideal candidate to fully integrate CAD and CAM. Use of CAD/CAM systems as they are known today, shows that generation of a 3-D model provides only geometrical support for machining yet will not yield NC part programs without extensive interaction with the user. Also, CAD/CAM systems are not sufficient to integrate with the other building blocks of CIM. The CAPP link will provide the time information necessary for planning and scheduling by calculating machining times."

Anthony K. Mason of California Polytechnic State University and Andrew Young of Northern Telecom wrote in their SME Technical Paper *Computer Aided Process Planning for Printed Circuit Board Assembly*: "As the electronics industry moves toward greater automation of its assembly processes, computer aided process planning (CAPP) becomes more crucial. Issues associated with the design, development, and operation of CAPP systems for circuit board assembly—issues which can differ significantly from those in metal working and other processes—include parts coding, variant versus generative strategies, the relationship of the CAPP system to CAD and postprocessing for insertion machine operation, and related CIM issues.

"Interest in CAPP is also shown by the increasing number of commercially marketed software packages that are designed to assist in manufacturing planning. In addition to helping construct a process plan, some of these systems help in the production of shop aids, postprocessor instruction sets, and test routines. Moreover, several firms in military electronics have constructed rather elaborate CAPP systems to meet, among other things, a need for exhaustive, traceable documentation of the process used on each and every copy of a product produced. Indeed, any firm producing electronics which must be supported in the field over a period of years, needs to know the production process that was used to manufacture individual units so that decision dealing with recall and modification can be made in the most economical manner.''

In comparison to metal working and many other manufacturing processes, circuit board assembly is a relatively simple process. One begins with a bare printed circuit board and attaches to it various types of mechanical and electrical components and subassemblies. While some components are mechancially attached, most are first temporarily affixed to the board by inserting the leads of the components into holes drilled in the board or by the use of adhesives. The attachment is made permanent by a soldering operation that can take various forms depending on the type of component that is to be attached. The soldering operation must also provide the connections required to make the circuits on the board electrically functional. The bare PCB can be viewed as a component itself.

The author discussed generative and variant CAPP systems. ''In the variant approach, one selects a design from a library of prior designs using one or more attributes of the product. The old design is then edited to account for differences between the product on which it was based and the new product. As edited, the old process plan becomes the new process plan. Under a variant systems, attributes must be assigned to each product so that a similar process plan can be retrieved. If the attributes of two products are the same, they are likely to have the same general process plan. In the context of PCB assembly planning, one could assign a process attribute to each board according to the types of components that were used to populate the board. However, a useful discrimination between boards would be difficult to achieve. Alternatively, one can assign a process attribute to the components.

''There are two approaches for assigning these attributes. First, one can use a predetermined coding system. Second, one can store rather arbitrary attributes in a relational database which can be expanded whenever one perceives the need for a new type of attribute. Either way, variant CAPP systems for process planning are basically data base managers coupled with an appropriate editing system.

''Unfortunately, the variant approach is not supportive of many of the features we would like our CAPP to have. It is interesting to note, however, that a common electronic spreadsheet, set up with a template of the ordered list of operations can be effectively used as a preliminary aid in process planning for electronic assembly and initial cost estimating. This very basic variant approach assumes that all PCBs are a variant of one very comprehensive PCB and is possible because the operation sequence is strictly ordered and the labor standards are in a relatively simple format.

''In a generative system, the process plan is produced using the facts presented by the new design and without direct reference to an existing process plan. Generative systems normally rely on a set of rules which can be referred to as a process planning 'knowledge base.' If we define an 'expert system' to be a computer program that performs a task normally performed only by a human expert, then generative systems for process planning may be viewed as a sort of expert system. It should be noted that under this definition, and 'expert system' is quite temporal, and that which we call an expert system today will just be another computer program tomorrow.

''There are strong arguments for variant systems and for generative systems. In choosing between them, consideration may be given to the following issues:

''*Group Technology (GT)*: There may be a strong argument for variant systems if one's manufacturing environment benefits from the application of GT concepts. The process of retrieval of similar designs and processes mitigates the costs associated with the proliferation of unneeded new products and process plans.

"The electronics industry applies GT concepts in several ways: (1) in the design of the component data base so as minimize the number of different components that have to be procured, stocked, and accommodated in production, (2) in analysis and procurement of capital equipment, and (3) in configuring the factory. However, GT for purposes of reducing the number of printed circuit board designs may not be a useful concept.

Innovation: Process innovation takes place in quite a different manner under variant than under generative process planning schemes. Since the variant approach is one of editing an existing process plan, innovation can be accomplished while editing. Innovation may also occur while editing a process plan produced by a generative scheme. Otherwise, however, innovation under a generative CAPP is a matter of changing the rule base that underlies the CAPP. The skill and effort needed to change the rule base may be substantial, and unless this issue is dealt with, innovation may be inhibited.

Response time: Generative process plans accommodate fast and automatic changes in process plans. This is the main reason they will dominate printed circuit board assembly process planning, particularly in highly automated environments.

"All commercial CAPP systems support the variant approach, with some requiring a predetermined, structured coding of parts, and some permitting the *ad hoc* creation of attributes. Recently, some of these systems have been designed to support a generative approach. To do this they must provide a programming language and data structure which supports at least primitive inferential logic."

Also see: Computer Aided Planning.

Computer Aided Testing. Computers were originally used to analyze data after it was put into the system manually. With the advent of interfaces between test instrumentation and computers the need for data input is no longer necessary Computers today can initialize the test, monitor test conditions, collect test data and analyze the facts.

Increased precision is realized through precise data collection techniques. The computer will begin testing at selected times as designated by the program. The data handling and analysis becomes error free since there is no chance of error due to manual data transposition.

The newer rule of the computer is to test models under various conditions to assist engineering in choosing particular design parameters. Reliability testing has been greatly enhanced since a computer program can randomize test conditions so that the test reflects years of actual use. The highest degree of accuracy and precision is being developed. A testing computer will retrieve the necessary engineering data from a CAD/CAM database and use it to perform the necessary tests.

Also see: Computers.

Computer and Automated Systems Association of the Society of Manufacturing Engineers (CASA/SME). The Computer and Automated Systems Association of the Society of Manufacturing Engineers (CASA/SME) was founded in 1975 to provide comprehensive and integrated coverage of the field of computers and automation for the advancement of manufacturing.

As an educational and scientific association, CASA/SME has become "home" for engineers, managers, and other professionals involved in computer-based technologies and automated systems. CASA/SME is applications-oriented and addresses all phases of research, design, installation, operation, and maintenance of the total manufacturing enterprise.

Specific CASA/SME goals are: to provide professionals with a focus for the many aspects of manufacturing which utilize computer systems automation; to provide a liaison among industry, government, and education in identifying areas of further technology development; and to encourage the development of the totally integrated manufacturing facility.

Computer Grade Tape. Computer grade tape is high-quality magnetic digital recording tape that is rated 1600 FCI (flux changes per inch) or greater.

Computer Graphics. Computer graphics is the generation of graphic images by a computer. Computer generated graphics are usually bit-mapped, meaning that the image is composed of a collection of picture elements (pixels), however, with raster display screens, the image may be a drawn (calligraphic) image. Computer graphics may be stationary or they may be animated (dynamic). Computer graphics are anything other than font-generated text on a computer screen. Even characters may be graphically generated as a bit-map, instead of calling a stored font (which is also a bit-mapped image, but is called by ASCII code). The advantage to describing text by their ASCII code as opposed to describing them as bit-mapped graphic images is the dramatically reduced memory needed to store the representation and the significantly decreased time required by the system to interpret the symbol and generate the display. With the advancing technology in video driver chips, computer graphics is becoming simpler and less expensive to implement. Several standards have emerged defining data file formats for graphical images. IGES (Initial Graphics Exchange Specification) is one of the data file standards.

Also see: Computers, Graphics, Initial Graphics Exchange Specification.

Computer Graphics Aided Three Dimensional Interactive Application (CATIA). The computer graphics aided three dimensional interactive application software program runs on IBM mainframe computers and defines three-dimensional parts which must be manufactured or machined. The CATIA program is three dimensional but can accommodate two-dimensional part programs by simply operating in one plane. The basic software module is the 3D Design Module, can then be coupled with other modules for developing the part programs for lathes, mills, or other NC machines. Essentially the two types of design data files are NC sets or lathe sets. Any machine that rotates the work-piece on a spindle can using the lathe set, which includesmachines which work by translating in different axes use the NC set program information. Normally the output of the CATIA program is then processed by a program such as APT to create the actual NC tape information or other form of input for the production machine.

Also see: CAD/CAM.

Computer Graphics Augmented Design and Manufacturing (CADAM). Computer graphics augmented design and manufacturing is a computer aided design software program which was developed by CADAM, Inc. CADAM allows a designer to design a part using the computer and to transfer that information to a CNC or NC machine to manufacture the part. The CADAM system can support from 2 1/2 axis to five-axis NC programs. The latest version of CADAM can handle three dimensional wire frame drawings and surface geometry definitions. The CADAM software does not directly work in three dimensions, but provides some of the information and commands necessary for the post processing software which defines the three-dimensional surfaces.

The CADAM software runs primarily on IBM mainframe computers and was one of the first well-known and widely used CAD software packages. CADAM can interact with other CAD/CAM software such as the APT (Automatically Programmed Tools) postprocessor. CADAM remains as one of the more widely used CAD software packages, especially in the environment of mainframe supported CAD workstations.

Also see: CAD/CAM.

Computer Integrated Manufacturing (CIM). Computer integrated manufacturing is one of several advanced manufacturing concepts that comprise modern manufacturing technology. As with other modern manufacturing concepts (e.g., Just in Time, Total Quality Assurance), CIM has a pervasive technological component. But it is more that just a new technology, it is a philosophy of operation. The philosophical basis of operation revolves from the definition of CIM which requires an understanding of the current concepts of manufacturing and integration.

The modern view of manufacturing encompasses all activities necessary to transform purchased materials into product, to deliver product to customer, and to support the performance of the product in the field. This concept of manufacturing starts with the development of a product concept, which may exist in the marketing organization; includes design and specification activities which are usually the responsibility of an engineering organization, and extends to delivery and after-sales activities which are usually part of the sales organization. This definition of manufacturing is far broader than the responsibilities that are traditionally associated with the manufacturing arm of a corporation.

The modern concept of manufacturing recognizes the importance of information. Performance of each activity within manufacturing requires information. Some of the required information is generated within the area that requires it; some comes from other areas within the manufacturing realm, and some comes from areas outside of manufacturing. For example, the activity "plan production of new product" within product planning requires inventory information (from within product planning), a bill of materials (which is produced by product design), and a sales forecast (which is provided by marketing). Activities also supply information to areas outside of the manufacturing realm. A new product is designed to meet specifications provided by marketing and the product's bill of materials and process design information must be provided to accounting for cost calculations.

The term "integration" must be viewed in light of the informational needs of the manufacturing activities. Integration means that the information required by each activity is available on a timely basis, accurately, in the format required, and without asking. Data may come directly from their sources, which include both manufacturing activities and other functional activities, or from an intermediate database. This definition implies that information needs for each activity must be identified and planned for in advance.

Computers are the tool of choice for automating activities. With their exemplary data collection, storage, and handling capabilities, they are also the tool of choice for automating integration.

Viewing the computer as the means, and using the above definitions of manufacturing and integration, CIM is defined as the use of computer technology to integrate manufacturing activities.

History. Over the years, each of the major functions of manufacturing followed individual courses of automation. Until the 1970s, the most aggressive and successful automation was seen in production operations. Discrete parts manufacturing made use of highly mechanized machines that were-driven and controlled by cams and complex mechanisms, such as automatic screw machines and chuckers. Process manufacturers made use of cam driven controllers and limit switches for operations such as heat treating, filling and canning, bottling, weaving and the like.

Design functions were enhanced by the use of computation aids such as slide rules, calculating machines, and analog computers. Designers of discrete products and parts permitted major portions of their designs to be done by people of less skill and experience than the senior designer.

Planning functions used fewer automated techniques than production and design. However, the use of tools such as planning tables, PERT charts, and network diagrams determined which activities were critical to the maintenance of a schedule.

Communications among manufacturing functions were largely paperbased and usually involved an exchange of drawings and specifications. Microfilm techniques improved different locations being updated and current with the latest changes in product or process design or specification.

As the development of computerized applications continued for each of the individual manufacturing functions, it became apparent that, for the first time in the hist separate manufacturing functions. Digital technology and the media for storing, transmitting, and processing data in digital format were central to the automation of common basis of automation, direct communications among these functions became technically feasible. This opportunity for automatic communication of information and Manufacturing, or CIM.

The concept Computer Integrated Manufacturing was coined by Dr. Joseph Harrington in his book *Computer Integrated Manufacturing* published in 1974.

Discussion. The historical approach to automation has been focused on automation of individual activities of functions. The result is a large number of stand alone islands of automation incorporating computerized technologies. The benefits of the new computerized technologies are severely diluted if the integration is still manual. When information is manually transferred from one automated piece of equipment to another, the data transfer is error prone and inefficient. More problematic, the same data may reside on multiple pieces of equipment. It can be changed in these different places by different people and reconciliation becomes nearly impossible. It has been estimated that 40% of the effort for data collection, verification, and entry is for creating and verifying redundant data.

CIM is the integration of the islands of automation. The challenge of CIM is both technical and cultural. The technical challenge is the breadth of technologies involved. Though they are all computer based, they are intended for widely varying applications. The challenge is further complicated by the number of different vendors, the incompatibility among systems (from the same vendor as well as different vendors), and the lack of standards for data storage, formatting, and communications. Specification of CIM technologies requires an understanding of the applications, the technology, the communications and integration requirements that will be required of that technology by other systems, functions, and people, and data communications and integration technologies.

CIM poses a cultural problem to most organizations. The modern definition of the scope of manufacturing typically includes activities within many different functional units: engineering design, manufacturing engineering, process planning, marketing, finance, operations, information systems, materials control, field service, distribution, quality, and production planning. CIM requires organizations to view their operations as a whole and not as fragmented functions.

Because CIM is the integration of technologies and information flows, it requires planning. Only if integration needs are planned for, can they be provided for in a cost-effective manner. Furthermore, computer integrated manufacturing is concerned with the automation of non-direct activities—not only manufacturing overhead, but also indirect activities in other functional units (but within the definition of manufacturing). As such, economic justification of CIM cannot be performed by traditional methods (which usually emphasize direct labor savings).

Despite the fact that traditional cost justification techniques show little or no benefits of CIM, CIM results in many benefits that translate directly to a corporation's bottom line. Quality and yield are increased as mechanisms such as adaptive control are put in place. Lead times are reduced as the time consuming steps involved in transferring information manually are eliminated.

Direct costs are reduced with automation. The indirect and overhead costs of expediting information are reduced with integration.

Product development times are shortened as fewer engineering changes are necessary, and parts designed are retrievable for leveraging the design effort. Manufacturing lead times are not compromised by unnecessary waiting for parts or information.

Developed by the Computer and Automated Systems Association of the Society of Manufacturing Engineers, the CIM Enterprise Wheel (see figure below) was designed to provide industry with a common vision of computer integrated manufacturing.

The remainder of the discussion portion of CIM and the CIM Enterprise Wheel is taken from the book *Introducing the New CIM Enterprise Wheel* (copyright CASA/SME, November, 1985). It was written by Daniel S. Appleton of DACOM.

The CIM Wheel is a top-down perspective of CIM—a view from the office of the business executive rather than the manufacturing technologist. It is the only view of integration that makes sense, because integration is more of an imperative to those who must derive economic benefit from it—that is, to the integrators—than

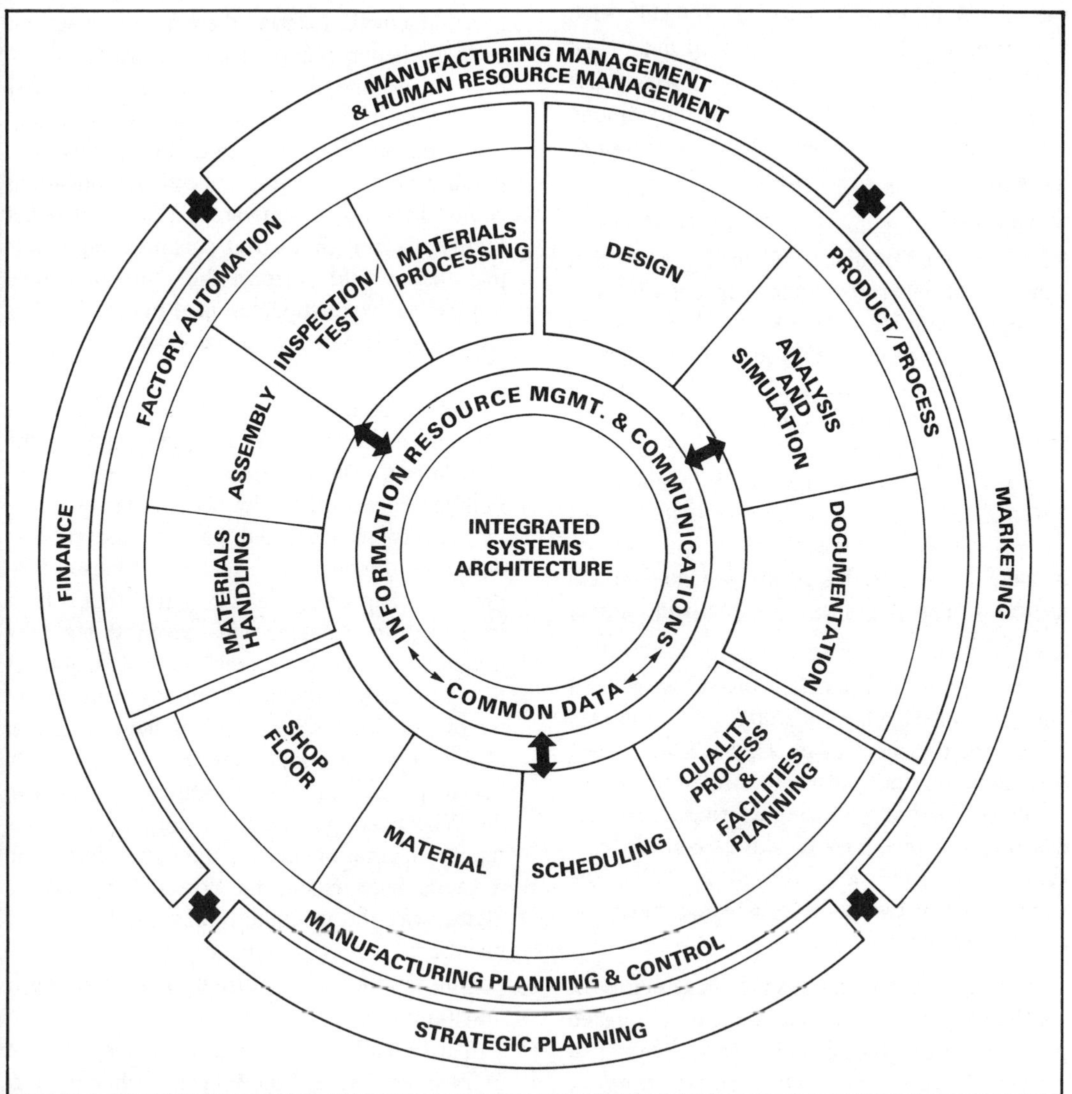

The CIM Enterprise Wheel.

it is to those who must submit to its rules and controls—the integratees.

The CASA/SME Technical Council decided that there needed to be five fundamental dimensions to the CIM Wheel: 1) general business management, 2) product and process definition, 3) manufacturing planning and control, 4) factory automation, and 5) information resource management. Each of these five dimensions is a composite of other, more specific manufacturing processes that, through the years, have shown a natural affinity for each other. As a result, each dimension is seen to be a family of automated CIM processes, and *archipelago*, which has emerged naturally because of an affinity among what used to be independent, stand-alone *islands of automation*.

The totality of the new CASA/SME Wheel represents the Technical Council's view that CIM is an enterprise-wide concept. It is a concept that reflects what General Electric so eloquently called the " Factory-with-a-Future," and in some cases what many manufacturers are searching for under the icon, the "Factory of the Future." In either case, the Wheel represents the infrastructure that the Technical Council believes is necessary if an enterprise is to think of automation as its competitive weapon and to manage its employees as knowledge workers instead of managing them as direct and indirect cost elements.

The Technical Council arrayed the general business management family of processes around the periphery of the CASA/SME Wheel (see figure on the next page). Although seen as an integral part of the manufacturing enterprise, the family of processes known as general business management was viewed as being the primary link between the rest of the enterprise and the outside world.

The general business management family of applications includes a wide range of automated processes from general and cost accounting to marketing, sales, order entry, human relations, decision support, program scheduling, cost status reporting, labor collection, and so on.

Most of the current general business management software was originally written in-house by data processing. However, in many companies, in-house written applications are gradually being replaced by purchased packages. General business management software usually resides on mainframe computers, and in some cases, it operates on smaller general purpose business machines, especially mini and microcomputers.

General business management applications can be bought in suites of software modules, but rarely will all of the modules be implemented by a manufacturer. Further, there are often serious overlaps between applications in this family of manufacturing applications and the applications in the manufacturing planning and control family. For example, accounts payable applications, purchasing, or labor data collection applications could show up in software packages in either family. Even if they do, it is almost impossible that they would perform the same functions. Similar software modules in different families are not constructed with the exact same functionality.

The second, third, and fourth families have been arrayed as thirds of the inner circle of the CIM Wheel. The processes represented in these families, however, do not occur in series, moving clockwise from the product and process definition dimension. In reality, all these activities are happening at the same time. It is important to recognize that even though the individual processes in each family of manufacturing processes individual process in each family of manufacturing processes have a natural affinity for one another, they also are closely tied to process in other families represented in the Wheel. As we shall see, most of the interfamily integration occurs through information resource management, the Wheel Hub. Because of the way the technology has grown in most manufacturing environments, *interfamily* integration is much more difficult than *intrafamily* integration.

The first CIM family of processes represented in the Inner Circle of the Wheel is called product and process definition (see figure on page 134). We have been conditioned to think of this as

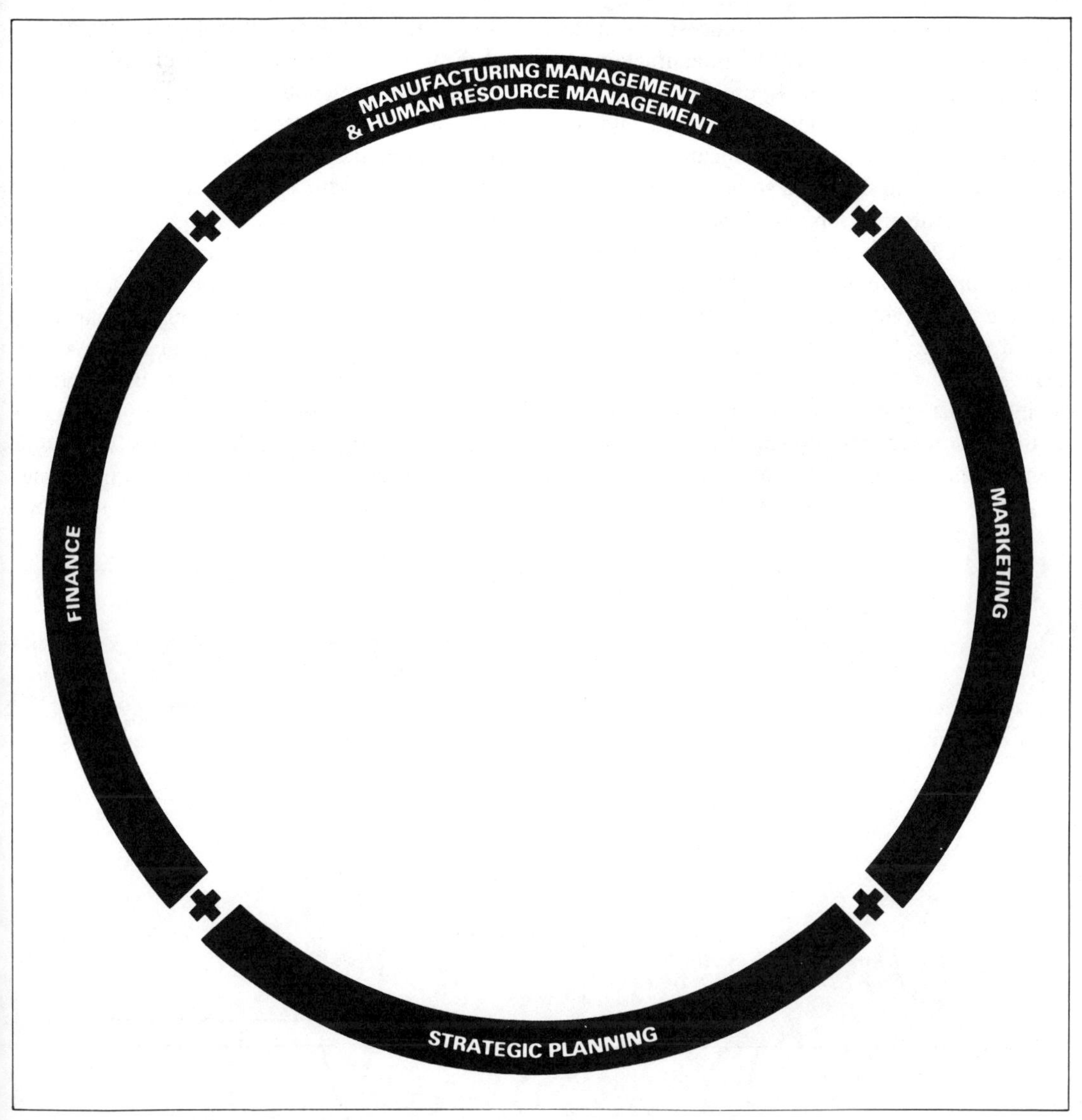

The general business management family of processes portion of the CIM Enterprise Wheel.

CAD/CAM. The islands of automation here include such functions as computer aided engineering, group technology, computer aided design, computer aided manufacturing, configuration management, modeling, simulation and optimization, computer aided process planning, and so on.

Of all of the data tracked and management in manufacturing enterprises, the yield information data are in the most disarray. It is not unusual for a manufacturer to have five (and up to 35) different types of part numbers; two group technology coding system; three (and up to 10) different bills of material; two, three, or four different types of CAD systems, each with its own different geometric representation scheme; and so on. Rarely will the software supporting these islands run on the same types of computers; in fact, each of these islands may well have its own computer, each computer being manufactured by a different hardware vendor.

The second family of CIM processes in the Inner Circle of the Wheel is manufacturing planning and control (see figure below). This family consists of islands such as inventory control, shop loading, capacity planning, master production, purchasing, and so on. The applications that generally support these islands run primarily on IBM computers; however, the software is rarely developed by IBM. Instead, the software is developed by various software housed, such as XCS, MSA, and COMSERVE; and it is sold as suites of ''integrated'' software modules. Rarely does an enterprise implement all of the software modules in a suite. In fact, even though most of these suites consists of from six to ten modules, rarely do we find more than three to five of the modules actually implemented. Since the suites cannot be effectively integrated with one another, software modules are often written in-house to perform manufacturing planning and control functions beyond those being performed by installed purchased modules.

Modules within a suite are integrated by the vendor in that they generally, but not always, are written in the same programming language, and often they will share the same database management system. However, functions performed by modules outside of the suite are often very difficult to integrate with modules within the suite because vendors hardly ever provide customers with their ''source code'' see that custom-

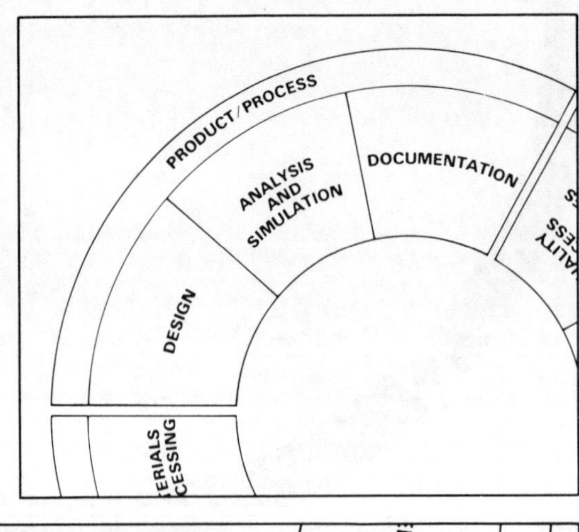

The product and process definition portion of the CIM Enterprise Wheel.

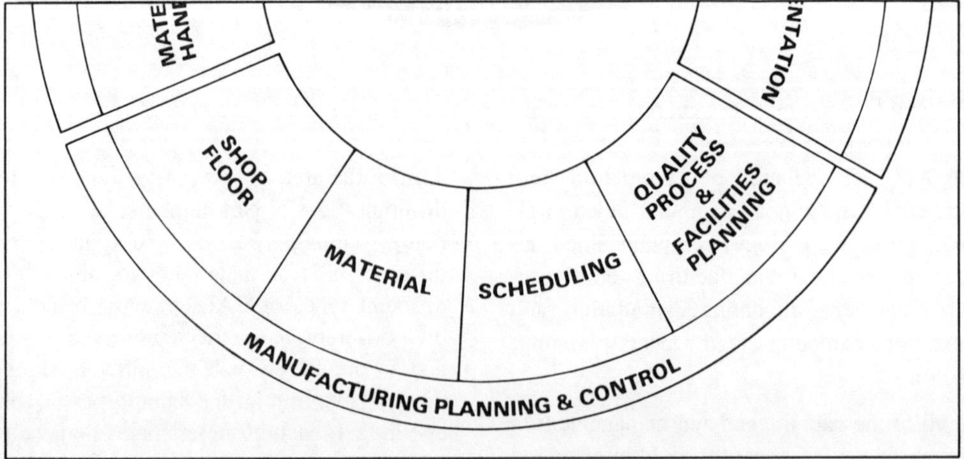

The manufacturing planning and control portion of the CIM Enterprise Wheel.

 Computer Integrated Manufacturing (CIM)

ers can modify database structures or programming routines to fit their own requirements.

The third family of CIM processes in the Inner Circle of the Wheel is called factory automation (see figure below). Robots, NC/

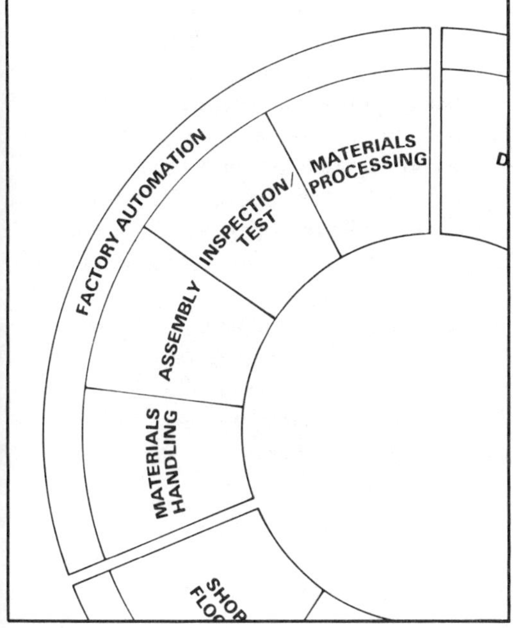

The factory automation portion of the CIM Enterprise Wheel.

DNC/CNC, flexible machining systems, automated material handling systems, automated test equipment, and process controllers are all applications within this family. These applications are more heterogeneous than the product and process definition applications, being supported by a mind boggling array of computer types and styles, each of which is focused on executing very specific algorithms, usually within a scheme of hierarchical control.

The last thing of interest to computer vendors who sell factory automation would be a requirement to integrate one stand-alone machine with another (unless they are all from the same vendor and operating within a flexible manufacturing system or a production cell). Usually each stand-alone machine is dedicated to performing a very specific task.

However, the need for hierarchic control is rapidly becoming recognized as a standard for factory automation, and many businesses are concerned about maintaining what they call an open system architecture, that is, a communications environment wherein any hardware or software product can be integrated. The National Bureau of Standards' Advanced Manufacturing Research Facility (AMRF) is a national laboratory wherein a factory automation control structure is being developed, tested, and evaluated.

The gulf between the factory automation family of processes, the manufacturing planning and control process family, and the processes in the product and process definition family is so wide that we call the territory in between "No Man's Land," to reflect the idea that few, if any, vendors participate in all three families. As a result, not much work has been done by the vendor community to interface, much less integrate these areas. This problem is, perhaps, why manufacturing enterprises have been forced to take on the integration problem themselves, calling it computer integrated manufacturing.

In many ways, the Hub is the most difficult part of the Wheel to comprehend. It has to do with a set of functions called information resource management (IRM) (see figure on the next page).

In the 1980s, information has emerged to be a major management issue in the manufacturing enterprise. In previous decades, information management was thought to be a problem that had to be solved by each organization for itself. In the age of computer integrated manufacturing, information is being viewed as a management opportunity for the enterprise as a whole. A tremendous amount of time and attention is being devoted to understand exactly what manufacturing information is, and to developing new and better techniques and tools for managing it.

Most CIM managers believe that there is a direct relationship between the efficiency of information management and the efficiency and the effectiveness of the overall manufacturing enterprise, especially since the percentage of total manufacturing costs attributable to non-

 Computer Integrated Manufacturing (CIM)

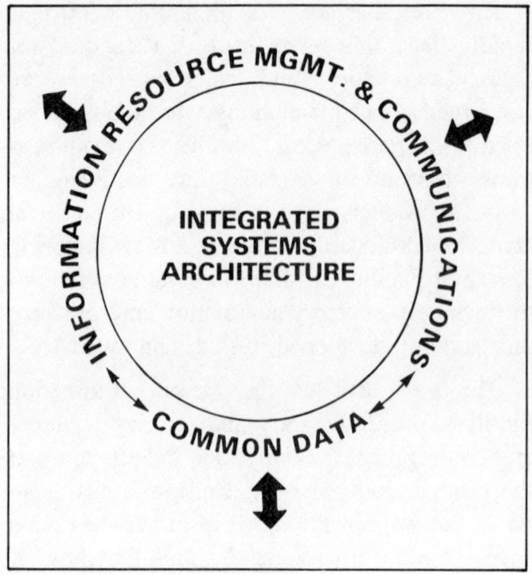

The information resource management portion of the CIM Enterprise Wheel.

touch labor has escalated so rapidly. Many CIM programs focus more attention on information management problems than they do on the problem of developing new and more sophisticated manufacturing machines, material transformation process, manufacturing management processes, and production facilities.

At the heart of CIM information resource management is the concept of data management. Manufacturing productivity is believed to be linked to the notion of shared or common data, especially between engineering and manufacturing. One of the objectives of CIM information resource management is to break down the walls that have traditionally existed between those two organizations, rendering them, in effect, into one organization through a common database.

There are basically two aspect to CIM information resource management. One aspect is intangible and the other is tangible. (Sometimes these are referred to as the logical and physical aspects of IRM.) The intangible (or logical) aspect is the information itself. The tangible (or physical) aspect includes the tools of information, that is, the computers themselves, and also the disc storage devices, printers, plotters, tapes, communication devices, operating systems, terminals, and so on, that are used to store, exchange, and process the information.

The tangible aspect of information resource management can be thought of as including four distinct layers of computer technology: communications technology, network transaction management technology, data management technology, and user interface technology. Each automated application within a family of CIM processes must be supported in some way by technology in each of these four technology layers; however, they are generally implemented differently in different applications. The integration of information within a CIM process family is accomplished by linking its applications together by mapping complementary technology layers. The same strategy can be used to physically integrate among the families. However, as we have said, *interfamily* integration is vastly more difficult than *intrafamily* integration.

Because of the increasing urgency to integrate the tangible information resources both within and among the CIM process families, there has been a tremendous push to establish standards for data communications, data exchange, data management and user interface. These standards activities are growing in strength and popularity, and today include efforts such as MAP, TOP, and IGES.

Establishing standards for the tangible aspects of information resource management solves only part of the problem. The other part of the problem must be solved by setting standards for the intangible aspects of information resource management. Unlike the standards for tangible information resources, these standards for intangible resources must be developed by each and every manufacturer. The objective is to control what data are stored in the environment, how to ensure that the stored data are accurate and consistent, and how the data can be manipulated.

Digital data are stored in two forms: as documents and as raw data that must be manipulated before they make sense to anyone. The form of the data has a direct bearing on what can be done with the data. Data stored in document form are not as flexible as data stored in their raw state. On the other had, raw data are more difficult to control because they require special tools such as data dictionaries, database management systems, and special techniques such as data modeling, just for understanding. Data standards are necessary to manage the quality and integrity of both forms of digital data.

In the manufacturing industry, these problems of standardization and integration of information management resources are difficult within an enterprise. But the problems of integrating intraorganizational systems pale when compared to those of integrating interorganizational systems, the manufacturing structure of the future. The CASA/SME Wheel is even more eloquent when applied to the problems that attend building networks of interdependent customers and suppliers.

One of the primary motivating forces behind computer integrated manufacturing is the fact that manufacturers have traditionally implemented technology, creating stand-alone islands of automation. Since there are over 150 distinct manufacturing functions, it is conceivable that without some strategy for integration, a manufacturer could end up with 150 stand-alone applications. The CASA/SME CIM Wheel provides a conceptual framework for describing how those applications relate to one another, and how they can be integrated into a useful mosaic. There is no surefire technique for determining which of the 150 manufacturing applications go into which CIM process family; indeed, different CIM program managers will undoubtedly come up with their own logic as to what the appropriate groupings of automated processes are and how these groupings inter-relate, both within a manufacturing enterprise and within a network of manufacturing enterprises.

The objective of the CASA/SME CIM Wheel is to reflect trends and currents in manufacturing automation and to stress the importance of "top down" planning. We believe that it accomplishes its objective. But, it is by no means carved in stone. The CASA/SME Technical Council is committed to support CASA/SME's goal of helping manufacturers understand the technical world around them and manage the introduction of new technology. That world is in a constant state of change. We are committed to keeping the CASA/SME Wheel aligned with those changes.

Applications. CIM is applicable to any manufacturing organization. However, the technologies used to automate the various manufacturing activities are dependent upon the industry and the manufacturing process employed by the company. Similarly, the technologies used for integration and communications are determined by the automation technologies and geographic considerations.

The concept of CIM can be applied to portions of an organization—e.g., the engineering function or the production function—on a stand alone basis. When approaching CIM in this manner, planning is still critical, both to assure integration of the various activities within a function and to assure that the hooks for interfunctional integration are put in place.

The concept of CIM can also be applied to nonmanufacturing organizations by redefining the term "manufacturing" to apply to a service, and integrating the informational needs associated with the design/specification and delivery of that service.

Some of the benefits of CIM are outlined in the figure on the next page.

Also see: Computers, CAD/CAM, Solid Modeling.

Computerized Simulation. Computerized simulation is when a computer is used to perform the simulation process. A software program which contains the necessary algorithms for simulating the desired process is utilized. Several simulation languages exist for performing simulation. GPSS (General Purpose Simulation System) from IBM was one of the first readily available and highly used languages. Many languages today have been designed after GPSS, with modern enhancements and greater ease of use. SIMON, SLAM, SIMULA, and SIMSCRIPT are some of the modern programming

Reduction in engineering design cost	15-30%
Reduction in overall lead time	30-60%
Increased productivity of production operations (complete assemblies)	4-70%
Reduction of work in process	30-60%
Increased product quality as measured by yield of acceptable product	2-5 times
Increased capability of engineers as measured by extent and depth of analysis in same or less time than previously	3-35 times
Increased productivity (operating time) of capital equipment	2-3 times

The benefits of CIM.

languages used today. Other programs such as ''Witness'' and''See Why'' by Istel are application languages which are actually written in FORTRAN.

Computer simulation allows much greater capabilities than what can be accomplished mentally or than what can be done using pencil and paper. Computers simulate days, weeks, or even years equivalency within a few minutes or hours. Computers can also allow the calculation of a large number of possibilities to determine the optimum scenario.

Also see: Simulation.

Computer Languages. Computer languages are means by which a user can create a description of a desired sequence or events to be performed, or simply a description of the desired results (a program) which is translated into machine language and executed by a computer. While a computer can only comprehend ones (1s) and zeros (0s), computer languages allow a user to create a program without being restricted to programming in ones and zeros. High level languages allow a person to be farther removed from the tedium of binary data representation while low-level languages allow a person to directly control many of the hardware functions of the computer and allocate the use of memory. Some languages are developed for specific purposes and may run under the control of another language such as an operating system or a ''shell'' program. Some of the current trends in computer language development are that they continue to become more structured, where lines of code are grouped into logical units and subfunctional modules rather than simply being strung together in a sequential fashion. Computer languages continue to become more sophisticated. Although von Neuman computer architecture can only do the basic functions of addition, subtraction, comparison, and movement of data between locations, high-level languages can use complex combinations of these basic functions to achieve the solutions to extremely abstract problems. The variety of computer languages continues to increase and languages are constantly becoming less machine-dependent by the establishment of international standards for versions of computer languages.

History. Computer languages began as first-generation languages during the 1940s with low-level languages such as machine language and assembly language. Machine language is literally the data and instruction representation which is used internally by the computer, but really isn't considered to be much of a language since it does not provide for much freedom of expression by the programmer. Assembly language was designed to allow a more logical and easier method to create programs.

During the 1950s and the early 1960s, early versions of FORTRAN and COBOL were created as second-generation languages. These languages began to incorporate syntactic conventions which expanded the command sets. Also, these were the first languages which could in principle deal with any type of problem, hence they were the first general-purpose programming languages, although FORTRAN was created with science and engineering applications in mind and COBOL was developed with business applications as an intended usage.

In the later 1960s to the present, third-generation languages such as standardized and modernized versions of FORTRAN, Pascal, PL/1, ALGOL, and Ada were brought forth. These third-generation languages were all high-level languages, and most did not allow any lower-level functions to be performed by the programs.

During the 1980s, fourth-generation languages emerged such as Query-by-Example (QBE), Intellect, Themis and Clout. Typically, these languages were nonprocedural. They describe the desired end result instead of meticulously instructing the computer in how to achieve the result. Many of the more general-purpose languages use this nonprocedural structure.

The 1990s will provide fifth-generation languages which will incorporate artificial intelligence type capabilities. True and total natural-language programming is a goal of many development activities in fifth-generation languages. These languages will be capable of utilizing the enhanced capabilities of the latest developments in computer hardware.

Discussion. There are many characteristics which define and describe computer languages. One of the simplest is whether the language is a low-level or high-level language, but some such as C, Forth, and Modula-2 are essentially high-level languages which allow low-level functions to be defined from within the program. True low-level languages such as assembly languages are machine dependent and may have significantly different command sets from machine to machine, even though the basic hardware functions which the assembly instructions are mnemonics for are basically the same in most computers. Low-level languages require a programmer to be familiar with the inner workings of the machine while high-level languages do not require any familiarity with the particular target hardware by the programmer.

High-level languages theoretically can be run on any machine without altering the program source code. The source code is translated by the computer before execution, either all at one time before execution, as a compiler does and then produces executable object code (compiled machine language), or line by line when executing the program as an interpreted language does. An interpreted language must convert each line of code into machine language immediately prior to execution each time the line is solved, even in a loop. Compiled code, is already translated into machine language and therefore the program executes much faster. Compiled code also protects the source code since the compiled code cannot be disassembled to determine the original program statements. This feature is desirable for developers who provide code to customers. Interpreted code, however, is much easier to debug since run-time errors can be directly related to the line of code which created the problem. Compiled code can sometimes produce run-time errors which result from perfectly allowable source code statements which either are not allowable in the particular combination, or in some cases, are not compatible with the particular compiler.

Computer languages can currently be divided into six distinct methods of problem solving. The first approach is imperative or algorithmic. This approach uses instructions to tell the computer what to do to achieve the desired result. The second approach is functional or applicative languages where mathematical or other types of functions are applied in order to solve a problem. The third method of problem solving is object-oriented where an object is defined as an area in computer memory that serves as a basic structural unit of analysis. When writing programs, users create new objects from existing objects. The fourth is using a small set of units or words to construct new words such as the language FORTH. The fifth method is logic programming which is a formal system of relationships between assumptions and conclusions. This method includes both propositional logic and first order predicate logic. The sixth method is using query languages where the user specifies the problem to be solved and any known relevant information. The language translator determines how to solve the problem.

Usually, when new languages are developed, an effort is made not to rebuild the same functional parts of a program. New languages are developed the ability to accommodate basic functions. Then, in other cases, the thinking is that all available languages are too specific, so a person desires a more flexible, general purpose. Some languages are considered to be application-specific languages while others are more general-purpose languages.

A major characteristic of computer languages is the structure of the language, not only whether the language is structured or unstructured, but what the structural unit actually is. Some languages, by the structure and organization of the language, are structured languages. Structured languages are those whose syntax breaks programs into components such as blocks, modules, packages, or subroutines. Unstructured languages follow a single linear sequence of commands, rather than being broken into components. Regardless of the existence or lack of structure in the language, the user can develop and write a program using structured format. The structured unit of a language is the basic unit of analysis in a given computer language, such as a line of code or a group of lines of code. Some languages such as MUMPS can contain an entire program in a single line of code.

Another classification used to describe computer languages is the purpose which they are used such as business for languages such as COBOL or QBE. Applications languages might be dBase, Lotus 1-2-3, or Multiplan. Science/Engineering languages such as FORTRAN, ALGOL, and Pascal are another category, with a specialized application such as robotics utilizing FORTH, while embedded systems might use Ada, and real-time programming might utilize Modula-2. Artificial intelligence includes LISP, Logo, PROLOG, and Smalltalk, while simulation languages might be DYNAMO or SIMULA. Educational languages are ones such as BASIC, COMAL, and micro-PROLOG.

Applications. Computer systems use many different types of programs. Some programs are used as operating systems, text editors, graphics editors, debugging tools, tracers, documentation and annotators, and others. These are all in addition to, and sometimes in support of, the translators and compilers which are being used by the programmer to create an applications program.

Applications of programming languages vary greatly. Some uses, especially for real-time programming languages, are in imbedded systems, where the computer itself is part of the machinery that has other primary functions. Ada was developed and adopted by the military as possibly a single programming language that could be used in multiple types of systems without having to select different languages.

Simulation is another area where computers and application programs provide capabilities which would be essentially impossible to accomplish without the use of computers. Huge volumes of data can be absorbed and processed by the computer, with projections being done over an extended length of time, without having to wait until the real time passes to determine the results. Many different scenarios can be programmed and some languages are very forgiving to the programmer since they can anticipate and prompt for missing data or suggest possible solutions to scenarios.

Early computing was done primarily with mainframe computers, but during the late 1970s the advent of lower cost minicomputers and the introduction of personal computers allowed individual users to use computers for new applications. The creation of Visicalc (and its successor, Supercalc), Multiplan and Lotus 1-2-3, which are spreadsheet programs, allowed the computer to be used as an electronic spreadsheet with virtually unlimited formula relationships being defined for each ''cell'' location. Boeing has developed a three-dimensional spreadsheet called Boeing Calc which allows thousands of two-dimensional spreadsheets to be linked with no restrictions on a cell at one level being linked by formula to a cell or cells in spreadsheets at other levels.

Another application of programming languages is in database management. Database management systems, as they are called, provide the framework for inputting, processing, and outputting data. The structured lines of data, or records as they are termed, can be sorted, combined, altered, and used to generate output such as mailing lists and labels. One of the best known and most widely used microcomputer database management software is dBase. An early standard, dBase II was developed for eight-bit CP/M and DOS based machines and later dBase III Plus, which is currently one of the most popular database development software languages was created for 16-bit DOS machines. dBase does not currently have SQL capabilities, which are highly desirable by some. These would allow a more substantial interface with IBM's mainframe database management software.

Word processing is an application which used to be done primarily with typewriters. Word processing is currently using computers which are either dedicated devices for word processing, or more recently, personal computers which can be used for additional task such as spreadsheet analysis or database management. Applications which previously were done using dumb terminals connected to a mainframe computer can now be accomplished using general purpose personal computers with appropriate software.

Education is an area where computer languages continue to play a major role. Computer aided instruction or CAI was a field where an instructor had to be proficient in the use of a programming language to develop the teaching tools. A recent type of programming known as authoring languages can construct CAI programs without a need to understand how the programs operate. The users can use more natural type language commands to instruct the program how they want their lessons and exams to be taught. These languages, such as PILOT, then write the computer aided instruction which in turn is executed on the computer by the student.

Software, in the form of computer language programs, continues to be the "glue" or critical link which causes hardware to perform useful functions. Intelligent, microprocessor-based devices must have software programs to operate. As the complexity of devices increases, software becomes an even greater part of computer systems.

Glossary. *Ada:* A language intended to be used primarily for military imbedded systems. Single language can be used in many applications.
Algol: An algorithmic language used for numerical analysis in scientific computing.
APL: A Programming Language, developed by IBM for data processing.
Assembly: Abbreviated, low-level mnemonics for machine language.
BASIC: Beginner's All-purpose Symbolic Instruction Code, a general-purpose, easy to learn language available for most microcomputers.
C: Revision of a language called B, which was a revision of a language called BCPL. Used for system's programming such as writing compilers, operating systems, and editors.
COBOL: COmmon Business-Oriented Language, developed by Grace Hopper for business usage.
FORTH: Shortened form of the word FOURTH, for fourth-generation language, which is a general-purpose language useful for astronomy, robotics, and graphics.
FORTRAN: FORmula TRANslation, a high-level language developed for science and engineering applications.
LISP: LISt Processing, an artificial intelligence application language.
LOGO: Comes from the Greek word *logos*, meaning "word" or "thought." Used for education, especially geometry, with potential as a general-purpose language.
MODULA-2: MODUlar LAnguage, a high-level language with low-level functions used for systems programming, including real-time programming applications.
MUMPS: Massachusetts General Hospital Utility Multi-Programming System, developed in the mid-1960s to be used in a time-sharing environment for the management of clinical data.

PASCAL: A general-purpose, numerically oriented language primarily developed for teaching programming.
PILOT: Programmed Inquiry, Learning or Teaching, developed for CAI (computer aided instruction) applications.
PL/1: Programming Language One, developed by IBM in the 1960s to use for science, business and "special purpose" programming.
PROLOG: PROgramming in LOGic, developed for artificial intelligence and database management systems programming.
QBE: Query-by-Example, a language developed to use for organization and query applications in database information handling.
SMALLTALK: A language developed for teaching and simulation uses. The language is intended as a flexible system of communication.
SQL: Systems Query Language.

Also see: Application Program, Assembly Language, BASIC, COBOL, Compiler, Computers, FORTRAN, Programming, Software.

Computer Numerical Control (CNC). Computer numerical control controls a numerically controlled (NC) machine. The NC program is entered into the computer, and the computer executes the program. Before CNC, NC programs were written on punched paper tape (later mylar tape for durability) and had to be read by a paper tape reader. The complexity and time involved with that method made it vastly inferior to today's methods of either carrying a floppy disk or data cassette to the CNC machine, or as is becoming more the case, simply downloading directly to the CNC from the programming computer. Often the source of the CNC program is the same computer system which provides the CAD and CAE for the parts which the CNC machine is to manufacture. CNC allows the ability to define more complex shapes and profiles than was possible with NC.

In computer numerical control, a numerical control system with a dedicated stored-program computer performs some (or all) of the basic NC functions. These functions control programs stored in the computer. Computer numerical control and soft-wired NC are synonymous terms. Computer numerical control is the dominant type of machine control being manufactured today.

The key element of the soft-wired controller is a microprocessor or minicomputer. Because a computer is involved, there is a tendency to confuse the soft-wired controller and its application to CNC with direct numerical control (DNC). There are, however, several differences between the two.

Today's hard-wired NC uses integrated-circuit (IC) digital-logic circuit packages—usually medium-scale integration (MSI)—which are mounted and wired in a fixed and permanent arrangement on plug-in printed-circuit boards (PCBs). The PCB connectors that receive these boards are also wired together permanently (this is usually called "back-plane" wiring) in a prescribed manner that is preengineered and tailored to accomplish the desired operation. In addition, changes in the operation of a hard-wired numerical control require wiring changes as well as the addition or omission of related hardware devices such as operator control buttons and PCBs. Customer-selectable options are feasable in a hard-wired NC provided their incorporation was planned for the original design.

A soft-wired CNC system incorporates a programmable control unit—usually a general-purpose computer (mini or micro) with a read-write memory. The computer and memory replace much of the general-purpose fixed-logic circuitry of the hard-wired NC with programmable logic that is stored in the computer's memory. This stored logic, together with stored computer instructions, is called an *application software program*.

This software program can be thought of as the mechanism for converting a general-purpose computer to a machine control system for a specific machine tool. The term "soft-wired" is applicable because the functions created to control the specific machine tool result from the application software program rather than from any physical wiring of a group of logic elements.

 Computer Numerical Control (CNC)

Soft-wired controls generally have offered additional features that conventional hard-wired numerical controls did not originally provide. Among these is part-program storage (the tape reader is used for input only; the machine is operated from the computer's memory). It is technically feasible to add memory to hard-wired controls to provide these features, but any change in the operation of these features requires some rewiring."

The ease of changing the operation of a soft-wired control after original manufacturing is its greatest difference from hard-wired NC.

An important inherent advantage of the soft-wired NC concept is that once a basic control capability has been provided, additional options can be added for a smaller incremental cost than is possible with a hard-wired control. This advantage lends itself to NC machine types with more sophisticated requirements, such as machining centers, boring machines, and special multiaxis machines. Thus NC users who have special requirements may find the soft-wired NC particularly advantageous.

The power of a software approach as it applies to NC is that a library of modular software programs can be developed. This library can grow in time so that eventually most NC functions can be applied by using existing program segments and less special software will have to be developed for each new or different requirement. The capabilities of this feature have been amply demonstrated in the application of computer systemsin other fields.

The basic elements of a computer numerical control system are shown in the figure below. The control is the heart of the system. It processes information received from the operator and machine interface. This information is interpreted and manipulated with hardware logic and computer programs (software). Memory provides the means to store programs and manipulate input data. Based on the information received, the control outputs database to the operator interface and machine.

The operator interface consists of devices which send, receive, and interpret information. Since the operations performed by NC systems are defined by the software, interface devices are needed to input the various programs from memory. Paper tape input is the most common. The operator station(s) is the other major operator interface element. It contains all the switches, pushbuttons, displays, etc., required to operate and monitor machine activities.

Machine devices are regulated by the control. Based on information supplied by operator interface devices and feedback from various machine devices, the control turns on and off machine outputs and controls machine motion.

The control performs "real-time" decisions on a process that is in operation at the same time. There are several types of control systems; however, each can be broken down into the same functional units. Each unit performs specific functions, and all units function together to execute the programmed instructions. The figure on the next page shows the five major functional units of a control. The dashed lines with arrows represent the flow of timing and control signals. The solid lines with arrows represent the flow of data.

All instructions and data are fed into the control through the input unit. Software, such as the system operating program, part programs, and diagnostics are input by means of paper tape, magnetic devices, etc., and are stored in memory until needed. The status of machine and operator station devices are input in the form of AC, DC, and analog signals. Analog signals are converted to digital signals (A/D converter) in order to be understood by the control. Input signals are sent to memory, where they are used by the control and arithmetic units to arrive at output decisions.

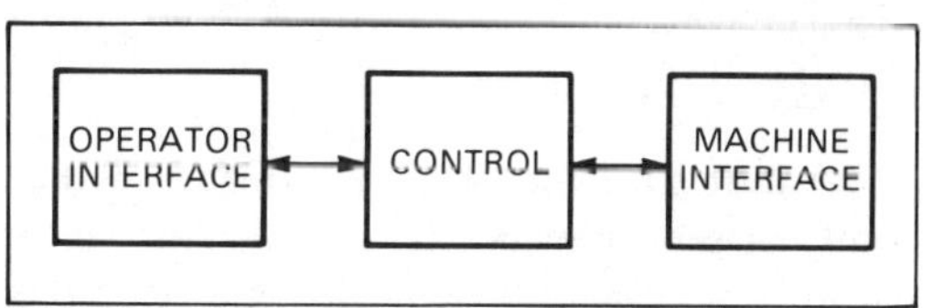

Numerical control interface devices.

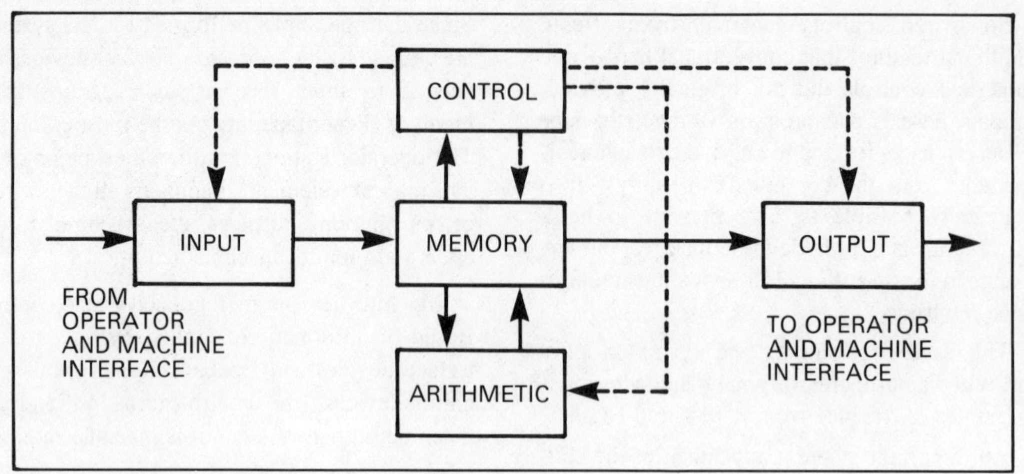

Five major functional units of a computer numerical control system.

For systems with many AC, or DC inputs, a scheme called multiplexing may be used. With this scheme, the state of many devices can be monitored on a single channel. This reduces wiring without restricting real-time operation.

The memory unit stores instructions and data received from the input. It also stores the results of arithmetic operations and supplies information to the output. The size of the programs and space required to manipulate data determines the amount of memory required. Basically, there are two types of memory—Random-access Memory (RAM) and Read-only Memory (ROM).

Random-access memory provides immediate access to any storage location point in memory. Information may be ''read'' or ''written'' in the same very fast procedure. Part programs are usually stored in RAM memory to enable editing. While there are many types of RAM, only certain types (i.e., coarse and bubble) are able to retain data during a power loss. Complementary Metal Oxide Semiconductor (CMOS) memory is retentive if it has battery backup.

Read-only memory stores information permanently or semipermanently. Information can be ''read,'' but cannot be altered. Only fixed programs such as the system operating program and diagnostics should be stored on ROM. ''Programmable'' ROMs are referred to as PROMs, and electrically erased PROMS are called EPROMs.

History. The history of computer numerical control is directly tied to the history of numerical control. With the advent of the microprocessors, computer numerical control was developed. NC machines were equipped with microprocessor-based controllers, containing local memory and input devices. The memory allowed the local storage of programs. This technology enabled the use of variables and real-time branching within programs based upon the value of the variables. CNC both eliminated the problems of punched tape and the dependency on the host computer.

CNC machines were quickly followed by distributed numerical control systems. With distributed NC technology, CNC machines are tied to a host. The host provides a central storage site for programs, which are down-loaded to machines for execution. The resident memory and processing power at the machine allows manufacturing to continue if the host fails. The communications link between the host computer also allows upstream communications. Information is available to management as to whether the machine is in use, if manual overrides have been instituted, and what expected lot completion times are.

Discussion. The key element of the soft-wired controller is a microprocessor or minicomputer. Because a computer is involved, there is a tendency to confuse the soft-wired controller and its application to CNC with direct numerical control (DNC). There are, however, several differences between the two.

Computers supporting DNC disseminate manufacturing data, and collect product information from, several machine controllers. Soft-wired controllers, on the other hand, generally support only one machine or a small number of machines. Also DNC computers may be remote from the machine tools, whereas soft-wired controllers are normally in close proximity. In addition, the software supporting DNC is usually written to support overall manufacturing activity. Software for soft-wired controllers, however, is written specifically for a particular machine/device and its required fixed sequences.

The arithmetic unit performs calculations and makes decisions. The results are sent to the memory unit to be stored.

The control unit takes instructions from the memory unit and interprets them one at a time. It then sends appropriate instructions to other units to cause instruction execution.

The output unit takes data from memory when commanded. Outputs are in the form of AC, DC, and digital signals. Digital signals used as axis drive commands are first converted to analog (D/A converter). Output signals are used to turn on and off devices, display information, and position axes.

The operator interface consists of all devices, exclusive of the machine, which send and receive control information. The figure on this page depicts some of the more common devices.

Punched tape is the commonly used input system for NC systems. Several different tape widths are available, but eight-channel, one inch (25.4 mm) wide tape is almost universally used.

Magnetic devices record and read magnetic spots on a moving surface of material. Each of these devices has a thin coating of magnetic material applied to a smooth, nonmagnetic material on plastic tape. Drums are thinly coated metal cylinders. The disk (floppy disk) resembles a phonograph record with magnetic material on both sides.

In the past, punch cards were used as input to NC systems primarily because of the availability of the equipment on which the programs could be prepared. Because keypunch operators and keypunch departments exist in many businesses for accounting purposes, the same equipment and trained personnel were utilized to assist in NC program preparation. Today, however, punch cards are used as input to relatively few NC machines.

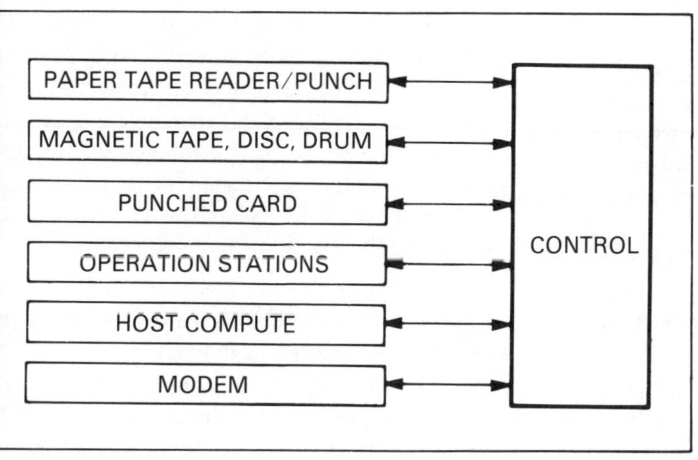

Operator interface devices used in a computer numerical control system.

The CNC operator station(s) consists of all the switches, pushbuttons, and displays required to operate the machine (unless the machine has been completely automated with numerical control, in which case the operator's attention is not required). Devices such as switches and pushbuttons command the machine to perform an activity, i.e., machine start and master stop. The commanded actions, machine member position, and state of machine devices are usually displayed for operator reference. Cathode ray tubes (CRT), light emitting diodes (LED), and plasma displays are some of the more common methods of displaying information.

The main purpose of the operator station is to initiate automatic operation, to input data, and to monitor activities using display devices. More and more, with the increasing intelligence built into the control systems, so-called ''manual data input'' (MDI) is employed. Manual data input is employed on NC systems, as an auxiliary means of providing an input. Manual data input requires that the operator set up a series of switches or controls to perform a given function and, when the operation is completed, set up the next step in the operation sequence. This may be a time-consuming approach for complex workpieces and is usually employed for relatively simple workpieces or for programs that are to be used only once or a very few times. The same kind of error is possible in manual data input NC systems as is possible in a non-numerically controlled system in which the operator sets up the machine for each operation.

The direct link of a general-purpose host computer to a machine tool is advantageous for certain applications. (A special purpose-design would be incorporated into the control system of the machine tool itself.) A direct numerical control (DNC) system can have certain characteristics which the computer can handle much more efficiently than an alternative type of system. For instance, computation may be required while running a part to correct for various machine conditions such as tool wear or errors inherent in the machine tool itself. An example of this is a precision leadscrew that has error throughout its length. The error in the screw is measured and stored in the computer memory in the form of a table. When a part is made on this machine, the computer system modifies the part program while it is running by making corrections based on the error table to obtain improved accuracy on the part.

Another application in which an on-line computer is advantageous is that in which the number of programs or different kinds of parts made from any one program are small, or changes in programs are very frequent and the same parts are to be made again a short time later. In this case, the computer has access to the programs, which are stored on either tape or magnetic disks, and it is merely necessary to enter the desired number of the program required for the parts to be made. The computer then searches for the data and has it ready at the proper time. In such applications, the computer is used as a source for rapid search and recovery of a specific program from a large number of programs. Several machines may be serviced by one computer operating in this mode.

A modulator demodulator (modem) converts data from the control into a form compatible with telephone transmission lines. The primary use of the modem for NC is diagnostics. For example, some controller builders can send and receive data from a customer's controller over telephone transmission lines to determine a control problem.

The machine interface consists of all devices used to monitor and control the machine tool. Extreme travel limits, miscellaneous position locations, and hydraulic and air pressures can be monitored. Additionally, solenoids for hydraulic and air control as well as motor control are provided. Outputs are usually a single DC and AC level or a DC output with remote AC switching devices.

Limit and proximity switches determine the location of a machine member. Proximity switches are located at defined intervals along the machine's travel. The control detects which switch is tripped to determine axis position. This method is no longer in common usage due to limited accuracy.

Computer Numerical Control (CNC)

There are usually two limit switches on each linear axis; one for plus motion, and a second one for minus motion. When the control detects that a limit switch has been tripped, machine operation is halted until the axis is manually moved off the limit.

Pressure and temperature switches determine system conditions. Oil and air pressure for the machine and temperatures of the control cabinet and lube may be monitored as needed.

Many machine functions are performed by applying air or oil pressure to devices. Power drawbars, turret indexers, toolchanger magazines, and coolant flow are but a few of the machine mounted devices and functions controlled by the numerical control unit.

Many machine tools use hydraulic or air operated cylinders to control spindle speed and axis feed transmissions. The control of these devices is programmed in the controller and activated by control codes.

A servomechanism (often termed a ''servo'') is a group of elements which convert the NC input into precision mechanical displacements. These elements include motors (hydraulic or electric), gear trains, and transducers (velocity or position).

The drive to spindles and slides in NC tools is usually provided by either hydraulic or electric motors.

Servomechanisms may be either open or closed loop as shown in the figure below.

In the case of the open-loop servo, there is no feedback signal to assure that the machine axis actually moved the distance programmed. For instance, if the servo is designed to move 0.0001 inch (0.003 mm) for each input pulse, and 100 pulses are programmed, the servo will move a table 0.010 inch (0.25 mm). The only assurance that the table actually moved 0.010 inch in this type of system is the reliability of the system.

The closed-loop servo, on the other hand, compares information feedback from the machine slide with programmed information to assure than the motion has actually been performed. The signal to the drive motor is modified by the feedback signal.

The basic or main elements of a numerical control system are shown in the figure on the next page. The principles here essentially are the same for positioning and contouring, although the principles of contouring are somewhat more complex.

From the machine control unit, the signals proceed via the *command-signal circuit* to the servomechanism and drive unit. Each machine slide or movement that is to be controlled by the system has its own servomechanism and drive.

A servomechanism amplifies the incoming signal and provides power to move the desired machine slide or carry out a mechanical move-

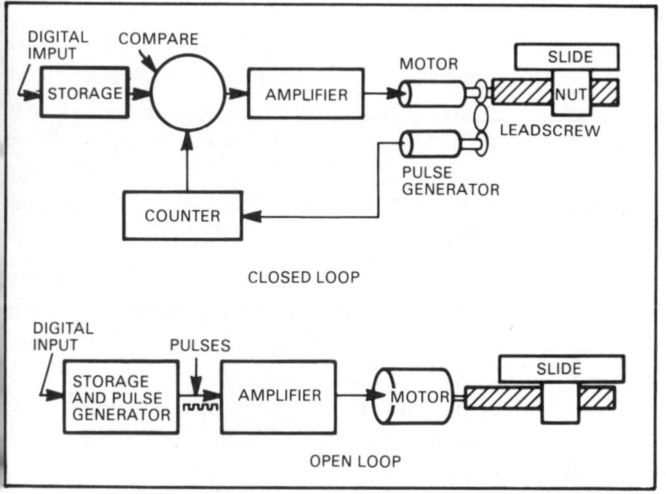

Open and closed-loop servomechanisms.

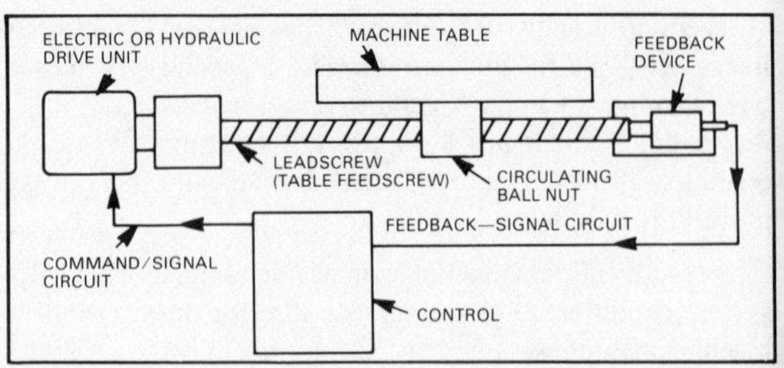

Mechanical elements of a computer numerical control system.

ment as required. Commonly, it can be electric or hydraulic. The servomechanism may be an electric motor which drives a machine table through a leadscrew; or the system may involve hydraulic motors, hydraulic rams, or other devices for moving the controlled machine elements or slides as required. Motors may drive the slides through low-friction leadscrews employing circulating ball nuts or through rack-and-pinion arrangements, or still other devices can be employed.

The controlled machine slide is any controlled part of the machine, or the controlled machine itself. It can be an addition to a machine tool, such as a retrofit positioning table or a conversion, but the controlled unit must be designed for numerical control in any case.

In the *electric or hydraulic drive unit*, the rotary hydraulic motor uses the pressure of fluid flowing through gears or against pistons to effect a shaft rotation. These motors are usually of the positive-displacement variety. The motor itself is very small the horsepower that it is capable of developing. Oil flow to the motor is controlled by a servovalve. A servovalve is typically an electrohydraulic unit using an electrical solenoid to actuate a small flapper which controls hydraulic pressure on a spool valve. The spool valve directs oil to the hydraulic motor, which produces the desired mechanical motion.

Both AC, and DC electric motors are commonly used for powering NC tools. Recently, printed-circuit motors have also become popular. These motors are high-torque devices which can be directly coupled to a leadscrew to drive a machine-tool slide. A stepping motor is a particular type of electric motor which is actuated by pulses and moves a fixed angular unit for each electrical pulse. The motor is usually clamped magnetically at fixed angular positions.

Precision slides are usually powered by a motor which drives the slide through a gearbox and *leadscrew*. The leadscrew is almost universally used to convert rotary motion from electric or hydraulic motors into linear movement of a machine slide. The recirculating-ball leadscrew is a precision screw with very low friction. The leadscrew is precisely made to provide a given linear displacement to the slide for each revolution. This distance can be very accurate.

Two kinds of transducers are commonly used in machine tools for *feedback devices*. Velocity transducers are used to measure spindle speed and slide velocity, and position transducers are used to measure slide displacement.

The most common velocity transducer is an electric tachometer. Electric tachometers provide a voltage which is proportional to the speed of a shaft. The voltage produced may be either AC or DC.

There are many types of position transducers in use. These include synchros and resolvers, digitizers, and linear induction types. Since measurement of position is one of the key elements of NC, this area has received a great deal of attention.

Synchros and resolvers are electric transformers which provide a voltage output that varies with the angular position of a shaft. This voltage measures of displacement. Coupled to a leadscrew through a gear train, the resolver or synchro provides position information on a slide.

A linear resolver is a precision-made scale which utilizes the resolver principle of induction. Signals are induced on the linear scale by an adjacent exciter scale.

A rotary digitizer is a device which provides a pulse for a given angular displacement at its shaft. For example a digitizer might generate a pulse for each degree of angular rotation of its shaft. These pulses can be counted and the relationship of the linear displacement to the digitizer angle can be used to determine the slide displacement.

The machine motion, as provided through the servomechanism, is recorded or monitored by a feedback (measuring) unit. *Feedback signal circuits* may be electrical, mechanical, or optical; corrections are automatically made by the machine control unit. Systems with feedback are generally classed as closed-loop types, whereas open-loop systems do not incorporate feedback.

When a machine slide and servo are coupled through a leadscrew; the leadscrew, the ways of the machine, and the accuracy of the servo (including the accuracy of the measuring system) all contribute to the machine error. A closed-loop servo can achieve a high degree of precision when a feedback element such as a linear system is attached to the machine slide. Such a system eliminates gearing, leadscrew, and backlash errors. In many cases, however, the drive and feedback units are attached to the end of a leadscrew, in which case the open-loop and closed-loop systems may have equal accuracy at the machine slide. The accuracy of the system must be designed for the jobs that the machine performs. Very high accuracy can be achieved—limited primarily by the ability to measure. The cost of the machine, however, is related to the machine accuracy and increases as the accuracy is improved.

Applications. CNC provides control to many manufacturing operations. Some example applications follow.

David J. Gayman, in an article in the January 1987 *Manufacturing Engineering* titled *Calibration, Compensation, and CNC* listed several features of CNC that make it attractive for retrofit to older machines. Among the strong points for new controls are the following:

- Increased accuracy and speed of the controls, seen at the workpiece as better dimensioning and tolerancing on the part itself.
- Circular interpolation, cutter compensation, and other built-in machining aides, again seen at the workpiece, this time as machining finish.
- "Canned" programming cycles, reducing time for some kinds of programming.
- Programmable machine/control interfaces, permitting closer mating of machine tool and control.
- Ports for a variety of input, including punched tape, digital tape, diskettes, and DNC.
- A gamut of program editing and visualizing aids, in some cases making prove-out an electronic process, rather than a cut-and-recut process.

Writes Gayman: "Plus, CNC has become widely accepted. Computers on all fronts, machine tools included, have progressed—and we humans have little by little accustomed ourselves to computers—until now computers are nothing special. The old, big, hot, unpredictable, and bizarre (especially bizarre when being programmed)...all have given way to compact, cool, and understandable."

In another application, switching to a CNC grinder spelled success for a Midwestern tool supplier. Hammill Manufacturing, Incorporated (Toledo, Ohio) has reduced the grinding time of a complex cold extrusion tooling part by about 40% while greatly improving concentricity and total indicator reading (TIR).

Company management attributes the production and quality improvement to a switch to a CNC cylindrical profile grinder manufactured by Weldon Machine Tool, Incorporated (York, PA) that grinds the part's four distinct features, including a hex shape, in one continuous operation.

The programmed grinding operation begins by plunge and traverse grinding the 15/16 inch diameter to a total tolerance of 0.0005 inch while creating the 1/32 inch radiused shoulder and holding the head length to ±0.001 inch. Using the same portion of the wheel, the one-inch diameter head is plunged to size. Next, a 15 degree angle dressed on the left corner of the grinding wheel plunges toward the workhead to create the 15 degree chamfer and to hold overall part length to ±0.001 inch.

Finally, the programmable workhead feature is brought into play to simultaneously grind the hex flats with their 3.32 inch (84.2 mm) blending radii at the points and the large 5/8 inch blending radius. The workhead generates the hex flats by indexing the part in precise, programmed coordination with wheel slide axis movement.

The part, which is bored 0.700 inch (17.78 mm) in an earlier procedure, is held during grinding by a hydraulic expanding stub arbor. The part programming, which was developed with computer assistance, is tape fed into the machine's CNC unit.

A future use of CNC was proposed by Edward M. Stiles of Ed Stiles Partners in his SME Technical Paper *Engineering the 1990s Quality Function*. He wrote: "Instead of a process distribution, and historical data on which to base control charts, process history will be made up of a record of actual measurements made on every unit, plus the listing of corrections made, part by part. Human error will be eliminated by the utilization of self-correction computerized numerical control (CNC) systems. The same CNC systems will track and report incipient equipment or tool failures for corrective action prior to actual break-downs and loss of product quality.

William Beeby and Phyllis Collier report in their book *New Directions Through CAD/CAM*: "At Lamb Technicon's F. Joseph Lamb Division about 10% of the machine tools in the company are NC—almost all CNC—and they perform nearly half of the machining done by the company."

Also see: CAD, CAE, Computer Program, Direct Numerical Control, Distributed Numerical Control, Numerical Control.

Computer Output Microfilm. Computer output microfilm (COM) refers to a microfilm printer that takes output directly from a computer, substituting for a line printer or tape output.

COM, which was originally developed for use as a graphics output device and later used for business or alphanumeric output to save space, is gaining in importance as a CAD activity to handle the blitz of graphics data being generated.

When drawings were created manually, as were the calculations for their design, one could keep up with the other. Then, when the analog and digital computers came into operation, drafters and graphmakers were no longer able to keep up. Just as the volume of data could only be created by computer, it also had to be managed by computer. The first efforts at marrying CAD to COM were unsuccessful. Addressibility and resolution were inadequate. These early systems could not emulate the smooth lines and curves of manual drafting. Although the graphics aspects of the new COM equipment could not be used, the machines excelled at reproducing letters and numbers. Thus, the love affair with COM use for alphanumeric printed data began.

In transferring drawings from the CAD database to production of an aperture card, a completed dataset is mechanically plotted. The full-size drawing is then taken to a COM center where it is photographed with a planetary camera. The frame of microfilm is then mounted on an aperture card. The original drawing is no longer needed.

Computer Part Programming. Computer part programming is the preparation of a manuscript, in an NC computer language, to define the necessary calculations to be performed by the computer.

Also see: Numerical Control, Programming.

Computer Program. A computer program is a list of statements which instruct a computer to perform a sequence of events and actions. Both the source listing (source program) and the compiled object file are examples of computer programs. A source program may be written in any of many different computer languages, which include assembly languages and macro programming languages. A computer program is considered original work by the person who writes a particular program for the first time.

The original author of a computer program has the same copyrights that the author of a book enjoys. Software piracy, or the unauthorized copying and use of computer programs has become a major problem of this era. The problem still exists of determining acceptable ways to protect a program, without hindering the legitimate user unnecessarily. The laws and penalties which apply to computer software programs are currently being changed to afford greater protection to the authors.

Also see: Compiler, Computer Languages, Computers, Programming, Source Program.

Computers. A computer is a term given to an automatic electronic machine which can perform calculations. Most computers consist of five major elements; (1) *input:* to get information into the system, (2) *storage:* to hold information and programs; (3) *processors:* to perform the required information processing; (4) *output:* to find out the results of processing; (5) *software:* to control the way all the elements operate in harmony to carry out the required function.

Today's computers range in size from being part of a wristwatch to large mainframes which require specialized rooms. Computers are used within many of the objects which we use daily, such as refrigerators, automobile ignition systems, alarm clocks, and even toys. The continual miniaturization and cost reduction of computers has allowed them to be the controller of choice in equipment and machines which either used to have mechanical controls or no automated control at all.

Computers have now become so prevalent in our society that the current generation of students is required to become proficient in the use of computers, and in some cases, required to own computers while at college.

Being able to communicate between computers becomes more imperative as computers proliferate our society. Local area networks and communications standards and connectivity are becoming a more critical aspect of selecting computers.

History. The history of computers can be traced back to thousands of years B.C. The Greeks and Romans used a kind of calculating device which consisted of stones manipulated on a flat surface to perform relatively sophisticated calculations. It was known by the Latin word *abacus*, which means "flat surface." In China and Japan, the abacus took the form of beads on a special frame. The beads of the abacus were called calculi, the origin of the word calculate.

William Oughtred, a mathematician, realized that logarithms could be represented as lengths on a rule. In the 1620s he used this idea to develop the slide-rule which was to be an intrinsic part of every engineer's working methods for about 350 years. It performs multiplication and division by sliding a moving rule against fixed calibrations. (The slide rule is an early example of an analog machine which employs continuous physical quantities, like length or voltage, rather than discrete numbers as in digital calculators and computers.)

The first calculating machine was developed by Blaise Pascal in 1642. In 1647, he produced his Arithmetic Engine to help his father, who was a presiding judge of the tax court at Clermont-Ferrand in France.

Around 1854 George Boole, a mathematician, developed a system for representing logical statements in terms of mathematical symbols. Using the symbols and some rules, one could determine whether a statement was logically true or false. Boole's purpose was to enable logical problems to be dealt with in algebraic notation. His methods were not widely accepted initially, and it was not until the next century that his ideas were applied. Boolean algebra established the basis of mathematical logic.

In the 1880s Herman Hollerith developed the modern machine-readable punched card and the associated mechanical card-processing equipment. Hired by the Bureau of the Census, he developed and used his inventions on the huge volume of data collected in the 1890 census. The presence or absence of a hole in particular positions on the card represented the presence or absence of particular characteristics about the individuals in the census. In order to read the cards, rods were passed through them. These rods made contact with a bowl of mercury to form an electrical contact to cause a counter to advance by one. The Hollerith tabulator was therefore the first computing machine which employed non-mechanical processing means. It proved to be so successful that a company was set up to develop his invention. This company eventually became known as International Business Machines (IBM), eventually the largest computer manufacturer in the world.

Vannevar Bush, a professor of electrical engineering at the Massachusetts Institute of Technology (MIT), put some of the earlier Boolean computing concepts into electronic practice by using thermionic valves. Claude Shannon was one of the operators of the MIT differential analyzer. Around 1938, with encouragement from Bush, he developed the technique of applying Boolean algebra to design electrical logic circuits constructed to perform arithmetic operations like add, subtract, multiply and divide. These developments, together with the publication of Alan Turing's description of a theoretical general purpose computer, had set the scene for the rapid development of the modern electronic digital computer.

The Electronic Numerical Integrator And Calculator (ENIAC) was designed by J.P. Eckart and J.W. Mauchly of the University of Pennsylvania, between 1943 and 1945. It was the first electronic computer (it used 18,000 electron tubes!), and was much faster than previous machines. However, the machine had no internal storage. It had to receive its instructions externally via switches and plugs. ENIAC was a special purpose computer, designed to handle mathematics problems in ballistics and aeronautics.

Introduced in 1951, the first electronic computer which was available commercially was the UNIVAC I (Universal Automatic Computer). Unlike its predecessors, it was used for non-scientific data processing. UNIVAC was built by Remington Rand, a company originated by Eckert and Mauchly. It used magnetic tape for input and output. Previously, computers had used the considerably slower punched cards and paper tape. Also, UNIVAC was the first computer to accept and process alphabetic data, as well as numeric data.

Discussion. Computers exist in many sizes and types, from imbedded application specific integrated circuit (ASIC) chips in devices such as wrist watches and pocket calculators to the expensive multiprocessor reduced instruction-set computers (RISC) machines which are designed for very large throughput. While computers can be designed as either analog or digital type processing machines, the vast majority of today's computers are digital type machines. The processing speeds and data size capabilities of today's digital computers easily handle digitized approximations of analog data to a significant number of decimal places such that the benefits of digital signal processing are much more desirable than those of an analog computer capable of continuously varying output. The instability and errors of analog circuitry actually make an analog computer using today's technology less stable and accurate than digital computers.

Computers are sometimes referred to as microcomputers, minicomputers, and mainframe computers. The earliest computers were mainframes since they were large, general purpose machines which usually were used as multiuser, timesharing centralized processing systems. As technological advances forced the price of computers lower, minicomputers, engineering workstations and finally personal microcomputers became a technological reality.

Minicomputers were first defined as a separate class of computers when large-scale integrated circuits allowed the physical size of computers to be reduced from the room-sized monsters to relatively smaller cabinets that did not require as stringent environmental control. The VAX line of minicomputers from Digital Equipment Corporation established the minicomputer as a viable solution for previously ''mainframe only'' tasks.

An engineering workstation is simply a micro or minicomputer which has appropriate applications software and is dedicated to a single user for the performance of specialized engineering activities such as CAD, CAE, document creation, or any other computer intensive activity. Engineering workstations usually utilize 32-bit sized CPUs with more than a megabyte of RAM memory. The 80286 (16-bit) and 80386 (32-bit) PC AT type machines are being used as workstations, sometimes with additional coprocessor or special function boards.

Microcomputers are the smallest computers, usually designed for single user or imbedded system applications. The affordable nature of personal computers has allowed many people to acquire and use personal computers for tasks which previously were accomplished using other means. Personal computers are microcomputers which have a display, keyboard and some type of data storage to provide all the basic computing functions needed for various single-user applications. The difference between an engineering workstation and a personal computer is that most engineering workstations have greater computing capability and higher resolution displays than the typical personal computer system. Also, most workstations are networked to host minicomputers which provide additional file serving and computing capabilities.

The capabilities of computers are sometimes described using two terms, millions of instructions per second (MIPS) and millions of floating-point operations (MFLOPs). Since the time required to complete an instruction varies depending on the type of instruction, and unless this detail is specified, the numbers can be misleading. For example, an NOP (no operation) instruction is the fastest type, since literally, no operation is required to execute the instruction, just read the instruction and then read the next instruction. If an application does not require floating-point operations, MFLOPs may be of no relevance. The capability to do large MFLOPs doesn't always compare precisely to the MIPS, especially if a numerical co-processor is employed to provide MFLOPs performance, which can be 5 to 20 times greater than the CPU chip alone could perform.

Since, most real applications require interaction with the peripheral devices such as communication ports, monitor screen, keyboard, and disk drives, total system performance evaluations must include measurements of these additional activities. In many computer systems which are designed with an input/output bus for interconnection of the different function cards and utilize main system memory which is not physically a part of the CPU chip, the data transmission speeds and data path widths may not be identical for all portions of the system. When the functional speed capability of a segment of the system is slower than the main system, such as the speed of access of the memory, NOP operations or wait states must be inserted between each instruction execution to slow the system speed to accommodate the slower device. In some faster systems, more than one wait state may be required for I/O bus access. Each wait state effectively cuts the speed of the system in half, so a 16 MHz system running with one wait state has essentially the identical performance and execution speed as an 8 MHz, no-wait state system. Another practice used to combine slower disks with faster systems is interleaving the data on the disk to allow the physical movement of the read/write head to occur without disrupting the speed of data flow. Some systems use cache memory to buffer the data between the faster CPU and the slower main memory.

Since the bottleneck in most computer systems becomes the CPU and the lack of true parallel processing, the use of multiple CPUs in a computer system provides parallel processing instead of time-phased multiprocessing. Reduced Instruction Set Computers or RISC machines have been developed to allow quicker execution of programs due to the reduced instruction set and the ability to optimize the CPU to perform the instructions which remain.

Applications. Computers are used in many different types of applications. The traditional data processing activities such as payroll, accounting, personnel records, and order processing are usually done on a centralized mainframe or minicomputer system with terminals being the main interface for the users. Engineering development and design work, due to its more specialized and varied use of computer resources and software is often done using dedicated workstations and departmental computing arrangements of networked minicomputers. Computers are also imbedded in numerous consumer and industrial devices. Televisions, stereo equipment, cameras, automobiles, even some greeting cards have imbedded microprocessors which provide some intelligent function. Many of these imbedded systems contain everything —CPU, memory, I/O, etc.—on a single chip. These custom chips can then be mass produced at the lowest possible costs. Usually when people refer to ''intelligent'' devices or sensors, they are describing devices with imbedded microprocessors.

Computers provide many capabilities which are unachievable or impractical to perform by other means. A software program can be executed to ensure repeatable operation without any human interference or complications. A computer can do simulations of accelerated scenarios, such as calculating the projected events over a time period of months or years in a few minutes. The sheer bulk of data which can be processed, maintained, and recalled by a computer is far superior to what any other device, including a human being, is capable of managing. Computers allow people to ''work smarter, not harder.'' While typing at a typewriter and shorthand stenographic skills used to be the primary methods by which secretaries handled correspondence, the minicomputer and microcomputer have provided electronic word processing capabilities which have obsoleted typewriters from most activities other than labels and pre-printed forms. When properly applied and used, computers can be a vital and integral part of modern methods and technology.

Computers allows electronic data interchange to be used for so many activities such as in electronic mail, facsimile transmission, digital voice communications, and ''paperless'' offices where all records and information are stored in computer systems in place of hard copy. The use of computers has spawned new areas of data management and security which were much different. Physical records were locked in a file cabinet, instead of being maintained on a computer network that is part of a dial up access system. Electronic funds transfer allows banks and other financial institutions to conduct business without using physical money or paper transfers.

Factory automation and Computer Integrated Manufacturing, CAD/CAE, and MRPII systems could not exist without computers. The common thread of connection and communication in many cases is simply the fact that a computer is utilized for each application. This does not mean that connection and integration is simple and forthright, but efforts such as the Semiconductor and Materials Institute (SEMI), the Electronic Design Interchange Format (EDIF), and the MAP/TOP committees are attempting to simplify the connection of computer systems.

Glossary. *abort:* The stopping of a computer at an irregular point in its program, usually before the normal completion of the executing sequence.

absolute address: The actual location number of a word stored in a computer.

accumulator: A location (register) in the ALU where arithmetic operations are carried out.

address: A character or group of characters that identifies a register, a particular part of storage, or some other data source or destination. A label or number specifying where a unit of information is stored in a computer's memory.

ALU: Arithmetic Logic Unit.

analog computer: A computer in which analog representation of data is mainly used.

archive: To store copies of software outside of main memory or tape, disk, or paper.
arithmetic logic unit: The unit of a computing system that performs logic and mathematical functions.
background processing: The processing mode for executing lower priority programs which do not require user interaction.
backplane: A printed circuit card located in the back of a rack or card cage which has sockets into which specific boards are fitted for interconnection.
batch processing: The technique of processing an entire group (batch) of similar or related jobs or input items on a system at one time without operator interaction.
BCD: See: Binary Coded Decimal.
binary code: Data which is represented by either ones or zeros.
binary coded decimal: A decimal notation in which the individual decimal digits are represented by a pattern of ones and zeroes. Each group of four binary positions represents a single decimal integer. For example: 2=0010; 3=0011; 4=0100.
bit: The smallest unit of information that can be stored and processed by a digital computer. Bits can be either on/off, yes/no, or one/zero.
boot: A program used to start the computer, usually by clearing the memory, checking devices, and loading the operating system into internal RAM from internal or external memory.
bootstrap: A technique for loading the first few instructions of a routine into storage, then using these instructions to bring the rest of the routine into the computer from an input device. This usually involves either the entering of a few instructions manually or the use of a special key on the console.
breadboard: Describes the rough model for constructing another device, usually referring to a circuit board that has a specific configuration on it.
bubble memory: A type of memory device using discrete, microscopic magnetic cells in an aluminum garnet substrate.
buffer memory: A section of memory used for temporary storage of data allowing transfer between the main CPU and storage devices, without slowing the CPU to the maximum transfer speed of the device.
burn-in: Operating a device at an elevated temperature to improve the probability that any device weakness will cause a failure.
bus: A circuit or group of circuits providing a communications path between two or more devices, such as between a CPU, peripherals, and memory, or between the functions on a single PC board.
byte: A sequence of adjacent bits, usually eight, representing a character that is operated on as a unit. It is usually shorter than a word. A measure of the memory capacity of a system, or of an individual storage unit.
cache memory: High-speed memory used as a buffer between a fast CPU and slower main memory in a computer system.
central processing unit: The portion of a computer system consisting of arithmetic, control, and logic elements.
chip: A small piece of silicon impregnated with impurities in a pattern to form transistors, diodes, and resistors. Electrical paths are formed on it by depositing thin layers of aluminum or gold. The commonly used name for an integrated circuit.
CPU: See: Central Processing Unit.
digital computer: A computer in which discrete representation of data is mainly used.
EBCDIC: See: Extended Binary Coded Decimal Information Code
emulate: Hardware imitation of another system's data processing capabilities to initiate the same results or permit software compatibility.
fault tolerant computing: An arrangement of computer architecture and software which allows parts of the hardware and/or software to fail without causing any detrimental effect to the operation and the execution of the application program.
foreground processing: The automatic execution of programs that have been designed to preempt the use of computing facilities. Usually a real time program. A program which can be interacted with by the operator while it executes. A

program which has a higher priority for use of the computer to execute than other programs.
gate array: An integrated circuit characterized by a rectangular array of logic sites. These sites consist of identical collections of diffused or implanted transistors, diodes, and resistors. Surrounding the array are the input/output circuits for offchip connections. The final process in creating the desired circuit is the custom metalization of the logic sites and their subsequent interconnections. Gate-array design permits a standard set of logic elements to be used for a wide variety of IC applications.
host computer: The primary or controlling computer in a multi-computer network. Sometimes the host computer provides additional data storage and output capabilities for the other computers in the network.
internal memory: The part of memory where information that is being worked on is kept. Also referred to as main memory.
mainframe computer: Very large computer, generally used by many users at one time in large data processing operations.
microcomputer: A small computer containing a microprocessor, input and display devices, and memory all in one box. It may or may not interface to a host computer and/or peripheral devices. Sometimes referred to as a desktop computer or a personal computer.
microprocessor: A basic element of a central processing unit that is a single integrated circuit. A microprocessor requires additional circuits to become a suitable central processing unit.
minicomputer: A class of computer in which the basic element of the central processing unit is constructed of a number of discrete components and integrated circuits rather than being comprised of a single integrated circuit, as in the microprocessor.
multiprocessing: Executing multiple programs at the same time, however the CPU actually only executes a portion of each program at a time and rotates among each task.
nibble mode: A form of dynamic RAM addressing in which four sequential bits (a nibble, or half a byte) are accessed at a time.
nonvolatile memory: Memory which retains its contents when the power to it is removed.
parallel processing: Executing more than one program at the same time. True parallel processing requires multiple processors.
Qbus: A type of backplane bus popularized by Digital Equipment Corp. with their low end PDP-11, LSI-11, Micro PDP-11 and MicroVax lines of computers.
real-time processing: The ability of a computer to execute a program fast enough that neither the operator nor other operations perceive having to wait for any other task to complete within the CPU.
register: A portion of memory which temporarily stores data (usually one word).
response time: The amount of elapsed time from the initiation of some action until the completion of the same action.
Unibus: A type of backplane bus popularized by Digital Equipment Corp. with their high end PDP-11 and Vax lines of computers.
VLSI: See: very large-scale integration.
very large-scale integration–High-density ICs characterized by relatively large size (approximately 0.25 inch on a side) and high complexity (10,000 to 100,000 gates).
volatile memory: Memory which loses its contents when power that is being applied to it is removed.

Also see: Absolute Address, Ada, Adaptive Control, ALGOL, Artificial Intelligence, Assembly Language, Automated Assembly, Automatically Programmed Tools, Bar Coding, BASIC, Batch Processing, Baud Rate, Binary Code, Binary Digit, Block, BPI, Bus, CAD-/CAM, Card, Card-to-tape Conversion, Cathode Ray Tube, CD-ROM, Chip, Circuit, Circuit Card, COBOL, Code, Code Conversions, Coding, Computer Aided Design, Computer Aided Engineering, Computer Aided Line Balancing, Computer Aided Manufacturing, Computer Aided Planning, Computer Aided Process Planning, Computer Aided Testing, Computer Graphics, Computer Integrated Manufacturing, Computer Languages, Computer Numerical Control, Computer Program, Computer Aided Design and Drafting, Computer Graphics Augmented Design and Manufacturing, Computerized Simulation, Console, Conversational Mode Computing, Core Memory, Cost Estimating, Cursor, Data, Database, Data Processing, Dedicated Computer, Digital Computer, Direct Computer Control, Direct Digital Control, Direct Numerical Control, Disk, Distributed Numerical Control, DOS, Fault-tolerant Computer, EBCDIC, Edit, Editor, Electronic Mail, Emulate, End of Block, EPROM, Ethernet, Even Parity Check, Expanded Memory, Extended Memory, Extended Storage, Fifth Generation Management, File Management System, First Generation Computer, Fixed-block Format, Flexible Manufacturing Cell, Flexible Manufacturing System, Floppy Disk, FORTRAN, Fourth Generation, Group Technology, Hardware, Hardwired, Hexadecimal Numbering System, Hollerith, Human Factors Engineering, Indirect Address, Initial Graphic Exchange Specification, Input/Output Channel, Instruction Set, Joystick, K, Kernal, LIFO, LISP, Local Area Network, Machine Readable, Macro Instruction, Magnetic Tape, Mainframe Computer, Management Information System, Manufacturing Automation Protocol, Manufacturing Requirements Planning, Mass Memory, Master File, Megabit, Memory, Memory Address, Microcomputer, Microprocessor, Minicomputer, MODEM, Motherboard, Mouse, MS-DOS, Multiprogramming, Multiprocessing, Network, Node, Noise, Numerical Control, Off-line, On-line, Open System Interconnection, Open-loop Control, Operating System, Optical Disk Drive, Parallel Processing, PC DOS, Personal Computer, Preprocessor, Programmable Controller, Progammable Logic Controller, Progammable Read-only Memory, Protocol, Second Generation Computer, Security, Significant Digit, Source Language, Software, Storage Medium, System Development, Technical and Office Protocol, Third Generation Computer, Throughput, Timesharing, Token Bus, Utility Program, Winchester Disk, Word Address Format, Word Length, Workstation, Write.

Configuration. Configuration refers to the complete technical description required to build, test, accept, operate, maintain, and logistically support equipment. Also the physical and functional characteristics of the equipment.

Configuration Management. Configuration management is a discipline applying technical and administrative direction and surveillance to: identify and document the functional and physical characteristics of a configuration item control changes to those characteristics, and record and report change processing and implementation status. Configuration management is a product of the aerospace industry. Many of the concepts and features of successsful configuration management can aid manufacturing automation in other industries.

An effective configuration control system contains four major features.

First, it is tailored from an effective set of guidelines and standards to fit the nature of the program including hardware and logistics support elements.

Second, corporate or division policy recognizes the importance of proper configuration management in the development of a new program, and emphasizes the need to generate an adequate plan for implementation.

Third, a configuration management plan is streamlined, yet adequately encompasses the entire life cycle of the program, recognizing the requirements of each phase of the life cycle and the complexity of the system configuration.

Fourth, the configuration management plan establishes the mode of operation and interface relationships among vendors, subcontractors, contractors, and customer.

The standard reference pertaining to configuration are U.S. government publications. No other available published works deal with the subject so completely. The standard configuration references are: *Engineering Drawing Practices (DoD-STD-100)*; *Specification Practices (MIL-STD-141)*; *Configuration Control—Engineering Changes, Deviations and Waivers (MIL-STD-480)*; *Configuration Control—Engineering Changes, Deviations and Waivers (short form)*

(MIL-STD-481); *Configuration Management Practices for Systems, Equipment, Munitions, and Computer Programs (MIL-STD-483).*

Console. A console is a part of a computer or numerical control machine tool system that houses equipment used for communication between the operator and the computer or machine tool.

Also see: Computers.

Contacts. Contacts are the physical moving parts controlling current through a switch or relay. Contacts are also cathode ray tube elements for ON and OFF instructions which are similiar to open and closed relay contact symbols.

Contact Sensor. A contact sensor combines elements of tactile and force sensing. A contact sensor must make contact with an object in order to sense it and measure the amount of resistance force or torque applied from the object.

Continuous Path Control. Continuous path control is the ability for a controller to control the motion of a device in a continuous manner along a particular contour, as opposed to only controlling the device from point to point, where the motion of the device between points being uncontrolled. Continuous path control includes both the actual route in cartesian coordinates, as well as the velocity of the device during the motion. The lack of continuous path control in devices which contain more than one axis of motion allows each joint to move independently at its own random velocity from the beginning rest position to the ending rest position of the device. The actual path which the device travels between beginning point to end point can be any of an infinite number of possible routes.

Continuous path control is utilized in robots for tasks such as arc welding, applying sealants, and spray painting. Continuous path control is utilized by numerically controlled equipment to machine parts. Inspection equipment sometimes uses continuous path control when the motion of the part must be known and at a constant velocity to ensure the accurate functioning of the inspection system.

Contouring. Contouring is the process of using a master form (a pattern) as the basis for producing very accurate reproductions of complex shapes and forms. This type of work, often performed on Die-Sinking and profiling machines uses a tracer to accurately reproduce the complicated forms and contours of the master form. The tracer maintains constant contact with the master form throughout the operation, thus controlling the position of both the work and the cutter. These machines can be subdivided into two groups. In the first group the tracer is hand controlled and is guided over the master form by the operator. The second group consists of automatically controlled machines. In this case, the tracer, once the cycle is started, follows the master form automatically without requiring the assistance or intervention by the operator.

Contouring Control System. A contouring control system is a numerical control system that generates a contour by controlling a machine or cutting tool in a path resulting from the coordinated, simultaneous motion of two or more axes.

Also see: Axis, Numerical Control.

Control Chart. A control chart is a graphical method for evaluating whether a process is or is not in a "state of statistical control." A *control chart factor* is a factor, usually varying with sample size, to convert specified statistics or parameters into a central line value or control limit appropriate to the control chart. *Control limits* are used as criteria for signaling the need for action, or for judging whether a set of data does or does not indicate a "state of statistical control." Lower control limits are for points plotting below the center level. Upper control limits are for those points plotting above the center level.

Also see: Statistical Control.

Controller. A controller is a device that provides control over the operation and function of a mechanism. A controller may control the motion and sequence of operation of a robot,

AGV, NC machine, or other automated equipment. A controller may handle the operation of a Winchester disk drive. The sophistication of controllers has continued to increase as needs and desires push the limits of technology. The computer industry has been a large influence on the miniaturization and increased sophistication of controllers. The use of voice-coil guided servomechanisms for locating tracks on disk storage media is a characteristic of contemporary drive technology. There are emerging and defined standards for controllers such as the SCSI (Small Computer Systems Interface) and ESDI (Electronic Storage Device Interface) standards for personal computers.

Also see: Computers.

Control Unit. The control unit is a portion of a central processing unit (CPU) that directs the operation of the computer, interprets computer instructions, and initiates proper signals to the other computer circuits to execute instructions.

Also see: Computers.

Conversational Algebraic Language. A high-level language developed especially for time sharing in which a remote console typewriter connects its user to the computer for on-line solutions to a problem.

Also see: Computers.

Conversational Mode. Conversational mode describes the communication between a human operator and a computer via a keyboard terminal or other input/output device. Questions and responses are elicited from the computer by the operator and vice versa.

Also see: Computers.

Cooling. Cooling is the process of reducing the heat content of a part, assembly, or material. Cooling can be required for a variety of reasons. Electrical and computer equipment requires cooling to maintain a safe operating temperature. When heat treating metal parts, part of the process is cooling, either by air, water, or oil. Cooling may be effected by conduction, convection, or radiation. Cutting tools require cooling fluid to constantly pass over the point of contact between the tool and the part to prevent the tool from becoming heated to a state which changes its properties to a form which can no longer provide cutting capabilities. Aluminum heat sinks on electronic equipment utilize both conduction from the components to the heat sinks and convection from the heat sink to the air around it. Some of the advancements in fast ECL chip technology utilizes IC chips which are cooled by liquid nitrogen. Due to the thermodynamics of heat being created when energy is used, cooling will always be a necessity, unless a process is 100% efficient in using energy, in which case no energy is wasted as heat.

Coordinated Axis Control. Coordinated axis control is when a mechanism with multiple axes is controlled by a controller which causes multiple axes to move at certain speeds and in certain directions, relative to each other, so the endpoint of the gripper or tool prescribes a desired path of motion. Coordinated axis control is necessary to impart continuous path control when a device is moving along a path other than a single axis of motion. If a single axis of motion is being performed, all the other axes can simply be locked and not moved. Depending where the path of motion is relative to the coordinate system of the mechanism, the same continuous path may require a complex or a simple coordinated axis control. Computer Numerically Controlled (CNC) machines are capable of complex, accurate, coordinated axis control. The quality and shape of the machined part is a direct result of the ability of the CNC machine to perform complicated and precise coordinated axis control.

Also see: Computer Numerical Control, Continuous Path Control, Controller.

Coordinate Measuring Machine. With the advent of numerically controlled machine tools, demand grew for a means to support this equipment with faster first piece inspection, and in many cases, 100% inspection. To fill this need, coordinate measuring machines (CMM) were developed by modifying precision layout machines. These machines provide digital readout of the position of a moveable machine member for display, computer calculations of geometric measurement routines, or computer-controlled positioning of the machine member. In these instruments, the travelling member, or gage head with a mounted probe, contact sensor, or in some applications an optical device, is guided along straight line paths. These paths are contained in a common plane and are mutually perpendicular, representing the X and Y axes of the rectangular system of coordinates. In most models, a third axis movement is provided by axial or radial movement of the spindle along the Z axis, that is, in the direction normal to the plane established by the X and Y axes.

These machines are equipped with workpiece staging tables of substantial size, made of dimensionally stable material (usually granite), and lapped to precision geometrical flatness requirements. The guideways of the travelling head are, depending on the model, of a cantilever, fixed bridge, floating bridge, horizontal bore, or vertical bore configuration. In the most common configurations, a nearly frictionless movement along the guideways is provided by air bearings.

The measurement function is accomplished by clamping the workpiece to the staging table and aligning it manually, or by computer offset calculations, with the coordinate system of the mutually perpendicular X, Y, and Z measuring guideways. Each slide is equipped with a precision linear-measurement transducer whose signed (+, -) position is digitally updated every 0.0001 inch (0.0025 mm) (in some cases every 0.000050 inch (0.0013 mm)) as the measurement sensing probe is moved from point to point on the workpiece.

Coordinate measuring machines can be grouped into three basic levels of measurement control capabilities. The most basic design provides an LED readout of the position of each of the machine's axes. All machine movements and alignments are provided manually by the operator. In this configuration, the operator must do all the calculations to arrive at decisions for measurement requirements. The next level of design provides computer assistance for alignment compensation and geometric measurement results. The results of the manual measurement done by the operator are displayed on VDT or hardcopy is provided by a printer. Measurement routines, programs, and results may be stored on various types of media. The highest level of design provides all of the aforementioned capabilities along with computer-controlled machine movements which allows for the operator to do nothing more than to clamp and locate the workpiece on the machine and initiate the measurement software.

Currently, coordinate measuring machines are being used in one of three ways in a manufacturing firm. The simplest approach is to place the CMM at the end of the production line or in an inspection area. With this approach, the CMM inspects the first part of a production run to verify the machine setup. Once the setup is verified, it then measures parts on a random basis. For many applications, this permits the best approach to inspection.

Another approach is to incorporate the coordinate measuring machine between two workcenters and then measure 100% of the parts produced at the first center before any secondary operations are performed at the second workcenter. This approach is possible because CMMs are capable of measuring three-dimensional geometry and making many different measurements within a short period of time. When this approach is used, the CMM indirectly controls the production process. In this setting, however, the CMM must be "hardened" to perform in the shop environment or it must be completely enclosed to provide an optimum environment for part inspection.

A third approach integrates the CMM into the production line. This permits the CMM to directly control the production process. In operation, an integrated system would measure the workpiece, compare the measurements with required dimensions, and, if necessary, automatically adjust the machine controls so that the part would be manufactured within the required specifications.

Direct computer controlled (DCC) CMMs are equivalent to CNC machine tools. A computer controls all the motions of a motorized CMM. In addition, the computer also performs all the data processing functions of the most sophisticated computer-assisted CMM. Both control and measuring cycles are under program control. Most DCC machines offer various programming options, including program storage and, in some instances, off-line programming capability.

The functional capabilities of CMM software depend on the number and type of application programs available. Virtually all CMMs offer some means of compensation for misalignment between the part reference system and the machine coordinates by probing selected points; some are limited to alignment in one plane, while most machines provide full three-dimensional alignment. Once the designated points have been taken, the program calculates the misalignment and applies the appropriate correction to all subsequent measurement readings.

Many software packages also provide a means for evaluating geometric tolerance conditions by determining various types of form and positional relationships (such as flatness, straightness, circularity, parallelism, or squareness) for single features and related groups of features.

''The idea of having a coordinate measuring machine (CMM) integrated into an FMS is something that appeals to almost everyone in a manufacturing/assembly environment,'' wrote John O. Wilson and Scott S. Lenger (FMC Corporation). ''Along with the increases in productivity, one would have automated process control/verification.''

The remarks were made in their SME Technical Paper, *In-Line Inspection—The Integration of a CMM into an FMS*. ''At FMC Aiken's installation, the task of CMM integration has proven to be a greater challenge than expected. That the CMM would be integrated into the total ''system'' was understood from the beginning. Automatic program downloading and uploading of resultant data to history files was strictly a software function and would be accomplished by the FMS vendor. The FMC project team was responsible for generating inspection routines that would satisfy the requirements of MIL Q 9858A.

''The CMM used in the FMC installation is a vertical column gantry machine with a working envelope of 70'' x 70'' x 52''. In the center of the envelope is a pallet stump with 360 degrees of rotation in 1 degree increments. This rotating platform can be used in the program to facilitate inspection with any probe. The 4 inch x 4 inch column terminates in a five port head that will accept up to five probe bodies and each probe body will accommodate up to five touch fingers. Used in conjunction with probe extensions, the tool array can be formidable; but with 25 tool points, this configuration can be used to probe almost every conceivable feature and its relationship to datum planes and targets.

''There cannot be too much preplanning associated with CMM integration. The goals must be established, the details specified, the personnel selected and trained and the project completion dates realistic. Given enough preplanning, a successful integration effort will be realized.''

Also see: Computers.

Coordinate Transformation. Coordinate transformation is the mathematical processing of robotic positional information from one coordinate system to another. Coordinate transformation is necessary when processing positional information from one type of robot to the next. Although robots vary widely in configuration, mechanically, most fall into one of the four basic motion defining categories: jointed arm, cartesian coordinate, cylindrical coordinate, and spherical coordinate.

The jointed arm consists of several rigid cantilevers attached by rotary joints and most resembles a human arm. A robot arm of this type is usually mounted on a rotary joint whose major axis is perpendicular to the robot mounting plate.

Cartesian coordinate robots consist of prismatic joints and a nonrotary base axis. The end effector is positioned within a Cartesian coordinate system. The Cartesian coordinate system is one whose axes or dimensions are three intersecting perpendicular straight lines and whose origin is the common intersection.

Cylindrical coordinate robots are constructed from a number of orthogonal slides and a rotary base axis. Additional rotary axes are often used to allow for end effector positioning. The cylindrical coordinate system consists of one angular dimension and two linear dimensions.

Spherical coordinate robots consist of a rotary base, an elevation pivot, and a telescoping extend and retract boom axis. Up to three rotary wrist axes may be used to control the positioning of the end effector. The spherical coordinate system has two dimensions that are angles, and the third dimension that is a linear distance from the point of origin; these three coordinates form a sphere.

Transforming mathematically between the cartesian coordinate system, the cylindrical coordinate system, and the spherical coordinate system to determine robotic positional information is referred to as coordinate transformation.

Coordinates. Coordinates are labelled points which define the location in three dimensional space of an object. Coordinates are used to give global directions such as longitude, latitude, and the points on a compass. Coordinates are also used when programming numerically controlled machines to describe the path locations in three dimensional space along which the tool point must move. Coordinates are usually numerically designated. Coordinates are sometimes referred to as absolute or relative. Absolute coordinates mean that a location description is being given relative to an established base origin point. Relative coordinates mean that a location description is being given relative to another coordinate system, which in turn is relative to the established base origin point. Instruments such as LORAN (LOng RAnge Navigation) equipment can triangulate based upon signals received from known locations on the globe and in turn can interpolate the longitudinal and latitudinal location of the LORAN unit.

Also see: Cartesian Coordinates, Polar Coordinates, Rectangular Coordinates.

Core Memory. Core memory is a programmbable high-speed random-access data storage device used to store information in ferrite cores. Usually employed as a working computer memory, the core memory retains information in the event of a power failure.

Also see: Computers.

Corrugating. Corrugating involves making parallel bends across either the length or width of a sheet to increase the section modules of the sheet. The bends may be "U," "V," or any desired shape. Corrugating is performed with a punch and die in a brake press.

Cost. The *Tool and Manufacturing Engineers Handbook* (Volume 5, Fourth Edition), notes that costs generally fall into the categories of fixed, variable, or semifixed costs. Among the manufacturing costs are direct labor, direct material, indirect labor, indirect manufacturing, general administrative, and tooling costs. Also described are fixed costs, variable costs, and semifixed costs.

Fixed costs are generally independent of the production quantity that is being built. Indirect labor and indirect manufacturing costs are generally fixed. Setup costs for machine tools are fixed costs.

Variable costs are those incurred on a per-unit basis of the quantity that is being produced. Variable costs increase with each additional unit that is produced. Per-piece direct labor and direct material costs for assembled or machined parts are examples of variable costs.

Semifixed costs are sometimes known as step variable costs. These costs are somewhat independent of quantity and vary with specific groups of units that are produced. The cost to change cutting tools and the completion of scheduled maintenance operations after a specific number of production units are examples of semifixed costs.

Direct labor is the cost of "hands-on" work related to the manufacturing of a product. Typical direct labor activities include machining, assembly, testing, and sometimes inspection and troubleshooting. Direct labor activities are characterized by the presence of some physical contact between the worker and the workpiece. This contact usually adds value to the product being produced. Direct labor is a variable manufacturing cost.

Direct material is the cost of all components included in the end product being produced. To be consided as direct material, the components or raw materials must be a permanent part of the end product being manufactured. Examples of material costs that are not direct include raw material from which tooling is fabricated, test equipment and packaging materials. Direct material is a variable manufacturing cost.

Indirect labor is the cost of all labor that is not directly associated with the product's manufacture. Examples of indirect labor include the salary costs of workers in the accounting, purchasing, and personnel departments, together with the salary costs of supervisors and managers. Indirect labor costs tend to increase in both absolute and percentage terms as the organization increases in size.

Indirect manufacturing costs (IMC) often is used synonymously with overhead costs. It includes all costs for rent, heat, electricity, water, and expendable factory supplies, together with the annual costs of building and equipment depreciation. Expendable factory supplies are often indirect materials that are consumered during the manufacturing process.

General and administrative (G&A) *costs* are those incurred at the plant or interplant level that are not easily associated with a specific workcenter or department. Examples include the costs of top executives' salaries, plant mainframe computer procurement/operation costs, and technical library facilities. Most G&A costs are fixed. The primary variable component is sales commissions.

Tooling and test equipment costs are those costs incurred in the fabrication of jigs and fixtures for machining. They also include costs for the programming, generation, and checkout of numerical control tapes, and costs for the design and fabrication of special-purpose test equipment. Tooling and test equipment costs are generally fixed.

Cost Analysis. Cost is an expense. It effects everyone as a cost of living, and businesses as a cost of goods sold. In general cost is something that is surrendered intentionally, unintentionally, or unconsciously, to obtain an end result. Understanding cost is a key to living a desired life style or making a profit on a manufactured product.

Analysis is a method of determining or describing the nature of something by resolving it into its parts. Analysis is used by anyone trying to understand what makes up a complex system or aggregate. A cost analysis then is the systematic determination of all the parts that make up the incurred expense of the whole.

An example of cost analysis is the breakdown of a plant's cost of goods sold data. Material, labor and overhead are all components of cost of goods sold. When a dollar value is placed on each component then you can determine which item adds the most cost to a product. Each component can be broken down further to understand what makes up its dollar value. A cost analysis allows businesses and people to understand expenses incurred in order to better control them.

Cost Center. A cost center is a numerical way of designating different parts of an organization. Many firms use three or four-digit numerical codes to identify departments, divisions within departments, and branches within divisions. Each unit of direct labor that is charged is usually associated with a customer order number. This order number enables the labor charges to be identified with a specific customer and/or product.

Most organizations usually associate an account number with the customer order number for labor charges. The account number usually contains all or part of the organization code of the individual performing the labor. A cost center is an identifiable accounting number that corresponds to this code. The purpose of the cost center is to identify the organizational segment performing the work on the customer order.

Cost Estimating. Developing cost estimates is essential to any company that manufactures a product. Estimates of time and cost are developed in a variety of ways for a variety of purposes. The accuracy of an estimate is strengthened by the awareness of the possible results of an erroneous estimate. Different reasons for preparing estimates include preparing budgetary estimates, deciding if a new product should be produced, preparing estimates as temporary work standards, cost control, determining make or buy decisions, and determining a selling price. Most people usually think that more time spent on an estimate ensures a more accurate estimate. For the most part, this is true. The amount of time spent, the level of detail, the skill of the estimator, and the estimating system, itself, all influence the level of accuracy. A skilled estimator sometimes approaches an estimate by asking the question, "How good does the estimate have to be?" The basis for this question is a recognition of the fact that the accuracy demanded from an estimate depends on what the estimate is to be used for and the consequences of possibly being wrong.

Computer cost estimating will not solve all estimating problems; but if properly planned, it can reduce problems. Before attempting to write or buy cost estimating programs for the computer, a series of management decisions must be made to guide the engineer who will be writing the programs.

Cost estimating has been a costly and time consuming problem. Attempts have been made to reduce the amount of engineering labor that is required to develop an estimate. Most companies attempt to establish guidelines that represent a ratio of work quoted to work booked. As example, if a company books only 2% of what it quotes, it suspects that the shop rate is too high or the work being quoted does not fit the plant's equipment. On the other hand, if the company books 20% of what it quotes, it suspects the work being quoted is priced too low or perhaps the estimating function has underestimated the amount of labor or material required.

Cost estimating forms are designed to produce an estimate as fast as possible, sometimes weighing speed over detail.

The increased use and popularity of standalone microcomputers has had a profound impact on all engineering fields. Spreadsheet software lends itself to automating the process of preparing cost estimates and reviews.

Spreadsheet software is useful only in automating the mathematics associated with compiling the cost estimate and cost review grids. This task is a small fraction of the total time required to complete a detailed cost estimate or review.

The computer can play a more valuable role in aiding the estimator in the process of data retrieval. Actual historical labor costs might be retrieved from a database and compared with similar parts that have not been fabricated before. Metalcutting formulas can be programmed. Thus, the computer can speed the process of manually estimating labor hours. The tabulated data and cost estimating software may be purchased commercially. One sample printout is located on the next page.

The computer can also play an instrumental role in the forecasting of production costs.

Cost Forecasting. There are basically two types of cost forecasting systems. The first seeks to forecast the total cost of building a group of end items prior to the time production actually begins. Parametric estimating systems often complete this task. They check conventional estimates to determine whether the obtained estimated total cost is accurate. Parametric estimating systems partially automate the procedure of making the initial cost estimate.

The second type of forecasting system seeks to predict the potential job costs of projects that are currently in process. Expended costs are compared with completed work. Revised cost totals, together with throughput time and workload forecasts, can be obtained.

```
##########################################################################################
SUMMARY OF THE ESTIMATE

CUSTOMER NAME   CURTIS MACH
PART NAME  KNURL HOLDER   PART NUMBER  62236T4
DATE   3/27/85

    MACHINE                  TIME/PC.   PC./ CYCLE     S/U TIME   OFF STD. W.C. RATE

HACK SAW - AUTO              1.384861        36          23          .2        $ 35

V MILL-PR FD & TAB           .370648         4           77          .2        $ 35

H MILL-PWR FD & TAB          .7437716        1           107         .2        $ 35

H MILL-PWR FD & TAB          .7916162        1           107         .2        $ 35

H MILL-PWR FD & TAB          1.888845        1           111         .2        $ 35

HORZ MC W/TC                 3.07977         1           37          .2        $ 45

N/C DR AUTO TC               2.197737        1           71          .25       $ 40

DEBURR                       .4905884        1           5           .25       $ 25

ASSEMBLE                     1.55            1           12          .2        $ 25

-----------------------------------------------------------------
    TOTAL                    12.49784                    550

ODD LOT PRODUCTION OPERATION AND SETUP COSTS
LOT NO.         LOT SIZE        TOTAL COST PER PIECE

 1               1000            $ 9.658146
 2               2000            $ 9.458861
 3               3000            $ 9.392431
OUTSIDE SERVICES
DESCRIPTION OF THE OUTSIDE SERVICE          HEAT TREAT
LOT NO.         LOT SIZE        COST         MARKUP           TOTAL COST
 1               1000            $ .5         $ .05            $ .55
 2               2000            $ .45        $ .045           $ .495
 3               3000            $ .4         $ .04            $ .44

DESCRIPTION OF THE OUTSIDE SERVICE          BUY 2 CARBIDE ROLLS
LOT NO.         LOT SIZE        COST         MARKUP           TOTAL COST
 1               1000            $ 1          $ .1             $ 1.1
 2               2000            $ 1          $ .1             $ 1.1
 3               3000            $ .8         $ .08            $ .88

DESCRIPTION OF THE OUTSIDE SERVICE          BUY 2 PINS
LOT NO.         LOT SIZE        COST         MARKUP           TOTAL COST
 1               1000            $ .4         $ .04            $ .44
 2               2000            $ .4         $ .04            $ .44
 3               3000            $ .4         $ .04            $ .44
ESTIMATE WITH OUTSIDE SERVICES - UNIT COSTS
LOT NO.         LOT SIZE        OPER COST       OSS COST        TOTAL

 1               1000            $ 9.658146      $ 2.09          $ 11.74815
 2               2000            $ 9.458861      $ 2.035         $ 11.49386
 3               3000            $ 9.392431      $ 1.76          $ 11.15243

ESTIMATE WITH MATERIAL COST - PRICES ARE PER PIECE

LOT NO.         LOT SIZE        OPER COST       MAT COST        TOTAL

 1               1000            $ 9.658146      $ 3.02401       $ 12.68216
 2               2000            $ 9.458861      $ 2.99101       $ 12.44987
 3               3000            $ 9.392431      $ 2.987178      $ 12.37961
```

A cost estimating for producing a part. This estimate, displayed several different ways is produced by using a personal computer.

```
##############################################################################
MATERIAL ESTIMATE
CUSTOMER NAME CURTIS MACH DATE 3/27/85

PART NO. 62236T4 NAME KNURL HOLDER
MATERIAL AISI 1018 CD FLATS SIZE 1 1/2 X 3 1/2

LOT SIZE  1000  WITH SCRAP  1031
PART LENGTH  3.75 WITH CUTOFF  3.875
PARTS PER BAR  37  BAR LENGTH  12  FEET
WEIGHT PER FOOT  17.85  WEIGHT PER BAR  214.2
TOTAL NUMBER OF BARS REQUIRED  28
PART WEIGHT  5.578125

TOTAL MATERIAL COST                          $ 2699.1
QUANTITY EXTRA                               $ 25
DELIVERY EXTRAS                              $ 25
MARKUP                                       $ 274.91
------------------------------------------------------------------------------
TOTAL COST                                   $ 3024.01

PART COST PER PIECE    3.02401
COST PER POUND  .45

LOT SIZE  2000  WITH SCRAP  2061
PART LENGTH  3.75 WITH CUTOFF  3.875
PARTS PER BAR  37  BAR LENGTH  12  FEET
WEIGHT PER FOOT  17.85  WEIGHT PER BAR  214.2
TOTAL NUMBER OF BARS REQUIRED  56
PART WEIGHT  5.578125

TOTAL MATERIAL COST                          $ 5398.2
DELIVERY EXTRAS                              $ 40
MARKUP                                       $ 543.82
------------------------------------------------------------------------------
TOTAL COST                                   $ 5982.02

PART COST PER PIECE    2.99101
COST PER POUND  .45

LOT SIZE  3000  WITH SCRAP  3091
PART LENGTH  3.75 WITH CUTOFF  3.875
PARTS PER BAR  37  BAR LENGTH  12  FEET
WEIGHT PER FOOT  17.85  WEIGHT PER BAR  214.2
TOTAL NUMBER OF BARS REQUIRED  84
PART WEIGHT  5.578125

TOTAL MATERIAL COST                          $ 8096.85
DELIVERY EXTRAS                              $ 50
MARKUP                                       $ 814.685
------------------------------------------------------------------------------
TOTAL COST                                   $ 8961.534

PART COST PER PIECE    2.987178
COST PER POUND  .45
```

A cost estimating program.

Parametric estimating systems attempt to correlate product costs with a number of easily identifiable features of the part being built. For parts that require machining, some of these features include weight, volume, tolerances, material, geometry, surface finish, and the amount of material removed during the metal-cutting process. For electronics products, some example part features include the number of active/passive componets, number of input/output terminations, weight, volume, and power dissipation.

Work progress forecasting algorithms seek to forecast potential costs, completion times, and remaining workload or projects on which work has already begun.

The algorithms require two key sets of input data. The first is the repository of historical cost curve data. A cost curve relates cumulative total job expenditures in dollars to time periods for the duration of the project. The model presented in the references uses a total of 142 historical cost curves that represent a cross section of the products produced by a job shop manufacturing facility.

The second set of data is used in conjunction with a work progress function. The work progress function is used to analytically determine with considerable accuracy the percentage of the project that may be considered to be complete. The function associates one of three parameters with each operation of each process routing of each job for all active jobs. Separate parameters are defined for routing operations that are not yet started, are in process, or are already completed. These variables describe the percentage of labor effort on a project that is complete at the time a new forecast is to be made. The obtained labor completion percentage is averaged with a percentage completion on the job with respect to purchased material. The result is a detailed assessment of the percentage completed work activity on the entire project.

The process begins by forecasting the potential total job cost. The completed work is compared to the expended total costs. Optimistic and pessimistic cost bounds are defined. The new cost forecast is obtained as a midpoint between these two cost extremes.

The next step requires the use of the cost curve repository. A curve from the repository is selected on the basis of the relative sizes of the cost percentages for purchased material, machined parts, assembled parts, and inspection costs. The goal is to select the curve whose shape is most likely to accurately represent the shape of the cost curve of the job being analyzed. Data normalization and transformation procedures take into account the effects of different total costs and completion times for the cost curves in the data repository.

The selected cost curve is transformed to fit through the previously forecasted values of total cost and throughput time. The data points are transformed so that the obtained curve will also pass through the point described by the costs expended to date and the time at which the forecast is being made. The curve shape and percentage job completion are used to define a time increment that is either added or subtracted to the current throughput time forecasts. Jobs with projected overexpenditures have increased throughput time forecasts. Jobs with projected underexpenditures have decreased forecast values for throughput times.

The model has provisions to define separate forecast values for both labor and material costs. Separate cost curves for labor costs can be defined. When cost values are divided by an aggregate labor overhead rate, a curve describing labor hour expenditures versus time can be defined. This curve can be used to define remaining preperiod forecasts of the labor hours to be expended over the job's remaining duration.

Counter. A counter is a device such as a register or storage location representing the occurrences of an event as incremented or decremented in response to an input signal.

Also see: Computers.

Counterboring. Counterboring is the process of enlarging one end of a drilled hole. The enlarged hole, which is concentric with the original hole, is flat on the bottom. Counterboring is used primarily to set bolt heads and nuts below the surface.

The counterbore is a rotary, pilot guided, end cutting tool, which has two or more cutting lips and usually has straight or helical flutes for the removal of chips and the admission of cutting fluid. Counterbores are used in conjunction with a pilot that guides the tool in the previously prepared hole. Two types of interchangeable pilots are generally used—a solid pilot and a roller pilot. Solid pilots are from 0.002 to 0.007 inch (0.05 to 0.17 mm) in diameter under the nominal size of the hole. The hole must be reasonably straight and should have a finish of 125 microinch (3.17 micrometer) or better for a successful operation. Roller pilots will yield improved counterbore performance if the hole is properly prepared.

Pilot size must be equal to or greater than the cutter-root diameter because chips can clog in the flutes if the area of the cut is too great for the chips to flow properly up the flutes. Chips can also clog if the cutting depth is too great. Too long of a pilot will tend to bind in the hold and may hit the bottom of the hole, the fixturing, or the work table. Roller pilots should be used if the hole finish is 64 microinch (1.62 micrometer) or better and the clearance between the roller pilot and hole is maintained between 0.0005 and 0.0001 inch (0.013 and 0.002 mm) . Holes should always be rough-reamed prior to counterboring.

CP/M (Control Program for Microcomputers). CP/M or Control Program for Microcomputers is an operating system which, prior to MS-DOS, was the most famous and widely used operating system. CP/M was written by Digital Research and originally designed to run on the Zilog Z80 microprocessor. CP/M can now be run on the Intel 8085, 8086, and 8088 microprocessors, essentially because they happen to implement a superset of the Z80 command set. Some variations to CP/M have been developed, such as Concurrent CP/M, which allows up to four virtual terminals and sessions to operate concurrently, and MP/M, which is a multi-user version of CP/M. CP/M has some limitation in the amount of memory it can address and the hardware processors which it can control. While CP/M is not widely used today, it paved the way for MS-DOS, PC DOS, and OS/2, which are the prevalent operating systems for microcomputers today.

Also see: Operating System.

CPM. See: Critical Path Method.

CPS. See: Characters Per Second.

CPU. See: Central Processing Unit.

Crater. In arc welding, a crater is a depression at the termination of a weld bead or in the molten weld pool. Crater is also frequently applied to the impressions left by the individual spark discharges in electrical discharge machining. It is also used to describe massive depressions resulting from short circuits in electrical discharge or electrochemical machining.

Also see: Arc Welding, Electrical Discharge Machining.

Critical Path Method (CPM). Critical path method is the term which describes a method of project management which consists of identifying the path of critical events, which means the events which if a longer time is taken to complete, would cause a longer overall time for the entire project. A task which is not on the critical path is said to have "slack time," which means the task can take longer time than is planned, up to a certain limit, without impacting the overall project time. The concept of critical path method of project management uses the principles that if the tasks which can cause direct and immediate impact upon the overall project are identified, these tasks can receive priority with regards to additional resources being applied to try to maintain the original schedule. CPM does not consider the impact of cost with regards to the actions taken, but only seeks to maintain the time schedule.

Also see: Project Management.

Cross Assembler. A cross assembler is a program which is run on a large computer for translating instructions for a second computer into machine language for the second computer.

Also see: Computers.

Crosstalk. Crosstalk represents electrical interference between machine servosignals causing one slide to jump whenever another slide is commanded to move.

CRP. See: Capacity Requirements Planning.

CRT. See: Cathode Ray Tube.

CSA (Canadian Standards Agency). The Canadian Standards Agency is the official body of the Canadian government that approves and establishes standards for the country of Canada.

Also see: ANSI, Standards.

CSMA/CD. A popular form of contention protocol is CSMA/CD Carrier Sense Multiple

Access/Collision Detect). Its origin is the University of Hawaii's Aloha network. The Aloha network used a radio-based packet scheme in which the secondary stations independently transmitted to the master without regard to the other station's signals. The master station broadcasted on one band and all secondary stations on another. Since the secondary stations transmitted at random, packets often "collided" when transmitted from different stations at the same time. After such collisions, the stations waited a random time before retransmitting. The Aloha scheme yielded only 18.4% maximum channel utilization. The slotted Aloha scheme provided for more effective use of the channel. With this approach, each station was synchronized on a master clock and any transmitted packet began on a specific clock interval. Collisions still occurred, but slotted Aloha provided for 36.8% maximum channel utilization.

CSMA/CD uses some of the Aloha concepts. However, before transmitting a packet, a station "listens" for a signal on the path and does not transmit until the signal (i.e., another station's packet) has passed through the cable. The sender then transmits its packet. Collisions can still occur when two or more stations sense an idle channel and transmit at the same time. However, the CSMA/CD protocol monitors the channel for a collision during transmission. If a station's output does not match the signal on the channel, it knows a collision has occurred. The protocol then ensures that all other stations know of the collision. Some carrier sense protocols (like Aloha) provide for a central site to transmit a busy signal on a separate subchannel when it is receiving data. The CSMA/CD does not work this way because it has no master station.

After deferring to a packet, the station transmits its signal in one of two methods. Under the persistent carrier sense approach, stations transmit immediately after the busy signal goes off. Most CSMA/CD protocols use a nonpersistent carrier sense where stations transmit with a random delay to avoid repeated collisions. Since each station generates its own randomizing variable, retransmissions among the stations rarely occur at the same time. Under heavy workloads, CSMA/CD increases this delay (with an exponential backoff algorithm) to prevent channel saturation. A central site carrier sense protocol can achieve 80% effective channel utilization; CSMA/CD achieves better than 90% utilization rates.

Curing. See: Electron Beam Curing, Radiation Curing.

Current Loop. A current loop is a communication line on which the presence or absence of electrical current is used to represent transmitted data.

Cursor. A cursor is the visual location which shows where the next character will be displayed or deleted when a key is pressed on a terminal keyboard. A cursor is usually shown as a square or an underline. A cursor may be flashing or shown in a contrasting color to make it more distinguishable. When nontext activities are being done on a screen, such as graphics or using a painting or drawing program, the cursor may be shown as a different symbol. Sometimes the cursor shape is even user definable, in which case the cursor can become anything which can be created within a certain dot matrix.

Standard conventions exist for text cursors, some of which are a square cursor depicting "insert" mode for text while an underline is for "typeover" mode. When pressing the delete key, the character under the cursor is removed, but when pressing the backspace key, the character to the left of the cursor is removed. Upon pressing a key, the cursor always moves to show where the next character will appear.

Cutter Offset. Cutter offset is a numerical control feature enabling a machine operator to use an oversized or undersized cutter.

Also see: Numerical Control.

Cutoff. Cutoff is the cutting action along a line. It may involve one or more cuts where the line of cutting is straight, angular, jogged, or curved. It is performed in a die operated by a press, similar to blades in shears. The use of cutoff operations is limited to blank shapes that nest readily. However, it is more versatile because it is not limited to straight-line cuts, as is shearing. A small amount of scrap or waste sheet metal may be produced at the start or finish of the strip or coil of sheet metal.

Cybernetics. Cybernetics is the field of technology relating to the comparative study of the control and communication of information handling machines and living organisms.

Cycle. Cycle refers to a complete series of events or operations in which conditions at the end are the same as at the beginning of the series. A succession of operations of this kind is usually, though not necessarily, recurrent. For example, in two-stroke internal combustion engines the complete cycle of operations is performed when the piston has completed two strokes in the cylinder and has returned to the position, in relation to the cylinder part, which it held at the commencement. The procedure is then repeated.

In electrical terms, a cycle can be defined as the complete series of changes that take place in the value of a recurring variable quantity in the time between recurrences which are identical in magnitude, sign, and direction of change.

Also see: Cycle Time.

Cycle Time. Cycle time is the amount of time from the release of a manufacturing order to its final completion. The cycle time may be limited to only the cycle time of one operation or expanded to the cycle time of an entire finished product. Cycle time includes set up time, queue time, move and transportation time, run time and will be impacted by lot size parameters. Cycle time may also represent the machine cycle time which would be the required amount of time to run a lot through a machine and not include the other elements of cycle time such as move, queue, etc.

Cycle time is the actual time recorded as opposed to a standard time which represents the time that should be needed to set up a given machine or assembly operation. Measurement systems are designed to capture as input the cycle time or actual time and compare to the standard time and report differences.

Also see: Cycle.

Cyclic Redundancy Check. Cyclic redundancy checking is another approach which entails the division of the user data stream by a predetermined binary number. The remainder of the number is appended to the message as a CRC field. The data stream at the receiving site has the identical calculation performed and compared to the CRC field. If the two values are identical, the message is accepted as correct. CRC is capable of ensuring an accuracy greater than 99.99% in data transmission.

Cylindrical Coordinates. Cylindrical coordinates refer to a coordinate system having a central axis about which a plane radiates. Cylindrical coordinates describe items which rotate about a central axis, with the rotation always being in one plane. Cylindrical coordinates also are the only coordinate systems which can describe three-dimensional space with two parameters, the position along the central axis and the radial orientation. Points in space can be converted between rectangular coordinates and cylindrical coordinates and polar coordinates.

A periscope is an example of cylindrical coordinates with the central axis variable. A compass is an example of a cylindrical coordinate system where the value of the central axis remains fixed. Some robots are designed to operate in cylindrical coordinates. Cylindrical coordinates are not usually as popular as rectangular coordinates, which do not require the use of any rotary actuators.

Also see: Cartesian Coordinates, Industrial Robots, Rectangular Coordinates, Polar Coordinates.

D

D/A. See: Digital-to-Analog.

Data. Data is a representation of facts, instructions, concepts, numerical and alphabetical characters, etc., in a manner suitable for communicating, interpreting, and processing by humans or by automatic means such as NC systems.

Also see: Computers, Database, Numerical Control.

Database. The most common use of databases is in the use of electronic data processing or computerized systems. The objective of such systems is to process data automatically into information. This is normally done in a computerized environment which performs basic functions such as input, processing, output, and storage.

Inputs to a computerized system can come in a variety of media but normally are inputs to update the system with transactions. Transactions are frequently called from source documents, such as payroll cards, and converted to computer-readable form. Transactions are processed via the computer hardware and update storage. Further processing of the data results in output.

Output can take the form of reports, documents, inquiry via CRT display, and input to other systems via various output media. It is during the storage process that the use of a database becomes important in the update, storage, and output of transaction information.

In order to understand the storage process, it is first necessary to have a basic understanding of data organization. All computerized systems must have a logical manner in which to store and process data efficiently. Data are typically stored in a *file*. A file is a group of related records which are stored on physical computer hardware such as a magnetic tape or disk.

A payroll file, for example, contains records for all employees of a company. In turn, a payroll record contains related fields of data. A record would contain fields such as a name, an address, and a social security number.

In order to aid a user in searching for data, each record in a file will use a field for identification. This field is called a *key*. A key could be the social security number field of a record in a payroll file. With knowledge of a particular social security number, a user can search the payroll file and locate the record and associated fields for that number.

Records in a file can be organized either sequentially or at random. In a sequence, records could be stored alphabetically. If a random method is used the computer must track the location of each record in the file using a file key.

The traditional approach to data processing is to have separate files for each application with some files being used for more than one application. This proves to be very cumbersome in updating and maintaining files when common data exists in separate files. Thus, a level of data organization known as the database has become prevalent in today's computerized society. By utilizing a database, records and files are consolidated so that a common pool of data records serves as a single file or data bank with minimum redundancy.

History. Databases as we know of them today have come through four stages of growth since the 1980s. During and prior to the 1960s, files were typically organized in a serial manner. That is that the logical and physical structure were the same and batch processing was the normal mode. By the late 1960s access to files was in the serial and random mode. Early database systems began to occur in the early 1970s. During this third stage, redundancy was sharply reduced through complex data organization. Field and group level data rather than complete records could be retrieved. Data management systems were used to coordinate logical file and physical file arrangements. Today elaborate software provides both logical and physical data independence. The creation and maintenance of the database is the responsibility of the database administrator. Database management systems may provide a form of command language to retrieve data and present it to programmers in the form they require.

Discussion. Three types of data organization exist for a database. They are often called schema, subschema, and physical storage.

The *schema* technique is viewed as a model of the overall data. The schema consists of a definition of the various record types in the data model along with the data elements they contain and the sets into which they are grouped. Schema are often drawn in the form of a block diagram. The diagram will be used to represent the total requirements of all programs and users of the database.

The *subschema* is utilized for a logical view or submodel. Typically a more limited view is held of the data by the application programmer.

The physical layout or storage device is yet another view of the database. Physical storage details what method will be used for location of data, what fields are to be used for pointers, and what security measures must be employed around the layout.

The software that enables a database to be deferred, created, and used is called a *database management system* (DBMS). Any access request from a user is intercepted by the DBMS and performs the necessary operations on the physical database. The DBMS is also responsible for applying authorization checks and validation procedures. The *Database Administrator* (DBA) is the person responsible for overall control of the database system. Typically the DBA's responsibilities would include managing database information content, storage structure and access strategy, backup and recovery, and security.

Because of the complex organization of data within a database, a number of methods are used to link related entities and attributes together. A pointer is a field that contains the address of another record. Pointers are used to assist in organizing data into a specific structure. Inverted files are another way to organize data logically. An inverted file is one in which the key to the file is an attribute and the data fields are pointers to each of the records that contain that attribute. One or a combination of these methods may be used to organize data into a specific structure. Normally these structures are in the form of flat files, trees, and plex structures.

A *flat file* is a record which contains the same set of fields. The record may be viewed as a line in which fields are the columns. *Tree structures* are usually represented upside down. The tree is made up of a hierarchy of elements called nodes. Nodes are related to parents; lower level nodes are referred to as children. By utilizing this type of structure, data tends to fall into a hierarchy of categories. A third form of database organization is called a *plex structure*. A plex structure is more complex than a tree because any item may be linked to any other item.

Physical file organization is also required of direct access storage devices utilized for databases. Physical file organization may be in the form of indexed files, indexed sequential files, and random files.

In between the database and the users is the data model. The data model is simply the information content of the database as viewed by users. The data model is the basic means by which independence is achieved. There are currently three forms which the data model can take. The relational approach, the hierarchical approach, and the network approach are currently in use today.

The relational approach is based on the mathematical theory of relations. A *relational* database is constructed principally from files using a technique called normalization. *Normalization* is a step-by-step process for simplifying complex relationships between data with associations in the form of two-dimensional tables.

Many database systems today are based on the hierarchical approach. A hierarchical data structure is often referred to as a tree structure because it is composed of nodes and branches. Each *node* is a collection of data items describing a record. The highest node is the root. Dependent nodes are branches. Structures expressing a downward relationship, such as a corporate organizational chart, may be easily expressed in a hierarchical structure. It will normally consist of one-to-one and one-to-many relationships.

The network data structure is composed of record types and set relationships. Record types are composed of data elements which describe the records. Set types express a relationship between two or more record types. The network database supports one-to-one, one-to-many and many-to-many relationships. It is considerably more flexible than a hierarchical data model in that intricate relationships may be precisely represented.

Applications A database can be used as a tool to create an environment for optimal use of data. Without a database, it is not particularly easy to retrieve data directly and benefit from logical masking, data integrity, and data sharing.

As an example of a database, it is interesting to consider the data processing in a distribution company. The company needs to maintain data about each inventory item stored because inventories are ordered in varying quantities by customers. The company uses a simple re-order point system. If the stock of a particular item drops below a certain level, then new stock is purchased from a vendor.

The database will require three principal records: vendor, inventory, and customer. As a customer places an order for one or more different inventory items, a sales order is built up in the database. This would require a sales order record type. In addition to this record, a sales order line record would also be required. This would contain information on the inventory item being ordered and the corresponding quantity of each item. The sales order record would contain data on the entire order such as customer identification, sales order number and delivery terms.

In addition, a purchase order record and purchase order line record would also be required. The relationship would be similar to the customer record, except the purchase order would be corresponding to a particular vendor. The purchase order would be built up as the inventory item or items go below the re-order point.

As a result, the database now contains seven different record types. The three principal records are: vendor, inventory, and customer and the four subsidiary records are: sales order, sales line, purchase order, and purchase line. The database has now established eight various relationships among the record types. An automatic relationship exists between customer, sales order, and sales line as well as vendor, purchase order, and purchase line. A manual relationship exists between the inventory record and the purchase order and sales order line records.

These relationships can be further explained by example. The relationship between a customer and the customer's sales order is automatic. There is no reason to have a sales order without a customer in the database. Each time a sales order is entered it should automatically be connected to an existing customer record. On the other hand, the relationship between a sales order line and an inventory item is quite different. The sales order line is subordinate to two relationships. An automatic relationship exists between the sales order and the line while a manual one connecting the sales order line to an inventory order exists. The manual relationship exists because the sales order line cannot automatically be connected to the item being ordered at the time the sales order is stored in the database. At the time the sales order is being processed, it may be possible to have a shortage of inventory items on hand. It would then be necessary to indicate a back order on the sales order line in some way so that when new stock enters the warehouse, back orders can easily be found and processed in the database. Thus when a relationship exists between two record types which are conditional on circumstances that can be determined only at processing time, the relationship is called *manual*.

Database Management System. A Database Management System (DBMS) is a generalized set of computer programs which controls the creation, maintenance, and utilization of the databases of a computer-using organization. Such software is a vital ingredient in implementing database-oriented EDP systems. Such systems utilize databases with complex data structures requiring powerful software to automatically perform tasks including the following:

- Database Creation. Defining and organizing the data needed to support the information system.
- Database Maintenance. Adding, deleting, updating, correcting, and protecting the data in a database.
- Database Processing. Utilizing the data in a database to support various data processing assignments such as information retrieval and report generation.

The database management system is controlled by the operating software of a computer system. DBMS can be viewed as a group of processing programs controlling the use of a database. It works in conjunction with the data management control programs of the operating system which are primarily concerned with the physical input, output, and storage of data during processing. A DBMS removes the database from the control of individual programmers and computer users and places responsibility for it in the hands of a specific individual often referred to as a Database Administrator. This improves the integrity and security of the database.

Also see: Database.

Data Circuit-terminating Equipment. Data Circuit-terminating Equipment (DCE) is the interface equipment that couples data terminal equipment (DTE) to a transmission circuit or channel and a transmission circuit or channel to a DTE. The equipment is usually installed at a user's premises. It provides all the functions that are required to establish, maintain, and terminate a connection. It also provides the signal conversion and coding between the DTE and the line. It may be an integral part of another unit or it may be a separate piece of equipment. A simplified form of DCE, called a *network terminating unit*, may be provided in a specialized data network.

Data circuit-terminating equipment is the interface at the endpoints of the data circuit, which are not necessarily the end points or origin of the actual data path. For example, a modem represents a DCE, which in turn may be attached to a video display terminal or a personal computer, which are examples of data terminating equipment (DTE).

Also see: Data, Data Terminating Equipment (DTE).

Data Conversion. Data conversion is the changing of the format, method of storage, language, coding, or symbolism of information without altering the intelligence or logic content. Also known as conversion.

Also see: Data.

Data Error Handling. There are many different methods in use today for detecting and correcting errors in data transmission. Most methods used to provide for data error detection entail the insertion of redundant bits in the message. The actual bit configuration of the redundant bits is derived from the data bit stream.

Vertical redundancy check (VRC) is a simple technique which consists of adding a single bit (a parity bit) to each string of bits that comprise a character. The bit is set to 1 or 0 to give the character bits an odd or even number of bits that are 1's. This parity bit is inserted at the transmitting station, sent with each character in the message, and checked at the receiver to determine if each character is the correct parity. If a transmission impairment caused a "bit flip" of 1 to 0 or 0 to 1, the parity check would so indicate. However, a two-bit flip would not be detected by the VRC technique, which creates a high incidence of errors in some transmissions. The single-bit VRC is not suitable for most analog voice-grade lines because of the groupings of errors that usually occur on this type of link.

Longitudinal redundancy check (LRC) is a refinement of the VRC approach. Instead of a parity bit on each character, LRC places a parity (odd or even) on a block of characters. The block check provides a better method to detect errors across characters. It is usually implemented with VRC and is then called a two-dimensional parity check code. A typical telephone line's error rate may be improved by two to four orders of magnitude with the two-dimensional check.

The *hamming code* is a more sophisticated variation of the VRC. It uses more than one parity bit per byte or character. The parity bit values are based on various combinations of the user character, and the parities are inserted in between the bits of the character.

Cyclic redundancy checking (CRC) is another approach which entails the division of the user data stream by a predetermined binary number. The remainder of the number is appended to the message as a CRC field. The data stream at the receiving site has the identical calculation performed and compared to the CRC field. If the two values are identical, the message is accepted as correct. CRC is capable of ensuring an accuracy greater than 99.99% in data transmission.

Also see: Computers, Data.

Data File. Data File is a collection of related data records or application data values organized in a specific manner and stored after, and separate from, the user program area.

Also see: Data.

Data Flow Diagram. The data flow diagram is a description of system processes shown in a graphic representation. The diagrams show both data into and data out of a system (or process). The data flow diagram also shows the component processes within the system and external interfaces with the system.

The diagrams depict the structured flow of data and storage of data within the system. Major processes can be shown at a high level as well as very detailed processes within that major process. Specific points for storage are depicted as repositories of data which is at rest within the process. Each data flow diagram indicates the external sources of the system from which the process data originates. Data flow diagrams depict the data flow through the processes. Information such as relevance of data or data frequency is not shown. The data flow diagrams are effective as an aid to identify the information where further analysis or review is required.

Data flow diagrams use symbols to depict the process with the use of plain English written directly on the diagram to describe the actual data in the process. Consistent use of symbols is used throughout the diagrams and is labeled accordingly.

Also see: Computers, Data.

Data Link. A data link is the physical computer equipment, especially transmission cables and interface modules, which automatically transmits information to and receives information from a remote location, and vice versa. Also known as communications link.

Also see: Computers, Data, Data File, Data Manipulation, Data Processing.

Data Manipulation. Data manipulation is the process of altering and/or exchanging information between storage words through user programmed instructions to vary application functions. Functions include sorting, merging, input/output, and report generation.

Also see: Computers, Data.

Datamation. Datamation is a shortened term for automatic data processing formed by combining data and automation.

Data Processing. Data processing is a computer procedure involving one or more operations for collecting data and producing a specified result.

Also see: Computers.

Data Set. A data set is one of several prescribed arrangements of related records composing the major unit of data storage and retrieval in the operating system.

Also see: Computers, Data, Modem.

Data Tablet. A data tablet is an input device with a direct correspondence between positions on it and the display surface of the CRT. Typically, it is used in conjunction with a mouse or a stylus for indicating positions on the CRT, for digitizing input of drawings or for menu selection.

Also see: Computers, Data.

Data Terminal Equipment. The data terminal equipment is the equipment comprising the data source, the data sink, or both. The equipment, consisting of digital end instruments, also converts user information into data signals for transmission, or reconverts the received data signals into user information. A DTE is the functional unit of a data station that serves as a data source or a data sink and provides for data communication control to be performed in accordance with link protocol. The data terminating equipment (DTE) may consist of a single piece of equipment that provides all the required functions necessary to permit users to communicate, or it may be an interconnected subsystem of multiple pieces of equipment, which performs all the required functions.

Examples of DTE are dumb terminals, computers communicating through modems, or computers communicating directly to other DTE equipment. In communicating directly between two DTE, the transmit and receive lines must be appropriately crossed (pins 2 and 3 of a serial line) to ensure that the transmission of one device is physically connected to the receiving input of the other device and vice versa. This is the purpose of using a null modem cable when connecting two DTE without modems in-between.

Also see: Computers, Data Circuit-terminating Equipment (DCE).

Data Transfer. Data Transfer (DX) is the process of transmitting data from computer to storage areas or from storage areas to computer through a specialized user program.

Also see: Computers, Data.

Datum. A datum is a theoretical exact point, axis, or plane derived from the true geometric counterpart of a specified datum feature. A datum is the origin from which the location or geometric characteristics of features of a part are established.

DBMS. See: Database Management System.

DC. DC, which stands for *direct current*, refers to a type of electrical current which is continuous, as opposed to AC *alternating current* which alternates, usually at 60Hz (cycles per second). Direct current is what comes from batteries and capacitors. Direct current cannot be conducted efficiently over long distances, which is why high-voltage alternating current is what is transmitted by utility companies. Computer logic uses DC to power logic circuits. Typically, motors, heaters, and lights are the only major devices which directly use AC power. Most appliances and devices which plug into AC outlets convert the power into DC before actually using the voltage and current within the device. One problem which AC power has is that AC essentially cannot be stored such as in a battery or capacitor. One advantage of using AC power is in variable speed AC motors where the frequency of the alternating current can be varied to control the speed, which allows the supply voltage and current to remain normal and constant, which is less likely to burn out the motor when driving it at very low speeds. For instance, at one extreme of 0Hz, or DC, the motor would be stationary, but electrically locked so it wouldn't be driven by the load.

Also see: Hertz.

DCC. See: Direct Computer Control.

DCE. See: Data Communications Equipment.

DC motor. A DC motor uses direct current to power the rotation of the motor. A DC motor has a unique advantage over an AC motor in that to reverse the direction, one must simply reverse the positive and negative leads on the motor. Also, to vary the speed, a simple rheostat or potentiometer can vary the speed, although to cut the voltage isn't always as desirable as other methods since the available torque is a direct result of the voltage applied to the motor. DC motors can be built with two different types of magnets, electromagnets, and permanent magnets. The permanent magnet or PM electric motors are much heavier than electromagnet electric motors, since the magnetic field strength is directly proportional to the physical size of the magnet although certain materials such as samarium cobalt have considerably greater magnetic strength per unit size than ferrite core magnets. DC motors are used in many devices which are part of computer systems and near digital circuitry since having alternating current near these types of circuits could cause interference. DC motors are preferred for miniature applications where the smallest possible size is a primary objective.

Also see: AC Motor, DC Servomotor.

DC Servomotor. A DC servomotor is a combination of a DC electric motor and a servospeed control system. The DC electric motor may be either brush type or brushless and can have an electromagnet or a permanent magnet. The servocontrol system can use either transistors or thyristors (SCRs), although SCRs can cause problems during turn-off. DC servomotors have a significantly different torque-speed characteristic than variable speed AC motors. An AC motor's useful torque operating range lies in the region between 90% and 100% of synchronous speed (synchronous speed depends upon the excitation frequency). However, a DC servomotor has a linear relationship between shaft velocity and torque, which allows it to be used at any speed up to 100% of rated speed.

Industrial robots and numerical control machines usually function best using DC servomotors of the permanent magnet, brushless type. By being brushless with the windings as the stator, heat is dissipated more easily. Using a permanent magnet eliminates the need for sliprings in the rotor assembly.

Also see: DC Motor, Industrial Robots.

DCTL. See: Direct-coupled Transistor Logic.

DDA. See: Digital Differential Analyzer.

DDC. See: Direct Digital Control.

Dead Band. A dead band is the range of values through which input can be varied to the servo portion without initiating a response from a machine tool. Generally if the dead band is narrow, good response can be obtained from the machine tool system combination. Also known as a dead zone.

Debug. To debug, or debugging, is the act of finding the errors (or bugs) in a computer program and removing or correcting them. Some computer languages include debuggers, which are software routines which analyze the source code which has been written and indicates problems such as syntax errors, improperly constructed programming statements, and other errors which would cause the program to not execute properly. A well known debugger for

C source code is the *csd* (C source debugger) from Mark Williams. Most debuggers available today are symbolic debuggers. One problem when trying to debug compiled programs is that the program must sometimes be compiled and run to be able to debug the functioning of the program, but the bugs relate to problems in the source code and a direct indication of where the source is incorrect is difficult to ascertain from the output of the object code (compiled source code).

Also see: Bug, Computer Languages, Programming, Source Code.

Deburring. Deburring is the process of removing burrs by any of several techniques. Unless otherwise indicated, the Class III definition is normally implied. *Class I:* Complete removal of all burrs when viewed without magnification. *Class II:* Removal of all burrs except those of a specified or given size. *Class III:* Removal of all loose fragments and burrs so that no sharpness or loose particles are detected by an inspection method specified.

The promise of improved quality and reduced costs has lured many companies to the use of industrial robots in their deburring operations. In 1983, one of every 100 to 200 industrial robots was involved in either deburring or fettling. *Fettling* is the process of removing sprues, flash, risers, gates, and runners from castings, and of smoothing parting lines and wide areas of flash.

Robotic deburring is usually applied for long-running parts in a wide range of production rates. Unlike other processes in deburring, robotic deburring requires a considerable amount of planning to produce optimum results.

Deburring expert LaRoux K. Gillespie in his *Robotic Deburring Handbook* listed 12 considerations to be resolved before a robotic system can be finalized.

1. Is the process a deburring or a fettling operations?
2. Should the robot handle parts or tools?
3. Will programming be performed on or off-line?
4. What edge (or surface) condition is required?
5. What is the initial burr size or edge condition?
6. What geometric features must be deburred?
7. How consistent are the edge locations?
8. What cutting tools should be used?
9. What robot characteristics are necessary?
10. What safety requirements must be satisfied?
11. What are the alternatives to robotic deburring?
12. What are the economics of robotic deburring?

In addition, it is essential that burr sizes are maintained sufficiently repeatable that the automated process will remove them.

Gillespie also noted: "The inherent limitations of robotic design are often subtle. As an example, many of the radial or polar motion robots cannot move in a straight line. Attempts to approximate straight lines on precision castings require memories far in excess of conventional capabilities. At least one radial motion robot has a wrist that allows it to move in a straight line; unfortunately, it cannot yet move in circular motions without significantly extended memory. SCARA robots have obvious limitations of too-limited vertical travel for typical part deburring, but they may have application on shallow parts. For successful deburring, robots should have the following characteristics:

- Accuracy;
- Repeatability;
- Continuous path capability;
- Easy and quick tool or spindle change;
- Machine rigidity;
- Continuous operation between part and tool;
- Low inertia; and
- Self-calibration ability.

"The following are also desirable features: Off-line programming and translating movements to similar features on the part."

LaRoux Gillespie also noted: "Before robots are selected for deburring, potential users should review available capabilities of each of the 38 basic deburring processes. Robots are not the

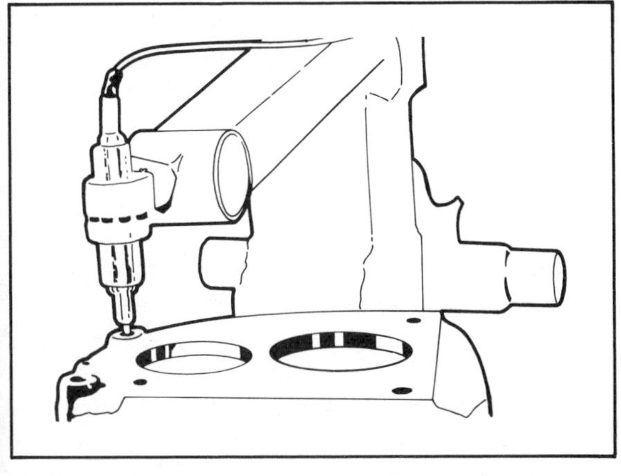

A robotic arm with live spindle for deburring gear box cases.

only answer to any deburring processes. They are a single answer to some needs. In addition, considerable savings result from users practicing burr prevention, burr minimization, and by closely evaluating deburring standards. In many cases, the most appropriate solution is to provide those performing manual deburring with the tools, training, and instruction necessary for them to do a better and faster job.''

It must be noted that the numerical control (NC) machine producing the part may also prove effective in deburring the part. In many cases, no additional fixturing, floor space or deburring machines are needed with NC deburring.

Considerations in NC deburring:

1. Rotary burrs, by their design, should be used at high velocities and low forces. In many cases, the recommended cutting tool velocity cannot be reached by the NC spindles.
2. Better control of tool changes is afforded when machining and deburring are on the same setup.
3. Collets or workholding fixtures may become clogged with fine abrasive particles if deburring and machining are done on the same setup.
4. Sticky machine way surfaces may result if coolants react with the rubber in some abrasive-filled rubber products.
5. During machining, thin flanges may become distorted. The distortion may cause the chamfering tools to not even touch edges.

In the future, automation (most specifically industrial robots) should have more appropriate tools. In the inspection area, automated systems will detect when burrs have not fully been removed. Sensing devices should indicate when a cutter change is needed.

Laser deburring stations are in the research stage although some robot-held laser systems are currently in production. In the future, robotic machining systems should exist that deburr one part face while a machining center machines the second face.

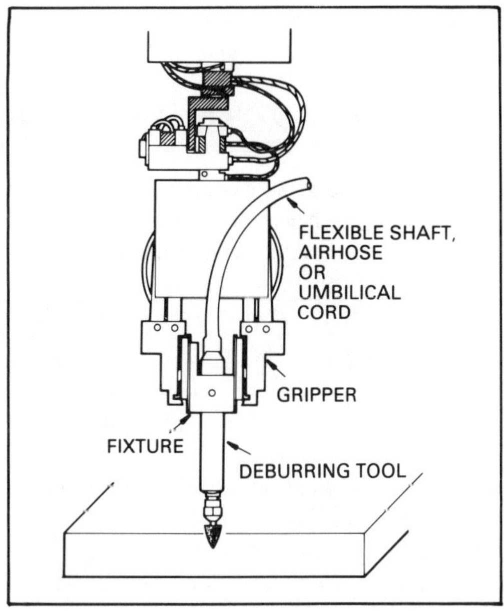

Deburring tool mounted in a robot gripper.

Also see: Fettling, Finishing, Industrial Robots, Lasers, Numerical Control.

Decimal Code. A decimal code, such as the conventional number system, in which each allowable position has one of 10 possible states. Also referred to as coded-decimal code.

Decision Support System. A set of computer hardware and software with appropriate communications to provide a manager or staff person with personalized computing capabilities by which to gather and analyze data.

Also see: Computers, Hardware, Software.

Decoder. A decoder is a circuit arrangement enabling the translation of data from coded form to a more easily recognized form without significant loss or information.

Dedicated Computer. A dedicated computer is assigned exclusively to one machine, application, or small group of machines.

Also see: Computers.

Default. Default refers to the state, condition, or value of some parameter if nothing is done to specifically cause that item to be something in particular. In some instances, default values may be random values which are unpredictable. In other situations a default state may always be the same predictable condition.

In computers, default conditions and values may refer to whether a circuit may be a one or zero, or indeterminate, (which means the voltage is floating (fluctuating) between 0 volts and 5 volts, but can't be decisively determined).

The notion of default values and conditions is a useful productivity tool when doing tasks such as data entry or other activities which have repetitive tasks. Fields which must have the identical data entered into them for each record can be programmed to have a default value unless the operator specifically changes the field. Also, default conditions are useful for safety situations to ensure a safe response is automatically presented, unless the information is specifically changed. A typical example of default values is when a photo copier automatically changes the number of copies back to one after a certain amount of time has elapsed with no activity.

Also see: Computers.

Degree of Freedom. Degree of freedom is a term which is used to describe the ability of multiple bar linkages and other objects to move in free space. Each direction (axis) that the assembly can move both forward and backward is called a *degree of freedom*. If a multiple-bar linkage has more than one joint which moves about the same axis, it still only represents one degree of freedom. Six degrees of freedom describe the maximum movement capable, unless a person considers time as a degree of freedom. The six degrees of freedom are lateral movement in the X, Y, and Z directions and rotation about the X, Y, and Z axes. An actuator is required for each degree of freedom, either a linear or rotary actuator. However, multiple actuators don't necessarily mean multiple degrees of freedom. The way linear and rotary joints are combined with each other has a determination on the maximum number of degrees of freedom.

Also see: Actuator, Industrial Robots.

Delete Character. A delete character is used to obliterate erroneous or unwanted characters; for example, on a punched tape, a delete character would consist of perforations superimposed on all other perforations rendering them indecipherable.

Demodulation. Demodulation is undoing or reversing the effects of modulation, that is removing the intelligence-bearing signal from a modulated carrier or reconstituting the signal that performed the modulation.

Also see: Modem, Modulation.

Demographics. Demographics is the statistical study of the human population. The revolution in graphics technology also has affected such disciplines as communications, agriculture and the creating of charts and graphs for demographics and scientific analysis purposes. Graphics systems have been implemented to perform analysis and graphics work for many agencies and companies, including the Bureau of Census, public school systems, and the National Bureau of Economic Research. Researchers have learned to adapt empirical and judgemental data to graphic form and to the computational capabilities of the graphics computer, thus taking advantage of the best of both forms.

Density. Density is the concentration of a group of like items within a particular space. The recording density of data on magnetic tape is expressed in bits-per-inch of digital data. The recording density of data on floppy diskettes is referred to as single, double, or quad density.

Naturally, the maximum density of a substance is limited by the particle size of that substance. Therefore, the newest advances in magnetic media storage technology include reducing the particle size of the magnetic recording substance. Instead of the long, slender tube-like particles, smaller hexagonal-shaped particles are used.

Density is also used when referring to the number of components within a fixed area of a circuit board or the number of pins on an IC or PGA (pin grid array). The density of a chip is determined by the quantity of logic circuitry within the IC itself while the density of actual logic circuitry within a particular space is determined by the relative density of the chips themselves.

Also see: BPI (bits per inch), Computers, Recording Density.

Delimiter. A delimiter is a flag that separates and organizes items of data. It also is a character (space, splash, asterisk, etc.) that provides separation of data segments within a continuous data string.

Depreciation. Depreciation is a decrease or loss in value. This can occur because of age, wear, or other causes. Accounting practices allow a business or individual to take deductions for certain items as expensed purchases. Items which cannot be expensed, such as real estate, computer equipment, machine tools, automobiles, and the like, are capitalized when purchased. The basis (purchase price) can be recovered by taking deductions for depreciation when filing income taxes. It is charged off against income as a business expense. This reduction in value is carefully calculated for all types of machinery, equipment, buildings, and other assets for business planning and for tax purposes.

There are several methods for calculating depreciation on equipment or other business assets.

Straight line depreciation involves assigning a given life to a piece of equipment and reducing its value by a set amount each year until the entire cost (less salvage value) is written off. For example, a $15,000 lathe could be depreciated in seven years and end up with a $1,000 salvage value. The owner would write off $2,000 each year for seven years.

With *accelerated depreciation*, there are several methods of increasing the deductions or write-offs during the early years, as follows:

- Double deduction for the first year, then straight line deductions thereafter.
- An extra 20% for the first year, then straight line deductions thereafter.
- Declining balance method, charging off a given percent of the remaining value each year. For example, 30% of the value remaining from the previous year would be written off each year.
- Sum-of-the-digits, charging off diminishing fractions of the value each year. The fraction is arrived at by assigning successive numbers to each year and totaling the digits. Thus, for a seven-year depreciation life the years are numbered 1, 2, 3, 4, 5, 6, and 7, which totals 28. On the first year 7/28 of the purchase price is written off. On the second year 6/28, the third 5/28, and so on.

While the simplest method of depreciation is straight line depreciation, the other methods of accelerated depreciation are very popular because they permit a capital asset to be depreciated faster in the early years of the investment.

The Tax Reform Act of 1986 generally changed the accelerated cost recovery system (ACRS) of depreciation for tangible property placed into service after 1986. The new system is referred to as the modified accelerated cost recovery system (MACRS) to distinguish it from ACRS. There are now three systems involved in the computation of depreciation. The depreciation system that applies to a particular piece of property is determined by the type of property and when it was placed in service. For tangible property, you use:

- MACRS if placed in service after 1986;
- ACRS if placed in service after 1981 but before 1987; or
- Straight line or an accelerated method of depreciation, such as the declining balance method, if placed in service before 1981.

No matter what method of depreciation is used, the total depreciation taken cannot total more than the basis (original cost) of the item. Within the different types of depreciation methods, there are different numbers of years used to recover the total cost of the item (become fully depreciated), depending on the type of property. For example, under MACRS, there can be 3-, 5-, 7-, 10-, 15-, and 20-year property. Under ACRS, there can be 3-, 5-, 7-, 10-, 15-, 18-, and 19-year property. Usually, a person can elect to use a longer recovery period than the timeframe listed for the property. For example, under ACRS, three-year property can be depreciated over 3, 5, or 12 years. The IRS (Internal Revenue Service) provides listings of the general classes of property and the allowable recovery periods.

There are other considerations that should not be forgotten when depreciation plays a part in planning. Depreciation can be allowed or allowable.

Allowed depreciation: The actual amount of depreciation claimed as deductions from income if it is within accepted accounting practices that are recognized by the IRS.

Allowable depreciation: In the absence of any actual depreciation claimed, the amount that *could have been* claimed is *added* to the value carried on the books until depreciation is claimed at a later time.

The personal computer has proven to be a useful tool for calculating depreciation and providing comparisons between depreciation methods. The printout in the figure on the next page is the result of a program that takes 3605 bytes. It was developed by J.E. Nicks for his publications *BASIC Programming Solutions for Manufacturing*.

Design. See: Auto-interactive Design, CAD/CAM, Computer Aided Design, Electronic Design Interchange Format, Software, Tooling.

Desktop Publishing. Desktop Publishing is a relatively new but potentially significant application of personal computer technology. It is called "desktop publishing" because most of the tools it requires can fit on top of a desk. The basic requirements are a personal computer (either a Macintosh or IBM PC or PC-compatible), page layout software, and a page printer such as a laser printer or a typesetting machine. UNIX-based workstations are considered to be professional typesetting systems, not desktop publishing.

The software required for desktop publishing includes word processing, graphics and page layout software. The word processor is used for basic text generation, editing, and merging. The graphics software allows the creation of illustrations and inclusion of prepared graphics called clip art. The page layout software permits the combination of the text and graphics and the entire page layout, including physical placement, font selection, and usually a WYSIWYG (What You See Is What You Get) preview of the finished document.

Also see: Laser Printer, Personal Computer, Postscript.

Deutsche Industrial Norman (DIN). The Deutsche Industrial Norman or German Institute for Norms is an official agency of the German

```
DEPRECIATION  SCHEDULE

CAPITAL EXPENSE IS $ 50000
FREIGHT IS $ 1000
INSTALLATION EXPENSE IS $ 3000
TOOLING EXPENSE IS $ 6000
OTHER EXPENSE IS ENGINEERING
DOLLARS FOR ENGINEERING ARE $ 5000

THE TOTAL EXPENSE IS $ 65000

ASSET LIFE IS 8 YEARS

STRAIGHT  LINE  DEPRECIATION

PROJECT  NAME          N/C  LATHE

YEAR           BEGINNING VALUE DEPRECIATION    ENDING VALUE

 1              65000          7250            57750
 2              57750          7250            50500
 3              50500          7250            43250
 4              43250          7250            36000
 5              36000          7250            28750
 6              28750          7250            21500
 7              21500          7250            14250
 8              14250          7250            7000
DOUBLE  DECLINING  BALANCE

PROJECT  NAME          N/C  LATHE

YEAR           BEGINNING VALUE DEPRECIATION    ENDING VALUE

 1              65000          16250           48750
 2              48750          12187.5         36562.5
 3              36562.5        9140.63         27421.9
 4              27421.9        6855.47         20566.4
 5              20566.4        5141.6          15424.8
 6              15424.8        3856.2          11568.6
 7              11568.6        2892.15         8676.45
 8              8676.45        2169.11         6507.34

THE DDB FACTOR USED IS 2
```

A computer-generated depreciation schedule.

government that approves, consolidates, and in some cases establishes national standards.

Also see: ANSI, Standards.

Device Driver. A device driver is a software program which controls a logical or physical device. A device driver is usually a software subroutine which provides the low-level control and interface to the device. For example, a device driver for a disk storage device would handle the functions such as the buffering of the data, the location of the required tracks and sectors, and any read/write verification. The low-level interaction with the servomechanism software is part of the device driver function. Once a device driver is written for a particular device, that portion of software functionality doesn't have to be rewritten for another application. Device drivers can be reused until either the device is changed, or a more desirable device driver is written. Certain devices such as disk drives, mice, and keyboards have become standards which have corresponding device drivers that can be used whenever a person includes that particular device in a new product. Using standard device drivers helps create uniformity in regard to the operation of certain devices.

Also see: Computers, Software.

DFW. See: Diffusion Welding.

Diagnostic Routine. A diagnostic routine is a maintenance test of key NC system components, performed by use of a special programmed tape and/or electronic instruments, to discover failure or potential failure of a machine element as well as the location of the failure. Also known as a diagnostic check, diagnostic subroutine, diagnostic test, or error detection routine.

Also see: Numerical Control.

Die Casting. A high production casting process, die casting is used for small-to-medium-sized parts. In die casting, molten metal is fed or injected into a die made of steel or high temperature alloy that often contains the impressions of several identical parts.

Industrial robots that have been on the market for any length of time have been tried in die casting operations. This application is a real test of the equipment's reliability, since the die casting environment is very hostile.

Also see: Industrial Robots.

Die Design. Die design is a critical aspect of the extrusion process and embodies both science and art. Optimum design is influenced by many factors, including the circle size of the shape to be produced, the maximum and minimum wall thickness, the press capacity, the length of the runout table, the stretcher capacity, the tool-stacking limitations, an understanding of the properties and characteristics of the metal to be extruded, and the press operating procedures and maintenance.

Computers are being used by some extruders to design and manufacture (with CNC machines) dies and to select process variables such as extruding speed and billet temperatures. Software employed is based primarily upon an analysis of metal flow. Various design stages are displayed on a CRT screen, and the designer interacts with a computer to change or modify the design, based upon experience.

Since all metals shrink upon cooling after hot extrusion, a shrinkage allowance must be provided in designing the dies. Deformation of the die under high pressures and expansion resulting from the high temperatures must also be considered in die design.

Another important consideration is the tendency for metal to flow faster through a larger opening than a smaller one. This must be compensated for in designing dies for use in extruding certain sections. For example, when a section to be extruded has both a thick wall and a thin wall, various means are employed to retard metal flow through the thick section and increase the flow rate through the thin section of the die.

Fine adjustments to the die for correcting or changing the rates of metal flow are made by varying the length of the land, also called the bearing, directly behind the die opening. By decreasing the length of the bearing, the rate of flow is increased; increasing the bearing length reduces the flow rate.

bearing, directly behind the die opening. By decreasing the length of the bearing, the rate of flow is increased; increasing the bearing length reduces the flow rate.

The geometry of the die aperture at the front and back of the bearing surface is known as the choke and relief respectively. If the die designer expects to encounter difficulty in filling sharp corners or completing thin sections of the extruded product, a choke may be provided on certain portions of the bearing surface. This slows the rate of metal flow and consequently fills the die aperture. Increasing the amount of back relief at the exit side of the bearing surface increases the rate of metal flow.

For the hot extrusion of some materials, such as brass, bronze, and other soft metals, the dummy block is made smaller in diameter than the billet. In extruding, no lubrication is provided between the bore of the container liner and the outer surface of the billet. As a result, friction prevents the outer surface of the billet from sliding and the undesirable skin of the billet is left in the container as the dummy block shears the metal during its forward stroke. An additional press stroke is required to remove this retained metal before the next billet can be charged into the container.

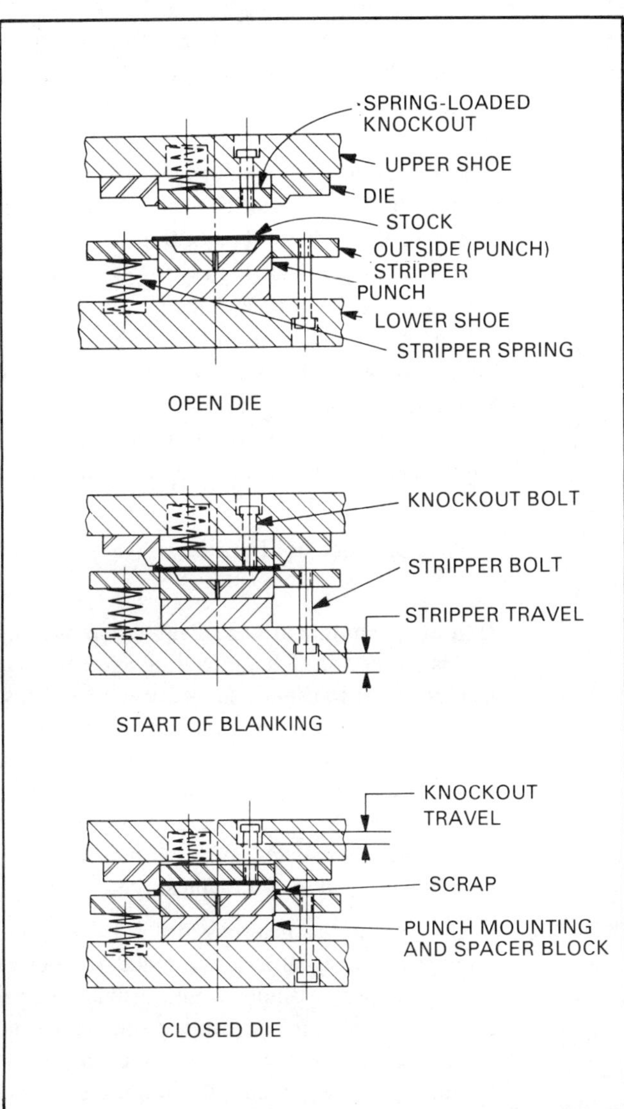

Return-blank blanking dies.

Dies. The term *die* is a generic term denoting the entire press tooling used to cut or form material. This word also denotes just the female half of the press tool. The female die steel works in opposition of the punch steel.

Blanking dies cut plain flat pieces of stock often without perforations, while *drawing dies* produce cylindrical, or cup-shaped work from flat material. The die is a fixture on the table, or sometimes on the ram of the press, and usually has a stripper plate or pins to pull the work off the punch as it ascends. The die is invariably wedge-shaped to provide clearance for the blank passing through the die. The stripper plate may also be a guide for the stock, to locate it directly under the punch as the operator of the machine feeds it forward.

More elaborate dies allow for a triple action of blanking, drawing or embossing; or where a large number of similar parts are required, gang or multiple dies with many similar openings allow duplicate punches to cut several blanks at each stroke.

A *die block* is a block (usually of heat-treated steel into which desired impressions are machined or sunk and from which closed-die forgings are produced on hammers or presses. Die blocks are usually used in pairs, with part of the impression in one of the blocks and the balance of the impression in the other.

A *die match* is the condition in which dies, after having been set up in the forging equipment, are in proper alignment relative to each other.

A *die set* is the assembly of the upper and lower die shoes (punch and dieholders), which usually includes the guide pins, guide pin bushings, and heel blocks. This assembly, which takes many forms, shapes, and sizes, is frequently purchased as a commercially available unit.

Die shoes are the upper and lower plates or castings which make up a die set (punch and die holder). Die shoes are also a plate or block upon which a dieholder is mounted, functioning primarily as a base for the complete die assembly. It is bolted or clamped to the bolster plate or the face of the press slide.

Differential Analyzer. A differential analyzer is an analog computer designed to solve differential equations by means of interconnected integrators.

Diffusion Welding (DFW). Diffusion welding is a welding process that uses solid-state diffusion to metallurgically bond joint surfaces that are held together under moderate pressure while being heated, generally in a controlled-atmosphere furnace.

Also see: Welding.

Digital. Digital refers to the representation of data or physical quantities by means of digits or discrete quantities. Strings of light pulses in an optical fiber are an example of a digital quantity. Digital products usually are much more impervious to noise and interference than analog type products. Sometimes digital devices don't require the adjustments or temperature compensations which analog devices do. This is usually due to the fact that digital quantities have greater variation between the mark and space or on and off values. The difference between the two states which make up digital values allows for greater fluctuation in the sensitivity and linearity than with analog products. For example, with TTL (transistor-to-transistor logic), a value of one is when the voltage is greater than 2 1/2 volts and a zero value is when the voltage is less than 1/2 volt. As you can see, the change of state between high (1) voltage and low voltage has considerable variance. With an analog circuit, any variance would change the perceived output value.

Also see: Analog, Data.

Digital Communications. Digital communication is the use of digital encoding and decoding techniques to transmit analog communications. Digital communication is far superior to analog because it is more error-free and is devoid of all periodic and sporadic noise and interference. A digital signal is reconstructed at the receiving end, which eliminates most of the degrading problems of analog communication.

In 1937, it was determined that the periodic sampling of a signal at a rate twice the highest frequency in the sample would provide all the components required to capture the signal and reconstruct it at a later time. A voice-grade line would, therefore, require a minimum sampling rate of 6,600 times a second since the voice grade band is about 300 to 3300 Hz. Channels actually occupy a 4 kHz bandwidth, so the Bell T1 carrier system uses a rate of 8,000 samples per second.

Common carriers transmit 24 voice channels together with time division multiplexing techniques. Bell's T1 carrier system provides the multiplexing by sampling the 24 channels at a combined rate of 192,000 times per second. An extra bit is transmitted at the beginning of each frame to function as a frame synchronization bit. This synch bit alternates values in each consecutive frame.

Digital Computer. A digital computer is a computer in which discrete representation of data is mainly used. All the data manipulated by the computer as well as the input and output must be in digital representation for a digital computer. Traditional von Neuman computing architecture is typical digital computing. A digital computer uses the digital transistor (PNP or NPN) as its basic building block. A digital computer can be constructed from basic gate array cells, or it can be an ASIC (Application Specific Integrated Circuit) type device. A digital computer, due to its use of ones and zeros or binary math, does not generate a build up of error as does an analog circuit. Digital calculations are precise, except when the origin of the input or the final form of the output must be in analog form. In such a circumstance, there will be some amount of error as the conversion is made from analog to digital or from digital to analog.

Also see: Analog Computer, ASIC, Computers, Digital Communications, Gate Array.

Digital Control. Digital control is the use of digital signals and communications along with a digital computer to provide the control for a device or system. Digital control is characterized by the data being composed of discrete bits instead of continuously variable voltages and currents such as is common to process control systems or something as elementary as the volume control on an audio amplifier, which represent analog control systems. Digital control uses discrete data as input and output to the control system. Digital control exhibits all the usual benefits of digital technology as opposed to analog technology. Being impervious to noise, drift, temperature, and no calibration requirements are some of the attributes of digital control. The simplest digital controllers are more complex and more expensive than the simplest analog controllers for the same basic job, such as the attenuation of a voltage or current value. But as the task becomes more complex, the digital controller becomes much more capable and cost-effective.

Also see: Computers, Controller, Digital Computer.

Digital Data. Digital data is data supplied in discrete, discontinuous form as digits, quantized pulses, or other coding elements.

Also see: Data.

Digital Differential Analyzer (DDA). A digital differential analyzer is a computer or logic circuit which uses numbers to represent analog quantities when solving differential equations.

Also see: Computers.

Digital Image. A digital image is an image that is composed of digitized information. The origin of the image may be analog creation or digital pixel creation, nevertheless, at this stage the image has been digitized. The purpose of digitizing an image is that it can then be stored in a computer as well as transmitted over digital data communication lines. A digital image can be edited and manipulated by a computer, to be expanded, reduced, rotated and cropped. If the resolution of a digital image is great enough, the viewer cannot discern the individual dots which make up the picture and it may appear to be a video image. One large dilemma with digital images is that because they are bit-mapped pixel images, they require a large amount of computer memory to store a single image, especially if the image is in color. With the continuing improvement in price performance of computers though, the availability and cost of memory is becoming much less of a concern.

Digital Signal. A digital signal is a nominally discontinuous electrical signal that changes from one state to another in discrete steps. A signal that is discontinuous time-wise, is also discrete, and can assume a limited set of values. The electrical signal could be changed in its amplitude or polarity. Analog signals may be converted to digital signals by sampling and quantizing.

Digital modulation is the process of varying one or more parameters of a carrier wave as a function of two or more finite and discrete states of a signal. An example is, the ternary signal coding modulation technique in which two independent bipolar coding sequences are interleaved. The digital modulation power spectrum nulls at a direct current (DC) level and again at one-half the bit rate. This requires less bandwidth than nonreturn-to-zero (NRZ) signals and provides better low-frequency response than return-to-zero (RZ) signals. The main disadvantage in using digital modulation is its low tolerance of poor signal-to-noise conditions, that is, low signal-to-noise ratios. Digital modulation can also be accomplished by coding the data, delaying it two-bit intervals, and then adding.

Digital-to-Analog. Digital-to-analog or D/A refers to the process of the conversion of a digital input signal or waveform to an analog output signal representing equivalent information. It converts the electrical pulses that represent digital data to analog voltages or electric currents thereby converting digital data (representation) to analog data (representation).

A common usage of digital-to-analog conversion is modern telephone transmission. Digitizing voice data allows the data to be transmitted over long lines and then converted at the receiving end back into analog data. This allows a high quality of data transmission, without excessive background noise, by using digital data transmission, digital switches, and digital transmission media. Digitizing voice data also allows multiple data streams to be transmitted concurrently over the same channel and then reconstructed at the receiving end and separated into the original individual data streams.

Operator output panels and other types of analog meters which are driven by digital computers require digital-to-analog conversion of the signals which are displayed by them.

Also see: Analog-to-Digital, Digitizer.

Digital Tube. A digital tube is a CRT that has only alphanumeric capabilities, with no graphics.

Digitizer. A digitizer is an input device that is capable of translating movement and point locations into digitized bit-mapped information. A digitizer is usually used to translate a pictorial drawing or line drawing into information which a CAD program can understand. Digitizers are usually tablets with a stylus or mouse type device to accurately move about the area to be digitized.

A digitizer may be required when transferring a graphical picture from one computer to another which has incompatible file definitions. Sometimes, a work of art which was created by something other than a computer may be input into a computer, and until the recent introduction of image scanners, digitizers were the only method of inputting nontext images. The digital scanners now on the market are capable of digitizing a page in a few seconds and appropriate software can translate the information into the proper format for a particular CAD program.

DIN. See: Deutsche Industrial Norman.

DINC. See: Distributed Numerical Control.

Diode. A diode is an electrical device such as an electron tube or semiconductor which allows current to pass in one direction only.

Diode Transistor Logic (DTL). A diode transistor logic is an integrated circuit logic employing diodes with transistors used as inverting amplifiers.

DIP. See: Dual-in-line Package.

Direct Access. Direct access is the facility to obtain data from storage devices or to enter data into a storage device in such a way that the process depends only on the location of the data and not on a reference to data previously accessed.

Direct Access Memory. Direct access memory is the capability of directly transferring data to or from memory thereby minimizing interruptions created by program-controlled data transfers.

Also see: Computers.

Direct Computer Control. Direct computer control is when a computer is used directly as the controller for a machine or device. Direct computer control allows the capabilities of the computer to be applied to the control algorithms, instead of having to use an electrical controller or an analog type control circuitry or even something as rudimentary as a limit switch. In a direct computer control system, any support device such as a CAD system or simulation and modeling system which is utilized to aid in the operation of the system is able to communicate directly with the controller of the machine or process. Some machines have imbedded microprocessors which means a microprocessor is integrated into the machine as a basic component and all functions are controlled by the imbedded microprocessor. Using computers to control machines allows greater flexibility and control capabilities which were unattainable when using less sophisticated control systems.

Also see: Cell Controller, Computers.

Direct-coupled Transistor Logic (DCTL). A direct-coupled transistor logic is an integrated circuit logic employing only resistors and transistors as active circuit elements.

Direct Digital Control (DDC). Direct digital control is when a machine is controlled directly by a digital computer (as opposed to the commands of the computer being translated by control device other than a digital computer). A numerical control machine uses numerical representations of motions and locations, created by a CAD system or some form of computerized programming workstation, but are translated into a numerical syntax for use by the machine. Direct digital control doesn't require the translation into numerical or analog signals and values. Usually, machines which can be directly controlled digitally have microprocessor controllers inside which accept the digital program and carry out the commands of the control program. Direct digital control provides more responsive control capabilities since machine control can be immediately influenced from a computer program. Some items such as digital encoders, digital cameras and test equipment with digital output enhance the ability to provide direct digital control.

Also see: Direct Computer Control.

Direct Material. Direct material is the cost of all components included in the end product being produced. To be considered as direct material, the components or raw materials must be a permanent part of the end product being manufactured. Examples of material costs that are not direct include raw material from which tooling is fabricated, test equipment, and packaging materials. Direct material is a variable manufacturing cost.

Direct Numerical Control (DNC). Direct numerical control connects numerically controlled machines directly to a programming or higher level computer or control system. Instead of loading a punched tape or floppy disk to download the program into the numerical control machine, a computer data communications link allows the control program data to be downloaded directly into the machine. Direct numerical control sometimes connects machines in remote locations to a central programming system. DNC allows the faster downloading of part program data for flexible manufacturing cells and decreases the setup time when changing from one part program to another. DNC eliminates the need for handling punched tape and the need to use punched tape readers, which sometimes do not function reliably. With the connectivity of current computer systems and the power of personal computer work stations and distributed computing, DNC allows programs to be developed wherever the programmer desires, and the resulting program to be easily downloaded to the CNC machine.

History. The original DNC concept was forwarded (circa 1965) as a means of reducing NC control costs through use of one powerful controller for a group of machines, rather than a separate controller for each machine tool. The cost of electronic control equipment was much higher in the early 1960's than it is today; so high that the DNC concept was driven by a need to reduce control cost.

Discussion. Proponents of the DNC concept believed that punched paper tape and tape readers at the machine tool could be completely eliminated by driving the NC machines directly from the memory of a central computer. Because NC machines of the day typically were not equipped with memory, the central computer would drive the machines in real time—that is, the computer would send NC data to the machine tools in sequence or pulse form during the actual machining operation. By operating in this manner, it was expected that maintenance costs of tape readers at the machine tool could be eliminated and that input errors caused by improper operation of tape readers at the machine tool could be avoided. In addition, the original DNC concept promised simpler management of NC programs and an elimination of a need for costly libraries of punched paper tape. It was believed that by storing NC programs and monitoring NC machines using a central computer, the optimization of NC programs and simulation of numerical control functions (verification of tool path using a plotter, for example) would be more easily performed. Also, it was thought that the collection and reporting of system operating data such as downtime, production, and maintenance information would be more easily accomplished.

The original concepts of DNC worked reasonably well in a few isolated applications; however, some of the promises of DNC never was realized on a broad scale. For example, the thought that DNC could eliminated the need for tape readers or other input devices at the machine tool proved unrealistic. Tape readers or other alternate input devices at the machine tool were found to be useful in the early DNC systems as backups to the computerized system. Such manual backup was required because occasional downtime of the central computer caused the entire DNC system to go down, sometimes idling a dozen or more expensive NC machine tools.

With the advent of computer numerical control came the availability of relatively inexpensive computer memory at the machine tool. The nature of the DNC concept was altered as a consequence. With computer memory at the machine tool, it is no longer required that the CNC machines of a DNC network be driven in real time. Instead, NC programs can be downloaded in total from the memory of the central computer to the memory of the computer at the machine tool; the connection between the central computer and the individual machine tools in the system need only be maintained for a short period of time—the time necessary to transmit the NC program. In this way, the uptime of individual machine tools is less dependent upon the uptime of the central computer and, because the machines are not driven in real time by the central computer, program editing at the machine is made much easier. This concept, known as distributed numerical control, is growing in usage. In fact, the acronym DNC, originally defined as direct numerical control, is now used by many experts in the controls industries to describe distributed numerical control.

Experience has shown that NC part programs which are to be stored and distributed via a DNC network should contain certain introductory information. Information of this type should include the part number of the workpiece to be machined, the drawing number, and any special processing-related information such as a list of fixtures or clamping devices that will be required to facilitate the machining of the workpiece. Special machining instructions should also be listed according to a standard format and a list should be included of the machine tools on which the part can be processed, including machine numbers and descriptions.

Part programs of a DNC system should be protected via an automated security system. Each part program should contain an authorization code at the beginning of the program which can be read by the central computer. Such a code can be employed, if necessary, to block the transfer of an NC program to a machine tool that is unauthorized for machining of the part. In this way, the use of specific machine tools for workpieces can be easily controlled.

In a DNC system, the part programs should be managed and classified in a logical fashion. Some of the classification breakdowns that have proven useful include access and reporting by program number or program name (workpiece name), machine tool authorization codes, date of program preparation, and date of change.

Experiences of users familiar with the operations of a DNC system indicate that the call-up of NC programs from the system should be sufficiently simple so that users who are not data processing professionals can call up programs without difficulty. Experience has also shown that a DNC system should be capable of downloading the same program to different machine tools simultaneously. This capability is particularly important and may have significant impact on the operating efficiency of a flexible production system, for example.

The system employed for data transmission is the heart of any DNC installation. System reliability and performance is often dictated by the viability of the data transmission network. Operating performance of the communication system in a DNC network is optimal if it does not present constraints causing one or both of the following conditions to occur:

- Machine is idled while waiting for transmission of data.
- Operator's time is wasted while waiting for responses from the DNC computer.

The operating baud rate or even the effective throughput rate is not of much concern when a part program is being downloaded if, for example, the CNC can accept the data to memory while another part program is running to produce parts.

The differences amount the various CNC machines in the DNC system must be accounted for in the design of the system. For example, some CNC units may have limited internal storage. In cases in which the NC program length is larger than the internal memory capacity at the machine tool, the DNC system must have the capability to automatically download only portions of the program that are within the capacity of the machine tool. This capability should be built-in from the start.

Often, NC programs are optimized by the machine tool operator during the first several production runs of a new part. An efficient DNC system should be capable of accepting optimized NC programs from the CNC machine tools in the network; however, a system should be established that prevents the revised program from being used in place of the original program before the changes are approved by the programmer.

The possibility of downtime of the central computer must be considered in the design of the DNC system. In many cases, short periods of downtime of the central computer will not stop production because current DNC systems generally do not drive machine tools in real time; a number of programs may be stored at the machine tool so that production can continue even when the central computer is down. In emergencies, NC programs can be downloaded to machine tools via a storage disk (a disk containing the programs for a day's production) and a portable disk reader. These and other provisions for emergency operation of the system are extremely important and should be considered in the design phase of any DNC system.

The printing of certain lists is an invaluable feature of DNC which should be designed to be compatible with current management style and reporting systems. The following are a few examples of the lists which can be printed using the capabilities of the central computer of a DNC system:

- Production schedules.
- Running times of programs.
- Tools required to machine a specific part.
- Instructions for the operator.
- NC programs contained on a disk.
- Block programs.
- Data on when each program was used last.

The ability of a DNC system to collect and report machine-related data is also important in some applications. Such data can be used to structure useful management reports. The following are examples of some data which can be collected using the central computer of a DNC system:

- Meantime between failures.
- Duration of downtime and causes.

- Machine utilization reports.
- Machine loading.

Applications. In general, two areas of application exist in which DNC has shown specific advantages. First, the DNC concept is often justifiable in applications that have large amounts of control information which must be managed, stored, and distributed—many NC programs or very complex programs. DNC facilitates the management of large numbers of NC programs and helps to sidestep the possibility of using the wrong NC program or using a program that is not the latest version. With DNC, lengthy NC programs can be loaded quickly, eliminating the costly nonproductive time often associated with the loading of complex programs via punched paper tape or other mechanical input media. The payoff is achieved in increased uptime and greater machine tool efficiency.

The DNC concept is also employed as the heart of the control system for so-called flexible production systems in which a number of numerically controlled machine tools are linked by means of electronic data communication and mechanical automation. Often employed to machine families of parts, such systems are equipped with a central computer which directs the flow of parts through the system and operates in a DNC mode, downloading NC programs to the member machine tools as required. In such systems, the central computer is also used to collect operating data.

Direct numerical control systems of the future are expected to perform more and more sophisticated tasks which will aid in production management and help increase machine tool utilization and productivity. In the coming years, the use of DNC will not be limited merely to tasks of monitoring but will play an ever-increasing role in controlling the manufacturing operation and optimizing production efficiency. Future DNC systems will be more fully integrated with data processing computer-aided design systems. More smaller NC shops will move to DNC in the years ahead as package systems continue to proliferate and further standards are developed.

Also see: Data, Numerical Control.

Discrete Components. Discrete components describes electrical components contained within an electrical assembly wherein each component usually performs a single logical function. Examples of these types of components are devices such as single transistors, capacitors, resistors, chokes, and similar devices. Discrete components have been replaced largely by integrated circuits, which combine multiple discrete function elements within a single device.

Discrete components are still widely utilized in various applications. Some instances of using discrete components are cases of when a single or few function elements are required, much less than the quantity provided by an integrated circuit (IC). Other situations which require discrete components are applications in which the voltages and currents to be handled by the circuitry are greater than the capability of integrated circuits.

The current designs and shapes of discrete components enable automated assembly by utilizing standard shapes and sizes.

Also see: Integrated Circuit, Printed Circuit Board.

Discrete Job. A discrete job is movement of a selected axis of an NC machine toolslide, etc., in a specific direction for a predetermined distance. Used for checking purposes.

Disk. A disk is a flat circular object which has a smooth magnetic coating on its surfaces. A disk can either be as flexible as the media used in a floppy disk, or it can be as rigid as the media used in a hard disk (Winchester disk). The disk in a hard disk drive can either be fixed or removable. A hard disk drive media must be operated in a sealed environment. Removable hard disk cartridges have sealing covers for when they are removed. When they are inserted into the drive, the system purges the air from inside the cartridge before allowing the head to approach the media. Larger capacity hard disk drives have multiple platters within the same unit to create greater storage capacity.

Also see: Computer, Density, Eraseable Optical Disk, Floppy Disk, Hard Disk, RAM Disk, Recording Density.

Diskette. See: Floppy Disk.

Display. See: Gas Plasma Display, Liquid Crystal Display, Refresh Tube Display, Resolution, Storage Tube Display, Touch-sensitive Display.

Distributed Numerical Control (DINC). Distributed numerical control uses a host computer to provide the parts program data to multiple numerically controlled machines. DNC or *direct numerical control* is when a host computer is directly connected to a numerically controlled machine instead of using punched tape. Distributed numerical control utilizes direct numerical control to control each machine. Distributed numerical control makes possible the direct control of machines by automatically generating programs using CAD/CAM systems.

In a DINC system, the host computer can download a program or group of programs, which then allows the machine or machine cell to function without continuous direct communication with the host computer. DINC allows greater flexibility in manufacturing systems.

Several well known products such as CADAM, CATIA, and APT provide the front-end intelligence for DINC. Other areas in addition to metal cutting and machining which use DINC are electronic circuit board assembly and electronic chip assembly.

Also see: Direct Numerical Control (DNC), Machine Cell, Numerical Control.

Distributed Processing. Distributed processing is when the computing and data processing is not done centrally on a single multiuser machine, but is done on multiple machines in different locations which are usually physically closer to the end users. Distributed computers allow the use of different types and capabilities of machines to better match the specific needs of the different users. Distributed computing allows a subgroup to have total control over the availability of a computing resource, while also allowing the computing resource to be available to share over a data network at their discretion. Usually distributed computing architectures utilize data networks to allow sharing of data and computing resources. The ability to network allows a distributed environment to not have to duplicate every support device at each of the distributed locations. Printers and file servers can be more centralized while the computing power is placed where needed and can easily be increased and added to as appropriate without disturbing the remainder of the existing environment.

DNC. See: Direct Numerical Control.

Document. Document is a collective terms for specifications, drawings, parts, lists, standards, and reports. Document refers to the information media for communicating ideas, descriptions, requirements, plans, and instructions related to a project, hardware or software.

Documentation. Documentation includes drawings, parts and wire lists, manuals, specifications, standards, reports, programs, and control information. Configuration and documentation are essentially the same thing. Configuration management is oriented toward change management. Documentation is oriented toward communicating the technical description of a product. *Configuration* is recognized to be the formal and disciplined approach to *documentation* and a major constitutent of the CIM Enterprise Wheel.

Also see: Computer Integrated Manufacturing, Configuration.

Documentation Management. Documentation management updates information on engineering documentation, including drawings, and provides a history of all revisions and change orders. Management of technical and engineering documents is vital to the operation of a manufacturing firm engaged in factory automation projects. For a factory to be truly automated, technical data as well as machining processes must be put on an information network.

DOS. DOS (disk operating system) usually is used as a shortened reference to either Microsoft's MS-DOS or IBM's PC DOS, which are both disk operating systems for the IBM PC and compatibles. A disk operating system can be other than those two products, such as Digital's RSX-11M for PDP-11 computers, or VMS for VAX computers or IBM's VM and VS. UNIX and CP/M are other examples of disk operating systems.

Any disk operating system performs the functions of providing for the basic control and interaction of the peripheral devices with the main computer system and interfacing the program (software) with the hardware. Application program software cannot operate without interacting with the operating system software, which in turn interacts with the hardware. A disk operating system is primarily distinguished from other types of operating systems by the fact that the operating system requires a disk to be present as part of the system. Typically, the disk operating system is physically resident on one of the disks associated with the computer system. In LAN (local area network) configurations, some types of nodes are capable of booting from the LAN and can be diskless systems, while still essentially running a disk operating system. In this scenario, the node uses the disk resources of other systems on the LAN. A computer that uses an operating system which isn't disk based usually stores the operating system in some type of nonvolatile read-only memory.

One of the significances of the introduction of disks with removable media for computers was that finally a computer could have access to theoretically unlimited data, although not at the same time. Prior to this, all the data used by a computer had to remain in the main memory of the system.

While the term DOS could be used to refer to any disk operating system, it usually is used to refer to either MS-DOS or PC DOS. While PC DOS and MS-DOS are not identical, the main difference lies in the vendor source. IBM commissioned Microsoft to develop a disk operating system for the IBM PC computer, which IBM sells as PC DOS. Microsoft also packaged the same disk operating system, with whatever minor changes were required to avoid copyright infringement, and markets the resulting product as MS-DOS. Most of the computers in the PC family, including compatibles can run either MS-DOS or PC DOS.

History. DOS began when PC DOS was developed by Microsoft for IBM for the IBM Personal Computer. DOS has gone through a number of changes and revisions. DOS version 2.0 incorporated many of the concepts of UNIX and XENIX when it was developed as the revision to the original DOS. DOS 2.0 and DOS version 2.01 were the first versions which accommodated a hard disk on the PC XT. When the AT was introduced, DOS 3.0 was provided to support the new machine. Version 3.0 has some bugs and 3.1 was introduced soon after to correct the bugs and other problems. Version 3.2 provides some networking support capabilities as well as support of 3 1/2 inch floppy drives. Version 3.3 has been introduced to support the PS/2 product line. Most people are waiting eagerly for the first multitasking or multi-user version of DOS. OS2, which is being developed for the newer Personal System/2 and AT class 80286 machines, is expected to support the 80286 and 80386 CPUs running in protected mode. All the versions of DOS introduced so far only support the real mode of operation of those two CPUs, which, even when running on an 80286 or 80386 machine, emulates an 8086 or 8088 chip. When DOS was first designed and developed, the 80286 and higher chips didn't exist.

Other vendors have developed their own versions of DOS for the PC, such as Compaq, who licensed a version of MS-DOS from Microsoft. This version can function on IBM machines and other clones, as well as the Compaq machines.

Discussion. DOS provides for the operation of all the basic parts of the personal computer, as well as many optional items which can be added at a later time. DOS supports different versions and configurations of some devices, such as keyboards with different layouts of keys or the optional selection of characters. Accommodations can be made for different country conventions such as how the time and date are displayed. Several types of monitors and disk drives are supported, as well as many memory configurations and quantities.

As with any standard, there usually are penalties which are paid for standardization. For example, any time a standard is frozen, all potential improvements from that time forward have to be forgone, unless the standard is updated. Some of the restrictions of the earlier versions of DOS, such as the 640K main memory addressing limit, were selected when the capability of the target hardware was much less than that of the the current machines. The original PC, which used the 8088 CPU chip, didn't offer more than the 640K of physical memory, but AT type machines which use the 80286 CPU can physically address 16Mbytes of memory. Current versions of DOS allow memory past the 640K limit to be addressed by using virtual disk capabilities using the VDISK function of DOS. This function treats a portion of RAM memory as if it were a physical hard disk, and the application software can address this portion of memory the same way disks are addressed.

When a personal computer is booted, it loads DOS from a floppy disk, hard disk, or network. Usually, the source is the machine's own disk drive (floppy or hard). To enable the machine to know how to load DOS upon power-up, and to also perform desired self-checks and self-tests, software referred to as the ROM BIOS is contained in a ROM chip in the computer. The ROM BIOS has the boot-up software, self-test software, and the Basic Input/Output System software. Once the DOS operating system is loaded into main memory, the system is ready for use by the operator.

DOS has a considerable list of commands, which are divided into two categories: internal commands and external commands. Internal commands are those which are resident in memory when DOS is loaded and the external commands are those which remain on disk and are always executed from disks.

Any disk operating system provides some basic functions which are not included in a nondisk operating system. Some of these basic functions are the ability to read and write to disks and the support functions of examining disks, setup and formatting, and using data from disks when executing application programs.

DOS provides standard interfaces between the different portions of the personal computer. DOS has a video driver which sends information to the screen for display, for example. When application software programs use the standard DOS interfaces, they can easily function with other programs which have been created likewise. But in some cases, the standard DOS handlers are too slow or don't offer the features desired by the application program. In these circumstances, the application program may directly take control of the device, circumventing DOS. As long as the program is used by itself, and provided the program adequately handles all tasks which it is preventing DOS from managing, there is no problem. However, if other programs are attempted to be run at the same time, or if transfer of data is desired between this program and others, problems may occur since the rules by which DOS manages resources may be violated.

Since DOS is, in fact, software running on the hardware, and application software must run under DOS, there is a penalty by running one software package under another software package which, in turn, interfaces directly with the hardware. On the other hand, if the desire is to eliminate the use of DOS, all functions needed by the application program, which DOS would otherwise provide, must be properly handled by the custom-written code. The large number of system support functions which must be coded is the reason developers use operating systems even though in many applications, the operating system may not provide the optimum performance. Using an operating system also provides a known standard interface for different people to develop applications which can run under the disk operating system when integrated at a target machine.

Applications. DOS is used on many personal computers as the basic operating system, although some people use XENIX, which is a PC version of UNIX. XENIX is also a disk operating system. Most applications which do not require exotic software or functions which cannot be provided easily by DOS use the standard DOS operating system for handling the interface between the different parts of the system.

In an attempt to use greater than 640K of main memory for program execution, Lotus, Intel, and Microsoft have defined a method of using expanded memory on an AT (80286 based) machine which uses pages of memory above the 1024K address location. The pages are swapped into and out of the memory under the 1024 address location. This Lotus/Intel/Microsoft (LIM) expanded memory specification (EMS) describes how the software must interact with the operating system and the hardware. Essentially, there is control software called the expanded memory manager (EMM), which resides in the memory above 640K and below 1024K, which swaps 64K pages of data from above 1024K into four 64K windows in this area between 640K and 1024K.

While the original version of DOS was very rigid, with predefined limits on the devices and disk formats which it could work with, the later versions allow for user-installable drivers and a configuration file which modifies and customizes the operating system each time the machine is started up. The configuration file, known as CONFIG.SYS, can modify things such as the number of disk buffers used by DOS and the maximum allowable number of files which can be open at one time. One of the often used special drivers is ANSI.SYS, which provides enhanced keyboard and screen control, consistent with the ANSI standard. With many applications which require the use of a mouse, the driver MOUSE.SYS or a similar mouse driver will be loaded at power up. Using a mouse on the PC allows the PC to operate similarly to a Macintosh, especially when using the Microsoft Windows operating system shell.

A program such as Microsoft's Windows or IBM's Top View are visual shells, which means they reside under DOS and, in a sense, wrap themselves around DOS to provide facilities which aren't available in DOS, and alter the visual display which would normally be available in DOS. Most of these visual shells provide icon-driven menus which make operating the computer much easier than trying to remember the DOS commands. However, they also force a user to use the menus or, in some cases, go through many levels without being able to take shortcuts and go directly to the desired command.

Essentially, DOS provides the service of command processing while the computer is operating. Keyboard commands are received, queued, and executed. Any peripheral devices which must be accessed are done transparently to the user and the user's program, and sometimes housekeeping tasks are done automatically, with no knowledge of the occurrence by the user.

One of the utility programs included with the DOS operating system is the program DEBUG. DEBUG allows a user to view and edit the contents of memory locations in the main memory of the computer. DEBUG can view the contents of the ROM chips as well as the RAM. Only the RAM can have its contents altered. One of the capabilities of DEBUG is to unassemble hexadecimal code machine language into slightly easier to understand assembly language.

Glossary. *BIOS:* The BIOS or Basic Input/Output System is software contained in a ROM chip which handles all the standard input and output of data between the computer system and its peripheral devices.

CP/M: Control Program for Microprocessors is an operating system developed by Digital Research, Inc. CP/M-86 was originally to be the operating system used on the IBM PC, but later DOS from Microsoft was selected.

personal computers: Personal computers are computers whose entire system is intended to be used by a single user and can generally survive in a normal office environment. A personal computer can easily fit on a desktop.

UNIX: UNIX is an operating system created by researchers at Bell Labs. UNIX provides things such as pipelines which allow the output of one process to be the direct input of another, without wasted overhead in unnecessary handling of the data.

Also see: CP/M, Expanded Memory, Extended Memory, UNIX.

Dot-matrix Printer. A dot-matrix printer contains a printhead which has wires fired by solenoids. The wires strike the back of an inked ribbon, which creates an impression of the end of the wire matrix on the paper. Dot matrix printers originally were configured with nine wires in the printhead, but now 12, 18, and 24 wires are becoming commonplace. Most 24 wire printheads are physically configured with two staggered columns of 12 wires, to provide overlapping printing and eliminate the "necking" between dots. Multipass and shadow printing are two additional methods of improving the quality of the print. Other advanced features of dot matrix printers are logic seeking and bidirectional printing, which improve overall output speed. Logic seeking printing performs line feeds on empty lines instead of printing a full line of spaces and performs a line feed and a direct move to the closest end of the next line of printing as soon as all characters are printed on a particular line. Bidirectional printing prints from left to right and then from right to left on the successive line to eliminate the wasted time of performing a carriage return.

Dot Pitch. Dot pitch is the distance between the pixels of a video display screen or the dots of a dot-matrix printer. The finer the dot pitch, the greater the visual resolution. However, the same bit-mapped image displayed on a monitor with a finer dot pitch will be physically smaller than the same image seen on a monitor with a larger dot pitch. The size of the dots is a different parameter than the dot pitch. Typical mid-range monitors of today used by personal computers use a 0.31 inch dot pitch. Some of the higher quality monitors use a 0.28 dot pitch, while the original IBM PC monochrome monitor uses a significantly larger pitch. The dot pitch is only part of the necessary parameters for a high-resolution graphics image. The other equally important requirement is to have a large number of total pixels in both the horizontal and vertical direction.

Double Precision. Double precision is the use of floating point numbers which have twice the number of digits that normal (single precision) numbers have. A double precision number is typically constructed by using a 64-bit floating-point data format where the high-order bit is used as the sign and the next eight bits are the exponent, with the remaining 55 bits used as the mantissa. Depending upon the bit size of the computer, successive bytes or words may have to be used to construct a double precision number. The purpose of double precision calculations and manipulation of numbers is to allow operations to be performed with greater accuracy than would normally be possible if only a single word length were used for numbers. As computers become larger in terms of the number of bits of the bus and CPU, double precision becomes less of a factor. A 32-bit computer's normal floating point number has twice the number of significant digits that a 16-bit computer's floating point number does, and eventually 128-bit and larger computers will allow a sufficiently large floating point number to be used.

Downtime. Downtime is the time in which a system or machine tool is not available for use due to failure or routine maintenance. Also known as cumulative lost time.

DRAM (Dynamic RAM). See: Dynamic Random Access Memory.

Drawing. Drawing is the name applied to the forming of a variety of steel parts such as automobile body panels, fenders, and doors.

Conventional drawing is the method used most frequently for producing cups in small and medium sizes. The cups can be drawn with or without a flange and in almost any desired cross section. The metal blank is forced to flow into the die cavity without any constraint other than against buckling. Internal resistance flow of the metal toward the punch and die sets up compressive stresses which, combined with tensile stresses created by the motion of the punch, give two-dimensional shear. This is a very desirable internal-stress condition which requires a minimum of applied forces to fan the metal.

In stretch drawing, the metal is relatively thin in proportion to the total area involved and would wrinkle badly if drawn in a conventional manner. Therefore, the edges of the sheet or blank are constrained in varying degrees to induce a predominance of tensile stress, thereby preventing buckling or wrinkling. Friction plays an important part in this operation and practically controls the severity of stress in any particular area. Consequently, details of the die as well as the shape of the part must be developed by the cut-and-try method. It is not unusual in the automotive industry to allow as much as a year for the design, construction, and try-out of large dies for stretch drawing.

Drilling. *Twist drills* are end cutting tools having one or more cutting edges, and having helical grooves adjacent thereto for the passage of chips and for admitting coolant to the cutting edges. Twist drills are used either for originating holes or for enlarging existing holes. They are used as the basic method of producing holes in a wide variety of materials.

The varying conditions under which drills are called upon to operate, as well as the wide variety of materials having different characteristics to be drilled, make it impracticable to lay down any hard and fast rules for drill speeds. In general, the permissible speeds depend on the hardness and the toughness of the material to be drilled. The harder or tougher the material, the lower the speed should be. Conversely, the softer the material, the higher the speed. Observation of the behavior of drills when operated gives an indication of whether the operating speed should be increased or decreased. If drills operate easily and without undue signs of heating, higher speeds should be tried. If, on the other hand, drills show signs of overheating, or if corners wear away rapidly, it is an indication that speeds are too high.

In production work, cooling of both drill and work is usually the most important factor to consider. The amount of heat generated is so great that tool life depends on the ability of the cutting fluid to carry away the heat at approximately the same rate it is generated. For that reason, lubrication must often be subordinated to cooling.

Some lubrication is desirable, particularly when drilling steel, as it will help to prevent chips from sticking to the cutting edges of the drill and causing clogging in the flutes. Cutting fluids should be selected on the basis of their heat-dissipating or lubricating qualities as conditions demand.

One of the most attractive production applications of laser beam machining (LBM) has been the drilling of cooling holes in jet engine components. Here, the laser must compete with the electrical discharge machining and electrochemical machining processes. Most jet engine manufacturers are using pulsed lasers for production or engineering development activities.

An application of automation to drilling was reported in the book *New Direction Through CAD/CAM*. In it, authors William Beeby and Phyllis Collier reported: "In 1980, Rohr Industries, Incorporated implemented a system that links together a Camsco computer system and a Trumpf CNC routing-drilling machine. An Auto-Trol machine digitizes part shapes from templates or mylar drawings, and the computer software automatically nests the flat shapes. When the nest has been arranged, the data are transmitted to a floppy disk that runs the router. At Rohr, the 30,000 shapes stored in the computer are aerospace components."

One application of drilling using numerical control was presented by authors L.F. Pau of the Technical University of Denmark and Dag Paus and Tom Stokka of Nordic Intelligent Manufacturing in an SME Technical Paper titled *Knowledge-Based Order-Specific NC Drilling System*. In this paper the authors noted: "A knowledge-based system generates hole positions and/or types, as well as the corresponding NC macros, on custom order beams for trucks.

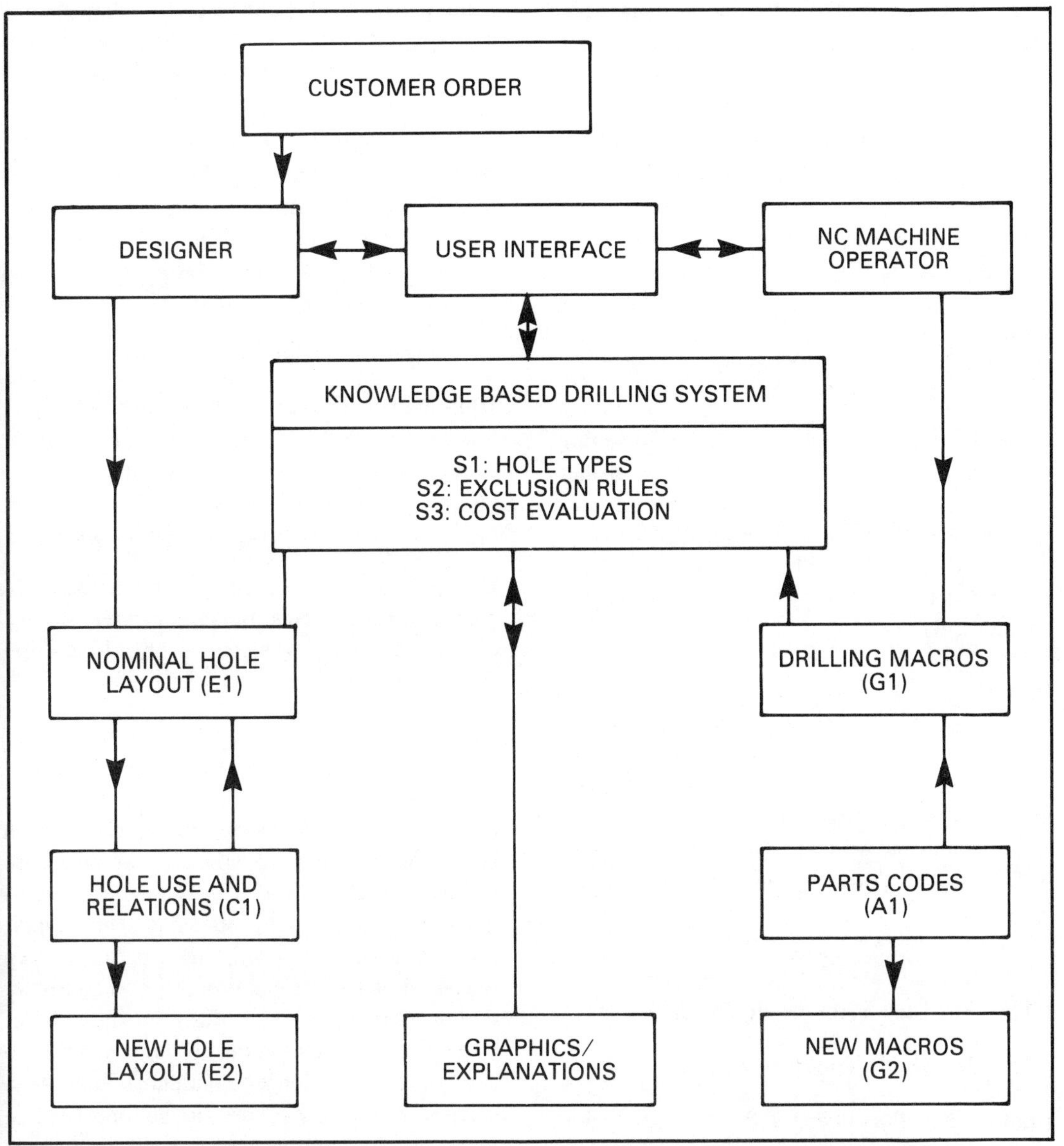

Knowledge-based order-specific NC drilling system.

This system exploits parts databases, CAD, and configuration procedures. It is implemented, partly in PROLOG, and in other languages. The main benefits are customization and reduced manufacturing times, for such orders."

Special emphasis must be put on the following techniques: parts description database, with attributes for all variants parts; specification procedures, with data structures encompassing all parts description databases; descriptions of the uses of each part, and of its relation to other parts; order verification procedures, for parts configuration and checks against parts in inventory; updates to CAD files; updates to production schedules at the flow control level; updates to NC macros, and distributed processing and data communication.

The project carried out for one company, has been focusing on one specific aspect of the above concept, which is knowledge based order specific NC drilling in the main beams onboard of medium and heavy duty trucks. The geometrical layout of holes is today available, both in drawings, and in CAD files. Custom-specific hole layouts must be generated automatically to produce ready-to-use NC command macros. In addition, the designer must have explanations about how the customized layouts were verified, and about related production costs per beam.

Previously, the generation of NC macros was carried out manually; the operator was including customized new holes on the basis of his/her inspection of parts descriptions. This process resulted in 3-10 erroneous holes being drilled per day; these errors had to be corrected by tedious welding, not to mention production delays and broken drills.

Driver. In PCs, a driver is a small program or routine that executes other programs or controls peripheral devices and the interfacing of these devices with the CPU. In shoe centerless grinding, a driver is usually a flat plate attached to the work spindle and against which the part is held. It drives, but does not actually support the workpiece.

Also see: Device Driver.

DTE. See: Data Terminal Equipment.

DTL. See: Diode Transistor Logic.

Dual-in-line Package (DIP). A dual-in-line package is an integrated circuit chip which has two parallel rows of pins. DIPs come in standard sizes with the distance between adjacent pins normally being 0.100 inch and the distance between the rows of pins 0.300, 0.400, or 0.600 inch. DIPs usually have standard pin counts of 4, 6, 8, 12, 14, 16, 24, or 28 pins per package. The DIP was developed to create a standard-size physical package for containing IC chips to facilitate automated insertion of ICs into circuit boards. The standard body allowed the development of automatic insertion equipment which could insert any chip of the standard dimensions. The first DIPs were made using ceramic packages, but as plastic DIPs were proven to be acceptable, they provided for cheaper product costs. As circuit designs became physically constrained from decreasing in size due to the requirements of the DIP packaging, smaller leadless chip carriers (DIPs are leaded chip carriers) and pin-grid arrays were developed and have become prevalent today.

Also see: Single-in-line Package.

Dual Power Supply. A dual power supply is made up of two EDM power supplies in a single cabinet which can be used to operate two machines simultaneously or can be connected to apply maximum amperage to one machine.

Dump. Dump is the process of removing all or part of the contents of a computer storage device. Also, the printed copy resulting from this operations.

Duty Cycle. The duty cycle of a product or device is the length of time the item is intended to be in continuous operation. The duty cycle of a motor is the length of time it can be run continuously without overheating or otherwise damaging itself. If a device can function continuously, the duty cycle is stated as being *continuous*. One of the main differences between industrial or commercial products and consumer or home products is the designed duty cycle. Typically, industrial products are designed to run for a much longer continuous time than consumer products. The span of time used to define a duty cycle varies from product to product. With laser printers, the duty cycle is relative to the number of copies components can reliably produce in a month's time. With some items such as pumps, the duty cycle may be stated in terms of a maximum length of time on and then a minimum length of time off.

Dynamic Random Access Memory. Dynamic Random Access Memory is computer memory which requires constant refreshing to maintain its data. The refreshing is a signal which must be provided at some minimum frequency to avoid the loss of data. The refresh signal is in addition to the main five volt supply. Since the refresh signal must be strobed, the CPU is required to do processing to refresh the memory, unless there is a separate automatic refresh generator. This type of memory contrasts to static random access memory, which does not require the separate strobed refresh signal and can maintain its data with a single five-volt supply line to the chip. The advantage of using dynamic memory is that it is capable of faster writes than static memory and is much cheaper to manufacture. With the current design of computer memory circuits, the refresh signal is a very insignificant portion of the design in terms of cost and complexity since most memory circuits have a built-in refresh signal from an IC chip.

Also see: Static Random Access Memory.

E

Earing. Earing is a phenomenon that occurs during the drawing of cup shaped objects from metal. Earing occurs more in deep drawing, due to it being related to the amount of working the metal blank receives. Earing in deep-drawn parts is related to planar anisotropy. When sheet metal is roll formed at the steel mill, a fiber structure is formed in the direction of rolling. The fibers actually are rolled-out impurities. The sheet metal is, therefore, stronger and has a greater elongation capability in the direction of rolling. This nonuniform strength causes four ears or lobes to occur, even though a circular blank is used. Earing becomes more severe when the sheet metal is cold worked to quarter hard or harder tempers.

In practice, enough extra metal is left on a stamped cup so that trimming is required to remove the wavy edge. Earing usually accompanies local wrinkling of the cup wall. Between the ears of the cup are valleys in which the material has thickened under the compressive hoop stress instead of elongating under the radial tensile stress. This thicker metal forces the die open against the blankholder pressure and allows the metal in the thinner areas around the ears to wrinkle. The die design, die radii, draw reduction, and lubricant are factors that affect earing. Earing can be expressed as a percentage calculated by: the average height of ears minus the average height of valleys, divided by the average height of valleys, times 100%. A good drawing material should exhibit less than 4% earing. Earing is very undesirable since it can lead to pinching or clipping in the draw dies because at the end of the draw, the full blankholder load is concentrated on the tips of the ears. These ear tips can be pinched off due to the high unit load, and an accumulation of the clippings in the dies can be a serious problem.

Also see: Drawing.

EBCDIC. The extended binary coded decimal interchange code (EBCDIC) is a set of 256 different characters, each character represented by a unique eight-bit byte (octet). The code is used for data representation in data storage and data transmission. There is no parity bit and letters are collated before numerals. The code is similar to the ASCII code except that the bit patterns that correspond to the alphanumeric characters are not the same for each of the characters.

EBCDIC is widely used in IBM architecture. When transmitting data between an IBM (EBCDIC data-type) system and a non-IBM (ASCII data-type) system, data conversion to/from EBCDIC from/to ASCII must be done by one system or the other. Magnetic tape units which are designed to emulate IBM standards store their data using EBCDIC representations. The current trend in popularity is toward ASCII, with its parity capability. All the normally used characters can be represented using the 128 character set of ASCII.

Also see: Computers.

Eccentricity. Eccentricity is the distance between the center of a datum circle and a datum axis of rotation.

Echo Check. An echo check is a method of checking the accuracy of transmission of data in which the received data are returned to the sending end for comparison with the original data.

ECL (Emitter Coupled Logic). Emitter coupled logic (ECL) is a type of transistor logic where the emitters are coupled together as part of the logic circuitry. A transistor has a base, collector, and emitter. ECL technology allows for faster logic functioning within a circuit than the traditional types of TTL, CMOS or HCMOS circuitry. ECL, while being very fast, is also very expensive, so it is only used in situations where the speed is truly needed and the resulting higher cost is not an impediment. Because of the markedly higher logic speeds, which means the electrons are more active within the ECL logic circuits, the heat generated is much greater. Since the limiting factor for operating speed initially is the amount of heat generated, liquid cooling is employed in some ECL IC chips to allow faster operating speed. ECL logic is used in larger mainframe computers and higher speed special-purpose processors.

Also see: CMOS, IC, TTL.

Economic Analysis. Economic analysis is in its most basic sense, an orderly study of the production, development or management of material wealth. An economic analysis can be done on one's personal financial/material wealth or it can encompass a national or world economy. Economists are those who are known to perform the scientific task of analyzing data pertaining to material wealth in order to make a prediction.

The prediction is a foretelling of trends that will effect the income of individuals and companies. Often companies use an economic analyst to set business strategies, such as capital spending, for the future. When determining a nation's economic trend, economists use standardized economic indicators. These indices include durable goods orders, housing starts, employment levels, etc.

Although an economic analysis is considered to be scientific in nature, human factors such as opinion can cause a variety of outcomes to an analysis. The results of an economic analysis then are only as good as the analyzer and, in themselves, only an indicator.

Also see: Cost Analysis.

Eddy-current Testing (ET). Eddy-current testing (ET) is a nondestructive testing method for detecting material composition, hardness, structure and surface or subsurface flaws in metals. In eddy-current testing, electric eddy-currents are induced in the test object by means of alternating current flowing in external electromagnetic coils. A probe measures a response and displays information on a cathode ray tube or meter for comparison with similarly induced responses of standards having known properties or characteristics.

Edging. Edging a method for deburring or chamfering sheet stock, tubing or rod ends, or workpiece edges by passing the stock across small grinding wheels or rotating angled cutters, or between pinch rolls.

EDIF. See: Electronic Design Interchange Format.

Edit. Edit is the function of using a software program to change or create a text or graphics data file in a computer. Any type of user modification of images, files, or programs is referred to as editing the source code or data.

A person may edit graphical images such as when creating or modifying CAD drawings. One of the advantages of editing using a computer instead of pencil and paper is that the drawing cannot be worn out no matter how many changes are made on a particular view. Also, the ease at which large-scale changes can be made is spectacular. Other functions which can only be done by computer are such things as scaling and rotating a large complex image. The ability to use color allows for complex drawings including three-dimensional images to be easily understood when viewing them on a two-dimensional screen. The state of the art in resolution in computer graphics continues to improve allowing for more complex graphics to be created.

Also see: Computers, Editor, Software.

Editor. An editor is a computer program used for modification and creation of data files. An editor allows an operator using a computer terminal to create and modify text or graphics. A text editor modifies text and a graphics editor creates and changes graphical images. Even though the grammatical meaning of the word editor means to change, an editor may also be used to create text or graphics when none exists.

A line editor is a text editor which is capable of displaying and modifying one line of text at a time, while a screen editor can display a screen full of data while manipulating text or graphics, by a full screen at a time. Most word processors are screen editors. A well known line editor is Edlin, which is included with the DOS operating system on personal computers.

A word processor is a text editor which has the ability to do many types of character attributes, formatting, text manipulating, merging, saving, and printing. Some word processors allow the user to sort lists in alphabetical order, while even more sophisticated text processors can merge text and graphics.

All computer source code must be created using a text editor. Some languages have a text editor resident as part of the compiler, while with others a separate editor must be used.

Many graphics editors allow a variety of functions such as panning and zooming throughout the drawing to facilitate its creation.

Also see: Computers, Edit, Software.

EDM. See: Electrical Discharge Machining.

Educational Robots. Educational robots are robots whose main purpose is to educate the user in how to program a robot and to better understand the requirements of a robot when developing applications. Some people have no feel for the limitations and capabilities of robots, which can be easily illustrated by using educational robots. When learning to write robot application programs, an educational robot can provide an inexpensive means to see the results of the programming effort. When developing peripheral or support tasks such as vision systems or inspection systems or grippers and end effector tooling, an educational robot can provide the means to test the functionality of the development.

Some robots such as the Heathkit Hero and the Rhino robot were designed primarily as educational tools. These robots don't have the accuracy or the durability to perform industrial applications as most commercial robots do. Several inexpensive remote controlled robots have been developed to allow people to visualize the basics of robotics.

Also see: Industrial Robots.

Effective Address. An effective address is the address of a memory location obtained by applying a memory reference instruction to a specific address. One instruction can go through several indirect addresses to reach the effective address.

Also see: Address.

Effective Draw. Effective draw is the maximum limits of forming depth which can be accomplished with multiple-action presses. As shown on a motion diagram for a typical double action or the upper action of a triple-action press, effective draw is the distance the inner slide is from the bottom of its stroke at the point at which the outer slide begins its dwell. For the lower action of a triple-action press, it is the distance the lower slide is from the top of its stroke when the inner slide begins its dwell—sometimes called maximum draw or maximum depth of draw.

EIA. There are two defintions for this term. EIA is a language for programming robots. The language was developed by Ikegai America. EIA is also the Electronics Industries Association, a trade group which establishes some interface standards.

Electric Motor. An electric motor is an actuator that uses electrical current as the power source of generating rotary motion. Electric motors are designed in many different types such as AC and DC motors. There are servo-controlled, stepper-controlled and constant-speed motors. A standard speed for electric motors which run on AC current is 1725 rpm (revolutions per minute). DC motors run at speeds from zero rpm (when using a motor as a brake) to several thousand rpm, although when extremely high speeds are needed, either a turbine is used or ratio drives are used to cause the rpm of the final drive to be greater than the speed of the source motor.

Electric motors essentially all work using the same principle, which is to produce a rotating electrical (magnetic) field in the stator which causes the rotor to rotate. The magnetic field is moved from pole to pole in sequence. The greater number of poles in the radial circle cause more constant torque output from the motor. If the poles are too few and far apart and the motor happens to stop in a certain location relative to the poles, the motor may have a problem starting because of the inability to create adequate force at the proper angle to rotate the motor.

Also see: Actuator, AC, AC Motor, DC, DC Motor, Industrial Robot.

Electrical Discharge Grinding (EDG). Electrical discharge grinding (EDG) is a process similar to electrical discharge machining except that the electrode is a rotating wheel, generally graphite but sometimes brass.

Electrical Discharge Machining (EDM). Electrical discharge machining (EDM) is the process of machining materials with sparks. The area where the sparking takes place is surrounded by a dielectric material. The cutting tool, called an *electrode*, does not physically contact the part being machined. Instead, the electrode remains the distance of the spark away from the workpiece. Thermal energy of the spark is used to machine the workpiece.

Only electrically conductive materials can be machined by EDM. Electrode material must also be electrically conductive. Dielectrics used in EDM are usually fluids, most commonly hydrocarbon oils. In some instances, distilled or de-ionized water is used. Dielectric fluid is an insulator until it ionizes. At the point of ionization, the fluid becomes an electrical conductor in the area where the spark occurs.

The dielectric fluid provides a path for the discharge as the fluid between the tool and the workpiece becomes ionized. Initiation of the discharge occurs when sufficient voltage is applied across the machining gap to cause the dielectric to ionize and current to flow. The tendency for the discharge to be initiated is increased if the spacing between the electrode and the workpiece is reduced, the applied voltage is increased, or debris from the previous discharges is suspended in the dielectric. The energy of the discharge vaporizes and decomposes the dielectric surrounding the column of electrical conduction. As conduction continues, the diameter of the discharge column expands and the current increases. The small area in which the discharge occurs is heated to an extremely high temperature so that a small portion of the workpiece material is elevated above its melting temperature and is removed.

With the high temperatures involved, it makes no difference whether the workpiece material has been heat-treated or not. The discharge temperature is high enough to machine the hardest and toughest of known metals, including exotic metals.

The size of the crater produced in the workpiece by the discharge is determined by the size of the discharge or, more accurately, the energy of the discharge. The amount of current is more influential than the length of time. It has been observed that, if the discharge current is doubled and the conduction time is reduced by one-half, the resulting crater is increased.

Often, EDM compares favorably with the combined cost of conventional machining and deburring. Delicate workpieces that are not strong enough to support the cutting forces of conventional tools can be processed by EDM without distortion. For instance, the electronics

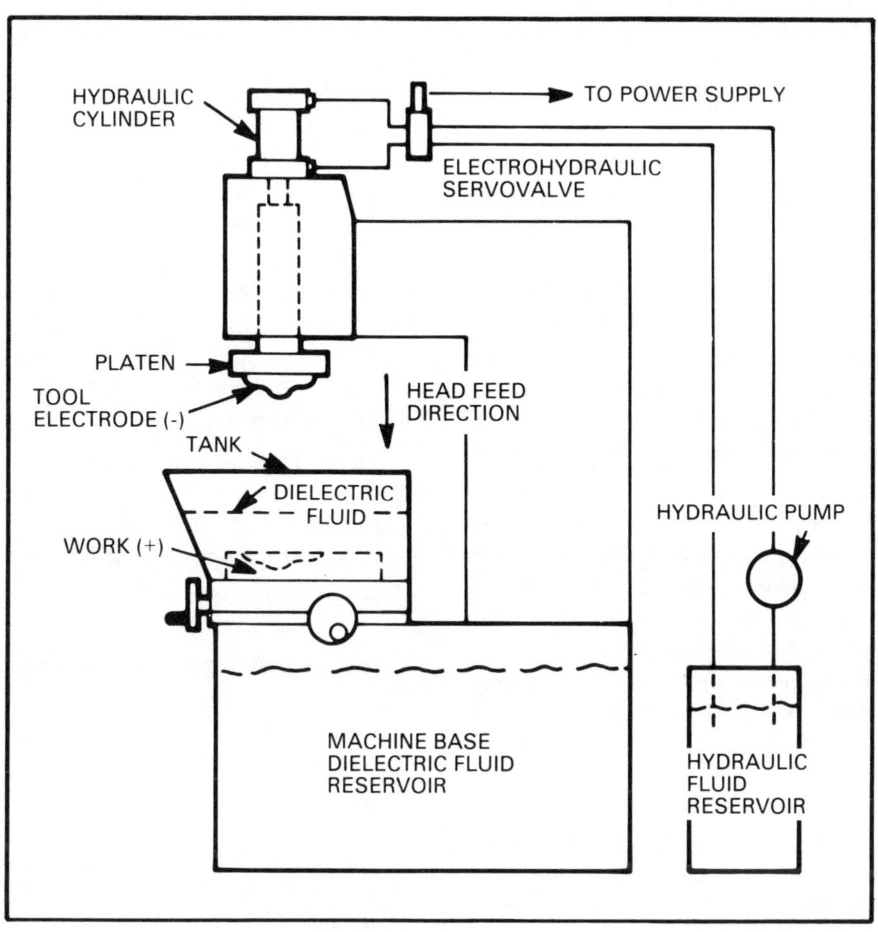

Components of an electrical discharge machine.

industry uses many EDM machines in the production of small, thin, copper, electronic parts.

Sometimes accuracy requirements dictate the use of EDM for two main reasons: when repetitive shapes are required, they can often be produced from an easy-to-make male electrode, and when machining accuracy must be maintained after heat-treatment of the part.

The various discussions of EDM applications are by no means complete nor could they be, because the successful use of EDM is the result of using this metal removal method to solve particular manufacturing problems.

Power supplies, control of the spark discharge, electrode materials, and dielectric fluids are being continually improved.

Recent developments in EDM power supplies have been concentrated on controlling the transistor circuits. The circuit represents the logical extension of microcomputer technology to EDM power supply control. All spark parameters, as well as machine tool functions, can now be controlled by a microcomputer. Control by microcomputers has provided another increase in cutting efficiency. Cutting conditions are monitored and then instantly changed by the computer, as required, for peak cutting efficiency.

The information and capability are available to document the steps necessary to produce both simple and complex parts by EDM. The computer serves as a repository for storing documented information and calling it forth for the use of others, or for commanding and directing equipment.

The use of numerical control with EDM machines is recognized as a method to increase table-positioning efficiency, such as that used in the case of multiple-cavity work, and to cut dies and punches with traveling wire. Numerical control is also used to orbit EDM machine tables, especially for large-mold workpieces. The addition of NC to the vertical ram movement makes it possible to EDM at different angles. Using an automatic electrode changer with the NC machines makes the EDM process completely automatic from roughing to finishing operations.

CNC-EDM equipment is designed to permit automatic, unattended operation under computer control. The computer is programmed by simple manual data input for conventional toolroom jobs or by tape for repeat or production jobs. System and equipment protection and warning monitors provide surveillance. Malfunctions and potential problems are reported locally or by telephone to remote locations. In some cases the equipment will fix itself. When a normally worn electrode is detected by the system, provisions can be made for a back-up electrode to be available in the tool changer. In other cases, where the fix is beyond the capability of the system to act, the computer will automatically shut down the machine and report. For example: a shutdown will be called for and reported in case of DC arcing, fire, or excessive dielectric temperature, among other problems. The system is also programmed to measure and compensate for electrode wear.

EVALUATING NC FOR EDM

1. What can I get for the dollar spent?
2. How do I keep from over or under buying?
3. Will something new appear soon after I buy, making the machine obsolete?
4. Should I buy a machine bigger, smaller, or the same size as my present equipment?
5. What type of tooling is required?
6. How do I build electrodes for this process?
7. How do I evaluate ease of programming versus burning speed?
8. Who will train and support me after the sale?
9. What future applications should I be aware of?
10. Should I consider purchasing a model with electrode changer?

Points to consider for evaluating numerical control and electrical discharge machining.

The machining center is of particular advantage when it is required to produce, in the workpiece, multiple cavities (alike or different) or where a number of different electrodes are required to produce an intricate workpiece.

Terry Bryce of the Mitsubishi International Corporation listed 10 points to evaluate numerical control usage for EDM. These points are presented in the figure in the left-hand column.

Also see: Numerical Control.

Electrical Discharge Wire Cutting (EDWC). Electrical discharge wire cutting, sometimes called traveling wire EDM, is a process that is similar in configuration to bandsawing, except in the case of EDWC, the "saw" is a wire electrode of small diameter. Material removal is effected as a result of spark erosion as the wire electrode is fed (from a spool) through the workpiece. In most cases, horizontal movement of the worktable, controlled by CNC on modern machines, determines the path of cut. However, some EDWC machines move the wire horizontally to define the path of cut, leaving the part stationary. On both types of machine configurations, the wire electrode moves vertically over sapphire or diamond wire guides, one above and one below the workpiece. The electrode wire is used only once, then discarded because the wire becomes misshaped after one pass through the workpiece. A steady stream of deionized water or other fluid is used to cool the workpiece and electrode wire and to flush the cut area.

Viewed from above, the electrode wire cuts a slot or "kerf." The width of the kerf is the wire diameter plus EDM overcut. Starter or threading holes are required. In steel or other material, a drilled hole suffices; for carbide, the hole may have to be produced by EDM.

As is the case with any electrical discharge process, EDWC requires that the workpiece be electrically conductive. The cut produced by the process is free of bellmouth or flaring and is controllable to produce small radii. Workpieces up to about six inches (152 mm) in thickness can be processed using standard equipment; workpiece stacking effects greater productivity.

Normally produced from hardened metals such as tool steel, stamping dies are routinely cut using EDWC. Using the EDWC process, complex-shaped blanks also can be cut quickly and easily. A distinct advantage of EDWC in prototype work is that NC programming used to product the prototype can be used for production parts, often with only minor modifications. The flexibility afforded by the use of NC in the process greatly speeds the production of blanks for test forming in blank development work.

Manufacturing of molds of all types is most often accomplished using EDM cavity-type sinking units. Expensive electrodes for these machines normally employ slight tapers and extremely accurate size requirements and provide an excellent application for EDWC.

Electrical Field Vector. An electrical field vector is the electric potential on a stationary positive charge per unit charge in an electric field, usually representing DC voltage or the instantaneous value of an alternating voltage.

Electrically Erasable Read-only Memory (EEROM). Electrically erasable read-only memory (EEROM) is a type of ROM which can be electrically altered (modified, erased, and reprogrammed). Unlike a ROM which is masked and cannot be altered, or an EPROM which requires UV light to erase and special circuitry to program, EEROM can be utilized in normal RAM memory locations. However, frequent alteration of the data in an EEROM can cause the device to exhibit unreliable functioning. An EEROM can maintain its data without continuously applied electrical power. The controls necessary to write an EEROM can be implemented in software, which makes the peripheral system requirements less than with EPROM. The greatest negative issue in regards to using EEROM is the relatively high cost of the device. A typical application of EEROM is to retain setup information in computer equipment. With the low total cost and reliability of CMOS RAM powered by lithium batteries, EEROM is not as widely used except in situations where a possible battery failure would be catastrophic.

Electrochemical Deburring. Electrochemical deburring is an adaptation of electrochemical machining that uses similar electrolytes and a stationary electrode to dissolve burrs from metal workpieces and flush them away with pressurized electrolyte. In addition to deburring the part, the process generates edge and corner radii.

Also see: Deburring.

Electrochemical Discharge Grinding. Electrochemical discharge grinding (ECDG) is a nontraditional machining process in which AC or pulsating DC current is passed from a conductive grinding wheel made of bonded graphite to a positively charged workpiece through an electrolyte. There is no mechanical contact between wheel and workpiece. Rather, workpiece material is removed by the action of electrochemical grinding, and oxides formed by the electrochemical action are removed by intermittent random-spark discharges as in electrical discharge grinding.

Also see: Grinding.

Electrochemical Honing (ECH). Electrochemical honing (ECH) is a process similar to electrochemical grinding in which nonconductive honing stones rather than a grinding wheel provide the abrasive action. Electrochemical action removes about 90% of the workpiece metal; the honing stones cut most aggressively on high or tight areas to maintain size and surface finish.

Also see: Honing.

Electrochemical Turning (ECT). Electrochemical turning (ECT) is a special application of electrochemical machining that uses noncontacting tools to make peripheral or face cuts on rotating workpieces.

Also see: Turning.

Electrode. An electrode is a conductor through which current enters or leaves an electrolytic cell at which there is a change from conduction by electrons to conduction by charged particles of matter, or vice versa. An electrode also is a component of the welding circuit through which current is conducted to the arc, molten slag, or base metal. It may or may not be consumable, depending on the particular welding process.

Electroforming. Electroforming is a process for making thin parts by electrodepositing metal on a mandrel or mold, which is subsequently stripped from the deposit. The process is similar to electroplating, except that the deposit is substantially thicker than an electroplated deposit.

Electrogas Welding (EGW). Electrogas welding (EGW) is a welding process for making 3/8 to 3/4 inch thick joints in one pass by depositing metal using either solid wire electrodes and gas shielding or flux cored electrodes without gas shielding. Molding shoes span the gap between parts being joined and confine the molten weld metal for vertical position welding.

Electrohydraulic Forming. Electrohydraulic forming is a method of shaping hollow tubes or preforms by discharging electrical energy inside the workpiece, which is filled with water or another suitable medium, producing shock waves that actually do the forming. Either of two methods—spark discharge or exploding bridge wires—is used to generate the shock waves. Deformation is controlled by forming into external dies, by regulating the amount of energy released, or by using shapers within the transfer medium.

Electron Beam Curing. Electron beam curing is a system for curing paint films using the energy of an electron beam. The process lends itself to high speed curing of paint on flat surfaces. Special paints must be used and personal shielding is required.

Electron Beam Cutting (EBC). Electron beam cutting (EBC) is a process related to laser beam welding, but utilizing beam spot intensities several orders of magnitude greater so complete vaporization occurs along the beam's path of travel.

Electron Beam Machining (EBM). Electron beam machining uses electrical energy to generate thermal energy for removing material. A pulsating stream of high-speed electrons produced by a generator is focused by electrostatic and electromagnetic fields to concentrate energy on a very small area of work. High-power beams are used with electron velocities exceeding half the speed of light. As the electrons impinge on

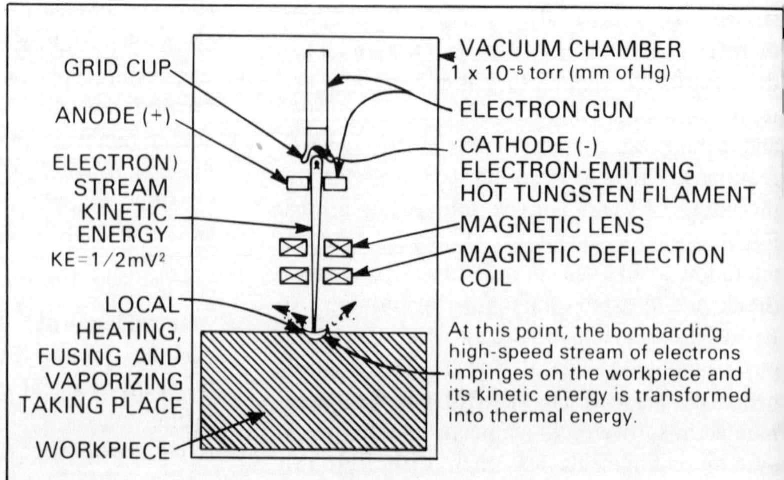

Elements of an electron beam machine.

the work, their kinetic energy is transformed into thermal energy and melts or evaporates the material locally.

Electron beams are concentrated on spots as small as 0.0002 inch (0.05 mm) in diameter. The process is usually performed in a vacuum, as shown schematically in the figure on the next page. A vacuum is used both to prevent collisions of electrons with gas molecules, which would scatter or diffuse the beam, and to protect the workpiece from oxidation and other atmospheric contamination. Lead shielding is required to protect the operator from X-ray radiation produced by the electron beam.

Any known material, metal or nonmetal, that will exist in high vacuum can be cut, although experience has shown that diamonds do not cut well. Holes with depth-to-diameter ratios up to 100:1 can be cut. Limitations include high equipment costs and the need for a vacuum, which usually necessitates batch processing and restricts workpiece size. The process is generally economical only for small cuts in thin parts.

The specific behavior of an electron beam, which is a beam of charged particles with very low mass but very high velocity, offers ideal conditions for the application of numerical control. The low-inertia beam can simply be controlled by electromagnetic fields, thus giving the possibility to economically adapt this extremely fast process to intricate machining geometries. A simultaneous control of the beam and of the workpiece movement is required in some cases.

An on-line computer control is applied to the workpiece movement, beam deflection, and beam focus on many EBM machines. This combination allows three-dimensional (X, Y, Z axis) movement of the focal point in combination with a defined operating speed on the workpiece. Furthermore, the computer controls the beam current as a function of workpiece position with time and also triggers the energy pulses. The advantage of this system is that the actual physical positions and electron beam parameters are continuously recorded in the computer and compared with the required positions and parameters which have previously been stored in the computer memory. The error signal is then used to correct any deviations.

The workpiece is moved continuously under the beam, and its position is recorded by the computer. When the position of a hole is reached, the computer triggers the electron beam pulse. During the pulse width, the beam follows the motion of the workpiece in order to impact on the same spot.

Most types of electron beam machining operations do not require expensive or complicated fixtures. In most cases, it is only necessary for the workpiece to be held so that it cannot move when the electron beam in impinged upon it. Movement of small pieces have been observed when being cut by the electron beam. When holes, slots, or patterns are to be cut through the workpiece, ample room must be provided under the workpiece so that electrons that pass through the orifice can be dissipated. If ample room is not provided, the electrons will strike the material located underneath the workpiece resulting in a vapor deposition or back splatter on the workpiece. For most cutting operations, a one-inch (25.4 mm) space under the workpiece is sufficient.

Electron Beam Radiation. Electron beam radiation is radiation generated from high-energy electrons that is used in crosslinking coating systems.

Electron Beam Welding (EBW). Electron beam welding (EBW) is a fusion welding process for making precise seams by directing a high intensity beam of electrons into the joint. Filler metal normally is not used, and welds that are 10 to 40 times as deep as they are wide are not unusual. Welding usually takes place in a full or partial vacuum.

Electronic Data Processing (EDP). See: Data Processing.

Electronic Design Interchange Format. The electronic design interchange format (EDIF) is a standard which has been developed by the manufacturers and users of CAD equipment to allow products to be designed and to have the ability to interchange design data with any other product which also communicates with the EDIF standard. Currently, only a few CAD and CAE companies' products have been implemented with the complete EDIF standard. One reason companies are reluctant to implement the EDIF standard is that customers would then have the ability to pick and choose different manufacturer's equipment for their CAD/CAE needs instead of being forced to stay with a single choice for certain related and consecutive CAD/CAE functions. This is because the design libraries and database formats of one vendor cannot be utilized by another vendor.

Also see: CAD, CAE.

Electronic Industries Association. Electronic Industries Association (EIA) is the association responsible for the promulgation of recommended standards pertaining to electrical connections and communications.

Electronic Mail. Electronic mail is the use of a computer and sometimes an electronic data network to provide mail correspondence between users. Messages are sent via the computer to the address(es) of the recipient(s). Some of the unique benefits of electronic mail are the ability to broadcast messages to multiple people without having to rewrite the letter multiple times. Transmission of the messages is essentially immediate. The actual time until the recipient receives the message may be a matter of seconds to a few minutes. Other activities which electronic mail lends itself to are things such as scheduling meetings with the ability to query the calendars of all the desired attendees to determine a commonly available date. Even the confirmation response can be made by electronic mail.

Some bulletin board services exist on subscription services such as CompuServe and The Source which allow subscribers to leave messages for anyone in the world who is also a subscriber.

Electronics. Electronics is the branch of science pertaining to the study, control, and application of currents of free electrons, including the motion, emission, and behavior of the currents.

Electroplating. Electroplating is a process for depositing metal on a conductive surface that the cathode in an electrolytic bath containing dissolved salts of the metal being deposited.

Electropolishing. Electropolishing is an electrochemical method of selectively removing surface layers or burrs from metal parts. It is similar to electrochemical machining in many respects, but is used merely to brighten or deburr metals rather than to remove stock and control dimensions.

Electroslag Welding (ESW). Electroslag welding (ESW) is a fusion welding process for making thick joints in one pass by depositing metal at very high melting rates using heat generated within an electrically conductive molten slag that covers the weld zone. Generally, filler metal is provided by a consumable guide tube that surrounds the welding electrode. ESW is not an arc welding process, although an electric arc is used to initially melt the slag.

Electrostatic Plotter. An electrostatic plotter is a type of plotter that utilizes electrodes to produce black dots which generate the image on various types of specially treated media. An electrostatic plotter doesn't use inked tipped pens, but instead uses a backplate or head containing a matrix of electrodes which places point charges on the desired parts of the media where the image will appear. The media then passes through a toner bath, which adheres to the charged areas of the media. A single bath is required for monochrome printing. For color printing, successive passes for each color are required past the head and through the toner bath for black, cyan, magenta, and yellow.

A familiar example of electrostatic plotters are the products made by Versatec, a Xerox company, that currently owns the majority of the marketplace. The Versatec plotters can produce output on paper, transparency, film, mylar, fan fold, roll chart, vellum, and bond paper. All of these types of media must be specially treated with a capacitive surface coating to be used.

Also see: Pen Plotter, Plotter.

Emissive Electrode. An emissive electrode is a filler metal electrode consisting of a core of a bare electrode or a composite electrode to which a very light coating has been applied to produce a stable arc.

Emulate. To emulate is to operate in such a manner as to appear to have all the characteristics of another device; to be capable of functioning as a substitute for another device, with functionally identical appearances.

It is also the hardware imitation of another system's data processing capabilities to initiate the same results or permit software compatibility.

An example is a hardware and software combination that enables one computer to execute programs written for another computer, or a device that produces the same set of outputs for a given set of inputs as does another device.

An example of emulation is the personal computer clones, which must appear to the software and to the user to be functionally the same as the original device. Due to patent rights, the operating system BIOS (Basic Input/Output System) ROM cannot be literally duplicated. Therefore, the clone manufacturers use several ingenious methods to fool the software. As far as the user attributes are concerned, some manufacturers simply decide to improve upon the original design, as opposed to trying to copy the original.

Also see: Computers.

Emulator. An emulator is a device or macroprogram which allows one system to imitate another enabling it to run programs written for the other system. The same data can be accepted, the same programs executed, and the same results obtained.

Encoder. An encoder is an electromagnetic or optical transducer used to produce digital data (code) indicating angular or linear position.

A numerical encoder generates the numerical value of the measured position by interpreting and transmitting a code based upon a printed disk pattern. There are dark and light areas printed upon a disk, which when evaluated along the radial axis of the disk correspond to a binary coded number. The coded areas can be read using photoelectric, magnetic, or electrical conducting systems. The pulse generating encoder is a device which provides an electrical pulse for a given angular displacement of its shaft. These pulses can be counted and the relationship of the linear displacement to the encoder angle can be used to determine the machine slide displacement.

An electromagnetic encoder can generate a force drag on a sensitive drive axis, while optical encoders provide no force drag and complete electrical separation between the drive axis and the position sensing apparatus.

Encoder accuracy is the maximum difference in position between the input to the encoder and the position indicated by its output. Encoder ambiguity is the inherent error resulting from multiple bit changes at code transition positions.

End Effector. End effectors or grippers can be thought of as the ''action end'' of the robot. Since they are so closely related to the robot application, they are often customized for that application. When the application to be performed by the robot is one which requires picking up an object, moving it, or positioning it—hand-like operations—the device that does the picking up and moving is called a gripper. It can be a claw-like device with two or three fingers or it can lift by suction or magnets.

End effectors (end-of-arm tooling) consist of grippers, hands, holders, and other tools for handling components to be assembled. They are not generally integral parts of robots and commonly have to be designed and built for specific applications. They should be designed to handle as many different components as possible. For example, multiple sets of jaws can possibly be used with a single end effector to handle several types of components. The end effectors can be provided with or without mechanical compliance or sensory devices and vision systems for more precise location.

Mechanical clamping of workpieces is most common, with pneumatic, hydraulic, or electric operation. Grippers apply surface pressure on the components to be assembled. They are available with jaws, fingers, expansion-contraction devices, and anthropomorphic hands.

Other types of flexible fixtures are coming into use for the robotic assembly of components have complex shapes, such as turbine blades, for example. Both modular fixtures (in which the robot assembles the required fixture in the work space) and phase-change fixtures (such as fluidized beds) have great potential for specialized assembly requirements.

The two designations (end effector and gripper), are parallel in that they both describe the action that takes place at the end of the robot arm. The designation end effector is more universal.

There are literally thousands of gripper and end effector designs available today. While most grippers are mechanical, some jobs require vacuum or magnetic grippers. There are generic end effectors available, and there are companies who just make end effectors and sell them to robot manufacturers or sophisticated end users. Still today, most end effectors are designed for a specific application.

Peter McCormick of EOA Systems, noted in the SME book *Developing and Applying End of Arm Tooling*: ''The type of end of arm tooling (end effector) a robot uses has a strong impact on the performance of a robotic application. End of arm tooling concept, design, and implementation are critical considerations in the ultimate success of a robotic cell.

''In conceiving an end of arm tool, it is useful to consider three variables: the level of existing technology in the overall facility, the degree of flexibility desired in the tooling, and the role of the end of arm tooling within the system.

''Coming to an understanding of the level of existing technology in the overall facility will enable the tool designer to create parameters within which the tooling design can be begun. Instead of approaching the tooling design question strictly as an electro/mechanical problem, consider the implications of the design on the ability of plant personnel to operate and maintain the tooling. If the plant environment is decidedly low technology, then it is reasonable to assume that the plant personnel will be more comfortable with (and resist less) a low-technology tool compatible with the surrounding environment. On the other hand, experienced users of flexible automation equipment better understand the nuances of automated equipment and also will likely feel comfortable with the technical capabilities of a highly sophisticated tool.

''Tooling flexibliity also is important. In creating the concept for tooling, as well as the overall system, try to understand the user's product as an economic entity which has a life cycle and whose fluctuations in demand have a very real impact on the economic viability of a robotic system. Increased systems/tooling flexibility will be needed as product life cycles continue to shorten and product demand becomes more unpredictable. Combine these uncertainties with the current disinflationary period and the result is more stringent cost/benefit analysis by management. Increased systems/tooling flexibility will likely mean more value added to the product per dollar of capital expenditure to create the system. This becomes very significant as product cycles shorten in the face of unpredictable demand which forces a prudent manager to amortize the system over less unit volume during a shorter time frame. Also, the inability to easily raise prices during disinflationary times places cost cutting at the top of any manager's agenda. Most managers look at the costs associated with inventory levels as the first area to be cut. In this economic environment, it becomes important to analyze when a product is produced as well as at what cost a product is produced. Systems which can demonstrate the flexiblity to remain economical at widely ranging volumes will certainly increase the likelihood of installation.

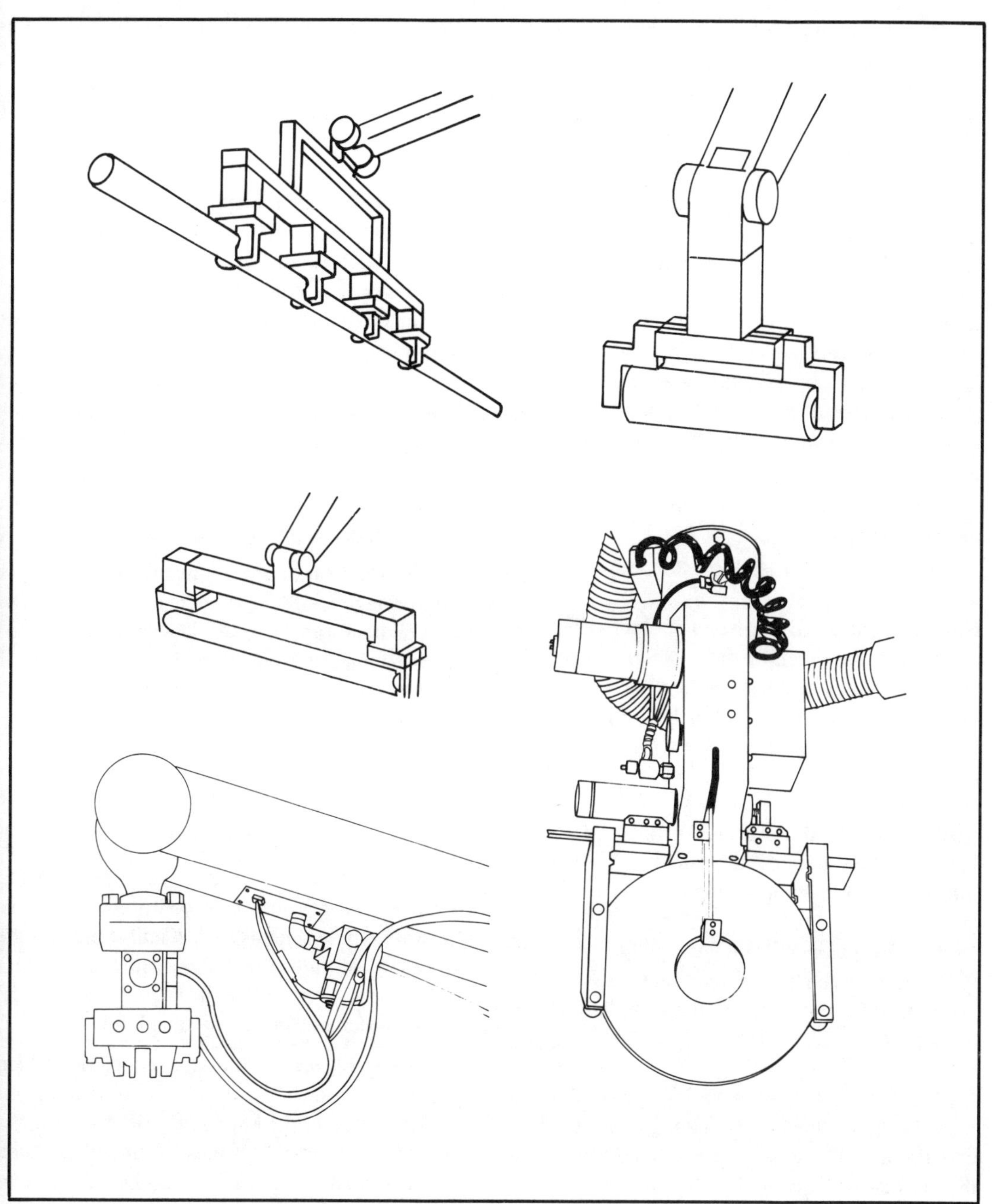

Sample end effectors.

"The remaining step in conceptualizing an end of arm tool is determining the role the end of arm tooling will play in achieving the system's requirements. At this point, the designer should try to understand the ability of the end of arm tool to perform a certain task as reliably as another component in the overall system. An example would be to incorporate a roll mechanism in the tooling to forego the expense of purchasing a six-axis machine. This type of examination leads to very substantive questions such as how can available technology be best applied, can other systems components more easily handle a perceived tooling function, how is sensory information to be generated and interpreted."

Some samples of end effectors are found on the previous page.

Also see: Gripper, Industrial Robots.

End of Block (EOB). An end of block (EOB) is the end of one block of data. An end of block character is a character or symbol punched on a tape denoting the end of a block of tape data; it is used to halt the tape reader after the block has been read.

End of Line Code. An end of line code is an indicator that defines the end of a line of information printed on a manuscript.

End of Program (EOP). An end of program (EOP) is a miscellaneous function signifying the last block of a program and the completion of a workpiece.

End of Tape (EOT). An end of tape (EOT) is a miscellaneous function signaling the spindle, coolant, and feed to stop after completion of all the commands in a block. This function is also used to reset control and/or the machine.

Endpoint. An endpoint is the coordinates on a display device to which a display writer is moved. An endpoint also is an extremity of a span, such as the last item to be processed before the completion of a process stage or the process itself.

Energize. Energize is a computer instruction setting a data table bit to 1, signifying ON if the pre-established conditions equate 1 with TRUE. The bit is reset to 0, signifying OFF if 0 is equal to FALSE.

Engineering. Engineering is the application of technology to provide useful functions for mankind. Engineering is the design of a product, the development of a process, or any other precise, calculated sequence of events which creates a desired end result.

There are several types of engineering disciplines, such as mechanical, chemical, electrical, industrial, and computer engineering. These scholastic disciplines of engineering are based on the type of knowledge which a person must be acquainted with to perform the work.

In an industrial environment, engineering is often categorized by the sequence of the design or manufacturing process. Design and development engineering refers to the development of product specifications. Manufacturing engineering and product engineering provide the expertise to support the production operations necessary to manufacture the product. Sales engineering and customer support engineering interface directly to the customer.

Leonardo Da Vinci and Sir Isaac Newton were engineers, not to be confused with those who were pure scientists. A scientist pursues the discovery of pure knowledge, while an engineer applies the knowledge in ways which useful work is then performed for mankind.

There are many professional societies which certify engineers and provide a common organization to further the knowledge of those practicing in the field. All states require those who provide certain engineering services for hire to be registered by the state, which have passing a state-administered exam as a prerequisite. This is done to protect the consumer by ensuring that those who offer their services are adequately qualified.

Engraving. Engraving is a process which cuts or forms lines or images into metal or wood. Die stampings, or chemicals may be used to engrave such parts as nameplates, or front panels.

Carl Anderson of Laser Fare Ltd. reports in his SME paper *Laser Engraving Using Vision Processing and Robotic Manipulation* that the laser, the robot, and the machine vision system have been tied together to manufacture class rings at Balfour Company.

The accompanying figure from his paper illustrates four zones in the work area.

The four zones are:

Zone 1: Laser front-end processor, laser computer for the beam steering, interface electronics, laser power supply and R.F. drive for Q-switch, and water to water heat exchanger.

Zone 2: Laser rail and all optical elements (including scanner head), mounted on a stationary platform doweled to the workstation surface.

Zone 3: Robot front-end processor. Robot and vision system associated electronics that is: power supply, controller and vision computer rack-mounted under the workstation.

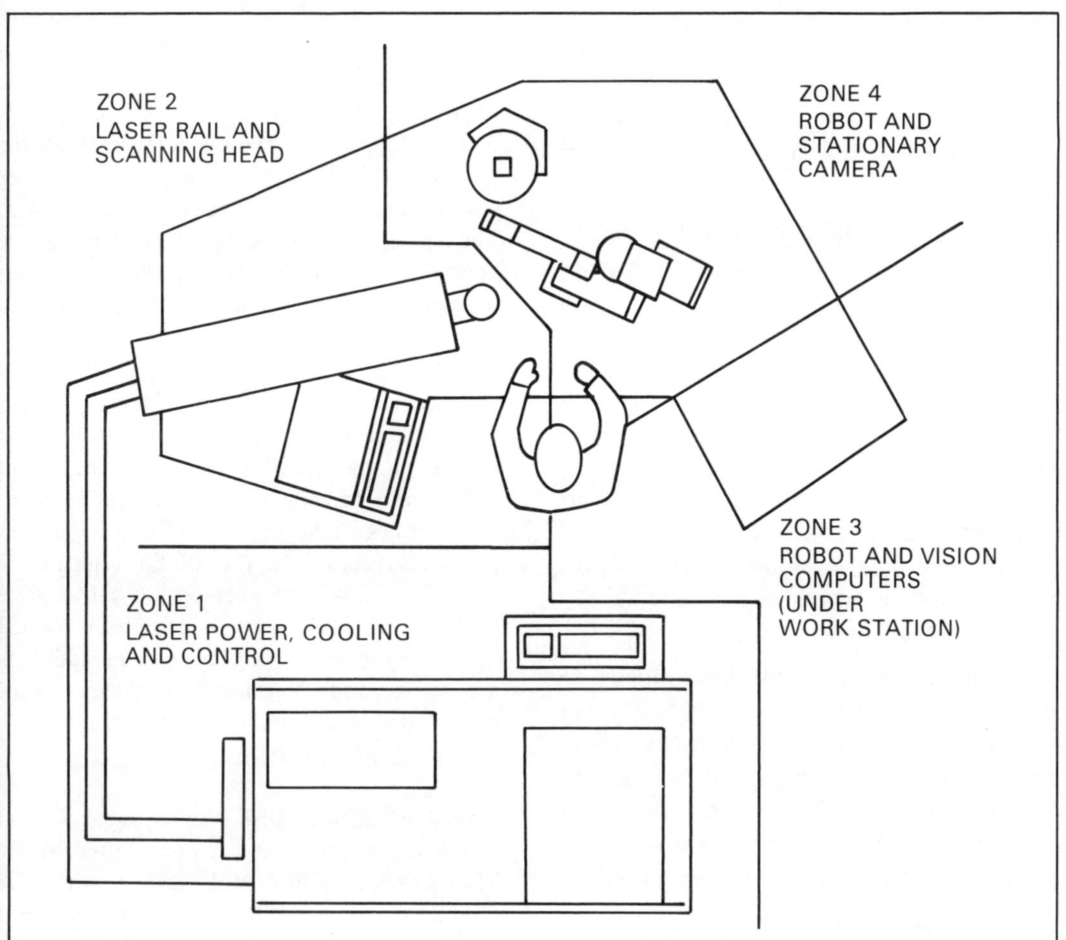

The four zones of a sample laser engraving system.

Zone 4: The robot and camera with light chamber. Pick-up and drop-off position/tooling located directly in front of operator.

All the components listed in zones 2, 3, and 4 are mounted on or in a structural steel framework/work surface. The robot and its controller are the decision makers for all system events. All other stations, the laser, pick-up tooling, as well as the camera/light-box, are mounted radially to the robot's primary axis.

Anderson explained: "The key to successful system operation was understanding and applying existing product terminology to programming efforts. The software was written to be operable by personnel of average skill found in the customer's manufacturing environment yet, it is comprehensive enough for the entire product range.

"To accomplish this, the operator has only to select from menu-driven screens guiding process choice and product configuration. As the choices further define the ring description, proper vision/object files are retrieved from memory. In addition, these choices govern the robot path used to present the ring to the camera due to differences in panel location and size. Type of processing is governed by this information as well. For example, processing may consist of external engraving on multiple panels exclusively, or may include internal engraving to identify material type and owner.

"Routines to introduce additional varieties or update families of objects is incorporated within the program, at a level normally inaccessible to the operator. This allows support personnel to quickly define object files (vision system definition of a panel) and store them to vision memory and/or teach new positions to the robot.

"After the operator (and us) had become skilled in operation of the system, the projected gains in productivity were approximately 8-20 times as fast, a favorable comparison to the projected estimate of 10."

Also see: Lasers.

Epitaxial Wafer. An epitaxial wafer is a thin disc or substrate of semiconductive material upon which a semiconductive layer having the same crystalline orientation as the substrate is grown by a vapor deposition process. This substrate is used as a source for all bipolar integrated circuits.

Also see: Wafer.

EPROM. An EPROM or erasable programmable read-only memory is a type of memory chip which can be programmed by electrical circuitry, and when programmed, the EPROM will retain the data without the continuous application of an electrical power source.

An EPROM can be erased by exposing the chip directly to ultraviolet (UV) light for approximately 30 minutes. The UV lights realigns the memory of the chip, eliminating all data. After erasure, the EPROM can be programmed again. The lens opening on the top of the chip must be kept covered with a light-blocking material, usually a piece of black electrical tape, to avoid decay and unwanted erasure of the data.

An advantage of EPROM over ROM (read-only memory) is that it can be reprogrammed and therefore used more than once. Also, since it can be programmed by the user, it has more utility, with different quantity scales of economical usefulness than ROM. EPROM is cheaper per unit quantity than EAROM (electrically alterable read-only memory), but unlike EAROM, EPROM must be completely erased before any programming changes or additions are made, and then must be totally reprogrammed as one continuous operation.

Also see: RAM, ROM.

Erasable Optical Disk. An erasable optical disk is one of three types of optical disk drives. The erasable optical disk differs from the CD-ROM and the WORM disks in that it uses a magneto-optical system that relies on an optically bi-stable medium that can shift between two states of reflectivity (which may differ only by a few percent) under the joint influence of the laser and a magnetic field. This technology utilizes the magnetic technology to write information to the disk and the laser technology for reading information from the disk, since, unlike a floppy diskette, there is no physical contact between the read/write mechanisms and the media.

While fully functional erasable optical disks have been successfully demonstrated for years, complete systems remain complex, expensive, large and elusive. The fact that erasable optical disks use magnetics causes them to not share the same invulnerability to damage and alteration as do CD-ROM and WORM.

Also see: CD-ROM, Optical Disk Drive, WORM.

Ergonomics. Ergonomics is the field of study of human motions and capabilities when performing tasks. While humans are capable of many varied tasks and motions, certain consistently repeated activities are less tiring and easier than others. An example of ergonomic conditions when typing at a keyboard would include the angle of the keyboard, changes in the angle of each row of keys, the distance which each key must be depressed, the feel of each key as it is depressed and the audible and tactile feedback. Chairs are a major area of ergonomic research with companies such as Herman Miller developing seemingly unusual and different chairs, but usually for reasons of ergonomics (as opposed to style or visual appeal). Five-star bases and hooded casters are some of the ergonomic additions to office chairs.

Error. An error is the discrepancy between a computed or measured value and an actual, specified, or theoretically correct value. *Error detecting* is the capability of detecting false results via a data code in which each acceptable term conforms to specific rules. Any variance from the rules during transmission or processing is detected as an error. *Error of form* is the deviation from the nominal surface that is not included in surface texture. The *error signal* indicates that a difference between output and input signals in a servosystem exists and that an alignment must be made between the controlling and controlled elements. The *error rate* is the ratio of the number of character substitutions for the total number of characters read.

Ethernet. Ethernet is one of the better known LANs and has one of the larger user bases. It was developed by Xerox's Palo Alto Labs in the 1970s. In 1980 the Digital Equipment Corporation and the Intel Corporation jointly published a local area network specification based on the Ethernet concepts.

Ethernet's primary characteristics include the CSMA/CD (carrier sense, multiple access/collision detect) protocol and baseband signaling on a shielded coaxial cable. The data rate is 10 mbit/s, with a provision of up to 1024 stations on the path. It uses a layering concept, somewhat similar to the low levels of the ISO model, and transmits user data in frames or packets. Ethernet uses Manchester encoding; at a 10 Mbit/s rate, each bit cell is 100ns long. Since the Manchester code provides for a signal transition within each bit cell, the Ethernet signal is self-clocking.

The Ethernet packet contains six fields, which are: Preamble, Destination, Source, Type, User Data, and CRC Field. Ethernet is designed around three layers which are the physical layer, the data link layer and the user layer.

Ethernet can support gateway interfaces into X.25 networks as well as IBM's SNA. Digital Equipment Corporation has integrated Ethernet into phase IV of its Digital Network Architecture (DNA). Users of the UNIX operating system can also utilize Ethernet. The IEEE standards committee has endorsed most all aspects of the Ethernet specification with their IEEE 802.3 specification.

Also see: CSMA/CD (carrier sense multiple access, collision detect), ISO, Local Area Network.

Even Parity Check. An even parity check is a check in which the sum of the 0's or 1's in a binary word is expected to be even.

Also see: Parity.

EXAPT. EXAPT is an acronym for extended subset of APT, a language processor developed in Germany and commonly used for point-to-point or lathe work.

Excess-three Code. An excess-three code is a number code in which the decimal digit **n** is represented by the four-bit binary equivalent of that number plus three.

Executive. An executive is a software routine which controls the execution of computer instructions or programs on the basis of established priorities and real-time or demand requirements.

Expanded Memory. Expanded memory is a special kind of extra or shadow memory which can be installed in an IBM PC, PC-XT, PC-AT, 386 PC, or a clone of any of these. The DOS operating system limits the addressing range of conventional memory to between 0K and 640K for the user and between 640K and 1024K for the ROM BIOS address. Expanded memory is up to 8 MB of memory on a PC and PC-XT, or up to 16 MB of memory on a PC-AT or 386 PC, which is accessed by paging four 16K windows into and out of a 64K area of unused reserved conventional memory in the main 640K to 1024K area of memory. Expanded memory consists of three parts: one piece of hardware (the bank-switched memory board), and two pieces of software (the expanded memory manager—known as the EMM—and the application program that uses the memory). Expanded memory can consist of anywhere from 64K up to 8 or 16 megabytes of memory, subdivided into small 16K pages that can be individually readdressed through bank-switching. Lotus Development Corporation, along with Intel and Microsoft, jointly defined a procedure for performing the bank-switching which is referred to as the Lotus/Intel/Microsoft Expanded Memory Specification or simply EMS. EMS can only be used to store data and to swap programs and graphics screens. An AST/Quadram/Ashton-Tate Enhanced Expanded Memory Specification or EEMS has also been defined to allow even more flexibility by not only allowing the storage of data and to be used as a swap area, but it can be used to run programs. Expanded memory is indirectly addressed by using the swap windows, while extended memory is directly addressable.

Also see: DOS, Extended Memory, Personal Computer.

Expert Systems. An expert system is usually a software program which may be written using artificial intelligence concepts. It is created to solve problems in a particular domain. A domain is a bounded area of knowledge and possibilities. While artificial intelligence is simply the capacity for a nonhuman system or device to exhibit and perform logical reasoning, an expert system is a programmed execution of events using information and knowledge which has been provided by people who are experts in the field. An expert system uses a compilation of the knowledge of one or more expert persons and through a computer program, performs the decision making as if the expert people were actually performing the tasks.

Most expert systems use a computer as the source of intelligence, however, hardwired logic can be utilized. Expert systems, while similar to artificial intelligence in that they utilize the basic concepts, are considered to be limited and bounded in their capability. An expert system is the application of artificial intelligence concepts to solve a specific range of functions. The intent in developing an expert system is to provide some level of nonhuman decision making, without having to completely utilize the on-line interaction of a human. An expert system may be complex or very simple such as a single decision based on the presence or absence of a single attribute. A *go*, *no-go* inspection machine is a simple expert system, such as an egg sorter which sorts according to size.

Expert systems didn't exist significantly until the widespread availability of computers provided the necessary intelligence, the ability to maintain large amounts of data and information, and the capability of searching and manipulating the data. An expert system may have to search a large decision table or a lengthy decision tree each time some input data is received.

Today expert systems appear everywhere from consumer products and automobiles to classrooms and hospitals.

History. Expert systems were not prevalant until computers became available in greater profusion. Since programming is the usual way expert systems are created, most of today's expert systems are computer based. Programming languages have been developed to allow easier programming of expert systems. The most accommodating languages provide a syntax and structure which readily adapt to the heuristic rules of decision making.

Attempts to create expert systems have existed for quite some time. Man has always been fascinated with the idea of creating a "smart" computer or machine. Companies have been founded solely to pursue the goal of developing expert systems. Learning Machines, Inc. and Symbolics are two companies that have designed and marketed computers optimized for artificial intelligence and expert system applications.

Discussion. Since a person trying to put the entire knowledge of any single person, or multiple people, into a computer would be a prohibitive task due to the vast quantity of information, an expert system which is focused upon a certain domain becomes the only viable option for creating a usable intelligent system.

The more an expert system can handle a scenario which would be cumbersome and difficult for a person to handle, the more critical that system is to enabling the task to be performed better. In many cases the expert system may only modify the program being executed by the main control system, so the value of the expert system could vary from nothing to a very great effect.

Since an expert system uses artificial intelligence to perform its calculations and manipulate the data, an expert system is really the implementation of artificial intelligence to perform a useful function in a particular domain or application.

Sometimes expert systems clash with other determining forces, especially when laws apply. Even though the statistical records support a decision in one direction, if statutes dictate another course of action, the system has no choice but to abide by the law. This is why even though law enforcement agencies may use expert systems to predict a crime or potential guilt, they must still operate by the law, and "blind" justice should always prevail.

Applications. The applications of expert systems in use vary widely. Some are simple uses of noncomputerized systems, while others are extremely complicated and intelligent.

One of the simplest forms of an expert system is the inertial seat belt retractors and seat back latches on automobile passenger restraint systems. The inertial locks allow the belt to extend and retract freely during normal acceleration and deceleration. However, when the inertia lock senses deceleration of the automobile greater than a predetermined amount, the mechanism mechanically locks to prevent the belt from extending and to provide the restraint of the passenger during a sudden stop. Once the deceleration forces have subsided, the lock returns to its previous operation. The inertia seat back latches work in a similar manner by allowing the front seat of a two-door car to be tilted forward without releasing any locks by simply pushing the seat back forward. During a sudden stop, the inertia lock holds the seat back in a rearward position, until the deceleration forces have subsided. This type of operation provides a minimum of inconvenience and intrusion for the user.

Another more sophisticated form of expert system is the intelligent stop light which many street intersections now use. The different lanes of the intersection have capacitance loops in the roadbed which sense the presence or absence of vehicles in the different lanes. Some lanes may have multiple loops to detect the presence of up to a quantity of vehicles. The knowledge of which lanes have vehicles and which do not influences the operation of the signal lights with regards to the length of time the lights remain in a certain direction, and whether or not certain lanes ever receive a green light. For example, if a left turn lane does not have any vehicles sensed, the left turn arrow will not receive its turn in the sequence of lights. This is sometimes apparent when a single vehicle is in the left turn lane, but either too far forward into the intersection, or too far back from the stop line, to be sensed by the roadbed loop. In this situation, the light for this lane will remain red, even though there is a vehicle waiting. The benefits of this type of signal light control is an improved flow of traffic by attempting to maintain a green light in the direction of maximum traffic flow. On some major streets and roads, the lights are programmed such that unless a vehicle is sensed on the side street, the light will stay green indefinitely in the direction of major traffic flow.

Another more sophisticated usage of an expert system is used by the IRS. It has programmed into the computers that process income tax returns, certain thresholds and "flags" which when matched by parameters in a person's return, generate a notice to the IRS examiner to possibly audit the return. These flags are usually based on combinations of deductions and lack of declarations that usually would be impossible to have. Other thresholds are based on certain maximum dollar amounts in different areas, especially in deduction and tax credit areas. For example, a person who listed charitable contributions which exceeded a certain percentage of his taxable income would probably draw an audit, regardless of what the contributions were stated to be.

An expert system which is relatively simple as far as the algorithms and calculations required, but is a constantly changing and growing database is Digital Equipment Corporation's order validation system. The system receives a field order as input and analyzes the items listed on the order. If a combination of options are listed which cannot "play" together or if certain basic and critical items which normally would be included in what seems to be the systems and configurations that have been ordered are not listed, the expert system will prompt the salesperson if the lack of the items is an oversight and whether or not the salesperson is aware the system cannot function properly as it is currently listed. This saves valuable time for customers by ensuring all the necessary pieces of a functioning system are ordered. It also saves Digital by avoiding the rush shipments in trying to correct a misplaced order.

Credit card purchase approval systems are another form of expert systems which contain all the information necessary to determine whether or not the card number is valid, if the card is restricted or reported lost or stolen, and if the amount of the purchase is within the allowable remaining credit limit. Because of the way the expert systems are set up, if a person makes several purchases and returns them in a short time, or if a sales clerk makes multiple attempts to get approval for the same purchase using the automated input system, the card holder may find successive purchase attempts being rejected, even though the customer is well below his credit limit. This is due to the expert system not clearing out the duplicate requests for the same purchase until a day or so later. Meanwhile the system has suspended charge approval since the system's total requests exceed the allowable charge limit.

The phone company computers which manage credit card calls are another form of expert systems that not only handle the input of data, but also analyze the numbers which are given to ensure that the combinations are actually a valid telephone credit card number, based on the allowable digits in each position of the number. Another use by the phone company of expert systems is in subscriber call routing. A computer system constantly monitors and analyzes the call volume and locations to determine the best routings for optimum performance. A call between two points may be routed three to one hundred times further than a direct point-to-point routing, simply to utilize the available capacity.

Several stock brokerage houses now use computerized programs to analyze the stock market and generate "buy" and "sell" signals to their clients. One of the biggest problems of expert systems plagues these expert systems. An expert system only contains the knowledge which was put into it. An expert system cannot reason and learn new rules and heuristics by itself. Therefore, if new situations arise or values go outside their expected boundaries, the expert system might give outputs which are not at all in line with what a human expert feels should be deduced. This is because the expert system is "playing today's ballgame" using yesterday's rules.

Some machining centers and flexible manufacturing systems use expert systems to determine the optimum routing of material within the cell or the most effective time to spend in each station. Compensation for tool wear and alteration of the working sequence when a machine or station in the cell fails is a major application for expert systems.

In CAD/CAM systems, expert systems may suggest materials to be used for a part, they may suggest cutting tool profiles and passes for machining a part, or the heat treatment of a part for a particular application.

In data communication systems, a use of expert systems is in statistical multiplexers which use the recent historical distribution of activity by the different nodes to assume that they will continue to have a proportional amount of activity on the network. Based on this assumption, the statistical multiplexer polls the more active points more frequently than the less active nodes. This changes as the historical information changes as less active nodes become more active and vice versa.

A very helpful use of expert systems is in hospitals where they are used by doctors to help raise all the possibilities when diagnosing a patient's problems. As the physicians carry out investigations and examinations, they feed the information into the computer, which in turn continues to search through a preprogrammed decision tree and suggests causes. This type of system works much in the same way as following the tree-structured lead through trouble-shooting sections in automotive repair manuals.

Glossary. *artificial intelligence:* Artificial intelligence is the ability for a nonhuman device to reason and make logical decision based on information and a known priority and hierarchy of decision making.

CAD/CAM: Computer Aided Design/Computer Aided Manufacturing.

data: A representation of facts, instructions, concepts, numerical and alphabetical characters, etc., in a manner suitable for communicating, interpreting, and processing by humans or by automatic means.

heat treating: A sequence of controlled heating and cooling operations applied to a solid metal to impart desired properties.

heuristic: An idea of information which aids in determining the solution to a problem.

machining center: A machine tool, usually numerically controlled, that can automatically drill, ream, tap, mill, and bore workpieces. It is often equipped with a system for automatic toolchanging.

Also see: Artificial Intelligence, Heuristic, Numerical Control.

Explosion Welding (EXW). Explosion welding is a solid-state welding process in which controlled energy from a detonating explosive causes impact bonding between similar or dissimilar metals. No diffusion occurs, no intermediate filler metal is required, and no external heat is applied. EXW is most often used to produce structural sheet, plate, strip or tubing clad with a corrosion resistant material that would be difficult to bond using another process.

Also see: Welding.

Extended Arithmetic Element. An extended arithmetic element is a CPU logic element providing hardware-implemented capabilities that include multiply, divide, and normalize functions.

Extended Data Comparison. An extended data comparison is an on-line application diagnostic routine that is user programmed to allow specific data files to be compared to the data table at various steps in the application cycle. A printed report describing the type of problem and its location is produced when a discrepancy is detected.

Extended Memory. Extended memory is a type of memory which is only available in the architecture of the IBM PC-AT, 386-PC, and clones of them. Extended memory is memory which can be assigned the address range of from 1 MB to 16 MB. Extended memory must start at 1 MB and be contiguous. Current versions of DOS only allow extended memory to be used for RAM disks, print spoolers, and disk caches.

Extended memory addressing by the Intel 80286 and 80386 CPU chips is done with the CPU being put into protected mode. Protected mode allows the operating system to erect barriers to prevent a program from interfering with the operation of another program or of the operating system itself. When PC DOS and MS DOS were originally developed, the 80286 processor and protected mode capabilities weren't being considered so therefore, current versions of DOS operate the 80286 processor in what is termed real mode, which is to emulate an 8088 processor. Versions of OS/2, the new IBM operating system, are expected to provide for protected mode operation of the 80286 and 80386 processors and allow the use of extended memory without special programming by the application software. Currently, some software uses extended memory by providing the necessary additional software capabilities.

Also see: DOS, Expanded Memory, Personal Computer.

Extended Storage. Extended storage is a storage medium or unit, such as a floppy disk or punched tape, that is not an internal part of the computer.

Extrusion. Extrusion is a plastic deformation process in which material is forced under pressure to flow through one or more die orifices to produce products of the desired configuration. In hot extrusion, heated billets are reduced in size and forced to flow through dies to form products of uniform cross section along their continuous lengths. With special tooling, stepped and tapered extrusions are produced. Cold and warm extrusion differ from hot extrusion in that the starting workpieces (often called slugs) are cold (at room temperature) or warm (heated to below the critical temperature of the metal). Metal is forced to flow by plastic deformation under compression around punches and into or through shape-forming dies to produce parts of the desired configuration. Since the temperatures of the slugs are always below the recrystallization temperatures of the metals to be formed, the process is essentially one of cold working.

The extrusion process provides a practical forming method for producing a limitless variety of parallel-surfaced shapes to meet almost any design requirement. Other advantages include improving the microstructure and physical properties of the material, maintaining close tolerances, material conservation, economical production, and increased design flexibility. Some shapes which cannot be roll formed because of their geometry can be readily extruded. Advantages of cold and warm extrusion include substantial cost savings in many applications, fast production rates, improved physical properties, close tolerances, energy conservation, and elimination of pollution problems.

Computers are being used by some extruders to design and manufacture (with computer numerical control machines) dies and to select process variables such as extruding speed and billet temperatures. Software employed is based primarily upon an analysis of metal flow. Various design stages are displayed on a CRT screen, and the designer interacts with a computer to change or modify the design, based on experience.

F

Fabrication. Fabrication normally transforms components or raw materials into a larger variety of products. The fabrication process includes operations which change the shape or dimensions of the part, its surface finish or material characteristics.

Typically, many fabrication processes include metal. Metal can be rolled out into sheets which can be sheared, stamped, cut or bent into desired shapes. The detailed manufacturing method to form the desired shape typically involves many detailed steps. The fabrication steps can be outlines on a routing or process sheet to indicate to the operator or fabricator the series of steps required to complete the process. In addition to the routing, additional documentation may be required to indicate the precise operations. This may be in the form of a detailed drawing which might include the lubricants, adhesives, and dimensions for the term.

The overall fabrication process initially requires a design of each component part or product. The list of elements must be generated for each product. For each listed item, a make/buy decision must be made. For those items whose production is justified in-house, specific process steps must be designed. Parts selected for purchase can be conducted by the procurement function.

Face. A face is a flat surface, usually at right angles and adjacent to the ground hole.

Face Milling. Face milling is a form of milling that produces a flat surface generally at right angles to the rotating axis of a cutter having teeth or inserts both on its periphery and on its end face.

Factory Hardened. Factory hardened is a term which is used to describe products that have been designed and manufactured to be able to survive while operating in a factory environment. Depending upon the industry, the equipment on the shop floor may be subjected to elevated temperatures, dust and chemical contaminants. When metal cutting is done, there is usually cooling fluid mist in the air, which permeates everything on the shop floor. Some machines emit strong electromagnetic interference which can damage devices with high levels of susceptibility. Even rough handling by operators is considered a hazard of the shop floor. Unstable and unclean power supplies are typical of a factory, with many periods of power fluctuation as machines start and stop.

Due to the hazards of the factory shop floor, typical designs for products aren't adequate. For example, fan cooling for computers would circulate contaminates into the system, so enclosed cabinets and cases are used with convection cooling of the internal devices. The National Electrical Manufacturer's Association has set standards for electrical enclosures with different designations for airtight, watertight, and explosion-proof devices.

Also see: NEMA.

Factory of the Future. The Factory of the Future has been the dream of many companies since robots and other forms of automation were introduced. It is a factory where, ideally, everything is done automatically with little or no human intervention (be it order entry, scheduling, warehousing, tool and fixture development, production, and so on. The factory will, or should, contain automated flexible manufacturing systems, programmable controls, machine tools, and unmanned material handling equipment, as well as sensor-loaded robots. Additionally, computers and robots will communicate and supervise other computers and robots.

The range of implementation of the Factory of the Future spans from "islands of automation" (that are custom-designed groups of computers and automated production machine and robots) to a factory nearly full of automated devices.

Richard Preston of the BDM Corporation discusses the United States Air Force Integrated Computer Aided Manufacturing Program Study *Conceptual Study for the Factory of the Future* in his SME Technical Paper *Factory of the Future Strategy*. He wrote: "Since the completion of the FOF project in late 1984, its results have been used successfully in the application of advanced computer and automated technologies in numerous companies. One of the more recent applications of FOF results occurred at Teledyne Microelectronics in the ongoing program for a totally integrated and partially automated facility for the manufacture of hybrid microcircuits."

This paper also sums up what appears to be the current feeling regarding FOF in that "the conclusion derived by the (USAF) project team is that there are no "pat" answers that will lead to instant solutions for current enterprise costs, productivity, and quality issues."

On the positive side, "the project results did provide identification of enterprise concepts and cost indicators that can be employed to analyze and compare existing functions to aid in their transformation into improved operations that take advantage of automation technology. The application of the FOF results provides a structured pattern or "road map" for the achievement of a cost-efficient, high-quality manufacturing enterprise.

Fail-safe. A fail-safe system is a system which if it fails, will always fail in a mode or manner which is safe. Examples are a relay which is powered on, so it will automatically open if it fails, or a motor which will default to a stopped position if it ceases to operate properly. The reason for fail-safe devices is safety of personnel and equipment. An automated device which would continue in an uncontrolled state upon failure of some part of the control system would be a hazard. Some computers and controllers even monitor themselves with a provide signal to ensure they are functioning properly (known as a *watchdog timer*). If the signal isn't detected after a certain maximum time, the system assumes a fault and shuts itself down or the particular portion which is being monitored. OSHA, the Occupational Safety and Health Administration, requires fail-safe pushbuttons and other operator devices on industrial equipment.

Also see: Default.

Fail-safe Work Method. Fail-safe work methods are methods of performing activities so that only correct actions can be completed. For example, a part with holes in the proper place can lock in a jig.

Also see: Jig, Just In Time.

Family of Parts. Family of parts describes a technique for grouping a series of similar parts into one family of parts for efficiency of engineering design and manufacturing. This technique in the design application avoids repetitive designs.

A common technique is to establish a drawing to cover a range of part numbers with minor variations. These variations can be referenced on an attached table, preventing the need to redraw pictures numerous times. Families of parts also use common routings where the specific operations are the same and only the specification changes are referenced on the common drawings.

Family of parts can be used in context with products categorized in families based on similarity. The product families are then placed with similar items and referenced using a bill of material structure to include all parts. The numbering scheme can be established to make a series (or family) of items easily recognizable.

Families of parts is also used in warehousing location systems to facilitate common equipment, racking and picking techniques. The commonality of parts within a family is the significant aspect in the effective use of the overall grouping.

Fan-in. Fan-in is the number of inputs available for connecting to a specific logic circuit, state, or function.

Fan-out. Fan-out is the number of parallel loads that can be driven by a circuit input.

Fastening. Fastening is the process of temporarily or permanently attaching two or more objects together. Items may be fastened in a rigid or flexible manner. Some of the major methods for fastening items include mechanical fastening, welding, brazing, cryofitting, crimping, swaging, soldering, adhesives, and shrink or force fits. Other fastening methods include stitching, stapling, riveting, and threaded fasteners. Selection of a particular fastening method for a specific application depends upon several factors, including the following:

- The chemical properties of materials to be joined.
- Size, weight, and geometry of the components to be assembled.
- Joint designs and accessibility.
- Functional requirements of the assembled product, including strength, reliability, environment, appearance, and whether it has to be dismantled for maintenance or repair.
- Production requirements (rate and total).
- Edge and surface preparations.
- Adaptability and compatibility of the joining method to automation, and effects on joint properties.
- Available equipment.
- Tooling requirements.
- Costs.
- Safety considerations.

Fault-tolerant Computer. A fault-tolerant computer is an arrangement of computer hardware and software. This arrangement allows parts of the hardware and/or software to fail without causing any detrimental effect on the operation of the application program. A fault-tolerant computer usually has multiple processors and triplicates of every critical component. Fault-tolerant computing usually employs three of every component, with "voting" occurring among the three independent but identical devices. The assumption is always made that the majority decision is the correct decision, either unanimous (three out of three) or majority (two out of three). When an opposing vote occurs, the other two processors raise a flag to indicate that the dissenter may be malfunctioning. Some systems automatically take the dissenter off-line until it passes a number of checks and tests at which time it may come back on-line.

Some of the major manufacturers of fault-tolerant computers are August Systems, Sequent, Tandem, and Triconex.

Redundant systems are different than fault-tolerant systems because a redundant system contains a backup or spare which may or may not be identical hardware, but is capable of duplicating the functions being performed by the primary device. A redundant system is only utilized when the primary system fails, while a fault-tolerant system uses all components of the triplicated system.

Also see: Computer, Off-line, On-line.

FCC. See: Federal Communications Commission.

Feather Burr. A feather burr is a very fine or thin burr.

Feather Edge. A feather edge is the same as a feather burr except that feather edge can also refer to the ends of a lead-in or lead-out thread, which is a very thin machined ridge. It is sometimes called a wire edge or whisker-type burr.

Features. Features are distinguishable characteristics of an object, device, or product. Features may be geometric shapes, colors, texture, odors, or other physical properties. Some well-known objects can be described by their features and may be easily recognized, even when seen in silhouette.

Features are especially important to machine vision systems and object recognition systems. Elementary features are the basic geometric shapes such as straight lines, edges, holes, and corners. The process of identifying features in an object recognition system is called *feature extraction*. In feature extraction, basic geometric shapes are identified and the relationships between the different features are compared to predefined patterns in order to identify an object.

When describing a product, the features are the notable functional and usable characteristics. For example, the features of a computer system might be the number of terminal ports, the speed of the processor, the system memory, or the file structure.

Also see: Machine Vision, Optical Character Recognition.

Federal Communications Commission (FCC). The Federal Communications Commission is a government body which regulates and licenses all public and commercial users of the airwaves. The FCC assigns radio frequencies for use within local areas for wireless transmission. The FCC also sets standards regarding the amount of ''noise'' and other useless emissions from electronic equipment.

The FCC has established different standards for emitted radiation and susceptibility for commercial, industrial, and consumer products. Any product which receives or transmits audio, visual, or data communication transmission, or which produces noise and emissions which interfere with or disrupt wireless transmission, is covered by FCC standards. Television sets, radios, telephones, CB radios, and personal computers are some of the products which are regulated by the FCC. Commercial radio and television are also regulated by the FCC. To ensure compliance with its regulations, the FCC routinely monitors the airways to detect offenders.

Also see: Standards.

Feedback. Feedback is the signal or data sent back to a commanding unit from a controlled machine or process for use as input in subsequent operations.

Feedback Control. Feedback control is closed loop control wherein a control system uses information regarding the output of the controlled process as part of the input information to the controller. The feedback may be an analog signal or it may be a digital signal. Sometimes the feedback is an opposite signal to the other primary input signal, in which case it is referred to as *negative feedback* because the feedback signal tries to force the controller in the opposite or negative direction of the primary input signal. Negative feedback is also referred to as *damping*.

Process control systems usually employ feedback control, as do most motion control systems, however, some motion control systems don't use any feedback, in which case they are referred to as *open-loop control* systems. An AC electric motor operating a common household appliance is an example of open-loop control. A servomotor is an example of a closed-loop control system while a stepper motor is an open-loop control system.

Also see: Closed-loop Control, Controller.

Feedback Device. A feedback device is the element of a control system that converts linear or rotary motion to an electrical signal for subsequent comparison to an input signal.

Feedback Loop. A feedback loop is the portion of a closed-loop system that provides controlled response information enabling a comparison of the information with a referenced command.

Feed Rate. The feed rate is the rate of movement between a machine element and a workpiece in the direction of cutting. It is expressed as a unit of distance relative to time or a machine function such as spindle rotation or table stroke.

Fetch. Fetch is a programming instruction used to locate a specified amount of data in storage and to load it elsewhere.

Fettling. Fettling is the process of removing sprues, flash, risers, gates, and runners from castings, and of smoothing parting lines and wide areas of flash. In 1983, one of every 100 to 200 robots was involved in fettling or deburring.

Industrial robots used for either fettling or deburring are selected based on tool and part weight, burr accessibility, forces involved, dexterity required, part complexity and the requirements of work space.

The removal of risers, sprues, gates, and runners from castings are currently tasks undertaken by robotic deburring. The casting-related finishing, however, is referred to as *fettling* or *cleaning* of castings.

There are some significant differences between deburring and fettling. Fettling requires cutting through thick sections of cast metal and smoothing wide sections of flash. Fettling requires high-power tools and large forces occur. Therefore, more powerful robots are needed.

Typically, fettling robots handle 45 hp tools weighing 50 lbs. Deburring, on the other hand, typically uses 1/3 hp with force of from 1 to 10 lbs. A large, rigid robot is required for fettling. Deburring, however, requires a soft touch and quick response so smaller robots are used.

Fettling is a process for working on or near surfaces. Deburring works only on the edges. Larger tools such as flap wheels, grinding discs, end mills, or rotary files are used in fettling. Deburring tools typically are small rotary files or brushes.

Fiber Optics. In 1956, Kapany first defined fiber optics as the art of the active and passive guidance of light (rays and waveguide modes) in the ultraviolet, visible, and infrared regions of the spectrum along transparent fibers through predetermined paths.

Fiber optics can also be described as the technology of guidance of optical power, including rays and waveguide modes of electromagnetic waves along conductors of electromagnetic waves in the visible and near-visible region of the frequency spectrum specifically when the optical energy is guided to another location through thin transparent strands. Techniques include conveying light or images through a particular configuration of glass or plastic fibers. Incoherent optical fibers will transmit light, as a pipe will transmit water. They will not transmit an image. Coherent optical fibers can transmit an image through perfectly aligned, small (10-12 microns), clad, optical fibers. Specialty fiber optics combine coherent and incoherent aspects.

Fiber Reinforced Plastic (FRP). Fiber reinforced plastic is a variation of plastic compound which has fibers of reinforcement intertwined in the plastic. The most common reinforcement material is glass fibers. The use of a fibrous reinforcement improves some of the properties of the plastic such as improving the flexural modulus and the heat deflection temperature in most compounds. In some types of plastics, the addition of glass reinforcement results in decreased impact resistance. In others, a loss of transparency occurs.

Fiber reinforced plastics can be used for structural members, sometimes replacing metal or wood, while providing imperviousness to caustic chemicals or moisture. As the high temperature characteristics of FRPs are improved, engine components and other similar products can be fashioned from plastics. Plastics are lighter in weight than some traditional materials, with the ability to mold in color and surface treatments. Complex parts can be economically molded from plastics, while machining or molding from metal could leave protrusions with too little strength.

FIFO (First In, First Out). FIFO is a term which refers to inventory control wherein the stock which was the first to be put into the storage area is the stock which is taken out first. This is usually done to allow the oldest product to be utilized first. If the product is foodstuffs which have a short usable life, this practice reduces the danger of product spoilage. The practice of FIFO doesn't hinder any other functions of a buffer storage or queue, although some problems may be caused in having to shuffle inventory to locate the oldest stock in the storage area during each session of retrieving inventory.

The concept of FIFO is also applied to inventory valuation in cases where the physical inventory is not segregated by age. When an item is removed from inventory (no matter what the age of the item), the value of the inventory is reduced by the original cost of the oldest item.

Also see: Inventory Control, Queue, LIFO (Last In, First Out).

Fifth Generation Management. The SME Blue Book *Fifth Generation Management for Fifth Generation Technology* provides a worthwhile discussion of this topic. It notes, "Fifth Generation Management (FGM) will be: The sharing of visions through communications and trust, based on honesty and integrity, accomplished by the integration of labor, the awareness of interdependencies, and management through the use of Program Management tools, which achieve further enlightenment of employees in order to manage competitiveness."

It lists as the components:

- Trust of superiors, peers, subordinates inside and outside of one's organization.
- Open communications between people and organizations.
- Removal of organizational barriers.
- Multiorganizational team work.
- Teams consisting of knowledge-based individuals.
- People and work are managed using program management techniques.
- Give people more responsibility and take advantage of their experience and knowledge.

Simply put, Fifth Generation Management (FGM) assumes a well-developed and flexible infrastructure of networked functions, together with their computers and applications, which are capable of referencing a common data architecture. Each of the functions becomes a node, or decision point, in the network. These nodes become reference points or knowledge centers, capable of teaming with other nodes in the support of the enterprise's business strategy. Rather than being held together by a "command and control structure," FGM is coordinated by its shared vision, values, and culture. This arrangement frees the organization from being locked into the management of "sequential hand-offs" and provides a way of coordinating the "iterative dialogue" which must go on between functions, such as "design for assembly." In short, FGM, makes it possible to effectively leverage the information infrastructure for competitive advantage.

FGM calls for a new leadership style and expects more of all employees in dealing with business variety so windows of business and technical opportunity can be more effectively met. Moreover, it recognizes that human interaction is as important as product quality, if competitive advantage is to be achieved. Finally, it taps into the power, wisdom and insight of its managers, professionals and employees who have, as a team in a nodal network, learned to use computer-based resources to enhance their decision making capabilities.

The following is a list of the five generations of enterprise management.

In the first four generations of enterprise management, raw materials and information are passed *serially* from one department to the next, much as in Adam Smith's pin-making factory with its division and subdivision of labor (*Wealth of Nations*, 1776), and reinforced by the work of Fredrick Winslow Taylor in his *Scientific Management*. Moreover, the hierarchical mode of organization predominates, even in third and fourth generation management. Matrix management does not fundamentally challenge the bases of power in second generation organizations. Fourth generation management does help to build the necessary infrastructure for a new way of doing business, although it does not overcome the habits and practices of second generation management.

Fourth generation management, digital interfacing, is what should be called *computer interfaced manufacturing* or *CIM I*. So much of what is called computer integrated manufacturing (CIM) is, in reality, computer *interfaced* manufacturing.

Fifth generation management, nodal networking, assumes the *computer integrative management of the manufacturing enterprise* or *CIM II*. First, its focus in not just on the manufacturing function, but the entire enterprise. Second, it is integrative, not integrated, because one is involved in a continually evolving integrative process. No one will awaken one day to find everything *integrated*. Third, each of the departmental functions and subfunctions become nodes, or decision points, on a network capable of bringing their accumulated knowledge to bear in an interactive mode at the functions work in parallel.

Interfacing is not the same thing as *integrating*, and yet the confusion of the two is leading to deep frustrations, and in some cases a rejection of the idea of CIM itself. CIM I will fail to solve the fundamental problems of the hierarchical organization, because it, like matrix management, is an overlay on second generation management.

File Management System. A file management system is software which performs the basic functions of manipulating data in a file structure within a computer. A file management system handles the creation, alteration, review, and outputting of file information. A goal of some file management systems is to manipulate data as well as to store and retrieve it. The report generation capabilities of most file management systems allow the creation of custom reports.

Commercially available file management systems may provide only basic functions such as address file type functions, or file cabinet storage and retrieval. More sophisticated file management systems provide extensive abilities to construct custom reports, and to interface with other popular application programs such as dBase III Plus and Lotus 1-2-3, and plain ASCII text files. They provide the ability to transfer data in both directions.

Typical applications for file management systems are name and address lists, mailing labels, reports, inventory, mail merge (form letters), calculations, and accounts receivable and payable.

Also see: Database Management System.

Fillet. A fillet is a rounded corner or arc that blends together two intersecting curves or lines. In three dimensions, a fillet surface is a transition surface that blends together two surfaces.

Also see: CAD/CAM.

Fillet Weld. Fillet weld is a weld of approximately triangular cross section using filler metal to join intersecting surfaces, such as a lap joint, T-joint, cornerjoint, or intersecting pipes or tubes. A fillet weld does not normally use a specially prepared groove. Filler metal is used to fill the flat or curved surface intersections.

Filter. A filter is a device that separates data according to a specific criteria. A filter also is an electrical device used to suppress unwanted electrical noise.

Finish Cut. A finish cut is the final cut made with EDM on the finished workpiece. The better the surface finish, the longer the finish cut takes. The finish cut should only remove the minimum amount of material necessary to achieve the finish required.

Finish Feed. Finish feed is feeding in small increments for finishing the part.

Finishing. There are many different processes employed for surface, edge, and corner preparation, conditioning, cleaning, and finishing. In recent years, there has been dynamic growth in the development and improvement of these processes, as well as the equipment, tooling, media, and compounds used.

Of the many different processes used for finishing, some simply clean contaminants from surfaces. Others remove or form the surface material to produce the desired results. The finishing processes can be broadly classified into mechanical, abrasive, thermal, chemical, and electrochemical methods, but some combine several methods.

Advantages, typical applications, and possible limitations of some of the more common finishing processes are presented in the figure on this page. While this data specifically applies to miniature precision parts made from aluminum, beryllium copper, and stainless steel, the comments are generally pertinent for other parts and materials.

Many different side effects that result from various finishing processes can affect subsequent processing, including machining, heat treating, assembly, plating, or painting. For example, inadequate finishing or incomplete burr removal can cause plating buildup, interference fits, excessive wear, or localized electrical or magnetic-field disturbances.

Most finishing processes remove material from all exposed surfaces and edges. As a result, part dimensions change, and radii are formed on edges. Provisions for surplus stock or minimization of burr formation in prior machining and care in finishing are essential for critical components. Most finishing processes also change surface finishes, often improving them.

Most mass finishing processes, such as barrel tumbling and vibratory finishing, reduce surface tensile stresses in workpieces. High-energy mass finishing, such as centrifugal barrel finishing, can induce compressive surface stresses. Compressive stresses are desirable for parts subject to

Process	Advantages and Typical Applications	Possible Limitations
Abrasive flow	Removes hard-to-reach burrs. Polishes surfaces.	Blind features not deburred.
Abrasive blasting	Good for hard metals.	Produces matte finish. Dust control required. Burr must be accessible.
Barrel tumbling	Low cost. Suitable for all materials.	Edges must be exposed. Slow and not effective for interior surfaces and edges.
Brushing and buffing	All accessible burrs and edges. Polishes surfaces.	
Centrifugal barrel	Fast process. Suitable for all materials. Residual compressive stresses.	
Electro-chemical	Removes hard-to-reach burrs.	Possible stray etching. Limited to conductive metals.
Electro-polishing	Good for thin burrs. Polishes surfaces.	Possible pitting and streaking.
Hand deburring	For hard-to-reach areas and small volume requirements.	Usually expensive and inconsistent. Burrs must be accessible.
Spindle finishing	Fast process. For uniform shapes.	Fixturing needed.
Thermal energy	For thin burrs. Deburrs blind features.	Covers part with oxide film. Burr area must be free of oil and water.
Vibratory	Versatile, economical process. Many deburring and finishing applications.	Not usually suitable for removing internal burrs in intersecting holes.

Advantages, applications and limitations of finishing processes.

fatigue loading because they tend to increase service life. However, compressive stresses can be a disadvantage for some applications, and can cause distortion, especially of thin, nonsymmetrical parts.

Some finishing processes change the color of surfaces, depending primarily upon the surface finishes, chemical changes, or contamination; for example, brushing operations and the compound used for tumbling control color; media types and compound control the luster produced.

A film is sometimes produced on the surfaces of parts subjected to incorrect mass finishing. This film is the result of using improper conditioners, agents, and abrasive particles. If allowed to dry, the film forms a barrier that inhibits proper plating and passivation. Films can be prevented by proper compound selection and solution use.

Different finishing processes vary with respect to their capabilities, possible applications, side effects, and costs. Capability factors include production rates, repeatability, and finishes produced. Total costs include those for equipment, consumables, tooling and fixtures, labor, and overhead. For a specific requirement, there are generally only two or three processes suitable. In some cases, especially for the removal of large burrs, more than one finishing process is required.

Several manufacturing and consulting firms are developing computer-aided selection systems for determining the optimum finishing process for a specific application. With one system, numerically coded input data entered into the computer includes workpiece materials, machining methods before deburring, the shapes of the burrs and classifications, the weight and volume of the workpieces, and deburring requirements. Output data includes the types of deburring methods that could be used, in order of their cost, and the deburring conditions resulting from the most economical method.

Specialized machines are being used extensively to remove burrs and finish parts when production requirements are high or families of similar parts are processed. Various machines available are manually, semiautomatically, or automatically operated, depending upon production needs.

Tools employed on the different machines include trimming dies, rotary cutters or pinch rolls, grinding wheels, brushes, and honing stones. Typical parts processed on such machines include stampings, castings, forgings, and extrusions; sheet and plate stock; many different types of workpieces; tubes and bars; gears; and cutting tool inserts.

Specialized machines for deburring and finishing include trimming presses, edgers, end finishers, gear tooth deburring/chamfering machines, and cutting-to-insert finishing machines. Also, when feasible, industrial robots and NC machining centers are used.

The NC/CNC machines that produce parts can often effectively deburr or finish the parts, or many features of the parts. These machines ensure the accuracy of features produced in the machining operations and can result in considerable savings by eliminating the need for subsequent operations. Because of more liberal tolerances on cast or forged surfaces, these machines may not be satisfactory. When wall locations vary from part to part, edges may be produced with too much or too little chamfering or rounding.

Deburring and/or finishing tools can often be stored in the magazines of NC/CNC machining centers that have automatic toolchanging capabilities. Macros are available for the NC/CNC units to remove burrs by automatic chamfering.

Several CNC machines are specifically designed for removing sprues, risers, gates, and runners from castings, a process called *fettling*. These are rigid, single-tool machines that often employ touch sensors to define surface orientations and then rotate the part to provide a straight-line path for the cutter. Such CNC machines are used on simple contours for nonprecision edge requirements.

For many companies, the use of industrial robots for deburring, fettling, and finishing has considerably reduced costs and improved quality. Robots can operate three shifts a day, virtually untended; they reproduce the same motions accurately; they can process parts faster than humans; they can use heavier, higher powered tools for faster finishing; and they can work in dusty, noisy environments that are unsuitable and/or hazardous for humans. Robotic finishing is not actually a process in itself, but is simply automation added to a mechanical operation.

Robotic finishing is typically used for parts that have production rates falling between job shop quantities and automotive industry requirements. Robotic methods are not normally economical for one-time runs of very low quantities because the engineering and programming time exceed savings. Unlike many off-hand processes, the use of robotic systems requires considerable planning to ensure optimum results.

For successful finishing, robots should have the following characteristics: high accuracy and repeatability, continuous-path capability, the capacity for easy and quick tool or spindle changes, rigidity, and low inertia. Other desirable features include easy off-line programming, circular interpolation, and the ability to translate movements to similar features on the same workpiece.

Additional features often required for robots are a large-reach envelope and a small wrist/arm configuration. Quick servo response, no backlash, and calibration are desirable for most applications. Load capacity of the robot is generally specified to be three times the expected working force. Five-axis capability is generally the minimum requirement; six axes are necessary when nonresidential, nonsymmetrical tools are used.

Depending upon the accuracy requirements for a specific application, the specifications and costs of a robotic system can vary greatly. The closer the tolerances and accuracies that must be maintained on the workpieces, the higher the cost of the system. For applications involving workpieces having wide tolerance allowances, lower cost robots with less accuracy but good repeatability can often be used. Such applications are especially used with compliance tooling or fixtures.

Some types of finishing are as follows.

Mass finishing refers to several processes for cleaning, deburring, deflashing, edge and corner radiusing, and surface finishing. In general,mass finishing means handling a quantity of workpieces in bulk. A more precise definition is: the edge and surface conditioning or workpieces in a mass of media, normally, but not necessarily, involving a mass of workpieces.

All mass finishing processes are based on loading parts into a container, usually holding abrasive or nonabrasive media, water, and a compound. While mass finishing commonly refers to workpieces loosely loaded into a container, equipment is available in which the workpieces are either fixtured or located in individual compartments within the container.

Action of the container causes the media to rub against the workpieces or the workpieces to rub against one another, thus producing the desired result. The major mass finishing processes are rotary barrel (tumbling), vibratory, centrifugal barrel, centrifugal disc, and spindle finishing.

Conventional *rotary barrel tumbling* was the original mass finishing technique. Ancient Chinese and Egyptians used tumbling barrels with natural stones as media to achieve smooth finishes on weapons and jewelry.

Barrel finishing is a low-pressure abrading process generally performed by the controlled sliding and rolling action of workpieces, media, compound, and water. Not all these ingredients are used in all applications. In a rotary or tumbling barrel (see figure on the next page), the upper layer of the workload has a sliding movement. As the barrel rotates, the load moves upward in the barrel to a turnover point. The force of gravity overcomes the tendency of the mass to stick together, and then the upper layer slides toward the bottom of the barrel.

Abrasive-flow machining (AFM) is a process in which a semisolid abrasive media is forced, or extruded, through a workpiece passage. The three major elements required for the AFM process are the machine, the workpiece fixture (tooling), and the media.

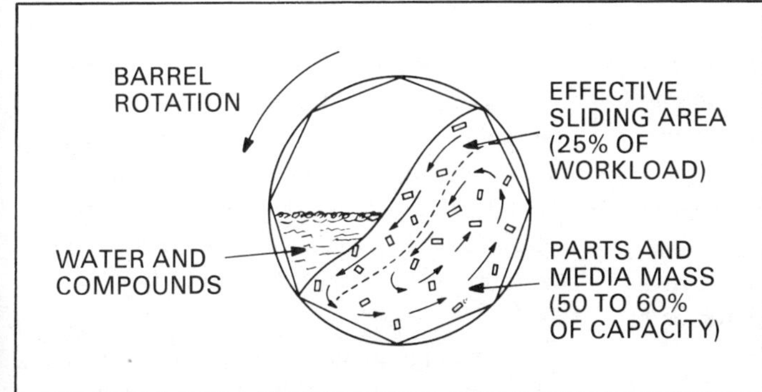

Cross section of finishing barrel showing flow of parts and media.

Abrasion in the AFM process occurs only in areas where media flow is restricted; other areas are unaffected. Inaccessible areas and any restricted places through which the media can be forced to flow can be finished, including complex internal passages.

Blast finishing refers to several versatile processes for the controlled cleaning and finishing of materials by the impact of media against the surfaces of workpieces. The original process, called sandblasting and often done outdoors, consists of introducing ordinary sand into a stream of high-velocity, pressurized air and manually directing it at the surfaces to be treated. Many different media and various processes are now being used for blasting indoors, using enclosures to contain the media.

The two major methods of modern blast finishing are dry and wet. Dry blasting is done with pressurized air or by airless processes. In air blasting, the media is propelled by a stream of pressurized air. In airless or mechanical blasting, the media is thrown against the work surfaces by centrifugal force, using a rotating bladed wheel. In wet blasting, the media is generally suspended in water that is forced by compressed air through nozzles to clean or finish the workpieces.

Shot peening is the cold working of a metal surface with a stream of spherical shot particles applied to the surface at high velocity under carefully controlled conditions. Shot peening cold works metals and improves stress characteristics in the parts blasted. This process also is used to form sheet and plate stock.

Shot peening systems may use programmable logic controllers (PLC) or computer numerical control (CNC). Either type of control provides repeatable programs or routines, tailored to the requirements of the individual parts to be peened, stored in the memory or on tape. These systems shorten setup time and lessen the possibility of operator error. When microprocessor-based systems are not used, operator training in the technology of shot peening is paramount to obtaining good results. Operators should be certified prior to using the process.

Other methods use bonded abrasives to finish parts. They include grinding with bonded wheels, coated-abrasive grinding, honing, and superfinishing.

Grinding with bonded wheels is a common process for both substantial stock removal and the fine finishing of flat, round, or curved surfaces to close tolerances and desired requirements. This process removes material by means of abrasive grains bonded into solid bodies that are referred to as wheels.

Advantages of grinding with bonded wheels include superior size control and the capability of varying the finish and lay of the surface produced. Since only a minimum force is exerted on the workpieces in finish grinding, there is little, if any distortion. The process is easily automated, and rough and finish grinding can often be combined in a single operation.

Coated abrasives are multipoint cutting tools available in sheet, disc, roll, belt, and other forms. They are used extensively for both heavy and light stock removal. Finishing operations performed include deburring, deflashing, blending, surface finishing, and polishing.

Honing is a controlled, low-velocity abrasive process using bonded-abrasive stones, sometimes called sticks, to remove stock and improve surface finishes. The process is used for both heavy and light stock removal.

There are several processes, in addition to mass finishing and blasting, that use unbonded, loose abrasives for finishing operations. These include lapping and free-abrasive machining (FAM).

Lapping is a low-velocity abrading process that removes controlled, very small amounts of material. It is accomplished with loose abrasive grains (usually retained in a viscous or liquid media, called the *vehicle*) between a tooling plate or wheel, called the *lap*, and the work surface to be finished.

The terminology is confusing, however, because lapping is also sometimes used to designate finishing processes that operate with fine-grained, bonded abrasives. Some machines are designed for operation with either laps or bonded-abrasive wheels. This section will primarily discuss lapping with loose abrasives.

Lapping and *free-abrasive machining* use the same basic techniques, but there are several different designs of machines employed for the processes. Some manufacturing personnel feel that the distinction between the two processes is as follows: with lapping, material is removed by abrasive grains that have been embedded in the lap; with free-abrasive machining, material is removed by rolling abrasive grains. However, this distinction is not always the case. For example, abrasive grains do not become embedded in hardened steel laps.

Basic lapping machines are usually of open-face design with pressure between the lap and work surfaces applied by hand or weight; these machines can be equipped with liquid cooling of the lap. Other lapping and free-abrasive machines are equipped for applying pneumatic pressure and with pumps to constantly replenish the laps with abrasives.

Burrs and flash, both internal and external, are rapidly burned away by the thermal energy method (TEM), also called *thermal energy deburring* (TED). Parts to be deburred or deflashed are placed in the chamber (pressure vessel) of a machine in which a gas mixture is ignited to create intense heat.

In addition to the thermal energy method, there are several other, less commonly used methods employing heat to deburr and finish parts. These processes include hot wire deflashing, and resistance heat, flame, electrical discharge, plasma, chlorine gas, and laser deburring. Most of these processes are used in very limited applications but are of interest for special requirements.

In *hot wire deflashing*, external edges and the surfaces of large cutouts of thermoplastic parts can be deflashed with an electrically heated thin wire that is moved parallel to the edges and surfaces. The heat in the thin wire melts and removes the flash.

In *resistance heating*, heat generated by current passing through a burr can be sufficient toburn away the burr. Workpieces are attached to a power supply that provides low-voltage, high-amperage current. The workpiece is attached to the negative side of the electrical source, and a positively charged electrode is placed in contact with the burr. The process can be used for any shape of part, provided electrical current connections can be made. In one application of this process, burrs are removed from gear teeth by using a meshing gear electrode.

With *torch flame deburring*, the flames from oxyacetylene or similar torches can be used to melt burrs and the edges of parts. This process has been used to trim flash, sprues, and gates from large castings.

Burrs can be removed by the *electrical discharge machining* (EDM) process previously discussed in detail. An electrode and the workpiece are submerged in an electrolyte, and the electrode is advanced to within 0.001 to 0.003 inch (0.03 to 0.08 mm) of the burr-laden edge. With power on, sparks melt the burrs, and the electrode continues to advance until it is below the burred edge. For circular parts with burrs projecting from their diameters, rod or tube electrodes can be used.

Burrs can be removed by placing the edges of parts in *plasma flame*. Thermal energy from the flame is concentrated at the corners and edges of the parts, and the high temperature quickly melts the burrs. This process is limited in that both small and thin precision parts may be distorted. Loose fiber ends resulting from machining fiberglass-reinforced products are being removed by laser beams and *flow-discharge plasma deburring*.

In addition, with *chlorine gas deburring*, on an experimental basis, burrs have been removed from heated steel workpieces by exposing them to a chlorine atmosphere in a closed chamber.

Electropolishing or, more correctly, *electrochemical polishing* has been used since the early 1930s, originally for metallography. The process is similar to electroplating with respect to processing, except that, in electropolishing, metal is selectively removed rather than deposited. Under proper conditions, metal is removed uniformly, thus causing smoothening and/or brightening. With proper techniques, the process can also be made burr selective—the burr being removed faster than adjacent stock. However, electropolishing cannot be confined to burr removal without stock removal unless the adjacent metal is electrically insulated or masked.

When the current is applied in electropolishing, a polarized film forms on the surface of the metal. The nature of this film is responsible for both the brightening and leveling actions. Film strength and viscosity of the electrolyte are responsible for the nature of this anodic film. The metal ions are converted to metallic salts and must diffuse through this film. The film is thinner over the projections and thicker over the depressions of the metal, allowing the projections to be more exposed to electrolytic action and to have lower electrical resistance than the depressions. The thinner film results in the projections being more rapidly dissolved by electrolysis than the depressions, which are more protected by the thicker film.

Atoms of the metal being removed in electropolishing pass through the polarized anodic film as metal salts and enter the electrolyte, either to be dissolved therein, deposited on the cathode, or to precipitate as sludge. Electropolishing solutions are therefore considered full sludging, semisludging, or nonsludging solution.

Over the years, the progression of various chemical surface treatments for metals evolved into what has become known as *chemical polishing*. During the industrial revolution, the pickling, or acid dipping, of ferrous metals became important as a means of removing oxide and scale. At the same time, an etch or open structure resulted in the surfaces of the metals, which provided better bonding or adhesion for galvanizing and enameling.

Copper and copper alloys, such as brass and bronze, required a pickle for similar reasons. The pickle most generally used was a mixture of sulfuric and nitric acids with various additives. It was found (accidentally) that, under the proper conditions, a brightening of the surface was attainable; hence, ''bright dipping'' became the accepted term.

Lastly, in *electrochemical deburring* (ECD), burrs are dissolved from metallic workpieces electrochemically and are flushed away by pressurized electrolyte. In addition to deburring, edge and corner radii are generated. The process is a specialized, static adaptation of *electrochemical machining* (ECM).

Also see: Barrel Finishing, Brushing, Buffing, Deburring, Feather Burr, Fettling, Finish Cut, Finish Feed, Finishing Teeth, Mass Finishing, Polishing, Spindle Finishing.

Finishing Teeth. Finishing teeth are the teeth at the end of a broach arranged at a constant size for finishing the surface.

Finite Element Analysis. Finite element analysis is a method for determining the structural integrity of a mechanicalpart by mathematical simulations.

Also see: CAD/CAM.

Finite Element Method. The finite element method makes use of approximating numerical functions to solve an increasing range of engineering problems.

Older, more traditional methods involve a model with extensive grid arrays and point difference equations of the desired problem. The model often has had its boundary overlapped, thereby including unnecessary data.

The finite element method more efficiently utilizes grid arrays in a model by defining its actual boundary as points as well as those points in its interior. These points are used in the approximating functions to find solutions for the many individual elements that make up the model.

Firmware. Firmware is imbedded software contained in a ROM (read only memory) or EPROM (erasable programmable read-only memory) chip and contained within the computer. Software which is loaded from external media, such as a tape or diskette, is not firmware. Normally firmware comes with the hardware it resides in and cannot be obtained separately without the accompanying hardware. Firmware usually will execute faster than software which does the same function, because firmware is optimized to work with the hardware and isn't as general-purpose as downloaded software.

An example of firmware is the ROM BIOS chip contained within all personal computers. The BIOS is the basic input/output system for the computer, which means it contains all the basic rules and reference data to allow the computer to function. The BIOS contains the instructions for how to access the peripheral devices and all of the I/O ports. Other examples of firmware are imbedded software such as servodrivers in hard disk drives and device controllers which are a part of the device.

Also see: Hardware, Software.

First-generation Computer. A First-generation computer is the computer design of the early 1960's characterized by vacuum tubes, electronics, off-line storage on drum or disc, and programming in machine language.

First Piece Inspection. First piece inspection is product acceptance of the first piece produced from a new machine setup or a new or charged process.

Also see: Inspection.

Fixed Automation. As opposed to flexible automation, fixed automation describes automation equipment which can only accomplish the one function it was designed and built to perform. Fixed automation was once the way automation was designed, with no thoughts about performing multiple functions, or being able to be reprogrammed and reapplied to another function. With the desire to automate manufacturing activities which are variable, the application of nonfixed automation such as robotics began to surface. Initially, the trend was more a fad. People tried to apply a robot to any and every manufacturing task. As the historical results of applications became available, people began to realize that there is a place for every type of manufacturing system. Certain manufacturing operations are still performed best by fixed automation, usually when the task is simple and the speed of accomplishment cannot be achieved by a robot. Fixed automation is less expensive than applying a general purpose robot or using flexible automation.

Also see: Flexible Manufacturing Systems.

Fixed-block Format. Fixed-block format is an arrangement of data in which the number and sequence of words and characters in successive blocks, as determined by hardware requirements or the programmer, are constant.

Fixed Cost. Fixed costs are generally independent of the production quantity that is being built. Indirect labor and indirect manufacturing costs are generally fixed. Setup costs for machine tools are also fixed costs.

Also see: Cost.

Fixed-cycle Operation. A fixed-cycle operation is an operation in which a preset series of steps direct machine axis movement and/or bring about the completion of such actions as boring, drilling, and tapping or a combination of these actions.

Fixed Functional Gage. Fixed functional gage is a general term describing a gage that is a direct or reverse replica of the dimension being measured. Gages made to nominal dimensions are termed master gages, and are used to set up manufacturing tools or other measuring instruments. Gages made to tolerance limits are normally used for determining functional acceptability. The latter are often referred to as GO,NO GO gages.

Fixed Head. Fixed heads are reading and writing transducers rigidly mounted on bulk memory devices.

Fixed-sequence Format. Fixed-sequence format is a means of identifying a word by its position in a block of data. Every word must be stated in a specific order, and all words preceding the last desired word, including those repeated from a previous block, must be present in the block.

Fixed Storage. Fixed storage is a storage device used to store data that is not changeable by computer instructions, such as magnetic core storage with a lockout feature.

Fixture. A fixture is a device or apparatus which holds or positions parts for processing or assembly. While a jig is considered to be a stationary aid, a fixture is usually moveable or portable. A fixture holds parts in precise orientation while operations are being performed. Fixtures ensure repeatability when performing the same operation on multiple parts, or when reworking a part to recapture the original properties.

Some automated equipment utilizes rotating turntables with duplicate fixtures on each side, allowing parts to be loaded and unloaded from one side while the other side is being processed by the machine. Transfer lines use pallets with fixtures upon which parts are located and moved throughout the different stations for processing. The use of the fixturized pallets ensures accurate location and presentation of the parts at each station, with predictable repeatability.

Numerical control (NC) machines can be used in place of complicated and costly fixtures. Universal and multipurpose workholders are now being used rather than single-purpose tools.

The versatility and possibilities available using NC machines opens the door to invention in fixture design. NC machine capabilities include machining of various surfaces at one time, initial referencing, and relocation of qualities from the fixture to the machine tool. In addition, precision, speed, and repeatability allow the tool designer to design with new insight.

Prior to the introduction of the computer numerical control (CNC) machining center, the amount of time devoted to finding ways to decrease machining time took priority over time spent trying to reduce setup time. Once CNC was introduced, however, the sharp reduction in machining time threw into focus the tremendous amount of downtime actually involved in setup.

The advent of the CNC multipallet shuttle machine marked a major breakthrough in reducing the time required to change parts, but there still remained the problem of the initial setup. In other words, the benefit of the multipallet CNC machine could be realized only after the initial setup was completed. Hence, the engineering problem became one of designing fixturing systems which could decrease the time involved in initial setup.

The four main objectives in designing a workholding fixture for NC are accuracy, rigidity, accessibility, and speed and ease of clamping and changing the workpiece. To be effective, these systems have to:

- Mount onto the machining center pallet precisely and quickly with no loss in location accuracy when moved from one pallet to another.

- Guarantee quick and accurate part location, and ensure accurate repeatability with multiple part changes.
- Facilitate one completed part per machining cycle.
- Virtually eliminate the need for single-purpose fixtures.

Since NC can perform many operations simultaneously, such as milling, drilling, tapping, and boring, fixtures must be designed for all these operations.

This means that every machined feature on a part must have cutter accessibility. The accompanying figure (page 241) shows an example where milled and drilled pads on a casting are accessible. Keep in mind that the closer a clamp is placed to a machined feature, the more it restricts the machining sequence. It can affect tool diameter, cycle time, finish, and accuracy.

Probably the most important consideration for NC tooling is keeping the fixture to a low profile. Often clamps must interfere with the ideal programmed pattern, and the programmer is forced to raise the tool above a clamp, move it over, drop it back down, and then proceed. This clamp, for two reasons, must be as low a profile as possible. First, it will allow a tool to be chucked as short as possible (see the second figure on page 241), and second, it reduces the amount of travel required to jump over a clamp (see the final figure, page 241), resulting in direct, cycle time savings.

Most NC machines have the capability of holding extremely close tolerances. Fixtures for NC play an equal role in allowing this accuracy to be transferred to the workpiece. While closer fixture tolerances can mean a more expensive tool, it can also mean a more accurate, consistent product.

Costly, intricate tools are not always necessary for workholding since the main purpose of NC machine workholders is for the holding and location of a part.

In limited production parts, strap clamps and blocks are favored as workholding devices because of their adaptability and low cost. In numerically controlled machine tool applications, strap clamps are hand-operated as opposed to power-operated. On automated lines, power clamps lower clamping time and increase production.

Low-melt alloy and plastic resin cast workholders, modular tools, and chucks and vises with altered jaws are now often used. Workability and simplicity are the major advantages included in NC workholding design. Designers should return to the fundamentals of tool design to incorporate these qualities.

A workholding fixture for NC consists of a base, locating, supporting, and clamping components.

The fixture base performs two functions. First, it allows mounting to the machine subplate or worktable, and second, it allows mounting of all fixture components.

There are four major methods of mounting fixture bases to the worktable:

- Banking a fixture against a stop (see figure on the top of page 242). This is a common method on machines using a rotary table or having a fixed zero.
- Keying to T-slots (see the bottom figure on page 242). Using this system gives accurate alignment and position in one axis only.
- Pinning a worktable (see figure on page 243). Often the T-slotted worktable has had a subplate bolted to it which has a consistent pattern of threaded holes and reamed or bored holes. These holes can effectively locate the clamp fixtures in both axes.
- The programmer has the option of calling out a specific table location for the fixture or allowing the operator to place it wherever it is convenient. For more accurate alignment, with parallelism less than ± 0.0005 inches (0.013 mm), one can remove a pin and pivot the fixture about the existing pin, using a dial indicator in the spindle.
- An indicating edge is a system used by many smaller job shops to simply have an edge on the fixture ground parallel to the locating pads. The operator uses available

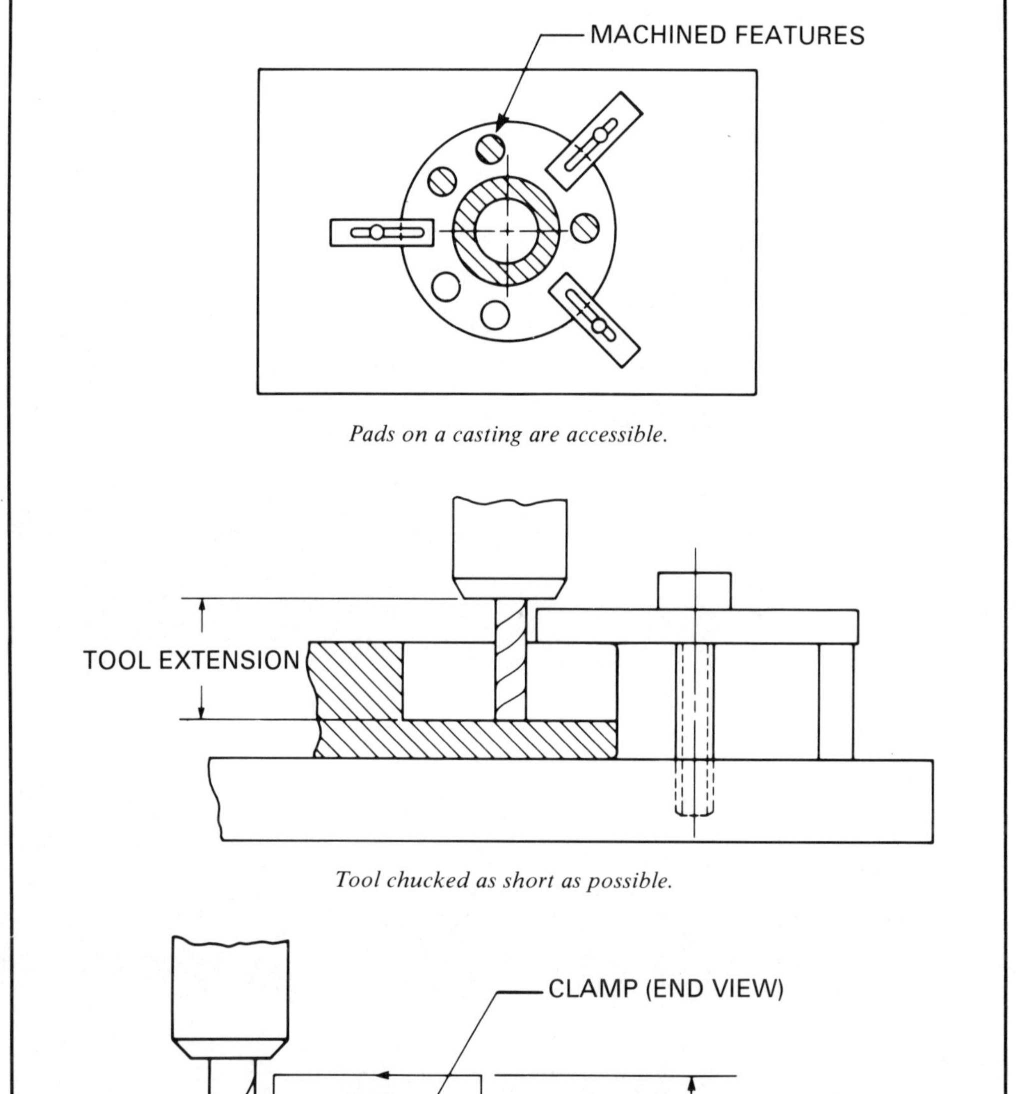

Pads on a casting are accessible.

Tool chucked as short as possible.

CLAMP (END VIEW)

KEEP THIS DIMENSION SHORT
BY USING SMALL CLAMP

CASTING

Minimize tool travel time over clamp.

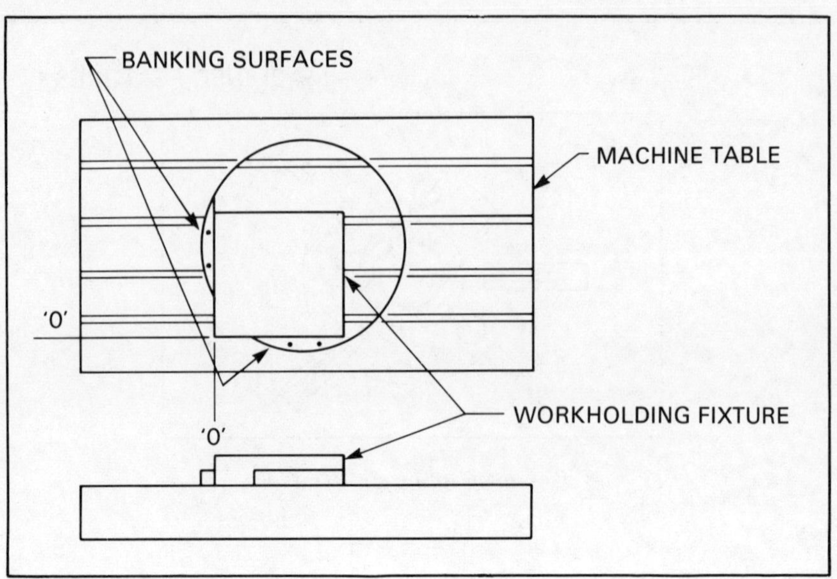

Banking a fixture against a stop.

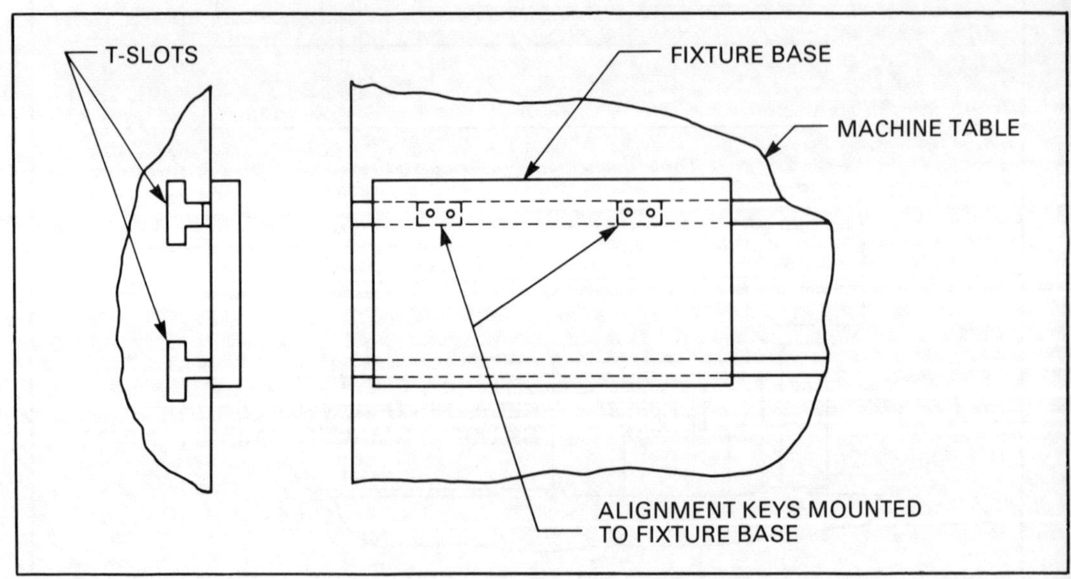

Keying to T-slots.

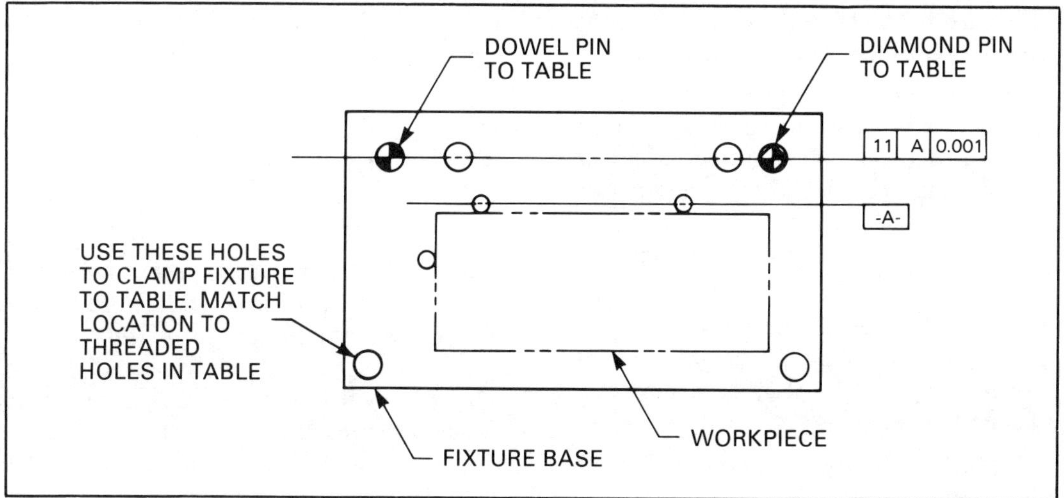

Pinning a worktable.

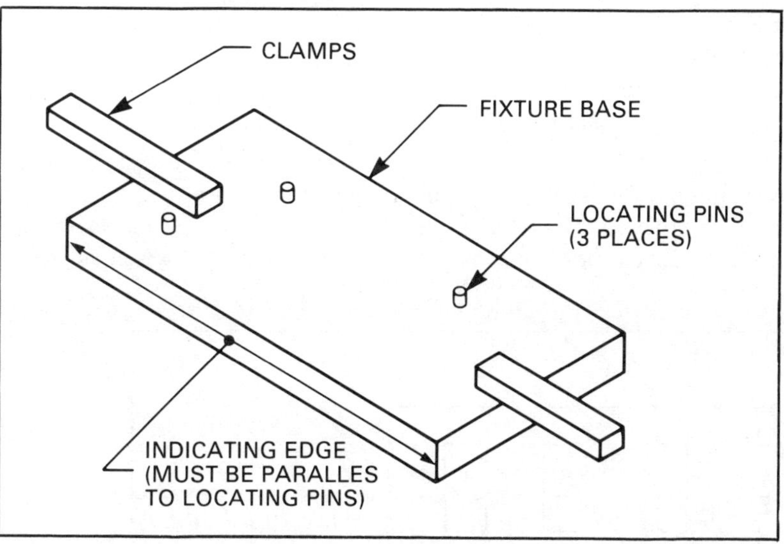

Indicating edge.

clamps and holes and secures the fixture while tramming across the indicating edge.

Position the indicating edge toward the operator, preferably without clamps along it (see the figure on page 243). The operator will clamp one end tightly and gently tap the other end with a soft-faced hammer, while an indicator is in contact with the edge and moves along the parallel axis.

It should be noted that this type of fixture is the least expensive to build, but takes more idle machine time to set up. Also, it can move easier if bumped or jarred.

Production fixturing can be accomplished by using multiple stations (figure below). The main advantage is the tool change time savings.

The cost of a fixture of this type is directly related to the number of stations; therefore, only long-run jobs can normally justify the added tooling expense.

Usually, it is desirable to hold the relationship from one station to another very close, ±0.001 inches, (0.03 mm); otherwise, parts could be produced out of tolerance and require reprogramming.

A way to avoid this problem and to save some tooling expense is to program the part after the fixture is built and the dimensions of the various stations are known. Time limitations, unfortunately, do not always allow this practice.

Numerous ideas become part of basic machine designs because NC machines offer many options in fixture design.

Also see: Jig.

Flag. A flag is an indicator used to signal the occurrence of a particular condition, such as the end of a word or boundary of a field.

Flag Bit. A flag bit is a processor memory bit which indicates the type or form of special condition that has been reached and which can be monitored by user-programmed instructions.

Flame Cleaning. Flame cleaning is a method of cleaning metal parts by playing flames from an oxy-acetylene torch over the surface to burn off oils, dry water residues, or heat scale rapidly and cause it to flake off.

Flash. A flash is a thin web or fin of metal on a casting which occurs at die partings, around air vents, and around movable cores. This excess metal is due to necessary working and operating clearances in a die. A flash also is the excess material squeezed out of the cavity as a compression mold closes or as pressure is applied to

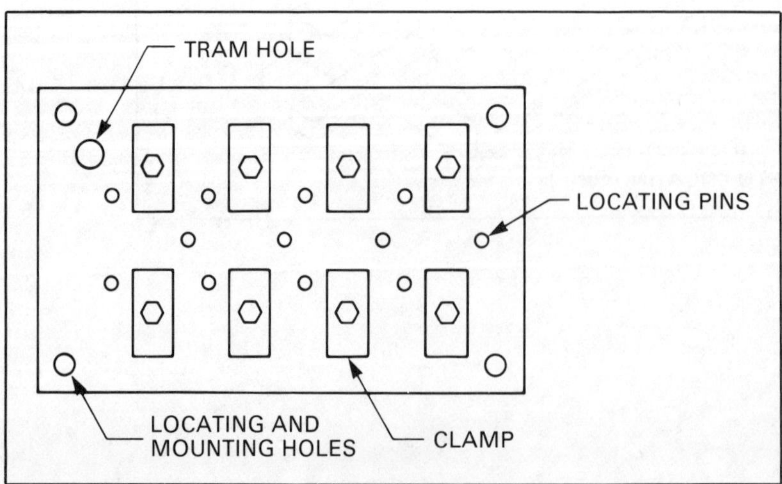

Multistation fixturing.

a transfer or injection mold. Flash includes both the fin and fragments that are left in the mold. Flash is also a thin electrodeposit that is less than 0.1 mil.

Flash Welding. Flash welding is a resistance welding process for making burr or miter welds in which electric current is passed across a loosely fitted joint, producing electric flashing (arcing) to heat the surfaces to temperatures at or slightly below the melting temperature, then pressure is applied to forge the surfaces together.

Flattening Die. Flattening dies are used to flatten hems; that is, dies that can flatten a bend by closing it. It consists of a top and bottom die with a flat surface that can close one section (flange) to another (hem, seam).

Flexible Assembly System. Flexible assembly systems utilize separate islands or cells of automation where the destination of output from each cell can be programmed. The material transportation system can consist of power roller conveyor systems, automated guided vehicles, and automated storage and retrieval systems (AS/RS) for moving magazines containing parts or printed circuit boards.

A printed circuit board manufacturer with a high volume of a single type of board having a low component mix might best use a dedicated in-line production arrangement. However, a company that makes many types of boards using many components will be most efficiently served with a flexible assembly line.

Whether to include dedicated machines in flexible automated assembly is determined by the volume of components and the mix of axial and radial leads, DIPs, SMDs, number of pins, and sizes. A robot-only flexible system is proper for a highly mixed, low-volume (thousands of boards per month) operation. An assembly line setup with robots may be quickly reconfigured with an existing program to handle several different product mixes.

A flexible assembly system offers the following production benefits:

- Each machine operates at its maximum output potential.
- Routing of the assemblies through the system is software-controlled and can be modified based on what is actually occurring in the process.
- There are no production line balancing problems.
- As production requirements increase, the manufacturer can expand capacity by adding new production cells.
- The assembly system can utilize different vendor's equipment in the various process operations.

Software is implemented on four hierarchical levels: machine control, supervisory, factory, and corporate. At the lowest level, the assembly machines generate production, consumption, and quality information. The data are collected and stored in the assembly supervisory system that monitors the process and communicates with a traffic management system. In the factory management system, all the various processes (inspection, storage, CAD, repair, and ATE) are integrated. Development of this system has been difficult because of the proliferation of computer systems, networks, and protocols. At present, no standard ''off-the-shelf'' factory management system exists. The highest level is the corporate management system, in which summary data collected through the factory management system is utilized by the financial and accounting systems.

Flexible Automation. Flexible automation is an adaptable, multipurpose design capable of being redirected or used for a new purpose or function.

Flexible Manufacturing Cell. A flexible manufacturing cell, sometimes referred to as an FMC, may be a part of a flexible manufacturing system (FMS). An FMC is a group of related machines which perform a particular process or step in a larger manufacturing process. A cell may be segregated due to noise, chemical requirements, raw material needs, operator requirements, or manufacturing cycle times. The flexible aspect of a flexible manufacturing cell indicates that the cell isn't restricted to just one type of part or process, but can easily accommodate different parts and products, usually within families of similar physical properties and dimensional characteristics.

Examples of flexible manufacturing cells are an automatic component insertion machine which can assemble standard components into circuit boards or a tracer lathe or CNC milling machine which can fabricate many different types of metal objects with the appropriate pattern or program tape. When implementing computer integrated manufacturing in a step by step manner, designing and implementing individual cells is the first step toward a totally integrated manufacturing environment.

Some of the benefits of flexible manufacturing cells were spelled out in the article *FMS: Too Much, Too Soon*, which appeared in *Manufacturing Engineering*, March 1987. The article was authored by Diane Palframan. In it she wrote: ''Being cheaper, FMCs are more easily justified. Without doubt, the inability of many companies to justify the cost of an FMS has held up sales.

''Another advantage of FMCs is that they are easier to control. There is no need for a very large, centralized computer system. There is also usually no need for a skilled computer operator. People are more likely to understand what is happening in a flexible manufacturing cell.''

In the SME Manufacturing Insights videotape *Flexible Manufacturing Cells*, Dr. Michael C. Burstein of the Industrial Technology Institute described an FMC as ''an automated machine for manufacturing, which includes some automated material handling, and—typically—some automated tool changing capability so that it is capable by itself of unattended operation, at least for some limited time.

''At one extreme, the CNC machine which hasn't its own material handling capability beyond that, you have a flexible manufacturing module which has material handling capability—limited capability—and tool handling capability. So that one, in fact, can have unattended operation of the piece of equipment.

''On the other hand, we go to a flexible cell, we generally either have a combination of machines—all of which are under common control, and all of which have common material handling. Now, the individual machines within the cell can, in fact, have their own tool changing capability, but in any case the important aspect is that there is common control and there is common material handling.''

In addition to disenchantment with FMS, Dr. Burstein listed other major factors contributing to the rapidly increasing popularity of FMCs: ''...there are really three major factors which have contributed. One of them is the intensity of international competition. Another is the major industrial restructuring—particularly within the U.S. And the third is the rapid pace of technological change.

''Which--both in product and process—had the effect of cutting the length of product life cycles to the point where management of a product market mix for a company requires them to manage carefully a whole collection of related product life cycles...in a sense a family of products represents a portfolio of product life cycles.''

Also see: Flexible Manufacturing System.

Flexible Manufacturing System (FMS). A flexible manufacturing system is one manufacturing machine, or multiple machines which are integrated by an automated material handling system, whose operation is managed by a computerized control system. A flexible manufacturing system can be reconfigured by computer control to manufacture various products.

Flexible manufacturing systems have been widely acknowledged as the cornerstone of the factory of the future. The FMS concept promises manufacturers a host of significant advantages. Among these are much more efficient use of capital equipment, increased productivity, improved product quality and consistency, reduction of work in process, reduced direct labor costs, and reduced floor space.

About 50 full-scale flexible manufacturing systems were running in the U.S. in 1985 and about 200 systems had been installed worldwide. These systems have sometimes been found to be very expensive and incompatible with a company's other manufacturing operations. Much of the difficulty in implementing the FMS concept has been attributed to the unnecessary complexity and inefficiency of the function being automated, and to the failure to design or redesign products for manufacturability in an FMS environment. Recent efforts have concentrated on simplification of the manufacturing process and tighter coordination of design with production processing in preparation for automation.

History. The first true flexible manufacturing system was designed in the mid-1960s by Molins Ltd., a British firm, and was called System 24. The system was ahead of its time and was dismantled before it was ever really finished primarily because of shortcomings of control technology at the time. However, a U.S. patent was granted for the design in 1983. This could require that current FMS designers having to make licensing agreements and to pay fees to the developers.

In the United States, the most notable early installations of FMS were those put into operation at Caterpillar Inc. by Kearney & Trecker. A very close working relationship developed between these firms. The systems were designed for specialized applications with very specific goals in mind. Most of these goals were met or exceeded.

The automotive industry (including truck, construction, and farm equipment) was the first industry to utilize FMS, and remains the major user in the United States and Europe. In Japan, the major user is the machine tool industry.

Most of the early installations were in very large companies—Caterpillar Inc., John Deere, and General Electric. The reason was not related to the larger manufacturing volume or the greater financial resources of these firms as it was to their greater sophistication in applying computers to various business and manufacturing functions, including numerical control.

Discussion. The driving force behind the development of flexible manufacturing systems is the desire to extend the benefits of automation to mid-volume and low-volume manufacturing and to the Just In Time management of materials. The competitive pressures that called for higher manufacturing efficiency, lower costs, and faster response to market demands cannot be met with current job lot oriented manufacturing. What is needed is the advantages of the automated production line combined with the flexibility of the job shop.

The rapid advances in computer technology over the past few decades provided the technology to add flexibility to automation. A major step was the development of programmable controllers for numerical control machines. With these devices the machines can be reassigned to new tasks quickly. The addition of robots to the arsenal of production tools was another major step towards achieving flexibility through computer control.

Of the 200 systems in use today (excluding the Eastern bloc nations), about half are in Europe and the rest are split between the United States and Japan. In terms of value the distribution is more even among Europe, the United States and Japan. The average investment per system (about $2.5 to 3 million) in Europe is lower than in other parts of the world.

There are four major suppliers of flexible manufacturing systems in the United States and a handful of potentially significant suppliers. The major suppliers are:

- Cincinnati Milacron—one of the strongest and most important machine tool manufacturers in the world. Cincinnati makes a wide variety of machine tools, FMSs, robots, and controls.
- Giddings & Lewis—holds an excellent position in machine tools—primarily large horizontal boring mills, machining centers, and vertical turret lathes.

- Kearney & Trecker—the largest and most successful U.S. systems builder.
- White Sunstrand—another pioneer in flexible manufacturing systems in the United States. The company boasts a very sophisticated electronics capability and is a leader in distributed numerical control.

Applications. Onan Corporation, a subsidiary of McGraw-Edison Company, manufactures electric generator sets, alternators, switch gear, and other power generating and controlling equipment. Trumpf America, Incorporated is producing a flexible manufacturing system that, when fully operational, will be the largest FMS sheet metal fabrication system in the United States. Trumpf is a logical source for this type of equipment since it is well-known for its metal fabrication equipment, particularly punching and nibbling equipment.

The Onan FMS consists of a CNC punching and laser cutting machine with a robotic tool changer, three tool towers with storage for 300 tools and an automated storage and retrieval system. This equipment, under DNC control produces automatically nested parts. As incoming sheet metal (on specially designed pallets) is weighed, stored, punched, and laser cut in an appropriate nesting pattern. The DNC system will continuously monitor inventory, tool status and work in process.

At General Electric's Erie, Pennsylvania locomotive plant, Giddings & Lewis installed a nine machine tool FMS that also includes a setup station, a transport system, a chip removal system, and a supervisory computer. The machine tools are: two vertical milling machines; three horizontal machining centers; three heavy horizontal boring mills, and one medium horizontal boring mill.

All of the machine tools are equipped with automatic or robotic tool changers. The transport system is a 212-foot rail tracked, chain-driven automatic transporter that moves work pieces between 21 load/unload stations. The executive computer directs and tracks the entire system.

Among the proven results of this installation are: a 240% overall increase in productivity; a reduction in cycle time from 16 days to 16 hours; a 38% increase in capacity, and a 25% reduction in floor space.

In his paper entitled *Real-time Operational Control of Flexible Manufacturing Systems*, and written for the *Journal of Manufacturing Systems*, Oded Z. Maimon of Tel Aviv University, outlined a real-time operation control scheme for an FMS. The FMS was made up of a set of machines, a material handling system, and service centers. All of these components work together for a common production goal.

The FMS integration is carried out from the production requirements to the machine's operation. In the system depicted in the figure on the next page, Maimon writes: "different levels of the control operate at different time scales and determine different operational decisions, from aggregated parameters to more detailed ones. Thus, the control system can efficiently handle the enormous amount of static and dynamic data and decisions required to automatically operate a complex system, such as an FMS."

An example of FMS is provided in the SME Technical Paper *Expert System Schedules Automated Cell*. This paper was authored by Vernon E. Lott and Daniel L. Manack of Texas Instruments, Incorporated.

The TI FMS is made up of four horizontal machining centers each with a 10-pallet carousel, two robot deburr stations, one dual position wash station, a coordinate measuring machine, a load/unload carousel, an AGV for material handling, and the hardware to support a tool management system including tool set and a TI-developed AGV with robot arm for tool delivery. The figure on page 250 depicts the FMS layout.

Lott and Manack noted: "A unique feature of the FMS is the 10-pallet carousels attached to each machine. The additional hardware cost was easily justified by the benefits provided which include:

- Reduced the number of AGVs from three to one. (Control system significantly simpler for single versus multiple AGVs.)

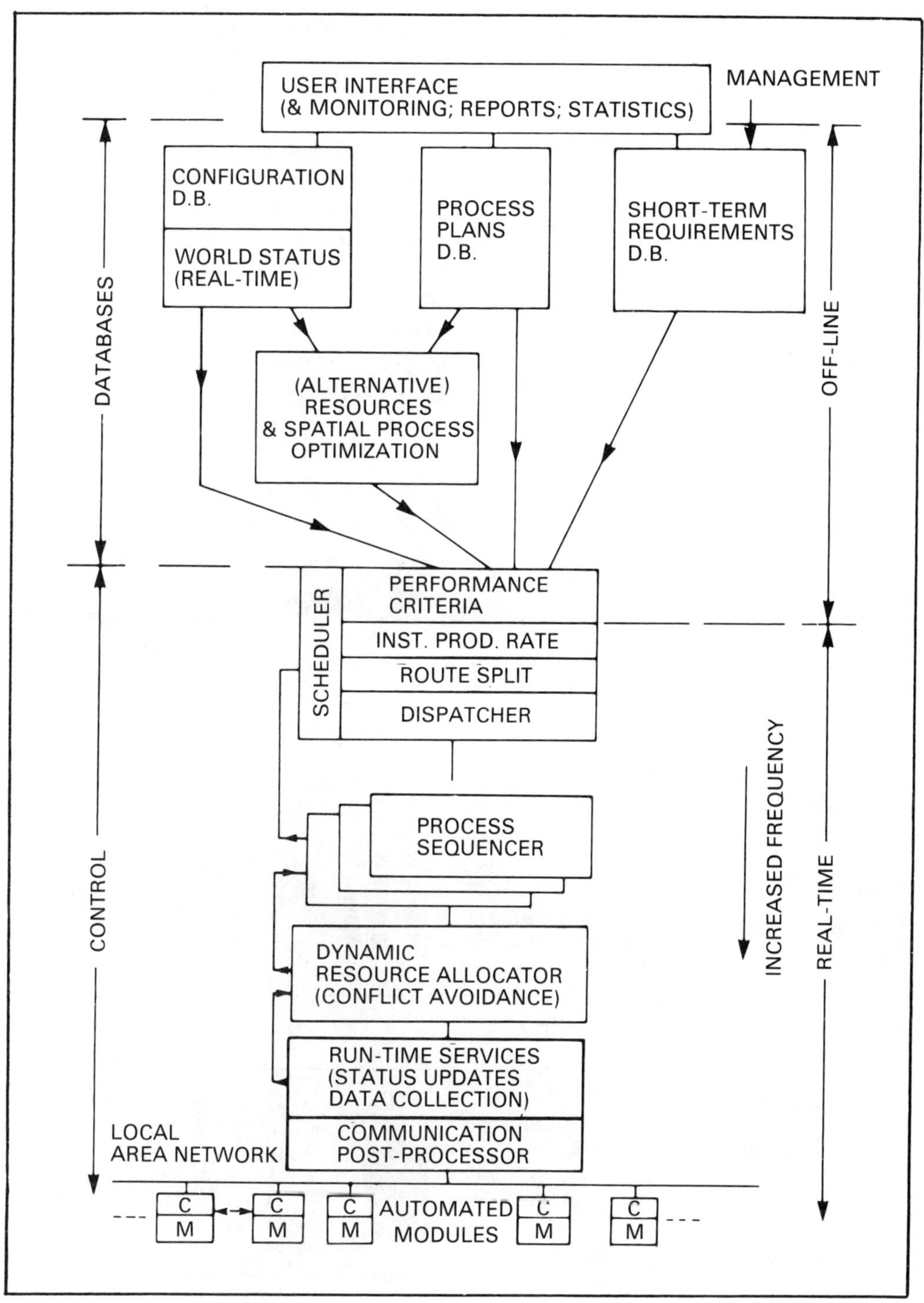

Flexible manufacturing system controller scheme.

- Eliminate risk of material handling system problems affecting machine utilization.
- Provide additional flexibility to run any one of 45 different operations without requiring an AS/RS.''

The horizontal machining centers are of four-axis configuration with 90 tool storage capacity and 6000 rpm spindles.

The robot deburr stations were developed at TI using standard equipment. Tooling concepts, refined over several years of experience with robotic deburr systems, facilitate efficient burr removal while maximizing tool life. Process consistency and elimination of manual deburr costs are the main advantages of this approach.

A coordinate measuring machine (CMM) verifies part features based on statistical sampling plans and process characteristics. The CMM operates within the shop environment and utilizes the production fixtures for part positioning.

The load/unload station is identical to the 10-pallet carousels on the machine. This station provides the operator with a central location for loading parts to the fixtures based on instructions from the scheduling system. Fixtures with completed parts also move through this station for unloading by the operator.

Transport of the pallets in the FMS is handled by the AGV. The AGV is designed to interface to varying equipment heights using floor mounted positioning cones.

The tool management system includes a tool set machine, storage area, and queue station.

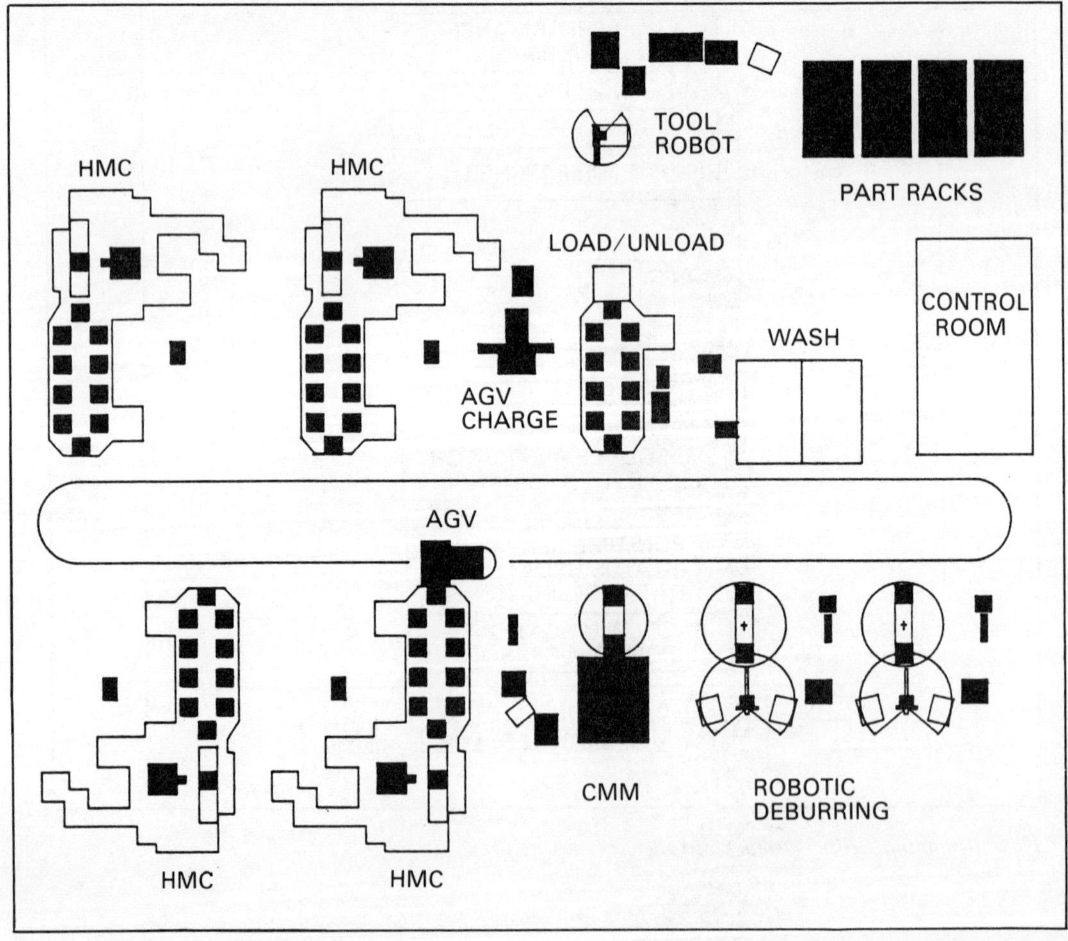

A sample FMS layout.

The operator receives information via a terminal as to which tools are required to support the production schedule as well as replacement tools that will exceed their predetermined life within certain time periods.

The FMS control system combines mature TI-developed manufacturing systems for real-time factory control and NC program support with artificial intelligence (AI) for planning, scheduling, and dispatching.

AI tools were selected over conventional programming approaches because of the significantly reduced development costs.

The authors continued: "The NC program support is provided by a CAD/CAM graphics system, a mainframe, and a terminal interface at the NC machine. Together, these systems provide a means of developing and maintaining NC programs in a central database under configuration management control and downloading to a specific machine upon request.

"The cell controller provides the real-time interface to automated material handling equipment. It receives point-to-point move commands from the Dispatcher and coordinates the operation of all material handling devices. This includes the automated guided vehicle which transports pallets between machines and secondary operations.

"Instructions to the operator for part loading and tool replacement are conveyed using TI Professional Computers (PCs). Application specific programs in the PCs provide instructions upon command from the dispatcher to drive a highly visible message center.

"The early phases of software development concentrated on creation of a test bed capable of modeling each machine and the material handling system. This model, also implemented on the TI Explorer, was used to test the dispatching subsystem before integration with other systems.

"The emulator was constructed to interact with the dispatcher, providing a simulated real world environment in a laboratory setting. In this configuration, the test bed exactly duplicated the message protocol of the cell controller. This approach allowed the scheduling and dispatching system to be tested before installing the FMS in production. Since the machines that make up the FMS were already producing parts as stand-alones, a smooth transition was essential.

"The FMS has satisfied or exceeded all expectations. The following represents key performance criteria.

- Machine utilization has doubled.
- Consistent processes capable of holding positional tolerances to ± 0.001 inch.
- Unmanned operational capability in excess of one shift.
- Satisfies complete machining requirements for components.
- Total manufacturing costs cut in half as compared to previous approach.
- Technologies developed for FMS currently being applied in related areas (i.e., robotic deburr and AI scheduling system).
- The integration of the FMS into the existing CIM architecture substantially reduces response time to part changes and new product introduction.

"The Texas Instruments' FMS is a demonstration of what can be accomplished when clear objectives are established and a methodical approach is followed. Concentration of the fundamentals of machine capability, sound fixture and tooling concepts, and flexibility through the AI scheduling and control system are the foundation for success."

In another case study, Kwok-sang Chui of the Cummins Engine Company reported in his SME Technical Paper *Case Report on Integrating FMS and Traditional Machine Tools*: "It is a recognized fact that to an FMS implementor a thorough understanding of the limitations of an FMS is more critical than the knowledge of its benefits. This is particularly true in the case in which an FMS is to be merged with conventional tools. Known to many in the manufacturing society, an FMS, aside from providing the flexible ability, also possesses other benefits such as inventory reduction, quick changeovers, floor space and quality optimization, and rapid reaction to product and market changes. The list can go on. The truth of the matter remains though that no system can deliver all these promises simultaneously. Some of these benefits must be sacrificed in order for others to be materialized. In other words, the objective or priority of an FMS application must be preserved even at the expense of other FMS benefits. As a project or manufacturing engineer, one must weigh each and every benefit, no matter how attractive it may seem to be, against the preset objective function throughout the life cycle of an FMS implementation. The interest of different projects may vary; the approach to protect the objective function can be generic.''

Dr. Chui reviews a case in which an FMS by Kearney & Trecker was built to produce a variety of diesel engine components of which the cast iron brake housing is the major driving force. After evaluating various options such as utilizing rebuilt machinery alone, supplementing rebuilt machinery with robotic and material handling systems, and strictly new capital equipment, a decision was made which favored the deployment of a flexible manufacturing system composed of six work centers and an automated guided vehicle. The original design of such system was projected to possess the capacity of up to 150 pieces/day with considerable savings in labor cost (216% over rebuilt machinery option). The figure on the next page represents a schematic layout of the FMS. The system is designed for two-shift operation; a 12-hour shift was proposed.

In the paper's conclusion Dr. Chui noted: ''the truth remains that the value of an FMS rests on its application. Furthermore, its value can be extended or optimized if such a system is properly integrated with conventional machinery which constitutes the main-stream of today's manufacturing environment.''

The author continued: ''Misapplication is often a result of lack of business vision. In the case of an FMS deployment, the project definition which represents the goal of the system must be clearly understood. Such definition should remain as the absolute measuring index throughout the project cycle. Redefining the project is permitted and encouraged, but one must realize that such redefinition often means additional expenditures.

''To conserve the project team's energy and to guard against any compromise on the project goal, a robust, or even rigid at times, boundary must be established. It is often tempting to have a flexible boundary so that all situations can be handled; yet one must realize that a flexible boundary can also be widened to a point where the value of the project is diluted, and its effectiveness is reduced or minimized.

''Engineering projections on the performance of any system, FMS in particular, are often too optimistic, as witnessed by the gross overrun of the direct labor costs. To obtain an accurate cost estimate, shrewd industrial engineering skill must be applied. Task elements must be refined, grouped, simplified, and balanced to yield an optimal workflow scheme.

''Too much energy was expended in ROI calculation and defense, in comparison to the engineering work such as part processing, workflow design, and balancing, equipment consideration, and installation, etc. Furthermore, many companies are still using the same ROI calculations method for FMS as well as for conventional tooling. An algorithm was presented at the 1987 Eastern Technology Conference (November, 1987, Springfield, Mass.) which suggested the use of an evaluating index called Efficiency Quotient or EQ to consider the total cost of a part. All financial impacts are to be included.

''For project goals, it is suggested that a single-purpose one is probably the best. The effectiveness of the system hinges on how much compromise the project engineer is willing (or not willing) to make on the project goal. This certainly does not preclude dual or multiple-purpose goals, but one needs to keep in mind that as the number of the system objectives goes up, the system's effectiveness goes down.

''Computerization is becoming an epidemic; computers today find their way everywhere, and FMS is no exception. Given the fact that almost all manufacturing systems are somehow connected to computers, it is imperative for the

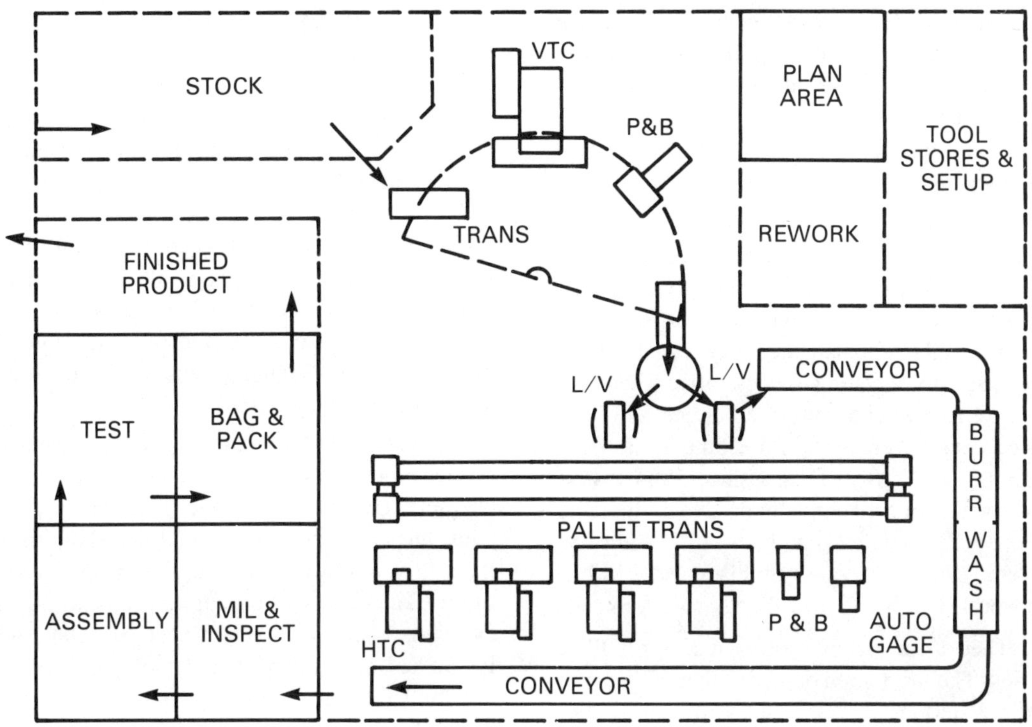

An FMS layout.

engineer to keep in mind that the decision-making responsibility rests on him or her. Computers are most effective and productive when they are used, not worshipped.

"Similarly, one needs to maintain an open mind on tooling utilization. True, by definition, FMS are best fitted for mid-volume mid-variety while transfer lines are more suitable for high volumes. But in fact, the value of a certain type of tooling is in its application; with 10% additional expenditure, this FMS, supplemented by conventional machinery, is capable of handling volumes above the so-called mid-volume level. Conversely, a transfer line can be engineered to produce mid-volume parts. This is particularly true with companies trying to introduce JIT practice. In short, it is much more productive to match the machine capability with the job requirement than to routinely follow rules of thumb.

"In closing, we observed that an FMS can be integrated with conventional tooling. In fact, such a harmonized system brings out the best of both and is more productive. In the course of the integration, it was demonstrated that fundamental engineering skills and knowledge are the paramount requirements to the success of the endeavor. It certainly is no easy task, but it can be accomplished with a combination of disciplined engineering, fundamental skills and analytical planning and execution."

There is another side to FMS and it was discussed by author Diane Palframan in an article in the March 1987 *Manufacturing Engineering*. In the article titles *FMS: Too Much, Too Soon*, she wrote:

"Many manufacturers, having long considered FMS technology, are now opting for something smaller, simpler, and cheaper. They are buying flexible manufacturing cells and even stand-alone computer numerical controlled (CNC) machine tools.

"Explains Tim Chapman, marketing manager for the manufacturing systems division of

Cincinnati Milacron (Cincinnati), 'Companies are going back and redefining their manufacturing environments. They thought they wanted a sophisticated FMS, but now they see that it could be more cost-effective to go for a more simplistic approach.'

"The trend away from complexity is widespread but not in every industrial sector. Aerospace companies, the defense industry, and, of course, many machine tool builders have already installed highly sophisticated FMSs that for the most part are operating successfully. These organizations will continue to automate and integrate. Vought Aerospace, for example, has been operating an FMS in Dallas since 1984 that is now capable of economically producing lot sizes of one. Cincinnati Milacron, which built the system, calls it the most sophisticated FMS in the world. The system has already paid for itself.

"Then there are the not-so-successful FMS stories. The most often quoted is that of Deere & Co.'s Waterloo, IA, tractor-making factory. For many years Deere operated a costly, extremely complex FMS at the Waterloo site. In a recent interview with *Forbes* magazine, Jim Lardner, vice president of Deere's components group, said of its FMS: 'Given the history of manufacturing industries that have embraced computers and computer systems, there is substantial anecdotal evidence that we computerized utter confusion and inefficiency.' Deere has now opted for flexible manufacturing cells (FMCs), and Lardner believes, 'What with the startup costs, the support system, the programmers, the automatic transfer devices, and so on, I don't think the economics support FMS, except in special cases.'

"Deere is not an isolated experience. FMS, like many new technologies including computer-aided design, may have been oversold in the early days by over-eager vendors. The performance of many systems never matched the promises. 'Intuitively, I think this probably was a contributing factor in the move toward simpler systems,' says Chapman.

"There were also unrealistic expectations on the part of the potential users. 'The Americans often expect an FMS line to perform miracles,' says Rudy Colombi, president of Mandelli, Inc. (Farmington Hills, MI), a subsidiary of Mandelli SpA (Piacenza, Italy), a machine tool builder. 'The Europeans have a more realistic understanding of FMS capabilities,' he adds. Another problem that Colombi sees is impatience. 'Whatever companies in this country want, they want it now. But what they don't realize is that FMS is not just a matter of capital investment and of building software and hardware but of training people. Training must be taken seriously with these systems.'

"Observes Peter Hall, manager, technical services, of the Swedish Machine Group's U.S. operation (Elk Grove Village, IL): 'Many U.S. companies thought they could leapfrog from their 30-year-old machines to the factory of the future. The factory of the future is a concept, an evolving process. It is achieved step by step.' Of FMS, he says, 'Many companies purchased a system that did not perform as anticipated. FMS is flexible but within definable limits.' "'Automate or liquidate' was the slogan in the early '80s, and the message is relevant to American industry today. Steady, ongoing investment in state-of-the-art FMCs is required to ensure continued prosperity and, indeed, survival. Today, companies are urged to simplify, to automate, and to integrate. In terms of FMS, this implies a modular, cellular process. According to Frost & Sullivan, Inc. (New York), the U.S. market for FMC is growing much faster than that for FMS and flexible assembly systems. The market research company says cells will grow an average of 40-50% per year to the end of the decade and that the market will be worth $540 million in 1989.

"'The obvious advantage of a cell is that it is cheaper. There is no great financial risk to the customer,' explains Shigeyuki (George) Yamane, marketing of Mazak Sales & Service, Inc. (Florence, KY). Mazak is owned by Yamazaki Mazak Corp., the Japanese machine tool builder. While many customers like to talk about FMS and, indeed, to dream about having their own custom-made system, often they simply cannot afford them, notes Yamane. Unfortunate ly, for many machine tool builders the talking can last for a year or two and never result in an order.''

Also see: Computers, Flexible Manufacturing Cell, Numerical Control.

Flexowriter. A flexowriter is an automatic typewriter incorporating an eight track tape reader and punch for preparation of punched tape.

Floating Point Representation. A floating point representation is a number representation system in which each number, as represented by a pair of numerals, equals one of those numerals times a power of an implicit fixed positive integer base raised to the exponent represented by the other numeral.

Floating Zero. A floating zero is a characteristic of an NC machine control unit allowing the zero reference point of an axis to be established at any position over the full travel of the machine tool.

Floppy Disk. A floppy disk is an inexpensive data storage device for small computers. It is a thin flexible circular object with a hole in the center which is contained inside a rectangular jacket. The floppy disk spins inside its jacket, driven by the center spindle and the drive hub. Elongated slots on the top and bottom of the jacket allow a read/write head to contact the media. A notch on one side of the jacket is sensed as a write enable indicator. Lack of the notch, or covering the notch with tape, indicates a write protect status and protects the disk from being written to.

Floppy disks, originally eight inches in diameter, were later developed as 5 1/4 inches, and more recently as 3 1/2 inch media. The 5 1/4 size has been commonly called a floppy diskette, indicating its smaller size than the original 8 inch product. A unique type of floppy disk is an 8 inch floppy inside a rigid cartridge. It is spun at a higher speed than the typical floppy disk. This product is known as a Bernoulli cartridge. The floppy disk was invented by a Japanese inventor, who holds over 1,000 patents more than Thomas Edison. IBM licensed the design of the floppy disk from this Japanese inventor.

Also see: Computer.

Flow Chart. A flow chart is a sequential listing of consecutive events, indicating the optional routes which can be followed in processing data or material. Flow charting was popularized by IBM to determine the logic flow of computer programs before writing code. IBM provided the first standardized symbols and thousands of templates which have been used by programmers.

A flow chart is a simplified and brief method for presenting logic flow, without spending time on peripheral details. Flowcharting has been taught as an integral and required part of computer programming, but isn't stressed as much today due to the emergence of powerful new programming languages. An emerging area of computer assisted engineering is CASE (computer-aided software engineering), in which various tools for structured software development aid the software engineer.

A commercially available software package known as Interactive Easy-Flow allows a person to use a computer to graphically draw flowcharts, with automatic placement and connection of the symbols.

A sample flow chart is shown in the figure on the next page.

Flow Rate. Flow rate is the inverse of cycle time. For example, a cycle time with 120 units per hour would have a flow rate per minutes.

Also see: Just-In-Time.

time. For example, a cycle time with 120 units per hour would have a flow rate of two per minute.

Also see: Just In Time.

Flux-cored Arc Welding (FCAW). Flux-cored arc welding is a fusion welding process that uses a tubular wire electrode filled with fluxing agents. Arc shielding is obtained from vaporization or decomposition of the flux, and may or

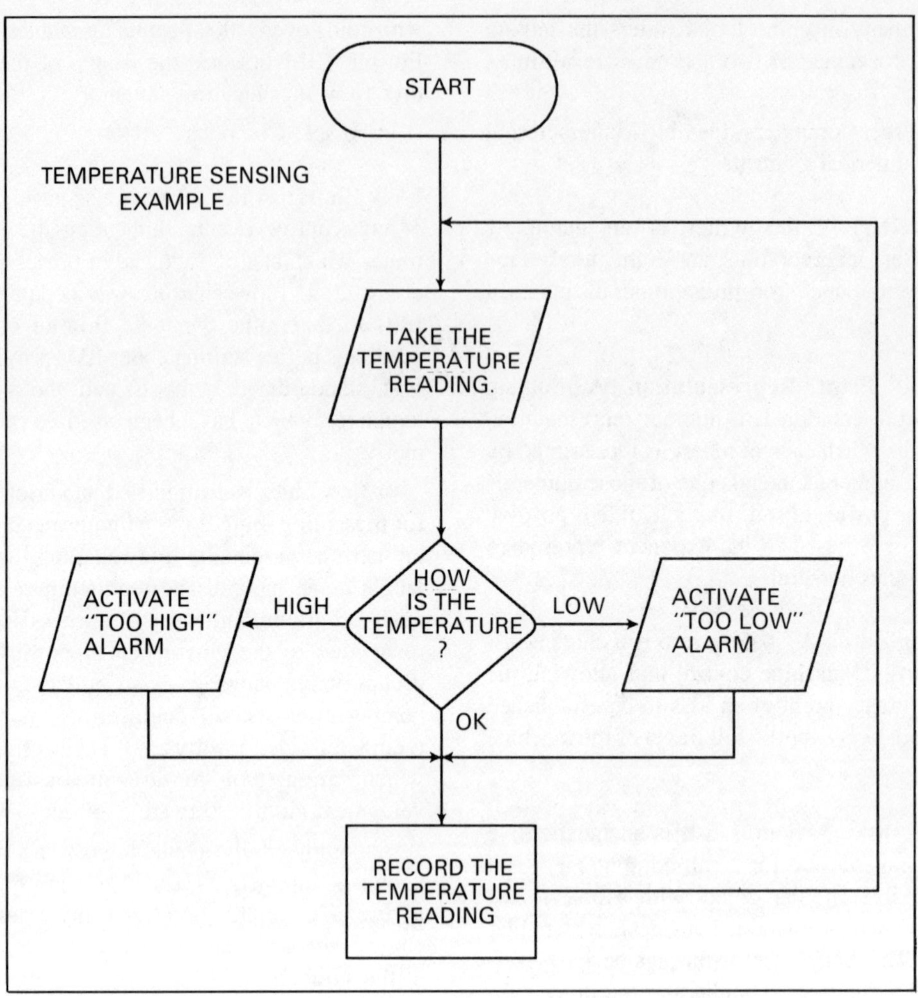

A sample flow chart.

may not be supplemented by carbon dioxide or inert gas.

FMC. See: Flexible Manufacturing Cell.

FMS. See: Flexible Manufacturing System.

Foam Molding. Foam molding is a molding process whereby heat-softened plastics containing a foaming (blowing) agent are injection molded into a cavity where they harden to produce a product that has a solid skin contiguous with a foam core.

Font. Font is a term defined within the field of typography as an individual set of characters of a particular typestyle in a unique size. A typeface, on the other hand, is the entire group of fonts in all the various typestyles and sizes. The typestyle is a collection of fonts of various sizes, with a unified style, such as bold or italic.

Typography is the art of using printed characters. Typeface refers to the general design of a group of characters, such as Times Roman, Helvetica, Palatino, Courier. The original design of a typeface is protected by copyright.

Some people, in an attempt to avoid paying royalties, design similar, but legally different typefaces. For example, TMS RMN, which is a Hewlett-Packard LaserJet typeface is similar to the original Times Roman.

The users of the Macintosh computer employ the word font to describe a general character design in a range of sizes and styles. What is normally called a typeface is called a font in Macintosh terminology.

Also see: Desktop Publishing.

Force Sensor. Force sensor is a device that is capable of measuring or detecting forces and torques exerted on a robots' wrist or end effector.

Foreground/Background Technique. Foreground/Background technique enables the automatic execution of programs according to the priority of the task to be performed. Low-priority programs can be executed when high-priority programs are not utilizing the system.

Foreground Processing. Foreground processing is the automatic execution of real-time or high-priority programs designed to preempt the use of computer facilities.

Forehand Welding. Forehand welding is a welding technique in which the welding torch, gun, or coated electrode is directed toward the progress of welding. With this method, the weld puddle is kept in front of the torch, gun, or coated electrode.

Forge Welding. Forge welding is a solid-state joining process in which the components are heated to a high temperature below the melting point and then forged together using dies, hammers or rollers. Once used extensively by blacksmiths, forge welding has largely been replaced by other welding processes.

Forging. Forging is the plastic deformation of metals, generally at elevated temperatures, into desired shapes by compressive forces exerted through a die. The forging process is usually classified by the type of equipment used or by the geometry of the end product. The simplest forging operation is by *upsetting*. Upsetting isthe compression of metal between two flat parallel platens. From this simple operation, the process can be developed into more complicated geometrics with the use of dies. The number of major variables involved in forging are the properties of the workpiece and die material, temperature, friction, speed of deformation, die geometry, and dimensions of the workpiece.

The basic principle in forging is the fact that the material flows in the direction of least resistance. Therefore, a part will expand more in its short dimension than in its longest dimension.

Forgeability is related to the material's strength, ductility and coefficient of friction. Because of the great number of factors involved, no standard forgeability test has been devised.

There are basically four types of forgings—design blocker type, commercial, close tolerance, and precision. Forging costs are usually broken down into four basic groups. They are: material, die, forging, and machining costs. The cost of forgings usually depends on items such as the alloy to be forged, metallurgical specifications, quality desired, dimensional tolerance, surface finish, and number of parts desired.

Automation of hot forging operations involves more difficult problems than those encountered in automating cold forging processes. The handling of the hot workpiece within a press, the lubrication of the dies, and the transporting of heated billets from furnace to forging machine requires rapid handling of the billet and synchronization of the furnace and press operations. In hot forging, the method and types of automation are determined primarily by production requirement. Four main types of automation, based on the level of sophistication and on the size of the production series, can be considered.

Automating forging systems involves the mechanization of operations such as heating, transporting the billets to be forged from furnace to forging unit, and feeding the billets to the first forging station. The workpiece is transported manually or automatically from one station to the next within the forging unit, whether press or hammer. The finished forging is transported automatically to the trimming press, where it is handled again by the trimming press operator.

Automation within a forging machine consists of equipping a standard forging machine, in general a mechanical press, with mechanical or pneumatic transfer devices. Thus, the billet or the forging is transported automatically from one forging station to the next within the same machine. The transporting mechanism usually consists of a walking beam system which is synchronized with the press operation. Thus, the high speed of the press, in terms of deformation rate and contact times, is maintained.

In automation by linking standard machines, the means of transporting workpieces outside, as well as inside, forging machines such as presses or hammers is mechanized using robots, hydraulically or pneumatically operated arms, and mechanically activated transfer systems. The operator is required only to supervise the sequence of operations and to stand by to stop the machines in the event of a malfunction.

In some cases, the heated workpiece is transferred toward the press by a conveyor. When the workpiece arrives at a predetermined position, the robot rotates to pick up the workpiece and carry it to the press die area. If the die area is clear, the robot positions the workpiece in the first die station and then activates the press. The upset workpiece is then transferred to the second die station, and the press is activated. Mechanical ejectors in the die remove the forged workpiece, and the cycle is repeated.

Machine vision systems can be used to inspect forgings. One system was described by James T. Barczak and Jeanne Temesan Merchant of General Motors in their SME Technical Paper: *Using Machine Vision to Inspect Automobile Forgings*.

This paper described two identical machine vision systems installed at the General Motors New Departure Hyatt Tonawanda Forge Plant to inspect six different families of automobile forgings. The inspection task requires multiple scenes to inspect features such as inner/outer diameters, part height, and concentricity. Custom systems designed by General Motors Machine Intelligence used high-speed processing hardware and a menu-driven, color graphics user interface to provide a flexible, model-based inspection systems.

The authors wrote: "The selection and installation of new equipment at General Motors New Departure Hyatt Tonawanda Forge provided the opportunity to be innovative...on the leading edge of technology. The new equipment (see figure on the next page) included four hotformers, each capable of producing forgings at the rate of 80 to 100 pieces per minute, an automatic guided vehicle system for material handling, and a machine vision system to inspect the forgings.

"The vision system is necessary to locate any nonconforming parts that may enter the system for any reason. This technology was selected based on the logistics of 100% inspection of millions of parts.

"When the decision was made to add vision to the material control system, it provided a challenging project to develop a system based on Tonawanda's needs. Since what was proposed had not been done before in the vision field, it presented the problem of finding a supplier capable of building such a system."

Parts are inspected as a batch process, at a rate of two parts per second. The parts travel on a conveyor at 20 inches per second. Thirty different parts are currently in production.

For years forging machines have been utilized with industrial robots. Upsetters, roll forges, programmable hammers, impacters, and forge presses have been interfaced with robots in single die to three die configurations. Both forge die and hot trim die operations have been installed. Parts range in size from a few ounces to several hundred pounds.

Other installations feature powdered metal technology in conjunction with the forging process.

Format. Format is a term used to describe the configuration of a medium which can be written

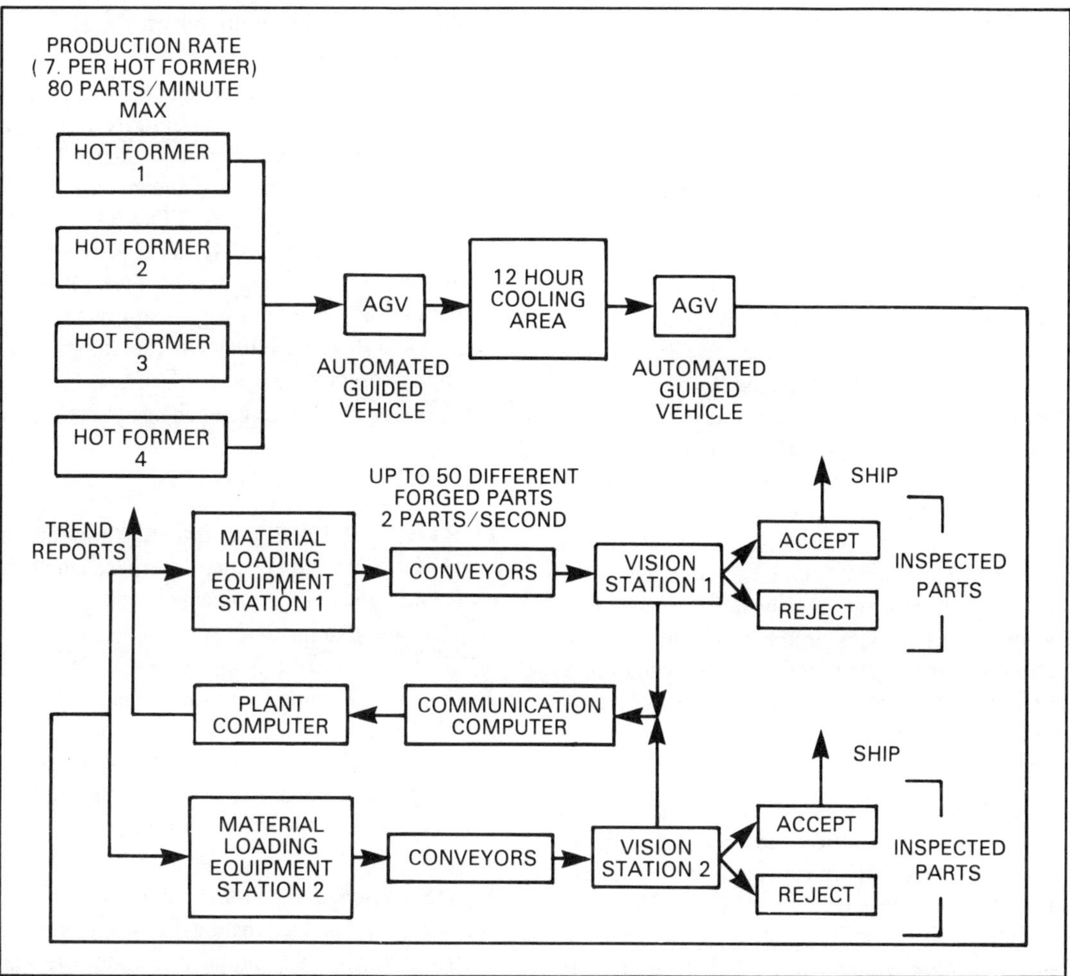

Material handling and inspection system used in a forging operation.

upon or reproduced. Format traditionally referred to printed substances such as books, papers, and forms. With reference to electronic data as medium and the function of "writing" data to magnetic media, the term format is now used to describe the configuration and layout of the magnetic medium.

Magnetic disk storage media is formatted by dividing the tracks (which are physically created when the disk is manufactured), into *sectors*. Data is located on the disk media during a read operation by using a directory table which indicates which tracks and sectors contain the desired data. Normally, reading and writing to a disk is done using one or more sectors at a time. Therefore, when a disk is "full," it means all the sectors have been written to, but not all the sectors are necessarily full of information. Using smaller sectors reduces wasted space, but might slow down the read and write operations. Disks have a maximum number of allowed sectors per type of media.

A hard disk is formatted with an interleave factor which means a sector is read and the next consecutive sector(s) is skipped. This is done to allow the CPU to have time to read or write the information. A faster CPU can operate with a smaller interleave factor, which means less time is spent by the read/write head passing over ignored sectors and more time spent actually reading from or writing to the media.

Also see: Asynchronous Communication, Asynchronous Operation, Disk, Fixed-sequence Format, Floppy Disk, TAB Sequential Format.

Forming. Forming is any of the processes of changing the shape of an object, usually without removing material. When forming sheet metal, the process is termed *braking*. When forming steel or cast iron, it is called *forging*. When forming plastics or other materials which can be made to flow easily, the process is called *molding*. Forming, such as cold working of a metal, improves certain properties such as the flow of the grain of the material. It also creates a work-hardened surface. Other methods of forming include drawing, fluid or pressure forming, peening, rolling, and stretching. In most forming operations, there is some elastic spring-back after the initial operation. This must be compensated for in determining the dimensions of the forming operation.

Automated equipment exists for some types of forming and fabrication. For the simpler processes, general-purpose equipment can be set up to perform the repetitive activities. For more complicated forming activities, specialized or single-purpose equipment might be required. If the forming includes varied activities, multiple machines may be necessary.

Also see: Brake Forming, Extruding, Forging, Machining.

FORTRAN. FORTRAN (Formula Translation) is a high-level, compiled language which was one of the first scientific computer languages developed. FORTRAN was written by John Backus of IBM around 1954, with the first compiler developed in 1956 and commercial distribution starting in 1957. FORTRAN was designed to simplify the task of using computers to perform computations on numerical information. FORTRAN can accept algebraic language as input. That is, the mathematical formulas may be written as the program statements and the FORTRAN compiler will convert the algebraic statements into machine code.

There have been various versions and dialects of FORTRAN including FORTRAN I, II, III, and IV. The currently accepted standards for the FORTRAN programming language are ANSI FORTRAN-77 and ANSI FORTRAN-66. FORTRAN is the most widely used language in science and engineering today. FORTRAN is available for virtually every mainframe computer sold today, as well as most minicomputers and microcomputers.

Also see: Computer Languages.

FORTRAN IV. See: FORTRAN.

FORTH. FORTH is a shortened form of the word FOURTH and is general-purpose language used for rbotics and graphics.

Also see: Computers.

Forward Chaining. Forward chaining is a form inference in which the antecedent or 'if' portion of a rule is matched against facts asserted in the global database to establish the truth or consequence or ''then'' portion of a rule.

Fourth Generation. Fourth generation, in the NC industry, is the change in technology of control logic so that computer architecture and core memory are included.

Frame. Frames are substantial collections of integrated knowledge about the world.

Free Machining. Free machining is a term used to describe metals having alloying additions, such as lead, manganese, or sulfur, that reduce the tool force required in machining operations.

Frequency Domain. Frequency domain refers to the analysis of control problems using algebraic equations that arise after applying Laplace and/or Fourier transformation.

Frequency Shift Keying (FSK). Frequency shift keying is a form of frequency modulation in which the modulating signal shifts the output frequency between predetermined values and in which the output signal has no phase discontinuity. The instantaneous frequency is shifted between two discrete values that are often called *mark* and *space* frequencies. FSK is a signaling method in which different frequencies are used to represent different characters to be transmitted. An example is the use of one frequency to represent a binary zero and another frequency to represent a binary one, with a smooth phase transition between them, if practical or necessary. The method is characterized by continuity of phase during the transition from one signaling condition to another. *Frequency change signaling* is similar to frequency shift keying but may involve a discontinuous change in frequency and change in phase with signaling condition transitions. FSK is synonymous with the terms *frequency shift modulation* and *frequency shift signaling*.

Also see: Modulation.

Friction Welding (FRW). Friction welding is a solid-state welding process that uses frictional heat generated by rubbing joint surfaces together under load to produce the desired bond. Normally, the faying surfaces do not melt. Filler metal, flux, and shielding are not required, although shielding gas is sometimes used when welding reactive metals.

FRP. See: Fiber Reinforced Plastic.

FRW. See: Friction Welding.

FSK. See: Frequency Shift Keying.

Full Duplex Mode. Full duplex mode, also known as *duplex mode*, can refer to either the mode of the communications which are transmitted over a channel or the physical configuration of the connection.

Related terms to full duplex or duplex are half duplex, and simplex.

A *simplex* transmission provides for the movement of messages across the path in one direction only. The sender cannot receive and the receiver cannot send. A commercial radio broadcast is one example of a simplex transmission.

Half duplex transmission provides for movement of data across the line in both directions, but in only one direction at a time. Human-operated keyboard terminals commonly use this approach. Typically, a message sent to a terminal, requires the operator to decipher the message, enter an appropriate response, and send the reply.

Full duplex (or duplex) transmission provides for simultaneous, two-way transmission between stations. Multipoint lines frequently use this approach. For example, Station A sends to the central computer at the same time the computer sends a message to Station B. Duplex transmission permits the interleaving of sessions and user data flows among several or many stations.

The physical lines or paths are sometimes described as half duplex or duplex circuits. A half duplex configuration provides for two conductors but only one is used for message exchange. The conductor is a return channel or common ground to complete the electrical circuit. The half duplex circuit is also called a *two-wire circuit*.

A full duplex circuit provides four conductors which provide two data transmission channels, one in each direction, and two return channels, respectively. The full duplex circuit is also called a *four-wire circuit*.

It should be noted that a half duplex circuit does not necessarily mean that the message flow across the circuit is half duplex. A modern telephone circuit is an example of a half duplex (two-wire) circuit which provides full duplex (concurrent bidirectional) communication.

Full Indicator Movement. Full indicator movement is the total movement observed with the dial indicator (or comparable measuring device) in contact with the part feature surface during one full revolution of the part about its datum axis or in traverse over a fixed noncircular shape.

Full-mold Casting. Full-mold casting is a sand-casting process that uses an expanded polystyrene foam pattern instead of a traditional mold cavity. Metal is poured directly into the embedded foam pattern, vaporizing the polystyrene and leaving a casting that duplicates the original pattern.

Full-range Floating Zero. A full-range floating zero is a feature on some NC systems permitting the zero point on an axis to be shifted over a specified range. The control retains data on the location of the original position of the zero. Full-range floating zero is also known as full-range zero offset.

Function Key. A function key is a terminal control key that causes the transmission of a signal not associated with a printable character. Detection of the signal usually causes the system to perform some predefined function for the user.

Functional Diameter. The functional diameter (virtual condition per ANSI Y14.5M) of an external or internal thread is the pitch diameter of the enveloping thread of perfect pitch, lead, and flank angles, having full depth of engagement but clear at crests and roots and of a specified length engagement. It may be derived by adding to the pitch diameter in the case of an external thread, or subtracting from the pitch diameter in the case of an internal thread, the cumulative effects of deviations from specified profile, including variations in lead (uniformity of helix) and flank angle over a specified length of engagement. The effects of taper, out-of-roundness, and surface defects may be positive or negative on either external or internal threads. A perfect internal or external GO-thread gage having a pitch diameter equal to that of the specified material limit and having clearance at crest and root is the enveloping thread correspondents that limit.

Functional Specification. Functional specifications are a description of the essential characteristics of a product, required to enable the product to serve a specific function, that relate to and define an end use. The reason for a functional specification is to ensure performance for a product's intended use, to ensure a long useful life, to define safety requirements, provide for compatibility with other equipment in the field, and generate a competitive sales advantage.

Functional specifications can also be called *engineering specifications* or *test specifications*. An example of a functional specification is a requirement that a plastic plug snap into a defined hole size with a certain amount of force.

There are functional characteristics which define functional specifications. These characteristics are classified as critical, major, minor, and incidental. The critical classification generally is safety-related while a major characteristic pertains to product failure. A minor characteristic is one which is required so that a product will not fall short of its intended function. An incidental characteristic is one which has no effect on product performance.

Fusion. Fusion is the melting together of filler metal and base metal (substrate), or of base metal only, which results in coalescence (joining) when the molten metal solidifies during cooling.

Fusion Welding. Fusion welding is any welding process or method that uses fusion to complete the weld.

G

Gaging. Gaging is the act of precisely measuring a dimension, quantity, color, capacity or other physical characteristics of an object with an instrument or test device. Gaging devices range from relatively simple plug, snap, pin, thread, optical, and air pressure gages, surface plates, gage blocks, and micrometers to much more complex equipment such as radiographic equipment, ultrasonic testing devices, and a wide range of reliability testers. Important quality assurance considerations in making any gage measurements (gaging) include:

- Gage Accuracy. Assurance that the gage was designed to be accurate within defined limits.
- Gage Application. Assurance that the design of the gage is appropriate for the measurement to be taken.
- Gage Restoration. Assurance that the gage measures within the original design accuracy limits.
- Operator Training. Assurance that the gage operator has the knowledge and experience required to properly use the gage.

Modern process equipment usually has capabilities for gaging parts as they are being manufactured, or at least without having to remove the part from the fixture or tooling.

Automatic gage systems often include statistical process control features to ensure that corrections are made to the process only when the corrections are warranted. For example, if one workpiece in 100 measures oversize because of some periodic malfunction in the process, statistical controls will cause the system to disregard the measurement and not provide a correction to the workpiece.

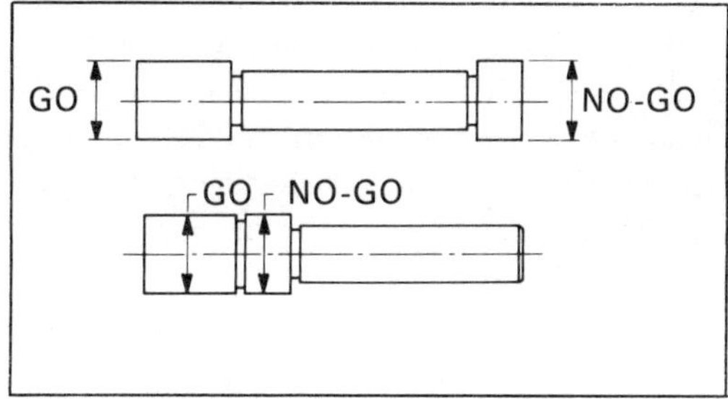

Typical plug gages.

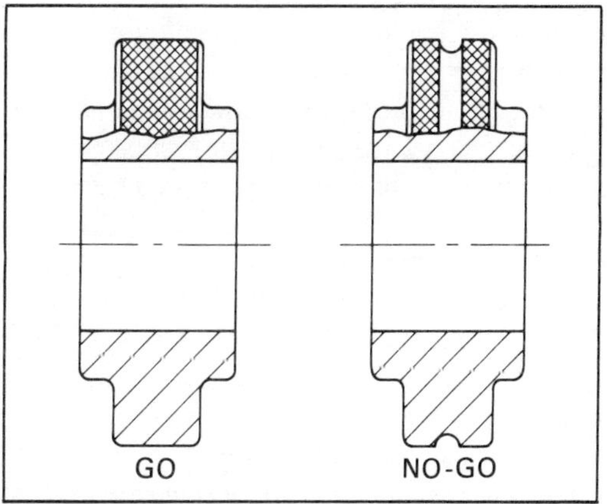

Ring gage set used to inspect the diameter of shafts.

Among the variables to be considered in automatic gaging are part size and shape, material, finish, production rate, tolerances, part handling, cleanliness of part, type of gage element, and inspection rate. Interchangeable and adjustable tooling enable a gage to handle different parts, sizes, and tolerances. To accommodate slightly misaligned holes or locations, automatic gages are sometimes designed with floating-type gaging elements. If a part's size or shape prevents it from being gaged in one operation, it is better to inspect the close tolerances first. This prevents wasting the expense of checking the broader tolerances first only to have the part rejected later.

Usually, locating surfaces for the gaging operation should be the same as those used in the machining operation. However, for final or assembly inspection they may be based on the function or end use of the part. A large variety of part transfer devices are used in automatic gages, such as walking beam devices, endless belts, gravity feeds, or robots.

With the advent of the computer, it has become practical to design automatic gages that can separate size tolerances from location tolerances at one gaging station. To obtain increased (faster) speed, this type of gage may employ laser gaging or vision (video) gaging. These gages do not rely on a fixed location of the part, but will determine the location of the datum surfaces and inspect in relation to them. Although air, air-electric, or electronic automatic gages require interchangeable or adjustable gage heads or tooling to handle different part sizes or tolerances, video systems need only to be programmed to inspect a wide variety of parts. With tutorial, menu-driven (software), user friendly programming, video automatic systems are easy to change from one part to another. Once the program is written, it is only a matter of calling up the new program from the system memory. Flexible, programmable gaging systems using CNC logic and LVDT transducers are also available. These systems can provide higher accuracy than optical systems or coordinate measuring machines and are designed to operate in a harsh shop environment. In addition to providing totally automatic parts measurement, these systems can directly control one or more machining process by sending statistically based compensation commands directly to the machine tool.

Gain. Gain is the amount of increase in a signal passing through a control system. It is also the sensitivity and ability in a control system to raise the power of a signal to a specified output.

Gantry Robot. A gantry robot is essentially a Cartesian coordinate manipulator configurated with three translational axes and up to three rotational axes. It gets its name from the type of support mechanism that allows the manipulator to move above the workpiece.

The gantry design allows the processing of all sides of a workpiece. Overhead mounting of the motion system frees floor space for part processing and for free material movement through the machine. The work area is open, providing good access for the operator. Many maintenance operations are carried out on a catwalk around the gantry, away from the hazards of the shop floor.

Multiarm/multihead systems are possible that share the same basic structure. One system currently uses two lasers, two welding heads, and two clamping heads.

These robots require a large, rigid foundation to ensure proper machine alignment. Due to the distance between the support columns, the foundation required may be larger than for a jointed-arm type of robot.

Gantry robots are made in many sizes and proportions. They carry heavy loads, and have the same characteristics at all points within their work envelopes. They occupy no floor space within the work envelope. The robots can be equipped with sensor-based active homing guidance and with axes of permissiveness for passive homing guidance, providing zero error positioning.

Gantry robots are controlled by point-to-point or continuous path NC systems and sometimes use a microcomputer or a programmable controller for complex sequence control.

The robots can be used for loading and unloading machines, material handling, inspection and gaging or for the movement of welding heads or machine tools.

These robots have proven to be a good configuration for a large work envelope laser machine tool, in terms of accuracy and reliability in a production environment. Planned applications (in 1987) for "laserobotics" include cladding, cutting, and welding of a variety of materials. Work will be performed for a number of industries including automotive, aerospace, and plastics.

Gantry robotic systems prep and paint fully assembled aircraft. One system automatically cleans, sands, polishes, laser mills, strips, paints and inspects aircraft.

Four of the world's largest gantry robots have been built to clean and refurbish the space shuttle's solid rocket booster (SRB) nonmotor components. Standing nearly 30 feet high, these robots are the first of their type to be used by the National Aeronautics and Space Administration (NASA) in the shuttle program. The robots have the capability of performing 55 separate motion routines to accomplish nine tasks on the SRB structures. Using a variety of end effectors, the robot arms move around the booster hardware performing operations inside and out. The figure on the next page shows an overhead gantry robot system at work.

Also see: Industrial Robots.

Gap Checking. Gap checking provides a check of the minimum or maximum distances between the boundaries of two design entities.

Also see: CAD/CAM.

Gas Plasma Display. A gas plasma display is a video display screen employing gas plasma or discharge technology, characterized by an exceptionally clear, flicker-free image.

Gas Tungsten Arc Welding. Gas metal arc welding is the predominant process used in robotic arc welding. Frequently, a special interface is required to connect the power supply to the robot so that the weld parameters can be set under program control. Shielding gases used most frequently for robotic gas metal arc welding of steel are pure argon, argon with 2 or 5% oxygen, and argon with 10 or 25% carbon dioxide.

Joint designs for robotic arc welding are most frequently single-pass fillets, with partial-penetration welds. The reason for this is that parts small enough to be quickly moved in and out of a robotic workcell do not generally require more weld metal to perform their intended function. The ideal joint design for robotic arc welding consists of intermittent fillets in varying locations around the assembly.

Gas metal arc welding is essentially a direct-current process, with alternating current rarely used.

Gate. A gate is a circuit, device, or element which blocks or passes a signal depending on one or more specified inputs. It is also the end of the runner in a mold or die where molten metal enters the casting or mold cavity. It is sometimes applied to the entire assembly of connected channels and to the pattern part which form them.

Gate Array. An integrated circuit, a gate array is characterized by a rectangular array of logic sites. These sites consist of identical collections of diffused or implanted transistors, diodes, and resistors. Surrounding the array are the input/output circuits for off-chip connections. The final process in creating the desired circuit is the custom metalization of the logic sites and their subsequent interconnections. Gate array design permits a standard set of logic elements to be used for a wide variety of integrated circuit applications.

Also see: Computers.

Gateway. A gateway is the node at which two or more networks are connected. It has the fundamental role of serving as the boundary between the internal protocols of the connected networks. A gateway is the total collection of

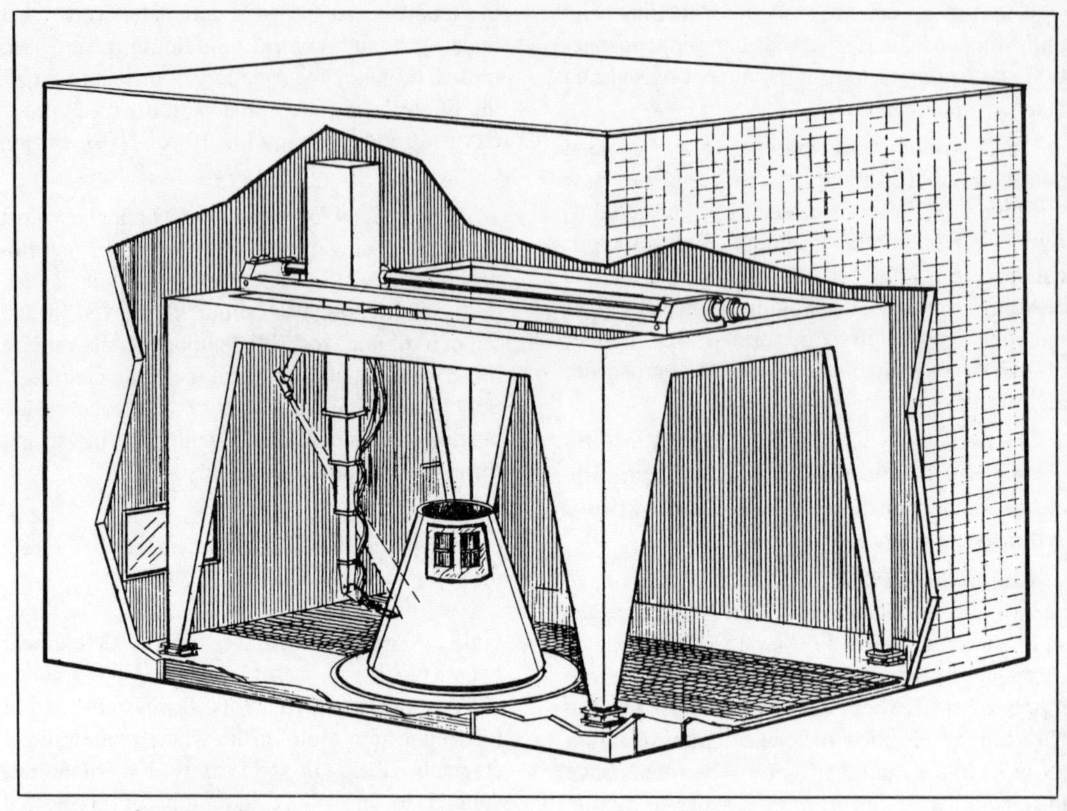

A gantry robot at work.

hardware and software that is required to effect an interconnection between two or more networks to enable users in different networks to communicate with each other. A gateway may be placed between two different types of networks or between identical types of networks.

One of the more popular gateways is between Decnet networks and X.25 (IBM mainframe) networks. Other gateways allow personal computer networks to interface to minicomputer and mainframe networks.

A gateway differs from a bridge which allows a program intended to run on a certain machine to be executed by a different type of machine. A bridge generally always includes hardware, while a gateway may only consist of software and the appropriate cables and connectors.

Gauging. See: Gaging.

G Code. A preparatory numerical code in a program addressed by the letter G indicating a special function or cycle type in an NC system. Also known as G function.

General and Administrative Cost. General and administrative costs are those incurred at the plant or interplant level and not easily associated with a specific work center or department.

Examples include the costs of top executives' salaries, plant mainframe computer procurement/operation costs, and technical library facilities. Most of these costs are fixed. The primary variable component are sales commissions and utilities.

General Processor. A general processor is the "executive" program containing the basic intelligence for NC work that is stored in computer memory before individual part programs can be processed. It can be a computer program for converting geometric input data into cutter path data for NC machines or a fixed software program designed specifically for logical manipulation of data.

Generation. Generation is a term used to refer to the general level of technology incorporated into the design of a computer or other evolving device.

Geometric Modeling. Geometric modeling refers to a complete and geometrically accurate description of a part, shape, or other entity. The geometric model is designed on a computer aided design system and is stored in a database. The geometric model in electronic format is the basis for subsequent engineering analysis, programming of the part for processing by a numerically controlled machine tool and, in some cases, transferring of the geometry to an on-line automatic inspection or gaging system within production operations. The geometric model itself may be a mathematical or an analytical model of a moving system. Geometric models can be relatively simple for use primarily as aids in creating sketches or drawings. Conversely, they may be sufficiently complex and complete so as to allow sophisticated thermal and structural analysis. Geometric models are increasingly utilized within coding and classification systems where group technology concepts are employed to exploit geometric commonality among parts within both design and manufacturing operations.

Graphical Kernal System. A graphical kernal system (GKS) is a standard which has been defined for describing the basic attributes and standards for generating and displaying computer graphics. The GKS specifies the interface to be used by the video display portion of the system. The GKS was devised in West Germany in the late 1970s and is becoming an international standard.

By using the GKS standard, software that is device-independent, can be written which will be capable of producing graphics when connected to other equipment which also abides by the GKS standard. The GKS hasn't been adopted by all developers of graphics-related software, because some had already established their own graphics standard and because only they want to maintain proprietary designs. The GKS is used more on high resolution CAD systems than on personal computers.

Digital Equipment Corporation has developed its own graphics standard known as ReGis graphics. The IBM personal computers have a CGA (Color Graphics Adapter), EGA (Enhanced Graphics Adapter), and PGA (Professional Graphics Adapter). More recently a VGA (Video Graphics Adapter) standard has been introduced.

Graphic Input. Graphic input is the input of symbols to NC systems that comes from lines drawn on a cathode ray tube or information obtained from drawings by a scanner on a digitizing tablet.

Graphics. The introduction of computer based applications for business brought with it an information explosion. The large volume of data produced by computer applications is frequently not used effectively by decision makers. They are inundated with voluminous amounts of data so quickly that by the time they have time to read, digest, and analyze the information, it has often been superseded. The decision maker is, in a very real sense, being engulfed by a kind of information pollution.

Management's ability to comprehend and act upon computer-generated information has been evolving at a much slower rate than the ability to generate the data. Unless the data generated by a computer application can be used by a manager on a timely basis, it is of little value to the business.

Studies have shown that we can communicate with text at 1,200 words per minute and with pictures at the equivalent of 40 million words per minute. The graphic or pictorial representation of data can be a quick and easy means of communicating large amounts of data effectively.

Graphic representation of information enables management to obtain, analyze, and interpret large volumes of data, and to make decisions on the basis of the information digested. Graphics for management decision making has received considerable attention since the introduction of inexpensive PC systems with graphics capabilities.

History. Computer graphics capabilities have been around since the early 1950s. Early applications were largely in the military field. The SAGE system, developed in the mid-50s was used to analyze radar images. The commercial graphics evolution began in the 1960s with the introduction of Sketchpad, a doctoral thesis at MIT. During the '60s and early '70s the major application and growth area for graphics was in computer aided design (CAD) and computer aided engineering (CAE).

Initially the automotive and aerospace industries were the leading users of CAD systems, and soon they became common features in most engineering environments. With the introduction of the microcomputer, graphics entered a whole new phase. Low-cost graphics systems, frequently incorporated into other business software such as spreadsheets, became commonplace. Much of the current growth in graphics is in this low end of the market. The Apple Macintosh, in particular, has had a great impetus on graphics development.

Discussion. The necessary elements of a graphics system are a processor and its memory, disk storage with a database or file management system for handling the stored data, graphics software, and a suitable display or output device.

Graphics software provides the interface between the hardware (e.g. host computer and the output device) and the application software or data files to be displayed. The graphics software will handle much of the data manipulation as well as the generation of the graphic output. Graphics software can either come as a stand-alone utility, or it can be embedded in application software. Graphics software for PCs tends to be menu-driven, while mini and mainframe graphics software make use of high-level languages. Graphics is primarily user oriented. It is therefore essential that graphics software is extremely user friendly.

Interactive computer graphics involves two-way communication between the computer and the user. The computer, upon receiving signals from the input device, can modify the contents of the frame buffer by creating a new pattern of pixels. To the user it appears as if the picture is changing instantly in response to commands.

The alphanumeric keyboard is not particularly suited for input to interactive graphics systems. The limitation imposed by the use of alphanumeric input and the limited cursor movement available with a normal keyboard do not lend to ease of use. A number of alternate input devices are available for use with graphics. These include the following devices:

Joysticks, which are similar to the devices used in TV games, move a cursor around on a display for entry of data at a specific point. Some keyboards come equipped with a joystick. Joysticks have largely been replaced with mice.

A *mouse* is a small device which is moved across any flat surface. A large ball in the body of the mouse is used to translate the movement of the mouse into information which is sent to the display. A screen cursor moves in relation to the movement of the mouse.

Data tablets are sensitized surfaces with a special pen which allow drawings to be created on the tablet. The cursor on the screen moves in relation to the pen on the data tablet.

Special pens can be used in conjunction with light-sensitive or touch-sensitive screens to input data directly on the screen. The main advantage of such input devices is that eye-to-hand coordination is facilitated by drawing directly on the screen.

Digitizers use a cross-haired device to input charts, maps, and photographs. Documents are translated into binary digits which represent the shades of white, black, or grey in much the same manner that photocopiers make copies of documents.

Digital video cameras record an image as a set of binary digits. The digital information can be used as input to a graphics system where the video image can be displayed and even changed.

Graphics output can be provided by a large number of general-purpose and dedicated devices:

Standard character *printers* produce graphics of limited quality. They can produce reasonable quality black-and-white graphs, pictures, or maps with varying tones of grey by overprinting. They are limited to producing two-dimensional graphics. Because of the very low resolution, they are not suitable for anything but the simplest of graphs, maps, or pictures. More expensive printers have a multicolor capability.

Pen plotters can produce monochrome or multicolor graphics of high quality. A typical flat-bed pen plotter has a large flat drawing area. The pen is driven by a mechanical arm in much the same manner as an artist or draftsman would move a pen or pencil. The pen widths and colors can be changed to produce color drawings and graduations in tone. Plotters are particularly suited for graphs, drawings, and maps, and can be used to produce three-dimensional plots.

Electrostatic plotters operate like photocopying machines, and produce a full page at a time. They can combine text and graphics data to produce high quality color graphics. Laser printers are a widespread example of a device of this type.

Spatial plotters are used to develop three-dimensional graphics by ''placing'' wires on the plotting surface. The location of the wire corresponds to the location of each point sampled in a survey or study. The wires are ''cut'' at the appropriate heights to indicate the density per measured unit of volume or area. Spatial plotters are also suited for model making.

Cathode ray tubes, or displays, are frequently used output devices for graphics. Generally speaking, there are two types of displays: (1) the alphanumeric display, which displays only letters and numbers, and, (2) vector displays which can also display lines and drawings. Most alphanumeric displays can be used for low resolution graphics with the appropriate graphics controller card. Displays do not produce permanent, hardcopy output.

Camera systems produce color photographs of a color CRT screen. The resolution depends on the display's resolution. These systems are growing in popularity because they provide a convenient way of producing 35mm slides in-house when used with a high resolution color terminal.

Computer output to microfilm (COM) produces one high-speed photograph per second. Equipment is available with a choice of 16mm, 35mm, 70mm or 105mm film in single exposure or movie format. Because of their very high speed, high resolution and versatility, COM plotters are gaining acceptance.

An important consideration in computer generated graphics is resolution. The resolution refers to the number of points, or pixels, per unit of area that convey the pictorial information. Screens can only display a finite number of horizontal or vertical lines. A display with 512 points per horizontal line and 512 horizontal lines would thus have a total of 262,000 individual pixels. In monochrome screens, each point would be stored as one bit—either white or black. Therefore, a screen with 512 x 512 pixels requires at least 33k bytes (262k bits) of memory to store the information displayed on the screen.

Color, on the other hand, usually requires much more memory, because additional bits are required. A four-color graphic display would require twice the memory of a monochromatic display (i.e. 66k in the example above).

The memory overhead for graphics can be greatly reduced by reducing the resolution. As the resolution decreases, the size of each pixel becomes larger, thus reducing the number of pixels displayed on the screen. With lower resolution, curved lines or shapes have less clear boundaries. An example of low resolution is the stairstepping effect on curves. The horizontal and vertical pixels that are used to display the curve are large enough to be seen, and the curve takes on a step-like shape.

Some producers of color graphic displays have attempted to solve the limited memory by offering two modes of operation. With one popular display, the user chooses two colors from a palette of 4096 colors in the high-resolution mode. Operating in this mode the system is able to provide a resolution of 754 by 482 points, or 370,000 pixels. In the second mode, a total of 16 colors are allowed, and the resolution drops to 377 by 240 points, or about 90,000 pixels.

Resolution can also be affected by the time taken to refresh the screen. Displays that are refreshed less often than 1/30 to 1/60 of a second appear to flicker. Many vendors add a fresh memory to their system to handle this problem. Graphic systems without refresh memories, for example, host-based systems using dumb terminals, must use some of their main memory for refreshing the screen. This further reduces resolution.

Color in a high-resolution multicolor graphics system has three principle attributes or dimensions: Hue, intensity and saturation. *Hue* represents what most people thing of as color—green, cyan (blue), magenta (red), yellow, etc. *Intensity* combines the hue with varying degrees of white or black. *Saturation* determines the ''richness'' of the color, for example, from pale red to fire engine red. Some systems allow a very large palette of possible colors from which the user chooses a few for use in one display. Other systems have a fixed number of hues with a few intensities and varying saturation for each.

The distinction between low, medium and high-resolution graphics systems is somewhat arbitrary. As a generalization, displays with less than 140,000 pixels (375 x 375) show very marked stairstepping on diagonals and curves, and are considered to be low resolution devices. Displays with over 250,000 pixels (500 x 500) show less evident stairstepping, and are known as normal or medium-resolution devices. High resolution starts at about 750,000 pixels.

Applications. Computers and their associated peripherals are capable of producing a wide range of graphic representations of data, ranging from graphs and pictures to maps and models. It has often been suggested that computer graphics is a solution in search of a problem. The potential application of graphics is restricted only by the lack of creativity of users.

Most graphics applications can be grouped into one of the two major categories: presentation graphics and decision support graphics.

Presentation graphics covers the wide area of visual aids and printed graphics. The graphic image is the end product. Computer aided drafting is an example. Other examples of presentation graphics would include solid modeling and analysis, printed circuit board design, process simulation, and topographic mapping.

Because the computer graphic generally replaces a manually prepared graphic, cost justification computer graphic systems for presentation graphics is fairly easy. The benefits of computer graphics can be directly measured in terms of reduced labor, shorter turnaround time and increased flexibility.

Decision support graphics convey information for decision making. The purpose is to convey information in a concise and easily comprehendible manner. It is very difficult to quantify the benefits that are gained from decision support graphics. It often makes sense to adopt the attitude that if better decisions can be made, the computer graphics can be justified.

An informal survey of 144 U.S. sites using business graphics showed the following uses:

Management Presentation	85%
Financial Planning	59%
Sales Presentation	45%
Scheduling Charts	37%
Product Planning	26%
Forms Generation	23%
Sales Targeting	22%
Inventory Monitoring	21%
Portfolio Analysis	13%
Other	35%

Of those surveyed, 75% intended to expand their use of business graphics.

One of the most unusual applications of graphics and COM plotters is in preparing computer animations or movies. The Walt Disney Studios' production, *Tron*, was produced entirely by computer graphics. Other applications include the observation of atomic motion or airplane flight. A computer modeling system will simulate the motion of an aircraft over time, and the "motion" of the plane will be recorded by a COM plotter. The use of movies greatly facilitates the understanding of the motion of the airplane over time.

Graphics Terminal. A graphics terminal is a terminal which is capable of displaying graphic images as well as text. A graphics terminal may use a composite video monitor or it may use an RGB (red, green, blue) monitor. The RGB monitor may be a TTL or an analog monitor. A TTL monitor can display two tones of a color, full intensity and half intensity, while an analog monitor can display as many as 8, 12, or 16 shades of the same color. A graphics terminal may use a bit-mapped display in which each pixel (picture element) has an address and description, or it may be a raster type display where each horizontal line is defined and displayed. Another type of monitor draws the image with the use of a raster. When the image is drawn, it is referred to as a calligraphic image.

Also see: Graphics, Terminal.

Gravitational Balancing Machine. A gravitational balancing machine is a balancing machine that provides for the support of a rigid rotor under nonrotating conditions and provides information on the amount and angle of the static imbalance.

Gray Scale. Gray scale is a term used to describe the intensity of shading between white and black. Plain white is one extreme and solid black is the other extreme. In between are many shades of gray.

In CAD systems which utilize a monochrome monitor, gray scale refers to the relative levels of brightness which are used to display various levels of detail within the drawing. In a monochrome monitor, the gray scale applies whether the phosphor color is white, green, or amber.

In a machine vision system, gray scale describes the discernible levels of brightness which can be perceived by the camera and associated computer circuitry to determine features of the image such as edges, holes, and surfaces. A machine vision system is limited by the resolution of the AD (analog-to-digital) conversion capability of the system. A typical system may provide 64 levels of gray in its gray scale.

In typesetting and publishing, gray scale refers to the practice known as *halftones*. With halftones, shading is accomplished by use of various densities of dot patterns to represent shading.

Grid. Grid defines the set of points (in X/Y coordinates) that determines the minimum accuracy at which basic entities are described or connected.

Grinding. Grinding is the process of removing metal by the cutting action of a solid, rotating, grinding wheel. The abrasive grains of the wheel perform a multitude of minute machining cuts on the piece being worked upon. Grinding is generally considered a finishing process used to obtain a fine surface funish and extremely accurate dimensional tolerance. Grinding may also be used as the complete machining operation on a surface. From .005 to .020 inch (.13 to .50 mm) of stock is commonly removed in a grinding process depending on the operations.

The grinding process is used in machining a wide range of metals in addition to cemented carbide, marble, stone, and a variety of ceramics. The commercial grinding abrasives are many times harder than the metals to be machined, hence the grinding process is used on metals too hard to machine through another process.

Grinding wheels are composed of abrasive grains on grit plus a bonding material. Grinding machines are designed to hold a workpiece rigidly and to feed it smoothly through the cutting path of a rotating grinding wheel.

Over the past decade, grinding machine builders have introduced a wide variety of new machine control features aimed at boosting the operating performance and overall productivity of new grinders. These new control features are generally classified as one of two different types: (1) machine controls designed to control grinding process variables, such as speed, infeed pressure or workspeed, in response to changes in the grinding operations, and, (2) machine controls designed to control positioning or tool path.

By far, the use of NC or CNC has excelled in tool path control but, to date, has not been as widely used for adaptive control of grinding process variables. Adaptive control of the complex relationships in grinding, and other machining processes for that matter, has lagged behind applications of tool path control, probably because adaptive control is significantly more complicated and is less easily developed for general use.

On modern grinders, position control or tool path control is characterized by the widespread use of manual data input (MDI), allowing the operator to program the grinder on the shop floor without the assistance of specially trained NC programmers. On many new machines, a variety of program storage methods are available, such as tape cassettes, floppy disks, and traditional paper tape, to facilitate increased flexibility in programming and changeover.

The latest generation of electronically controlled grinders feature ''canned cycles'' to assist in programming—a convenience that speeds programming and helps eliminate costly programming errors.

In the future, grinding machines will feature increased capability due to still more sophisticated controls. Advances are expected to be made in both tool positioning and the controlling of the physical process of grinding. It is expected that control capabilities now found only on NC lathes, machining centers, and sophisticated milling machines will carry over and be available on grinders as well. This phenomenon is evidenced by the fact that some machine builders are now offering equivalent grinding machines that closely resemble, with respect to controls, certain machining centers and turning machines in their product line.

For example, toolchangers have been standard equipment on CNC and NC machining centers for many years, and some observers believe that features such as these (namely, wheel changers) will be developed for automated grinders in the future. This trend is now evidenced by a number of grinding machines that are sold with an auxiliary wheel arbor as standard equipment. This second wheel can be mounted, balanced, and dressed off the machine while the grinder is in operation. When a wheel change is required, the second wheel is ready for installation with minimum downtime. An automated wheel-changing feature would be particularly effective in speeding changeover in job-shop grinding operations.

The control of grinding process variables through the use of adaptive control will come more slowly than with more sophisticated positioning controls.

Gripper. A gripper is a robotic end effector which grasps an object to allow the robot to pick and place an object. Many types of grippers exist from special-purpose grippers to simple basic grippers. Some grippers have parallel action jaws while others have scissor-type action. Simple grippers may have only two opposing jaws while others may have three or more jaws. Grippers can contain test apparatus or scales to perform other operations while the part is moved by the robot. Some grippers may use electromagnets and others have suction cups. These types of grippers may actually be end effectors, but not grippers in a pure sense since they don't pick up objects by gripping the object. Designing grippers is a specialty field. Sometimes a substantial amount of design time is required to design and develop a suitable gripper.

A sample gripper is illustrated below.

Also see: Industrial Robots.

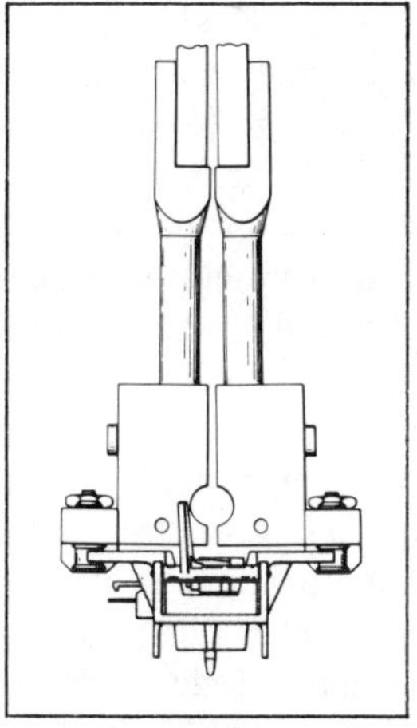

A sample gripper.

Ground. Ground is a conducting connection between an electric circuit or equipment and earth or a conducting body serving in its place.

Group Technology. Group technology (GT) is a manufacturing philosophy that identifies and exploits the underlying similarity of parts and manufacturing processes. By grouping similar parts into part families based on either their shapes or their manufacturing process, it is possible to reduce costs through effective design rationalization and design data retrieval. It is also possible to lower stock levels and purchase quantities and simplify and improve production planning and control. Further, GT reduces tooling costs and setup times; in-process inventory; throughput time, NC programming costs and allows for a more efficient NC machine utilization. Group technology is also an essential element in the implementation of computer integrated manufacturing (CIM). The basic philosophy behind group technology is to use (as much as possible) the existing work in the design and manufacturing of products.

Dr. Nancy Lea Hyer of the University of North Carolina at Chapel Hill served as Editor for the book *Capabilities of Group Technology*. Dr. Hyer noted in the book's Preface: "Broadly defined, group technology (GT) is the application of knowledge about groups. More specifically, GT implies the notion of recognizing and exploiting similarities in three distinct ways: (1) by performing like activities together; (2) by standardizing similar tasks, and (3) by efficiently storing and retrieving information about recurring problems. In the past several years, the interest in GT among the U.S. manufacturing community has increased dramatically. U.S. industry's fight for survival in a highly competitive domestic and international marketplace, advances in computerized data storage, and the emergence of the computer integrated manufacturing (CIM) philosophy for company operations all contribute to GT's increased attractiveness."

Converting to a more flow-oriented operation is fundamental to just-in-time manufacturing. Group technology is a key step in the conversion process from standard batch production—emphasizing a functional plant layout in which machines are grouped according to specific operations—to a layout based on the concept of grouping products according to their design and/or manufacturing similarites. Fundamentally, GT is a manufacturing philosophy. It is an organizational principle of identifying and bringing together closely related components and processes to benefit from their similarities.

The principle of grouping parts and processes can have a profound effect on most aspects of the manufacturing cycle. Because job shop operations typically involve an extremely high variety of parts and processes, the positive effect is particulary significant in this type of manufacturing. Here, GT can produce great benefits for product engineering, manufacturing engineering, production control and procurement by helping to identify patterns of similarity that would otherwise be difficult to discern. Through the selection of similarities, simplification and waste elimination, opportunities come into clear focus and can be accomplished.

History. Group technology is not a new principle. As early as 1920, Frederick Taylor advocated the idea of grouping parts requiring special operations. Also in the 1920s, the Jones & Lawson Machine Co. built machine tools using principles of departmentalization by product rather than by process, product standardization and minimal routing paths. It is likely that this approach would today be referred to as group technology.

Years ago, group technology embraced coding and classification systems that allowed engineers to retrieve drawings to see what had been done before and to avoid redundancy. That was on the design side. Then manufacturing observed early group technology principles in the formation of families of parts. This meant that the foreman would look at the parts awaiting processing and would arrange then into groups according to tooling and setups required. The idea for parts grouping, began to find its way into American industry. As a result of this innovation, design of the factory floor is changing as machine tools are grouped according to the type of end product rather than the functions they perform.

Discussion. Group technology can have a profound effect on virtually every aspect of manufacturing. However, the most immediate and significant impact is felt on the factory floor. The development of group technology cells results in the elimination of work in-process inventory, the promotion of visibility, quality improvement, and greatly reduced material handling and tracking.

A good system is an important ingredient for a properly functioning GT system. The first step in many group technology applications could be the classification and coding of the elements of production. Several developed classification structures have been based on the E. G. British concept of hierarchical classification, which assumes that all elements of the company are subject to classification, from the product itself to the controls over production. In this hierarchical ordering, each level is dependent on the previous levels and ultimately allows a great deal of information to be stored in a relatively small space. This kind of coding system is particular adaptable to computer applications.

When using the classification system to store and retrieve information, a design engineer requires different information than that required by a purchasing agent or a manufacturing engineer. Therefore, the various characteristics required by each user are retained in such a way that the user can retrieve only the required information. For example, though a coded database might carry essential characteristics for a product—such as its shape or geometry, dimensions, function, chemistry, and other specifications—the attributes stored for the disparate user would vary for this same product. Whereas customer support might require management, pricing substitutes, and maintainability information, a finance user would need cost, price, estimating, and vendor information. A production worker needs process plans, tooling, in-work inventory, raw material, and schedule data. Raw material families include material form and chemical makeup.

The initial reason for the development of group technology retrieval systems for piece part designs was to avoid redesign. These benefits have been appreciated, however, only by organizations whose management and technical staffs used the system diligently.

In a large organization with thousands of design engineers and drafters as potential users, the administration, management, and control of group technology classification systems becomes a difficult task. Experience has demonstrated that design retrieval from a centrally located design retrieval center may not always be beneficial. Some functions of design, however, proved opportunities to realize enormous optimization for group technology systems. For example, an electrical/electronic (E/E) design group can request that a limited number of designs particular to their requirements be extracted from the huge part design library. This specialized minifile contains the preferred designs for electrial housing, bracketry, and similar parts appropriate to electrical designs, and has been classified to satisfy specifically the needs of the E/E design group. As much as 95% of the piece parts required for a new design can be retrieved from the system.

For maximum efficiency and benefit, manufacturing engineers have traditionally worked with design groups to advise design engineers concerning the producibility of a proposed design while it is in the definitional stage. Sound economic and production management principles can thus be incorporated into a design while it is developed. This design/shop interface works especially well in complex and sophisticated areas in which the use of exotic materials and special forgings is frequent.

Group technology is not an easy achievement. Its development in a company can require an extreme effort in setting up a logical coding system. What may seem logical at first glance may turn out to be inappropriate. Trials, errors, and revisions must be endured. In addition, because it means changing procedures and methods that may have been used for decades, people must reorient to new ways of doing their job.

Much of a piece part design represents the arbitrary decisions of designers or drafters. After the basic criteria are established and the design is determined, the requirements can be specified as a set of dimensional and other relationships. The balance of the design, even for the simplest of parts, is often a result of habit or personal preference of choices made from a series of options. As a result, a number of parts that are otherwise equal might vary by such noncritical differences as the bend radius of an angle or type of corner relief, for example.

When piece parts are identified by shapes into families of similar parts, one of the most obvious benefits from such information is the grouping of like articles for production. To create processing cells which include a variety of machine tools that work and which are dedicated to the efficient production of a family of similar parts requires a great deal of data. For example, piece part classification for large structures, such as airframes, demonstrates that a large variety of parts are designed and manufactured from formed and extruded cross sections. A machine called a *partsmaker* is a highly efficient machine tool for the fabrication of this type of part. The problem in the manufacturing and industrial engineering communities is to ensure that all parts configured from extruded cross sections lend themselves to fabrication by numerical control. This will allow a partsmaker, which operates on the bar feeding principle, to be utilized to the maximum degree. An optimum equipment load for this type of equipment can be readily obtained by analyzing the piece part drawing file and changing the processing on all applicable part configurations to NC processing. Through this procedure, a company can obtain optimum processing for large number of part families, and determine the requirements for a partsmaker-type equipment.

Another example of equipment loading includes the selection of forming equipment, by the shape characteristics of the part to be formed. For example, a particular part might be formed on a drop hammer, hydraulic press, bag press, or by the electroform process. In nearly every instance, one process in preferred over the others. Using the characteristics defined in a classification system, industrial engineers and process planners are able to determine the best processing solutions.

Coding and classification can be tool for a concept called *generative process planning*. Generative process planning is adopted by many companies, to help make the classification processes workable. This generative method of parts identification is derived from the following realizations:

- Hierarchical classification structures can be defined in decision-tree logic.
- A unique path through a decision tree can be represented as a specific code character.

- A combination of unique decision-tree paths can identify a specific engineering or manufacturing decision.
- Characteristic codes, used as a shorthand, define a combination of paths that lead to a prescribed optimal design or manufacturing decision.

The purpose of the generative planning system model is to demonstrate that uniform manufacturing process plans can be generated directly from engineering design information. It is universally accepted that classification benefits are potentially greatest when they are implemented in the design process. Only in this manner can the advantages from a classification system cover the whole business spectrum.

Applications. A major benefit of GT comes from reduction in nonmachining time due to the combination of tooling, setups, and so forth, of small lots. Scheduling and process planning are simplified, eliminating duplicate efforts for similar parts. The use of GT in part design decreases redundancies and helps to standardize parts.

In traditional batch/job shop manufacturing, parts that may require only a few minutes of actual machine time can spend days or even weeks on the factory floor. There is an almost random sequence of parts to be dealt with; machine tools have to be set up for every new run; tooling has to be chosen and delivered; and long, frequent setups are usual. The overall effect is constant queues of work in-process inventory, which is very costly.

The primary effect of applying group technology on the shop floor is to reduce lead times. Other benefits include elimination of costly WIP inventory, reduction of delays and setups; elimination of unnecessary materials handling and communication; product quality improvements, and improved manufacturing engineering. From a broader perspective, positive effects of the grouping principle go far beyond the factory floor. The approach has a significant application in the areas of product engineering, production control and procurement.

By helping to identify selected similarities, group technology can reduce the proliferation of parts. Furthermore, it promotes design standardization and provides feedback from manufacturing. It also makes cost estimating easier.

The benefits of applying group technology are clearly substantial. They include reduction in part design; the number of drawings through standardization; new shop drawings; industrial engineering time; production floor space required; raw-material stocks; setup as well as production time, and work-in-process inventory.

To achieve these benefits requires the handling of a great deal of information. As stated previously, design processes and parts need to be grouped according to select features. The computer is best suited for these complex tasks.

Because many companies perceive that GT requires extensive upfront commitments, GT has not been used extensively in the U.S. The documented benefits would seem to warrant a popular perception that the ''up-front'' work is too demanding. However, this perception is often not justified. Rarely does every single part in an operation's inventory need to be rigorously classified and coded before GT benefits can be realized. It is usually sufficient to start by classifying and coding only those parts currently in active use. In most manufacturing operations, the number of new components released begins to level off shortly after a classification/coding system has been implemented.

But nonetheless, the key to group technology, particularly in a highly diversified environment, is the classification and coding of parts, processes and facilities according to appropriate characteristics. Then, a meaningful code must be assigned to reflect those characteristics. Thereafter, the computer performs step-by-step prompting and interrogation, then automatically assigns the proper code.

It is important to note that although the Japanese just in time manufacturers have generally formed manufacturing cells to handle families of parts, these cells do not necessarily need to involve physical rearrangement of a facility. Significant benefits can accrue simply through the application of the GT principle to achieve administrative improvements. However, to achieve full competitive status vis-a-vis the Japanese and other foreign competitors, usually requires the physical conversion to manufacturing cells.

A few examples of GT applications may be in order. William Beeby and Phyllis Collier report in their book *New Directions Through CAD/CAM*: "At Deere's (Deere & Co.) Hydraulic Division facility, more than 20 software programs can group parts according to type, manufacturing processes, demand history, cost, and so forth. The Hydraulics Division organization was subdivided according to families of components. The formation of part families includes analysis of part codes production flow, starting machines used, production requirements, and unique processes. Adjustable cell arrangements based on practical use are made. Then these cells, which may or may not be located together, form departments.

"As the result of a five-year program, the Rocketdyne Corporation became deeply involved in the implementation of group technology, using the MICLASS system for coding and classification. Rocketdyne activities included common coding between divisions on machine tools implementing interactive graphics—a system called MI-GRAPHICS—as well as a CADDS 4 Computervision 3-D system. These links also communicated with the management information system, or MIS, throughtout the plant on a standalone basis.

"The MICLASS group technology system at Rocketdyne was in batch manufacturing operations to classify and code drawings, as well as in analysis activities. Such activities include the retrieval of engineering drawings, standardization of drawings, analysis of raw material requirements, production group and machine tool requirements, and the optimal routings, machine tool use, and so forth.

"Rocketdyne has used group technology as a powerful tool for identifying and extracting the similarities and differences present in product mixes and across geographically distant manufacturing plants. Successful utilization of group technology requires a thorough understanding of the product mix and production requirements for planning, scheduling, and manufacturing decisions. Successful utilization should suggest alternative ways of meeting such goals.

"Rocketdyne discovered that the parts must be coded accurately and uniformly to load the master data file to permit analysis for identifying manufacturing work cells, shop load, and standard manufacturing process plans. The coding programs must also provide a standardized capability and discipline so that all manufactured parts can be uniformly coded. Such precautions make parts coding easier and more accurate for less experienced personnel, thereby reducing time and costs incurred for the coding effort.

"Management discovered that group technology is a powerful tool for accessing data and permits more effective decision making as a result of the family-of-parts capability and equipment information."

The efficient use of group technology should result in simple part geometry descriptions that can be stored on a digital computer, accessibility of part information using the part geometry or other attributes, and families of similar parts. Group technology systems work well where programs exist that have built these part families; analyzed manufacturing costs; and defined and calculated machine loading, equipment utilization, and procurement. Where programs have identified machine-tool requirements so that group technology cells can be found, the systems also work well.

Though group technology and classification and coding schemes have been around for a long time, it has not been until recently that companies have shown much interest in seriously adopting the technologies. This increased interest can be attributed to the number and varieties of products deluging industry, creating smaller lots for each product; a growing need for more economical ways to produce a quality product; increases in product varieties and therefore, the kinds and properties of materials required; the need for less waste of material; and a rapidly growing need for better communication between design and manufacturing.

Group technology emerges as one of the prime forces that will integrate the engineering and the manufacturing process together. Group technology lends order to what has usually been a highly unordered collection of processes. It drastically reduces the negative phenomenon of the needlessly recreated design and improves production scheduling and control processes.

Glossary. *automated guided vehicle (AGV)*: The method of a programmed carrier used to transport goods from storage to point of use.
automated storage and handling system: A computerized material handling approach that utilizes programmed pick-and-place equipment.
computer aided design: The use of computers to aid in designing.
computer aided manufacturing: The use of computers to aid in various phases of manufacturing.
fixed automation: Automation specified for only one product line, process or part.
flexible manufacturing system (FMS): Programmed system of fully integrated machines, or machining centers, capable of performing a series of machining operations on different workpieces, within a product group, flowing through the system one piece at a time—no setup time required between each piece.
intelligent automation: Automation that includes process feedback to adapt to specified variables, that can include artificial intelligence.
job shop manufacturing: The production of discrete units planned and executed in irregular-sized lots, usually for a customer order.
manufacturing cycle: The length of time between order release and shipment.
numerical control (NC): A technique for controlling the actions of machine tools and similar equipment by direct insertion of numerical data at a given period.
programmable automation: Automation that uses the same machine and potentially the same tools, but uses variable application software (product test, program, etc.).

Gundrilling. Gundrilling is a variation of the drilling process that uses a single-lip, self-guiding tool to produce deep, precision holes. High-pressure coolant is fed to the cutting area, generally via a passage down the length of the tool.

Gundrilling has been incorporated into high-production transfer equipment in many industries. Automatic equipment, with up to 352 gundrilling spindles on one machine, is being used in production. Numerically controlled gundrilling consistently produces tube sheets with up to 17,000 holes per part.

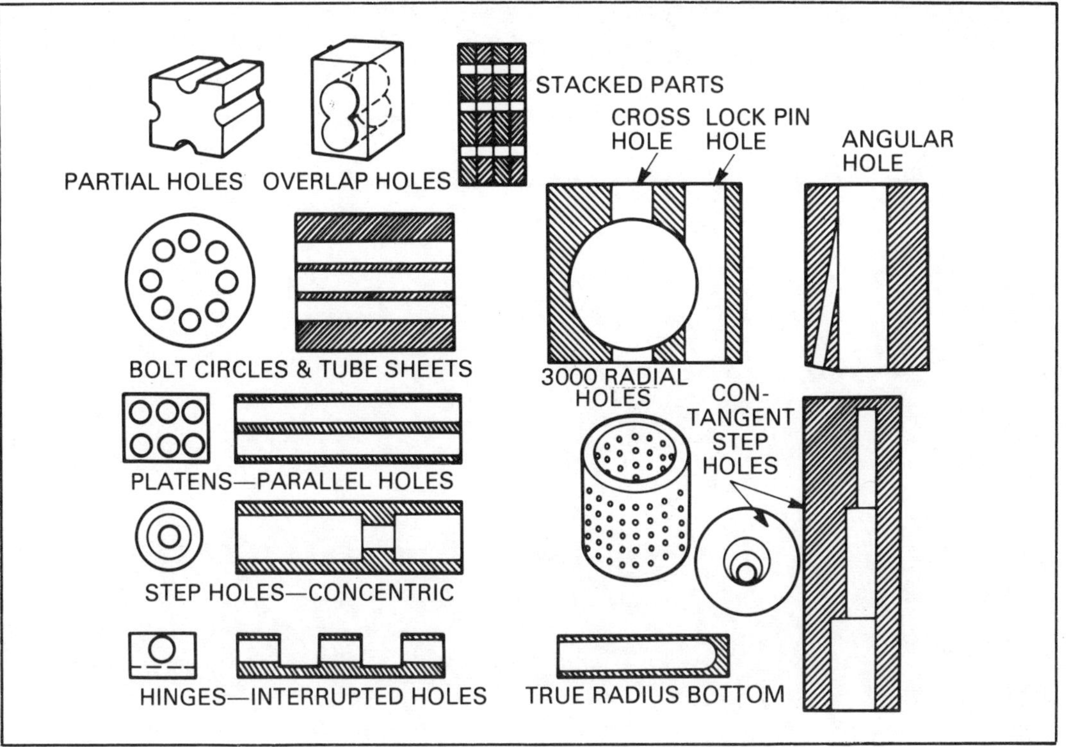

Various types of holes produced by gundrilling.

Half Duplex Operation. Half duplex operation describes an alternate, one direction at a time transmission mode of operation. It is synonymous with one way reversible operation or two way alternate operation. Sometimes this involves a circuit that is designed for duplex operation but can only be operated alternately in each direction because of the limitations of the terminal equipment. In some circumstances, the bandwidth of the transmission media doesn't allow full duplex transmission of the type of data in question. Current technology allows most communications applications to be carried out in a full duplex mode when desired.

A walkie-talkie is an example of a device which operates in half duplex mode only; push-to-talk. Some data terminals can only be operated in half duplex mode. This is why terminals have a feature called *echo*, which transmits (echos) the data to the terminal as it also transmits the data over the communication line. If the echo attribute is selected when a device is not transmitting half duplex, double duplicate characters will appear on the screen.

Halftone. Halftone is a term used to define a gray scale which is created by the use of dot patterns. Darker tones have a correspondingly denser dot pattern. When viewed from a distance, the dots in a halftone area blend smoothly and appear to be an even shade of gray. The greater the dot per inch resolution, the better the ability to create halftones. Desktop laser printers with their 300 dots per inch capability are the minimum resolution needed for useful halftones. Typesetting equipment with its minimum 1200 dots per inch provides much better halftone capability.

Newspapers which are printed entirely in black and white utilize halftones in pictures and headline backgrounds. Magazines and other types of media also utilize halftones. With the large usage of color, halftones aren't used as much as they once were. However the advent of the readily available laser printer has caused people to start using halftones in documents which previously had text only or simple single-shade black-and-white drawings.

Also see: Gray Scale.

Handshaking. Handshaking is an exchange of predetermined characters or signals between two communicating stations. The exchange provides for control or synchronism after a connection is established, particularly during the information transfer phase of a call.

Typically, the following information is exchanged during a BSC (binary synchronous communications) handshake: Message available for transmission; Start of text transmission; Acknowledge or rejection of the text; Detection of errors; Retransmission after error detection, and End of transmission.

To ensure error-free data transmission, handshaking is a requirement for most communications protocols.

A rudimentary form of handshaking is when a person dials a modem line, and upon hearing the tone, connects the computer to the phone line allowing the computer to respond and establish the connection.

Also See: Binary Synchronous Communication.

Hard Automation. Hard automation is automation which is composed of physical components as opposed to being implemented in software or by simply programming a general-purpose, flexible machine. Hard automation can only be used for the purpose for which it was designed. When high-volume automation is needed which will probably use up the working life of the equipment, hard automation usually provides the least expensive and most reliable alternative. When the system is something new which hasn't been developed before or when the technology hasn't been proven, it is sometimes desirable to first develop the methods and processes using flexible or soft automation to be able to modify and reprogram the system until all bugs are eliminated. After the system development and prototyping stage has been completed, hard automation can be built for the production use. When referring to molding, hard automation would be steel molds, whereas soft automation might be wooden molds which have a very short life but are much easier and quicker to create.

Also see: Fixed Automation, Flexible Manufacturing System, Soft Automation.

Hard Copy. Hard copy refers to copy which is readable by a human without the use of any special device. The fact that something is readable doesn't mean it is necessarily understandable by the reader. Hard copy may also be readable by machines such as optical character readers (OCRs) and digitizers. Copy in this context means any information which is printable. Hard copy might be printed by a computer or reproduced from microfilm.

Hard copy is usually generated by some type of printer if the source is computer data. Hard copy can also be created from a computer screen by means of a Polaroid Palette. Another method of creating a hard copy of the image on a CRT screen is to use programs known as "print screen" utilities which capture the image on the screen and output the data through the printer port.

Also see: Soft Copy.

Hardening. Hardening is the process of increasing the surface wear resistance of a part. It is accomplished by heating a piece of steel to a temperature within or above its critical range and then cooling (or quenching) it rapidly. In any heat treatment operation, the rate of heating is important. Heat flows from the exterior to the interior of steel at a definite rate. If the steel is heated too fast, the outside becomes hotter than the inside, and the desired uniform structure cannot be obtained. If a piece is irregular in shape, a slow heating rate is all the more essential to eliminate warping and cracking. The heavier the section, the longer the heating time must be to achieve uniform results. Even after the correct temperature has been reached, the piece should be held at the temperature for a sufficient period of time to permit its thickest section to attain a uniform temperature.

The hardness obtained from a given treatment depends on the rate of quenching, the carbon content of the steel, and the workpiece size. In alloy steels, the type and amount of alloying elements influence only the hardening ability (generally the depth-hardening ability) of the steel, and does not affect the hardness itself, except in unhardened or partially hardened steels. A very rapid quench is necessary in order to obtain the desired martensitic structure which provides the hardened characteristic of steel. For low and medium plain-carbon steels, quenching in a room temperature water bath is a method of rapid cooling which is common practice. For high-carbon and alloy steel, oil is generally used as the quenching medium, because its action is not so severe as that of water. Certain alloys can be hardened by air cooling. But for ordinary steels, such a cooling rate is too slow to give an appreciable hardening effect. The temperature of the quenching medium must be kept uniform to achieve uniform results. Any quenching bath used in production work should be provided with means for cooling.

Hard Facing. Hard facing is the general term describing deposition of an overlay that is wear resistant, abrasion resistant, corrosion resistant or tough, or possesses antiscuff or antifriction properties. Most hard facing deposits are alloys, but ceramics and cermets also may be used. Hard facing by welding is common. Plasma-arc spraying is generally used for depositing ceramics or cermets.

Laser hard facing is a method in which a high-intensity beam of light is focused on the surface of the part, then rapidly scanned over the area while injecting the powder into the interaction zone. This method allows for the hard facing of irregular shapes and deposits powder only where it is needed. Deposits of 0.010 to 0.10 inch (0.25 to 2.5 mm) are attainable in a single pass. Due to the speed of the process and the localized heat, distortion and dilution are minimized. Rapid solidification of the hard facing permits the formation of previously unattainable microstructures.

Hardware. Hardware is the portion of a computer system which has physical attributes and can be touched and seen. The remainder of a computer system consists of firmware, which is the code imbedded in the system in ROMs, and software. Software is electrically represented data loaded from magnetic media and stored in the memory of the computer system. Hardware is named as such because it is composed of physical materials. Software is program control information. Software is stored by the alignment of magnetic particles on a storage medium. Hardware is discrete and finite and takes up inventory space. Hardware is capital inventory. Purchases of hardware are subject to investment tax credits, and can be depreciated. Design, development, and manufacture of hardware receives different consideration by the IRS (Internal Revenue Service) than the development and reproduction of software.

When referring to computer products, the term hardware is used generally to refer to anything other than the software data. The monitor, system unit, and keyboard are all examples of computer hardware.

Also see: Firmware, Software.

Hardwired. Hardwired is a term used to describe a device which is electrically connected to another device with wires that cannot be quickly disconnected, such as by pulling a plug or disconnecting a connector. Usually a hardwired device requires the use of a screwdriver or soldering iron to disconnect its wiring. Some of the reasons for hardwiring devices are to have a connection that won't accidentally become disconnected and vibrate loose. Also, a hardwired connection usually isn't susceptible to corrosion or oxidation. If a device isn't intended to be portable, it is usually hardwired to discourage casual moving of the device. Some devices require special electrical requirements and are hardwired to avoid inadvertently connecting the device to an inappropriate source. Most quick-connect devices such as plugs cannot handle the large current loads that a hardwired connection can within a limited cost and space. Some states in the U.S. have an electrical code requiring devices such as smoke detectors to be hardwired instead of battery powered, to ensure they will always have power.

A hard-wired system is an NC system with a fixed wired program built in when manufactured and not subject to changes by programming. Changes are possible only through altering the physical components or interconnections.

Also see: Numerical Control.

HDLC. See: High-level Data Link Control.

HDM. See: Hydrodynamic Machining.

Head. A head is a device, such as a small electromagnet on a storage medium, that reads, records, or erases information on the storage medium.

Headchanging Machine. Headchanging machines, like machining centers are a relatively new class of multifunction, numerically controlled machine tools. They differ from machining centers in that single or multiple-spindle heads, rather than tools, are transferred to a single workstation in proper sequence to perform the required series of operations. The single workstation is equipped with a spindle drive and slide feed unit; the workpiece remains in a fixed or indexable position. Additional workstations can be added on some machines if required.

Advantages of headchanging machines are similar to those of machining centers: increased productivity and versatility, reduced capital equipment and labor costs, and less material handling. Headchanging machines are generally used for larger lots of similar workpieces and for faster production requirements than are usually obtainable with machining centers. The use of modular heads and pallets on headchanging machines, however, permits quick changeover to suit various workpiece requirements. These machines are being integrated with other NC machines in flexible machining systems.

Programs can be loaded into the CNC unit by punched tape, magnetic tape cassette, or manual data input (MDI). With MDI, the control can be programmed by the operator when the first workpiece is being machined or during dry-run cycling. Programming normally only involves the Z axis with cutting tools preset to the required length in each head.

Header Card. A header card is a card placed at the beginning of a batch of cards and containing information about data on all cards in the batch.

Header Record. A header record is a file record containing the file name and other information to aid in locating and retrieving the file.

Heat Treatment. Heat treatment may be defined as a process which combines the application of controlled heating and cooling operations on metals or alloys in their solid state in such a way that they will produce desired properties. Heat treatment can be applied to a variety of commercially used metals—iron, steel, aluminum, copper and many others. However, because of the versatility and wide use of ferrous alloys in industry, the treatments applied to steel are by far the most widely used.

Heat treatment may be used to improve tensile strength, ductility, toughness, wear resistance, machinability, formability, bending properties, corrosion resistance, magnetic properties and many other properties. Internally, a metal or alloy consists of one or more types of atoms packed together in orderly three-dimensional arrangements called crystals. The crystals, in turn, are bonded together in diverse ways which are described in terms of microstructure or grain structure. Any given structure can be altered to some extent by plastic deformation from compressive, tensile, or shear forces, but the available time-temperature combinations obtained by heat treatment provide a greater variety of properties. Heat treatments are carefully controlled combinations of such variables as time, temperature, rate of temperature change, and furnace atmosphere. In actual practice, the heat treatment process may use temperatures that range from as low as -325 degrees F (-198 degrees C) to as high as 2,350 degrees F (1287 degrees C), depending on the type of metal being processed. There are various types of heating environments such as molten salt baths or controlled atmosphere heating furnaces. The selection of a specific treatment must be based upon knowledge of the properties of the work material and of the finished part.

Heat treating is the manual loading and unloading of heat treating furnaces which represents an environment where labor turn over is usually high. This in itself provides a good basis for an intitial robotic application since labor acceptance of this type of device has been quite good.

Robots are capable of handling the payloads and cycle rates associated with most heat treating operations. The increases in production are most rewarding in heat treating applications, since operator fatigue is quite high when handling hot parts.

HERF. See: High-energy-rate Forming.

Hertz. Hertz (Hz) is the unit of frequency equivalent to one cycle per second. It is the frequency of a periodic phenomenon which has a period of one second. It is the Systeme International d'Unites (SI) unit for frequency. Standard alternating current power is 60 cycles per second or 60 Hz in the United States and 50 Hz in Europe.

Also see: Baseband, Broadband.

Heuristic. Heuristic is a label used to describe an iterative problem solving approach. A heuristic idea or information aids in, or provides direction for, the solution of a problem. Heuristic problem solving may not always utilize actual pieces of the final solution in the early stages of the iterative problem solving process.

Expert systems are most often programmed to process in a heuristic manner. The iteration can take place in one or two forms. In the first form, the problem solving logic is iterated with each iteration resulting in a refinement of the solution. A physical example might be the bracketing of a target by artillery before firing for effect.

In the second form of heuristic reasoning, the collective intelligence gained from the solution of multiple problems is analyzed to develop new and refined problem solving rules. This is how a chess player defines the game. Early application of heuristic reasoning in computers was directed at development of a chess playing machine.

Also see: Artificial Intelligence, Expert Systems.

Hexadecimal Numbering System. The hexadecimal numbering system is a numbering system which is based on the number 16 (hexadecimal meaning six and ten). Since there are no single digits other than zero through nine, the hexadecimal numbering system uses the letters A through F to supplement the single digits, with the letter A coming next after 9. After the letter F, the next number carries a one (1) into the second column and begins over with a zero in the least significant digit's column. Since it would be very confusing if a person unknowingly switched between numbering systems without indication, hexadecimal numbers are usually followed by the designation Hex or a small "h". As is apparent, the number 10 Hex, if converted to the decimal numbering system would become a 16, and if converted to the binary numbering system (base 2), would become 10000. Likewise, ABh is equivalent to 171 in decimal. Hexadecimal numbers are used within a 16 bit computer for processing efficiency. The numbers are converted from and to the decimal numbering system during input and output operations (see figure on next page).

HFIW. See: High-frequency Induction Welding.

HFRW. See: High-frequency Resistance Welding.

Hierarchy. Hierarchy is a data structure consisting of sets and subsets such that every subset of a set is a lower rank than the data of the set.

High-energy-rate Forging. High-energy-rate forging is a forging process carried out on equipment having extremely high ram velocities resulting from the sudden release of compressed gas against a free piston.

High-energy-rate Forming (HERF). High-energy-rate Forming is any of several processes that apply force to a workpiece at rates well in excess of a similar conventional process. Most of the high-energy-rate forming processes involve the sudden conversion of chemical, electrical or mechanical potential energy to mechanical kinetic energy.

High-frequency Induction Welding (HFIW). High-frequency induction welding is a process for making a continuous weld using radio-frequency alternating current and an induction coil to induce electric eddy currents in the joint area, which heat the metal. There are three principal variations: continuous seam, used to produce welded tubing up to 6 inches (152 mm) diameter and 3/8 inch (9.5 mm) wall thickness; butt end, used to weld tubing end-to-end; and magnetic pulse, used to make sleeve or lap joints between two tubular components.

DECIMAL	HEXADECIMAL	DECIMAL	HEXADECIMAL	DECIMAL	HEXADECIMAL
0	0	16	10	32	20
1	1	17	11	33	21
2	2	18	12	34	22
3	3	19	13	35	23
4	4	20	14	36	24
5	5	21	15	37	25
6	6	22	16	38	26
7	7	23	17	39	27
8	8	24	18	40	28
9	9	25	19	41	29
10	A	26	1A	42	2A
11	B	27	1B	43	2B
12	C	28	1C	44	2C
13	D	29	1D	45	2D
14	E	30	1E	46	2E
15	F	31	1F	47	2F

The hexadecimal numbering system.

High-frequency Resistance Welding (HFRW). High-frequency resistance welding is a process for making a continuous weld using radio-frequency alternating current applied across the joint through contact shoes. There are three principal variations: continuous seam, used to produce welded pipe and tubing; finite length, used to join coil ends or sheet edges of carbon or ferritic stainless steels; and melt welding, used to produce a small cast nugget within the joint.

High-level data link control (HDLC). High-level data link control is a high level control that can be applied to a data link. In essence, it is the level of control above the data link layer (above level 2 of the OSI model). The high levels include the application, presentation, session, transport and network layers, while the data link and the physical layers are considered to be the low levels.

HDLC protocols eliminate much of the hand-shaking and time-consuming line turnarounds. They operate in both full duplex and half duplex modes.

All HDLC protocols are not identical, but are generally compatible. For example, within each protocol, it is possible to add control characters to an information field, and have those characters recognized only by that protocol. The control characters are transparent to other protocols.

While other data link controls such as binary synchronous communications (BSC) are character oriented, HDLC controls are bit oriented.

Synchronous Data Link Control (SDLC) is IBM's product offering of an HDLC.

Also see: Binary Synchronous Communications, Full Duplex Mode.

High-level Language. A high-level language is a language which can generally, at least in principle, be used on any computing machine. High-level language is an appropriate term when discussing the class of computer languages known as imperative/algorithmic languages.

An advantage of high-level languages is that they don't force extreme restrictions on the programmer. A negative feature of high-level languages is that they don't provide much explicit control over the basic functions of the machine, such as data memory location and data manipulation. High-level languages usually allow a programmer to write a program in human logic patterns using logical and mathematical statements. The more recent developments in high-level languages are described using terminology such as natural-language programming, which is meant to indicate that the command formats are closer to normal conversational speech.

FORTRAN was the first widely developed and utilized high-level language. Other languages such as C and Pascal try to provide both high-level capabilities and low-level abilities when necessary.

Also see: Low-level Languages, Programming Languages.

High Noise Immunity Logic (HNIL). High noise immunity logic is a type of bipolar logic that possesses a noise rejection capability of approximately 4 volts versus 0.4 volt for transistor/transistor logic (TTL).

Histogram. A histogram is a graph or diagram which illustrates the history of some activity over time. The image created by a strip chart recorder is one example of a histogram. Any data which represents some activity over time can be made into a histogram. Histograms facilitate comprehension of data. As the old saying goes, "A picture is worth a thousand words..." An electrocardiogram or EKG is a histogram of the beating of a heart over a period of time. Most process control systems have a method of providing data histograms because for some processes the continuous status of the process must be known. The charts showing the performance of a stock in the stock market over some period of time is another example of a histogram.

HNIL. See: High Noise Immunity Logic.

Hobbing. Hobbing is a gear tooth generating process consisting of rotating and advancing a fluted steel worm cutter past a revolving blank. In the actual process of cutting, the gear and hob rotate together. The speed ratio of the two depends upon the number of teeth to be generated on the gear, and on whether the hob is single or multithreaded. The hob cutting speed is controlled by change gears that vary the speed of the hobbing machine's main drive shaft.

In gear hobbing the cutter is called a *hob*. Most hobs are cylindrical in form with cutting teeth following a helical path around the outside of the body corresponding to the thread of a worm screw. Flutes are cut across the threads, forming rack shaped cutting teeth.

At the start of the operation, the gear blank is moved into the rotating hob until the proper depth is reached, with the pitch-line velocity of the gear being the same as the lead velocity of the hob. The action is the same as if the gear were meshing with a rack. As soon as the proper depth is reached, the hob cutter is fed across the face of the gear until the teeth are complete. Both gear and cutter rotate during the entire process.

Hobbing is a continuous process in which the hob and blank both rotate in timed relationship with each other. In addition to the rotary motion, the hob and gear blank are fed relative to each other to produce the spur gear, helical gear, or worm gear. In the hobbing process, the cutting action is continuous in one direction until the gear is completed.

Hobbing machines with electronic control are available, using programmable controllers or CNC. Some machines with programmable controllers feature electronic lead control, set with digital thumbwheels, eliminating the need for lead-change gears. Other features include digital selector switches for setting travel limits and cutting cycles, infinitely variable feeds and speeds, automatic feed and speed changing, power tailstock positioning, and inch/metric digital readouts.

On CNC hobbing machines, the microprocessor and software control all moveable axes. The number of teeth; helix angle; swivel angle for the hob head; amount and sequence of hob shifting increments; radial, axial, and tangential motions of the slides; crowning and taper cutting; and cutting parameters (hob speed and feed) can all be preset.

The conventional gear train is replaced with individual drives and encoders for the work and hob spindles and the radial, axial, and tangential slides. The rotary encoders for the hob and workpiece spindles generate impulses which are processed by the microprocessor to obtain the programmed number of teeth. To produce helical gears, additional pulses generated by the axial rotary encoder are added to or subtracted from the hob spindle rotary encoder, in relation to the worktable rotation.

Eliminating the need for change gears permits faster setup, rapid changeover, and increased productivity. Other advantages include improved accuracy and reduced downtime, noise, and wear. Another important benefit is increased flexibility, since programs for a number of different gears can be stored in the control memory.

Hobby Robots. Hobby robots are robots designed for entertainment or educational purposes. This type of robot is distinguished from an industrial robot in that the industrial robot has definite work to perform and it does so for long periods of time. An industrial robot is usually used in a commercial or industrial application. The hobby robot is just for the pleasure or education, or satisfaction of the creator. Some types have been constructed to investigate the practical limits of putting intelligence into a machine of some type.

Hobby robots usually are intended to do human-like nonrepetitive types of chores. A hobby robot may include a microprocessing unit, vision sensors, touch sensors, mobility and other robotic features found in industrial robots. Some hobby robots also possess remote control features.

Some of the well-known hobby robots are the Hero and Rhino. Many toy stores sell battery powered robots which are not considered hobby robots as much as they are considered toys.

Hollerith Code. Hollerith code is a 12-bit code representing letters, numbers, or special symbols punched in 80-column cards with 12 rows per column.

Holographic Nondestructive Testing. Holographic nondestructive testing (HNDT) or holographic interferometry combines holography with interferometry, permitting the detection of material defects and impending fatigue failure, the measurement of residual stress, and vibration mode analysis. Holography is a two-step process that permits the reconstruction of three-dimensional images. A hologram of an object is formed in the first step of the process. In the second step, the original image is recreated, thus permitting the observer to see an exact replica of the image in size and position. The recreated image also displays depth of field and parallax.

When producing a hologram, light emitted by a laser source is split into an object beam and a reference beam. The object beam is expanded and directed toward the object, where part of it is reflected toward the recording medium, which is usually a high-resolution photographic film. In recent years, thermoplastic emulsions have become available. Although the reference beam travels directly to the film, it is diverted so that the optical paths of both beams are approximately equal. Because the two beams are from the same coherent source, they interfere with each other as they pass through each other. The film emulsion receives exposure wherever constructive interference occurs and virtually no exposure where destructive interference occurs. The varying exposure results in a very fine pattern in the emulsion of the film. This fine pattern diffracts the beam to form a reconstruction of the object beam when the film is illuminated after development by a replica of the reference beam. The replica of the object can be viewed or photographed.

Two holograms can also be made of the same object in a single photographic emulsion before and after a certain deformation. Then the hologram is reconstructed, the two wavefronts are simultaneously recreated. This results in a set of fringes superimposed over the image of the object; the spatial frequency of these fringes provides information regarding the deformation of the object. Because each fringe relates to the deformation of the surface, variation in the geometry of a fringe is directly relatable to the topology or three-dimensional shape of the surface. The variation in the shape of the surface depends on the manner in which the object is deformed. Deformation is commonly achieved by acoustic, thermal, or mechanical stresses. The type of stress used is determined by the part, its accessibility, and the type of flaw.

Holographic nondestructive testing has made the precision of interferometric measurements—previously restricted to regular, specularly reflecting surfaces under laboratory conditions—applicable to generalized objects in industrial environments. As a result, the technique has been utilized in a wide range of nondestructive testing applications. For example, in the area of structural analysis, HNDT has been used for nondestructive testing of aircraft panels and honeycomb structures; aircraft wing assemblies and fins; detection of cracks, delaminations, and disbonds; and rotating structures such as turbine blades and naval propellers, small arms barrels, vibrating vehicle components, rock mechanics, and computer components. Holographic techniques have been used to record flowfields associated with turbine blades, projectiles, wind tunnels, rocket engines, and plasmas, as well as to analyze particle fields associated with aerosols, liquid droplets, pollutants, and spacecraft waste tanks.

The advantages of HNDT are many. It is, first of all, a noncontact technique; thus objects may be studied without disturbing their motions by adding accelerometers or strain gages. Furthermore, sharply curved surfaces or other areas where the addition of such devices is awkward may be readily studied. It is a full-field technique; thus displacements may be studied across an entire surface rather than at discrete points. It offers interferometric precision; therefore, although a variety of techniques can be used to study much larger motions, HNDT can resolve displacements of less than one micrometer.

Pulsed lasers can provide sufficient energy to record double-pulse holographic interferograms of large objects. Furthermore, extremely short exposure times (approximately 30 nanoseconds) and pulse separations (approximately 1 microsecond) permit studies of fast transient events in the presence of relatively large general motions.

The limitations of HNDT fall into three general areas. First, the technology is different and relatively new. The equipment required is expensive, and while a high skill level is not required to operate a holographic laser, some training is required. Secondly, the technique measures only displacements. Strain must be calculated from displacement data and material properties. Finally, the holographic interferogram does not give quantitative data directly. While a number of applications such as crack detection can be conducted with only visual inspection of a hologram, a variety of others rely on quantitative evaluation of the fringes in the hologram. Great progress has been made in development of techniques that produce quantitative results.

Holography. Holography is the process of making or using three-dimensional images on a two-dimensional surface. A hologram is created by using an interference pattern from a split beam of coherent light, such as from a laser beam, to create a three-dimensional image. To view the image, a beam of coherent light must be used to backlight the image. Lasers are used to provide the light source because they are the best known source for coherent (parallel focused beam) light. A coherent beam of light can maintain the relationship and integrity of an image at any point that the beam of light may be intersected and viewed. Holograms are used to create art forms and pictures, as well as for testing materials and products to find internal faults and flaws. Holographs are sometimes used on movie sets to create images which appear to exist, but contain no physical substance, only visual attributes. The original Star Wars movie used holograms to create some of the images in the film. Some future uses of holographs might be to add visual aids to operator manuals in factories or to assist surgeons.

Home Position. The home position is the fixed location in the basic coordinate axes of the machine tool. It is usually the point in the work process in which tools are fully retracted permitting any necessary changes.

Honing. Honing is a low-velocity abrading process. Material removal is accomplished at lower cutting speeds than in grinding. Therefore, heat and pressure are minimized, resulting in excellent size and geometry control. The most common application of honing is on internal cylindrical surfaces. The cutting action is obtained using abrasive sticks (aluminum oxide and silicon carbide) mounted on a metal mandrel. Since the work is fixtured in such a way as to allow floating and no clamping or chucking, there is no distortion.

For small diameter bores, a one-piece mandrel with a U-shaped cross section is used. Two integral wear shoes and the honing stick provide an unevenly spaced three-line contact with the work circle. The work is given a slow axial reciprocating motion as the mandrel is rotated, thus giving the characteristic cross-hatch pattern to the surface.

A characteristic of honing that makes it possible to generate geometrically accurate cylinders is that the interface between the work and abrasive stick is completely independent of the machine. The floating work fixture enables the tool to exert equal pressure on all sides of the bore independent of machine vibration or clamping distortion.

Host Computer. A host computer is the primary or controlling computer in a multicomputer network. It provides a coordinating function to the other computers or operations. Sometimes the host computer provides central data storage and output capabilities for the other computers in the network. In some networks, no computer has the responsibility of being the host, therefore, all nodes are considered peers of each other.

Also see: Computers.

Hot Pressure Welding (HPW). Hot pressure welding is a process similar to forge welding used for joining ductile metals in electronic circuits. Resistance heating is generally used, with heat being transmitted to the joint through an indentor tool.

Hot Wire Welding. Hot wire welding is a variation of arc welding processes (GTAW, PAW, SAW) in which a filler metal wire is resistance-heated as it is fed into the molten weld pool.

HPW. See: Hot Pressure Welding.

Human Factors Engineering. Human factors engineering is a design process which takes into account the relationship of the human individual to an engineering project, product design or facility design. Industry is beoming increasingly cognizant of human factors. All human elements impacting a design cannot easily be identified. Neither can they easily be put into an equation or set of rules, or otherwise quantified. Nor can they be accurately predicted at all times. Yet, these factors can often outweigh all other factors.

Humans are the benefactors, components and beneficiaries of the production systems they design, construct and operate. Although abilities and physical characteristics vary widely among individuals, many guidelines are available to help people function more effectively.

Critical components of any production system are the links between the equipment and human elements of which it is composed. A primary objective of human engineering is to ensure that when such links are designed, consideration is given to criteria such as human abilities, placement, safety functions/requirements, and range of motion/reach.

Hunting. Hunting is an unwanted oscillation of an automatic control system in which a machine component moves slightly back and forth on both sides of a desired position.

Hybrid Circuits. A hybrid circuit is a circuit which has both digital and analog circuitry combined into one overall circuitry or device. As technology continues to advance in the direction of larger-scale integrated circuit chips and more specialized chips such as ASICs (application specific integrated circuits), combining analog and digital circuitry become more sensible. Allowing the combination of the two types of circuitry makes it possible for the most appropriate type of circuit to be used for each respective function. This avoids suboptimizing when trying to do the entire range of intended product functionality with just one type of circuitry or the other, or having to utilize more than one circuit board or multiple chips. Because analog circuits have different requirements and properties from digital circuits, combining the two places a greater number of requirements on the resulting circuitry than if they were kept separate.

Hydraulics. Hydraulics is the science of using fluid as a transmitter of force and energy or simply moving fluids within restricted containers. Hydraulic fluids are used in many situations since they can provide high forces in a small space. Hydraulic robots provide the greatest weight handling capacity—much greater than any electrical robot of comparable size. One problem of using hydraulics in precision machines such as robots is that hydraulic fluids are compressible and the flexing of the conduits when carrying hydraulic fluid of several thousand pounds-per-square-inch (psi) pressure can cause noticeable inaccuracies in movement. Contaminants in hydraulic valves can cause severe malfunctions in hydraulic equipment, which can occur even when using small micron filters. Fluid control equipment and major equipment vendors such as Parker Fluid Power can perform most any type of activity to a reasonable degree of acceptability, depending upon the specific criteria desired.

Also see: Industrial Robots.

Hydrodynamic Machining (HDM). Hydrodynamic machining is a process that removes material and produces a narrow kerf by the cutting action of a fine high-pressure, high-velocity stream of water or water-based fluid with additives. Pressures up to 60,000 psi (414 MPa) are employed. Part profiles are created using optical tracing or numerical control systems.

Hz. See: Hertz.

I

IC. See: Integrated Circuit.

ICAM. See: Integrated Computer-Aided Manufacturing.

IEEE. IEEE is an acronym for the Institute of Electrical and Electronic Engineers. The Institute of Electrical and Electronic Engineers Standard Board reviews and approves functional descriptions of communication interfaces, which are then referred to by the standard document number.

Also see: IEEE 488, IEEE 802.3, IEEE 802.4.

IEEE 488. The IEEE Standard Board approved IEEE Standard 488-1975 "Digital Interface for Programmable Instrumentation," first published in 1975 and again published in 1978, with minor editorial changes as, IEEE Standard 488-1978. It is referred to as "IEEE 488." The major areas defined by IEEE-488-1978 are: Mechanical—The interface connector and cable; Electrical—The logic signal levels and how the signals are sent and received; Functional—The tasks an instrument interface may perform, such as sending data, receiving data, triggering the instrument, and so on; and the protocols to be used in such functions. Also, the standard bus defined by IEEE-488-1978 is commonly called the GPIB, or general-purpose interface bus. Today a wide variety of instruments include interfaces conforming to this mechanical, electrical, and functional standard.

Briefly, the basic hardware, characteristics and interface functions are briefly as follows. The system uses a 16-line bus to interconnect up to 15 digital instruments. Each instrument on the bus is connected in parallel to the 16 lines of the bus. All active interface circuitry is contained within the various instruments, and the interconnecting cable (containing 16 signal lines) is entirely passive. The cable's role is limited to that of interconnecting all devices in parallel, whereby any one device may transfer data to one or more other participating devices.

Eight of the 16 lines in the cable are used to transmit data. The remaining lines are used for data byte transfer control (or "handshake") and general interface management (bus management). Five of the bus management lines are bus control and the other three are handshake lines. Data bytes are transferred by means of an interlocked handshake, permitting asynchronous communications over a wide range of data rates.

Every participating digital device must be able to perform at least one of three roles: talker, listener, or controller.

A talker can transmit data to other devices via the bus, and a listener can receive data from other devices via the bus. Some devices can perform both roles.

A controller manages operation of the bus system primarily by designating which devices are to send and receive data. The controller may also command specific actions within other devices.

A minimum IEEE-488 system consists of one talker and one listener, without a controller. In this configuration, data transfer is limited to direct transfer between one device manually set to "talk only." Devices can be a talker; a listener; a talker and a listener, or a controller, a talker and a listener.

Also see: IEEE.

IEEE 802.3. The Institute of Electrical and Electronic Engineers (IEEE) Standards Board endorsed most all aspects of Ethernet with their Standard, IEEE 802.3.

Also see: Ethernet, IEEE.

IEEE 802.4. The Institute of Electrical and Electronic Engineers (IEEE) Standards Board endorsed most existing aspects of the Manufacturing Automation Protocol (MAP) with their Standard, IEEE 802.4.

Also see: IEEE, Manufacturing Automation Protocol.

IGES. See: Initial Graphics Exchange Specification.

Image Table. An image table is an area in memory dedicated to input/output data. The value 1 represents ON; the value 0 represents OFF.

Imaging. Imaging is the action of creating a viewable image. Normally the image is created on some image plane, however in the case of a holograph, the image appears to be suspended in mid-air. Images are created when light is reflected off an object and the reflected light is allowed to strike a device capable of creating an image. The device may be photographic film, a camera lens, or the human eye. Machine vision uses a camera and a lens system to capture the images that are then processed. If a CCD or CID type camera is used, the image is converted directly into a digital image. An image may be digital or analog. An analog image is what is usually considered as a video image. A digital image uses relative values of shades of gray for a monochrome pixel image, or predefined byte combinations to represent the color of each multicolor pixel.

Immediate Access. Immediate access is the ability to obtain data directly from a storage device or to place data directly in a storage device in a relatively short period of time and without serial delay due to other units of data.

Immediate Feedback. Immediate feedback is the principle of checking production right after a part is made to provide information on deviations from standards so that correction will be made. The concept extends to automatic shutdown of a process in case of many defects or a parameter drifting out of a desired control range.

Immediate Input/Output Instruction. An input/output instruction immediately transfers input data from a selected input module to the appropriate 16-bit work in the input image table without waiting for an input/output scan.

Impact Printer. An impact printer is a printer that creates the image of the characters on the media by physically striking the media and leaving an impression which is inked. Dot-matrix printers use electronic solenoids which fire tiny pins into the back side of an inked ribbon. The matrix of concurrently fired pins causes an impression to be created on the media, with the ink transferring from the ribbon to the media. Daisy-wheel printers are also impact printers that use a wheel of fully formed characters to strike the back of the ribbon. Other types of impact printers use cones and balls with fully-formed characters. Impact printers are very noisy and faster printing speeds only increase the noise. Putting the printer inside a sound absorbing enclosure only eliminates a portion of the noise.

Examples of nonimpact printers are laser printers, ink-jet printers, thermal printers and pen plotters.

Also see: Dot-matrix Printer.

Impedance. Impedance is the total opposition a circuit offers to the flow of alternating current, including resistance and reactance. A circuit which verifies that a transmission line is the correct electrical load for its communication devices.

Also see: Circuit.

Inbound Stockpoint. Inbound stockpoint is used with a pull system, a defined location next to the place of actual production to which materials are brought as needed and from which material is taken for immediate use.

Incremental Dimension. Incremental dimension is the dimension from one point of departure to the next. In an NC program, if the entire job is incrementally dimensioned and the tool programmed to return to its start point, the algebraic sum of all the intervening plus and minus motion is zero.

Also see: Numerical Control.

Incremental System. An incremental system is an NC system in which each coordinate or positional dimension, whether input or feedback, is taken from the last position instead of from a common data position, as in an absolute system.

Also see: Numerical Control.

Indexing. Indexing involves the movement of one axis at a time of an NC system part to a precise position through numeric commands.

Also see: Numerical Control.

Indirect Access File. An indirect file is a file which is protected by being accessed only through an intermediate copy. I/O affects the copy, not the file itself.

Indirect Address. Indirect address is the address, or identification, in a computer instruction that indicates the location of an address of a referenced operand rather than the operand itself.

Also see: Computers.

Induction Brazing. Induction brazing is a process for producing a brazed joint in which the parts being joined are induction heated to brazing temperature.

Industrial Robots. The Robotic Industries Association (RIA) adopted a definition for industrial robot which reads: "A robot is a reprogrammable multifunctional manipulator designed to move material, parts, tools, or specialized devices through variable programmed motions for the performance of a variety of tasks."

Implicit in this definition are two requirements. The first is that the manipulator be a multiaxis machine capable of moving parts or tools to a location in space. The second is that there be a control system, including memory, that suitably drives the manipulator through the programmed motions and is capable of interacting with and providing stimuli to external devices related to the working environment. The variety of geometric configurations of robots are illustrated in the figure on the next page and are identified as:

- Cartesian Coordinates.
- Cylindrical Coordinates.
- Polar Coordinates.
- Revolute Coordinates.

Industrial robot control systems range widely. In this discussion, attention is focused on the more sophisticated robots.

The figure at the bottom of the next page illustrates a number of geometric executions, all of which provide three articulations to locate a part or tool (hereinafter referred to as the *end effector*) in space. The figure at the bottom of the next page is representative of a wrist assembly. This wrist assembly also incorporates three articulations to achieve any orientation of the end effector. Thus, six articulations are required to position and orient. In fact, however, many jobs are accomplished with less than six articulations if proper consideration is given to layout arrangements and part configuration.

It is worthy of note that, while each of the configurations illustrated offers certain advantages, the consequences of selecting any one design become less significant with sophisti-

Industrial Robots

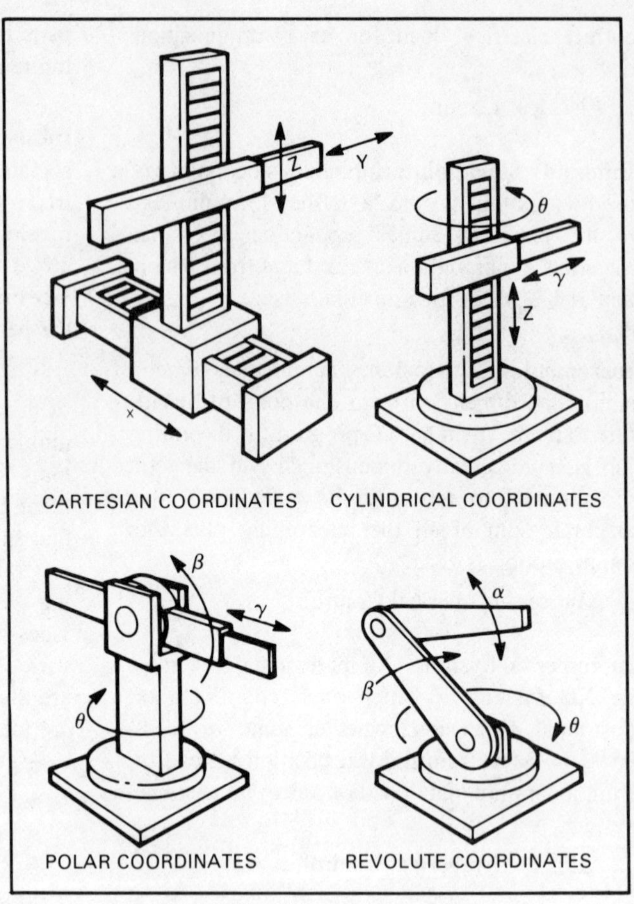

Industrial robot arm configurations.

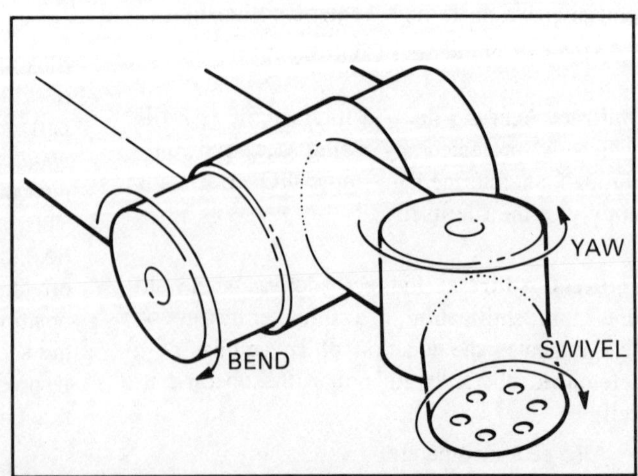

Wrist articulations on an industrial robot.

Industrial Robots

cated computer-controlled systems. This derives from the fact that such control systems are capable of driving the manipulator in machine, world, or tool coordinates irrespective of the machine coordinate system. The figure on the this page illustrates this concept.

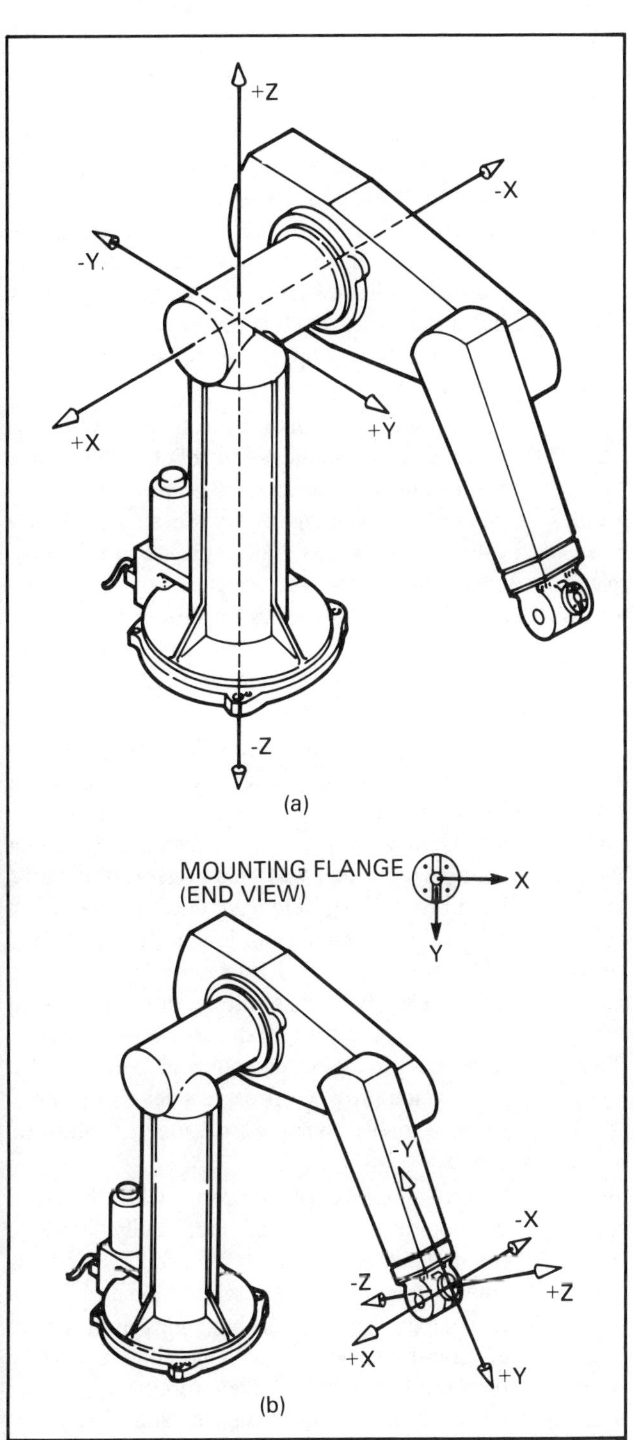

Coordinate systems: (a) World coordinate system and (b) tool coordinate system. Sophisticated computer control systems are capable of driving modern robots in world or tool coordinate systems regardless of the specific machine coordinate system.

Industrial robots, contrary to popular belief, are not highly intelligent, humanoid type devices which can perform dexterous, logically deduced actions. Industrial robots are actually computer numerically controlled machines which execute programs, sometimes complex with branching functions, and are as complex as the programming allows. Since the program of a robot is a numerically defined set of motions, smooth continuous path actions are sometimes difficult to impossible for robots to perform. High-speed point-to-point motions are easy to execute, but the accelerations and decelerations of the motions sometimes are perilous to the items being moved as well as the robot itself.

Current research and development in robotics is in the areas of developing robots capable of greater mobility, but able to fix themselves to a location, determining their orientation and performing a highly precise task. Another aspect of current development is in the area of expert systems and artificial intelligence. The typical time intensive task of programming a robot to every point in its program is being replaced by the ability to interface a CAD description of the workpiece or the workplace. Points in the robot's path of motion are calculated from the CAD information. These calculated paths are then modified or "touched up" by manually stepping the robot through the entire program, which requires much less time than manually generating the program "from scratch."

History. Although the image of a robot varies widely among the technical population, it is generally accepted that modern day industrial robots originated with a programmable material handling apparatus by George C. Devol in 1954.

The industrial robot created by Devol had its origin in two technologies: numerical control for machine tools and remote manipulation. Numerical control is a method of generating actions based on previously stored data. The data could include data points to which the machine is to move, timing signals to start and stop tasks, and logical statements for decision control sequences. All operations and variations thereof are stored in the memory so that a programming change will not require major hardware changes.

Flexible automation has been a central issue in flexible manufacturing innovation, and numerical control has played an important role in improving and increasing flexibility. Many manufacturers produce a variety of products in batches rather than one product in a repetitive fashion over an extended period of time. Batch manufacturing requires frequent product and schedule changes, so a special-purpose production line or some robots are not the most efficient solution. Flexible manufacturing systems require flexible automation. Numerical control has facilitated the evolution of the contemporary industrial robot which is programmable and can perform different operations by simply modifying data stored in memory.

The second technology responsible for the origination of the industrial robot is the remote manipulator. A teleoperated or remote manipulator is an apparatus that performs a task at a distance and can be used in unsafe or inaccessible environments. The structure often resembles a human arm and allows for dexterity in a wide range of work space. This makes the remote manipulator highly suited to modern manufacturing operations.

The father of modern industrial robotics is Joseph Engelberger, who founded Unimation, Incorporated, now a division of Westinghouse. The Unimate robot, which is essentially a polar coordinate device, was used largely for material handling and spot welding applications. The first robots, by Unimation and Cincinnati Milacron, a CNC machine tool builder, were powered by hydraulics. Hydraulics still provides the greatest force relative to the size of the actuators. At this time, electric motors had not been perfected to a state which provided necessary levels of torque to be a viable source of motion for industrial robots.

ASEA, a Swedish company, was the first to develop a well-received electrically powered robot. The two ASEA robots, one large and one small, represented a new state of the art in robot technology. This was the beginning of the age of electric-drive robots, which has continued to the present. Electrically driven robots were more adaptable to computer controls and highly pre-

 Industrial Robots

cise movements. Electric-drive, pneumatic actuators and hydraulic-drive will be discussed later.

Discussion. The manipulator configuration to be used for a specific assembly application will usually be determined by one or more of the following requirements. The first is the needed work envelope. The envelope is the three-dimensional volume of spatial points in which the end of the robot arm can be moved. The second requirement is the weight (payload) of the components to be assembled.

The most popular manipulator configurations for robotic assembly systems include Cartesian coordinate, articulated arm, SCARA (Selective Compliance Assembly Robot Arm), and gantry.

With Cartesian coordinates, manipulator movement is linear along one of three perpendicular axes X, Y, and Z. Robots with this type of manipulator are most popular for handling small parts in pick-and-place applications. When properly supplied with parts that can be easily oriented, the robots can be successfully integrated into assembly systems, providing a high degree of accuracy, repeatability, and speed.

Robots with cylindrical coordinate manipulator configurations are among the most popular in use today. They can be supplied with three, four, or five axes of motion. For the assembly of small parts, tabletop units are available.

For larger weight requirements, floor-mounted robots with greater payload capacities are available in the same configuration.

Articulated arm robots are among the most versatile available, but usually with some sacrifice in repeatability. They are available with up to six axes of motion and payload capacities from 4 to 176 lb (2 to 80 kg). Such robots are generally more expensive than other types, and their complexity can be a disadvantage. However, they are well-suited for applications requiring simultaneous movement of all axes. Also, unlike the large cylindrical coordinate manipulators, they are more efficient with respect to space requirements. The SCARA (Selective Compliance Assembly Robot Arm) robot was designed primarily for assembly applications and is gaining worldwide acceptance for use in such applications. It has an asymmetrical, horizontal work plane and offers vertical axis insertion at the end of the wrist from any location within the work plane.

A gantry robot is essentially a Cartesian coordinate manipulator configuration with three translational axes and up to three rotational axes. It gets its name from the type of support mechanism that allows the manipulator to move above the workpiece.

Industrial robots are available with electric, pneumatic, or hydraulic actuators to move the robot arms through desired paths. Servoelectric drive mechanisms are the most popular drive for assembly applications.

Electric-drive robots commonly have an electric motor servocircuit with a potentiometer, transducer, optical encoder, or resolver to communicate the position of the arm. They are fast, accurate, quiet, adaptable to sophisticated controls, and relatively inexpensive. Possible limitations include: drive axis hunting (overshooting), power limitations, and, sometimes, the need for a gear train or other means of transmitting power. However, direct drives that eliminate gear trains are becoming increasingly popular.

Pneumatic actuators for robots are clean and relatively inexpensive. They can develop moderately high forces at high speeds, but the compressibility of air limits the accuracy attainable. Positional accuracy and repeatability also are sensitive to changes in load. They may be noisy in operation because of the discharge of air. Pneumatic manipulators are most generally used for light-duty, two-position, pick-and-place applications.

Hydraulic-drive robots can develop high forces at moderate speeds and can be accurately controlled. However, they cost more than electric actuators and are subject to fluid and noise pollution. They are usually used for moving heavy parts.

End effectors (end-of-arm tooling) consist of grippers, hands, holders, and other tools for

Industrial Robots

handling components to be assembled. They are not generally integral parts of robots and commonly must be designed and built for specific applications. They should be designed to handle as many different components as possible. For example, multiple sets of jaws can possibly be used with a single end effector to handle several types of components. The end effectors can be provided with mechanical compliance or sensory devices and vision systems for more precise location.

Mechanical clamping of workpieces is most common. Grippers apply surface pressure on the components to be assembled. They are available with jaws, fingers, and expansion-contraction devices. Other types of flexible fixtures are coming into use for the robotic assembly of components having complex shapes, such as turbine blades. Both modular fixtures (in which the robot assembles the required fixture in the work space) and phase-change fixtures (such as fluidized beds) have great potential for specialized assembly requirements.

Various sensors, both contact and noncontact, are used on robot grippers or workholding pallets for one or more of the following purposes to:

- Locate and identify parts to be assembled.
- Ensure that the part is correctly oriented.
- Verify the presence of a part.
- Guide a robot.
- Determine forces and/or temperatures.
- Compensate for position errors.
- Control gripping forces.
- Provide overload protection.
- Recalibrate a robot.

The use of sensors is desirable for many assembly applications, especially those requiring precise positioning and inserting, because they can increase reliability, minimize cycle time, reduce costs, and prevent damage through incorrect assembly. Sensors can provide direct digital input to robot, work cell, or system controls.

Machine vision systems can also provide input. Machine vision systems are being increasingly applied for object recognition, guidance, and inspection operations. Advantages of these systems include reduced tooling (gripper) and fixture costs, increased flexibility because of their reprogrammability, reduced scrap and/or rework of assemblies, and improved product quality.

In terms of control methods, most precision robots are point-to-point machines, continuous-path robots, or controlled-path robots.

Point-to-point robots are programmed by moving each axis of the robot to a position that yields the desired location of the robot end effector, and then recording the individual position of each of the robot's axes into memory. In replaying these stored points, each axes runs at its maximum or some limited rate until it reaches its final position. As a result, some of the axes reach their destination prior to others. Because there is no coordination of motion between axis, the path and velocity of the end effector between points is not easily predicted.

Continuous-path robots are generally programmed by the operator who physically grasps the robot arm, or a similar teaching arm, and leads it through the desired path in the exact manner and velocity desired for the robot's motions. During this programming, the position of each axis is recorded into the robot's memory on a continuous-time basis, thus generating a continuous-time history of each axis position. The sampling frequency is typically in the range of 60 to 80 Hz.

Controlled-path robots are programmed by manipulating the robot arm to the desired tool center-point location and tool orientation. Instead of moving each axis individually, computer control allows coordination of all robot axes when teaching the program. The position of the robot's arm is recorded in the controller's memory at each desired program-point location. During program replay, the robot's computer control generates a mathematically definable path between the program points, including the control of velocity along the path and acceleration/deceleration for the start/stop points.

With the controlled-path type of robot motion control, several modes of generating the robot's

Industrial Robots

path are possible, depending on the type of motion interpolation employed. These interpolation types include joint, linear, and curvilinear. Joint interpolation results in controlled-path motions in which a computer calculates the individual robot axis motions so that each arrives at the program-point location at the same time. Linear interpolation results in straight-line motion of the robot's tool center point. Curvilinear interpolation results in tool center-point motions that follow a desired curved path. Robots are available with one, two, or three interpolation systems in a single unit. Selection depends upon specific requirements.

The control functions for a robotic assembly system may be broken down into levels or areas of responsibility: the robot controller and the workcell controller.

A robot controller contains the circuitry and memory to operate the robotic system. The control circuitry directs the operation and motions of the robot and allows communication with external devices. For programming convenience, the robot controller should have a handheld teach pendant, and a CRT display is a desirable option.

The memory of a robot controller stores the operating system software and the information entered into the programming and operating systems. An erasable programmable read-only memory (EPROM) is a fixed memory whose contents can only be read and cannot be altered by the user. A random access memory (RAM) is used as the system's working memory. User programs that are stored in the robot controller must be loaded into the RAM to be executed. Because the RAM is a volatile memory, a battery backup should be provided to prevent loss of program information if power is interrupted. Batteries should be checked periodically to ensure that they have adequate power when needed.

Bubble memories sometimes store programs and data that are not being executed by the systems. These memories are nonvolatile, and stored programs and data are retained when power is lost. Off-line memories provide backup and bulk storage for user programs and data in devices away from the robot controllers. Such storage devices may be digital cassettes or floppy disks.

The work cell controller has the responsibility for controlling sequencing operation and integrating all components of the cell. Depending on the purpose and scope of the cell, the components can include conveyors, automation equipment, vision systems, and other peripheral devices.

Programmable logic controllers are used extensively as the hardware for performing the control functions. More recently, personal computers (PCs) are being increasingly applied to shop floors for this purpose. Advantages of using a personal computer as a workcell controller include low cost, space savings, simplified programming, and many options.

Industrial robots are programmed by either on-line or off-line methods.

One method of on-line programming consists of lead-through teaching in which the programmer manually moves the manipulator through the desired cycles. In another method, motion sequences are recorded in memory by depressing buttons on the teach pendant or control console. Disadvantages of on-line programming include the need to stop production for programming in some cases and, complex procedures may be required for some applications. When a CAD/CAM database exists, on-line programming often duplicates information already in the database and is not readily compatible with such databases.

Advantages of off-line programming include eliminating the need for stopping production, the capability of moving programs from one robot to another of similar accuracy and programming language, and the ability to easily incorporate programs into automated assembly systems. A number of programming languages are being used. Some of these languages are complex, and most require a language processor on the robot, which is costly and generally used infrequently. Development work is being done with simulation and modeling (automatic programming) in which a robot is assigned a task, develops a plan, and programs itself.

Lastly, a word about pick-and-place units. Pick-and-place units are the simplest form of mechanical manipulators. They are either pneumatically operated or cam-controlled and are equipped with up to five or six axes of motion; or they can lift and lower, reach and retract, wrist turnover, body rotate, grip and ungrip.

Significant disagreement exists throughout the robotics industry concerning the definition of characteristics of pick-and-place units versus robots. For purposes of this discussion, pick-and-place units are somewhat arbitrarily defined as being lower cost mechanical manipulators that offer limited shop floor programming flexibility to the user. At the upper end of the manipulator spectrum are industrial robots which are servocontrolled and can be easily reprogrammed by the user to perform different tasks.

Applications. For years, robots have been used in manufacturing operations. The capabilities of robots are limited, and for a successful application, the proper selection of the robot itself is only one of the important application ingredients. It is also important to understand and review the possible applications of a robot.

Robots are typically used for applications which are dangerous, difficult or undesirable for humans to perform. Some of the earliest applications of robots were materials handling in the area of forging, as well as spray painting, and spot welding.

In forging, for example, the means of transporting workpieces outside as well as inside forging machines may be with the use of robots. The operator is required only to supervise the sequence of operations. Upsetters, roll forges, programmable hammers, impacters and forge presses have been interfaced with robots in single die to three die configurations. Both forge die and hot trim die operations have been installed. Parts range in size from a few ounces to several hundred pounds.

Other installations feature powdered metal technology in conjunction with the forging process. Thus far, the parts are small and light with the emphasis on robot control of the complete sequence. In one application, an environmentally controlled rotary hearth furnace heats the parts. The robot loads and unloads the furnace thereby assuring orientation. The time the part is in the furnace is automatically interlocked with the robot.

The robot places the part in the forge press where final compression and forming takes place. The part is automatically ejected from the forge die. Because placement of the part is critical, tooling is equipped with locators to assure proper placement of the part in the die. As in die casting, the exact repetition of the cycle through heating and forming operations assures consistent part structure and grain flow, and resultant high quality parts.

Palletizing heavy parts is another simple application which robots perform well. It is a monotonous task which requires constant availability of the device. Highly accurate assembly tasks are sometimes an excellent application for robotics.

For many companies, the use of industrial robots for deburring, fettling, and finishing has considerably reduced costs and improved quality. Robotic deburring, fettling, and finishing are typically used for parts that have production rates falling between job shop quantities and automotive industry requirements.

Most applications are for large parts and are dedicated to a specific part or close geometric family of parts. Some precision and intricate parts are being deburred by robots, but they are the exception because many robots are not technically capable of performing precision deburring.

For successful deburring, fettling, and finishing, robots should have the following characteristics: high accuracy and repeatability, continuous-path capability, the capacity for easy and quick tool or spindle changes, rigidity, and low inertia. Other desirable features include easy off-line programming, circular interpolation, and the ability to translate movements to similar features on the same workpiece.

Robots can be applied in metalcutting in two ways. The simplest form is one robot tending one or more machines. The more complex form integrates a series of robotic cells into a total manufacturing process.

One robot tending one or more machines can be thought of as an ''island of automation'' in which there is little linkage between what occurs within it and the events preceding and following it. The figure below is illustrative of such an island or work cell. One robot sequentially loads and unloads three machine tools—a vertical milling machine, a broach, and a drill. This is quite typical, cost effective, and very simple with respect to application technology.

Even with the robot doing most of the work, some of an operator's time is still required to move raw material into position and to remove finished pieces. Also, the work cell requires a certain amount of attention such as tool dressing, toolchanging, and the like.

In more complex, intergrated manufacturing systems, a series of robotic work cells are linked to form process continuity. This continuity could be called a flexible transfer line.

Industrial robots are used for the application of adhesives and sealants. The robots generally manipulate the dispensing gun, but occasionally the parts to be joined are manipulated by the robots. Advantages of using robots for such applications include reduced labor requirements and costs, faster production, consistently high quality, and reduced adhesive usage. Adhesives can be applied to a number of parts simultaneously by using multiple guns on a single robot. A possible limitation is that high volumes are generally necessary for cost-effectiveness. However, the flexibility of robots permits handling a variety of different parts in small batches.

One robot dispensing system being used by automotive manufacturers is of gantry design to allow the system to straddle production lines. The figure on the next page illustrates a car-door assembly system. A programmable interface integrates the unit with the parts handling system.

Industrial robots are increasingly being applied to arc welding applications, primarily because of their versatility. High-volume production is not required for economical robotic arc welding. The versatility of robots makes them economically justifiable for low to medium production requirements. They are well-suited for handling a variety of welding operations, small lot sizes, and parts having a variety of welds requiring different approach angles. Design changes are easily accommodated, and different parts are handled by changing the program and end-of-arm tooling.

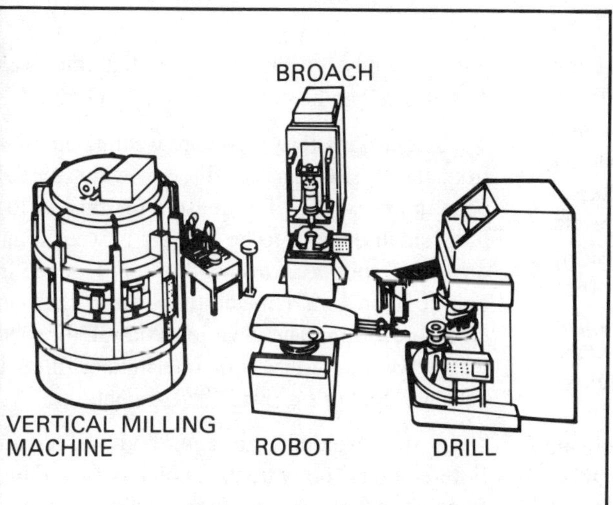

Robot-driven work cell—vertical milling machine, broach and drill loaded and unloaded with a robot.

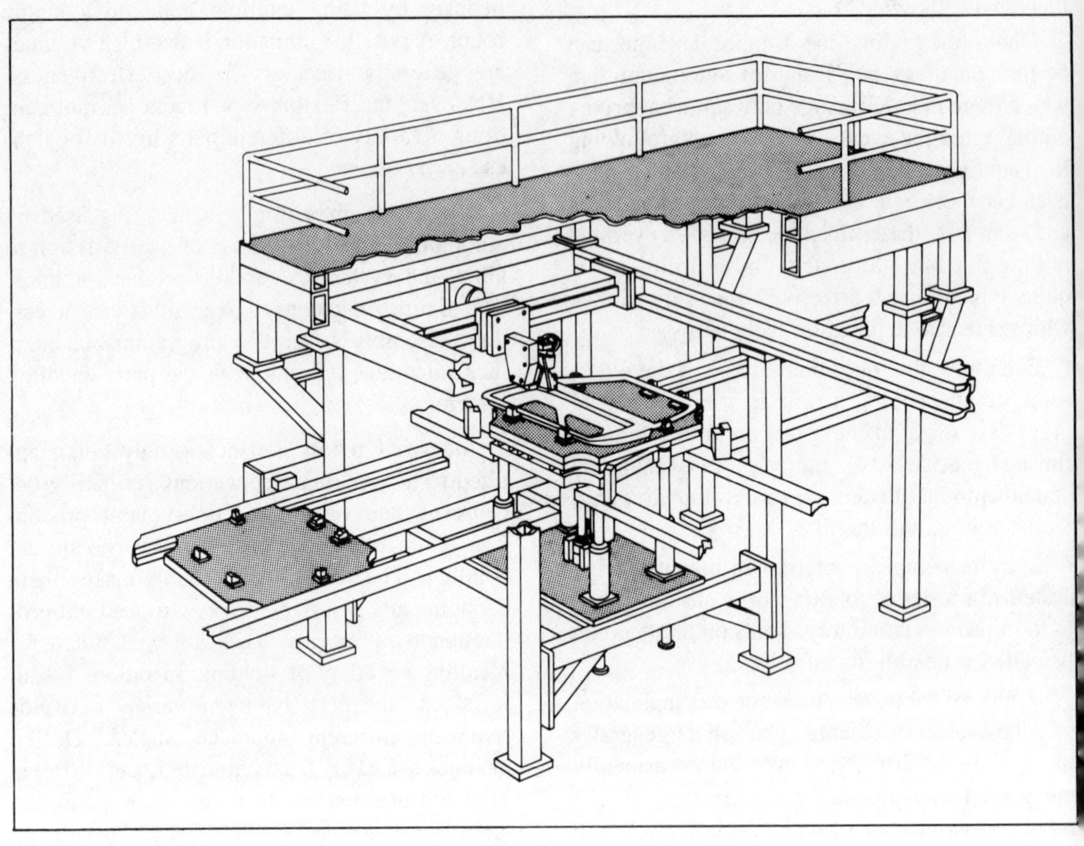

Robot door assembly system.

Robots, however, are not applicable to all arc welding applications. Assemblies requiring welds that are difficult or awkward to reach and that necessitate long reaches are not generally suitable for robotic welding. Also, unlike a skilled operator, robots made with the present state of the art do not have the ability to sense what is happening during welding nor do they have the dexterity of an operator. Rapid advances in improved sensors and software capabilities, however, are narrowing the differences between robots and skilled welders.

Gas metal arc welding is the predominant process used in robotic arc welding. Frequently, a special interface is required to connect the power supply to the robot so that the weld parameters can be set under program control.

Joint designs for robotic arc welding are most frequently single-pass fillets, with partial-penetration welds. The reason for this is that parts small enough to be quickly moved in and out of a robotic work cell do not generally require more weld metal to perform their intended function. The ideal joint design for robotic arc welding consists of intermittent fillets in varying locations around the assembly.

In spot welding, an auto body is already fixtured on a ''body truck'' which is moved by a floor conveyor. A rail system guides the body truck and maintains proper alignment to the robot. The welding operation is done around the rear "wheel well" while the body is moving. The body truck position is measured relative to the robot base. The ability to program the body when it is stationary, execute the program when the body is in motion, and a minimum of modification to the line has made this application economical.

In MIG (metal inert gas) seam welding, a MIG welding gun is attached to the end of the arm. By defining the tool center point (TCP) to be coincident with the weld wire tip, the operator orients the gun around the TCP making programming easier. The ability of the control to generate a straight line between programmed points results in straighter weld lines than could be made manually and the controlled velocity results in a smooth, slow and even motion of the weld tip, all resulting in excellent weld quality and consistency.

An inspection robot usually requires precise control. Robot-based inspection systems can provide the maximum in flexibility. A variety of sensors can be mounted on the robot wrist. Robots with articulated arms can reach inside of parts with cavities, such as car bodies and appliances. Robots can be programmed to measure reference datum points and planes on a part. Algorithms, using these reference point measurements, can calibrate the part coordinate system, which allows the CAD/CAM database to be integrated into the robotic measuring system. This integration eliminates the need for precision part fixturing, allows real-time adaptive robot path trajectory control, and provides off-line programming of robot path trajectories.

Inspection robots have a number of limitations. They can inspect parts only at moderate throughput rates; about 100 parts per hour would be a nominal rate. They are, therefore, not well-suited for very high-speed production lines where 1000 parts per hour are common. Gaging accuracy is another concern. At the present time, robot gaging systems rely on the repeatability of the robot for the gaging tolerance.

The main components in a robotic inspection system are the robot, sensors, part presentation device, computer/control system, and software.

An assembly robot is designed to fasten parts together. Assembly applications characteristically require a more articulate robot with high-level sensory feedback, control capability, complex tooling, and parts feeders.

Robotic assembly (the use of industrial robots for assembly) is being applied much more often. One major robot manufacturer has estimated that by 1990 more than 16,000 robots will be used in assembly systems. About 30% of these units will be dedicated to mechanical assembly applications, while the remaining 70% will be used in the electronics industry.

Most current applications of robotic assembly systems involve small products in medium volume requirements, families of products, and products or production mixes that are likely to change significantly. Predominant users of such systems include the automotive, electronic, electromechanical, and precision mechanical industries.

Robotic systems assemble numerous automotive components. The assembly of universal joints involves the insertion of needles in bearings. Systems for engine cylinder heads include the assembly of valves, tappets, and covers. Spot welding guns are positioned automatically by means of robots in the assembly of automobile bodies (see figure on the next page).

A vision-guided robot carrying pneumatic wrenches (see figure on the next page) is being used to tighten bolts on car underbodies being transported down a moving conveyor line at the Oldsmobile Div. of General Motors Corporation. The robot is mounted on a carriage that is temporarily clamped to the car and travels with it. Solid state cameras view the underbody and report the location of gage holes to the robot system, which then locates the position of the-bolts. Cycle time is 53 seconds to find and torque the 12 fasteners on each body and return to the starting position.

Westinghouse Electric Corporation, with funding from the National Science Foundation,

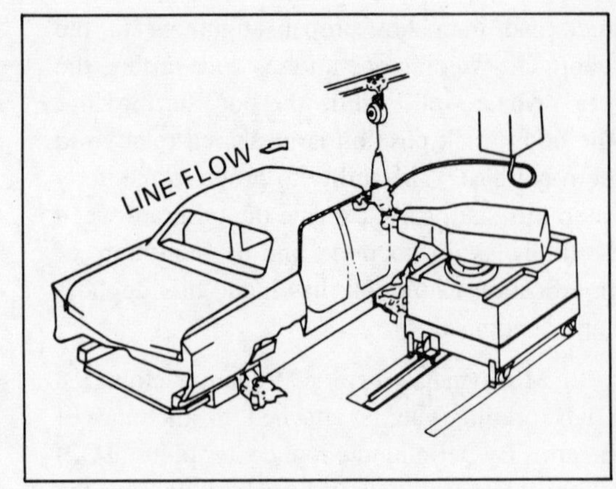

Spot welding guns are positioned accurately and automatically by means of robots in assembly of automobile bodies.

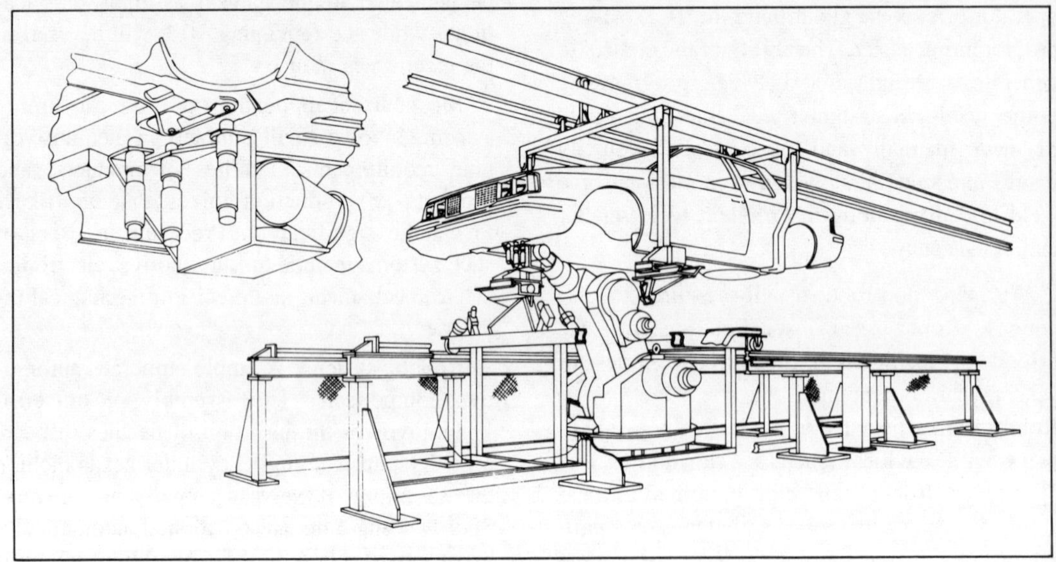

Underbody bolt securing using a vision-guided robot.

developed a programmable assembly system for the low-volume or batch production of small, fractional-horsepower electric motors having a number of variations. Called APAS (adaptable-programmable assembly system), the system consists of six computer-controlled workstations, robots at four of the six stations, part presentation equipment, fixtures and tools, transfer conveyors, and sensory devices in a complete assembly line (see the figure accompanying the term: Automated Assembly).

Industrial robots are used extensively for press loading. Robots are particularly suitable for low-volume production requirements and for operations in which large differences exist in size and geometry of the workpieces to be handled. Two basic types available are servo and nonservorobots. Servorobots have good accuracy and programmability, but are more costly. Nonservorobots, with fixed stops, have good positional accuracy and are less costly, but programming is limited and slower.

In investment casting, wax patterns are processed through a series of slurry and sanding operations to build a rugged shell to serve as a mold during the metal casting operation. Previously, manpower had been considered indispensable for the delicate manipulation during dipping and sanding operations. As many as 10 such operations are necessary to form a shell around the wax pattern. However, with the weight that can be processed by manpower restricted, other means had to be developed to increase the output of the shell building area. The introduction of robots capable of manipulating several hundred pounds at variable speeds and with carefully controlled acceleration and deceleration characteristics has increased the productivity of shell building by a factor of five to eight times. Even where individual shell weights cannot be increased due to characteristics of the product to be made, multiple shells can be manipulated at one time to increase the throughput at a location.

Several benefits have resulted to the investment caster. He/she has been able to penetrate new markets which previously had been restricted to sand casting techniques. The efficiency of the shell building areas has increased dramatically as more useful shell weight per linear foot of floor space is processed. Delicate parts are now being run under carefully controlled shell building programs by utilizing the robot as the manipulative device with a resultant decrease in defective shells.

Edward Mroz and Ralph B. Stevens of Engineered Systems & Development Corporation report in an SME Technical Paper titled *Robots Handling Magnetic Disks in Clean Rooms* that advancing technology in information storage on computer rigid disks is increasing the requirement for automated disk manufacturing operations in a clean-room environment. Clean room standards and practices demand that automated systems be designed to control contamination generated by equipment. Two robotic work cells developed for use in thin-film coating and certification of rigid computer disks were developed with attention to clean room, specific design decisions.

The figure on the next page shows a single-disk, internal-diameter gripper that will pick and place disks from Empak, Fluoroware, and Encore cassettes. The fingers are made of Delrin, a material which neither outgases nor wears substantially with use, and the design allows minimum contact with the disk inside diameter. A specially designed shroud around the finger joints contains any particles generated so that they can be removed via a vacuum.

Dave Kalitzki of the Organization for Industrial Research, reported in his SME Technical Paper titled *Using Group Technology to Justify and Implement Robotics in Batch Manufacturing Environments*: ''With approximately 60-70% of all manufacturing in the United States being batch manufacturing, the problems perceived as obstacles to the application of robotics in batch manufacturing are not unique just to robotics. In fact, the same problems (large varieties of parts) are considered hindrances to the introduction of other systems designed to improve productivity—including design standardization, process plan standardization, and manufacturing standardization through the use of either ''virtual'' or actual manufacturing work cells.

''Group Technology has allowed the application of advanced production and manufacturing systems in these batch manufacturing environments by taking advantage of the grouping of discrete parts by manufacturing character. Group Technology applications have provided practical and financially attractive solutions to these batch manufacturing problems.''

S. Jetley and J. Bunyea of the SUNY College of Technology reported in their paper titled *Expert System for Robot Maintenance* that an expert system has been developed to aid the maintenance personnel in the diagnoses of malfunctions developed in the Puma-600 robotic system. The Puma system consists of various separate components. The expert system starts by asking the user to identify the component with the fault. A series of questions is asked and test directions given to diagnose and correct the fault. The knowledge base is provided in the robot's maintenance manual and is limited to that information. However, the system is designed to be continually updated and the knowledge base increased.

These are just some of the many ways industrial robots are used in manufacturing. While the

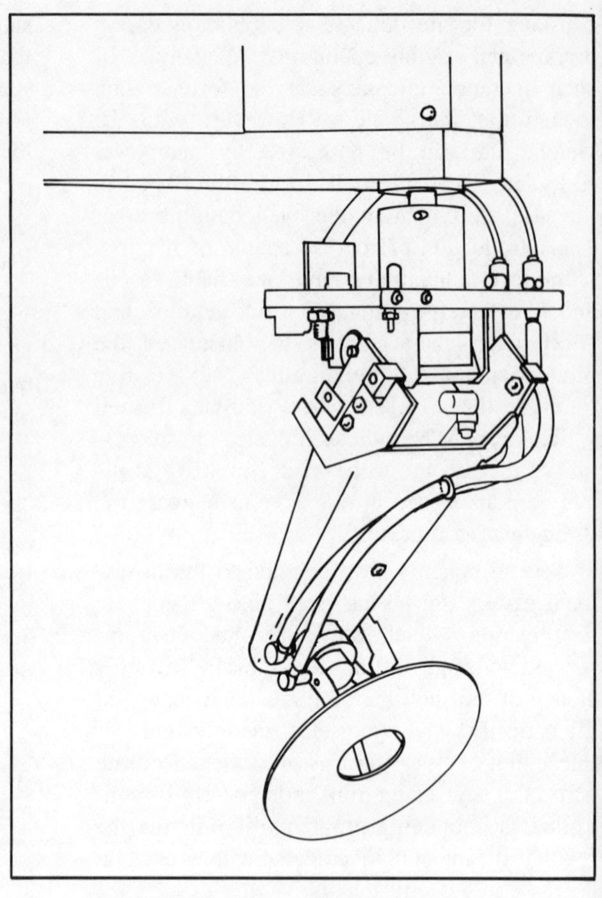

A single-disk, internal-diameter gripper that will pick and place disks in clean room applications.

industrial robot was originally conceived with a point-to-point mode of operation, it has been the continuous path mode of operation which has dominated most industrial robot applications. No longer do potential users need to wait for some adventurous industrialist to experiment with applications for the first time; most of this has been done and applications in the nonsensory handling and processing areas are well-documented. Further developments in robot structures and mechanisms now give theuser a superb range of devices/products from which to choose as any visit to a robot exhibition will demonstrate.

Glossary. *actuator:* The part of an industrial robots that converts energy (hydraulic, pneumatic or electrical) into power.
arc welding: Any fusion welding process that joins metal parts by melting at the point using the heat of an electric arc. Generally, the arc is sustained between a welding electrode and the parts being joined.
batch processing: A manufacturing operation in which a designated quantity of material is treated in a series of steps. Also, a method of processing jobs so that each is completed before the next job is initialized.
CAD: See: computer-aided design.
CAM: See: computer-aided manufacturing.
Cartesian coordinate system: A system of two or three axes that intersect each other at right angles forming rectangles. Any point within the rectangular space can be identified by the distance and direction from any other point. Also known as rectangular coordinate system.
clean room: A working environment with specially designed air filtration systems and highly regulated air pressure. Clean rooms are isolated

from contaminants that are present in the air and on surfaces with these environments.

computer-aided design (CAD): The use of computers to aid in designing products.

computer-aided manufacturing (CAM): The use of computers to aid in the various phases of manufacturing. Numerical control (NC) is a subset of CAM.

computer numerical control (CNC): A self-contained NC system for a single machine tool utilizing a dedicated computer controlled by stored instructions to perform some or all of the basic NC functions. Punched tape and tape readers are not used except possibly as backup in the event of computer failure. Through a direct link to a central processor, the CNC system can become part of a direct numerical control (DNC) system.

data bank: A complete collection of information such as that contained in libraries stored in drums, disks, or other storage media for computer processing. A library is a collection of files which is a collection of records. Records are a collection of items. Also known as database.

database: See: data bank.

die casting: A high-production casting process used predominantly for small-to-medium-sized parts. In die casting, molten metal is fed or injected into a die made of steel or high temperature alloy that often contains the impressions of several identical parts.

end effector: Grippers, hands, holders, and other tools for handling components to be assembled.

floppy disk: A flexible, magnetic-based disk used to store data input to NC machine control units.

forging: The process of deforming to the desired shape by forming in presses, hammers, rolls, upsetters, and related machinery. The product resulting from this deformation process.

hardware: As a food gatherer, man allowed nature to do all the manufacturing or processing of materials and foods he consumed. Man's first step towards manufacturing; such as food growing, making tools and weapons, dressing skins and animal husbandry were for his own use, therefore the entire cycle of manufacturing and quality activities were performed by the same individual.

interpolation: A function of control enabling data points to be generated between specific coordinate positions to allow simultaneous movement of two or more axes of motion in a defined geometric pattern. For example, in NC, curved sections can be approximated by a series of straight lines or parabolic segments. Also known as linear interpolation.

linear interpolation: See: interpolation.

numerical control (NC): A technique for controlling actions of machine tools and similar equipment by the direct insertion of numerical data at a given point. Data is automatically interpreted.

programmable controller (PC): A solid-state industrial control system with a memory which can be set to operate in a specified manner to store instructions that implement functions such as I/O control logic, timing, counting, arithmetic, and data manipulation.

random access memory (RAM): A type of memory that can be accessed (read from) independent of the time of the last access or the location of the most recently accessed data.

resistance spot welding (RSW): A resistance welding process which produces bonding at one spot. The size and shape of the weld nugget are influenced largely by size and contour of the electrodes that introduce low-voltage, high-amperage electric current into the assembly.

software: All programs, routines, and documents associated with a computer.

spot welding: See: resistance spot welding.

Also see: Adhesive, Arc Welding, Arm, Articulated Arm, Assembly Robot, Automated Assembly, Bin Picking, Computer Numerical Control, Degree of Freedom, Die Casting, Educational Robots, End Effector, Fettling, Forging, Injection Molding, Inspection, Intelligent Robot, Heat Treatment, Hobby Robots, Human Factors Engineering, Hydraulics, Joint, Jointed Arm, Kinematic Analysis, Laser, Machine Vision, Numerical Control, Pick-and-place, Point-to-point Control System, Rectangular Coordinates, Relative Coordinates, Remote Center Compliance, Repeatability, Servocontrol, Servomechanism, SCARA Robot, Teleoperated Robotics, Transfer Lines.

Initial Graphics Exchange Specification. The Initial Graphics Exchange Specification or IGES establishes standard information structures to be used for the digital representation and communication of product definition data. The primary purpose of IGES has been to exchange graphics-related information between different makes of computer-aided design (CAD) systems. The mechanism of exchange is a so-called ''neutral'' or intermediate file. Each sending and receiving CAD systems has a translator that converts its internal data formats to and from the IGES neutral file.

Unfortunately, CAD systems suppliers differ in their support for the IGES information structures. This often prevents full translation between CAD systems. And, as CAD systems evolve, their capabilities and data representations have extended beyond the scope of IGES. Three-dimensional solids modeling, nongraphic data management, electronic design, and other-forms of CAD were not addressed by the early versions of the IGES specification and have remained problematic in subsequent editions.

IGES Version 1.0 was embodied in ANSI Standard Y14.26M-1981. The IGES Steering Committee, led by the National Bureau of Standards, continues to upgrade IGES. Version 2.0 was published in 1983, and Version 3.0 in 1986. A version 4.0 is planned but will probably be the last IGES specification. The IGES Steering Committee has already formulated a new standard called PDES (Product Data Exchange Standard) to replace IGES and deal with its inherent shortcomings where solids modeling and nongraphic data management are concerned.

Injection Molding. Injection molding is a molding procedure whereby heat-softened thermoplastic or thermoset material is forced from a plasticating barrel into a relatively cool mold cavity for hardening.

The fully automatic injection-molding machines are designed to operate continuously without constant operator attention. Because of a uniform operating cycle, molds with fewer cavities can turn out greater production than the manually operated machines. Microprocessor control is a standard feature on most new units.

Many machines have a two-stage pressure arrangement on the mold-closing mechanism to prevent damage to the mold in case of incomplete ejection of parts or runners. Low pressure is used to close the mold; after the mold is completely closed, a limit switch actuates the high-pressure stage to lock the mold closed. If a piece of material is left in the mold, the limit switch cannot function and a warning light flashes.

The part design should lend itself to automatic or mechanical ejection upon opening of the mold. Self-shearing gates and a sorting arrangement to separate the parts from the runners can make this a more economical setup.

Fast mold changing systems are utilized for injection molding operations. Advantages include reduced time for tooling changes which contributes to productivity gains. Varying levels of fast mold changing are available—from automatic, push-button setups to less sophisticated semiautomatic systems that require more operator assistance. The end result is a capability for integrating injection molding operations into a flexible manufacturing system for producing a variety of plastic parts.

Input. See: Computers, Graphic Input, Lag, Light Pen, Manual Data Input, Remote Input/Output.

Input/Output. See: I/O.

Input/Output Channel. In an automatic data processing system, an input/output channel is a functional unit, controlled by the central processing unit. This channel handles the transfer of data between main storage and peripheral equipment.

Inspection. Inspection involves evaluating the quality of some characteristic in relation to a standard. The purpose of inspection is to determine product conformance to established/agreed upon specifications. Therefore, inspection is often referred to as "product acceptance inspection." Inspection of each quality characteristic consists of the following actions: interpretation of the specification; measurement of the quality of the characteristic; comparing the interpreted specification with the measured quality, and determining conformance to the specification. Titles of persons engaged in full-time inspection or "product acceptance" include: Inspector, Test Technician, Final Inspector, Quality Assurance Auditor, Quality Control Inspector.

History. As a food gatherer, man allowed nature to do all the manufacturing or processing of materials and foods he consumed. Man's first step towards manufacturing; such as food growing, making tools and weapons, dressing skins and animal husbandry were for his own use, therefore the entire cycle of manufacturing and quality activities were performed by the same individual.

As tribes evolved into communities, marketplaces were established. The marketplace was an ideal location for craftsmen to set up small workshops. The small size of shops allowed craftsmen to see all aspects of the design, manufacture and marketing of their products. The consumer usually purchased goods personally, allowing the craftsman to see and learn first hand the quality needs of the customer. The consumers quality needs define "fitness- for-use" as the proprietor examined project before sale, to assure that the apprentices were learning their trade and to protect the customer, but there were no "inspectors" as we know them today.

As sole proprietorships grew in size, worker's face-to-face contact with the consumer was virtually eliminated. Products were no longer sold to the ultimate user; they were sold to intermediate merchants ("middlemen"), and were shipped to other locations.

Growth also brought about a division of labor which resulted in the worker only making a part of the completed product. In making just a part, the worker's knowledge of the design, manufacturing process and marketing was reduced considerably and replaced with very limited knowledge, focused on the specific operation performed by the worker.

Armed with limited knowledge, gained impersonally from specifications rather than face-to-face contact with the consumer; the craftsman was now paid only by the piece of work produced. Compensation was based on quantity not quality.

The larger size of manufacturing faculties removed the control from the sole proprietor and shifted it to supervisors. Responsible for high volume and quality; the supervisors initially examined the huge quantity of product being manufactured. As the examination became full-time the supervisors created a new job; that of "Inspector."

Prior to World War I, inspectors reported to production supervisors. But after a series of serious quality failures and scandals, advocates of Frederick W. Taylor's school of "Scientific Management" encouraged the formation of inspection departments, independent from manufacturing.

During World War II, the tremendous requirements for high quality, reliable weapons, vehicles, ships, airplanes, and supplies forced the U.S. industrial base to adopt Taylor's "Ssientific" methods of management. The U.S. government required that U.S. suppliers set up independent inspection systems which would guarantee products fitness-for-use. The basis for inspection systems were outlined in military specifications (MIL-I-45208A *Inspection System Requirements* and MIL-Q-9858A *Quality Program Requirements*).

They identified specific statistical sampling plans to be used by suppliers' inspection departments. Almost overnight, formal inspection departments were formed in most manufacturing companies in the United States. Since World War II, the complexity of military hardware has increased hundred-fold and methods of inspection have had to keep pace.

In industries not related to defense, the recent trend has been towards implementation of JIT (Just In Time) manufacturing. JIT emphasizes operator awareness of quality factors which determine the product's fitness-for-use. In JIT, the responsibility for quality is shifted from an independent inspector operator/department back to the operator/department responsible for manufacturing the product.

The concept of training operators in what constitutes fitness-for-use is a throwback to the early days of manufacturing where the worker came in contact with the end consumer and was completely responsible for product quality.

Discussion. As products have grown in complexity, the job of manufacturing them has been divided among many departments. The job of inspection has also become complex and divided among various departments. Inspection always includes the interpretation of the specification; measurement of the product, comparison of the interpretation with the measurement, and the determination of conformance to the specification.

Inspection also includes additional elements depending on the purpose.

If the inspection purpose is to distinguish good lots from bad lots, then the terminology includes such terms as acceptance inspection, sampling, inspection vendor, (incoming) inspection, process inspection, and final inspection.

Sampling inspection or *acceptance inspection's* purpose is to determine the acceptability or nonacceptability of lots of product based on inspection of a sample randomly pulled from the lot. Results of the sample inspection are used to determine acceptance of the lots. Vendor (incoming) inspection is sampling inspection performed by the purchaser, on material procured from another company. *Process inspection* is sampling inspection of material or product between departments of the same company after a process or operation has been performed. *Final inspection* is inspection performed by the manufacturer prior to shipment to the customer after all operations have been completed.

If the purpose of the inspection is to distinguish good pieces from bad pieces, the terminology includes detail inspection, 100% inspection, and screening or sorting. Detail/100% inspection is the sorting of good pieces from bad when the process is inadequate to meet the tolerances of the specification or the process is adequate, but shop difficulties have created defective product.

If the inspection purpose is to determine if the manufacturing process is changing, the terminology includes *control chart sampling*. A control chart sampling is performed to see if the process is changing. Shewhart Control Charts are used to compare averages of samples to statistically derived upper and lower control limits. Control chart sampling detects significant, assignable causes of variation.

If the inspection purpose is to rate the quality of the product the terminology includes product auditing and product quality. The quality of the product is determined at a particular point in time by categorizing the seriousness of defects and assigning weights or demerits proportionally. Results are charted as defects per unit or demerits per unit of production.

If the inspection purpose is to rate the accuracy of inspectors, the terminology includes *inspection accuracy* and *overinspection*. The purpose of verifying *inspection accuracy* is to measure the effectiveness of inspectors in finding defects. A comparison is made between the defects found by the inspector, and the defects that should have been found. The ratio of one to the other is the accuracy of the inspector. *Overinspection* is the unnecessary inspection of conforming product or the mistaken rejection of conforming product due to poorly defined standards and inadequate training.

If the inspection purpose is to obtain product design and reliability information, the terminology includes qualification and testing and reliability testing. Qualification testing judges the service capability of the product when compared to the specification. Reliability testing establishes the mean-time-between-failure (MTBF) of the product by testing the product to failure. MTBF can also be calculated using complex formulas or computer generated models. Failure analysis is performed to determine the failure mode of the product. This information aids in the future design of more reliable products as measured by the MTBF.

If the inspection purpose is to measure process capability, the term is process capability measuring. Process Capability measurement is accomplished using control charts. Control charts contain subgroups which must have identified parameters measured (inspected). The measured data is then used to chart the process capability.

Fixed inspection utilizes multiple contacting or noncontacting sensors mounted in a test fixture that holds the part to be inspected. This approach lends itself to the inspection of parts at high throughput rates. However, to switch from one part to another requires changing the inspection fixture. Flexible inspection utilizes sensors that are moved along a programmed path trajectory of the part being inspected. This approach lends itself to processing at moderate throughput rates. Changing from one part to another can be accomplished quickly by downloading a new path trajectory program to the machine controller.

Industrial robot-based inspection systems can provide maximum flexibility. A variety of sensors can be mounted on the robot. Robots with articulated arms can reach inside of parts as well as car bodies and appliances. Robots can be programmed to measure reference datum points and planes. Algorithms, using these reference point measurements, can calibrate the part coordinate system, which allows the CAD/CAM database to be integrated into a robotic measurement system.

Inspection robots have a number of limitations. They can inspect parts only at moderate throughput rates. They are, therefore, not well-suited for very high-speed production lines.

For inspection applications, the Cartesian-style robot appears most appropriate. Because of its geometric design, the Cartesian robot can provide higher repeatable positioning accuracy than other articulated arm robot styles. Experience has shown that a fixed-axis robot is adequate for most inspection work. This is because coordinate transformation algorithms can correct sensor limitations imposed by a five-degree-of-freedom robot.

Robot positioning accuracy is determined by the ability to match its actual position in three-dimensional space to the command position called for by the program's position instruction. The error between actual and command positions is caused by a number of factors. These factors include: servocomponent design, structural natural frequencies, bearing friction, gear backlash and load torques.

Robot settling time is a very important performance consideration because it causes a direct tradeoff between measurement accuracy and process inspection time. This tradeoff is shown by the data in the graph on the next page. This graph compares robot positioning error distributions for two different settling times. This data was obtained with the robot executing an inspection path program representative of a typical process inspection cycle for a production application. Increasing the settling time by 117% reduced the variability in robot position by 36%; however, process inspection cycle time increased by 19%.

There are three types of sensors typically used in gaging systems: one-dimensional sensors, contour sensors, and array sensors. One-dimensional sensors give the range of distance from the sensor to a point on the object. These sensors use triangulation techniques, where the light source is usually a single-point laser diode set at a known angle from the pickup sensor.

A contour sensor analyzes a line of light across an object; the light is usually from a laser. The laser in the visible range (red) is a gas tube device containing helium and neon. More recently, these are being replaced by solid state lasers that operate in near-infrared range.

An array sensor can take area images of the part for locating features such as holes. This two-dimensional information can be further enhanced by adding a range sensor to obtain Z axis information, effectively creating a simple but limited three-dimensional device.

Sensor selection is determined by the geometry of the part.

A robotic inspection system may be composed of two to four small computers that are required to perform different distributed processing functions. These computers are interconnected by a data communication and control network. This configuration allows a host computer to coordinate and control the robot and sensor(s).

Lasers also are used in inspection.

The uses of the laser in metrology arise from the characteristics of laser light that differentiate it from ordinary light. Those characteristics are the extreme intensity, the highly directional, small collimated beam, the monochromaticity, and the coherent nature of the light.

One of the largest uses of lasers is by contractors for alignment and surveying. Here, the intensity and directionality properties make the laser a natural tool.

Applications. Most modern manufacturing entities have permanent, independent inspection departments. These inspection departments are usually part of a corporate Quality Control or Quality Assurance Department. The manager of

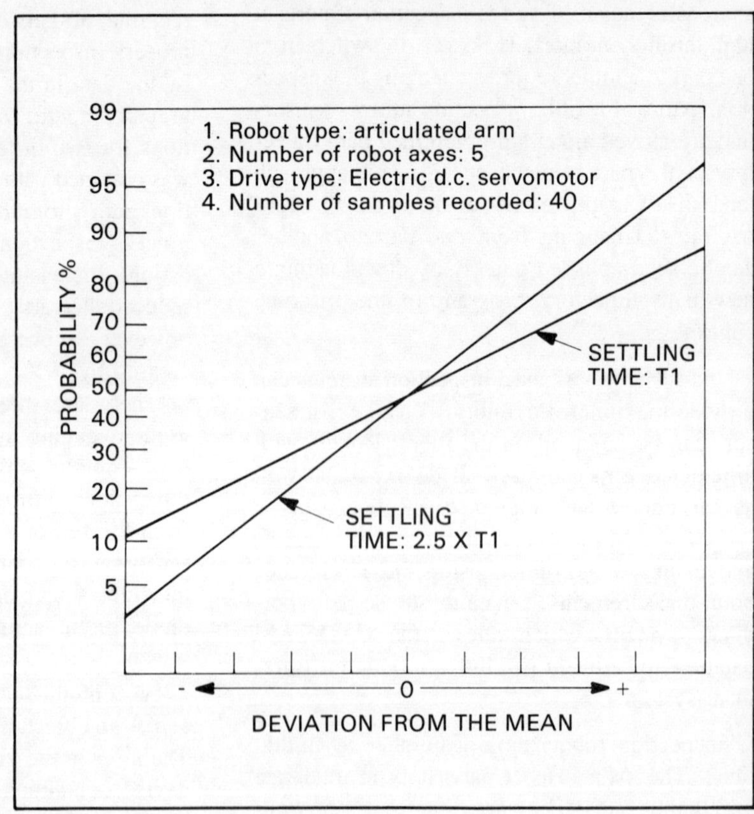

Comparison of robot position repeatability for two different setting times.

the Quality Assurance/Quality Control Department is typically at the same management level as the managers of engineering and manufacturing. With purchased material inspection, inspectors engaged inspect incoming raw materials, piece parts, and subassemblies. Incoming inspection is routinely located near the receiving area to facilitate product flow. Due to the variation of product inspected; sophisticated and highly specialized tools are located in incoming inspection areas. These include: coordinate measuring machines, optical comparators, colorimeters, electron-microscopes, frequency generators, Rockwell Hardness Testers, optical microscopes and a variety of mechanical measuring equipment such as calipers, height gages, micrometers, torque wrenches, bore-indicators, gage blocks, gage pins, etc.

Process inspection is performed at various stages throughout the manufacturing process to identify nonconforming material and to prevent value adding activities from being performed on discrepant product. Process inspectors are located on the shop floor where the production process or manufacturing operation is performed.

Final inspection and test are typically performed by a group separate from incoming and process inspection. Final inspection/test involves functional and visual/mechanical inspection of the finished product. Inspection of packaging may also be performed by the final inspection department. Inspection and test equipment required to perform final inspection are usually dedicated and calibration control is maintained.

Incoming, process and final inspectors require some or all of the following information.

1. The customer's order—defining what products the customer requested and any modifications he or she wants to the specification.
2. The product specification—defining the characteristics of the product.
3. Quality characteristics—identification of specific quality characteristics to be inspected, procedure and equipment to be used for same.
4. Pertinent industry standards—established industry standards.
5. Previous inspection results for product under consideration.
6. Information on past history of products from same source of supply.
7. General Information–knowledge of product application, prior complaints.
8. A detailed inspection/test procedure,which contains: (a) tests to be performed with step-by-step procedure; (b) measurements to be taken and equipment to be used; (c) data to be recorded; (d) sample size, for each test; (e) procedure for sample selection; (f) required accuracy of inspection/test equipment; (g) environmental conditions during tests; (h) criteria for "acceptance" or "rejection" decision on quality (including maximum and minimum test limits, allowable number of failures for "acceptance" of lots, and basis on which production is to be shut down.); (i) reports to be prepared; (j) action to be taken after "rejection" decision; (k) disposition of test specimens, and (l) requirements for "certification" of test results by an independent agency.
9. Responsibilities of the inspector in accepting or rejecting the product, notifying supervision of rejected material, and requesting or requiring that the line be shut down.

Due to the complexity of manufactured products, it has become necessary to engage in formal inspection planning. Persons responsible for inspection planning must understand what constitutes fitness-for-use of the product being inspected. Included are the inspector, inspection supervisor and quality control engineer in conjunction with the design/manufacturing engineer.

Steps required to properly perform the inspection planning function include:

1. Preparation of a flow diagram showing the various materials, components and processes which collectively or sequentially turn out the final product.
2. Identification of the quality characteristics to

be inspected.Interpretation of the specification to assist the inspector by:
   a. defining terms used
   b. providing supplemental information
   c. classify defects (major, minor)
   d. providing samples as required.
3. Preparation of written inspection instructions.
4. Instructions for inspection of nonproduction operations, i.e.
   a. internal handling (use of correct containers for ESD protection, etc.)
   b. internal storage (adequate identity and traceability)
   c. packing (product identification, lot numbers, traceability, protection against environment, damage from handling)
   d. shipping (special marking as required by customer, care in loading to prevent damage in-transit).
5. Inspection data planning to determine the data recording requirements for each inspection station, and inspection data to include recording of inspection measurements or results on control charts for frequency distribution.

Sound inspection planning requires periodic reappraisal of current plans and analysis of situations influencing the formal planning function. These are:

1. the education, experience and training of the work force,
2. the stability of the process, and
3. the stringency of product specifications.

Failure to properly apply inspection increases the risk of catastrophic product failure, while over application of inspection results in excess cost and much internal friction.

The inspection manual, usually called the Quality Control or Quality Assurance Manual, outlines the responsibilities of the Inspection Department and, in addition, includes much information on the responsibilities of most other company departments with respect to quality control and inspection.

The content of the Quality Control/Inspection Manual should include, but is not limited to:

1. A statement of legitimacy and purpose approved by senior management.
2. Table of contents.
3. The Quality Assurance Department organization, including inspection department organization charts, job descriptions and a statement of responsibilities.
4. Documentation of the quality system.
5. A description of the system for control of drawing documentation and design changes.
6. A description of the incoming inspection function.
7. A description of the in-process inspection function.
8. A description of the final inspection function.
9. Description of the sampling plans used by the company.
10. The description of standard plans for the use of control charts.
11. Description of the systems in place for the calibration and maintenance of inspection and test equipment.
12. Description on how nonconforming material is dispositioned.
13. Description on packing, storage, and handling.
14. Copies of all inspection forms used.
15. Descriptions of training and operation certification programs.
16. Descriptions of product qualification and reliability test plans.
17. A description of the system for inspection data feedback to production.
18. A description of the corrective action system.
19. Product identification procedure.
20. A description of the quality reporting system and information for upper management.
21. A description of the cost related to quality system.
22. A description of workmanship standards.

Automated inspection and testing are used to reduce costs, improve precision, reduce inspection time and to alleviate manpower shortages.

Automated gaging and testing is used extensively in mechanical, electronic and chemical industries.

The advent of numerically controlled (NC) machines has required development of new inspection equipment that uses the NC principle. NC machines are controlled by tape and validity of the tape is critical. Inspection of, and certifying the tapes stabilizes the process capability of the NC machines. Coordinate measuring machines (CMM), also tape controlled, are used to inspect parts made on NC machines.

Automated, infrared inspection of solder joints is now commonplace in the defense and computer industries.

The use of automated inspection is expanding rapidly.

Laser scanning instruments can be used in a broad range of manufacturing operations and a variety of industries. Some of the potential areas of application are wire manufacturing, centerless grinding, plastic extrusion, metal product fabrication, and nuclear reactor metrology.

By modifying the techniques by which the detector output is digitized and interpreted by the processor unit, measurements can be made on translucent materials such as fiber-optic cables or transparent material such as glass tubing.

In addition to simple diameter measurement, product position, gap size, and multiple dimensions, measurements are possible by examining the detector output in different ways. More elaborate scanner geometries can achieve dual-axis inspection. In these applications, the laser beam is alternately swept across the measurement field in two axes 90 degrees apart.

By stacking individual scanners back-to-back or detecting only the edge of a product and relating it to the position of a reference edge, products much larger than the range of an individual scanner can be measured. Extra-high-speed scanners, which measure at four to six times the normal rate, allow detection of smaller defects, such as lumps or neckdowns, in moving product applications.

Glossary. *AOQ:* average outgoing quality, i.e., the quality of the product leaving the inspection department after acceptance sampling and any sorting required.

AQL: acceptable quality level, i.e., the lowest quality of incoming product which is to be accepted regularly.

ATP: acceptance test procedure.

calipers: mechanical tool for measuring parts.

colorimeter: meter used to test color of painted product to standard.

cost of quality: those costs associated with finding, correcting and preventing defects, both internal and external.

CMM: coordinate measuring machine, often numerically controlled.

defect classification: categorizing defects a major or minor depending on their importance to fitness-for-use.

destructive testing: inspection or testing which results in product being unfit for use.

diode: an electrical device such as an electron tube or semiconductor which allows current to pass in one direction only.

disposition: to determine the fate, in the case of rejected material.

first piece inspection: product acceptance inspection on first piece from new machine setup, new or changed process.

inspection error: failure of the inspector to identify discrepant parameters.

JIT: Just In Time manufacturing.

MIL-I-45208A: military standard titled *Inspection System Requirements* outlining basic system for production of military/defense hardware.

MIL-Q-9858A: military standard titled *Quality Program Requirements* outlining comprehensive quality requirements for complex military weapon and communications systems.

MRB: material review board, i.e., for disposition of discrepant material; consists of members from design, engineering, production and quality control. It sometimes includes members from purchasing.

MTBF: mean-time-between-failures.

NC: numerically controlled (as applied to machine tools and inspection/test equipment).

nonconformance: parameter identified as not in compliance with the specification.
nondestructive testing: inspection and testing which does not affect product fitness-for-use.
product acceptance inspection: final inspection prior to shipment to customer.
quality characteristic: parameter defined by the specification.
roving inspection: first piece, in-process, final inspection performed at production station at predetermined intervals.
sensory qualities: those qualities for which we lack technological measuring instruments.
tollgate inspections: lot-by-lot product acceptance inspection.
X axis: the axis of motion that is horizontal and parallel to the workholding surface.
Y axis: the axis of motion that is perpendicular to the X and Z axes.
Z axis: the axis of motion that is parallel to the principal spindle of the machine.

Instruction. Instruction can be defined as a set of bits which cause a computer to perform a specific prescribed operation and which may also indicate the values or locations of its operands. An instruction set is a list of machine language instructions that a computer can perform. Instruction storage is the area of storage media containing coded instructions.

Also see: Computers.

Integrated Circuit. An integrated circuit or IC is an electrical device. An IC contains a miniaturized circuit on a wafer substrate, which is encased in protective packaging, with connection pins or leads protruding externally to allow connection to the remainder of the circuit. Integrated circuits, sometimes referred to as chips, are manufactured in larger circular wafers, which contain multiple circuits in a grid, which are later cut apart and tested. The acceptable chips are mounted in the chip carriers, with the connections made between the pads of the IC chip and the pins, using tiny gold wires or other means.

Advantages of ICs are the much greater reliability of an IC circuit verses the same circuit built of discrete components. ICs are much less susceptible to damage, except in the form of static electricity, in which case ICs may be more susceptible than some types of discrete components. As technology continues to advance, more complicated circuits are being created as integrated circuits such as (application specific integrated circuits) ASICs and hybrids. In large quantities, ICs are much less expensive than discrete component circuits. The initial cost of creating an IC is large, but the manufacturing costs of making the production parts are relatively low.

An integrated-circuit diode-matrix memory is an integrated circuit containing a matrix of diodes which may be individually open-circuited or short-circuited to represent a program.

Also see: ASIC, Circuit, DIP, Hybrid, LSI, SIP.

Integrated Computer-Aided Manufacturing (ICAM). ICAM is an acronym for Integrated Computer Aided Manufacturing, a program that was initiated by the United States Air Force in the late 1970s and was concluded in the early 1980s. The program was a part of the technology modernization effort that the Air Force has been subsidizing throughout its U.S. industry supplier network. The ICAM program was comprised of dozens of projects that were carried out by contractors in the private sector over approximately eight years.

Beyond the manufacturing modernization that was initiated and executed within aerospace design and production facilities, the program left a permanent imprint on U.S. manufacturing technology by fostering the concept of integration in addition to process improvement and automation. Moreover, the ICAM program was one of the first modernization initiatives to recognize the importance of data and information to the integration manufacturing enterprise. Under the aegis of ICAM, data and information flows were analyzed for manufacturing operations by uniquely applying structured analysis techniques used by software designers. These techniques were embodied in the standardized formats of IDEF methodologies that were used to analyze activities and information flows necessary to the manufacturing process. By supporting the use of standardized structured analysis techniques among all of its contractors in the ICAM program, the Air Force assured enhanced communication and interchangeability of architectures. The ICAM program also resulted in a generic architecture of manufacturing that has been used to represent and analyze the manufacturing enterprise.

The program also supported three key implementation programs (referred to as wedges) that concerned sheet metal fabrication, electronics manufacture and composites fabrication. The composites fabrication was pursued most vigorously because of significant performance advantages and strategic merit. This resulted in some of the advanced methods of manufacture that are successfully used today for production of aircraft parts from composite materials.

The ICAM program results are being carried forward by continuing modernization and improvement programs supported by the Air Force. In many instances, these new programs utilize the integration concepts and structured analysis techniques that were developed under ICAM.

Integrator. An integrator is a device or circuit that integrates an input signal, usually with respect to time.

Integrity. Data integrity exists when data does not differ from its source data, and has not been accidentally or maliciously altered, disclosed or destroyed.

Also see: Data.

Intelligent Robot. Intelligent robots have sensory perception, making them capable of performing complex tasks which vary from cycle to cycle. Intelligent robots are capable of making decisions and modifications to each cycle.

Also see: Industrial Robots.

Intelligent Terminal. An intelligent terminal is a terminal that can execute certain fixed or variable programs for its user without continuous intervention by the user during their execution. An intelligent terminal may offer such capabilities as dual session, which is the ability to maintain a communications link with two host computers simultaneously, and allowing the operator to toggle between the two sessions on the screen, or allowing the user to configure a split-screen, with both sessions viewed concurrently. An intelligent terminal may also offer capabilities such as a printer port, integral modem, or a device input port. Other features of intelligent terminals are horizontal scrolling, software setup, and multiscreen memory. Some more sophisticated controllers offer intergral text editors and memory for local activities. Some terminals have various integral fonts for quicker display of text in various styles.

Also see: Computers, Modem, Monitor, Terminal.

Interactive. The term interactive refers to a process of function which cannot occur by itself, but requires the interaction of a person, typically, or another process to continue and complete. An interactive mode of operation is such that a sequence of alternating entries and responses takes place between the device and a user similar to a dialogue between two persons.

An interactive data transaction is a single (one way) message that is transmitted via a data channel, and to which a reply is required in order for work or a conversation between two persons, a person and a machine, or two machines, to proceed logically.

An interactive display system can be used in a conversational mode; i.e., a user can enter display commands or inquiries and the system can respond as in a conversation by changes in display elements, display groups, or display images in the display space on the display surface of a display device that is a part of the system. An interactive display system is usually supported by a computer or data processing system, computer programs, a menu, display commands, and the display device, such as a CRT.

Interactive graphics, interactive database queries, interactive routing, etc. are other examples of typical interactive processes.

Interchange Station. An interchange station is the position in which a tool of an automatic toolchanging machine waits for automatic transfer to the machine spindle or the appropriate coded drum station.

Interface. An interface is a shared boundary between two identifiable entities. For example, the boundary or a point that is common to two or more similar or dissimilar systems, subsystems, devices, or other entities, across which information or control flow may take place. Another example is the shared boundary between two pieces of equipment, such as the interconnection between two electronic devices, two layers of a layered communication system, or between data circuit-terminating equipment (DCE) and data terminal equipment (DTE). In open systems architecture, an interface is the connection between adjacent layers at the same node or between corresponding layers at different nodes in a network. The boundary is defined primarily by functional, interconnection, and signal characteristics in communication systems.

The interface must allow the entities to be interconnected and to interoperate. In order to adequately define the interface requirements, the type, quantity, and function of the interconnecting circuits, the type and form of signals that are to be interchanged via those circuits, and the mechanical details of plugs, sockets, pin numbers and the like, must be specified.

In layered systems, the interface is the set of protocols that is applicable to the interaction between any two layers within a single system, such as a computer, data processing, or communication system.

Interferometer. Interferometer is a visual reference gaging procedure using split and recombined light beams to accurately compare the surface geometry of a part against a master and to measure surface texture.

Also see: Gaging.

Interlock. Interlock defines a circuit arrangement that prevents additional operations from taking place until another operation is completed first.

Also see: Circuit.

Intermediate Transfer Arm. An intermediate transfer arm is a mechanical device in automatic toolchanging that automatically grips and removes a programmed tool from the coded drum station and places it in the interchange station. It also grips and removes a used tool from the interchange station and returns it to a coded drum station.

Internal Memory. Internal memory is the part of memory where information that is being worked on is kept. It is also referred to as main memory. Internal memory is usually RAM (random access memory), which means data can be written to any portion of the memory at any time. In a computer, the main memory usually has a basic structure in terms of what types of functions are located at relative location of memory.

Also see: Computers, Memory, Personal Computers.

Internal Setup Time. Internal setup time occurs when elements of a setup are performed while the machine is not running.

International Organization for Standardization (ISO). This is an international organization whose members are the national standards bodies of most of the countries of the world. The organization is responsible for the development and publication of standards in various technical fields, after developing a suitable consensus. It is affiliated with the United Nations. Member bodies of the ISO include, for example, the Association Francaise de Normalisation (AFNOR), the British Standards Institution (BSI), the American National Standards Institute (ANSI), and the Standards Council of Canada. The headquarters of the ISO is in Geneva, Switzerland.

Interpolator. An interpolator is a device that defines the path and rate of travel of a cutting tool when provided with a coded mathematical description of the path. All points between programmed end points are defined resulting in smooth curves or straight lines.

Interpretive Language. An interpretive language is a computer programming language which, during execution, the computer interprets each line of code and converts it into machine language as it executes the line. If lines of code are executed multiple times, the lines are interpreted (converted into machine language) multiple times. The sequence is to interpret a line of code, execute that line of code, and then interpret the next line of code to be executed.

A compiled language has the entire source code converted into executable object code by a compiler, before any execution of the code occurs. The advantages of interpreted languages are the lack of having to compile the code before execution can occur to obtain sample output. Another benefit of an interpreter is the ability to run a trace during execution, which will identify the exact line of source code which is causing an error in execution. Compiled code, while capable of providing error messages, cannot relate the error to the exact line of source code in some instances, since the source code has been transformed into object code. The ideal situation is a language, such as BASIC, which can be executed as interpretive language for quick program creation, and then compiled and executed as compiled code for speed and source code protection.

Also see: Compiler, Computer Languages, Computers, Programming.

Interrelated Datum Reference Frame. An interretated datum reference frame is one that has one or more common datums with another datum reference frame.

Interrupt. Interrupt defines a break in the normal process of a system or program that enables high-priority work to be done and then normal processing resumed.

Inventory Control. Inventory includes raw materials, purchase parts, work in process, and finished goods in different stages of the manufacturing or distribution process.

Inventory Control simply stated is the management function which has the responsibility of having the right material in inventory at the right stage to meet production needs. This is balanced with the objective to attain a minimum level of inventory while maintaining high service levels.

Inventory control sets inventory investment goals which may include: return on investment, customer service, facility and equipment utilization, inventory investment limits and inventory turnover ratio. After the goals are established the execution of the MRP output is also the responsibility of inventory control management. This involves the release of planned material orders to initiate the ordering of raw material.

Inventory control may also involve the management responsibility of the physical control of inventory. Physical control of inventory includes the movement of material within the manufacturing facility, storage of inventory and the accurate reporting of inventory balances for planning and financial reporting.

Also see: Materials Requirements Planning.

Inventory Level. In stockless production, inventory level applies to work-in-process inventory levels.

Investment Casting. Investment casting is a type of casting process which involves the formation of an expendable pattern in a die or mold and the use of the pattern to form a mold in an investment material. When the mold of investment material (refractory particles and a liquid) has set, the pattern is melted, burned, or dissolved out and the part is cast.

The steps in the investment-casting process consist of sometimes beginning with a full-sized model of the finished part, which may have been machined from some material other than the material of the production parts. The full-sized model is then used as a guide to create a metal die. The metal die, actually a split mold, is used to create wax patterns of the finished device. The wax patterns are attached to a mold cluster, which is dipped (invested) into the mold material. When the mold cluster has hardened, it is heated to cause the wax patterns to melt and flow out. The mold is then fired to harden it and molten metal is poured into the mold. After the metal has hardened, the mold material is removed to expose the finished components.

Parts made by the investment-casting process require a minimum of finishing and can be held to close metallurgical specifications. Parts for valves, sewing machines, locks, rifles, golf clubs, and aerospace and military equipment components are a few of the applications that have demonstrated the versatility and economy of the investment casting process.

The majority of investment castings are built in small lots. The intelligent robot, possessing decision making capabilities, may be used to identify and process each lot individually. Process controllers and computers could provide the backup information to the robot to enable it to construct a good shell around the mold. This would enable each investment caster to decide what technique is to be employed to build a shell, and to exercise complete control over the method chosen. The robot provides a building block approach which allows the user to determine the extent of automation used at any phase of his/her expansion plans.

I/O. An acronym for input/output, I/O describes the data which is input into a process or computer or other device as well as the output which may come from a computer or other type of device. I/O can be internal within a machine between different processes or I/O can be between a device and the external world. The placing of an imaginary boundary defines the interface across which input and output traverses.

I/O also can refer to the physical portion of a computer or communication device which handles the input and output of data. A serial or parallel communication port is an I/O port. A keyboard is primarily an input-only I/O device, although the indicator LEDs on a keyboard are output devices. A monitor is an output-only device, except when fitted with a touch screen, in which case it becomes an I/O device. A printer is usually an output-only device, although some have optional keyboards and some plotters can function as a digitizer input device.

I/O Address. An I/O address is the unique logical identifier given to an input or an output (or combination input/output) physical location. Most software executing in the computer must have some identifier to direct output to the appropriate device on the I/O bus. Having a logical address allows different physical devices to be designated as the respective I/O device, as well as having the ability to reconfigure the I/O if the desire is to rearrange the relationship between physical devices and logical addresses.

In some products such as programmable controllers, different types of I/O have different ranges of valid addresses. Having a convention of reserved address values allows different physical devices to be communicated with, as if they were some other device. For example, a cartridge tape unit can be installed and interfaced to a floppy disk controller. Most computer products are provided with accompanying documentation which lists all the I/O addresses.

Ion. An ion is an electrified portion of matter of atomic or molecular dimensions. *Ion exchange* is a reversible process by which ions are interchanged between a solid and a liquid with no substantial structural changes of the solid. *Ion implantation* is a process by which atoms of virtually any element can be injected into the near-surface region of any solid. Unlike discrete coatings, this process alters the chemical composition near the surface of the solid.

Ion Vapor Deposition (IVD). Ion vapor deposition or IVD is a method of applying a corrosion-protection coating (aluminum) on high-strength steels, titanium, and aluminum workpieces. During coating, the workpiece is placed in an evacuated chamber and held at a high negative potential with respect to the vapor source, thus becoming the cathode of a high voltage circuit. An inert gas (usually argon) is then introduced into the chamber to raise the chamber pressure. By maintaining the proper gas pressure in the chamber, a DC glow discharge is established about the workpiece. A portion of the evaporated plating metal is ionized in this region and accelerated toward the workpiece, thus forming an adherent aluminum coating on its surface.

Island of Automation. An island of automation is a single piece of automation that is working independently of any additional process.

Also see: Automated Assembly, Industrial Robots.

ISO. See: International Organization for Standardization.

Iteration. Iteration is a process of problem solving in which the same calculations are repeated over and over with gradually narrowing results until a precise solution is found. The iteration process is a process of repeating computations in which output of each step is input to the following step.

J

Jig. A jig is tooling usually considered to be a stationary apparatus. Jigs assist in the assembly or manufacture of a part or device. This differs from a fixture, which may be movable or may rotate. A jig for arc welding is usually constructed of electrically conductive material so the ground clamp of the arc welder can be attached to the jig instead of directly to the workpiece. A jig used in subassembly or final assembly might provide assembly aids such as alignments and adjustments.

Some activities cannot be performed without the aid of jigs. Tasks such as gluing, welding, and other fastening and joining may require jigs and fixtures. A stationary mold might be considered to be a type of jig.

Since a jig is a fabrication aid, it is desirable to minimize the cost of creation of a jig. Reusable or general-purpose jigs provide some decrease in cost. Using inexpensive materials can reduce the material cost, but the design and construction time might still be considerable. If a jig is to provide alignment and repeatability for assembly, the quality of construction is a required attribute.

Also see: Fixtures.

Jig Boring. Jig boring originally pertained to tool (jig and fixture) manufacturing, but the continually increasing demands for accuracy within many branches of metalworking has extended the application possibilities for jig boring machines. The importance of the jig boring machine in manufacturing has been firmly established. Without its aid, the present day state of the art in precision metalworking could never have been achieved. The modern jig boring machine brings into close agreement the professional disciplines of the machinist and the metrologist.

Jig boring machines are used for a wide range of applications. The locating and measuring features of the machine are employed for establishing the dimensional detail of workpieces, including: jigs used for the production machining of multiple parts; press tools such as lamination dies; and gages used to qualify parts produced on other machines.

These machines also are used for the production of:

- prototype parts needed before custom tooling can be designed and manufactured;
- parts for which the required accuracy of hole location and surfaces, as well as the quality of the surface finish, cannot be otherwise obtained;
- parts calling for the ultimate in dimensional integrity, such as mating components in an assembly;
- delicate or complex parts with a minimum of distortion, and
- parts, including die components, machined prior to hardening to allow for the more efficient application of jig grinding for finishing.

In general terms, the jig boring machine employs a precision spindle to drive the cutting tool and a table to support the workpiece. The table and spindle are movable and are fitted with built-in measuring devices that provide the means for establishing X, Y, Z, and A coordinate positions. The machine is designed to locate and bore holes and to generate surfaces to the highest level of accuracy. Three basic designs of jig boring machines in common use are: open-sided (C-frame), adjustable-rail, and fixed-bridge construction.

Open-sided construction jig boring machines of the C-frame design employ a single column for supporting the machine's vertical spindle and housing assembly (see the figure below). Guideways in the column control the perpendicular alignment of the spindle centerline throughout the full range of its adjustment along the Z axis.

The machine table is supported on a compound slide and is movable along the X axis. The compound itself is supported on the machine base and is movable along the Y axis. Coordinate settings locating the table under the spindle's vertical centerline are controlled by the linear positioning system for each axis.

Next, consider the adjustable-rail construction. On planer-type jig boring machines (see the figure on the next page), the crossrail is supported and adjusted vertically on two columns. The rail serves to carry the vertical spindle in its housing along the Y axis. The table is supported on the base of the machine and is movable along the Y axis.

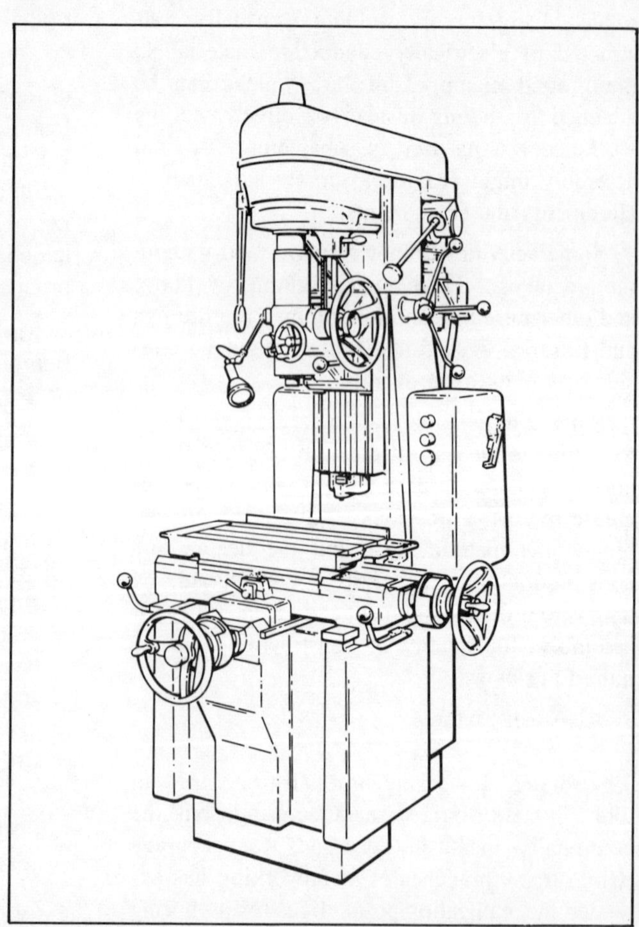

Jig boring machine of open-sided construction.

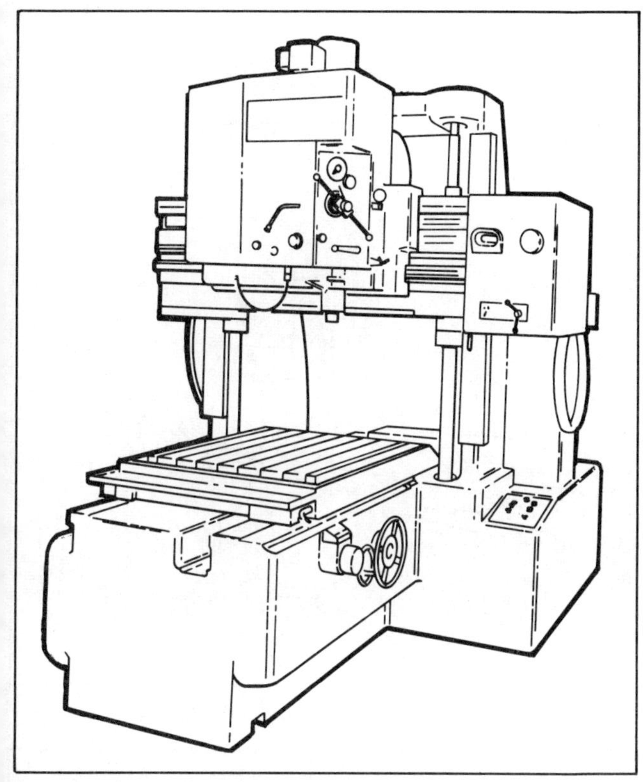

Adjustable-rail or planer-type jig boring machine equipped with graduated-scale measuring system.

On fixed-bridge construction jig boring machines of the design illustrated on the next page, the worktable is mounted on the base guideways and transverses in the longitudinal (X axis) direction. The spindle is supported on the cross-slide carriage and travels with it in the transverse (Y axis) direction on the guideways of the fixed bridge. Vertical guideways, an integral part of the cross-slide carriage, support the spindle housing and guide its vertical adjustment.

Jig boring machines equipped with numerical or computer numerical control systems (see the figure on the next page) are effectively employed when the job process can be preplanned. Machine functions for coordinate positioning and contouring operations are automatically controlled, thus relieving the operator of the need to attend to tedious, repetitive setting of machine dials and other control devices.

Production output of NC machines can be predicted with greater certainty since their operation is less dependent upon the operator. The precision machining of curvilinear details in cams, templates, and press tool components can be developed efficiently. Many jobs exist that would be impractical to process on a manually operated jig boring machine. One job, for example, permits precise, irregulary curved forms to be generated on cams or master templates without operator involvement.

Also see: Boring, Boring Machines, Numerical Control.

JIT. See: Just In Time.

Job. A job is the smallest definable increment of scheduled work. While a job can be further broken down into tasks, the job represents some

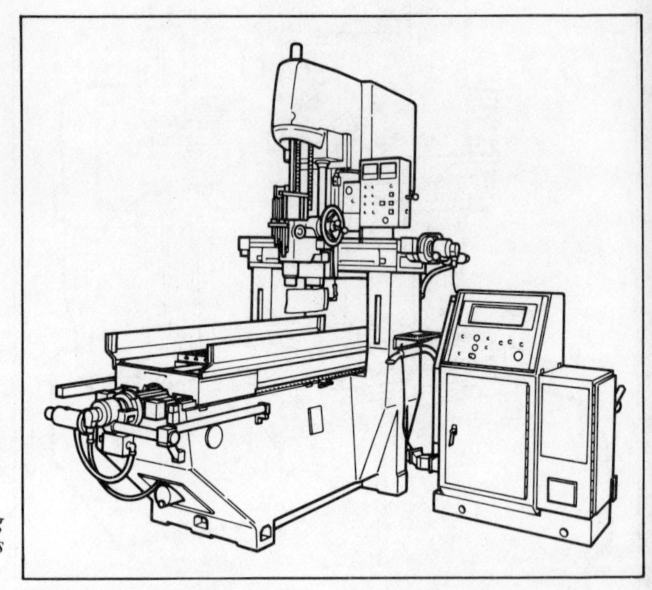

Fixed-bridge design of a jig boring machine on which the worktable traverses in the longitudinal direction.

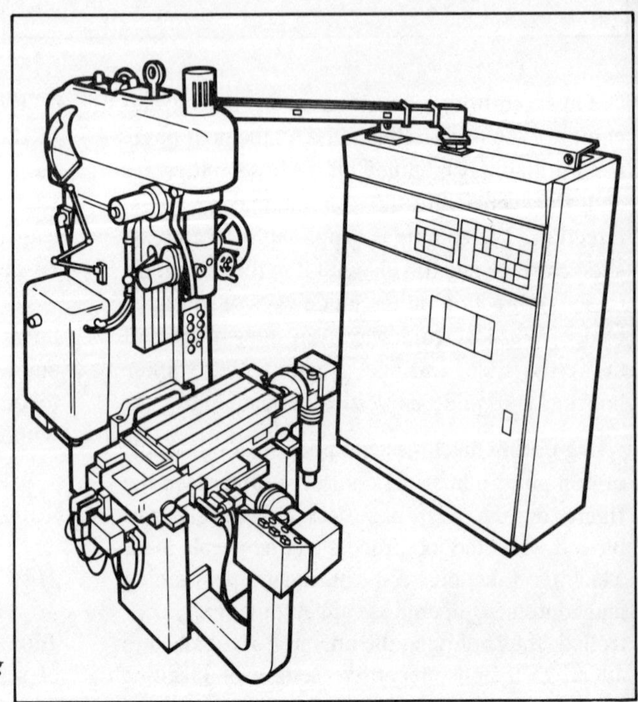

Numerically controlled jig boring machine of the open-side type.

amount of work which must be maintained and tracked as a single reportable unit. A task is usually not independently scheduled by an MRP II system. Time and cost accrual is usually done in terms of a job. A job may be assigned to a single person or department. The precise definition of a job is generally agreed upon by the particular people involved, so there is no universally identical definition. It may be given a unique identifying number for tracking purposes.

Concepts such as JIT influence smaller job sizes and quicker flows through the production cycles. MRP and MRP II systems use the job as an increment of work which can be scheduled and planned. Both the content of the job and the relationship among jobs is controlled by an MRP II system.

Jobs are also often called *lots*.

Job Control Language. Job control language (JCL) is a problem-oriented language designed to express statements in a job that are used to identify the job or describe its requirements to an operating system.

Job Control Statement. A job control statement is a computer statement identifying the job in a *job stream* and describing the job's requirements to the operating system of the computer.

Job Shop. A job shop is a business that provides services in small production quantities, single quantities, and prototype services. A job shop usually does not have the higher overhead costs of a mainline business company. A job shop sometimes specializes in certain types of services or processes. Job shops usually expect a highly unpredictable flow of work, sometimes employing people in a flexible manner. Job shops specialize in quick turnaround of jobs, which sometimes costs a premium over longer lead time establishments. The concept of a "job" being the smallest production quantity is the origin of the term job shop.

Job shops usually form around large industries such as the automotive industry, the electronics industry, and the metal fabrication industry. Job shops are sometimes the first to try some new processes, since they do not have to necessarily make a long-term commitment to the change, the way a major corporation might. The advantages of being small also are disadvantages since most job shops have a smaller cash flow than a larger firm and less expendable income for research and development.

Also see: Job.

Joint. A joint in robotics is a point at which two members of the manipulator's mechanical structure are united. These juncture points may be designed for two separate links of the manipulator to intersect for transverse actions to each other; to telescope for extension or rotate or pivot radially at a single point.

Also see: Industrial Robots.

Jog. There are three definitions for this term. The first is a control function enabling the momentary operation of a drive to accomplish a small movement of the driven machine. Secondly, a jog is the manual movement of a selected axis to accomplish a small movement of the axis. Movement can be in either direction at varying feed rates. Lastly, a jog may be an intermittent motion imparted to the slide by momentary operation of the drive motor after the clutch is engaged with the flywheel at rest.

Jog Switch. A jog switch is a switch for manually advancing machine members in small increments or for actuating functions that are part of a full cycle.

Jointed Arm. In robotics, a jointed arm is distinguished by a series of links and joints that rotate on a base to allow horizontal (X) sweep but achieve vertical travel (Z) and reach (Y) by rotating its joints.

Also see: Axis, Industrial Robots.

Joystick. A joystick is a data entry device employing a hand-controlled lever used to manually enter X, Y, and Z coordinates.

Also see: Axis, Computers.

Jump. Jump refers to a conditional or unconditional instruction which, when executed, causes the control unit to terminate one sequence of instructions and begin another sequence at a different memory location.

Just In Time. Just In Time, (or JIT) has been the source of great excitement in manufacturing over the past few years. Some feel that JIT is the golden ring to seeking manufacturing efficiency. While that mystique may be justified to some extent, many companies go into JIT with erroneous perceptions of what it is.

JIT is not an inventory program, an effort that involves suppliers only, a cultural phenomenon, a materials project, a program that displaces MRP, or a panacea for poor management. Rather, it is an enterprise-wide operating philosophy that has as its basic objective the elimination of waste. Under JIT, waste is considered anything other than the minimum amount of equipment, materials, parts, space and workers' time essential to add value to a product. The key operational phrase thus becomes ''adding value.'' JIT strives to identify activities not adding value and to eliminate them.

The simplest way to identify nonvalue adding activity is to apply a few tests to each step of a routing: Does an activity (for example, inspection, transport, receiving, purchase ordering and expediting) add cost without changing the physical or chemical characteristics of the item? Does a piece stop or pause at any point during its journey through the manufacturing site? Is an operation performed to compensate for anything that was not done properly the first time? If the answer to any of these questions is ''yes'' then the activity becomes a candidate for creative problem solving and thus, elimination. This philosophy points to two basic tenets of JIT: Identify and eliminate the ''cause for pause'', and if you can't use it now, don't make it now.

History. It is a belief among some that Just In Time was developed and refined in Japan. In reality, JIT originated in the United States back in the 1920s.

Henry Ford, in an effort to establish his company without external financing and to combat the depression of 1920, conceived and executed the principles that are the foundation of JIT as we know it today. The key JIT principles the Ford Motor Company employed were reducing production cycle time and eliminating waste. One of the most noteworthy accomplishments in keeping Ford product prices low was the shortening of the production cycle. During the business depression of the 1920s, the Ford production cycle was reduced from 21 to 14 days. The book, *Today and Tomorrow*, by Henry Ford provides an exciting commentary on how Ford successfully applied the principles we know as JIT at the Ford Motor Company.

However, for many reasons, the use of JIT did not expand in the United States until the early 1980s. It took the Japanese, who achieved worldwide supremacy using a JIT operating philosophy, to awaken anew JIT interest in the United States. Japanese adoption of a JIT operating philosophy is both clear and predictable.

Most of us know that Japan is small, cramped and a resource poor country. The combination of enormous human resources with few natural resources may help to explain the Japanese adoption of JIT. The Japanese make do with little and avoid waste. The modern Just In Time approach, featuring ''hand-to-mouth'' management of materials with total quality control seems in character with the Japanese historical penchant to conserve. To the Japanese factory worker, JIT objectives seem reasonable, proper, and easy to accept. JIT attempts to control costly sources of waste of idle inventories, defective parts and non-value added operations.

In contrast to Japan, Western countries have abundant space, energy and material resources. High-performing manufacturing companies learned to cultivate consumer demand for change and variety. Thus they held goods and parts in inventory in order to be responsive to changing consumer demand. With low interest rates, inexpensive materials, and plentiful storage space, the strategy was affordable. As Western consumers became more accustomed to annual style changes and ''planned obsolescence,''

ESSENTIALS OF JIT
Stable final assembly production
Limited batch production
Short supply and production lead times
Most reliable supply and quality
Integration with Materials Requirement Planning or equivalent system.

IMPLEMENTATION STRATEGIES
Make all problems visible
Incremental progress-start simple
Involve people (action groups).

STAGES
Set up time/lead time reduction activities
Layout changes (Industrial engineering involvement)
Quality Assurance Program (preferable using process control)
Balance production
In house JIT Inventory Management
JIT using supplier.

VENDOR/JIT PHILOSOPHY
Fewer suppliers
Long term commitments
Responsibility at source for quality of product
Cost rather than price
Component rationalization
Stream line documentation procedures

An overview of Just In Time.

a "throw-away society" replaced earlier generations of careful quality-conscious buyers.

Turning out new goods and keeping well-stocked shelves of finished goods and components became acceptable. Waste in the form of defective parts or shelves full of "passable" ones, was not a dominant concern. Hence JIT was not a necessary philosophy for success in Western countries. The situation changed dramatically as Western consumers increased purchases of the lower priced, higher quality goods produced by JIT methods in Japan. As demand for American goods declined, the competitive reality set in and created the incentive for American companies to adopt JIT.

Discussion. JIT requires a comprehensive program that is carefully planned and far-reaching in scope. While every manufacturing operation is slightly different, JIT implementation typically involves 10 distinct steps, as indicated below.

Step One: Plan Opportunities. JIT shakes the very foundation of manufacturing as it is currently practiced and challenges most of our long-standing operating principles. Such a profound changes does not lend itself to a cherry-pick approach that is championed by an individual functional manager or supervisor. Significant change management requires a carefully conceived plan that provides a clear strategic advantage in the marketplace. Such a strategy should become manufacturing's contribution to the business plan.

The first phase provides the base and direction of the program. It involves gathering facts, assessing the opportunities and delivering an enterprise-wide, time-phased plan that is reflected in a multiyear pro forma improvement projection.

The next phase, implementation preparation, is essential for the achievement of continuous improvement. It is here that the productivity organization is trained in improvement leadership and that performance tracking mechanisms are installed. The third phase launches the actual improvement program by establishing directed task groups with specific project assignments.

As the program matures, there is a need for re-evaluation of the priorities and, occasionally, for the rededication to the principles of JIT.

Step Two: Organize For Success. A productivity organization identifies and implements operational improvements. This organization typically consists of four levels:

Chartered Organization. Each and every person in the organization should understand where the company is going and what his or her specific contribution is to the improvement efforts. The input for the charter comes from the first three phases of the planning and task initiation methodology.

Implementation Steering Committee. This group is charged with the execution of the JIT plan. It prioritizes and assigns problems that need to be remedied.

Task Groups. Numerous task groups are established. These are multidisciplinary and trained in specific JIT techniques such as variation research, setup reduction and procurement strategies. Each group, once assigned a task, recommends a specific solution to the steering committee. Once accepted, the line organization and the task group are assigned a new problem.

Improvement Vehicle. After a recommendation from the task group has been implemented, there is a need to continuously refine the improvement.

Step Three: Awareness/Education. There is a significant need for all personnel to gain both awareness of JIT technologies and receive education. The latter involves applying techniques introduced during this activity. Everyone should be involved in trying (and usually failing at first) to improve operations. In the long run, it becomes the only means to assure continuous improvement.

Step Four: Housekeeping. Housekeeping in the JIT sense means more than just a clean workplace. It is an effort to establish an attitude that each person is responsible for his or her equipment, ensuring that necessary tools are in the right place and that everything is clearly visible for all to see and have access to.

Step Five: Quality Improvement. This activity should commence early in the JIT journey. It takes more time to realize ''zero defects'' than any other element of JIT. There are some differences between JIT quality programs and standard ones.

In JIT, the emphasis is on continuously reducing variance for improved manufacturability. Staying within accepted or engineered tolerance limits is not the goal of JIT; rather JIT strives for elimination of variance.

The operator, not the inspector, becomes responsible for ''zero defects''. It is incumbent upon management to provide the necessary tools and mechanisms to enable the operator to manage the quality control process.

Whenever an error or problem is discovered, the operator should be allowed to stop the process and take immediate corrective action.

Proactive maintenance of equipment becomes the operator's responsibility. The quality principles above apply to preventive maintenance activities.

A policy of all functions being responsible for a new product launch not only assures proper quality from the start, but also should result in total manufacturability. Operators and supervisors require extensive training in quality assurance techniques. The long-range goal is to eliminate inspectors because they add no value to the product and they cause queues.

Extensive use of fail-safing techniques is employed.

Step Six: Uniform Plant Load. A significant contribution from the Far East practitioners of JIT was the concept of uniform plant load (UPL). The idea is simple: If you sell daily, then make it daily. This means that every model within the product group is manufactured in small lots on a daily basis. Whenever possible (and it is almost always possible), the end items are manufactured to demand and not to stock. In order to foster a make-to-demand environment, most companies have to dramatically reduce the manufacturing lead time.

UPL is not the master schedule or the final assembly schedule; it is the cycle time required to meet—but not exceed—demand. It is not how fast a machine or process can operate; it is the production rate for all components and assemblies that is synchronous with the demand rate.

The primary benefit of UPL is to eliminate indirect labor costs for managing or transporting excess inventory and to allow direct labor to operate multiple machines because these machines have typically been slowed down to the UPL rate.

Step Seven: Redesign Process Flow. To achieve rapid throughput and the production opportunities afforded by UPL, the process flow or functional layout usually requires rethinking. The objective is to eliminate any operations that do not add value and then group dissimilar but dedicated equipment together in a cell configuration. Wherever possible, cells should be dedicated to a product group (focused facility) so that maximum flexibility to UPL demand cycle time can be achieved. The UPL rate will vary as demand varies. Once the cell has been optimized for direct labor application (''Keep the man busy, not the machine''), the UPL rate determines the operation's rate in the cell; the higher the demand, the shorter the cycle time and the higher the number of operators assigned to the cell. The goal of a good cell is to always maintain the same labor hours per unit, regardless of the demand rate.

Step Eight: Setup Reduction. By dedicating equipment to product groups, setups are minimized. Where multiple components must be shared by the same manufacturing source, then it becomes necessary to significantly reduce the changeover time in order to economically make just enough for each day's demand.

The techniques of setup reduction are well documented, so they do not require repeating in this overview. Experience has shown that the most effective setup task groups are those that have the actual operators as members. The operator knows the problems, and by participating in the solution, he or she will make the improvements work.

Step Nine: Pull System. Once the UPL has been established and the cells put in place, it is usually time to establish the pull system. Occasionally, it is advisable to select a few critical part numbers and institute the pull system prior to cell completion. However, the full benefits of negative feedback (self-regulating mechanisms) will not be realized until the UPL and cells are in place.

There are two types of pull systems in JIT. The first type is ''overlapped'', which is when continuous flow has been established on a line or cell, then the empty space previously occupied by a part is the signal to make one more part. The second type is ''linked'', which is when parts compete for the same resource and cannot be made one at a time, or when they have to travel significant distances in a lot mode, then it is advisable to use a pull card or signal (Kanban) to authorize the manufacture of the components. When a higher-level work cell requires more parts, it goes to get the replacement container. When picking up the parts from the point of manufacture, a card or signal is left behind. This card becomes the authorization to make one more fixed quantity container's worth of parts. Priority control is the sequence in which the authorizations were received (first in, first out) (FIFO). Violations of the sequence cause the entire system to collapse. Furthermore, if there are no authorizations to make, nothing is made.

The pull system causes some changes to MRP II but does not negate its overall usefulness. MRP II is still required for long-range parts and capacity planning. The pull system manages the short-term parts requirements and execution.

Step Ten: Supplier Networks. The last activity of the JIT cycle is to get the supply continuum involved in delivering only when needed. Quality improvement commences at the beginning of the JIT program, but the pull system is not implemented until the flow and demand have been sufficiently smoothed out. The key criterion for specifying JIT deliveries from the supplier is a smooth demand at a commodity level, which is the level where co-op contracts (JIT supplier agreements) should be created (rather than at the part number level).

To become flexible to the customer's demand, the supplier must aggressively implement JIT also. The customer plays a significant role in the supplier's JIT implementation.

Applications. Just In Time in a general sense applies to any kind of production. Improving process capabilities, reducing setup times, and similar activities are useful to any manufacturer.

JIT promotes repetitive manufacturing. Over time, manufacturers have transformed a job shop operation to repetitive production as their volume increased. Henry Ford made the transition with his first assembly line. This conversion means reducing the inventory between operations, transferring material in a flow, and decreasing the throughput time. One way to think of it is as an attempt to make an entire industry perform as if it were part of a single assembly line.

There is, however, a great deal to be learned from JIT which can be applied to a job shop. JIT has not been applied correctly to any industry, so it is a question of how smoothly a job shop can be expected to perform. Many manufacturers who operate by job shop methods may not need to do so. It is worth a review to determine if job shop production is potentially repetitive. Even if that review turns out negative, most of the principles of JIT apply except for uniform plant load.

There are valid reasons that require job shop methods. Some of them are: unique customer orders which require custom engineering; operations performed are "on the edge of the state of the art," which suggests that yields are poor and sometimes unpredictable; quality control requires inspection or holding in lot sizes; and small-volume orders arrive at irregular times. While a valid condition exists that demands a job shop operation, it may not always be true. A review of JIT principles may show a way.

JIT applies to a substantial part of the operation. JIT may apply to a combination of types of manufacturing. It is most fully developed when used to perfect production that is already repetitive. As with almost anything that is difficult, it is easy to disclaim applicability.

JIT requires a positive attitude, a determination to seek out problems and solve them, and nonacceptance of the belief that barriers to a smooth flow of production are permanent. Companies which have embarked on JIT have rarely attained all they wished to do, but they are much better off than if they had decided not to try it at all.

Glossary. *attachment:* A die, tool, adapter, or extension that may be exchanged at the time of a setup. The machine, remaining unchanged, usually provides frame, power and master control.

backflush: The deduction from work in process inventory. The component parts used in an assembly or subassembly by exploding the bill of materials by the assemblies produced.

balancing operations: The activity that tries to match actual output cycle times of all operations to the cycle time of use for parts as required by final assembly.

broadcast system: A sequence of specific units to be assembled which is communicated to supply and assembly activities so that each part merges with the correct assembled unit.

cause-and-effect diagram: A statement of a problem or symptom with a branching diagram leading from the statement to the potential causes.

continuous flow manufacturing: Production of nondiscrete material without interruption as in extrusion, distillation or rolling operations.

control chart: A statistical device primarily used for the study and control of repetitive processes. It is designed to reveal the randomness or trend of deviations from a mean or control value.

cycle time: The time between completion of two units of production. For example, the cycle time of computers assembled at a rate of 20 per hour would be one completed computer every three minutes.

external setup time: Elements of a setup performed while the machine is running.

fail-safe work methods: Methods of performing activities so that only correct actions can be completed. For example, a part with holes in the proper place can lock in a jig.

flow rate: The inverse of cycle time, for example, a cycle time of 120 units per hour would have a flow rate of two per minute.

immediate feedback: The principle of checking production right after a part is made to provide information or deviations from standards so that correction is immediate. The concept extends to automatic shutdown of a process in case of many

 Just In Time

defects or a drifting out of a desired control range.

inbound stockpoint: Used with a pull system, a defined location next to the place of actual production to which materials are brought as needed and from which material is taken for immediate use.

internal setup time: Elements of a setup performed while the machine is not running.

inventory level: In stockless production, this term applies to work-in-process inventory levels.

job shop manufacturing: The production of discrete units planned and executed in irregular-sized lots, usually for a customer order.

lead time: The span of time required to perform a group of activities. It includes time for order preparation, receiving, inspection and transport.

level schedule: A schedule such that the use of all parts in materials for a given period of production is as evenly distributed over time as possible.

line balancing: Reassigning and redesigning work on an assembly line to make work cycle times at all stations approximately equal.

linear production: Actual production to a level schedule, so that plotting actual output versus planned output is even.

machine: Those basic elements of equipment that do not change during a setup.

move card or move signal: A card or other signal indicating that a specific number of units of a particular part number are to be taken from an outbound stockpoint and taken to a point of use or inbound stockpoint of another cell.

outbound stockpoint: Used with a pull system, a defined location near the point of production on a plant floor to which produced material is taken to await pickup.

pipeline stock: The amount of stock between the point where it is made and the point where it is used.

process capability: the basic physical capability of production equipment and procedures to hold dimensions and other characteristics of products (such as electrical resistance) within acceptable bounds for the process itself. Not the same as tolerance required of the produced units themselves.

process control: The ability to maintain a process within a given range of capability by feedback and correction.

process manufacturing: Production which adds value by mixing, separating, forming or chemical reactions in either batch or continuous mode.

production card or signal: A card or other method of notification that a specific number of units should be made in a pull system.

production network: The complete set of all cells, processes and inventory points, including all suppliers, from the point of completion of the product all the way back to the raw material.

production rate: The number of units completed per period of time; same as flow rate.

production time: The time for a production card to make a complete circuit through a production cell allowing for all the operations necessary for the production (i.e. setups, inspection, material placement).

pull system: The production of items only as demanded for use or to replace those taken for use.

push system: the production of items as required by a given schedule planned in advance.

queue time: The time a job waits before processing at a work center.

repetitive manufacturing: Production flow or similar units planned and executed by a schedule, or by arate.

safety stock: A stock of items known to be of good quality; kept in reserve and used to replace items found defective in further processing.

setup time: The time required for a machine, assembly line, or work center to convert from production of one specific item to another. It is the elapsed time from production of the last good unit of Item B until production of the first good unit of Item A.

signaling system: A set of signals to perform specific actions in production at specific times or in response to need for material.

small group improvement activities: A name for small groups of 5-12 people who normally work as a unit for the purpose of seeking and overcoming all kinds of production problems associated with their work.

stockpoint: A designated location into which material is placed and from which it is taken. Not a stockroom; rather a way of organizing material for easy flow-through production.

throughput time: The elapsed time from which material starts into a process until it is finished, typically through a work center.

tools: Items (wrenches, for example) used during a setup but which are not part of the machine.

time bucket (or bucket): A length of time summarized in a production plan or material requirements plan; usually in days, weeks, months, or quarters.

variance (production): The deviation of actual occurrences from plan for all kinds of reasons stemming from both known and unknown causes.

variance (statistics): The arithmetic mean of the squared deviations of individual items from their arithmetic mean or from a forecast value.

waiting time: As used in stockless production, it is move time or transport time, and idle time spent waiting to be used.

Also see: Manufacturing Resource Planning.

K

K. In data processing, K is used to define storage capacity, two to the tenth power, 1024 in decimal notation. For example, information can be stored in 1024 locations in 1K computer memory, in 2048 locations in a 2K memory, etc.

KANBAN. (Pronounced kahn-bahn) KANBAN is a Japanese word which literally translated means "visible record" or "visible plate." The word more generally is meant to be taken to mean "card" or "billboard" or "sign." KANBAN is principally a scheduling approach which uses standard containers with a card attached to each. The system of using KANBAN was developed in Japan and has been used in conjunction with a Just In Time manufacturing system.

The KANBAN system is a pull rather than a push planning system. A push system is principally a schedule-based system. A pull or Just In Time system is based on providing parts when they are needed but without guesswork. Authorization is provided by alerting the work center that material is needed. KANBAN is a tool by which this requirement is communicated to the previous operation or vendor which fills the need. KANBAN can be used with most discrete part manufacturing systems, but many process systems must be run continuously to maintain proper and adequate quality. A steel mill, for example, may require days to start up and reach steady-state production.

Karnaugh Map. A karnaugh map, used to design logic, is a truth table rearranged to show a geometrical pattern of functional relationships for gating configurations. It facilitates recognizing essential gating requirements by detailing similar logical expressions and thereby allowing duplicate logical functions to be combined.

Kernal. Kernal is a functional subset of a computer system such as the graphics processing portion of the input/output handler. A kernal is a logically separated segment which can be modified and enhanced separately from the remainder of the system, as well as being able to be incorporated into larger systems. A kernal is typically only useful to a software programmer who will incorporate the kernal into the remainder of the system. A subroutine is not the same as a kernal, unless the subroutine comprises some useful purpose.

Keyhole. Keyhole is a technique of welding in which a concentrated head source penetrates completely through a workpiece forming a hole at the leading edge of the molten weld metal. As the heat source progresses, the molten metal fills in behind the hole to form the weld bead. Keyhole welding cannot be accomplished with welding processes using exposed electrodes.

Keypunch. A keypunch is a keyboard-activated device which punches holes in cards to represent data.

Kinematic Analysis. Kinematic analysis is a process for plotting or animating the motion of part in a machine or structure order design. Kinematics simulation allows the motion of mechanisms to be studied for interference, acceleration, and force determinations which are still in the design stage.

Knife-edge Bandsawing. Knife-edge bandsawing is a variation of bandsawing that uses a sharp-edged blade to cut soft or fibrous materials that tend to tear or fry.

Knockout. Knockout is a mechanism for releasing workpieces from a die; it is also called ejector, kickout, liftout, or shedder. Also, a removable piece on an electrical panel to allow cables to enter.

Knowledge Engineering. Knowledge engineering is the process of eliciting knowledge from individuals who are deemed to be experts in their field and structuring that knowledge into a compatable form.

L

Label. A label is an ordered set of characters used in a program to identify the location of an instruction, routine, item, file, message, or record.

Ladder Diagram. A ladder diagram is a description of the programming technique used to program many U.S. designed programmable logic controllers (Europeans use statement-language type programming, generally) which consists of describing the logical elements of the sequence of actions and boolean functions as rungs on a ladder where there is a limit to the number of activities per rung and some limit to the number of rungs. While the ladder logic program is written from left to right and secondarily from top to bottom, the programmable controller performs the logic solving by solving the first activity in each rung from top to bottom, and then solving the second element on each rung, etc., until reaching the last element in the last rung. Because of this type of sequential solution, an element may be solved before another because its containing rung occurs first, even though the elements are at the same location within their respective rungs. Furthermore, the first element of the last rung is solved long before the last element of the first rung. This sequence can obviously be altered with skips and jumps, however. The process of sequentially solving each element in order is part of the scan of the programmable logic controller, which also includes any diagnostic routines and reading and writing to the I/O.

Also see: Programmable Logic Controllers, Programming.

Lag. Lag describes the difference in real time between a response relative to an input. Lag is the tendency of the dynamic response of a passive physical system to respond later than desired. Lag can also be thought of as the time parameter characterizing the transient response of a first order exponential system to a step input.

An example of lag is in a servocontrol system where the command (input) to the motion control system is given at a specific instant in real time, the signal is then obeyed by the motion actuator, with the resolver or encoder position indicating the attained physical position of the axis (output).

In product marketing usage, lag is used to describe the elapsed time between which a trend may begin or a market window may open, and the response by a company to take advantage of the opportunity.

Also see: Lead, Servocontrol.

LAN. See: Local Area Network.

Lancing. Lancing is cutting along a line in the product without freeing the scrap from the product. It is performed using a progressive die operated on a press. Lancing cuts are necessary to create louvers, which are formed in sheet metal for venting functions.

The figure at the top of the next page illustrates lancing.

Land. Land is the peripheral portion of the drill body between adjacent flutes. Land may also be

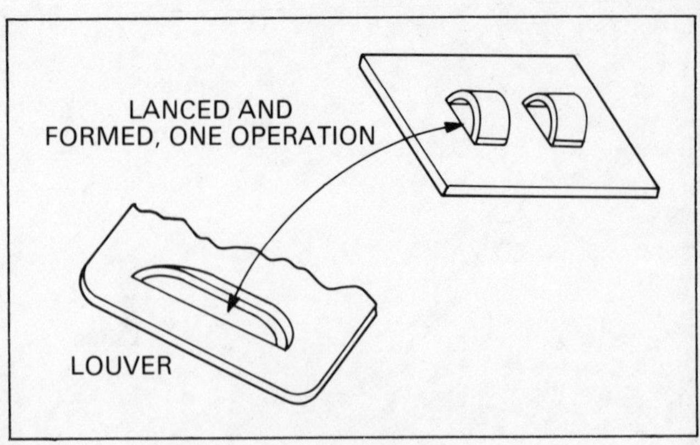

Lancing is cutting along a line in the product without freeing the scrap from the product.

the section of the reamer between adjacent flutes or the narrow surface of a profile-sharpened cutter tooth immediately behind the cutting edge. Land may also be one of the threaded sections between the flutes of a tap or the thickness of the top of the broach tooth.

Also see: Broaching, Drilling.

Language. Language is a set of symbols combined with specific rules necessary for their interpretation.

Also see: Ada, Algorithmic Language, APL, Assembly Language, BASIC, C, COBOL, Computer, Computer Languages, Conversional Algebraic Language, FORTRAN, High-level Language, Interpretive Languages, Language Translator, LISP, Low-level Language, Machine Language, PL/1, Problem Oriented Language, Procedure Oriented Language, Source Language, Symbolic Language.

Language Translator. Language translator is any assembler, compiler or routine that converts statements from one language into equivalent statements in a different language.

Lapping. Lapping produces exceedingly smooth and highly accurate surfaces, either flat, cylindrical, tapered, shouldered or contoured. A lap, usually of cast iron and charged with fine abrasive, is rubbed against the surface of the work either by hand or by machine. In many instances, a lap bonded together has been used to replace the metal lap, charged with loose abrasives. Abrasive-coated paper laps are also used. Laps are used with kerosene and lard oil lubricants.

Formerly limited in use to the production of precision surfaces in gages and measuring devices since 1930, lapping has been generally adopted as a production process in many industries, particularly the automotive. In this industry, it is used for the finishing of crankpins, bearings, piston rings, ball bearing races, value tappets and other wearing parts. Lapping tolerances of 0.00005 inch (0.0063 mm) in diameter and 0.000025 inch (0.0013 mm) in straightness and roundness are attainable.

Stock allowances in lapping should never exceed 0.0005 inch (0.013 mm); preferably not over 0.0002 inch (0.005 mm) for rough lapping or 0.0001 inch (0.002 mm) for finish lapping from a rough lapped surface. Preliminary operations should be grinding or honing and should be as accurate as possible relative to geometrical and location accuracy because lapping provides little correction for these factors.

Large Scale Computer. A large scale computer has a large (500 bytes or more) memory and multiple I/O cha:_els. Such computers process multiple programs concurrently.

Also see: Computers.

Large Scale Integration. Large scale integration occurs when integrated circuits are designed and built with complexities that require more than 10,000 transistors to duplicate. Specific sizes of large scale integration (LSI) are very large scale integration (VLSI) and ultra large scale integration (ULSI).

Also see: Integrated Circuit, Metal Oxide Semiconductor, Very Large Scale Integration, Ultra Large Scale Integration.

Laser Printer. A laser printer represents the latest technology in printers. Laser printers use a laser to create an electrostatically charged image on a drum. Toner which has adhered only to the electrostatically charged areas of the drum creates a visible image on plain paper or transparency.

Most all desktop laser printers use one of three manufacturer's laser engines; Canon, Kyocera, or Hitachi. Other manufacturers of laser printer engines are Fuji, Xerox, NEC, Ricoh Sharp, Konishiroku, Kyocera, and Casio. Canon has the greatest market share with manufacturers such as Hewlett-Packard and Apple using the Canon engine. Most laser printers are designed differently than photo copiers in that the imaging drum is part of a replaceable unit along with the toner cartridge. The reason for this was to avoid the need to have to periodically recondition or replace the drum as must be done with photo copiers. The typical duty cycle (number of pages a laser can print per month) ranges from 3,000 to 40,000 pages per month.

Laser printers for use with personal computers usually offer one of two types of interface protocol. One is the Postscript page description language developed by Adobe Systems, which uses mathematical equations to describe the images for the page. A second is the Hewlett-Packard Printer Command Language (PCL) which uses a bit-mapped representation of the entire image.

Currently, Canon and the others have introduced their second generation engines, which are better quality and less expensive than the first generation engines. A significant quality feature of the second generation engines is the ability to produce solid black in large areas, with no variations in color.

The resolution of laser printers is typically 300 by 300 dots per inch, which is significantly better than dot matrix printers. Furthermore, the resolution of the dots is extremely sharp and well-defined, with no fuzzy edges. Typeset resolution is usually 1200 or more dots per inch, but to the untrained eye, the 300 dots per inch laser printer output seems the same as typeset quality. The advent of affordable laser printers has spawned the desktop publishing revolution.

Also see: Desktop Publishing, Printers.

Lasers. A "solution looking for a problem," was how the laser was described in its early days and those words have proved prophetic. Every day a new use is discovered for laser technology and lasers are now applied to processes which had always seemed adequate until the laser highlighted their limitations. The laser is illuminating not only the frontiers of scientific, nuclear and space research but also providing an invaluable tool in everyday activities. Even when they are used simply for display purposes, lasers can be impressive. As the lasers are switched on, the night sky is crossed by shafts of light, so pure in color and so straight and well-defined that they seem like tangible rods.

All these properties stem from the process of stimulated emission which gives the laser its name–Light Amplification by Stimulated Emission of Radiation. Although the first laser was not built until 1960, the concept of stimulated emission had been proved theoretically by Albert Einstein as long ago as 1917.

Discussion. Light is energy in the form of electromagnetic radiation, emitted by atoms. To dispatch a photon of light (a packet of radiation) the atom must be excited by an input of energy that boosts its energy level in a series of definite steps called transitions. The wavelength of the photon of light emitted as the atom falls back toward the ground state (the unexcited state) depends upon the transition the atom goes through. Transitions at high energy levels produce long wave infrared light while those at low levels produce short wave ultraviolet light; visible light comes in-between.

In a conventional light source, such as bulb filament, the emission of light is a rather haphazard affair, as atoms fall back spontaneously towards the ground state-radiating photons with a tremendous variety of wavelengths in all directions and at random intervals of time. However, Einstein showed that an atom can be stimulated to emit a photon before it would do so spontaneously if it is hit by a photon whose wavelength corresponds exactly to the transition the atom will go through.

When this happens, the atom will emit a photon of precisely the same wavelength as the photon that hit it and the two will move off travelling in the same direction at the same time (in phase). This process can be repeated on many other atoms, to produce an identical photon each time, each of which can contribute to further stimulations, stimulated emissions can thus amplify one photon into a whole stream of photons of exactly the same wavelength, travelling in phase and in the same direction. This is known as coherent light.

When Theodore Maiman produced the first working laser in the spring of 1960, he used a solid rod of ruby. It has since been shown that almost anything, from krypton to carbon dioxide can be made to operate as a laser.

With the ruby rod laser, the ''lasing'' material is not the ruby itself but the chromium impurities it contains. The idea is to flood the rod with light energy from a flashtube to ensure that the majority of chromium atoms are in an excited state—in fact, at the third transition level. The atoms soon loose energy and decay to the second level and, for a while, there are more atoms in level two than level one (the ground state.) Sooner or later, one of the atoms in level two decays spontaneously to the ground state, emitting a photon which can stimulate other transitions to the ground state, as Einstein predicted.

This process of stimulation begins not just with a single atom, but with many thousands of atoms. The light, although of the same wavelength, shoots off in all directions. However, light that is not travelling parallel to the laser axis, or close to it, is lost through the walls of the crystal before it can stimulate much emission. On the other hand, light that is parallel to the axis is trapped within the rod. Mirrors at each end reflect the light back and forth, and as it traverses the length of the rod it stimulates more and more emissions, all moving in the same direction. By making one of the two end mirrors less than perfectly reflecting, the amplified light can escape from the rod in a thin, parallel coherent beam of great intensity and a single wavelength. This is referred to as a ''raw beam'' in the accompanying figure. Different lasing materials from ruby will produce light of different wavelengths. The raw beam is then sent through a focusing lens to machine or measure the workpiece. (See the accompanying figure).

Many mediums have lasing capability, but only a few lasers are powerful and reliable enough to be practical for machining operations. Lasers can be classified by their lasing medium. They are, solid state, liquid, or gas. The two types most used in machining are the discharge-pumped, carbon dioxide gas laser and the optically pumped solid state laser.

Gas lasers usually consist of an optically transparent tube filled with either a single gas or a gas mixture as the lasing medium. A typical commercial carbon dioxide gas laser contains carbon dioxide, helium, and nitrogen. Carbon dioxide supplies required energy levels for laser operation, nitrogen keeps the upper energy levels populated through collisions, and helium provides intracavity cooling. The pumping source is some form of electrical discharge applied by electrodes.

The carbon dioxide laser operates at a wavelength of 10.6 microns either pulsed or in a continuous wave (CW). In high-power gas lasers, the available output is limited by two factors: ability to cool the gas, and ability to properly stabilize the gas discharge.

Thermal energy upsets the lasing equilibrium of the gas. Cooling is necessary primarily to keep the lower laser level depopulated.

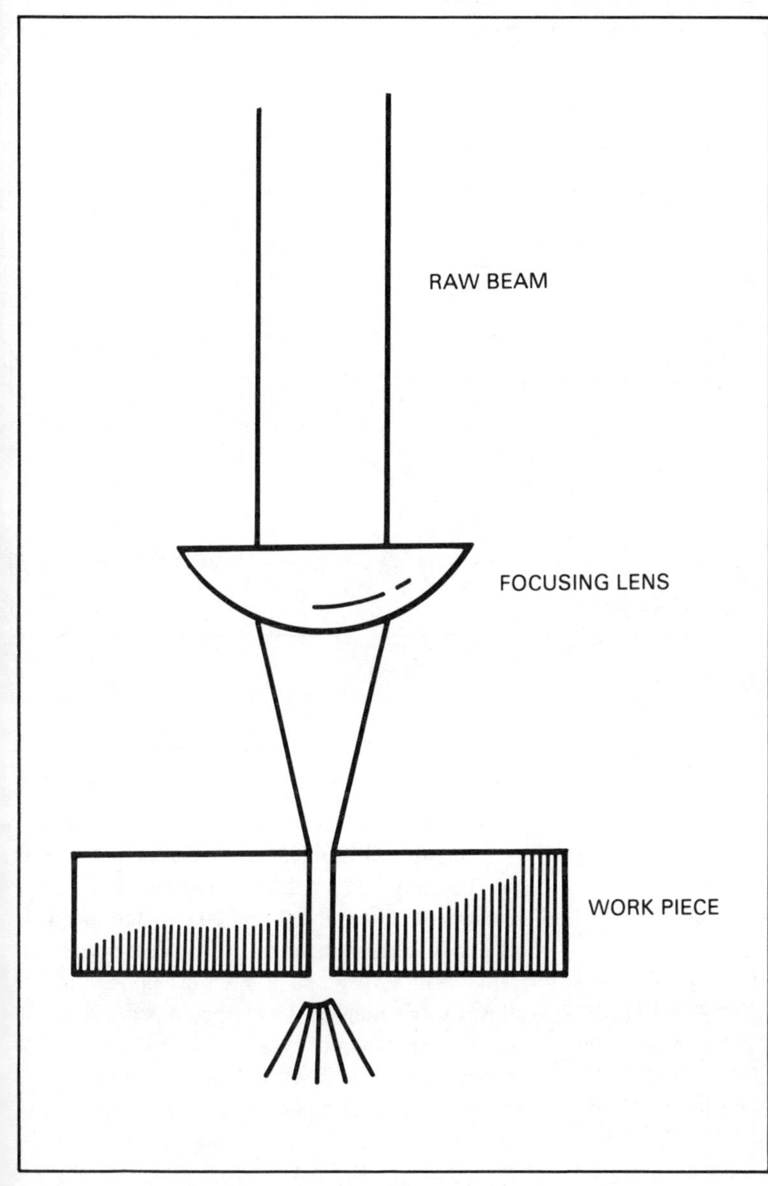

The basics of a laser material processing.

Solid-state lasers consist of a crystalline or glass host material and a doping additive to provide the reservoir of active ions needed for the lasing action. The original solid-state lasers used ruby (with approximately 0.05% chromium dopant) as the lasing medium.

Another common solid-state laser uses a single crystal of yttrium aluminum garnet (yag) doped with neodymium as the lasing medium. The neodymium-doped yag (Nd:yag) laser is relatively efficient, allows for high pulse rates, and can be operated with a simple cooling system. Both ends of the rod (lasing medium) are parallel and polished to high flatness. The Nd:yag solid-state laser operates at a wavelength of 1.06 microns. Most solid-state lasers operate only in the pulsed mode; however, the ND:yag laser may be operated either pulsed or in a continuous-wave mode.

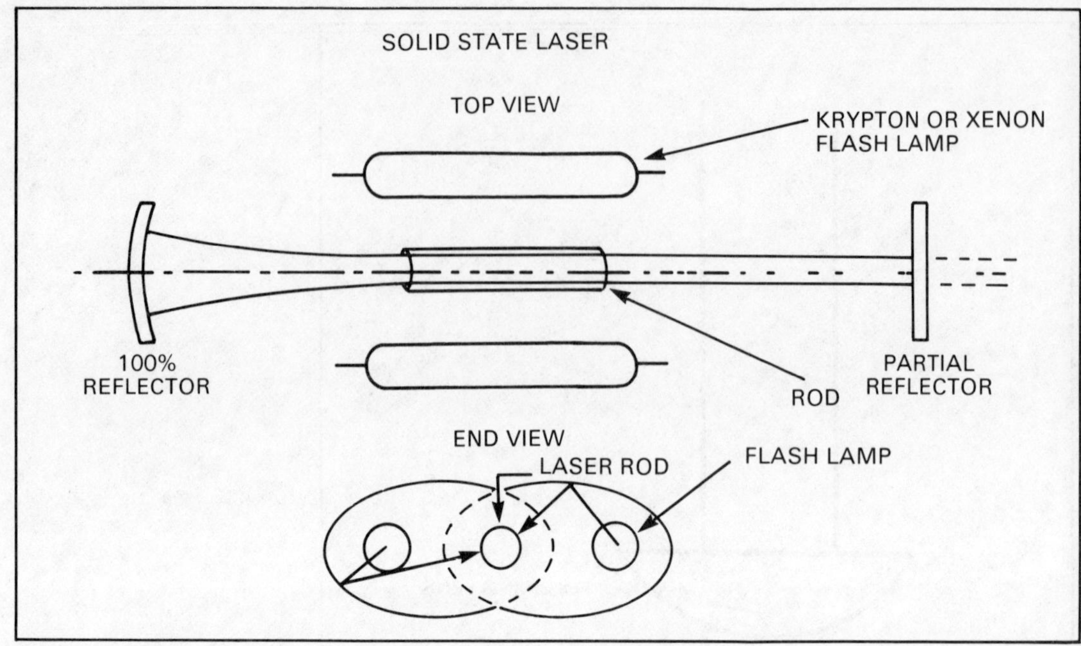

Top and end view of a typical Nd:YAG laser.

of 1.06 microns. Most solid-state lasers operate only in the pulsed mode; however, the ND:yag laser may be operated either pulsed or in a continuous-wave mode.

Since the Nd:yag laser material is electrically insulating, it must be powered by a means other than simple electrical excitation. Energy is injected into the laser medium to generate an intense light flux. The light is absorbed by the medium and collimated into a laser beam. The pump which optically excites the laser material is usually a krypton or xenon filled arc-discharge lamp. The figure above is a diagram of a solid-state laser system.

To efficiently use the light produced by the lamp, the laser rod and arc lamps are mounted in a reflective cavity. The cavity may be in the form of an elliptical or circular cylinder to focus the light from a linear arc-discharge lamp onto the laser rod as is illustrated in the next figure.

The energy source for the pulsed flashlamps usually consists of a DC power supply and a bank of energy storage capacitors. The capacitors are charged to a predetermined voltage. Their stored energy is then discharged through an inductor to limit current peaks, then through the flashlamp to create the necessary pump light.

In solid-state lasers, removal of waste heat is a fundamental problem (i.e., the excess energy not usefully converted into laser radiation). The radius of the rod is limited by the need to conduct surplus heat to its cooled periphery. This requirement sets a practical upper limit to the power which can be extracted per unit length of rod and thus per system. Waste heat may be taken up by a heat-sink in relatively small systems. Larger systems may use streams of fluids for carrying waste heat away from the laser cavity. Typical fluid cooling systems are water, water to air, water to water, and refrigerated recirculating water.

Laser systems require maintenance, of course. The majority of the system is conventional electronic and cooling equipment which requires little maintenance. Pumps, blowers, and mechanical equipment used to position workpieces and heads require conventional preventative maintenance. Equipment suppliers provide recommended service guides covering cleaning of optics, refilling of sealed plasma

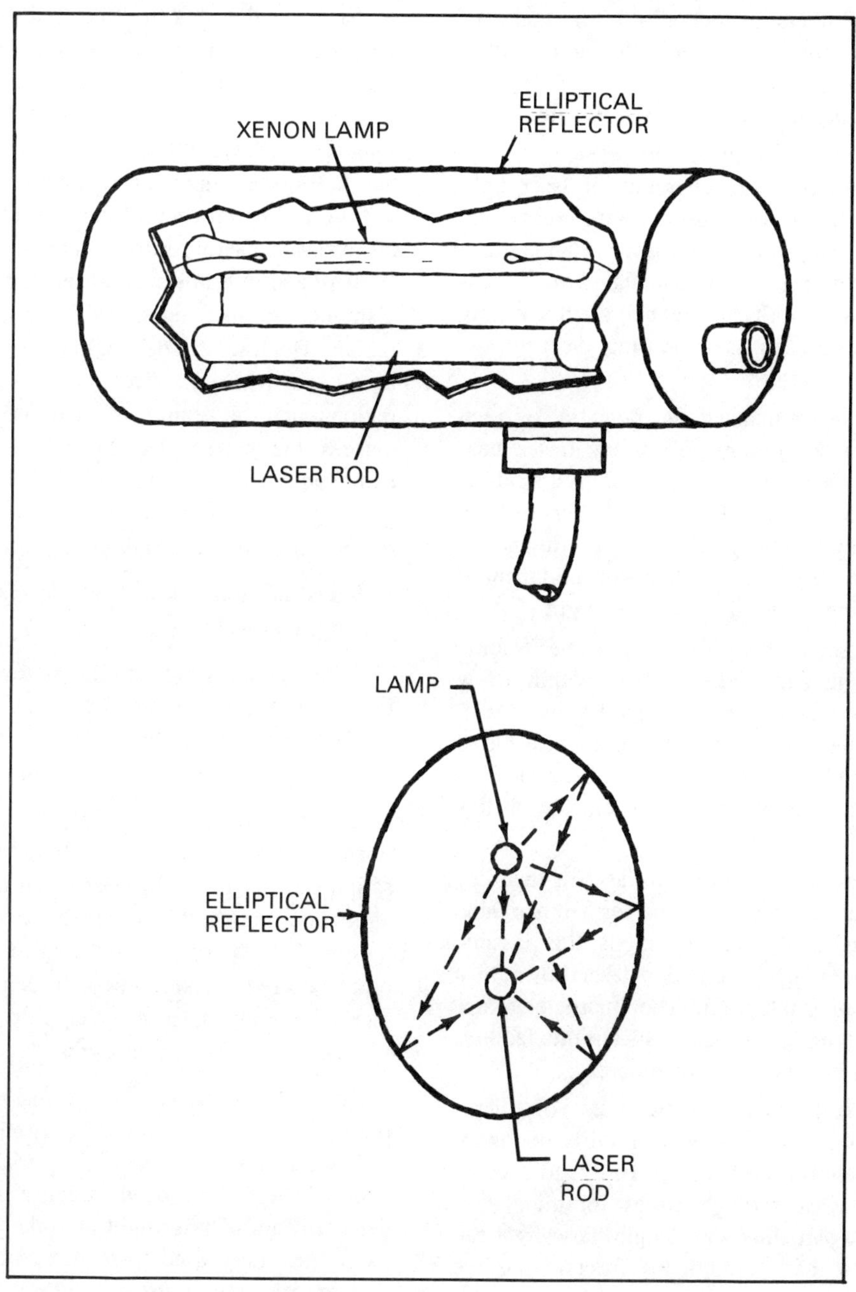

Eliptical focusing structure using linear lamp.

tubes, changing and maintaining of filters, recommended lubrication, and materials and schedules for mechanical components. Depending on size and operating cycles, the demands for gas mixtures for carbon dioxide lasers may be less than one cubic foot per hour (cfh) or could be up to thousands of cfh or more.

Gas is recirculated in systems which require large quantities of gas. In the case of flashlamps, some suppliers recommend changes after a given number of hours of operation, for example, "Change lamps after 200 to 400 hours." Other suppliers indicate expected number of "shots" between changes. Output is measured, and lamps are changed when output drops.

Applications. The coherence of laser light and its parallel beam endow it with great value as a measuring tool. If the laser is aimed at a target its distance can be calculated very accurately from the reflected light through a variety of techniques such as measuring the time the pulse takes to return.

Armies have realized the potential of such systems for accurate range finding. Laser measuring has been adopted in several civil fields as well, and the variety of people who benefit from its accuracy increases all the time. Engineers, for instance, use lasers to check the alignment of aircraft wings or to dig tunnels straight.

With lasers, geologists can plot the location of continents and land masses to within a few meters—a hundred times as good as with conventional techniques. It will soon be possible to detect whether continents are indeed drifting relative to one another at a rate of a few centimeters every year.

Another valuable feature of lasers is that they can concentrate energy to perform remarkable delicate and small-scale operations. It is possible now to focus the beam of a laser down to a diameter of 0.025 mm. The microelectronics business in particular benefits from this facility, through it is valuable in many areas.

Perhaps the most beneficial use of microwelding is in eye surgery. It is fairly common, for instance, for the retina of the eye to become partly detached through disease or injury. By choosing a particular wavelength, laser light can be made to pass through the front of the eye without any effect and focus on a small spot on the retina. The concentration of energy will then produce scar tissue to "weld" the retina back into place firmly and with minimum damage to the retina.

The laser has many important advantages that have motivated engineers to keep working on ways to use it in the manufacturing environment, primarily in the area of "materials working." The laser is a noncontact machine tool: nothing touches the part being worked on, and there are no saw blades or drill bits to break or keep clean. This also makes the laser attractive for drilling tiny holes or cutting patterns complex for a saw to follow. The absence of physical contact also makes lasers useful for machining materials that are hard and/or brittle such as ceramics and even diamonds, or soft and easily deformed such as rubber. The laser can deliver energy precisely to a tiny spot with little effect on the surrounding regions, and the beam can reach into otherwise inaccessible places. The laser is also proven compatible with computer controls, which are growing increasingly important as factories continue to become more automated.

Industrial applications include cutting, machining, and welding.

Laser beam cutting (LBC) is a process related to laser beam machining that severs metals by melting and vaporization when high intensity laser light is directed along a cutting line.

Laser beam machining (LBM) is a nontraditional process for cutting, drilling, slotting or scribing metal parts. In laser beam machining, an intense, coherent beam of single-wavelength light is focused on a small area of the workpiece, where the optical energy is converted into thermal energy intense enough to melt and vaporize the workpiece material.

Laser beam welding is a fusion welding process for making structural, assembly, sealing, and conduction welds in material up to 3/4 inch (19 mm) thick by directing a high power density beam of laser light onto the joint. Inert gas is generally used to reduce oxidation, but filler metal is rarely used.

Drilling holes was one of the early uses of the laser beam. Not very useful, but one which was used to attract attention to the process, was the "punching" of holes in razor blades. Since that time, laser drilling has become more sophisticated. Holes continue to be drilled in hardened steel and are now drilled in all metals (including many high-temperature alloys), plastics, paper, rubber, ceramics, composites, and crystalline substances such as diamond.

Several methods are used for cutting holes. The technique used depends on hole size and shape.

Laser systems are available for cutting round holes, welding, and making perforations in a circular pattern. These devices typically rotate a focusing lens in a horizontal plane (or the plane of the lens) on an axis coincident with the incoming stationary beam. The focused spot will always be on the focal axis of the lens, and will be rotated in a circle with the lens. The rotating lens assembly typically is driven by a variable speed motor and may be equipped with a gas jet. The effective radius of operation is limited by the lens size used. For holes larger than one inch (25.4mm) diameter, it is usually better to rotate the mirror.

Lasers will cut hard or soft, tough or brittle, stiff or resilient materials. It does not matter whether the structures are strong or weak.

Laser cutting is usually assisted by a flow of gas. The flow of gas performs several functions such as cooling the area around the cut and blowing away swarf and slag. Reactive gases increase cutting speed. Self-burning of the material, however, can cause poor kerf quality.

The oxygen supplied in gas jet-assisted cutting enhances the cutting of steel for several reasons. Oxygen reduces reflectivity, helps initiate the exothermic reaction once the metal reaches a high temperature, cools the material around the working area, directs molten metal and vapor away, and sweeps away molten slag from the bottom of the cut.

It is important that the proper gap between the gas jet and the workpiece be maintained. This may be accomplished with self-adjusting, height-sensing units that control the gap automatically regardless of surface uneveness.

Motion control of the laser beam in relation to the workpiece is usually accomplished by an optical tracer or a numerical control system. Numerical control systems vary widely in their sophistication. Some simply control X-Y axes of motions, while others may be coupled with a computer system which assists in part layout on a sheet in order to maximize material utilization. Numerical control systems also establish the cutting path based on dimensional and geometric input, determine stock status or availability of most economical sheet stock, establish shop loading schedules and report job status, maintain cost data, and prepare shipping and billing papers.

For the die shop, considerable savings are envisioned in making steel rule die blocks. Not only can very intricate shapes be accurately cut, but there is also the opportunity to use NC to permit production of these die blocks without requiring time consuming layouts.

In addition, depending on laser power density, cutting is accomplished by vaporizing or burning the wood. Vaporization is preferred, because wood from the kerf is then removed so rapidly that there is no time for heat to be transferred to surrounding material, and no charring is visible. It has also been found that when cutting wood, much more efficient cutting occurs when sufficient power density is available to cause vaporization. With inadequate power density, cutting is slower, and two to four times more laser energy is required per unit volume of wood removed. For example, in cutting ponderosa pine, 0.005 kW hours per cubic inch of material removed was used when the power density was sufficient to cause instantaneous vaporization. When insufficient power density was used, the energy value more than doubled to 0.011 kW hour per cubic inch.

Ever since its introduction, military establishments have been watching the progress of the laser closely and examining the feasibility of a death ray. At first it was thought that lasers would be too weak to be effective over anything but a very short range. Indeed, an Austrian scientist provided theoretical proof that death rays were never really likely. However, the development of powerful carbon dioxide lasers changed the picture. These lasers could soon punch a hole through virtually every substance.

The short wavelength (that is, high frequency) of a laser makes it ideal for communication. The relatively low frequencies used by radio waves can carry only limited amounts of information, but lasers can make it possible to use the higher frequencies of light to carry much more information. The problem is that, unlike radio waves which can be broadcast in all directions and be picked up out of sight of the transmitter, lasers travel in a single straight beam. They are also badly affected (like all light) by atmospheric conditions, so they are unlikely to be used for general broadcast; but for rapid communication between two points their value is escalating each year. Many telephone companies are beginning to replace conventional cables with fibers carrying laser light which carry far more calls of much higher quality.

Lasers have combined to work with robots in some manufacturing firms.

D.P. Soroka and R.L. Swensrud of the Westinghouse Electric Corporation reported in their SME Technical Paper *Application of the Westinghouse Laser Systems*: "The combination of lasers and industrial robots is an effective marriage and one which is expected to play a prominent role in factory automation schemes. The attributes of these systems offer the potential user a level of flexibility and cost effectiveness which can easily justify the transition from present method to an advanced manufacturing process. The large productivity improvements realized by laser robotic system users typically results is shortened payback periods. Many users have also found that a laser robotic workstation enables the product designer to exercise greater latitude in the part configurations, as manufacturing operations and sequencing may change from present methodology. An example of such an opportunity would be to replace formed sheet metal part die trimming with laser cuts. The user can eliminate the expense of design, manufacture and maintenance of the dies, and can accommodate design changes by simple robot reprogramming, whereas die changes are much more difficult to accomplish.

"The area of laser robotics is a mature technology, however there is an ongoing evolution upward toward new levels of expanded capability. Near-future developments are expected to include product offerings which continue to expand the overall application base. Such developments would include the following:

- A production-reliable Nd:YAG fiber optic beam delivery system for jointed arm robots. For small parts processing with lower power levels, this arrangement is expected to result in an overall system cost reduction.
- Advanced sensory devices for improved process control.
- Special applications and new laser robot arm configurations.
- Special optic configurations such as laser beam steering mechanisms and beam shaping for unique applications requirements.
- Improved control technology for enhancements to process and motion control.

"Further productivity gains are possible by configuring these 'islands of automation' into totally integrated manufacturing systems. By combining today's automatic material handling, lasers, robotics, and sophisticated computer and software technologies, the efficiency of a flexible manufacturing system can be realized now."

One important development in automating the machining process was described by Nicholas O. Akinkuoye, Northern Illinois University, in his SME Technical titled: *Megatrend in Automated Manufacturing: The Design and Development of a Laser Beam Machining Robot*.

"The laser machining robot is a unique, fully integrated adaptive system, designed with two- or three-axis coordinates. Other designs allow for up to five-axis systems, covering every plane and angle depending upon their applications. The robots are driven by controllers based on a microprocessor, can usually be interfaced with a teach pendant control panel, and can interlock with other machines in the industry, factory or shop. The laser machining robot is designed to cut specialty items such as plain carbon steel. It can perform jobs such as drilling tiny holes in aerospace composites and diamond or carbide wire drawing dies.

Notes Akinkouye: "In order to cope with the escalating overheads and the fast ever-changing requirements of the manufacturing industry today, modern industry needs automatic machinery which can increase production. The rationale for the design of the laser beam machining robot centers around the usefulness of robots in hot, heavy-duty, hazardous, and boring work environments. To this end, laser beam machining robots are useful for high-speed cutting and profiling deep penetration welding, microtrimming, microwelding, microdrilling, scribing, engraving, heat treatment, and surfacing. With the use of laser beam machining robots all these techniques can be carried out rapidly and uniquely on a vast range of materials including metal, plastics, timber composites, and even diamonds.

"A laser beam machining robot, like any other robot, is made up of basically two most important features: The robot manipulator arm which in the case of this application culminated in a drilling, welding, or scribing torch and the robot controller, which is typically the brain of the laser beam machining robot. The controller operates the manipulator arm which provides the necessary commands and other data needed by the support equipment. The robot controller memory contains a program of a series of programs which contains data to complete a specific laser operation such as drilling, welding, etc."

The laser has also been teamed with the robot at an Austin Rover trim line in the United Kingdom. The production process was the topic of an SME Technical Paper titles *CO_2 Lasers in the Automobile Press Shop* written by Gerard H. Parsons of Control Laser Limited.

In his paper, Parsons explains: "The purpose of the laser robot was to provide a flexible panel trimming substitute during the development phase so that the trim die could be manufactured once and once only after the trim line is established. The simple economic justification was contained in the cost savings. In 1983 Austin Rover was spending $350,000 annually in trim die adjustment and a similar sum on panel hand trimming. The Swindon machine paid for itself in the first six months."

Parsons said: "The production experience with the Austin Rover system had been good. Continuous two-shift working had resulted in over 4,000 hours of laser running by the time the working party met to discuss development. The first conclusion was that the basic machine concept was sound. The design for the Telco machine therefore retained the four braced pillar gantry structure with stress relieved X-axis slideways. The Z-axis slideway suspended between twin Y-axis cross beams had proved to be very stable and the Dexter recirculating bearings on all three major axes had performed without problems."

Parsons listed the following future body shop applications:

- Trim die cutting
- Draw tool hardening
- Laser panel blanking
- Laser pressed panel trimming

In the SME Technical Paper *Automated Laser Robotic Workcell for Jet Engine Component Rework*, author P.A. Lovoi of InTA, describes an automated 20-axis laser cutting and welding robotic workcell. The system utilizes an adaptive 3-D vision controlled end effector. The cell is designed to be a general purpose shop tool that has also been specialized to repair combustion chambers for turbo fan jet engines. The system consists of eight major subsystems: laser, laser beam delivery, robot, end effector, 3-D vision, handling, control, and software. The workcell is being designed and built for the Navy's Norfolk, Virginia Naval Air Rework Facility (NARF).

Wrote Lovoi: "The design philosophy consists of two thrusts: understanding the welding and cutting requirements before designing the hardware and designing a general purpose machine before designing the specialized machine.

"The workcell (see figure on page 351) is designed to be operated by the control computer and not the individual subsystem controllers, but no standard protocol, digital interface was avail

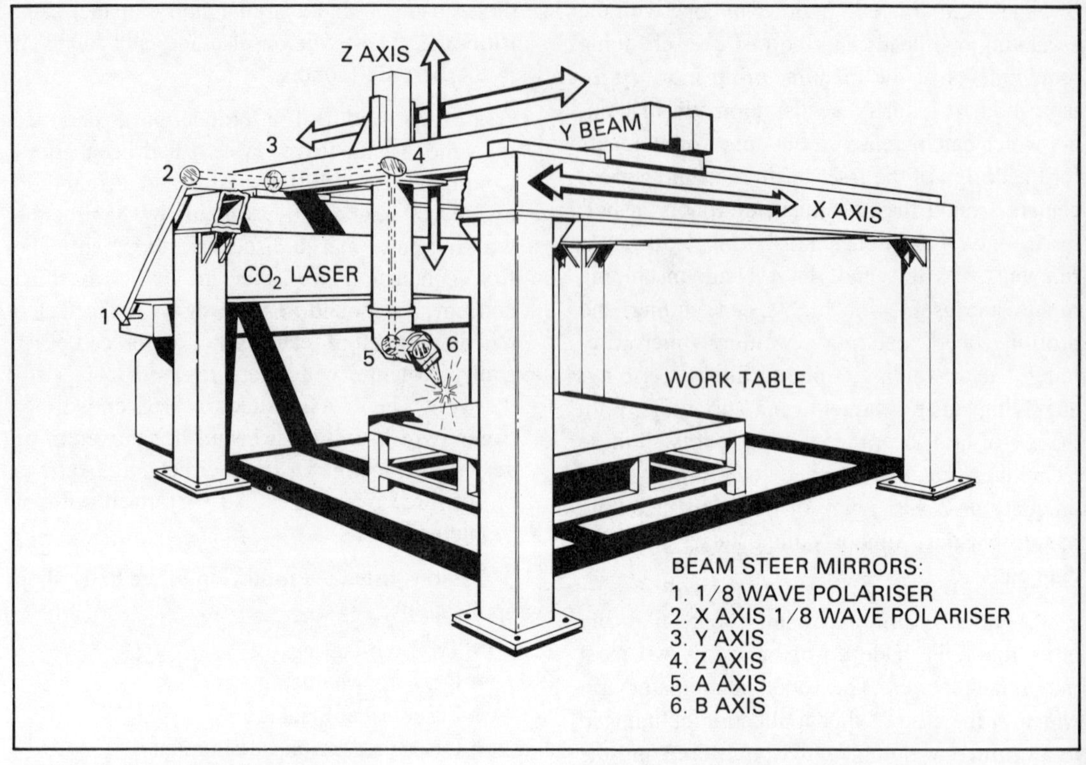

A sample laser/robotic system.

able for the laser functions including mirror tuning, laser gas control, cutting and welding gas control and all interlock indicators.

''Temperature variations within the laser cavity effect output power. To increase the stability of the output power, a controlled temperature chilled water system is used to cool the CO_2 laser. Additional control of the laser power is provided by feedback control using a thermal detector on the rear laser mirror. This thermal detector has a slow (10-20 s) response time so fast control over the laser output power is not possible. Though presently this mode is open loop, InTA is designing an optical sensor to provide the fast feedback necessary to close the loop.''

In the May 1987 *Manufacturing Engineering*, a manufacturing engineer (Randy J. Horning of Harley-Davidson) discussed the pluses and pitfalls of machining deep drawn stamped parts with a five-axis laser. In the article titled *Successful Five axis Laser Machining*, Horning wrote: ''The advantages of five-axis laser machining vary depending on the application but generally include reduced part variability, consolidation or operations, and outstanding process flexibility. Reduced part variability is attributable to several factors, the most important being the noncontact nature of laser machining. This eliminates the part variability that results from cutting tool wear and results in lower amounts of stress induced on the part being machined, thereby generally producing less deformation of the part. If cosmetic appearance of the part is a consideration, the noncontact nature of the laser process can lessen variability in metal finish by eliminating surface defects such as dents and scratches that result from performing secondary operations with some type of die.

''Many times, there are opportunities to consolidate multiple operations into one operation using a five-axis laser system. Secondary operations such as grinding and deburring can be

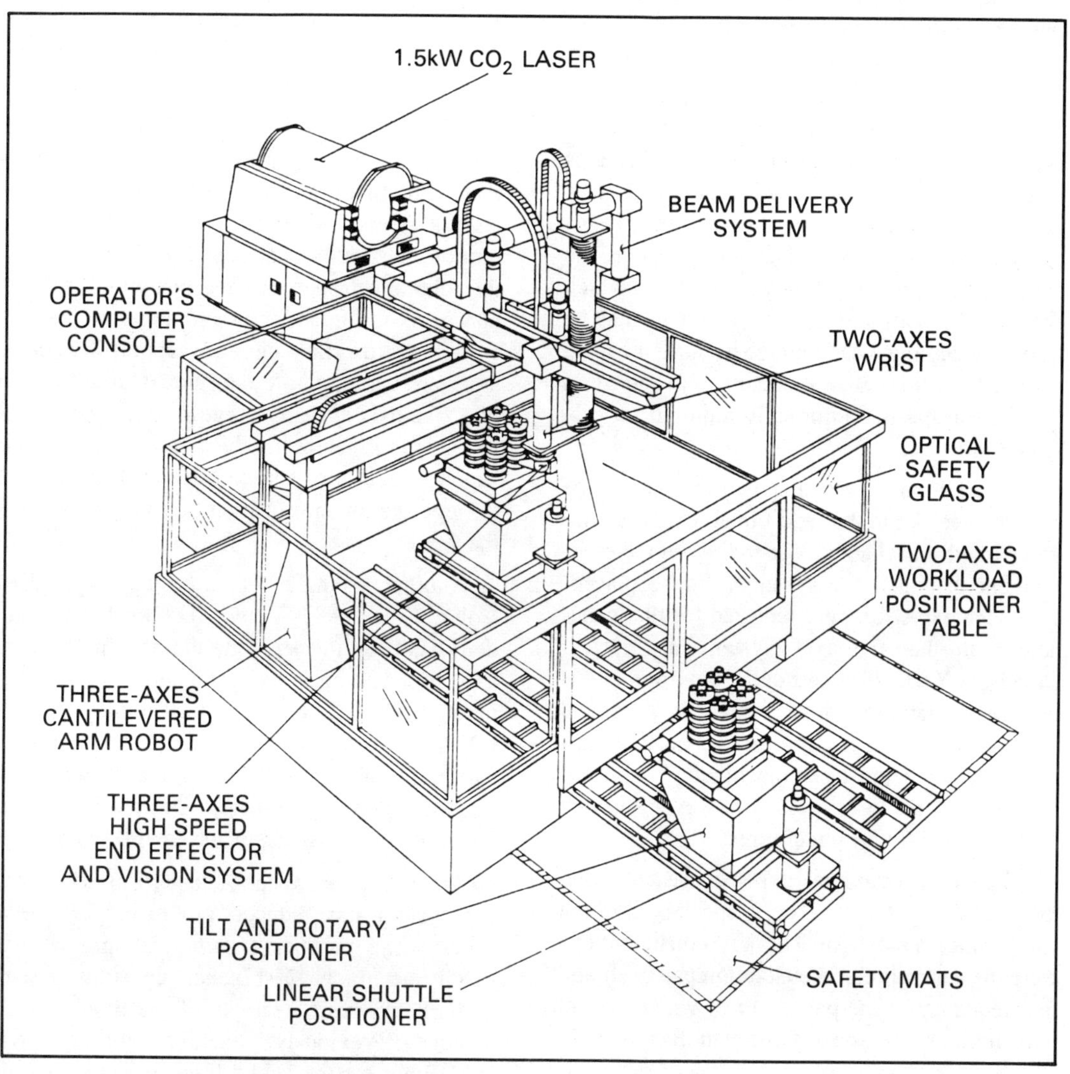

Automated laser cutting and welding work cell.

reduced or eliminated because laser cut edges are generally smooth and relatively burr-free. Parts that require multiple setups for conventional trimming or machining offer an even greater benefit potential. This is because, in alarge majority of these cases, the part can be completed in one operation with a five-axis laser.

"The final benefit of the process is one that can be difficult to attach a cost savings number to, and that is its outstanding flexibility. It can be difficult to attach a dollar figure to this benefit because the full impact of this flexibility may be fully realized only after the system has been in use and its capabilities are better known. There are some areas of process flexibility, however, that can be analyzed beforehand. For example, with a five-axis laser system, piece part design changes can be reacted to very quickly and usually with lower tooling costs. Lead time requirements for customer orders can be significantly reduced. And, for those interested in reducing piece part inventory levels, shorter runs can be justified because of lower setup times compared to conventional tooling.

''There may be more benefits that can be realized—again, depending on the application—but there are also disadvantages with five-axis systems, including cost, cycle times, and complexity. In a large majority of cases, the greatest disadvantage of a five-axis laser system is the initial acquisition cost of the system itself, which can approach half a million dollars. This may be an overwhelming disadvantage because it can be very difficult to realize the short-term return on investment that is traditionally required for justification.

''In addition to the acquisition cost of the system, there can be additional expenses involved with installation, depending on the environment in which the system is to be placed. Two such expenses were required for the installation of the five-axis system at Harley-Davidson York. The systems' environment is the metal stamping area, which is a logical location because it is deep drawn metal stampings that are being laser machined. Within 20 feet (6 m) of the laser system, however, are two 1000 ton (8896 kN) punch presses.

''The first additional expense was incurred through the installation of a special vibration dampening foundation. The key considerations were the deflection of the isolation pad itself and the frequency of the pad, both of which should have limits set by the system manufacturer.

''The second additional expense, again because of the environment, was the construction of an enclosed room to house the laser system. It was felt that this was needed mainly because of the noise level in the area, the quality of the air, and the high heat and humidity during the summer months, all of which have the potential of leading to problems.

''Another possible disadvantage of using a five-axis laser to machine deep drawn metal stampings is that for simple shapes, or for parts requiring many inches of machining, the laser may actually be slower than conventional machining. This depends on the part and many times will be found to be of minor significance relative to the overall benefits of the process. All the same, it can still be a disadvantage that must be dealt with.

''The final disadvantage of five-axis laser systems is the complexity of these systems compared to other processes or even to other laser systems with only three or four axes of motion. The additional complexity results from the relationship between the operation, maintenance, fixturing, and programming of a five-axis system. Extra training and support for these areas are paramount to the successful utilization of a five-axis system. Because of this complexity, lack of resources in the areas of training and support could prove to be a major disadvantage through the underutilization of a very expensive machine.''

Another example of laser use is applied Volvo. Since 1984, Volvo AD (Goteborg, Sweden) has been investigating the benefits of five-axis laser cutting in the production of prototype parts. Volvo's five-axis system employs a one-kW CO_2 laser and has been used to cut axle carriers, dash panels, sunroofs, inner hoods, and hinges, among other parts. The parts are produced in small batches of up to 100 units.

According to Project Manager Lutz Hanicke, the system has reduced processing and lead times while producing parts with production-quality accuracy. Dash panels constitute a particularly dramatic example of the time savings achieved. Previously, punching, nibbling, sawing, filing, cutting, and drilling were required to make the part. To produce a run of 35 panels took 700 hours. Producing 35 panels using the five-axis laser system requires 20 hours of programming and 7 hours of production time, for a total of 27 hours.

With an approach that integrated metalcutting, fabrication, and design, Gladd Industries cut part costs by 40%. In an article in the February 1987 *Manufacturing Engineering*, titled *Cutting Metal and Costs with CNC Lasers* author Gregory T. Farnum noted that: ''In addition to manufacturing a line of industrial processing ovens, Gladd also operates a contract manufacturing division. Called the Specialty Products Division, it utilizes a CNC laser fabricator as well as other production machines. The laser fabricator, from U.S. Amada, Ltd. (Buena Park, CA), employs a Spectra Physics 1200 W CO_2 laser.

Farnum wrote: "The air movement equipment company's original belt guard consisted of angle iron, sheet metal, and expanded metal, with the angle iron forming the frame to which the sheet metal and expanded metal were attached. The expanded metal provided cooling air for the enclosed drive components. The original design called for 12 pieces of angle iron, 4 pieces of sheet metal, and 6 pieces of expanded metal. A bandsaw, shear, and nibbler were required for cutting.

"Drawing on its laser, design, and programming capabilities, Gladd came up with a design simplification that significantly reduced fabrication costs. The new design completely eliminated the need for angle iron. Rather than build a frame, the sheet metal is now laser cut, formed, and welded together. The new sheet metal body provides sufficient strength and durability, and the product is much lighter and easier to handle. Due to the laser's ability to cut accurate contours, fewer pieces of expanded metal are required to fit cutouts in the sheet metal. The new design uses just four pieces of sheet metal and four pieces of expanded metal, reducing the number of parts required by almost two-thirds and significantly cutting material costs as well as production time.

"The laser's accuracy provides an added advantage in that there is less material waste. With the original design, the pieces overlapped considerably. Now, pieces are cut to fit exactly.

"Important time and cost savings are also provided by Gladd's computerized part programming capability. Because every end user's application is different, the belt guard must be built in virtually an unlimited number of sizes. To minimize design engineering time, Gladd Engineering developed a master program on their mainframe computer, enabling them to accomplish size variations quickly and easily."

To generate a new belt guard size, only a few key specifications—sheave sizes, shaft diameters, and the distance between shafts—need to be entered into the master program. From these, the master program automatically develops the dimensions needed to produce the new size. Once the new program has been generated, it is downloaded to the laser's CNC unit.

"The word *integration* often conjures up visions of extremely high-level, sophisticated systems—systems that frequently exist only on paper. But in this application, Gladd's laser operation integrated design, programming, and metalcutting expertise to slash direct labor and material expenses, reduce design time, and significantly cut the cost per part."

Ron Sanders of Laserdyne describes other laser applications in his featured article *A Brief Guide to Multiaxis Laser Machining*. This article appeared in *Manufacturing Engineering*, June 1987.

In it, Sanders wrote: "With low volumes characterizing many aerospace applications, formed parts are often trimmed with hand tools or manually operated milling machines. For short runs and low volumes, these labor-intensive methods are often more economical than hard tooling such as trim dies. The high cutting rates associated with laser cutting make it an attractive alternative to these conventional machining and routing processes.

"During prototype development, laser machining offers a significant advantage over die trimming and piercing operations. First, with laser machining, setup involves mounting the minimal amount of fixturing required to locate the workpiece relative to the cutting head. For many prototype parts, this is an epoxy replica of the part extending up to the trim line. Once this is done, the NC program describing the machining parameters (laser process conditions) and sequence (motion path) is loaded into the CNC unit. Part programs are readily generated by downloading design information through a CAD/CAM postprocessor into the CNC. The entire setup process can be accomplished in minutes, in contrast to the hours required for setup of hard tooling. Lower trimming rates associated with laser cutting are more than offset by the almost negligible tooling setup time.

"High cutting rates and the absence of tool contact and the accompanying distortion make laser machining a cost-effective solution to notching, piercing, slotting, and cutoff of bent tubing. Multi-axis laser machining is also being used in producing cutoff and cutouts for intersecting tubing assemblies in aerospace applications."

Also see: Drilling, Gantry Robot, Industrial Robots, Inspection, Laser Beam Cutting, Laser Beam Machining, Laser Beam Welding.

Laser Beam Cutting (LBC). Laser beam cutting is a process related to laser beam machining that severs metals by melting and vaporization when high intensity laser light is directed along a cutting line.

Also see: Lasers.

Laser Beam Machining (LBM). Laser beam machining is a nontraditional process for cutting, drilling, slotting or scribing metal parts. In laser beam machining, an intense, coherent beam of single-wavelength light is focused on a small area of the workpiece, where the optical energy is converted into thermal energy intense enough to melt and vaporize the workpiece material.

Also see: Lasers.

Laser Beam Welding. Laser beam welding is a fusion welding process for making structural, assembly, sealing and conduction welds in material up to 3/4 inch (19 MM) thick by directing a high power density beam of laser light onto the joint. Inert gas is generally used to reduce oxidation, but filler metal is rarely used.

Also see: Lasers.

Latching Relay. A latching relay maintains a set position by mechanical means until it is released electrically or mechanically.

Latch Instruction. A latch instruction is an instruction for a programmable controller which causes an output to stay in the ON position.

Layers. Layers is a method of portioning a file so that categories (such as all drawing notes) may be established, to allow selectable viewing of drawing portions.

Layout. Layout is the representation in viewable graphic format of a design logic or concept. When referring to electrical component and circuit board design, the layout is the translation of the logical schematic circuit into a physical circuit utilizing the available product shapes and sizes. The layout is used as a model to ensure the physical components can be properly assembled. The layout is further used by translating into the actual artwork which is a pattern used to construct the printed circuit board (the layout of the circuit is printed onto the PCB material). Printed circuit board design layout is more than simply placing all the components physically in an array with a minimum distance between objects.

When discussing architectural objects, the layout usually refers to a floor plan which shows the locations and configurations of all the physical objects in the room. A layout drawing is usually included to show the assembly of components of any product assembly to assist in the reassembly when making repairs or alterations.

Also see: CAD, CAD-D.

LBC. See: Laser Beam Cutting.

LBM. See: Laser Beam Machining.

LCD. See: Liquid Crystal Display.

Lead. Lead is the tendency of the dynamic response of a passive physical system to respond sooner than desired. Lead is a term normally used when referring to servocontrol systems which exhibit dynamic responses. It is an attempt to anticipate some of the responses which may occur during a movement of an axis by compensating for the undesired motion with a countering or complementary motion. This may be done to neutralize the effects of gravity, or the differences of moving with and without a load, or sometimes the differences in response when moving at different velocities.

Lead can usually be implemented by knowing where the endpoint of the move is to be and altering the move command accordingly, which can be a variable value. Lead can also be implemented by generating a positive bias voltage in the servocontrol circuitry, which becomes a fixed value. Lead is more difficult and limited to generate than lag. Theoretically, lag can have a value from nearly zero to infinity (no response).

Also see: Lag, Servocontrol.

Lead Through. Lead through is a term usually applied to programming a robot or similar type of device whereby the operator physically ''leads'' the device through the path of motion while the device ''records'' sufficient points along the path to recreate the moves. Lead through programming obviously cannot be done with larger type devices, but may be much more expedient for smaller robots. Lead through is also useful when programming a complex motion or orientation which, while not a problem to repeat as part of a programmed move, it is difficult to manually servo the robot into the desired configuration. Lead-through programming avoids the problem of accidentally moving the robot arm into an object and damaging either the robot or the object. With the increasing ability to model motions on three dimensional CAD systems, the need for manual programming of robots is decreasing. The part and workplace envelopes can be defined in the CAD system and the motion program generated by the computer off-line and downloaded into the robot.

Also see: Industrial Robots, Off-line Programming, Programming.

Learning Curve. During the cycle of producing a product, an organization's ability to perform work will vary based on its ability to learn or acquire further competency. As people learn or acquire this competency, the capacity of the organization can increase. Thus is possible to anticipate a certain degree of improvement over a period of time as learning occurs.

The learning curve can be calculated based on the relationship of product output to resource input over a period of time. Knowledge about the curve and the expected rate of improvement can help an organization project out estimated costs.

Although the learning curve can be successful in project oriented industries, it is rarely used outside of these industries because limitations exist. One difficulty pertains to the introduction of new products which are very similar to prior models. Because prior knowledge is easily applied, the cost reduction estimates from the prior experience are not realized. In addition, savings are typically realized in the form of direct labor. With the proper automated equipment in place, learning skills do not apply.

Least Significant Digit (LSD). The least significant digit is a digit representing the smallest value. The rightmost digit.

Least Squares Circle (LSC) Center. The least squares circle center is the center of a circle from which the sum of the squares of the radial ordinates of the measured polar profile, has a minimum value. It is established by mathematical analysis of the profile and generally requires a computer to be practical.

LED. See: Light-emitting Diode.

LIFO (Last In, First Out). LIFO or last in, first out is terminology used primarily to describe inventory management and control procedures. LIFO simply means that inventory is placed into a queue or storage area, usually a dead-end configuration of storage space, and the last items placed into the area, which are usually the closest accessible to the mover, are removed first. LIFO-type inventory management is not practiced when the product has a shelf life which is relatively short compared to the time the item may be in the storage. Additionally, if the entire queue or storage is not routinely emptied in the course of normal usage, LIFO material movement practices would leave some items in storage indefinitely, long past the usable life. The only time LIFO would be a desirable practice is when the product, such as wine, is required to be consumed, but the desire is to leave the bottles lying down as long as possible to improve the flavor and quality. In this type of situation, the demand hopefully never completely empties out the supply and the original bottles can remain for a significant length of time.

Also see: Inventory Control, Queue.

Light-emitting Diode. A light-emitting diode or LED is an electrical device which has a lens and light generating capability. An LED is a direct current solid-state lamp. An LED can be illuminated by 5 volt DC. LEDs are especially useful in digital computer circuitry because they can be driven with the same voltage level as the logic circuitry, thereby not requiring an additional power supply, provided the current requirements can be met by the base power supply. LEDs are available in a variety of lens colors, with red and then yellow being the most prominent colors. LEDs are used frequently as pilot lights to indicate the operation of a circuit. Due to the nature of an LED, as it operates on and off, the disturbance to the impedance and power consumption of the circuitry is low enough that no noticeable negative impacts are created. LEDs can be assembled into arrays of individual elements which can then be used to represent a bit-mapped display. Sometimes, characters and numbers are constructed using 7-segment or 11-segment displays, which may also include a decimal point. LEDs are probably one of the brightest types of displays, and they do require substantial current when driven at the high levels of illumination.

Light Pen. A light pen is a pen-like device which is connected by cable to the computer. All light pens function in the same manner. A CRT monitor ''writes'' to the screen by firing an electron beam at the phosphor-coated back of the display screen at intervals of 1/60 second. Starting in the upper left hand corner and traveling left to right down the face until it reaches the lower right hand corner of the screen, whereupon the process starts over again. A light pen uses a lens that senses the glowing state of the phosphors. The light pen determines the X-Y screen coordinates of the detected area by measuring the time it takes for the electron beam to reach the light pen. The pen then passes the coordinate information back to the computer through the color/graphics adapter, and the software determines the next events to occur. Light pens can be used like touch screens or they can be used like mice or tablets for drawing.

Linear Circuit. A linear circuit is an analog circuit having output that is an amplified version or predetermined variation of its input.

Linear Interpolations. This type of interpolation is a function automatically performed in the control that defines the continuum of points in a straight line based on only two coordinate positions. All calculated points are automatically inserted between the coordinate positions upon playback.

Linearization. Linearization is a mathematical procedure whereby a path defining a curve is subdivided into many small linear segments that closely approximate the desired path, such that tool motion resulting from the consecutive subdivisions will machine the shape to a specified tolerance.

Line Printer. A line printer is a printing device that prints an entire line of characters at one time.

Line Receiver. A line receiver is a device used in conjunction with a line driver to detect signals at the receiving end of a long line.

Liquid Crystal Display (LCD). A liquid crystal display is a type of display which uses a medium known as a liquid crystal to create the viewable elements. An LCD functions similarly to a CRT display in that it is capable of being provided with information which can be mapped onto the display screen for viewing. A liquid crystal display has voltage applied at locations on the back of its grid-like area of pixels, which thereupon become viewable. An LCD display is not as fast as a CRT video display and does not provide the high resolution possible with a CRT. LCDs are monochrome, with the color usually being grey. The viewing angle of an LCD display varies from product to product but typically can be distinguished within an included angle of 90 degrees. Higher than normal ambient operating temperatures affect the display by reducing the viewing angle, with extreme temperatures (50 degrees to 60 degrees C) causing the viewing angle to be reduced to a few degrees of included angle. An LCD display is a reflective display which means the image is distinguished by the reflection of ambient light, therefore it cannot be seen in the dark without some form of lighting being projected on the display screen. An advantage, however, of LCD displays over LED (light emitting diode) and florescent displays is of the lower power requirements to maintain the display.

LISP (LISt Processing). LISP, which stands for LISt Processing, is a programming language which was developed by John McCarthy at MIT in 1959. LISP has derivatives from languages such as FORTRAN and IPL, which in turn was the major evolutionary precursor to Logo. LISP is used mostly in artificial intelligence (AI) and expert systems programming. The original standard was LISP 1.5, but a possible new standard is COMMON LISP. LISP programming is exceptionally memory-intensive, while the runtime code is considerably reduced in size. In principle, LISP is not exceptionally difficult, but recursive programming can be confusing, which could be blamed on the previous lack of clearly written and informative textbooks. LISP is available to run on mainframes, minicomputers, and is resident on certain special-purpose machines from manufacturers such as LISP Machine, Symbolics, and Tektronix. LISP was created to process symbols in a way recognizable to speakers of the human language. The emergence of LISP has been in concert with the emergence of artificial intelligence.

Also see: Artificial Intelligence, Computer Languages, Computers.

Load Carrying Capacity. In robotics, the load-carrying capacity is the maximum weight that an end effector can effectively handle. Load carrying capacity can be affected by the end effector's position in the robot's working envelope.

Also see: Industrial Robots.

Local Area Network (LAN). During the past few years, local area networks have become one of the most publicized and controversial topics in the data communications industry. LANs are not really as new as some people might think, but with the increasing capability of personal computers and the desires to share data, peripherals and computing resources, LANs have become an item of which the general computing public is more aware. The publicity stems from what the LANs are purported to do for an organization; the controversy comes from how they are to do it. However, LANs will continue to become a greater and more important part of the computing environment.

The LAN is distinguished by the area it encompasses; it is geographically limited from a distance of several thousand feet to a few miles and is usually confined to a building or a plant housing a group of buildings. In addition to its local nature, the LAN has substantially higher transmission rates than networks covering large areas. Typical transmission speeds range from 1 Mbit/s to 30 Mbit/s. LANs do not ordinarily include the services of a common carrier. Most LANs are privately owned and operated, thus avoiding the regulations of the FCC or the State Public Utility Commission. LANs are usually designed to transport data between computers, terminals, and other devices. Some LANs are capable of voice and video signaling as well. Switching, digitizing schemes, data link controls, modulation, and multiplexing are often found in local area networks.

Local area networks have become popular for a number of reasons, the primary one being that most businesses transmit over 80% of their data and information locally, that is, within the local office or branch. This locality of data flow requires a transport system to move the data between the local machines. Moreover, many local applications (such as computer-to-computer traffic) require high transmission rates—certainly higher than the voice grade technology (300-56,000 bit/s) of the local common carrier. LANs are seen by some organizations as a means to bypass local loops and all the problems inherent in the common carrier's end office connection.

History. Initially, LANs were developed as a means of tying together expensive resources for backup and sharing. For example, in the early 1960s vendors built channel adaptors to join CPUs and others developed interface boxes to allow smaller computers to act as ''peripheral devices'' to large mainframes. In the early 1970s, LANs were used to share memory and printers in order to expand the life of an organization's systems. Lately, LANs have been used to tie together components that have outgrown the centralized computer room.

As distributed data processing (DDP) is now becoming a prevalent technology in the industry, local area networks are one way to implement it.

LANs are seen as a path for increased office automation. Integrated voice and data is a recent trend in the ''office of the future.''

Discussion. A LAN usually contains four major components which serve to transport data between end users.

First, the LAN path most often consists of coaxial TV cable or a coaxial baseband cable. Cable TV (CATV) coaxial cable is used on many networks because it has a high capacity, a very good signal-to-noise ratio, low signal radiation, and low error rates. Twisted pair cable and microwave are also found in many LANs. Baseband coaxial cable is another widely used transmission path, giving high capacity as well as low error rates and low noise distortion.

Thus far, optic fiber paths have seen limited application, mainly due to their higher cost, but their positive attributes virtually assure their widespread use in the future. The immediate use of lightwave transmissions on local area networks is point-to-point, high speed connections of up to 10 miles. A transfer rate of over 44 Mbit/s can be achieved on this type of path. Infrared schemes using line of sight transmission are also used on the LAN path. Several vendors offer infrared equipment for modem and local loop replacement. Up to 100 kbit/s over one mile distances are possible with infrared schemes.

Second, the interface between the path and the protocol logic can take several forms. It may be a single CATV tap, infrared diodes for infrared paths, microwave antennas, or complex laser-emitting semiconductors for optic fibers. Some LANs provide regenerative repeaters at the interface; others use the interface as buffers for data flow and/or simple connections, like that of RS232-C.

Third, the protocol control logic component controls the LAN and provides for the end user's access onto the network. LANs employ different methods and techniques to provide this function.

The last of the four major components is the user workstation. It can be anything from a word processor to a mainframe computer. Several LAN vendors provide support for other vendor's products and certain layers of the OSI model.

Typical topographies for LAN are either ring or bus, with the ring using token passing and the bus using CSMA/CD. It is possible, however, to have a logical token ring on a physical bus topography. Another type of network is a star network, where a central host has individual lines to each of the other stations on the network.

Local networks can employ most of the concepts of data link control to manage the flow of data on a communications path (link) such as polling/selection, hub polling, contention, and time slots. Sliding windows and cyclic redundancy checks are also employed. Two other popular approaches are carrier sense multiple access/collision detect (CSMA/CD) and token passing.

Another confusing and controversial issue regarding LANs is whether to use broadband or baseband. With regards to LANs, each technology has advantages and disadvantages; neither is clearly preferable to the other, rather the specific application and usage leads to a preference one way or the other. Broadband networks have greater capacity; baseband LANs are capable of 1 to 10 Mbit/s rate, whereas a broadband LAN can operate at speeds well over 150 Mbit/s. However, the broadband subchannels operate at substantially slower rates, so the individual user is getting only the RF modem speed supporting the user work station. Broadband permits multiple channels, providing for multiple types of protocols and subnetworks on one system. While baseband does not have this feature and is restricted to one protocol, nothing precludes the baseband net from supporting subnetworks and multiple applications through time division multiplexing (TDM) techniques. Broadband can accommodate more users than baseband; RF modems and amplifiers can provide for more extended coverage. Moreover, its technology is based on CATV and is highly reliable. For example, mean time between failure (MTBF) of its coupler taps is 30 to 40 years and 18 years for its amplifiers.

Some broadband proponents say the costs are comparable. However, a baseband operation is simple and has small start-up cost. All that is needed (excluding software) is a cable, two baseband transceivers, and two interface units. The broadband components can be considerably; more expensive.

Applications. Many local area networks are available today. All use either baseband, broadband, or both technologies, and the majority of the networks use some form of CSMA/CD, token passing, or the conventional polling/selection protocol. Several of the vendors have announced support for selected layers of the ISO Open Systems Interconnection model such as X.25 and X.21. A unique approach by one vendor, Ungerman-Bass, is with its Net/One LAN that permits the interconnection of both baseband and broadband. The RF modems and the baseband transceiver can be exchanged; the resolution is made by altering an ''encoder/decoder'' board in Net/One's protocol and network logic boards.

Some of the installed bases of LAN include:

Apollo Computer, Chelmsford, MA–name: DOMAIN; maximum distance: 3,250 feet; path: baseband coaxial; type: ring; protocol: token passing; speed: 10 Mbit/s; maximum connections: >100.

Corvus Systems, San Jose, CA–name: OMNINET; maximum distance:4,000 feet; path: twisted-pair wire; type: bus; protocol: CSMA; speed: 2.5 Mbit/s; maximum connections: 255.

Gould, Inc., Andover, MA–name: MODWAY; maximum distance: 15,000 feet; path: coaxial; type: bus; protocol: token passing; speed: 1.544 Mbit/s; maximum connections: 250.

IBM (GS), Atlanta, GA–name: Series 1/Ring; maximum distance: 5,000 feet; path: coaxial; type: ring; protocol: CSMA/CD variation; speed: 2 Mbit/s; maximum connections: 16.

Network Systems, Minneapolis, MN–name: Hyperchannel & Hyperbus; maximum distance: 3,000 feet; path: coaxial; type: bus; protocol: priority CSMA; speed: 50 Mbit/s & 6.3 Mbit/s; maximum connections: 256.

Prime, Framingham, MA–name: PRIMENET; maximum distance: 750 feet; path: coaxial; type: ring; protocol: token passing; speed: 8 Mbit/s; maximum connections: >200.

Ungerman-Bass, Santa Clara, CA–name: NET/ONE; maximum distance: 4,000 feet; path: baseband coaxial; type: bus; protocol: CSMA/CD; speed: 4-10 Mbit/s; maximum connections: 200.

Wang, Inc., Lowell, MA–name: WANGNET; maximum distance: 2 miles; path: broadband coaxial; type: bus; protocol: CSMA/CD & FDM; speed: 9.6 kbit/s to 12 Mbit/s; maximum connections: 512 to 1000's.

Xerox Corp., El Segundo, CA–name: ETHERNET; maximum distance: 1.5 miles; path: baseband coaxial; type: bus; protocol: CSMA/CD; speed: 10 Mbit/s; maximum connections: 100 to 1000.

One of the largest collections of LANs in private industry is the internal DECNET system within Digital Equipment Corporation. The entire system includes over 10,000 individual nodes on multiple networks.

Glossary. *baseband:* If pertaining to a radio signal, baseband or narrowband, pertains to a range of frequencies that is narrower than 1% of the midband value of the range. Another characteristic of baseband is that it generates digital waveforms and uses time division multiplexing (TDM) schemes. Baseband bandwidth contains the narrow band (0 to 300 Hz) and voice band or grade (300 to 4000 Hz). Therefore, any bandwidth of less than 4 kHz would be considered to be a baseband. Typically video images are unable to be transmitted on a baseband communications network.

broadband: If pertaining to a radio signal, broadband, or wideband, pertains to a range of frequencies that is broader than 1% of the midband value of the range. If pertaining to a voice frequency signal, a group of voice frequency channels that use a bandwidth greater than 4 kHz or greater than 20 kHz. Both 4 kHz and 20 kHz have been proposed as the limiting width between broadband and narrowband for voice channels.

CATV: The acronym for cable television and also implies any of the defined standards regarding cable television transmission, connectors, media, etc.

coaxial: A cable consisting of an insulated central conductor with additional insulation on the outside and covered with an outer sheath.

CSMA/CD: A popular form of contention protocol which allows multiple stations to exist on a network, with intelligent handling of data transmission collisions and retransmissions.

cyclic redundancy check: Cyclic redundancy checking is another approach which entails the division of the user data stream by a predetermined binary number. The remainder of the number is appended to the message as a CRC field. The data stream at the receiving site has the identical calculation performed and compared to the CRC field. If the two values are identical, the message is accepted as correct. CRC is capable of ensuring an accuracy greater than 99.99% in data transmission.

FCC: The Federal Communications Commission (FCC) is a government body which regulates and licenses all public and commercial users of the airwaves. The FCC assigns users radio frequencies to use within their local area for wireless transmission. The FCC also sets standards regarding the amount of ''noise'' and other non-useful emissions from electronic equipment.

mean-time-between-failures: The mean-time-between-failures or MTBF is the arithmetic mean value calculated from samples of actual times from beginning of use until failure of a part or component.

network: A network is a combination of nodes or stations which have the ability to communicate with each other in a predetermined fashion utilizing contention protocols and error checking capabilities. Networks can be local area networks (LANs) or wide area networks (WANs). A manually switched combination of lines between a shared resource usually is not considered to be a network, however if software handles the switching, then this would logically be a star network.

optical fiber: Fiber optics can be described as the technology of guidance of optical power, including rays and waveguide modes of electromagnetic waves along conductors of electromagnetic waves in the visible and near visible region of the frequency spectrum, specifically when the optical energy is guided to another location through thin transparent strands. Techniques include conveying light or images through a particular configuration of glass or plastic fibers. Incoherent optical fibers will transmit light, as a pipe will transmit water, but not an image. Coherent optical fibers can transmit an image through perfectly aligned, small (10-12 microns), clad, optical fibers. Specialty fiber optics combine coherent and incoherent aspects.

OSI model: The Open Systems Interconnection (OSI) model was developed by a 1977 subcommittee (SC 16) of the International Organization for Standardization (ISO)'s Technical Committee 97. The model is intended to assist the exchange of information between distributed systems. The function of OSI is to provide

protocols for different vendor's/manufacturer's products to connect with each other—thus allowing an open systems interconnection of user applications. Currently seven layers have been defined within the OSI model. From the bottom up, the seven layers are: Layer 1-Physical, Layer 2-Data Link, Layer 3-Network, Layer 4-Transport, Layer 5-Session, Layer 6-Presentation, Layer 7-Application.
RF modem: An RF modem operates at radio frequencies. A modem is a signal conversion device that includes both a modulator and a demodulator. It is a combination of equipment that changes the type of modulation, modulates an incoming signal, demodulates an incoming signal, converts digital signals into quasi-analog signals usually for transmission, or converts quasi-analog signals into digital signals, usually for further processing.
sliding window: A sliding window data link control is one which allows multiple messages to be in the process of transmission at the same time, which are usually numbered so a specific message can be retransmitted if an error is detected. The sliding window concept has much greater efficiency than the stop-and-wait alternating sequence of data link control.

Also see: Manufacturing Automation Protocol, Technical and Office Protocol.

Local Intelligence. Local intelligence describes the dispersement of a typically centralized intelligence capability such as the use of personal computers as DTE devices instead of ASCII terminals. The use of personal computers places considerable intelligence and computing capability locally at each terminal location. The uses of local intelligence include preprocessing input data before transmitting the data upwards in the hierarchy to the host computer, to reduce both the volume of transmitted data and the requirements for host processing. Local intelligence can provide much greater sophistication in the performance of DTE activities such as local editing and file storage of data. In a networked system, local intelligence can be made available to other locations via the network. With the continuing decrease in the price/performance ratio, local intelligence is becoming much easier and cost justifiable. Distributing intelligence locally allows greater flexibility in changing and modifying the configurations than when using a highly centralized architecture.

Also see: Distributed Data Processing, Personal Computer.

Logic. Logic is a method of solving problems through the repeated use of functions defining basic concepts. AND, OR, and NOT are the basic logic functions. Logic also is the systematic scheme defining the interactions of signals in an electronic system.

Logical Device. A logical device is capable of executing logic operations; i.e., operations that follow the rules of symbolic logic, such as the Boolean operations, of AND, OR, and NOT. Integrated electrical circuits can be designed and used to perform the logic operations.

Also see: Logic, Integrated Circuit.

Logic Element. A logic element is any one of a variety of forms of electrical switches within a computer which corresponds to a basic, elementary logical function.

Also see: Computers, Logic.

Logic Family. A logic family is a group of digital integrated circuits sharing a basic circuit design having standardized I/O characteristics.

Also see: Computers, Logic.

Logic Level. Logic level is the voltage magnitude associated with signal pulses signifying ones and zeros in binary computation.

Also see: Binary Code, Logic.

Logic Line. Logic line is a block used to construct the user's unique logic, usually ending with a coil. Types of logic lines are: relay, timer, arithmetic, counter, or special functions.

Also see: Logic.

Longitudinal Redundancy Check Character. A longitudinal redundancy check character is used for checking the parity of a track in the longitudinal direction on tape on which each character is represented in a lateral row of bits. Usually it is the last character in each block.

Lookup Table. A lookup table is a tabular arrangement of data such that on the basis of a multiple entry into the table, a single or unique location in the table will permit an identification of a particular item in the table (array). A single entry usually permits a set of items to be selected. Another entry permits the selection of another set, and an element common to both sets can then be selected, for example, a code table, a logarithm table, a periodic chemical elements table, a mileage table, or transportation system timetable. The table may be arranged in chart form, with guiding lines to entries and elements, such as the Smith chart that is used for transmission line analysis.

In the programming of such devices as programmable controllers and other computers with limited capability to represent or manipulate data, tables are often used to improve the effective processing capabilities and the ability to accommodate extended data.

Also see: Programming.

Loop. A loop is the repeated execution of a series of instructions for a variable number of times, but usually with address modifications changing the operands of each iteration, until a terminating condition is completed.

Also see: Computers.

Loop Table. A loop table is a short piece of punched tape with ends joined to form a loop for continuous reading of a program or operation.

Also see: Computers, Numerical Control.

Lot. A lot, in many cases is a batch. A *lot size* is the number of pieces or good ordered for manufacturing.

Low-level Language. The term low-level language refers to programming languages which are directly tied to particular machine hardware, as opposed to high-level languages which can, in principle, be used on any machine. Historically low-level languages were developed first and sometime later high-level languages followed. Low-level languages include machine languages and assembly languages. While low-level languages require greater learning and conformity by the programmer, they also are usually less bulky and execute faster than high-level languages.

Programming in low-level languages is the closest a person can come to programming in machine language. Low-level languages allow a programmer to exercise direct control over all aspects of data movement and manipulation within the machine. The negative aspect of using low-level languages is that significant reprogramming must be done to allow a program to run on another machine. For minimum code size and maximum execution speed, low-level languages have no equal.

Also see: Assembly Language, Computer Languages, High-level Language, Machine Language, Programming Languages.

LSC. See: Least Squares Circle Center.

LSD. See: Least Significant Digit.

LSI. See: Large Scale Integration.

Lubrication. Lubrication is the process of interposing a lubricant between two contacting and moving machine elements to reduce friction and wear. Liquid, plastic and solid materials are used as lubricants although sound design requires the application of liquid lubricants whenever possible. Common categories of lubricants include mineral oils, animal and vegetable oils, synthetic liquids, greases, solid lubricants, and miscellaneous fluids.

Under conditions of thick film lubrication, the viscosity of a lubricant is its only important property. The *viscosity* of a liquid is defined as the ratio of the shearing stress to the rate of shear. Since many applications of lubricants are where thick film conditions do exist, and since bearings designed for thick film lubrication will occasionally operate in the thin film or boundary condition, the oiliness of a lubricant becomes important. As a lubricant is quite often required to operate in a system for long periods of time and occasionally at elevated temperatures, its oxidation and its thermal stability are also of interest.

M

Machine Language. Machine language is the binary representation of data understood by a digital computer, the ones (1) and zeros (0) which comprise the binary digital data format. Machine language is what all higher level programming languages, and all lower level languages such as *Assembly Language*, are translated into before actual execution of the program by the computer. While the binary representation of the data partially characterizes machine language, the other aspect is that the higher level commands are translated into the basic functions of a computer which are moving, storing, and adding data values. A combination of these basic elementary functions are combined to create the more complex functions of multiplication, division, and advanced mathematical functions. Machine language operations employ the use of registers within the memory of the computer as temporary storage locations for the data as it is being manipulated.

Since most software programs are written in a higher-level language and then translated into machine language, the conversion process (compiler) creates machine language which is not necessarily optimized. Therefore, machine language translations of programs do not offer the same conciseness and efficiency.

Also see: Assembly Language, Computer Languages, Programming.

Machine Loading. Departmental or work center machine schedules are prepared from the information of the master production schedule. A *production scheduler* is the person responsible for the maintenance of the machine loading. Computerized scheduling is preferred to properly consider all loading factors and to model the interaction between machines. The three major considerations in placing a production order in a machine schedule are: when the order is required, whether and where the machine and labor capacity are available, and the sequencing of the operation required to produce the order. Scheduling options may optimize utilization, flow time and throughput.

For most applications, the orders are scheduled according to machine capacity, especially when the machines are independent of each other. This requires that the machines be listed according to types and operations, with their standard time values that can be performed on each.

The basic information to begin machine loading includes: amount of units to be produced per order, time required per piece per operation per machine, and time to set up a given machine or line to run a particular part.

Many of the current applications of robots in loading and unloading situations are relatively simple, straightforward installations best described as pick-and-place operations. Typically, one part is handled at a time, the robot moves are well-defined straightline motions, and a minimum of interfacing is required between the robot and associated equipment.

In one application, *Robotics Today* reported in its August 1982 issue that at the General Electric aircraft engine manufacturing plant in Evendale (Ohio), an air bearing-mounted sled moves an industrial robot to handle loading and unloading operations at four presses.

In the previous manual load/unload arrangement, one operator took care of two presses. The original cycle time was approximately 15 minutes. This has now been reduced to 13 minutes. The total cycle consists of a three-minute preheat, followed by a 10-minute forming operation.

Production runs on these parts are relatively short. Typically, 35 to 50 parts are run in a batch-type model. In the first year—the robot was installed in July 1981—about 24 different part families were handled.

The robot is programmed to load and unload four presses, traveling between the machines on a specially built sled that rides on four air bearings. During transport, the sled is lifted off the floor about 1/2 inch (12.7 mm). An air motor-driven wheel on the front of the sled pulls the vehicle along the floor. Two rollers on the sled bottom track in a U-shaped channel embedded in the floor.

Limit switches at each of the four press stations indicate when the sled is properly positioned. It is then lowered over two pins for positive location. In addition, four magnets on the bottom of the sled assure that it stays down over the pins.

Also see: Industrial Robots.

Machine Readable. Machine readable is the ability of some represented form of information to be absorbed and comprehended by a machine. Machine-readable information may also be bar codes or human-readable characters which can be read by an optical character reader (OCR). Typically, machine-readable information is simple, concise, and digital in format to prevent errors. One of the earlier forms of machine-readable input was paper with black lead pencil marks within defined boundaries with the machine being able to ascertain the existence or absence of a black mark in the defined space. With the advent of OCRs, the need to substitute cryptic representations for normal ASCII text representations of information was eliminated, and while the earlier OCRs required the use of a special font style, today's OCRs can recognize almost any legible font style and size. Machine readable can also be used to describe the use of magnetic cards which are encoded and deciphered by magnetic card readers.

Also see: Computers, Machine Vision, Optical Character Recognition.

Machine Tool. A machine tool is the tooling portion of a metal cutting, turning, or forming machine. Broadly interpreted, a machine tool is a powered device which is capable of material processing (i.e. machining, stamping, forging.) The term may include assembly or testing. The term usually does not cover fixtures, tools or gages. The term *machine tool* is used to refer to the entire machine which performs the operation and also to the cutting tip itself.

The cutting tool is the part that contacts the workpiece to perform the operation. The tool of a cutting or turning machine pierces the workpiece to remove material. By using a tool piece, the user can replace the tool instead of the entire support assembly when the original tool becomes broken or dull. The tool tip is usually made of a material which is much more expensive than the support portion, as well as usually containing physical properties which make the tool material too brittle to be used for the entire assembly. Cutting tools are constructed from many materials, depending upon the desired properties such as heat resistance, wear ability, and surface finish. Some materials can be used for a significant length of time, provided a rougher finish is acceptable. However, for a smoother finish a different material with a shorter duty time is utilized. Machine tools are sometimes formed into intricate shapes to provide the necessary profiles for the finished parts. Machine tools may contain turret-type storage devices to quickly change tools as part of the machining process (turret lathe), and/or may index or transfer the workpiece past a number of tools as in a screw machine.

Also see: Machining Center, Tool.

Machine Vision. Machine vision is a process in which information is extracted from visual sensors to enable machines to make intelligent decisions. Computers and cameras capture and process the data to perform useful activities such as machine control, inspection, sorting, and assembly. Machine vision is the front-end imaging portion of object recognition and robot guidance systems. Machine vision may be used to view infrared emissions, or to capture high-speed actions into perceivable viewable frames (a *frame* being a snapshot at a moment in time of the dynamic image). Machine vision has become a viable industrial tool with the usage of less expensive and better quality cameras and microprocessors. Machine vision-like systems are now being used as OCRs (optical character readers) and as graphic entry systems for desktop publishing systems. In general, machine vision is a sensory capability for machines and mechanical systems to achieve higher intelligence capacity levels.

Vision is one aspect of sensory input to machine tools and related automation. Other aspects include the ability to absorb sensory data and logically conclude preprogrammed algorithms for manufacturing operations.

Machine vision by itself doesn't always provide a complete application. Usually the vision system is only part of a complete system such as an inspection system or robotic guidance controller.

Trade organizations, such as the Society of Manufacturing Engineers (SME), have organized special interest groups to support machine vision. The Machine Vision Association of SME provides a forum for the exchange and dissemination of ideas and information.

History. Machine vision is a relatively new technology as far as its use and application. The lack of availability of enough of the critical components of a vision system at a cost-effective price prevented the advancement of usage for some time.

A machine vision system is usually made up of several elements. A required component is the camera or other type of image capturing device. The first types of cameras to be used in vision systems were television cameras and vidicon cameras. These cameras could capture a fairly detailed analog image, but the large task remained of translating the analog image into digital information which could be processed by the remainder of the computer system. The computing speed which the then state-of-the-art computer systems could provide, was much too slow for the real-time needs of a vision system requiring from one to five images to be processed every second. Also, the algorithms being used at that time usually processed every pixel of information in the picture, without any gross elimination of unnecessary data. The types of images which were handled during this stage were monochrome type images with each pixel location being designated as either black or white.

As the capabilities of computer technology advanced, array processors were designed to more quickly process the vast arrays of pixel information used to generate an image. More direct digitizing methods were employed using the vidicon type cameras. A new type of image definition began to emerge, which was called *gray scale imaging*. More sophisticated algorithms were developed to do the post processing, which could eliminate large quantities of useless data. Faster computers, such as the lower-priced minicomputers, provided greater computing horsepower.

The next significant development in machine vision was the introduction of solid state cameras. CID (charge injection device) and CCD (charge-coupled device) cameras began to emerge as the preferred front end imaging device.

At present, high-performance personal computers provide greater processing capability than the dedicated machines of a few years ago. The cost of 32-bit diskless minicomputers is so low, comparatively, that they can be used economically instead of developing special purpose array processors. RISC (reduced instruction set computer) computers provide even greater speed, provided all needed instructions still remain in the reduced set.

The current trend is to integrate expert systems and artificial intelligence capabilities into machine vision systems, both for enhancing the captured image, and for aiding in the post processing for whatever application the vision system is being used as a front end.

Discussion. Machine vision is the visualizing or capturing of an image and converting the information which describes the image into something which can be utilized by the remainder of the computerized system, such as a visually controlled robot arm, a visual inspection system, or a security system.

Most people attempt to make a machine visualize in the same way as a person does, but they do not realize that a grown person visualizes by relating what is seen to what is remembered by the brain regarding shapes and relationships and what certain images mean. This additional knowledge and relationship is what expert systems, and even more thoroughly, artificial intelligence, can provide for machine vision systems. Machine vision tries to recognize the basic attributes of an image, such as differences in contrast, and outlines of shapes and range.

The edge detection capabilities usually consist of recognizing the differences between the pixels of dark and light. In an abbreviated algorithm for edge detection, the computational routine simply starts at one edge of an image and evaluates the pixels one by one until there is a change in illumination value. All previous pixels are discarded except for the pixels that border the change in light value. The location and value of these pixels is maintained, and as row after row of the image is processed (as in processing raster lines), an outline including any holes, is generated in a two-dimensional plane.

Lighting is very critical in machine vision applications. Reflections from excessively bright light could mask a feature or totally wash out an image. Systems that use gray scale processing cope with the problem of lighting. In one implementation of gray scale processing, the threshold for declaring a pixel either black or white is set at a value close to either the maximum or minimum and then recalculated successive times while each time incrementing the value of the threshold. The number of times the threshold is changed determines the number of gray shades discerned.

Early systems required the camera to reimage for each gray scale level, but more modern systems can store the frame and do the entire gray scale processing in software. A common system today can process 64 levels of gray, which was decided mainly by the need for sufficient gray shade discrimination. Eight and 16 levels of gray were not sufficient for many applications, while 64 levels is usually adequate, even when the lighting causes the image to be skewed in the gray scale range.

Some manufacturers continue to develop custom designed and built processors to handle the vision processing chores for their systems while others use standard platforms. Use of standard platforms may allow easier upgrading when the standard platform evolves into newer generations of hardware. Manufacturers who have concentrated on inspection applications have been fairly successful. The application of machine vision to real-time machine control is not as widespread since the need for "real-time" machine control may require processing an image plus post-processing computations, all within no more than one fifth of a second. This type of performance is difficult for most systems to reliably maintain. Similarly, there is a need for an "expert" to understand the control parameters.

Algorithms used by many vision systems calculate several features and properties of the object(s) being imaged. Some of the usual, more basic geometric properties are the circumference, the major axis, the minor axis, and the included area. Any internal holes are located relative to the outer perimeter of the part, and if the entire perimeter is not clear, some systems provide the capability to complete the edge when obstructed or damaged. Additional properties which are calculated are the first and second moments of inertia and the centroid of the part. These attributes are useful in the object recognition applications where these might be the only unique properties of similarly shaped objects. Another use for determining these properties is inspecting parts using a rules-based system. These factors allow a determination if the entire part is present, without damage or missing attributes.

The methods used for recognition of parts can be by two methods. One method uses "taught" objects, which are usually multiple-stored "pictures" of a "good" object. When an object is imaged, the image is rotated and compared to the stored images to see if it matches. Depending upon the level and accuracy of match, the system will determine that the image does or does not match the stored picture. When the image is very simple, or if many different parts are to be inspected, or if a very large complex part which can be simplified into a large quantity of simple attributes is imaged, an alternative method of recognizing an image is to use a *rules-based system*. A rules-based system has programmed rules which describe attributes and relationships, which are used to analyze the image. When a sufficient number of rules are satisfactorily accepted, the image is deemed to match the object in question.

The problem with this approach is that if very similar objects are compared, with no major feature differences, the system may not be able to distinguish between the two. An advantage to the rules-based system, however, is that as soon as a match is made, processing on that particular image is suspended. This can sometimes speed the recognition process over the matching to a taught image. Also, if the object is an extremely large object, with thousands of distinct features, but with a small number of unique and different features, such as a printed circuit board, the rules-based method is much faster and allows new parts to be correctly handled by the system without ever having been taught the particular new object.

Applications. Today machine vision is applied significantly in manufacturing industries. Machine vision as applied to manufacturing extracts information from visual sensors to enable machines to make intelligent decisions. Such decisions are needed in quality control (detection of defects), process monitoring (prevention of defects), product routing (parts acquisition and sorting), and statistical reporting (performance evaluation).

The three main industrial application categories are inspection, identification, and machine guidance. Among the inspection tasks are the following:

- Gaging. Checking to make sure that dimensions fall within acceptable tolerance bands.
- Verification. Checking to make sure that a product is present, complete, or the right one in the proper orientation.
- Flaw detection. Checking for unwanted features of unknown shape anywhere on the observed portion of the product.

Among the identification tasks are the following:

- Symbol recognition. Deciding which one of many possible symbols is present in a given location. Examples of this application are reading serial numbers or bar codes.
- Object recognition. Deciding which of many possible objects is present by examining features of the object under test.

Among the guidance functions performed by machine vision are the following:

- Object location. Two or three-dimensional determination of position and orientation for purposes of part acquisition, transfer, and assembly.
- Tracking. Continuously updating the position of a feature relative to a tool to control continuous processes such as gluing or welding.

Machine vision applications include inspection and gaging tasks. These applications use both taught part methods and rules-based methods of recognition. In process control, machine vision is sometimes used to detect levels in tanks, especially when the substance cannot be contacted by a physical probe, either due to the substance being hazardous to the probe or the probe contaminating the substance.

Most large, complex computer printed circuit boards are visually inspected for the correctness and integrity of the etch. This is necessary in many cases due to the inability for humans to accurately inspect a panel, and the prohibitive time if manual inspection was attempted.

Some applications follow.

In his SME Technical Paper *The Automatic Tool Verification System*, Frank Czerniejewski described a process where, "...up to 400 lathe cutting tool assemblies are measured for positional accuracy and verified for proper type by a vision system. The heart of each station in the system is a high resolution camera mounted on an X-Y table. A tool is seated by a robot arm inside a hood. The table is moved first to a calibration circle, then to the expected position of the tool tip. A picture of the tool is taken and converted to line segments (X,Y locations) which outline the tool tip. The actual position is compared to the expected position by a single reference point and a series of small windows stepped around the outline curve. The reference point and outline windows are modified for the effects of station-to-station differences and actual tool location. A single calibration tool allows these corrections for station-to-station and wear variations without reteaching the tools. A tool is accepted as being correctly identified when it is in the right location and has the proper profile. Each tool is inspected in less than 2 seconds with an accuracy of plus or minus 2 mils."

In another example (*Detection of the Machining Problems by a Machine Vision System*), Ibrahim N. Tansel and Bruce Clark described a system which used texture analysis techniques to identify the machining problems (i.e., chatter, deep cuts resulting from by broken inserts) from the tool vibration marks on the machined surface. An example of a machine vision/inspection system is illustrated in the figure below.

Roger J. Schlichtig of R.A. McDonald Research Labs, presented another example of machine vision in his SME Technical Paper titled, *A Hardware Based Machine Vision System*: "In the recycling of used brake shoes, the first step is to sort the incoming parts into one of about 200 possible classifications. Presently, these incoming shoes are dumped onto a conveyer and a group of people grab them off and throw them into an appropriate bin. The throughput is approximately two shoes per person per second, with a high degree of inaccuracy. In addition, these people only do a presort of the shoes, breaking them into about ten major categories.

Schlichtig noted that the system "can quickly classify the shoes into their proper category, and supply the necessary position and orientation information to a robot arm controller, or gating mechanism to properly sort the shoes. In this application, the objects are recognized by applying a series of general-purpose templates to the image, and analyzing the results with the CPU.

"Let us assume that the camera is set so that up to five shoes can be in the image at one time.

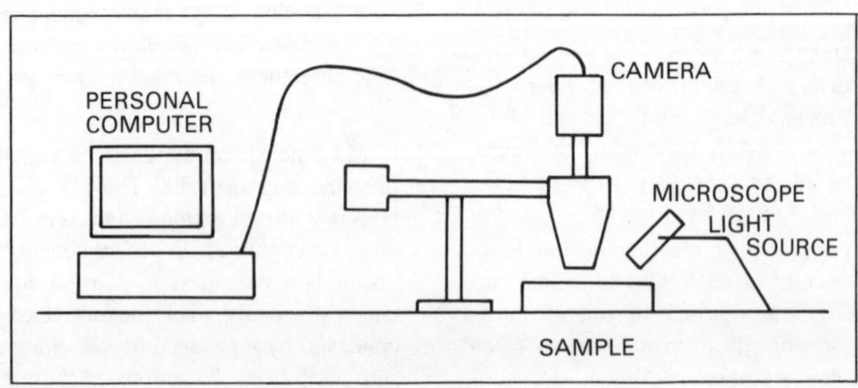

A sample machine vision/inspection system.

The actual number of shoes will be random, as will be their position and orientation. The only necessity in their placement is that they not be touching one another. The figure below shows a typical edge enhanced image of five brake shoes. All the shoes in the figure are the same type, so we will be able to see the effects of having a given shoe in a variety of rotations.

In the SME Technical Paper, *Automated Laser Robotic Workcell for Jet Engine Component Rework*, author P.A. Lovoi of InTA described an automated 20-axis laser cutting and welding robotic workcell. This workcell uses a vision system.

Noted Lovoi: "The vision system consists of four major components: laser, galvometer scanners, detector, and control computer. The vision system galvo scanners, detector, and computer ride next to the end effector, the vision laser is mounted in a remote cabinet.

"The vision laser is a 3W argon ion laser tuned to a single wavelength line at 488 nm. The laser is focused and coupled to a fiber with a GRadient INdex (GRIN) optical element. The fiber guides the laser light to the end effector where it is passed through a spatial filter, recollimated and directed into the two axis galvo scanners assembly. The galvos can direct the beam with a resolution of 0.175 mR in each axis. The beam intersects the surface to be measured and the scattered light from the intersection is imaged onto the lateral effect photodiode. The figure on the next page shows a layout of the vision system optics. This detector produces four signals, A, B, C, and D, that are proportional to the distance of the centroid of light falling on the detector from the center. Two differences are taken, A-B and C-D, to determine the position of the centroid from the detector center. The two sums, A+B and C+D, normalize the difference to provide true position signals, (A-B)/(A+B) and (C-D)/(C+D). The sums also provide intensity information that is used in grey scale operation of the system.

"The vision system is adaptive in that the beam pointing is determined by the vision computer based on the feature being scanned and previous scan data. For each scan position of predictive window is opened and the scan only covers that window. If the feature is not detected a knowledge based expert system is called that determines the most appropriate measurements to be made to regain tracking of the feature. Using this method very little data is collected that does not contain useful information. Fea-

Five brake shoes at random locations and orientations.

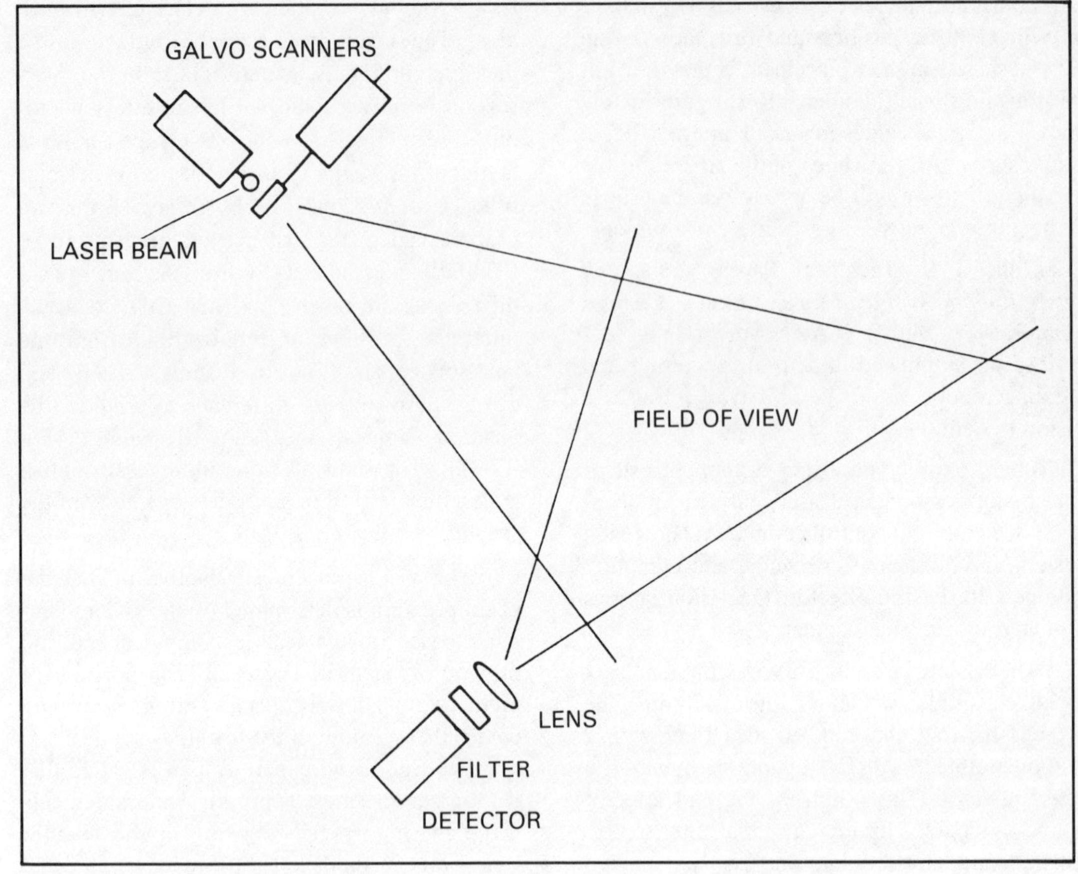

A machine vision system using a laser, galvometer scanners, detector and computer control.

Using this method very little data is collected that does not contain useful information. Features can include seams, steps, rows of holes, and marking pen lines.

''Directed high velocity gas jets across the detector and scanner ports protect the vision system optics from the welding and cutting debris and vapors.''

Also see: Artificial Intelligence, Expert Systems, Industrial Robots, Inspection.

Machine Vision Association of the Society of Manufacturing Engineers (MVA/SME). Founded in 1984, the Machine Vision Association of the Society of Manufacturing Engineers is a technical society that provides leadership in education and information exchange for the manufacturing, research, and academic communities. MVA/SME promotes individual professional development to effectively utilize machine vision technology for quality and productivity improvement.

MVA/SME members have access to the latest information and developments in machine vision technology through a wide variety of seminars, clinics, conferences, expositions, films, videotapes, and publications.

Members of MVA/SME are employed in one of four areas:

- Planning, selecting, and applying machine vision systems for manufacturing or processing;

- Designing vision systems and/or machines and equipment used in machine vision systems;
- Research and development leading to the creation of new or improved machine vision equipment or processes, and
- Administrative, educational, or governmental activities relating to machine vision.

The world headquarters for MVA/SME is in Dearborn, Michigan.

Machining. See: Abrasive Flow Machining, Boring, Broaching, Electrical Discharge Machining, Grinding, Drilling, Milling, Reaming, Sawing, Threading, Turning.

Machining Center. A machining center is a multifunctional, numerically controlled machine tool with automatic toolchanging capabilities that permit the programmed, automated production of a wide variety of parts and cutting operations.

Machining centers are available in a wide variety of types, sizes, configurations, capabilities, and costs. Machines having manual quick-change tooling arrangements are not considered to be machining centers. Some builders and users consider turret-type machines to be machining centers, but they differ in that their tools are permanently retained in their turret throughout the cycle. In multispindle or multistation machine tools, the spindles are successively indexed into the machining position, rather than the tools being automatically inserted and removed from a spindle. Typically such machines are used for higher-volume production, which does not require the flexibility of a machining center. Headchanging machines are also considered to be machining centers.

Machining centers are available with numerical control of two to nine axes. Three, four, or five-axis controls are the most common. Machines with five-axis control, used primarily for contouring complex workpieces, are available in several designs. One design employs a horizontal, rotary indexing table as the fourth (B) axis and a vertical rotary table (mounted on the B-axis table) as the fifth (A) axis. In another design, the B axis is combined with an A axis that automatically tilts the B axis. A third design combines the B-axis table with a tilting spindle (A axis).

There have been continuing evolutionary changes in the design and construction of machining centers and their controls and attachments. Improvements in design and the use of more sophisticated controls have resulted in increased productivity, versatility, accuracy, and reliability. Machining centers are now being increasingly integrated with automatic workpiece gaging, monitoring functions, fault detection, and work changing.

History. Machining centers are employed primarily for automating the production of relatively small volumes of a variety of workpiece shapes and sizes requiring multiple operations. While they may be operated as stand-alone machines, machining centers can be effectively integrated with other machines and material handling systems to form flexible manufacturing cells and systems (FMCs and FMSs).

Major reasons for the use of machining centers are the advantages of increased productivity and versatility, improved workpiece quality, and better process flow times. Productivity increases per work hour are significant, especially in applications requiring many tools and frequent changeovers. Reduced times for setup, toolchanging, loading/unloading, material handling, gaging, and troubleshooting all contribute to increased productivity.

Increased versatility results from the capability of machining centers to perform multiple functions (drilling, reaming, boring, milling, threading, contouring, and other operations) on a single machine, often completing parts with one setup and minimum operator attention. This eliminates the need for a number of individual machines, each with its own operator, thus reducing capital equipment and labor costs. Precision, operating range and workpiece size may be compromised however when compared to the single-purpose machine tools.

Discussion. Two of the major types of machining centers are vertical spindle and horizontal spindle models. Most machines have a single spindle, but they are available with multiple spindles. Right angle head attachments are available, and some machines have spindles that can be swiveled from vertical to horizontal positions. There are also machining centers with two spindles, one horizontal and the other vertical.

Vertical spindle machining centers are those that have vertical spindles at each station. These provide clear work areas for easy setup and loading/unloading. They are generally preferred for flat plate-type workpieces, for box-type parts requiring deep reach of the tools, or for three-axis machining on a single face of parts, as (mold and die work). An advantage is that most thrust forces are directed downward into the machine base and heavy tools can be used without concern about deflection from gravity. Vertical spindle machines are generally lower in cost than those with horizontal spindles and are sometimes less versatile. However, the variety of work positioning and indexing equipment available for processing multisided parts increases the versatility of these machines.

Horizontal spindle machining centers are those that have horizontal spindles at each station. These types of machining centers are generally more flexible and costly than vertical spindle machines and are available in a wider range of sizes. They are usually preferred for machining large, multisided, cube-type parts because they have no restrictions as to workpiece heights. With a simple rotary or index table, four sides of a workpiece can be machined in a single setup, and the addition of a right angle head permits machining the top surfaces.

Some manufacturers have introduced universal machines, which are machining centers with both horizontal and vertical spindles, automatic changeover from one to the other, a tool changer for both spindles, and an automatic pallet changer. With a CNC rotary/tilt worktable, universal machines can have up to five-axis machining capability. With this type of machine, five-sided machining is possible, fully automated and in a single setup. The benefits are a significant reduction in setups, part handling, fixturing costs, inventory requirements, and throughput time, in addition to achieving the high part accuracies inherent with single setup machining.

A variety of automatic toolchangers (ATCs) is available on machining centers offered by different builders. Most consist essentially of a chain, drum, or dial-type magazine or matrix for idle tool storage, and a device for interchanging these tools with one in the machine spindle. A desirable feature is random selection, bidirectional rotation that automatically selects the shortest route for the next tool, thus reducing changing time.

Many ATCs will accommodate multispindle, right angle, and other heads. The storage of large-diameter heads and tools often requires that adjacent pockets in the storage magazine hold smaller diameter heads or tools, or be left empty. On some multispindle machines, a single ATC system changes the tools in all spindles simultaneously. Some ATCs are integral with the machining centers, while others are free-standing, separate from the machines and with shuttle transfer systems. An advantage of a separate unit is that variations in tool weight and movements do not affect the machine accuracy.

A double-ended, swivel-arm mechanism was the original and is still the most common method of changing tools. A new tool is taken from storage with one end of the arm, and the other end of the arm removes the used tool from the machine spindle. The arm then swings 180 degrees, inserting the new tool in the spindle and placing the used tool in a storage pocket. Some ATCs provide direct transfer of tools from storage to spindle without the need for an intermediate exchange mechanism such as a transfer arm. Typical of this armless-type design is direct loading from a rotary drum to the spindle by axial motion of the quill. Grippers at each tool storage pocket around the drum periphery clamp and unclamp the tools at the required times.

The capacity of various storage units available is generally about 120 tools, but it may range higher.

Workpieces may be loaded onto and unloaded either manually or automatically. When the table of the machine is of sufficient size, one or more workpieces can be set up on the table while another is being machined, thus minimizing downtime.

Applications. Selection of a specific type of machining center depends primarily upon the application. Factors such as the size, complexity, and variety of workpieces to be machined, production requirements, the number and types of tools needed, tolerances that have to be maintained, and even personal preferences must be considered. Group technology aids in selecting the proper type and size machine, the required tooling, and desirable options for the most efficient and economical operation. Whenever possible, future requirements should be anticipated.

Machining centers can handle a variety of part sizes and shapes, and changeover from the production of one part to another can be done quickly and easily. Additional cost savings result from reduced material handling, workpiece inventories, and floor space requirements. Also, one operator can often attend two or more machining centers. Effective use of such centers often depends on the use of computerized scheduling and sequencing of the work.

Modern machining centers maintain close, consistently repetitive tolerances, resulting in the production of high-quality parts. There is also an increasing use of in-process or post-process gaging, integrated with the machines, that reduces inspection costs and the output of scrap parts or those requiring rework.

In February 1982, *Robotics Today* reported a 60% improvement in productivity and better quality was achieved with the use of a robot for part loading and unloading in a three-machine line at a Eaton Corporation's Fluid Power Operations facility.

Three special-purpose machines located in the machining cell are arranged in a "U"-shaped layout with one of two drilling machines first, the boring machine second, and a shuttle-type drilling machine third (see figure on the next page). The robot was located in the center of the cell.

Each of the three machines has dual fixtures for machining two parts per cycle. In the first operation, a hole is drilled on each side of the barrel of the casting. These holes are ultimately used to support a shaft in the completed differential assembly. The holes are bored to size in the second machine. The third machine drills eight holes, four on each end of the housing.

The operator loads parts onto an indexing conveyor which then moves the housings into the machining cell. Two parts are placed in each fixtured pallet where they are held in proper orientation for pickup by the robot. Parts are indexed automatically to the pickup position. A photocell signals the robot to remove incoming castings from the conveyor. The next two parts are then indexed into position.

The robot moves the housings progressively through the machining cell and ultimately discharges the parts onto a belt conveyor for transport out of the cell. An unload fixture is used to strip the parts from locator pins as the robot's dual grippers open. A photocell mounted on the fixture tells the robot that the parts have been removed from the grippers.

The operator is responsible for monitoring machine and robot operations and inspecting the completed housings. It is interesting to note the amount of manual effort the operator is spared in this installation. Each part is picked up and moved four times as it travels through the cell. This means a total of 160 lbs (72 kg) are moved in completing one part, or over 12 tons (10.9 t) per shift.

The robot grippers are designed to hold two housings, with the parts oriented end-to-end. This permits the robot to transfer two housings simultaneously from the indexing conveyor to either the first drilling machine or the boring machine. The housings must be loaded individually into the third machine (drill) since they are located side-by-side in the fixtures. This is accomplished by having the grippers mounted

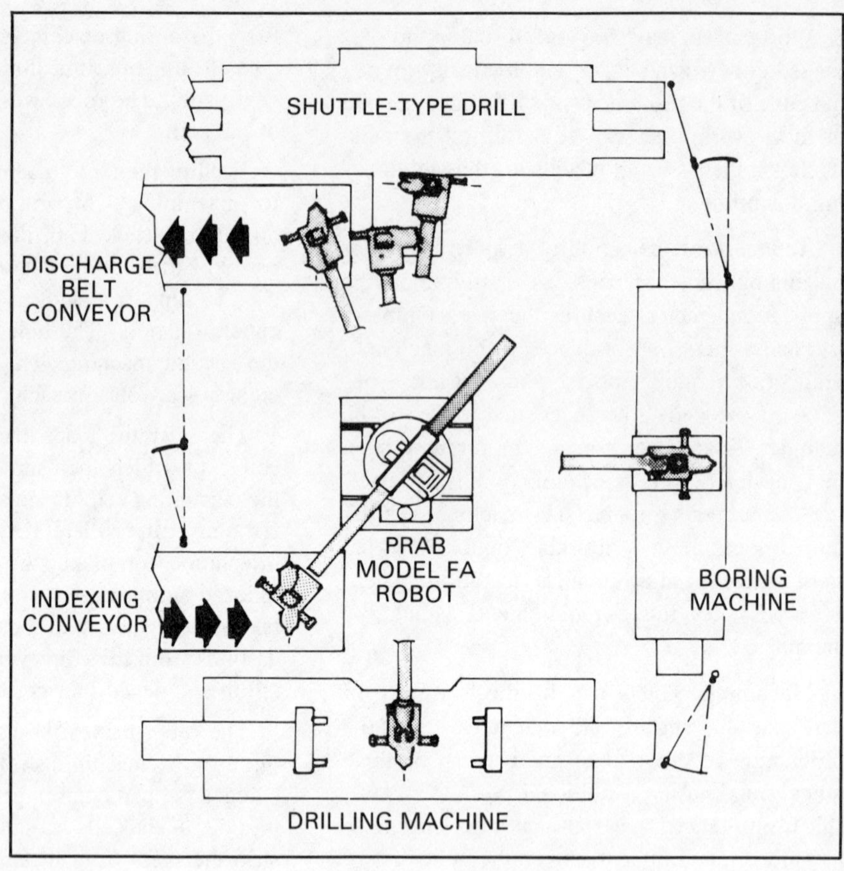

Machining cell layout.

on a rotary cylinder which permits the parts to be rotated 90 degrees in the robot arc.

Gripper jaws are opened and closed by means of a hydraulic cylinder. Part orientation is maintained during transport by the robot using locator pins which are inserted into the housing flange holes as the gripper jaws close. When the parts are unloaded from the jaws, they must be stripped off the pins to allow the gripper to move away.

Macro Instruction. A macro instruction is an abbreviated command which when executed, in turn calls another instruction or set of instructions, usually more complex and extensive than the calling instruction. A macro saves time when frequently executing the same instruction, by being able to execute an abbreviated keystroke sequence. Macros, when constructed as programs, typically execute a sequence of subroutines or programs, sometimes with program logic interspersed between the calling statements. These macro program are called *scripts*. Macros may also be simply a collection of program statements, with no calls to subroutines or other programs. The definition of what constitutes a macro is the typing of a statement as a substitute for a longer sequence of typed characters. Macros are sometimes used as a method of substituting more familiar commands for unfamiliar commands or to make another software package appear similar to a familiar package.

Also see: Computer Languages, Programming.

Magnetic Conveyor. A magnetic conveyor is a conveyor which uses a translating magnetic field on the back side of a steel surface to cause magnetically attractive parts to move along the top surface of the conveyor. A magnetic conveyor can move parts up or down an incline and at whatever speed the magnetic field is translating. In addition, a magnetic conveyor holds the parts in fixed locations relative to each other once they are on the conveyor surface. A magnetic conveyor is an excellent way of separating magnetically attractive material from nonmagnetic material such as removing steel chips and shavings from the coolant troughs in a machining environment. The lack of any projections on the top side of the conveyor allows the liquid coolant to freely run down the conveyor surface back into the tank while the shavings are securely removed to the desired location. When the parts to be conveyed are nonmagnetic, vibratory conveyors can sometimes provide similar performance characteristics as magnetic conveyors.

Magnetic Disk. A magnetic disk is a rotating circular plate that is coated or permeated with magnetic material on which information is recorded and stored for subsequent use.

Magnetic Tape. Magnetic tape is a medium which consists of mylar tape on which an iron oxide or similar type substance is deposited and which can store the signals of recorded information. Magnetic tape originally was constructed in a form of 1/2 inch in width which was divided into multiple parallel tracks, each track which could accommodate a recording density of 200, 300, 600, 900, 1800, 2400, and as high as 9600 bps (bits per inch). Magnetic tape was the first easily portable means to move large amounts of data from machine to machine or to store data off-line for mainframe computers. Magnetic tape is nonvolatile, as long as the tape does not come in contact with a strong magnetic field, which could distort or erase the data. Magnetic tape drives and form factors have been compressed with the advent of tape drives for microcomputers. With the increasing size of the hard disk drives being used by the microcomputers of today, a need for a large capacity backup storage device emerged. Magnetic tape, with its long serial record capability, is suited for this purpose. Other types of magnetic media are better for random access applications.

Also see: Bits Per Inch.

Magnetic-three Feed Rate Coding. Magnetic-three feed rate coding is a method of feed rate coding using three digits of data in the F word. The first digit defines the power of 10 multiplier and determines the positioning of the floating decimal. The last two digits are the most significant digits of the desired feed rate.

Mainframe Computer. A mainframe computer is a very large computer, generally used by many users at one time in large data processing operations. A mainframe is characterized by being a multi-user system, capable of batch-job mode or on-line operation. A mainframe may support other minicomputers and microcomputers with its disk storage capacity and processing capability. The distinction between mainframes, minis, and micros, as well as super computers, has blurred considerably as the price/performance of all computers has dramatically improved with continuing advances in integration and processing capability.

Also see: Computers.

Main Memory. Main memory is memory that is directly associated with a central processor. It may be by supplemented other peripheral storage. Other types of memory storage are bubble memory for conditions of shock and vibration when nonvolatile storage is needed. Nonvolatile memory can maintain its contents when the power to the memory is removed, while volatile memory loses its contents when power is lost. Compact disks (CD) are one of the newer forms of memory used today. CD memory uses a laser beam to read the data. Read-only memory is written to, using special equipment which then can only be read, while read/write memory can be erased and rewritten to. Some memory can only be accessed in a sequential manner such as streaming tapes (this is not the same as sequential data files versus random-access data files) while hard disks and start/stop tape drives allow random access to data.

Also see: Mass Memory.

Maintenance Management. As productivity in manufacturing seems to be progressively approaching its growth limit, maintenance has become one of the last cost savings frontiers to management.

Until most recent times, maintenance management programs were available only on mainframe computers. However, as personal computer prices dropped, many maintenance departments found that they could dedicate a portion of their budget to the development of a PC-based maintenance system.

In the SME Technical Paper *Microcomputer-aided Maintenance Management and Inventory Control*, Kishan Bagadia of UNIK Associates lists six parts of a basic maintenance management system. They are: equipment data, preventive maintenance, work order systems, maintenance (repair) history, and spare parts inventory control.

Equipment data or equipment history should provide the user with important information regarding the company's equipment by punching a few keys. This information would include the equipment's manufacturer, serial number, cost, location or purchase date.

Preventative maintenance is the periodic checking of existing facilities to discover any potential problems which could lead to equipment breakdown or early depreciation. Computer-assisted scheduling comes into play in this area. The accompanying figure presents a sample preventive maintenance weekly schedule.

Work order systems provide a documented authorization for the expenditure of material or labor. The computer may also generate a work order number at this time.

The use of a computer in maintenance management also allows a manager to maintain an inventory of spare parts and can provide a system for controlling those parts. The logical follow-up in this area is one of purchasing parts to maintain inventory. The computer in this case provides the accounting, purchasing, and record keeping function.

Bagadia points out 21 advantages of computerized maintenance. They included:

1. Increased productivity of the maintenance work force.
2. Reduced maintenance costs.
3. Reduced overtime.
4. Reduced outside contract work.

PREVENTIVE MAINTENANCE WEEKLY SCHEDULE 04/23/83

MACHINE	JOB #	CRAFT	DESCRIPTION	FREQ	
445001	1	01	CHECK THE BATTERY	52	OVERDUE 2 WEEKS
445001	2	01	CHECK SPARK PLUGS	52	
445001	3	11	CHECK OIL	26	OVERDUE 1 WEEK
445002	1	03	CHECK HYDRAULIC PRESSURE	52	
445002	3	09	INSPECT THE WELD AT LEGS	52	
445002	4	07	CHECK THE INSULATION	26	OVERDUE 1 WEEK
445003	2	01	CHECK ALL SWITCHES	52	
445003	3	01	CLEAN THE TERMINALS	52	
T57001	2	01	CHECK THE WATER LEVEL	52	
T57001	3	09	CHECK WELD ALL MOUNTINGS	26	OVERDUE 1 WEEK
T57001	4	01	CHECK RELABY THE WEAR	52	
778001	2	F11	LUBRICATE DRIVE MOTOR	52	
778001	3	02	CHECK BELT & PULLEYS	52	

A computer-assisted maintenance management schedule.

5. Reduced maintenance backlog.
6. Reduced downtime.
7. Reduced cost per repair.
8. Reduced employee frustration which results in improved morale.
9. Better service to other departments.
10. Reduced inventory costs.
11. Paperwork will be significantly reduced to help make possible the most productive use of employee time.
12. Access to maintenance information is dramatically expanded.
13. Maintenance and supervisory personnel can use their time more effectively.
14. The computer system is effective in showing individuals and crews what they contribute to the operation.
15. Longer useful life for production equipment.
16. Reduced systematic planning, scheduling and control by making material and special tools available when needed, providing clear instruction, and providing up-to-date history records.
17. Use of historical maintenance information to make better capital decisions in relation to older equipment.
18. Reduced follow-up role required of the skilled trades supervisor.
19. Stored inventories can be made more effective by expanding recorded information. As a result, the right parts or materials are more likely to be in the right place at the right time.
20. Increased overall plant productivity.
21. Increased overall money saved.

Bagadia notes in his paper: "The microcomputer has already invaded almost every aspect of industrial activity with a varying degree of success. Now it is taking its first steps into attacking the problems of maintenance management."

Management Information System. A Management Information System (or MIS) is an information feed back system in which data is recorded and processed for use by management personnel in decision making.

Mandrel. A mandrel is the core around which fiberglass that is impregnated with plastics resin is wound, as in filament winding. It is the portion of an extrusion die that forms the hollow center in an extruded tube; a form used as a cathode in electroforming; a mold or matrix; the portion of a blind rivet that is preassembled in the rivet body. During rivet setting, pulling or driving the mandrel or pin forms the secondary (blind) head. The mandrels can be smooth, serrated, or threaded.

Manipulation. Manipulation is the controlling and monitoring of selected data upon which action can be taken to vary application functions.

Manipulator. A manipulator is a device which can move and place objects in a dexterous manner. The most common type of manipulator is a robot. Manipulators are usually programmable, and can be simple devices capable of only basic moves, or sophisticated systems with sensory support.

Descriptions of the capabilities of manipulators includes such properties as degrees of freedom, axes of motion, repeatability, and accuracy. Some manipulators can only move freely between endpoints while others are capable of following a defined path and maintaining programmed speeds when moving between points.

Also see: Industrial Robots.

Man-Machine Interface. By definition, a man-machine interface (MMI) is the link between a human and a machine. A terminal, with video display monitor and keyboard is the MMI for a computer. Often MMIs are customized designs or proprietary interfaces on specialized equipment such as programmable controllers, robots, and numerical control machines. MMIs can provide only information output as in display-only devices such as CRT monitors and indicator panels. MMIs can also allow input only as in start/stop buttons on control panels. The ease of providing input, interpreting output, and the quality of comments and error messages determines the degree of user-friendliness in the

MMI, and often the success of the system. The advent of bar code equipment, magnetic cards, OCRs, voice input and voice response devices, and other alternative-media devices has provided for a large selection of forms and formats for man-machine interfaces.

Also see: Programmable Control, Industrial Robots.

Manual Backup. Automated industrial control systems require alternate manual controls to accommodate failure of the automated control system. Most automatic control systems can be operated in a manual mode. A manual backup system is a hardwire control linkage which entirely bypasses the automated control system. An industrial control system has three modes of operation—automatic (programmed), manual (nonprogrammed using the computer), and manual backup (using hardwired controls). A manual backup system is required by most process control installations to ensure the ability to maintain safety. Manual backup usually is used to augment an automatic control system, as opposed to being a redundant option of an automatic control system. When systems used to be primarily manually controlled, manual backup did not exist as an obvious need.

The manual backup system for a process control system usually consists of a hardwired potentiometer and a display mechanism to provide feedback to the manual controller.

Also see: Manual Control, Process Control.

Manual Control. Manual control takes place when the direct control of a device is being performed by human intervention, as opposed to automatic control. Automatic control is when a system operates without any direct human interaction. Manual control allows a human operator to have direct interaction with the controlled devices of a system. A manual control system usually requires human readable indicators to facilitate the proper adjustment of the control elements.

In an electronic system, manual control is when a person is able to adjust potentiometers directly, not by having to effect changes through control circuitry. In the process control industry, manual backup control is a required safety precaution to ensure the ability to safely maintain a process if the automated control system fails.

Also see: Automatic Control, Computers, Manual Backup.

Manual Data Input. Manual data input is a means of manually inserting commands and other data into an NC control.

Manual Lead-through. This system physically manuevers the robot end effector through the desired sequence of actions that will be automatically repeated in playback.

Manual Mode. Manual mode is a mode of operation in which NC machine tools are controlled manually through axis jog switches.

Manual Part Programming. Manual part programming is the preparation of a manuscript in machine control language and format to define a sequence of commands required to accomplish a given task on an NC machine.

Manufacturing Automation Protocol. Manufacturing Automation Protocol or MAP is a communications protocol specification which is currently being developed. MAP was originally proposed by a committee of potential manufacturing equipment users and manufacturing equipment manufacturers, as a means of reducing the existing problem of connectivity between many different manufacturers' machines.

MAP is a response to two preeminent technological and one stark economic trends. First, there is growing recognition of the need to coordinate production facilities and automation equipment with corporate operational policies. One example is just-in-time scheduling, which minimizes work-in-progress inventory and working capital. Secondly, there is a need to combine automation technologies provided by a widely divergent set of vendors into a single integrated plant control system. This trend has precluded the typical office automation solution of buying all required computing resources from a single vendor.

The economic trend stemmed from the General Motors Corp. recognition in the early 1980s that its U.S. operations were not as competitive with those of many overseas auto manufacturers. GM adopted a strategy of automating its plants in order to compete in the world market. Company officials realized that the required multivendor factory automation solutions could only be accomplished on the basis of a widely supported nonproprietary factory communication specification. This economic reality, felt first by GM and now becoming widely accepted, led the company to create MAP.

GM selected Open Systems Interconnection (OSI) standards as the basis for MAP because of the worldwide acceptance these standards command. This support has helped MAP achieve the broadly based constituency that is necessary for any effective standard. Computer manufacturers and the Corporation for Open Systems have enthusiastically supported the OSI basis of MAP because this will allow them access to a much broader and larger market than is available to any proprietary or U.S.-based solution. The emergence of Technical and Office Protocols (TOP), a complementary suite of OSI protocols created under the guidance of Boeing Computer Services (BCS) of Seattle, WA, cemented this worldwide computer vendor support.

International MAP users groups have been formed to allow proponents of MAP to have a forum and organizational base for professionals and supporting corporations involved with factory floor communications. The Users Group's objectives include generating a standard set of communication specifications based on the OSI model, and accelerating acceptance of industry communication standards. Other goals are to improve manufacturing industry productivity by installing MAP technology. The Users Groups also provide marketplace feedback to computer and support system manufacturers. The Secretariat to the World Federation of MAP/TOP Users Groups is the Society of Manufacturing Engineers, of Dearborn, MI.

History. The MAP specification, first published in 1982 as Version 1.0 and revised in 1985 to Version 2.1, is primarily comprised of subsets of the OSI communication standards. In the event that an OSI standard does not exist for a communications feature, the MAP and/or TOP technical subcommittees have written interim definitions which have been included in the MAP and TOP specifications. However, these interim definitions will be replaced with appropriate OSI standards as they emerge.

The MAP specification has been through several versions. MAP version 3.0 is the current version. Until recently, vendors and manufacturers of MAP and TOP-compatible equipment have a dilemma of deciding whether to develop products for the momentarily current version or to wait for the next version to evolve. Solutions currently exist that will integrate MAP 2.1 version within the newer 3.0. In addition, the MAP/TOP Users Group released a Stability Statement at its September 1987 meeting that provides a six-year upward compatible agreement that begins as of June 1988.

MAP was first demonstrated at the National Computer Conference (NCC) in July 1984. This demonstration was orchestrated by General Motors and Boeing and featured multiple vendors' equipment. The purpose was to demonstrate the interaction and functionality of using a common communication protocol. The National Bureau of Standards (NBS) participated in the demonstrations as conformance testing is a major part of designing and producing true MAP compatible products. Similarly, a demonstration that conveyed MAP 2.1 and TOP 1.0 functionality took place during AUTOFACT 1985 in Detroit.

Several vendor companies offer MAP compatible products, which implement the MAP 2.1 specification at the beginning of 1988. Most of these products are communications oriented and require integration to larger manufacturing equipment.

Discussion. MAP uses a token bus, broadband communications network as its physical link, as defined in the IEEE 802.4 standard. TOP (Technical and Office Protocol) uses a baseband (IEEE 802.3) as its network specification with a provision for broadband in certain circumstances (i.e. where an existing broadband network currently exists).

MAP is based on the seven-layer OSI model. The seven layers are: Level 1-Physical, Level 2-Link, Level 3-Network, Level 4-Transport, Level 5-Session, Level 6-Presentation, Level 7-Application. Messages and files are exchanged by communication between two devices using the application layer services and protocols. ISO divides these services and protocols into two categories called Common Application Service Elements (CASE) and Specific Application Service Elements (SASE). The MAP specified protocols for each of the seven ISO/OSI layers are listed in the figure on the next page.

CASE provides services which are, by definition, common to distributed application processes wishing to exchange information among themselves. ISO has thus far defined three types of CASEs: association control, context control, and commitment concurrency and recovery. The association control CASE permits applications to establish associations among themselves, transfer messages over those associated connections, and either gracefully terminate or abort the associations. The context control CASE allows application processes to define and manipulate the meaning of the information they wish to exchange. Finally, the commitment concurrency and recovery CASE permits application processes to coordinate the activities of separate associations. MAP Version 2.1 provides only the association control CASE. For this reason, the meaning of transferred messages and the coordination of associations must be defined by the programmers of the distributed applications.

The second category of application services and protocols, SASE, provides information transfer capabilities to satisfy the needs of particular applications such as file transfer and distributed job processing. MAP Version 2.1 provides limited support of the OSI File Transfer Access and Management (FTAM) SASE.

OSI FTAM contains many features that can be used by peer file transfer applications to access and manipulate, via telecommunications means, files, or bodies of information treated like files. As a subset of the OSI FTAM, MAP Version 2.1 FTAM provides services and supporting protocol that permit applications to create or delete files, read the characteristics of a file, and transfer the entire contents of a file in a binary or ASCII text file format. MAP Version 2.1 FTAM, by providing its own association and connection services, differs from the OSI FTAM standard that assumes the use of association control CASE.

In addition to association control CASE and the FTAM SASE within the application layer, MAP Version 2.1 also defines a network management application, directory service application, and a standard for constructing manufacturing-related messages. MAP 2.1 components were defined by MAP technical subcommittees since OSI standards for these features did not exist.

MAP Version 2.1 network management includes features that permit performance management and event processing within an operational MAP network. The technical subcommittees have divided network management into two major components: the network management agent and the network manager. The network management agent is an application responsible for collecting statistics and accepting event notifications generated by CASE and the session, transport, and network layers. In addition, the network manager by using association control CASE and a special agent-to-manager protocol.

The network manager consists of two applications that collect performance statistics and network events from the agents in the network and then display this information to a network administrator.

MAP Version 2.1 provides vendors with some implementation latitude that will result in network management product variations. For

LAYERS	FUNCTION	MAP SPECIFICATION
USER PROGRAM	APPLICATION PROGRAMS (NOT PART OF THE OSI MODEL)	
LAYER 7 APPLICATION	PROVIDES ALL SERVICES DIRECTLY COMPREHENSIBLE TO APPLICATION PROGRAMS	CASE, FTAM, MMFS/EIA 1393A DIRECTORY SERVICE NETWORK MANAGEMENT
LAYER 6 PRESENTATION	TRANSFORMS DATA TO/FROM NEGOTIATED STANDARDIZED FORMATS	NULL AT THIS TIME
LAYER 5 SESSION	SYNCHRONIZE & MANAGE DATA	ISO SESSION KERNEL
LAYER 4 TRANSPORT	PROVIDES TRANSPARENT RELIABLE DATA TRANSFER FROM END NODE TO END NODE	ISO TRANSPORT CLASS 4
LAYER 3 NETWORK	PERFORMS PACKET ROUTING FOR DATA TRANSFER BETWEEN NODES ON DIFFERENT NETWORKS	ISO CONNECTIONLESS NETWORK SERVICE
LAYER 2 DATA LINK	ERROR DETECTION FOR MESSAGES MOVED BETWEEN NODES ON THE SAME NETWORKS	ISO/DIS 8802/2 LINK LEVEL CONTROL CLASS 1 OR CLASS 3 ISO/DIS 8802/4 TOKEN ACCESS ON BROADBAND MEDIA
LAYER 1 PHYSICAL	ENCODES AND PHYSICALLY TRANSFERS BITS BETWEEN ADJACENT NODES	10 MBPS AMPSK ISO/DIS8802/4 OR 5 MBPS CARRIERBAND PHASE COHERENT FSK

PHYSICAL LINK—(COAXIAL CABLE)

MAP specifications summary by layer (2.2 Reference Spec).

example, MAP Version 2.1 does not specify the network manager's end user interface or use of the collected statistics and events. Additionally, MAP Version 2.1 only requires the support of the network management agent; therefore, vendors may decide not to provide a network manager at all.

In MAP 2.1, network management is an interim solution that requires future extensions to provide facilities for complete administration and control of MAP systems, their applications, and the connections among those applications. This is due in part to the fact that MAP Version 2.1 assumes the physical and link layers are managed by non-MAP-specified mechanisms. In addition, the statistics and events required by the 2.1 specification will limit how much useful information can be derived about the state of the network. Consequently, many vendors may elect to provide proprietary extensions to their network management facilities. Unlike MAP Version 2.1, MAP 3.0 provides a clearer definition of Network Management and therefore eliminates many of the issues surrounding Network Management in 2.1

The MAP Version 2.1 directory service provides application process names to network address translation. A full MAP network address is in excess of 50 octets. Rather than burden end-users with knowing the full address of an application with which they want to establish an association, directory service allows the use of a symbolic name that represents that address. As with network management, directory service is divided into two components: an agent called the *directory client service agent* and a server called the *directory service agent*. The directory client service agent accepts symbolic names and looks in a local directory for the appropriate network address, or, if necessary, by using association control CASE and a special inquiry protocol, they request the directory service agent to provide a translation. The client service agent then passes the full network address back to the user. The directory service agent processes requests for name to address translation and maintains a directory information database that contains the mappings between the symbolic names and the full network addresses.

MAP Version 2.1 also specifies a language for machine independent, process-to-process information exchange between manufacturing related of applications and programmable devices such as robots and programmable controllers. This language in version 2.1, called Manufacturing Message Format Standard (MMFS), defines a syntax for messages and a semantics for the components of a message. For example, MMFS defines the bit encodings and the meaning of those bits for commands such as reading and writing I/O registers in programmable controllers. In general, MAP Version 2.1 implementations of MMFS will provide ease of use application programmer tools for encoding and decoding MMFS messages.

MMFS is currently being enhanced by an Electrical Institute of America (EIA) committee and defined as an SASE called the Manufacturing Message Service (MMS). This SASE will be published as EIA Standard RS511. MAP 3.0 will require support for MMS instead of MMFS.

Underlying the application layer are the presentation and session layers. These layers provide the services required by applications to format their data exchanges and manage their dialogs in an orderly manner.

The presentation layer offers applications and a set of data representation services. For associated applications to understand the information transferred among them, they must first agree to use a common set of encoding rules to represent information such as character codes, real numbers, and file formats. The purpose of the presentation layer is to negotiate these common encoding rules called the *transfer syntax*. In essence, by negotiating the transfer syntax, the presentation layer decouples the syntax or encoding used for the information exchanged by the applications from the semantics or meaning of that information. As a consequence, the presentation layer eliminates the need for application programmers to agree prior to application development on the transfer syntax of the exchanged information. In addition to negotiating transfer syntax, the ISO presentation working group is also considering alternate encodings for the purposes of data compaction/decompaction and encryption/decryption.

The session layer provides services for connection establishment and termination, token control, and data synchronization. Connection establishment and termination provide session users with the ability to make and break connections between themselves. Token control determines which session user has the right to invoke certain services. For example, the data token controls the right to send data when a two-way alternate session connection is used. Synchronization services provide a means for the session user to identify points in the data stream and to resynchronize to a given point.

MAP Version 2.1 supports only the basic functions of the session layer called the *session kernal*. These functions permit CASE and FTAM to establish a session connection, transfer normal data without first requiring the right to send (called two-way simultaneous or full duplex operation), and release the session connection. Synchronization services are not provided in the current MAP 2.1 specification.

Below the presentation and session layers are the transport, network, link, and physical layers. Together, these layers provide the communication services required by applications to reliably exchange information among themselves.

The transport layer provides functions for error-free delivery of messages passed from the session layer. This layer includes functions for message acknowledgment, sequencing, flow control, and error recovery. ISO has divided the transport layer protocols into five classes numbered 0-4. Class 0 provides basic facilities for connection establishment, data transfer, and protocol error reporting, while Class 1 adds the ability to recover from network disconnects or resets. Class 2 permits connection optimization by providing multiplexing of several transport connections over a single network connection, and Class 3 adds flow control and network disconnect recovery. Class 4 adds the ability to detect and recover from message error, and provides for increased throughput capability by allowing a single transport connection to use multiple network connections. An appropriate transport protocol class is chosen based on the reliability of the lower three OSI layers. However, regardless of the protocol class used, the transport service remains the same.

MAP Versions 2.1 and 3.0 require the use of transport protocol Class 4. This is due primarily to the fact that the probability of residual errors (those errors that remain undetected) is high for the lower three layers of MAP. Therefore, the error detection and recovery capability of the transport protocol Class 4 ensures reliable delivery of messages across the network.

The network layer provides transparent routing and relaying of messages exchanged between two transport entities. The network layer is rather complex since one of its major features is the routing of messages across various subnetworks ranging from private local area subnetworks to public wide area data subnetworks. At last count, ISO had prepared eight standards to describe the variety of services and functions offered by the network layer.

MAP Version 2.1 requires support for the connectionless network protocol often referred to as the *internet protocol*. This protocol provides network layer services for routing, addressing, and optimizing the size of data units passed within a subnetwork. The use of the internet protocol does not, however, preclude the interconnection of various subnetworks. In fact, by supporting other network protocols in addition to the internet protocol, systems can route data units received from one subnetwork to a destination system on the same or different subnetwork. These systems, called routers or intermediate systems, were used during the AUTOFACT '85 MAP demonstration to interconnect the three subnetworks at the demonstration in downtown Detroit to another MAP subnetwork located at the GM Technical Center in Warren, MI.

The link layer provides services to the network layer for transfer of data units between two adjacent systems linked by a physical communications medium. For local area networks, the Institute of Electrical and Electronics Engineers (IEEE), whose local area network link and physical layer standards have been endorsed by ISO, divide the link layer into two sublayers: logical link control and media access control. Logical link control provides addressing and control services to the network layer for transmission of data units. Media access control accepts data units from the logical link control, breaks those data units into smaller units called frames, and, when permitted, transfers those frames to the physical layer for transmission.

Since the IEEE has defined three types of logical link controls and two major means for controlling access to the physical medium, MAP Version 2.1 defines the specific logical link control (LLC) and media access control (MAC) which must be used in the MAP link layer. First, MAP Version 2.1 specifies the use of the LLC Type 1. This type of LLC, noted for its minimal protocol complexity, permits network layers to exchange data units without the need for a specific link connection. However, LLC Type 1 does not ensure error-free and in-sequence delivery of those data units. As mentioned earlier, this is the major reason why transport protocol Class 4 is utilized in MAP 2.1. Secondly, MAP requires the use of the token passing MAC method. Through this method, a token is passed from one system to another in a defined sequence. Only the system holding the token is permitted to transmit. Since the sequence and token rotation time can be adjusted, this MAC method lends itself to environments where the maximum time required for a system to access the network must be deterministic and relatively constant. LLC Type 3 is used in MAP 3.0.

The final layer of the OSI reference model is the physical layer. The physical layer provides the electrical interface to the transmission medium. This layer includes both the mechanical connection to the medium as well as the means for activating the medium to transfer bits of information. The IEEE has defined three different physical layer implementation techniques with corresponding media characteristics suitable for use with a token passing MAC. Since each is rather technical to describe, suffice to say that MAP Version 2.1 recommends the use of a physical layer that supports 10 megabit per second raw transfer rates over 75 ohm broadband coaxial cable. In MAP Version 3.0, the layer supports 10 megabit over 75 ohm broadband coaxial cable and five megabits over carrierband. This medium, identical to that used for cable television, is suited to factory environments due to its high noise immunity and ability to be installed over long distances.

The definition of MAP is continually being refined and enhanced. Current work is aimed at such activities as improving MAP FTAM, network management, and directory services. In addition, extensions to MAP to support real-time device communications requirements are being developed.

Applications. MAP is intended to eliminate the "islands of automation" which currently exist in many factories. These islands are due to different manufacturers' equipment having different communication and interface capabilities. MAP will allow a multiuser environment in factories with standard messaging for factory floor devices. Users have a common, standardized network specification to which machines and equipment can be designed to and can be simply connected together to "plug and play". One of the goals of MAP is to provide access to any computer on the MAP network from any other device or terminal on the MAP network, even if they are different machines running different operating systems.

Benefits of using MAP are a decrease in implementation time and the cost savings associated with it. The connection of a machine or subsystem into a larger system can be done while focusing on higher level interface considerations.

As VLSI implementations of MAP provide less expensive connectivity, the installations of MAP can be expected to increase dramatically in years to come. Large scale implementations for

MAP and TOP are currently (as of early 1988) limited. But, with the advent of Version 3.0 products comes a broader acceptance and hence more widespread installations.

Deere & Company is one of the first companies to put MAP to work. David C. Scott (Deere & Company) described these systems in his paper, *Making MAP a Reality a User's View*. The paper appeared in the SME book, *A Program Guide for CIM Implementation*.

Wrote Scott: "What turned out to be the first manufacturing production MAP-compliant network was incorporated in a CIM pilot project in 1984. This pilot was initiated in the sheet metal facility of the John Deere Harvester Works, in East Moline, Illinois. The Harvester Works covers over 260 acres (95 under roof), and is located on the Mississippi River. The facility performs complete design and manufacturing from punching and welding of sheet metal, machining of bars and castings, through assembly and inspection for a full line of grain harvesting combines.

"In January 1985, the first NC programs were downloaded from the DNC system to the laser punch and the shear. Now, NC programs generated and nested on two Computervision Designer 5X systems can be downloaded to the Numeritronix DNC system for storage. Programs can be uploaded across the network, as well, so if a program does not run properly on the shop floor, it can be modified in the memory of the controller and uploaded to the DNC system in the new usable form.

"It took considerable communications expertise and about three months to progress from getting equipment in-house to using the DNC system for production. Most of the problems lay in not specifying all of the communications features needed at the time of equipment acquisition, partly because this pilot project came together as the culmination of three separate evaluations of need within the company at the same time.

"In summary, the project helped gain practical experience in networking and MAP, and in developing specifications for equipment for this environment. At a minimum, the communications section of a machine process specification must include consideration for:

- cable type and pin-out
- connectors
- numbers of bits per character
- parity
- protocols and
- speed.

"Consistent protocols and speed difference compensation are advantages of broadband networks. New CDS and other full MAP products provide those services now. MAP was in its infancy in 1984; Deere and Concord Data Systems were pioneers.

"Other existing equipment was also connected to the network. A TIM (Token Net Interface Module) was added in a room where a Visual 50 terminal acts as a console device for the DNC system. Several Lear Siegler terminals used by engineers to access the CV CAD/CAM Systems for data management and file manipulation also were connected. With these terminals on the network, a user at any terminal can access either system. An Okidata printer also was added when new software which accommodated hardware handshaking became available.

"A VAX 11/750 that is used as backup on another system in the plant also was added. As long as the system it backs up is operational, it is used for development and software testing. It is now also being used for network management. A program was written for the VAX that allows statistics on all of the devices on the network to be collected with one command.

"Four Wiedematic punch presses with GE 7500 NC controllers have also been added to the DNC network. For this retrofit installation on less intelligent controllers have been added to the DNC downloads menus and NC programs to operators on the 1502 terminals. The 1502 then supplies the machine controller with one block of the program at a time as the NC machine requests it, similar to reading a tape. To add these NC machines, on four-port TIM was needed, as was some coaxial cable to drop from the main broadband; the initial backbone utility cable topology was designed to permit future additions such as this.

"The LAN also has been extended for a separate project being installed in the welding operations. This system links a VAX 11/730 and two cell controllers each controlling up to 16 controls on manual spot welders. The system, designed mainly to improve the quality of spot welding, is being supplied on a turnkey basis by Medar. The total network configuration is shown in the figure on the next page.

"Since completion of this project, a team was formed to update the strategy to address CIM for the factory. Most of their proposals will not require large capital investments, but instead involve finding better ways to use equipment already in-house. Integrating systems already in use, finding common formats for related software programs, and optimizing equipment usage are some key portions of the plan. Based on the success of the MAP pilot, the team recommended expansion of the network throughout the entire facility to provide an essential communication utility to support these CIM concepts.

"The prices of MAP equipment are dropping, but even now, the LAN would not be cost-competitive if there were no side benefits. For example, terminals that were once dedicated to one of two computer systems can now access either system. The status of the machines and computers connected to the LAN are all available from one point on the network. Network experience allows exploration into newer and wider uses of the utility.

"Perhaps the largest challenge remaining, on a company-wide basis, is how to plan for and manage the shop floor communications, computing, and automation directions to facilitate CIM in a large decentralized company such as John Deere. Experiences such as those in the pilot need to be shared and incorporated in future efforts if they are to be worthwhile. To address this challenge, Deere has taken several steps.

"At the corporate office, a Computer Aided Manufacturing Services organization was formed. Their responsibilities include factory automation planning, factory local area networks, and MAP activities, among others. Under this group's leadership, a Shop Floor Computer Communications and Control (SFCCC) Users Group has been formed. A steering committee with members from corporate Manufacturing Engineering, Computer Systems, Quality and Reliability Engineering, and CAM Services was formed to coordinate and provide input to the overall group. This general group is composed of factory representatives from the same disciplines to form a core shop floor communications team with members from Plant Engineering, Computer Systems, and Manufacturing Engineering Systems. Similar to the GM MAP/TOP Users Group experience, the response to the SFCCC has been very enthusiastic.

"The role of the SFCCC is three-fold: education, coordination, and direction. At the meetings, educational segments on topics such as MAP are included with unit updates on experiences, current problems, and issues being faced. Members are asked to share lists of vendors, contacts, informational sources, and other such data which may be useful to others. Based on discussion of the issues surfaced by the group, plans for their resolution to provide future direction are formulated.

"At the first general meeting, the users decided to form three interim working groups to develop guidelines for the three primary issues identified during the meeting. The first developed boilerplate specifications for inclusion in shop floor equipment specifications which will deal with the desired capabilities of the computer control. It covers physical and electrical characteristics, protocols, and general capabilities required to integrate the equipment into an overall CIM system. Basic examples are upload and download commands and terminal mode. The application layer of MAP using the EIA-1393 specification will be the ultimate answer when complete. The second group worked on guidelines for the communication utility and physical cabling schemes, including plans on how to evolve from current systems (e.g. point-to-point) to the ultimate broadband for MAP in a cost-effective way. The final group

 Manufacturing Automation Protocol

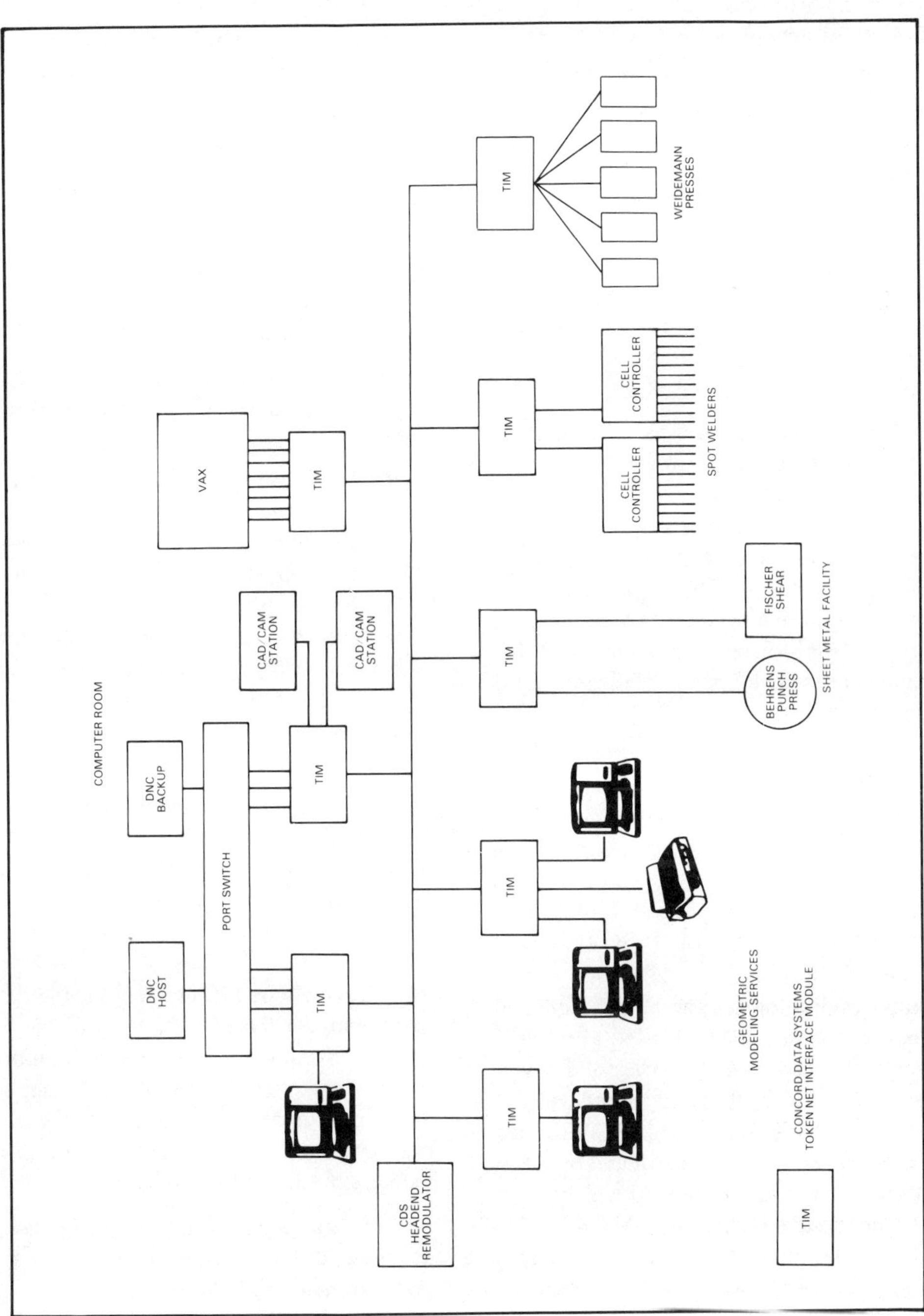

The Deere Harvester MAP Network.

addressed flow and placement of data in the evolving distributed hierarchy on the shop floor.

"Another obvious need identified is training at all levels. To address this, outside training vendors are being contracted to provide courses ranging from overviews of LANs, to MAP, to detailed broadband engineering instruction. Extensive contacts are being maintained with vendors of LANs, controllers, and computer equipment to obtain education on their directions and products and encourage their participation on future pilot projects, stressing MAP.

"CIM will become a reality on a broad scale within the next few years. However, to be ready, the foundations must be laid now. The communications schemes for the shop floor are essential and the MAP approach offers a very attractive alternative, with an expanding support base from both users and vendors. But rather than just waiting for MAP to happen, it is important that those companies which hope to realize the CIM potential begin planning and learning now through activities such as pilots and users groups. The benefits are tremendous."

Another MAP application was reviewed by Douglas Richardson, ITP Boston, Inc. in the January 1988 *Manufacturing Engineering*. The article titled *Implementing MAP for Factory Control* explained: "As MAP matures and comes into widespread use, it will bring substantial savings in development costs and increased flexibility for the developer. It is still in a very early stage of acceptance, however, and developers should consider all options and develop a clear picture of the objectives MAP is intended to meet. A number of questions should be asked at the very outset of the project: At which levels in the overall system is the use of MAP feasible and appropriate? Should it be used solely for communication between devices? Between devices and cell control? Between cell control and higher level factory control? Among different application components of factory control?

"Hardware choices usually establish fixed constraints. For example, specific devices needed for a project may not support MAP. Constraints also arise from the existing MIS environment into which the project must fit, and from the use of software packages for specific areas.

"Before attempting to sort out the many technical options, the developer must have a clear understanding of the basic requirements for the project as a whole. This is far easier said than done, since the underlying issues and tradeoffs are fuzzy, hard to quantify, and hard to communicate to management.

"There are two guidelines of overriding importance for those considering implementing MAP:

- Do not force MAP into a software architecture where it does not fit. If the project's constraints are such that MAP cannot be the backbone of its communication scheme, abandon it.
- If MAP is the backbone of your network, use it wherever possible. Any boundary between MAP and another communication protocol will be a headache, and there are big returns in minimizing these boundaries.

"Designing the communication portion of any system is one of the most complex problems the application designer has to face. The development team is likely to have only a few members who are skilled in communication, while the other developers must focus on the application itself. For this reason, the development team for the FOF (Factory of the Future) project developed a set of transaction-oriented routines that made it easy for the application developers to send and receive messages. These transaction interface routines are the only part of the system that call MAP functions.

"The transaction interface can be thought of as an eighth layer situated between the application program and MAP itself. This layer allows the applications to work with a simplified picture of the communication network. It also isolates the use of MAP, which is important because future revision to the MAP specification may require changes to existing programs.

"In the FOF project, the transaction interface does the following:

- Manages all the details of establishing the appropriate MAP connections.
- Appends a unique transaction number to each message and generates an acknowledgement message for recovery purposes.
- Buffers messages if a connection is down and handles the details of re-establishing the failed connection.

"The disadvantage of machine independence is that the MAP application interface is not integrated with the host computer's operating system. For example, there is no way to wait for an operating system event (such as the expiration of a timer) or a communications event (such as the receipt of a message). The real effect of this limitation may not be clear until a project is well into the design stage.

"Applications that control real-world systems must be interrupt-driven to some degree. In some cases, an interrupt may represent a physical condition, such as a limit switch closing. In other cases, it may represent a more abstract condition, such as an unanticipated inventory shortage or a time-dependent event. The computer's operating system provides the tools necessary to deal with these conditions.

"The problem is that the MAP application interface only allows the programmer to wait for a MAP communication event. This causes difficulty when a program must respond to MAP messages or other external events."

Author Richardson also noted: "An automated factory is vulnerable to many kinds of failures. The problems of limiting failures and designing the overall application to minimize their effect is outside the scope of this discussion. One point should be clearly understood in designing a MAP-based system, however, MAP is not failure-proof.

"Despite the extensive error checking protocols embedded within its layers, MAP is vulnerable to the problem that all communication systems face: When a host computer or application process crashes, messages can be irrevocably lost. If a crash occurs while the message is in the pipeline between sending and receiving processes, all messages in the pipeline are lost.

"One approach to this problem is to reduce the incidence of failure by using highly reliable, or fault-tolerant computers. This approach, which was used in the FOF project, can reduce the number of failures due to the computer hardware. Unfortunately, it cannot eliminate the possibility of software errors, either in the system software or in the application software.

"Therefore, even when designing a system for a fault-tolerant computer, the possibility of failure must be taken into account. This has several implications for the application designer.

"First, the application has to deal with the fact that messages may be lost if a failure occurs. In such applications as monitoring, this may be an acceptable risk. In a control application, on the other hand, it is unacceptable; the application must provide a protocol on top of MAP to detect and resend the lost messages. This generally involves uniquely numbering the messages, checking the sequence of incoming messages, and generating ACK/NACK messages in response.

"Application using a database management system (DBMS) must deal with an even more complex issue: how to make sure that consistency is maintained between MAP messages and the information contained in the DBMS. Consider an application that processes material handling commands, and suppose that a command involves updating the inventory status of a unitized load (in the database) and sending a message via MAP to an AGV to move the load. Further, suppose that a failure occurs while this command is being processed. Depending on the order in which the actions take place, one of the following inconsistent states may result: The message is sent and processed, but the inventory status is not updated...or the inventory status is updated, but the message is never sent. Either situation results in serious real world problems. The designer must ensure that an inconsistency of this sort is harmless or come up with a way to eliminate the inconsistency.

"Failure must be anticipated. The design effort required to deal with system failure and recovery is substantial.

"The widespread interest in MAP and the substantial standards effort that MAP represents illustrate both the hope being placed in CIM and the potential problems that must be overcome. The MAP crusade has the potential to foster unrealistic expectations, which can only be followed by disappointment. However important it may be, MAP is only one ingredient for a successful CIM application. Caution must be used when anticipating the benefits derived from its use.

"One problem with communications standards is that they are written by communications specialists who often view English as a second language to Computer Science. These specialists sometimes overlook the fact that communications standards will be used by nonspecialists—application developers. Most of the literature on MAP deals with details of the lower layers that are completely irrelevant to the user. More attention should be focused on how MAP can be used, rather than on how it was built. Such a shift in emphasis will foster greater use of MAP and move it closer to becoming a communications standard.

"Another problem with standards efforts is the design by committee syndrome. In the effort to satisfy conflicting points of view, committee efforts often result in an awkward, cumbersome design. The description of the application interface in version 2.0 of the MAP specification occupied 62 pages. In the first draft proposal for 3.0 spec, it occupied 393 pages.

"Futhermore, users often abet the tendency toward cumbersome standards by pushing for every conceivable feature, enhancement, and frill. The success of UNIX lends credence to the argument that the virtues of simplicity far outweigh the inconvenience of limitations. As the well-known software engineer Pogo once said, 'We have met the enemy, and he is us.'"

Also see: Computers, Local Area Network, Network, Open Systems Interconnection, Technical and Office Protocol.

Manufacturing Requirements Planning. MRPII is a system which supports the operations planning and execution of a manufacturing firm. MRPII includes high level planning, operations planning, operations execution and operations reporting.

High level planning is planning which is performed by senior management and includes business planning, sales planning and production planning. Operations planning includes the material requirements planning, capacity requirements planning and master scheduling. Operations execution and operations reporting are those areas where the plans are acted upon. These areas include purchasing, inventory control, shop floor control and performance measurement reporting. There are additional data requirements to perform MRPII and these include definitions of bill of material, item masters, process definitions and product costing.

Business planning defines the long-term direction, objectives and strategy, taking in consideration the environment and business market and the internal analysis position of the firm. The business plan is the driver for the marketing plan, operations plan and the financial plan.

Capacity planning is the need to determine current and forecast facility requirements to meet planned production volumes.

Sales planning establishes the rates of all product lines defined in dollars and in units to meet the business plan. The sales plan should not be confused with a forecast which is further defined by time.

Production planning is the function of establishing the overall level of manufacturing output. The purpose of a production plan is setting a production rate that will meet management's objectives with respect to raising or lowering inventories or backlog balanced with maintaining a production force relatively stable.

Master production scheduling defines the quantities of goods to be produced by time period based on the product mix, models/features/options, material availability, capacity availability and time fence policies. The master production schedule is a statement of all production plans in terms of what, how much and when. A master production schedule to be valid must be detailed, time-phased, realistic/believable, inclusive to cover all demand, written and approved. The master production schedule is a statement of what the firm intends to do and not what the firm wishes to accomplish.

Material requirements planning is a management tool that plans material and production in a priority sequence, in a time-phased manner to meet a firm's requirements. Material requirements planning uses the bills of material, inventory data and the master production schedule and calculates and maintains orders and priorities for purchase and manufacturing requirements. Material requirements planning produces a plan of action for production and purchase requirements which includes information for the what, how much, and the when. Material requirements planning maintains valid schedules, satisfies dynamic requirements of changing orders, highlights pitfall shortages, minimizes inventory levels and maximizes customer service.

Purchasing in an MRPII environment supports the execution of the plan by ensuring that the right material arrives, in the right quantity, at the right time, with the highest quality. Purchasing uses the net requirements, part information and vendor data to facilitate order placement, monitor orders, and monitor vendor performance. Purchasing is responsible for the vendor; this includes status of tooling and overall capabilities, financial and labor stability, warranty policies, product quality, service level reputation and people/equipment capabilities.

Inventory control management is the process of setting inventory objectives, planning the inventory flows, and controlling the process to achieve the objectives. Inventory control is responsible for cycle counting, inventory support, design for personnel accountability, and audit history.

Shop floor control in the MRPII setting is a set of procedures, approaches, techniques and policies used to control shop capacity and priorities, report work order status, monitor capacity, maintain delivery, quality and productivity performance. Shop floor control determines operation due dates, establishes operation priorities, projects, short range shop load, and maintains input/output control.

MRPII performance measurement measures the planning and control functions in manufacturing, values inventory and determines income, and supports the decision-making process.

An item master is a file which defines each stock keeping unit. The file includes a part number, part description, lead time, lot sizes, costing information, inventory status, requirements, planned order, and open orders. One item master file is organized and maintained centrally for company-wide use. A change to the item master is maintained through a formal change procedure.

A bill of material or product definition defines each product on a level-by-level basis to service the needs of all the firm's functions, marketing, engineering, manufacturing and finance. One bill of material system would be organized and maintained centrally for company wide use. The bill of material is a "model" of how the firm chooses to manufacture the product. A bill of material is maintained (historical, current and future) through a formal engineering change procedure.

The process definition or routing is a "road map" of how to produce a specific subassembly or manufactured item. The routing includes the operations to be performed, their sequence, the work centers to be involved, and the standards for setup and run. The routing may also include tooling information, operator skill level required, inspection operations, testing requirements, and special instructions.

Product costing is the association or assignment of the cost of doing business with purchased and manufactured parts in order to support performance measurements, inventory analysis, make/buy decisions and pricing decisions. Product costing establishes a standard cost of each item which is a predetermined cost, and explicit statement of what cost should be under efficient methods of operation which can be attained and sustained. It is the product of two factors determined separately—quantity and price.

History. There were four steps in the evolution of MRP. These steps include principally two methods of ordering material, the order point and material requirements planning. The four steps are: a better ordering method, priority planning, closed loop MRP, and manufacturing resource planning: MRPII.

The order point method establishes a point to order based on an average usage for the expected replenishment lead time plus a safety stock to guard against greater than average demand or lead time. Material requirements planning determines a time to order based on time phased schedules for the parent items that use the material.

Before a better ordering method was in place, the order point method was used to order material. The depletion in the supply of each inventory item was monitored and a replenishment order would be issued whenever the supply drops to a predetermined quantity. This quantity was determined for each inventory item separately. In the order point system some form of an economic order quantity computation normally determined the size of the replenishment order.

A better ordering method was the manual calculation of the requirements for a 13-week period. The typical scenario was that a finished product list was developed and a manual calculation of the material required was performed netting out items in-stock and on-order. A shortage list was generated for those items which would not satisfy this total requirement. This process was long and time-consuming and was performed approximately four times a year.

In the late 1960s and early 1970s computer capabilities increased and the task previously performed manually began to become automated and was performed more often.

The priority planning system became popular as material requirements planning became more sophisticated and it became possible that additional benefit could be gained by not only ordering material at the right time, and to establish the due date on orders when they were issued, but also to keep these dates correct and in line with the latest requirements.

Closed loop MRP is a system of material requirements planning which includes additional planning functions of production planning, master production scheduling, and capacity requirements planning. This system also includes the execution function of the planning process that is the shop floor control functions, dispatching and vendor scheduling. The critical element of the closed loop MRP system is the feedback for the planning system from the execution systems so that the planning can be kept valid throughout all the planning systems. When the planning aspect has been completed, the plans are reviewed for reasonableness to check for a fit.

MRPII evolved from the closed loop system and integrated additional characteristics. The operating system which is used to run manufacturing was integrated with the financial system. This also included integrating the business planning with the manufacturing resource planning system. It incorporated the ''what if'' capability. This allows for simulation of various situations of what would happen if different policy decisions were to be implemented. Finally, the system became the system for the company as a whole since it integrated all the major functions of company sales, finance engineering, and manufacturing. The nontechnical difference was the manner in which management uses the system.

Discussion. Manufacturing resource planning or MRPII is the effective planning and controlling of resources, priorities and performance to get the right products to the right customers in the right quantity, at the right time and at the highest quality.

The role of manufacturing resource planning (MRPII) is to take inconsistent supply and demand factors as part of the nature of manufacturing. These obstacles are a basic characteristic of the manufacturing environment, and the manufacturing resources available still need to be balanced against demand.

The MRPII planning process moves from broad objectives to detailed activities. The detailed activities include a control phase which requires follow-up action for both material and capacity. The tools available for the support of this control function are input/output control, shop dispatching, and purchase order status reporting. A plan is needed before control can be established.

Typical benefits of MRPII are improved inventory turns, reduction of part shortages, improved customer service, increased labor productivity, and more accurate gross margin analysis.

MRPII logic basically balances supply and demand. The demand is derived from the master production schedule. The supply includes open manufacturing and purchase orders, finished goods, and raw material. The bill of material is used to explode the demand at each level of production. Lead time is used to determine the required date based on an established time to purchase or manufacture an item. The netting process forecasts potential shortages. If supply and demand are not balanced, the computer recommends a specific action.

Applications. Manufacturing resource planning can be successfully implemented in many different types of manufacturing environments. The following is a list of potential applications for MRPII:

- Conventional manufacturing which includes fabrication and assembly.
- Fabrication only with no assembly.
- Assembly only with no fabrication.
- Process manufacturing.
- Repetitive manufacturing.
- Low-speed manufacturing.
- High-speed manufacturing.
- Make-to-order.
- Make-to-stock.
- Engineer-to-order.
- Job shop
- Flow shop.
- Manufacturers with distribution networks.

MAP. See: Manufacturing Automation Protocol.

The Manufacturing Automation Protocol and Technical and Office Protocol Users Group of the Society of Manufacturing Engineers (MAP/TOP of SME). The Manufacturing Protocol and Technical and Office Protocol Users Group of the Society of Manufacturing Engineers became an official technical group of SME in 1985. It was formed as an industry group, addressing communications integration issues within a multivendor environment, focusing on engineering, office, and manufacturing.

Realizing that common communications are essential for integrating these processes, the MAP/TOP Users Group focused its development activities on existing and emerging international communications standards, and accepted industry operating practices. One of the group's ongoing activities is developing and publishing a generic, nonproprietary set of communications specifications.

To provide a forum and organizational base for those professionals and supporting corporations who are involved with communications integration, the MAP/TOP Users Group has become a technical group of the Society of Manufacturing Engineers. SME serves as the secretariat for this group.

MAP/TOP Users Group of SME activities are directed at educating users, and encouraging industry-wide adoption of standards-based communications technology. The objectives of the MAP/TOP Users Group of SME are:

1. Generate a standard set of communication specifications using current industry standards, user-driven MAP and TOP specifications, and accepted industry operating practices.
2. Improve the productivity of manufacturing and associated industries, by implementing a common communication specification referenced in various manufacturers' hardware processor and factory floor devices.
3. Promote interoperability of multivendor devices, to improve the integration of engineering, office, and manufacturing functions.

4. Using the Open Systems Interconnection (OSI) model, assist in the definition of industry standards where they are lacking, and provide MAP/TOP Users Group involvement to accelerate acceptance of communication standards.
5. Provide marketplace feedback to computer-device manufacturers, thus encouraging them to develop and support nonproprietary communications products implementing common specifications.

Mask Design. A mask design is the artwork which creates the mask used in the photo-etching process of an integrated circuit chip. An integrated circuit chip is made of many layers of different substrates, which are photoengraved onto the wafer. The registration of the different layers is critical to ensure the proper stack-up of the many different substrates. The mask design is not the same as the graphical representation of the circuit logic, since a single logic element may be composed of many alternating layers of different and similar substances. The mask design is the final step in the design process of the integrated circuit before fabrication. Manufacturing process constraints and performance criteria dictate certain restrictions on the minimum sizes of physical elements. The smaller the circuit's physical size, the greater the functionality which can be included on a standard size chip.

Also see: CMOS, Integrated Circuit, NMOS.

Mass Finishing. Mass finishing may be any of several processes for cleaning, descaling, deburring, deflashing, surface finishing or producing edge or corner radii by conditioning workpieces in a container partly filled with a mass of abrasive or nonabrasive material. Usually, parts are loaded loosely, and the finishing media is suspended in water or a water-based solution. Media may be crushed and graded stone, corncobs or nut shells, hardwood sawdust or pegs, or preformed ceramic, resin-bonded or metallic shapes.

Also see: Finishing.

Mass Memory. The mass memory, along with the main memory or internal memory, constitute the total memory available to the computer system. Mass memory, sometimes called off-line storage, represents the total peripheral memory available to a computer system, such as hard disk drives, floppy disk drives and cartridge tape units. The trend in mass memory is to constantly increase in size and speed. Hard disk drives are now available in Gigabyte sizes, with optical storage being developed with capabilities of hundreds of Gigabytes. As computer speeds and data transmission capabilities increase, information which used to be kept in mass memory can now be kept in main memory. Data organization and addressing are constraints imposed by large memory usage.

Also see: Computers, Internal Memory, Memory.

Master Code Relay (MCR). Master code relay is a mandatory hardwired relay that can be de-energized by an emergency stop switch. When it is de-energized, its contacts must open to de-energize all application input/output devices. Its function must never be replaced by computer software or user-programmed MCR codes. MCR instructions are user-programmed fence codes for MCR zones.

MCR zones are program areas in which all nonretentive outputs can be turned off at the same time. Each MCR zone must be delimited and controlled by MCR instructions (fence codes). These instructions, or fence codes, must never replace master control relay hardware.

Master File. The master file is generated by a computer, and usually updated during data processing, to retain relatively permanent information for use in subsequent processing of additional information of the same nature for the same application.

Material Handling. Broadly, material handling is the activity of transporting, storaging, and presenting of material. When a component or product is moved other than within a process of manufacturing or inspection or testing, the activity is generally termed automation. In a manufacturing environment, components and raw materials must be received, distributed to the necessary points of use, and the finished products moved to the next operation or shipped to the eventual customer of the manufacturing process. Material handling also includes the presentation of the parts to a particular process machine, such as machine loading. Material handling may utilize conveyors, fork-lift trucks, or simply humans carrying items. Material control can be a major task utilizing large computers to track many different products worldwide. In the case of air, rail and truck routing, material control requires the added function of traffic management.

Initially utilized for basic warehouse applications, automated guided vehicles systems (AGVS) have expanded greatly in recent years. In the factory, the AGVS can be a key element of an automated storage and retrieval system (AS/RS) and manufacturing resource planning (MRPII).

The first uses four automated guided vehicles AGVs were in material handling applications over 25 years ago. Under computer control, the driverless AGV can be dispatched automatically to pick up and position a load, and follow a route along the factory floor. Through computer interface, the system provides inventory control and status reports for management decision making.

Physically, the AGV may be a single carrier (mobile vehicle) or a tractor pulling unpowered trailers. Some AGVs resemble lift trucks, and many can be operated manually as well as under automatic control.

Typical load carrying capabilities are in the range of one to four tons. Typical speeds are from two to four hour mph with a velocity governed primarily by safety considerations.

Writes Richard Miller in his SME book, *Automated Guided Vehicles and Automated Manufacturing*: "Another significant application for AGV material handling is in the electronics industry for printed circuit board (PCB) manufacture. While automatic insertion equipment has been used by the industry for populating bare circuit boards for many years, these machines were only 'islands of automation' in the PCB manufacturing process. In many cases, conveyor systems were used to link these islands, however, this approach lacked the flexibility necessary for facilities where a high mix of boards are produced.

"Automated guided vehicle systems were introduced in 1983 to interface these islands of automation, replacing the complex conveyor schemes and providing significant improvements relative to machine access and machine deployment. Greater flexibility in product routing also was achieved."

In general, automated guided vehicle systems may be used in three classes of facilities: manufacturing/warehousing applications, manufacturing/assembly applications, and distribution application.

Also see: Inventory Control, Machine Loading, Industrial Robots.

Material Requirements Planning. Material Requirements Planning or MRP is the basic function of calculating the raw material or components required to manufacture a lot of a product. MRP absorbs the bill of material and the build schedule, along with customer and vendor order requirements, to generate a list of quantities of materials to order and the dates when material must be ordered and delivered to maintain the desired manufacturing schedule. The MRP system provides build schedules, manages inventory, and maintains procurement operations.

As the need and desire for more integrated management systems increased, MRP systems were expanded to incorporate the financial tracking and accounting aspects which evolved into MRPII. MRP can simply include the basic item cost and roll-up a total product cost based upon an indented or flat bill of material.

James W. Branam of Rockwell International discussed MRP in his SME Technical Paper *MRP and Schedule Execution in the Job Shop of the Future*. In his discussion of MRP, he wrote: The first basic MRP concept of Independent vs.

Dependent Demand was well understood by job shop companies. They know that they could not order component parts until they knew about the end item that they were building. Most MRP to old system comparisons were based on the old system being a reorder point system. However, the job shop people understood the concept of Independent vs. Dependent Demand and ordered components using a ''block ordering'' approach. This approach typically was to perform a manual bill of material explosion (and then later with computers) and order all components to be due on the same date. This date was based on an arbitrary or historical offset from the ship date.

''The second basic MRP concept of component Time Phasing was only partially practiced by the job shops. The time phased the jobs but not the components within the jobs. This lack was more due to a lack of data processing capability than a lack of understanding of the concept. MRP was touted as a system that would bring the components in on an as required basis rather than having them set in inventory for long periods of time. This idea sounds a lot like the Just In Time idea of today. However, the MRP approach never realized its full potential because move, queue and safety times were inflated to provide cushion in the lead times. This again highlights the weakness in the schedule execution phase of present MRP systems. The cushion provided by the inflated lead times was deemed necessary because of the weakness in the schedule execution phase. This weakness is more predominant or evident in job shops than in Make to Stock companies because of the less routine or regimented flow of work through the plant.

''One further note on MRP concepts. The original term MRP stood for Material Requirements Planning. Since then it has been expanded to mean Manufacturing Resource Planning. This term also seems to mean the same as the Closed Loop Manufacturing Control Systems. The origi-nal Material Requirements Planning concept was a good one. However, it did not give the expected results because of a lack of discipline or regimentation in other areas (that supplied data to MRP or were to execute the plans developed from MRP).''

Also see: ABC Inventory, Bill Of Material, Inventory Control, MRPII.

Matrix. A matrix is a two-dimensional array of circuit elements which can transform a digital code from one type to another. A matrix is an array of input and output leads with logic elements connected at some of their intersections.

Mean-Time-Between-Failures. Mean-Time-Between-Failures or MTBF is the average length of time between failures of a system. It is also a measure of reliability. The MTBF gives an indication of the expected availability of the product, as well as an estimated frequency of having to repair or replace the product. In systems which have continuous usage, a large value for the MTBF is desirable and required in some situations. The MTBF is usually calculated initially from accelerated tests of a product and relating the results to the performance of similar products. Once sufficient time in use has occurred, experimental historical data is available to substantiate the MTBF. MTBF often characterizes equipment maintenance activities with monitored data as initial input. This provided an opportunity for preventive maintenance and often leads to detection of chronic system faults.

Also see: Mean-Time-To-Repair.

Mean-Time-To-Repair. Mean-Time-To-Repair or MTTR is the average length of time required to repair a device. With MTBF it provides useful guidance in projecting equipment and system uptime. MTTR is useful to calculate both the downtime of a system from when it fails until it is operational again. MTTR also helps estimate the cost of the repair itself. A common practice in electronic systems today is to diagnose the problem down to the module level and simply replace the module or component, which dramatically reduces the total repair time and quickly returns the system to operation. The defective module is then repaired off-line and is available for a future replacement repair.

Also see: Mean-Time-Between-Failures.

Mechanical Drum Programmer. A mechanical drum programmer is a sequencer which operates switches by means of movable pins placed on a rotating drum. When the pins are changed, the switch sequence also changes.

Medium Scale Integration. Medium scale integration or MSI is a solid-state integrated circuit having between 12 and 100 gate-equivalent circuits.

Also see: Circuit.

Megabit. A megabit is equivalent to one million binary bits.

Megahertz (MHz). A megahertz is equal to one million cycles per second.

Megohm. A megohm is equal to one million ohms.

Memory. See: Associate Memory, Bubble Memory, Buffer Memory, Bulk Memory, Computers, Core Memory, Direct Access Memory, Electrically Alterable Read-only Memory, EPROM, Expanded Memory, Extended Memory, Electrically Erasable Read-only Memory, Internal Memory, Nonvolatile Memory, Photo-optic Memory, Random-access Memory, Sequential-access Memory, Static Random-access Memory.

Memory Address. A memory address is the sequential numerical identifier of a specific physical location in a computer's memory. The memory address allows data to be maintained in an orderly manner and ensures the storage and retrieval of any information from any place inside the system. The CPU and the computer system are designed to function using certain rules and procedures. Since the designer has no knowledge of the actual information which a user will place in a computer, and a general-purpose computer must be capable of being reprogrammed, the use of a memory address allows the designer to design a usable system. Since the memory address is a logical designation, the memory address location can physically be embodied in different type of components and even be rearranged within the same component. For example, when a hard disk is configured or a DOS check disk-fix (CHKDSK/F) command is executed, the hard disk is checked and physical segments are mapped into logical address locations, with any unusable segments of the disk being identified as such and skipped over.

Menu. A menu is a list of choices or options which can be chosen by a user or operator. Menus are especially helpful as they remind a user of all the available choices without the need to remember all the options. But once a user is more experienced, menus can become frustrating if they force the user to sequentially go through multiple menus to reach a frequent and known endpoint. Some software offers menus for the novice user and command key sequences for the experienced user.

Menus in today's software programs can be ''pull-down'' in nature in which, upon selection of a single command word at the top of the display screen, a menu drops down. Menus can also be ''pop-up'', whereby they appear upwards from the selected command. The Apple Macintosh computer was one of the first products to popularize pull-down menus. Lotus 1-2-3 solidly entrenched the concept of pull-down menus in IBM PC software.

Metal Oxide Semiconductor. Metal Oxide Semiconductor or MOS describes a class of devices that are constructed using metal oxide semiconductor field-effect transistors or MOSFET. MOSFETs or MOS devices are much more suitable for integrated circuits than bipolar circuits because MOS transistors are self-isolating and can have an average size of less than a millionth of a square inch. This has made it feasible and practical to design circuits which use over a million transistors per circuit. Because of this high density capability, MOS transistors are used for high density random-access memories (RAMs), read-only memories (ROMs) and microprocessors. Several types of MOS circuit devices have been developed including PMOS, NMOS, and CMOS. PMOS uses aluminum for electrodes and interconnections in a metal gate p-channel MOS. NMOS uses silicon gates in an-channel MOS, while CMOS uses complimentary MOS which employs both p-channel and n-channel devices. NMOS and CMOS have been the dominant technologies in use today with silicon gate CMOS becoming more attractive for new designs.

Also see: Bipolar, CMOS.

Microcomputer. A microcomputer is a small computer containing a microprocessor, input and display devices, and memory all in one box. It may or may not interface to a host computer and/or peripheral devices. Due to the size, microcomputer systems are sometimes referred to as desktop computers. Since the system may be dedicated to the use of a single person, they may be called personal computers. A microcomputer used to be identifiable by its relatively low speed and memory capacity, with modest capabilities. Today's microcomputers and super-microcomputers have capabilities exceeding early mainframes. Personal computers can be networked, used as network servers, file servers, multiuser hosts, and distributed computing resources. With the addition of optional hardware to provide enhanced performance, desktop computers can become typesetting publishing systems, CAD workstations, and multifunction engineering support systems. Microcomputers still provide the low end dedicated functions such as being imbedded in products and containing a complete microcontroller's functionality on a single microcomputer chip.

Also see: Computers, Microcontroller, Minicomputer.

Micro-instruction. A micro-instruction is an elementary instruction, usually a single add, shift, or delete operation command. It is the part of a microprogram that specifies the operation of related subunits, such as the main memory and input/output interface, as well as the operation of individual computing elements.

Microprocessor. A basic element of a central processing unit is a single integrated circuit. A microprocessor requires additional circuits to become a suitable central processing unit. A microprocessor chip may have some memory, especially if it is high-speed cache memory, and some other functions such as memory management resident on the same chip as the basic CPU. This cuts down on the chip count required for implementing a microprocessor, as well as ensuring greater compatibility between these particular components, since the user has no choice in what is combined. This arrangement usually optimizes the chip's capabilities and provides greater performance.

Some of the more well-known microprocessors are the Zilog Z8 and Z80, the Intel 8080, 8085, 8086, and 8088, as well as the newer 80186, 80286, and 80386, while Motorola has its M6800, M68000 and M68020. One of the interesting differences between the microprocessors is the Motorola series has the order of their bits reversed as compared to the Intel and Zilog CPUs. This means that the Motorola bytes and words have the least significant bit on the left side instead of on the right side. This bit reversal causes some additional processing when transferring data between a Motorola CPU and Intel or Zilog style microprocessors.

Also see: Computers, Microcomputer.

Milling. The milling process may be considered as a method of metal removal in which the portion of the work piece that is to be removed is projected into the space dominated by a rotating body carrying one or more individual cutting teeth. The profile of the finished surface produced on the work piece, therefore is determined by the intersection of a plane perpendicular to the motion of the work piece and the outline of the body of revolution generated by the rotating cutter.

Two main classes of milling operations are face milling and peripheral milling. In *face milling*, the finished surface is mainly parallel to the face of the cutter. In *peripheral milling*, which includes slab and form milling, the finished surface is mainly parallel to the periphery of the cutter.

Because of the limited period of engagement of each tooth with the work piece, the removal of metal in milling is accomplished by the separation of small individual chips. The rotary motion of the cutter and the translating motion of the work piece combine to produce a chip whose cross section varies from instant to instant as the direction of motion of a tooth changes with respect to direction of motion of the work piece. Chip thickness is maximum at the instant that these motions are perpendicular to each other, and substantially zero when they are parallel.

Programmable milling machines feature pushbutton programming by means of manual data input (MDI). These machines bridge the gap between standard and NC milling machines. They provide the flexibility of many standard machines with the production capabilities of NC machines, without the need for NC coded part programming. Major applications of these machines include tool and die, prototype, and short-run operations, but they are used less frequently in shops that have programming capability for other machines. Programmable milling machines are available in horizontal and vertical-spindle models.

The horizontal-spindle model has an electronic control system that includes an MDI panel and a portable control box. Programming is done by the machine operator in shop language. As the operator machines the first workpiece, each move, with its corresponding speed, feed, and dwell, is entered into the system memory by pushbuttons. An alphanumeric screen displays the events for checking. Then, subsequent workpieces can be machined without any settings. These machines can be equipped with CNC for three-axis contouring operations.

In addition to mechanical-electrical, mechanical-hydraulic, and mechanical-electrical-hydraulic controls, numerical control and computer numerical control (NC/CNC) are being applied to milling and other machine tools requiring complex cycles. In some cases, both tracer and numerical control are provided on the same machine, and some machines are provided with the option for NC or CNC retrofit at a later date. Automatic toolchangers are also available on some NC machines (particularly machining centers), with a punched tape establishing the programming, selecting cutters from the changer, and determining milling speed and feed. Machining centers are discussed in another section of this book.

An NC/CNC milling machine provides control for at least two axes of simultaneous motion, with three or four axes under tape control being quite common. Some machines are provided with five or more axes under NC. In addition to the conventional longitudinal, transverse, and vertical movements, some machines have the column and spindle carrier mounted in circular swivel ways to permit the spindle to swivel in the horizontal and vertical planes. An NC/CNC system with continuous-path capabilities is desirable for most applications, and circular interpolation is frequently used when curved surfaces and contours are to be milled.

A CNC milling machine for automatic three-axis operation with full contouring capability has a power-operated spindle-speed changer with a variable range of 60-4200 rpm. Traverse rates for all three axes are 120 ipm (3048 mm/min), repeatability is plus or minus 0.0005 inch (0.013 mm), and positioning accuracy is plus or minus 0.001 inch (0.03 mm). The machine can be programmed while producing the first workpiece, and the program can be stored in memory for subsequent requirements.

Universal milling machines, available with or without NC or CNC, have a vertical milling head that can be swung to a position alongside the ram to permit horizontal spindle operation. Rotary and tilt tables permit five-sided machining of a workpiece at virtually any angle. Pendant control offers digital readout in all three axes, and programs can be transferred to cassettes for storage.

An NC and CNC system can be designed for as many axes of motion as desired. A milling machine with five-axis control is shown in the figure on the next page. A major advantage of such machines is that very complex workpieces can frequently be completely machined with one setup. Multiple-spindle machines are also available for increased productivity by simultaneously machining identical workpieces.

NC can be applied to a five-axis profile miller with a horizontal spindle. This machine can cut in five directions simultaneously. Milling an elliptical part with sloping walls, using five axes of the machine at one, and machining complex dies are examples of operations the machine can perform automatically under NC or CNC. An important fact is that the number of axes used to designate a particular system for NC or CNC does not refer to the number of functions that can be programmed, as these can be numerous.

In a general sense, a special milling machine is any machine designed and built to machine a specific part or family of parts. Usually a basic standard machine that is specially modified is considered a special milling machine if the modification results in approximately 50% or more change in the basic design.

Thread milling is a process that produces threads with a special thread-form cutter by either conventional milling or planetary milling. With the availability of NC/CNC machines having three-axis contouring capability and advanced controls, thread milling is undergoing a resurgence in popularity. Ease of programming with canned routines, longer cutter life, and high-quality threads are among the advantages of this milling method.

Millions of Instructions Per Second. Millions of Instructions per Second or MIPs is a term used to describe the relative speed and computational capability of the CPU of a computer. Since there are many different instructions that are part of the instruction set of a CPU, some of which execute in a shorter time than others, the maximum MIPs can vary depending upon which instructions are utilized. If the simplest instruction, which is a ''no operation'' (NOP), is used, the CPU will perform its maximum MIPs. While MIPs is used as an indicator of processor speed, the performance of useful activity by a computer includes much more than just executing CPU instructions. Functions such as disk access, keyboard input, display data management, and other utility tasks are all part of the normal activities required when executing a useful program. Unless the instructions used for a MIPs benchmark are disclosed, the results may be misleading.

Also see: Computers.

Minicomputer. A minicomputer is a midrange computer which is not as large and expensive as a mainframe, but larger and more capable than a microcomputer. Originally, when the term minicomputer was first used, it referred to a computer within a certain range of speed (MIPS and FLOPS), with a particular user capability. Generally, the distinction between micro, mini, and

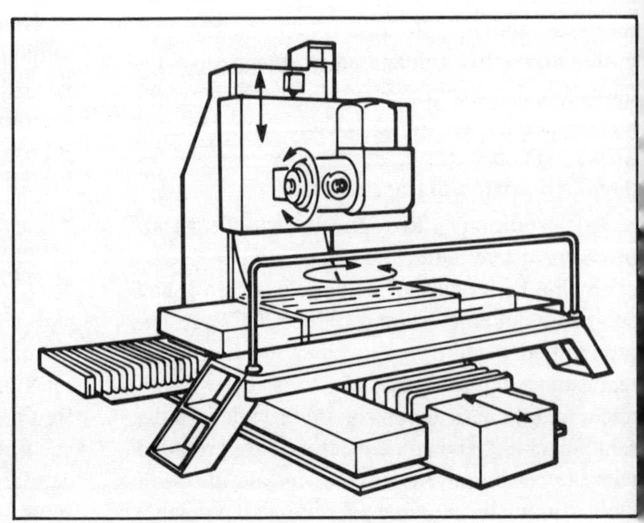

Milling machine having five axes of motion that are tape controlled.

mainframe computer involved the bit size of the CPU. Micros were 8-bit machines or less, minis were 16 to 32 bits and mainframes were from 32 or 64 bits and larger. As technology has progressed, the original definition of a minicomputer has become invalid as microcomputers now have capabilities far surpassing the original minicomputers, and even surpassing many mainframes of yesterday. The capabilities of today's minis and superminis, are on a par with many mainframes. Sometimes the only way to categorize a computer is to assign it the category it was originally defined before it evolved, or simply consider it to be in the same class as machines of similar functionality and vintage.

The VAX computer by Digital Equipment Corporation is probably the most popular minicomputer, with the top end of the product line, both as single machines and when used in clusters, competing against mainframes. The CRAY computers are one of the few mainframes that continue to represent the top end of computing capability, as well as the top end in price for a certain performance.

Also see: Computers, MIPS.

MIPS. See: Millions of Instructions Per Second.

Mirror Image Programming. Mirror image programming is a machine control unit feature enabling the reversal of all instructions programmed for a specific axis, usually the X axis. One result of this feature is that two mirror-image parts can be produced by a single program tape.

Mislogic. Mislogic is the incorrect relay panel wiring, incorrect PC programming, or electrical noise in a transmission line resulting in improper application operation.

MMI. See: Man-Machine Interface.

Mnemonic. A mnemonic is a two-, three-, or four-letter code that describes the function of each instruction of a binary computer. Each mnemonic has an equivalent hex (or binary) number to represent the function. A mnemonic is a simple code, usually alphabetic, that is representative of the function of the instruction it represents. Since computers only understand binary language, ones (1) and zeros (0), and binary notation is too difficult to program with, the manufacturer of the computer CPU defines a mnemonic code for the instruction set of the processor. Essentially, a mnemonic is a macro instruction for the binary code.

When programming in assembly language, a programmer may write the entire program source code using mnemonics, which is then assembled into binary coded machine language. Several standard microprocessors such as the Z80, M6800, and 8085 are widely used for assembly language programming. Examples of the mnemonics for the M6800 range from "ABA" which means "add accumulator A to accumulator B and place the sum into accumulator A," to "WAI" which means "wait for an interrupt." Mnemonic code is a programming code that facilitates recall because it is written as meaningful notation resembling the original words.

Also see: Assembly Language, Macro Instruction.

Model. A model is a representation of a device, that is created to allow the performance of some activity in which the model can be used instead of the actual device, with the results being essentially the same as if the actual device were used. The reasons for creating a model may be to build a facsimile of a device in a shorter length of time than if an actual device were built. Usually, models are less expensive to build than the actual device. Models usually do not include all the features and functionality that the actual device does. A model is usually constructed by more manual methods than a production object, mostly because the production systems are not in place and available during the time when the models are needed. Sometimes, various designs cannot be visualized or simulated without creating a physical device. When referring to electrical design, a model is constructed in the computer and is a computer-defined representation of the functional attributes of the device. A preproduction model of a device is called a prototype.

Also see: Breadboard, Simulation.

Model File. The model file is a computer file which simulates a three-dimensional definition of the tool or part. It may contain textual information to specify material, weight or other information which is not part of the geometric definition. The model file is used as a source of views for the drawing file and for geometric information for animation sequence, as well as the data necessary for numerical controlled machining operations.

Modem. A modem is a signal conversion device that includes both a modulator and a demodulator. It is a combination of equipment that changes the type of modulation, modulates an outcoming signal, demodulates an incoming signal, converts digital signals into quasi-analog signals usually for transmission, or converts quasi-analog signals into digital signals, usually for further processing.

A modem is normally part of a terminal installation and is connected to a data channel. It is used for the transmission and reception of data at either or both ends of a circuit or channel. It may include clocks and signal generators but normally does not contain error control equipment. It may be considered as a signal processor. Many additional functions may be added to a modem or may be considered to be part of a modem in order to provide user (customer, subscriber) services or communication system control features.

Modems can operate at many different data transfer speeds, with 300, 1200 and now 2400 baud being typical for personal computers. Nine thousand six hundred baud and higher are used for commercial application, however the higher baud rates sometimes require conditioned or direct connect lines to ensure functional integrity of the transmission.

Also see: Modulation.

Modulation. Modulation is the variation of a characteristic or parameter of a wave in accordance with a characteristic or parameter of another wave. For example, modulation is a variation of the amplitude, frequency, or phase of a carrier wave in accordance with the wave form that represents intelligence by means of superposition or mixing. The carrier may be a continuous direct current signal or a continuously alternating signal such as a sinusoidal wave. The carrier is used as a means of propagation. The superimposed or mixed signal is used as the intelligence bearing signal. The variation of the modulated carrier is detected at the receiver. The information or intelligence frequencies are normally called the baseband.

Modulation is also the process, or the result of the process, of varying certain features, characteristics, or parameters of one signal by means of another signal. For example, in continuous carrier modulation, the modulated signal is always present. In pulsed carrier modulation, there is no signal between pulses.

In lightwave communications, modulation is the variation of a characteristic or parameter of a lightwave in order to superimpose an intelligence-bearing signal on a carrier wave. An example is a variation of the amplitude, frequency, or phase of a lightwave by an analog or digital signal that is first coded to bear intelligence, then transmitted, and finally recovered by a photodetector at the receiving end of an optical cable.

Also see: Carrier Frequency.

Monitor. A monitor is a video display unit. Monitors can be bit mapped such as the EGA (Enhanced Graphics Adapter) compatible monitors used with personal computers; or they can be raster-scan devices such as the ones used with high resolution CAD terminals. The function of the monitor is to display the information sent by the host computer, as well as to display input information such as keystrokes or other device input.

Monitors may be monochrome or color. Typically, a color monitor cannot display simultaneously all the colors and shades which it is capable of generating. To generate colors requires additional memory dedicated to each pixel location, which is the limiting factor of some video display systems. To supply the additional memory is only part of the solution, since unacceptable times would be required to repaint the screen with each picture change.

A text monitor, monochrome or color, is capable of only displaying text (ASCII characters). Usually 24 lines by 88 characters is a typical screen display size. Some monitors display a 25th line for use as a status line. With smaller characters, monitors can display up to 132 characters wide, useful for spreadsheets and wide documents.

Some of the latest advancements in monitors include flat screen displays, smaller dot pitch, and greater viewability.

Also see: Characters, Graphics Terminal, Monochrome, RGB.

Monitoring Controller. A monitoring controller is used in an application to continually check a process and alert an operator to application malfunctions.

Monochrome. Monochrome refers to a single composition of color. With regards to a video display terminal or monitor, monochrome may be black and white, which is actually the presence or absence of a single color. The opposite of monochrome would technically be termed polychrome, but when referring to monitors, most people simply call a polychrome monitor a color monitor.

The color of the phosphor used within the cathode ray tube of the monitor determines the display color, such as green or amber. More recently, white phosphor, high-resolution terminals have been developed, which when used in reverse video mode, appear to be black print on white paper.

Another visual attribute sometimes noticed with a monochrome monitor is the bleeding of the image as a smooth scroll is performed. This phenomenon is due to the persistence of the phosphor. Short persistence phosphor disappears quickly while long persistence phosphor maintains its image much longer after the signal has been removed.

Also see: Monitor, RGB.

Monolithic Integrated Circuit. A monolithic integrated cirucit is an integrated circuit with at least one element formed within a silicon substrate.

Also see: Circuit.

MOS. See: Metal Oxide Semiconductor.

Most Significant Bit. The most significant bit is the leftmost bit of a word.

Most Significant Digit. The most significant digit or MSD is the digit representing the greatest value, usually the leftmost digit.

Motherboard. A motherboard is the large main board in a computer system, which has other smaller boards attached to it (known as daughter-boards). A motherboard usually contains all the major elements and components of the system, with the daughter-boards usually being optional or expansive portions of circuitry. A motherboard has connectors on it for the daughter-boards to be plugged into. A system which uses a motherboard does not use a backplane, but it may contain a bus structure which connects the expansion board sockets.

The IBM Personal Computer uses a motherboard with expansion slots for optional cards. By using a motherboard instead of a backplane, the designers were able to reduce the cost and improve the reliability, since when using a backplane, every card must be plugged into a connector, even the main CPU. Any connector is a possible failure point, plus the degradation of signals and increased signal path in using a connector.

Also see: Backplane.

Motion Control. Motion control is a method of delivering the end of arm tooling to a point or through a trajectory as prioritized by the system controller. In robotics, two methods of prioritizing manipulator movement are: end of axis control and coordinated axis control.

End of axis control controls the delivery of tooling through a path or to a point by driving each axis of a robot in sequence. The joints arrive at their preprogrammed positions in a given axis before the next joint sequence is activated.

Coordinated axis control controls the delivery of tooling through a path or a point by driving all joints of a robot simultaneously to reach programmed points and paths.

Motion Hold. Motion hold is a means of externally interrupting continuance of motion of the robot from any further sequence of action steps without dissipating stored energy.

Motors. See: AC Motor, AC Servomotor, Air Motor, DC Motor, DC Servomotor, Electric Motor, Servocontrol, Servomotor, Stepper Motor.

Mouse. A mouse is a computer input device. It is a small hand-held device with one, two, or three buttons on top and a cord extending from the rear of the device to the computer. The small size and shape coupled with the long cord resembles a mouse, hence the name. A mouse is moved on a flat surface in an X or Y direction, or combination thereof, and the movements are interpreted and converted into cursor movements on the terminal monitor. The buttons on the mouse may be programmed and interpreted to have various meanings. Some programs only use the left button, while others also use the right button, as well as simultaneous pressing of the left and right buttons. The buttons on top of the mouse may either be pressed (held down) or clicked (pressed momentarily and quickly released), as required by a particular application program's needs. Mice typically either have a ball on the underside which physically rolls (provided the surface has sufficient friction) or they use an LED with a sensor and are moved across a reflective pad which has a grid of nonreflective lines. The reflective pad type mouse is capable of more consistent and finer resolution of movement than the ball type mouse.

MRP. See: Material Requirements Planning.

MS-DOS. MS-DOS (Microsoft Disk Operating System) is a computer operating system technically known as a BDOS (Basic Disk Operating System). A BDOS is a control program which forms the basis of the operating capabilities for a computer that uses disk drives. Diskless computer systems do not require the portion of the BDOS which handles the reading and writing of data to and from the disk.

MS-DOS was developed by Microsoft Corporation to use on the IBM Personal Computer. IBM commissioned a version of DOS to be provided for their PC to which they would own the rights, called PC DOS. MS-DOS is the version of DOS which can be licensed directly and solely from Microsoft. Functionally, PC DOS and MS-DOS are interchangeable, as far as application software and programming languages are concerned. Since MS-DOS runs on the Intel family of microprocessors, any computer which uses the Intel 80XX through 80XXX processors can theoretically run a version of MS-DOS.

Also see: Computer, Operating System, Personal Computer, PC DOS.

MTBF. See: Mean-Time-Between-Failures.

MTTR. See: Mean-Time-To-Repair.

Multifunctional Robot. A multifunctional robot is a type of robot with the capability to perform a variety of work duties in a single operation location.

Multimedia Terminal. The term multimedia terminal refers to industrial input devices which offer multiple methods of input and multiple methods of output. These terminals may be intelligent or may even perform some data processing.

Some of the types of data input methods include magnetic card, bar code, infrared scanner, laser scanner, touch screen, ASCII keyboard, custom keypad, voice, light pen, and digitizing tablet. A combination of input methods is usually employed.

Displays might include CRT monitor, LCD and LED displays, hardcopy output, and pilot lights.

Multimedia terminals are designed to survive in harsh industrial environments, with such attributes as membrane keypads and sealed displays. Other features might be the ability to survive rough handling, high temperatures, or irradiated perils.

The more common multimedia terminal products available offer communications capabilities such as RS232, and RS422 multidrop, among other interconnect methods. In large quantity applications, low cost and reliability are a much sought after attribute.

Multiple-rung Display. The multiple-rung display is the feature permitting more than one rung of program logic to be displayed on a CRT at one time.

Multiplexing. Multiplexing pertains to the use of a single or common channel or path in order to make two or more channels, such as by sharing the time of the channel (as in time division multiplexing). Multiplexing can also be superimposing many frequencies at the same time (as in frequency division multiplexing) in order that many signal sources and sinks may communicate during a given time period. In addition, multiplexing means to use one channel for connecting two or more communication source and sink pairs.

Multiplexing in communication systems is the creation of data links, circuits, or channels by using a given set of equipment and applying various means of increasing the capacity of the physical equipment to carry messages. For example, multiplexing may be accomplished by timesharing, frequency division, time division, space division, phase shifting, or other means such as asynchronous time division multiplexing (ATDM), color division multiplexing, heterogeneous multiplexing, homogeneous multiplexing, pulse code modulation multiplexing and statistical multiplexing.

Also see: Channel.

Multiprocessing. Multiprocessing occurs when more than one program is executed, seemingly all at the same time, however, the CPU actually only executes a portion of each program at a time and rotates among each task. Multiprocessing allows more efficient use of a computer by being able to queue processes and execute them as soon as the CPU is able, without the operator being required to interact and begin the process at that time. Multiprocessing allows the use of batch files, which are sequential lists of executable instructions for the computer. Sometimes each line of the batch file is the command to execute another separate computer program.

A common example of multiprocessing is use of a print buffer on a computer to send data to a printer while the computer continues to execute other programs. Multiprocessing may be used to run a compiler in the background while doing other tasks in the foreground.

Also see: Background Processing, Computers, Foreground Processing.

Multiprogramming. Multiprogramming is a technique for handling two or more routines or programs by interleaving in succession the execution of a few instructions from each program.

N

NAND. NAND is a combination of the Boolean logic functions NOT and AND combined. A logic operator having the property that if P, Q, R..., are statements, the NAND of these statements is true even if only one statement is false, and false if all statements are true.

A NAND gate is a component that implements the NAND function. Output is produced under all input conditions except when all inputs are energized.

Nanosecond. Nanosecond (ns) is one thousandth of a microsecond.

National Electrical Manufacturers Association. The National Electric Manufacturers Association (NEMA) is an organization comprised of manufacturers of different types of electrical equipment. The organization establishes industry standards and specifications for the functionality, design, and performance of different electrical products.

An example of NEMA designations for electrical enclosures are NEMA 5, a water-tight enclosure designation and NEMA 12, which is the most common type of electrical enclosure used industrially. NEMA 12 is an indoor-only use enclosure that provides protection against dust, falling dirt, and noncorrosive dripping liquids. The NEMA designations define such characteristics as the allowable penetrability of air, dust, and water into the enclosures. The method of sealing the door and the type of material used to construct the enclosure are also defined. Explosion proof housings for electric motors is another type of NEMA defined enclosure. NEMA designations establish design standards.

National Institute for Occupational Safety and Health. The National Institute for Occupational Safety and Health (NIOSH) is a research institute established by the Occupational Safety and Health Agency (OSHA) to conduct research on the health effects of exposures in the work environment. NIOSH also develops criteria for dealing with toxic materials and harmful agents, including safe exposure levels. NIOSH is further charged with training an adequate supply of professional personnel to meet the purposes of the OSHA. In general, NIOSH conducts research and assistance programs for improving protection and maintenance of worker health.

Also see: Occupational Safety and Health Administration.

Natural Binary. Natural binary is a number system to the base 2, in which 1 and 0 have weighted value according to relative position in a binary word.

Also see: Computers.

NC. See: Numerical Control.

Necking. Necking and swaging reduce the diameter on the tube or drawn cup. Necking is accomplished between dies in a manner similar to other drawing operations. It can also be performed by spinning and swaging. Swaging is primarily used to reduce diameter, and it represents the squeezing operating in which part of the metal under compression plastically flows into contours of the die; the remaining metal is unconfined and flows generally at an angle to the direction of applied pressure.

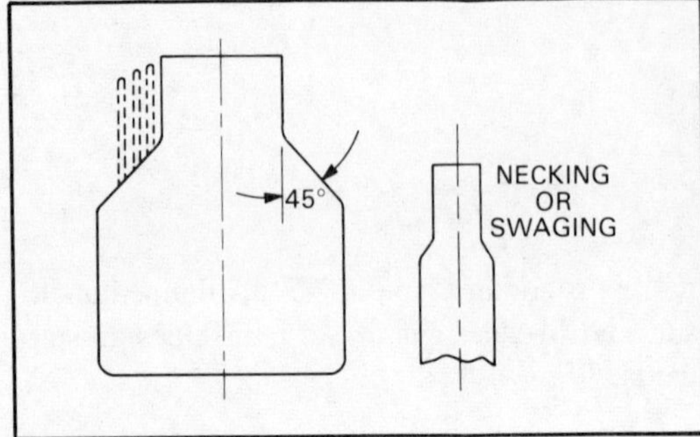

Necking and swaging will reduce the diameter on the tube or drawn-cup. Necking is accomplished between dies in a manner that is similar to other drawing operations.

Negative Logic. Negative logic is logic in which the 1 state is represented by voltage having a lower or more negative value that the voltage representing the 0 state.

Negative Metal Oxide Semiconductor (NMOS). Negative Metal Oxide Semiconductor technology uses p-type silicon substrates in which electrons are the active charge carriers. NMOS memory is characterized by fast access times and the potential for high density configurations.

Also see: Metal Oxide Semiconductor.

NEMA. See: National Electrical Manufacturers Association.

Nesting. Nesting is a programming technique in which a segment of a larger program is executed iteratively (looping) until a specific data condition is detected, or until a predetermined number of interactions have been performed. The nesting technique allows a program segment to be nested within a larger segment and that segment to be nested within an even larger segment.

Also see: Computers, Programming.

Net. A net is an organization of stations capable of conducting communication operations in accordance with a protocol and usually on the same channel.

Also see: Channel, Network.

Netlist. A netlist is a listing of the electrical logically related connection points (pins) comprising a particular signal in an electrical or electronic design. The netlist is used by electrical CAD and CAE systems as a basis for continuing operations such as circuit simulation and generation of the physical layout of a circuit board. The netlist can be obtained once the schematic is entered into the system. A netlist is a universally understood method of defining an electronic circuit design. A design represented by a netlist in one CAD system can be entered into another system and understood by a different CAD system. A netlist does not include any geometrical or graphical representations and therefore does not create any compatibility problems when transferred from system to system. A netlist exists before any physical layout design has been created but a physical layout is required for the fabrication of a printed circuit board. The netlist is an intermediate representation of a design which can only be used for evaluating the electrical logic functionality of a design, but not the timing analysis or other physical design dependent operations.

Also see: CAD, CAD-D, CAE, Schematic.

Network. In communications, a network is an organization of stations that is capable of interstation communication but not necessarily on the same channel or circuit. A minimum of two or more interrelated circuits can be a network. A network is a combination of switches, terminals, and circuits in which transmission facilities interconnect user stations directly (that is, there are no switching, control, or processing centers in the combination). In topology (geometric configuration; an interconnected group of nodes), the interconnections are accomplished by branches.

Examples of specific uses of data networks include the public information networks such as Compuserve, The Source, and the Dow Jones News service. There are literally hundreds of information and news retrieval networks.

Two of the more well-known standards in network specification are the Open Systems Interconnection (OSI) standard and IBM's Systems Network Architecture (SNA). Another well-known network standard is Digital's DECNET.

One of the keys to success of any control system is its networking setup. In the automated factory, equipment must transmit data directly to other segments or the operation, such as storage and retrieval, transportation, and management functions through a control network. A reliable communications network provides administrative control and the ability to monitor the total operation.

During the past few years, local area networks have become one of the most publicized and controversial topics in the data communications industry. LANs are not really as new as some people might think, but with the increasing capability of personal computers and the desires to share data, peripherals, and computing resources. LANs have become an item which the general computing public is much more aware.

A LAN is distinguished by the area it encompasses; its is geographically limited from a distance of several thousand feet to a few miles and is usually confined to a building or a plant housing a group of buildings. In addition to its local nature, the LAN has substantially higher transmission rates than networks covering large areas. LANs are usually designed to transport data between computers, terminals, and other devices. Some LANs are capable of voice and video signaling as well. Switching, digitizing schemes, data link controls, modulation, and multiplexing are often found in local area networks.

Networking standards are in various stages of development by many organizations and companies.

Nibble Mode. Nibble mode is a form of dynamic RAM addressing in which four sequential bits (a nibble, or half a byte) are accessed at a time, instead of the normal methods of addressing a byte or a word at a time. Nibble mode is typically used when the needs are only to access a few bits of information at a time and memory is scarce, or when a higher continuous update of a few bits of information is desired. The actual time to access four bits is less than the time required to access eight bits.

Also see: Computers.

NIOSH. See: National Institute for Occupational Safety and Health.

NMOS. See: Negative Metal Oxide Semiconductor.

Node. In a data network, a node is a point at which one or more data transmission lines are interconnected. In network topology, a node is a terminal (end) of any branch of a network or terminal that is common to two or more branches of a network. In switched communication networks, it is a switching point that may include patching and control facilities. In a data network, it means the location of a data station where one or more functional units interconnect transmission lines or data links.

A node is functionally the interface between the network and the device connected to the network, both physically and logically. Typical references to nodes in a network include the host node, remote nodes, spare nodes, etc.

A communication network in which all connections are direct, point-to-point between all users is a nodeless network. Some networks physically are nodal networks while logically they are nodeless, since the logical connections are initially fixed and function as if they were directly connected point-to-point.

Also see: Channel, Local Area Network, Net, Network.

Nodule. A nodule is a rounded projection formed on a cathode during electrodeposition.

Noise. In a communications system, noise is the sum of unwanted or disturbing energy introduced into the system from natural or man-made sources. Unwanted light waves coupled into an optical fiber are one example of noise. Noise may be audible in voice or sound communication equipment and visible in visual systems, such as radar and television. In visual systems noise has been given various descriptive names such as snow, rain, stripes, and running rabbits. In the audio frequency range, certain frequencies and amplitudes of noise are referred to as pink noise, white noise, blue noise, black noise, background noise, static, or echoes.

Other types of noise include antenna noise, atmospheric noise, intermodulation noise, pseudorandom noise, and intrinsic noise.

Noise immunity is the ability of a device to reject noise signals that are unwanted and interfering. *Noise spike* is a voltage or current surge in an industrial operating environment.

Nonlinear Programming. Nonlinear programming is the act of creating a program for locating the maximum or minimum of a function of several variables which are subjected to constraints, when the function and/or constraints are nonlinear.

Also see: Programming.

Nonretentive Output. Nonretentive output is an output which is continually controlled by one program rung so that when the rung changes state, either from true or false, the output turns on or off.

Nonvolatile Memory. Nonvolatile memory retains its contents when the power to it is removed. Typically nonvolatile memory refers to some form of IC chip type memory such as ROM, EPROM, EAROM, and battery-backed CMOS RAM. However, the term can be equally applied to disk drives and magnetic tape storage media, as well as CDROMs. Older computers used ferrite core as nonvolatile memory, hence the name core memory to describe the main memory of the computer. With the lower power requirements of today's memory chips and the high energy content and long life of lithium batteries, CMOS memory backed by batteries has become an economical and viable form of nonvolatile memory. The life of some lithium batteries is in excess of five years and the power requirements of memory chips while inactive (standby or idle power usage) is very low. Also known as nonvolatile storage.

Also see: Computers.

NOR. NOR is a logic operator which gives a truth table value of true, only when all variables connected by the operator are false.

NOT. NOT is a logic operator having theproperty that if P is a logic quantity, then the ''NOT P'' quantity is true if P is false, and false if P is true.

Notching. Notching is a stamping operation that cuts a piece of metal from the edge of a strip. By making several notches, notching gradually makes a blank contour before the blank is detached by a cutoff or parting step.

Notching is also a cutting operation used for removing a piece of metal from the edge of a strip to a required blank. In some cases, notching is done on the product itself. Generally, progressive dies are used for notching. The term *seminotching* represents the same cutting operation if it is done at the central portion of the strip.

Null. The term ''null'' represents the absence of information, as contrasted to a zero indicating no information.

Numerical Control. Numerical control is a technology whereby control of production equipment in real time is through prerecorded, symbolic instructions. Instead of a person operating a machine, a set of prepared instructions are read and decoded by the machine, causing it to operate. The instructions are prepared prior to operating the machine and are usually not prepared at the machine. Numerical control technology was first developed for machine tools. Though the technology has since been applied to a wide variety of production equipment, the term numerical control generally is used to refer to machine tools only.

Numerical control machines are extremely versatile. They are also far more accurate, produce more uniform parts, are faster, and have lower tooling costs than the traditional machine tools. In addition to improving the performance of traditional machine tools, numerical control or NC technology has served as an enabling technology in the development of new machining technologies. Processes such as laser cutting, electric discharge machining, electron beam welding, and high-speed cutting would not be possible without the precision available from NC.

Numerical control, computer numerical control, and direct numerical control have given the manufacturing industry the capability to exercise a new and greater degree of freedom in the designing and manufacturing of products. This new freedom is demonstrated by the ability to automatically produce products requiring complex processing with a very high degree of quality and reliability. Furthermore, products which previously were impossible to manufacture economically can now be made with relative ease using NC machines.

History. NC machine technology was developed at the Massachusetts Institute of Technology's Servomechanisms Laboratory under contract from the U.S. Air Force. The technology was first demonstrated on a milling machine in March of 1952. Its development predicated upon digital control technology, which was also in the early stages of development in the early 1950s. The Air Force continued to sponsor further NC development at MIT through 1956.

Initially, control of NC machines was at a machine language level. Machine movements were assigned codes, and the code, together with numerical values for distances, were input to the machine. For a straight line this is relatively simple. For a curved line, however, it is quite tedious. The curve must be divided into short line segments, and the starting and ending point of each segment must be calculated and input to the machine. The number of segments is determined by the required tolerances. The difficulty and tediousness of these calculations helped fuel the development of an NC programming language. The U.S. Air Force sponsored research at MIT, and in 1959 the APT language for Automatically Programmed Tools was introduced. APT is an English-like language that defines the workpiece geometry, defines the cutting tool configuration, describes the tool motions, and records the sequence of machine functions. Similar to other computer languages, a program written in APT is translated (compiled) to numerical values that the machine can understand.

Initial machines used hardwired logic circuits with a small storage buffer (one or two blocks). Punched paper tape was the instruction media. A program had to be read through the tape reader each time the part was made. The inability of punched tape to withstand repeated use was a major source of problems. Advances to coated and mylar tape only partially relieved this problem.

Development of direct numerical control (DNC) technology eliminated the use of punched tape. In direct numerical control, each NC machine is connected to a host computer, and the part program is sent to the machine in real time, block by block. Offsetting the advantage of the elimination of punched tape is the fact that the NC machine is dependent upon the host, and manufacturing is shut down if the host fails.

With the advent of microprocessors, computer numerical control (CNC) was developed. NC machines were equipped with microprocessor-based controllers, containing local memory and input devices. The memory allowed the local storage of programs. This technology enabled the use of variables and real-time branching within programs based upon the value of the variables. CNC both eliminated the problems of punched tape and the dependency on a host computer. However, CNC introduced critical data management problems. The same part program could be stored on multiple machines, making the management of change difficult.

CNC machines were quickly followed by distributed numerical control systems (also abbreviated DNC, and thereby the cause of much ambiguity). With distributed NC technology, CNC machines are tied to a host. The host provides a central storage site for programs, which are down-loaded to machines for execution. The resident memory and processing power at the machine allows manufacturing to continue if the host fails. The communications link between the host computer also allows upstream communications. Information is available to management as to whether the machine is in use, if manual overrides have been instituted, and what expected lot completion times are.

The current challenge in the use of numerical control technology is its integration with design and process planning.

Discussion. Initially, numerical control machines were divided into two generic categories: point-to-point and continuous path. Point-to-point machines (e.g. drill presses) performed an operation at predetermined points, listed in the input instructions. Continuous path machines (e.g. routers and milling machines) simultaneously moved and operated on the part. Thus their speed as well as X-Y motion required control. Over time, the capabilities of the point-to-point machines became more complex, and today these two classes of machines have for all practical purposes merged.

The components of a numerical control machine are the machine itself, the machine control unit (MCU), the interface between the machine and the MCU, and job-specific accessories such as tools and tooling.

The MCU and its interface with the machine distinguish NC machines from traditional machines. The MCU contains the executive program, diagnostic programs, input devices, and for distributed NC configurations, communications devices. The interface between the machine activators and the MCU transmits the signals from the MCU to the machine causing machine operations to occur.

NC machine actuators may be operated in either a closed-loop or open-loop mode. Open-loop machines have no feedback mechanism; command signals are issued to the machine, but there is no check to determine if the command was successfully executed. Closed-loop systems include a feedback mechanism in which one or more control variables are measured, compared with the value input by the command, and corrected if in error.

Numerically controlled machines also have software modules—an executive program, diagnostic programs, and customized part programs. A customized part program instructs the machine to perform the operations necessary to machine a specific part. The executive program interprets the customized part program, causing the machining of the part. The diagnostic programs are used to diagnose operational problems in the machine.

Direct and distributed NC machines also must have communications links to a host computer. The usual configuration is for many DNC machines to be connected to one host, allowing sharing of part programs from a centralized storage location.

There are four types of NC programming methods: manual, digitizing, language based, and graphics based. Manual methods require that numerical values, giving tool position and direction, be formatted appropriately and input to the control. Digitizing, more sophisticated that manual, allows a print to be traced. The tracing is converted to geometrical data and tool positioning by the computer.

Language based programs, like other software languages, use English-like statements as input to a computer that translates them to instructions to the machine. Compiling language based programs is a two-step process. First, a cutter location file (CL) is produced that describes the shape of the end product and the path that the cutter will take to produce that shape. The second pass, or post-processing, creates a customized program for an individual model of machine tool, taking machine parameters such as speeds and feeds into account.

More sophisticated are systems that allow the NC programmer to use the geometric model of a part created during design on a CAD/CAM system to generate the NC program. This method has the additional advantage of enabling the engineer to test the program on the computer

Numerical Control

by overlaying the tool path on the workpiece geometry. This computer simulation of the process minimizes the need for on-the-machine testing.

In addition to the computer technology, NC machines often require sensors to provide feedback. Closed-loop machines require feedback to ensure that the command was executed. Both open- and closed-loop systems may use sensors in the form of vision systems to provide input on workpiece recognition and orientation, worn or broken tool detection, cycle counting, equipment status and redundant safety features. In unattended operations, sensors provide in-process quality control checkpoints.

The number of axes or machine motions to which numerical control is applied commonly ranges from two to five. In general, NC machines are grouped into two classes: positioning machines and contouring machines. The functional capabilities of both types of machines are explained in the following sections.

The two axes of a representative point-to-point or positioning system are the straight-line movements of the longitudinal and cross or transverse slides, these two machine motions occurring at 90 degrees to each other. They are respectively X and Y axes, and these motions position the workpiece by positioning the table or surface on which it is mounted according to rectangular coordinates. Two-axis control, if provided with contouring capability, could be used for two-dimensional contouring.

A third axis may be added by applying numerical control to the up and down movement of the spindle of a vertical milling machine or of an upright drill, for example. This becomes the Z axis. These axis designations are diagrammed in the figure on the next page.

In contouring systems, the third axis provides three-dimensional control —for milling cavities in dies or molds or for milling other contours in three dimensions.

In its simplest form, the positioning machine is provided with NC dimensional control of the slide position only. Slide feed rates and spindle-rotating speeds, for example, may be selected manually. However, most modern NC positioning machines provide tape control of feeds and speeds, coolant on-off, turret indexing, etc. The method of handling these functions varies considerably from one manufacturer to another and cannot be generalized sufficiently to depict in diagrams.

A point-to-point machine (sometimes called a positioning machine) is one that moves the slides until a specific point on the workpiece is at the exact position at which the machining operation can begin. In some machines, the table slides move the workpiece to a specific location under the tool so that machining can start. In other machines, the table and workpiece remain stationary and the tool is moved to the desired location in relation to the workpiece. Certain machines can position both part and tool simultaneously.

In the first instance, each slide attempts to move at its maximum traverse rate to the new location, ignoring the status of other slides in the system. Because the slides operate independently of each other, the tool path between operations can be predicted only roughly. The path is affected by the distance between points, acceleration or deceleration, and the maximum traverse rate of each slide. The lack of linearity of the tool path between locations is of little consequence since the tool is not in contact with the workpiece during the traverse sequence.

The positioning machine operating in the manner described previously would be useful as a ''hole-making'' machine, that is, drilling, tapping, or boring holes at different locations on a workpiece. Such equipment would be more useful if it could machine (i.e., mill) between adjacent points; however, milling cannot generally be accomplished at maximum traverse rates. Straight-cut systems are capable of moving the cutter at a controlled feed rate along paths parallel to one or more of the machine axes.

A milling machine, for example, might be designed for a maximum error of 0.0005 inch (0.013 mm) when making a cut. That is, the machine slide, which is stationary during the

Numerical Control

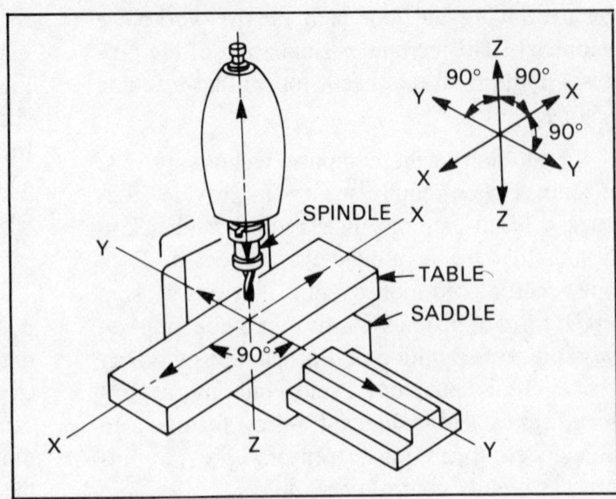

Diagram of a vertical spindle machine tool showing the axes—X, longitudinal; Y, transverse; and Z, vertical.

cut, would be required to hold its position to within 0.0005 inch (0.013 mm) as the maximum allowable error in order for the cut to be within tolerance. The slide would be required to have the rigidity necessary to maintain position under the loads imposed by the cutting tools. The moving slide would travel at the speed required for the cut. Many such methods of rate modification have been devised; thus, straight-cut positioning machines have been developed.

Because traverse rate is controlled on a per-axis basis and the path described by the tool between adjacent points in a multiaxis system is unpredictable, the straight-cut-positioning machine is usually limited to milling along a principal axis of the machine. Its path is very predictable because it is as linear as the guide surface (ways) of the machine.

A two-axis machine capable of drilling, milling, boring, and counterboring is illustrated in the figure on the next page. In a point-to-point machine of this type, the table and workpiece are moved in both the X and Y axes by NC and positioned beneath the spindle. Spindle feed is controlled manually by the operator because NC of the third (Z axis) slide is not provided. However, such a machine can be of substantial economic benefit because once it is set up, the operator is concerned only with spindle feeds and speeds and toolchanging.

When a two-axis machine is equipped with an automatic feed-cycle unit, several approach points and depths can be preset by the operator. They are then selected by tape command, and the machining operation (including machining to programmed depth) can continue under NC until a tool change is required.

A three-axis positioning machine equipped with a tool turret is illustrated on the next page. This machine requires an additional servoloop to control the Z axis machine slides, connected in parallel with the X and Y axes. Since the machine has a turret, which is indexed by tape control, a workpiece requiring as many as eight different tools can be machined without stopping the cycle for a tool change. This type of machine also provides tape control of feeds and speeds, and a tape-controlled dwell cycle.

This three-axis machine may also be equipped with a tool-length compensator. This permits the operator to index to each turret position manually, and advance the turret slide to the workpiece with the rapid-advance control.

Using a thickness gage, the operator can then preset each tool to compensate for variances in tool length and the programmer can program actual hole depths. Once a machining cycle is started, the cycle can continue under tape control without interruption.

The rapid-approach-to-depth operations involve movement of the Z-axis slide. Thus the analog principles of the servoloop section of

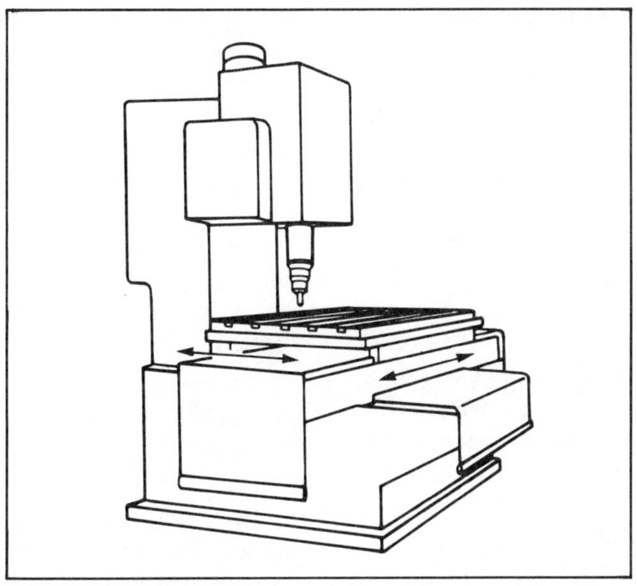

Two-axis, tape-controlled drilling machine.

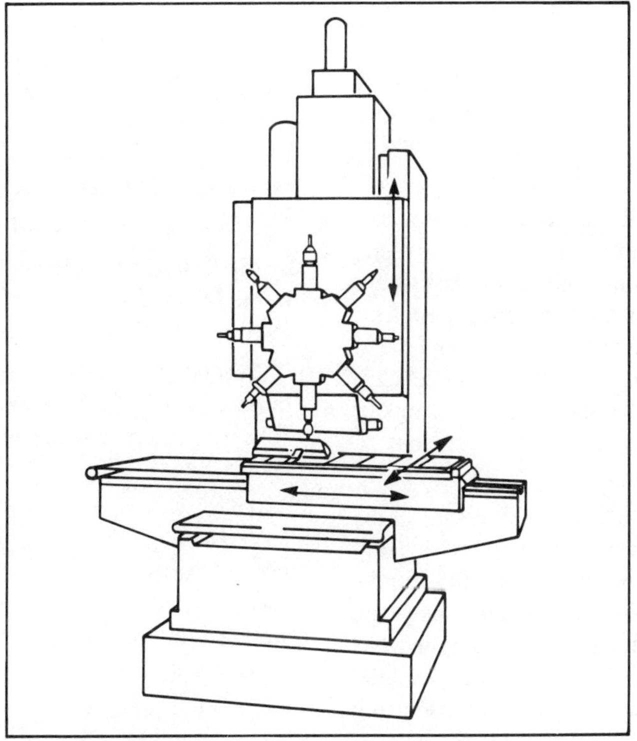

Three-axis, tape-controlled drilling machine with turret.

the NC system are applied in the same manner as for a single-slide system.

Some manufacturers of positioning NC systems have devised methods, sometimes proprietary, for controlling the slides simultaneously to produce angular movements. These methods have good economic reasons for existing in that they have greatly reduced the programming effort, have reduced the tape length, and have reduced the need for higher speed readers. Some of the angular cuts, or "slopes," produced by these methods are not exact; however, this "slope" potential has many advantages if close tolerances are not required. The machining of a chamber, when the chamber does not have to mate with another surface, is a good example of this application.

No matter how sophisticated a positioning machine is, it should not be confused with a contouring machine. A true continuous-path contouring machine must include both interpolation and buffer-storage elements. Positioning systems have been built with buffer storage in cases when the "read time" without buffer storage was considered excessive, that is, when the amount of data to be read or the rate at which the data should be read consumed too much of the total cycle time.

Since the advent of numerical control a new generation of tool holders, ranging from general to highly specialized use, has been developed. Varieties and combinations are almost endless. Most NC machine builders use their own type of spindle-to-tool locking and mounting devices. These are categorized as either straight shank or tapered shank tooling, and many variations of each exist.

Most toolholders have a special device in the back which allows for some type of locking drawbar to pull the tool firmly into the spindle and to release the tool automatically upon tape command to allow automatic tool changing. NC machining centers, probably the most versatile machine on the market today, can perform many tasks. As a result, the toolholders must be capable of holding a variety of tools. Almost all NC machines are more productive if all the tooling is set at approximately the same length. this allows the distance between the tool and workpiece to be set at a minimum, always using the longest tool.

While most NC machines select tools from a numbered pocket in the tool magazine, some magazines select tools by a set of code rings on the tool itself. NC lathes use a special tool holder for turning applications called a qualified tool holder. Qualified tooling is used with indexable inserts.

The four main objectives in designing a workholding fixture for NC are; accuracy, rigidity, accessibility, and speed and ease of clamping and changing the workpiece.

Evaluating numerically controlled facilities for the first-time buyer is generally encumbered by various fears. The expenditure is large, and the changes on the enterprise are anticipated as being large and disrupting. Frequently, engineers or managers evaluating the project are uncomfortable in examining a decision with so many unknowns and the perceived risks.

The entire process—from the project's conception through presentation of a proposal to buy—should be well thought out. Whether it is a one-person shop or a multibillion dollar corporation, the same general approach is needed.

A four-step approach should be considered.

First, define the need, the objective. What is to be accomplished? In this area, business objectives should be considered.

Second, consider why the objectives are required. This is most important because it establishes the criteria on which the alternatives are to be evaluated.

Third, develop a list of alternative methods to meet the objectives. These alternatives will have various strengths and weaknesses. The list should include the present production method as well as the production methods of the new alternatives.

Fourth, reevaluate the objectives in view of the alternatives. This step is most crucial to eliminate unreasonable objectives and solutions. By the same token, new opportunities may be identified which were not known or in proper perspective at the start of the study. It is important to realize that not every opportunity could or should be reduced to financial considerations at this time. Depending on the potential bias of the person performing the analysis or the bias of the company, the conversion from explanation to dollar value may be meaningless. This step should be saved for later in the analysis.

The evaluation to this point should be as unbiased as possible. One common bias often ignored is the unwarranted assumption that the future will be a continuation of the experiences of the past. There is no reason to expect that competitors will preserve the status quo, that changes in business will not occur, or that the work force and management will act in the future as they did in the past. These assumptions make a convenient basis for evaluation; but they are no more valid than any other well thought-out scenario.

At this point, conferring with management is wise. Refer to statements of business objectives to establish the environment under which the alternatives should be evaluated.

Evaluating only the costs associated with deteriorating current equipment should not be done. There is an additional cost of last value added—overhead absoption and profit—when current equipment fails to operate on a regular, dependable schedule. This element occurs when equipment is not available for production. The loss may occur when a machine is actively under repair or when the equipment is waiting to be returned to production. The total cost is the sum of repair costs and lost production value.

After each alternative has been evaluated based on technical characteristics, each alternative should be assessed based on value to the company as compared to the other alternatives as well as the current production method. The present method should be the benchmark against which the alternatives are compared.

The first step in such an analysis is to develop a sample of the workload to be processed across the new equipment. It must be representative of the work to be performed. From that sample, projections can be made of the effect on the total workload, eliminating the need for a very large, detailed study while preserving statistical accuracy. After the project has been adopted and implemented, this sample forms the basis for a performance audit, a critical but often omitted step.

The categories of benefit and costs should be identified next. Typically savings can be obtained in the following areas.

- Direct labor.
- Tool and fixture cost.
- Consumable tool cost.
- Inventory carrying cost.
- Tool setting cost.
- Programming cost
- Inspection cost.
- In-plant transportation cost.
- Maintenance cost.

It should be noted that the impact of these factors varies with the project. For example, if workpieces are costly, the inventory considerations may be more of an influence than the direct labor savings. The optimum balance of the costs is what is desired.

A detailed discussion of each of the just-listed factors can be obtained in the fourth edition of the *Tool and Manufacturing Engineers Handbook*, Volume I.

Several arguments have been advanced on the subject of consumable tool expenditure for conventional versus numerical control facilities. The heart of the matter is that, per operation hour, the tool cost for NC is higher because more actual metal removal occurs. Hence, per hour usage is increased. However, due to the greater number of hours required to achieve the same value of output, the overall tool consumption for conventional machining exceeds that of NC. One company audited its tool expenditures per direct labor hour and found the results shown in the figure on the next page.

Further examination showed that types of tools used did not change very much, but the effectiveness of their use did change drastically. Cutting parameters were controlled better with

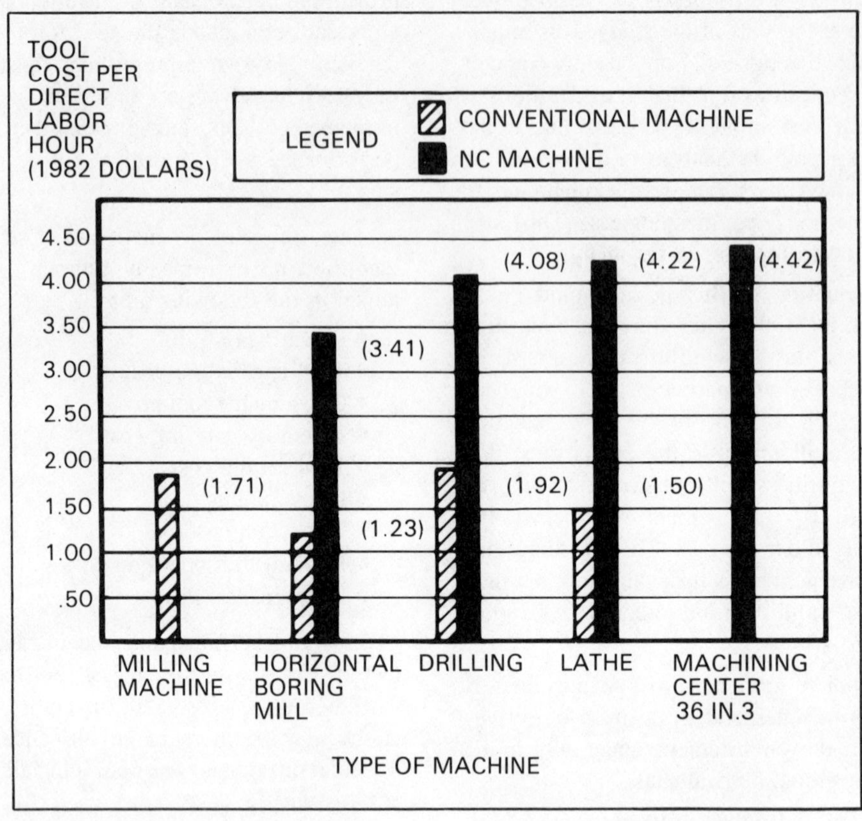

Sample consumable tool costs—typical as experienced by one major manufacturer in the United States.

NC, whereas operator prejudice and misuse were more common on conventional machines.

Another major area of concern in the justification of numerical control is programming costs. Normally, not recognized is the fact that there is such a thing as "conventional programming costs." These costs include the time that process engineering personnel spend setting standards, troubleshooting the conventional machine operations, processing the workpiece, etc., for all the conventional operations.

Regardless of the type of input applied to a numerical control system, the programming data must be put onto the input media—punched paper or mylar tape, magnetic tape, etc.—by programming.

Programming is the preparation of a detailed set of instructions for input to the NC machine, by which a series of machining operations are accomplished. Essentially, there are two principle types of NC programming—manual and computer-assisted.

In manual programming, part programmers work on a form sheet (program manuscript) and set up the sequence of operations to be performed by the machine in about the same mannner as they were physically setting up the machine for those operations. They first define the coordinates of the points to which the cutter is to be positioned, and then defines the feeds, speeds, etc. for the operation. They must know the characteristics of the machine being programmed and be able to enter the proper data into the form and sequence required to make the part to the drawing.

Manual tape preparation can be accomplished by an experienced individual familiar with machine operations. The complexity of the part and the features of the machine tool control determine whether or not manual programming is feasible. Most modern CNC controls feature part surface programming and advanced edit capability which make manual programming easier. Complex workpieces may require computer-assisted programming. A variety of considerations must be taken into account from the part programmer's viewpoint. Some of the considerations are: characteristics of the machine tool numerical control unit combination; axis nomenclature, and the tape preparation equipment, such as small computer systems that allow programming in terms of the input format of the machine control.

Other considerations include tape format, which in turn includes:

- Available letter addresses in a block.
- Maximum and minimum number of digits permissible for all codes.
- Maximum block length.
- Programming sequences within a block.
- Miscellaneous function coding and their specific functions.
- Preparatory function coding and their specific functions.
- Spindle-speed coding (usually direct rpm).
- Feed-rate coding (usually direct ipm or mm/min).
- Control features that aid manual part programming.

Special programming parameters are also a consideration. These parameters include:

- Selectable absolute/incremental; linear/circular interpolation, and inch/metric.
- Minimum programmable motion.
- Tape-reader limitations (in newer control, CNC buffered read eliminates such limitations).
- Cutter-path calculations (in newer controls, part surface programming eliminates such calculations).
- Special features or options that may affect part programming and tape preparation.

The coded information on tape (see figure on the next page) represents input data to the control unit and is acted upon in such a fashion that the control unit directs the machine tool through its various operations.

Manual Data Input (MDI) control is a type of computer-based numerical control designed for programming directly on the shop floor compared to standard CNC controls, which have a limited capability for manual entry of part program data, MDI controls are designed to be programmed manually. Some general characteristics of an MDI control are:

- Compact size.
- A compact, easy-to-use, powerful input data format. The operator directly enters a part program using the keyboard on the control. While a tape cassette or tape reader/punch may be used for storage and subsequent input of part programs, generally the operator creates programs at the machine. A professional part programmer is not required to prepare part programs.
- A CRT display that has a wide variety of easily read screen formats to interactively guide the operator. This conversational technique makes manual data input programming quite easy (the operator simply fills in blanks).

MDI controls are generally applied to low-cost two and three-axis milling machines, machining centers, and two-axis lathes. They can also be applied to higher-cost milling machines and machining centers.

In addition, some MDI controls have the capability to "learn" and memorize a sequence of operations.

With computer-assisted NC programming, a general-purpose digital computer usually assists the programmer in defining the operations to be performed by the machine tool in order to produce a part. The advantages of its use over complete manual programming are that (1) less time is required by the part programmer, (2) fewer errors occur in the final program, and (3) overall programming costs are lower.

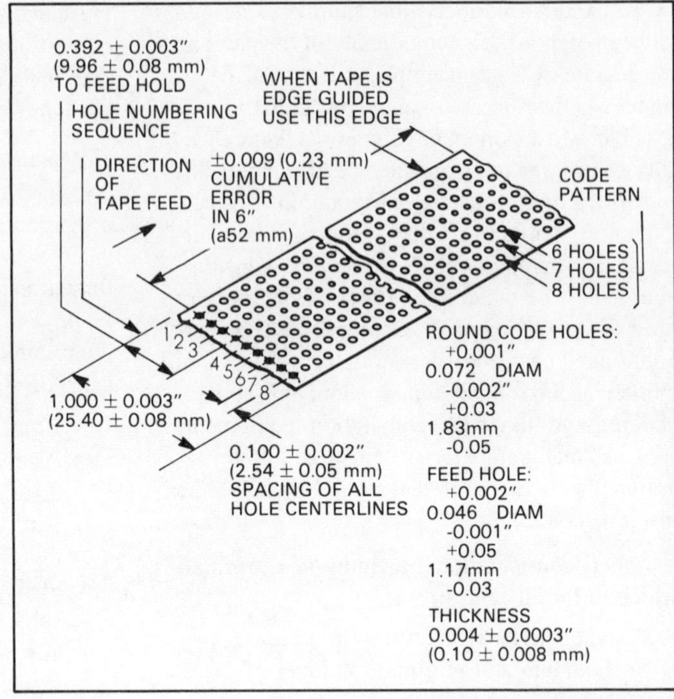

Standard eight-channel tape specifications.

Computer-assisted NC programming requires a programming language that allows the programmer to address the system using English-like commands. The programmer still works from the engineering drawing and enters data on part shape, machine parameters, and tool specifications. Then the computer does the calculations and generates instructions for parts programming and production. Still another program, the postprocessor, transforms the instructions into a parts program for a specific machine and NC unit. Often considered the industry standard, the automatically programmed tool, or APT, is the most common programming language.

The part programmer describes the operations to be done in work-like statements which are interpreted by the computer to produce hundreds of instructions as a result of very simple statements.

Computers that are more and more sophisticated are being built into CNC units. As a result, relatively complex workpieces can be programmed at the machine tool using so-called manual data input (MDI). Although MDI is generally not considered to be computer-assisted programming in the usual sense, it is increasing in use along with advances in CNC capability. Relatively complex workpieces which once required complex manual programming or off-line computer-assisted programming can, in some cases, be programmed via special commands right at the machine tool using MDI techniques.

Compared with manual programming, the principal benefit of computer-assisted programming is increased productivity. Increased productivity results from two factors: (1) a reduction in the user's part programming time and (2) improved utilization of the numerically controlled machine tool. Computer-assisted programming reduces part programming time in a number of ways.

First, it reduces and simplifies the mathematical calculations which the part programmer must perform. This is especially important for the more complex parts.

Second, it reduces the time spent correcting errors introduced by manual part programming and mathematical calculations made by hand.

Third, it provides a standardized input method that reduces the need to worry about varying tape formats or about forms of instruction accepted by different machine tools or machine tool controls.

Lastly, it allows for the use and reuse of stored sets of instructions and repeated machining patterns.

Reduced part programming time can result in lower part programming, labor costs and shorter lead times for part production. The degree of savings will depend on the capabilities of the selected computer-assisted system and the applicability of the system to the type of machining for which it is intended.

Computer-assisted programming systems also increase productivity by improving utilization of the machine tool. The NC machine tool is a major capital investment; the machine pays for itself by manufacturing parts.

Machine tool productivity in the period from 1975 to 1985, increased in relation to the number of NC machine tools sold. Although roughly half of the NC tools used in 1985 operated on manual programming the ratio was changing in favor of computer NC.

Because of the improved productivity in using computer-aided NC systems and often beginning with timesharing systems, shops that had owned limited programming systems are bringing in the newer systems even though, in some cases, their older systems might still work. Some of the NC units can be retrofit with CNC, a step that lengthens the productive life of an NC machine tool.

An unavoidable trend that will have to be reckoned with is the increasing cost of change itself. Aside from the rising research and development costs associated with new NC product development and the rising cost of acquiring new technology, there will be major, additional costs in training people to use this new technology. Computer technology will eliminate some jobs while creating new jobs for which people may be poorly prepared. Recognition of this ''human factor'' will likely be reflected in increased private and public training programs for computer users. Users will also benefit from the development of simpler ''user-friendly'' systems designed with increased attention to ergonomics, the man-machine interface.

In addition, computer-assisted NC programming systems will become increasingly more prevalent in manufacturing. Their use will help to improve productivity in an expanding array of manufacturing tasks. Future development of these systems will be limited only by the ability of developers to effectively use increasingly available computing power for more sophisticated and cost-effect applications.

There are a wide variety of general-purpose and specialized part programming languages from which to choose, and the number and types of these languages are increasing rapidly. There are a number of publications that are designed to inform users of the latest developments in NC part programming languages and that can be referenced for information on specific languages.

To close this discussion, presented is a word about computer numerical control and direct numerical control.

Computer numerical control (CNC) by Electronic Industries Association (EIA) definition is an NC system in which a dedicated stored program computer performs some or all of the basic NC functions in accordance with control programs stored in the read-write memory of the computer. Computer numerical control and soft-wired NC are synonymous terms. Computer numerical control is the dominant type of machine control being manufactured today. On the soft-wired control, the 110 volt a-c outputs are wired directly to the electromechanical devices on the machine (solenoids, for example). Likewise, switches and sensors on the machine are wired directly to the inputs. This contrasts with conventional hard-wired controls in which logic functions are wired together in a fixed, preengineered arrangement.

The key element of the soft-wired controller is a microprocessor or minicomputer. Because a computer is involved, there is a tendency to confuse the soft-wired controller and its application to computer numerical control with direct numerical control. There are however, several differences between CNC and DNC.

Direct numerical control (DNC) is a system of connecting a number of numerically controlled machines to a common computer memory. This connection is made for part-program storage, with provision for on-demand distribution of machining data. Typically, additional provisions are available for collection, display, or editing of part programs, operator instructions, or data related to the numerical control process.

The original DNC concept was forwarded (circa 1965) as a means of reducing NC control costs through the use of one powerful controller for a group of machines, rather than a separate controller for each machine tool. The cost of electronic control equipment was much higher in the early 1960s than it is today; so high that the DNC concept was driven by a need to reduce control costs.

Computers supporting DNC are used to disseminate manufacturing data to, and collect product information from, several machine controllers. Soft-wired controllers, on the other hand, generally support only one machine or a very small number of machines. Also, DNC computers may be remote from the machine tools, whereas soft-wired controllers are normally in close proximity. In addition, the software supporting DNC is usually written to support overall manufacturing activity, i.e. machine loading, productivity, and efficiency trends. Software for soft-wired controllers, however, is written specifically for a particular machine/device and its required fixed sequences.

Applications. Authors William Beeby and Phyllis Collier wrote in their book *New Directions Through CAD/CAM*: ''NC techniques provide users with greatly increased flexibilities in a variety of production activities: interactive graphics, programming, families of parts, sheet layout and nesting, machine cycle optimization, and parts verification. For large corporations and job shop operations, NC can lower prices for programming systems, and can enhance graphics capabilities and programming aids, direct and distributed numerical control and manufacturing data management systems. All shop activities will be affected by these technologies.''

NC machining applications continue to be focused on the manufacture of complex parts. Large volumes of simple parts are still more efficiently manufactured on specialized, dedicated machines. Initially, the time and effort involved in creating a part program offset the reduced setup time of NC machining, and limited its use to large volume, complex parts. With the development of programming languages and efficient programming methods, the time and cost of creating part programs has been reduced sufficiently to make NC technology cost effective for complex parts of all lot sizes.

Numerical control finally allowed the repeated manufacture of large quantities of part with high accuracy and repeatability from piece to piece. Provided the cutting tool did not wear significantly, the parts were identical to a high tolerance, as they came off the machine after the initial production run.

Numerical control is applicable to a wide variety of industrial tasks. In evaluating the applicability of NC to a particular job, the heaviest weight should be given to jobs which include:

- A long series of operations in which an error in the sequence would destroy the value of the operations.
- A wide variety of different sequences of operation which must rapidly and frequently be set up on the same piece of equipment.
- A relatively complex sequence of operations to be performed.
- An operation in which it is impractical for a human being to operate the environment required. Some NC machines run unattended by an operator.

- However, this benefit usually is associated with the use of industrial robots which load and unload NC and non-NC equipment grouped to form a robotized machining cell.

Numerical control has been shown to be one of the most significant advances in part manufacturing since the development of methods for interchangeable parts production.

Over the past 25 years, NC has demonstrated the ability to improve in dramatic ways such things as:

- Planning, flexibility, and scheduling.
- Setup, lead, and processing time.
- Machine utilization.
- Tooling cost.
- Cutting tool standardization.
- Accuracy, efficiency, and productivity.
- Material flow and workpiece handling time.
- Interchangeability of work, tools, etc.
- Safety.
- Cost estimating.

Numerically controlled machine tools provide an economic means for manufacturing management to make detailed plans of operation and, at that same time, retain documentary support for those plans.

When conventional machine tools are employed in job shop manufacturing operations, it is common practice to pre-establish only a general description of the operation sequence content. This broad planning approach makes it necessary for the machine operator to do the more detailed planning at the machine. This phase of operation planning frequently involves final determination regarding the type of setup and workholding devices to be employed, cutting tool configuration, and operating feeds and speeds. Often these factors are based entirely upon the skill of the operators and are dependent upon operator knowledge of workpiece requirements and capabilities of the machine and cutting tools. In shops engaged in repetitive part manufacturing, this detailed planning must be performed each time a job is assigned.

Fundamentally, the numerical controlled machine tool indirectly provides the user with great knowledge of the operation and provides the means for retaining all the detailed planning and skill required for any metalcutting operation.

NC has improved manufacturing flexibility. Because of the advantageous capabilities inherent in NC machine configurations, substantially more individual operation on a workpiece can be completed in a single setup than would be possible if conventional machine tools were employed. This concept is clearly demonstrated on NC machining centers on which a combination of milling, drilling, turning, tapping, boring, and reaming operations can be performed in a single setup on multiple sides of the workpiece.

NC can speed the introduction on new product lines and engineering changes and special workpiece configurations can be implemented in less time than required when conventional machine tools are employed. Factors influencing this savings of time in the multioperational capabilities of many NC machines, the reduced number of fixtures and, in some cases, reduced complexity of workholding devices, and the detailed operation-sequence-planning function carried out in the office.

In programming the numerical control operation and preparing tool charts, lead time is reduced significantly because of major simplification of tool design. Many workpieces can be completed in a few operations on NC machines as compared with three to four times as many operations for accomplishing the same work on conventional machines. Significantly, with NC, drastic reductions are often possible in the number of machines required. Also NC has proved to make short runs profitable.

Numerical control has made an impact in scheduling. Acceptance of the complete numerically controlled machining concept means that fewer setups will be performed on a given workpiece with less spindle idle time. Reduced lead time means that management can forecast for a shorter period into the future and thereby realize a proportional increase in accuracy of forecast requirements.

The ability to convert raw materials into finished parts in a relatively short time results in forecasting of raw material requirements only, rather than in forecasting of completely finished components. When many variations and special requirements are involved, better forecasting and scheduling techniques result.

Reduced quantities, reduced lead times, and fewer setups all lead to less work in-process. These factors permit management to do a better job of establishing net requirements and to plan machine shop loading and scheduling.

NC has lead to improved setup and lead time. It is practical to run smaller batch quantities on NC machine tools than would be economical with conventional machine tools performing the same operations. This is because of the minimal setup costs involved with these machines as a result of preset tooling techniques developed to support the machine operation to reduce nonproductive machine time; larger lot sizes (through group technology) can also be handled effectively with NC.

Capabilities of NC machines enable several conventional machining operations to be replaced with a simple NC operation. Because of the smaller quantities and fewer operations performed, the work in process for a given volume of output is less than that of equivalent production on conventional machines.

Lead time is a complex factor composed of many individual operational elements—market-forecast accuracy, number of variables involved, number of operations performed, number of jobs in process, total available machine capacity, and work-force capacity. Because NC machines provide a means for converting raw material to a useable finished part in a minimum number of setups, most of the factors influencing lead time can be reduced. Forecasting accuracy is inversely proportional to the forecasting period. Thus the closed-loop principle applies in that less stock-out and error-forecast protection is required because of faster response to changing sales requirements. Also, ability to quickly convert raw material to finished parts reduces the need to maintain a finished inventory of such parts.

Better control of processing and machining time is a result of implementing NC machine tools. When the human element is removed from the manufacturing process, as is the case on numerically controlled machine tools, manufactured parts are made in the manner prescribed by the manufacturing engineering department. Machining operations are done in the order that has proved to be most economical and desirable fro the manufacturing standpoint. This order cannot be altered without the approval of the manufacturing engineering department. The cutting tools and fixtures which are required for the manufacturing of these parts are programmed into the manufacturing process and are the same for all parts. On a machine tool that has automatic toolchanging capabilities and automatic tool selection, the tools which are called for on the original tape or other program control media are those which must be used in order for the machine to operate correctly.

Feed rates and cutter speeds may be made unalterable if so desired. Most numerically controlled machine tools, however, have a dial on the controller which allows the feed rate to be reduced in the event that the material being cut is harder than normal or the cutter is dull. The feed rates can, therefore, be reduced or increased if circumstances require it. Most NC controls feature override capabilities of 0-120%. Some controls can be overridden to as much as 999%. Many modern numerically controlled machine tools have adaptive controls in which feeds are reduced automatically by the controller as the load on the cutter is increased because of hard materials or dull cutters.

As a result of these control innovations, the time necessary for a part to be manufactured is known and can be controlled precisely by management. This time is predictable and is the same for all parts, with the exception of those parts which require a reduced feed rate. When a consistent reduction in feed rate exists, this reduction becomes evident to management because of the increased production time and steps can be taken either to correct the tape or to make necessary adjustments. Again, the important point is that the control of the manufacturing process is now in the hands of the management team and is removed from the judgment of the machine operator.

The utilization of NC machines, generally, yield a higher unit operating cost per hour than equivalent conventional machines. However, in many cases, reduced cycle time per job offsets much of the increase in operating cost per hour. In situations in which the operating cost per hour is higher than equivalent conventional equipment, multiple-shift operations are advisable. Lack of operator fatigue and minimal operator interference or opportunity to delay the machine result in substantially higher machine-utilization potential.

In terms of tooling costs, workpiece-holding devices generally form the most expensive portion of total tooling costs for job shop machining operations. A prime requirement of NC machine workholding devices is maximum accessibility to the workpiece for performing as many operations as possible in a single setup. This suggests considerable simplicity in the configuration of the workholding devices, which in turn means simpler design, manufacturing, and prove-out.

One of the prime advantages of NC machines is their ability to position accurately and repetitively. This capability eliminates the need for almost all cutter-guiding elements, such as bushings and set gages, in the design of the workholding device. This configuration eliminates those tooling elements most prone to deterioration: the cutter-guiding elements which come in contact with the cutter each time the surface is generated. As a result, tool-maintenance time is reduced, less interference with production is encountered, and tool-maintenance labor costs are avoided.

The reduced tooling costs have an indirect influence on time factors. The greater simplicity of the workholding device results in minimization of total lead time in tool engineering, in manufacturing, and in prove-out for production. Subsequently, when engineering design or customer requirements for modifications are encountered, fewer tooling modifications are necessary. With conventional tooling, such modifications are primarily related to the cutter-guiding devices rather than to the workpiece-holding portion of the tooling. On NC machines, however, the machine-control tape or other storage media acts as the cutter-guiding/location device, and virtually no time-consuming, costly changes need to be made to the workholding device.

In the area of cutting tools, with the programmer developing a detailed list of cutting tools required for the complete operation, it follows that standardization of cutting tools will be promoted. Such standardization provides the advantage of familiarity on the part of programmers, tool-crib setup men, and machine operators, thereby reducing the amount of time required in each phase of the work. Standardization by influence would include development of preferred tool sizes and configurations, which in turn influence the engineering-design phase, tool storage, inventory requirements, purchased-quantity prices, and tool-sharpening costs. Over a period of time the inclusion of preferred sizes at the engineering-design phase reduces the need for special cutting tools and virtually eliminates the extended lead time to precure such special items. Also eliminated are the procurement at a high unit cost (with the prove-out problems attendant to every new tool and the obsolescence of tools caused by engineering modifications.

Numerically controlled machine tools use the same basic tools as conventional machine tools, except that they are used for more operations. Thus the number of tools required in inventory is reduced. An example of this is an end mill. This tool can be used for cutting slots of various widths, for machining pockets and for internal and external contouring. It has indeed become a versatile tool in the hands of the programmer while at the same time reducing the need for special tools. This reduction leads to a reduction in tool inventory as well as to a reduction in costs to maintain tools and preset them.

To take maximum advantage of the savings in tooling, it is necessary to establish a standard tool list. As an example, drill sizes and the number of drills required can be greatly reduced by looking at the functional requirements of a part and having the engineering department design around standard tools, when possible, rather than using special tools. For example, a 1/2 inch drill is often called for to make a clearance hole for a shaft or hydraulic tubing. A 17/32 inch drill, which would provide an equal amount of clearance, is also required for clearance holes for 1/2 inch screws. Since the 17/32 inch drill can serve both functions, the 1/2 inch drill can be eliminated. Similar circumstances often occur when metric drills are used.

NC machines produce accurate results. Numerically controlled machines generally are designed from the ground up with a degree of sophistication that often is not encountered with conventional machine tools. The machine design elements provide inherently higher quality potential because of greater rigidity in the structural members. Closed-loop measuring systems used on many machine make the measuring devices independent of the feed-drive systems. This provides significantly greater accuracy potential than is commonly available on conventional machines.

The programmed sequence of NC operations also contributes significantly to increased accuracy with cutter-workpiece relationships established in response to the instructions on the tape or other media. For NC machines with automatic toolchanging capability, the correct tool is selected, feeds, and speeds are changed automatically to the optimum levels for the cutter-workpiece conditions, coolant is provided as required, the number of passes over the workpiece for roughing and finishing cuts is controlled by the machine, the sequence of cuts is optimized, tools are withdrawn to clean chips as required, and complete tool movement to full required travel is accomplished each time. Machining with NC equipment increases repeatability from piece to piece and from run to run as compared with repeatability attainable using conventional machine tools.

Material flow time has been significantly influenced by NC machines. Many workpieces can be completed in only a few operations on NC machines as compared with three to four times an many setups for accomplishing the same work on conventional machines. Consequently, reduced material-flow time results from the reduced number of operations. This means fewer assignments by the supervisor and less in-process work in storage in the plant at any one time.

Numerical control machine capabilities make it possible to convert raw material to finished parts in minimum lead time. This capability reduces the need to keep finished parts in inventory. It provides faster response to introduce engineering modifications without the possibility of high obsolescence costs for existing inventory and also permits faster response to customer requests for special designs.

Many workpieces can be completed in only a few operations on NC machines as compared with three to four times as many setups for accomplishing the same work on conventional machines. Consequently, reduced material-flow time results from the reduced number of operations. This means fewer assignments by the supervisor, fewer tool-kitting operations, and less in-process work in storage in the plant at any one time.

In the area of workpiece handling, the need to transport parts from operation to operation and internal machine-operator workpiece-handling efforts are both reduced with NC machine manufacturing methods. For example, on an NC machining center, numerous different operations are performed in a single setup. With conventional machines, the workpiece is often scheduled over several different machines to have the same work performed. There are fewer setups and requirements for the operator to load and unload the workholding device.

The manual time internal to the machine operating time during the normal NC machining cycle permits the operator to perform work such as deburring with the result that a separate operation for this purpose can be eliminated, thereby reducing workpiece handling.

In the area of safety and numerical control, specialization in planning the details of the operation, in setting up the cutting and work-holding tools, and in operating the machine all contribute to great operator safety. Under direction of the tape or other program storage media, the machine establishes the cutter/workpiece relationship. The operator is not required to interfere with the operation to adjust measuring devices in hazardous locations. These measuring devices are generally located near the cutter or workpiece on conventional machines for greatest accuracy, but at the same time in the most hazardous location for operator safety.

Transfer of the detailed operation-planning function from the shop environment to an office environment has direct influence on operator safety, since it removes the need for concentration on work for which the operator has not special training, in an environment that is highly distracting. When this detailed planning function is eliminated from the machine operator's scope of responsibility, full attention can be concentrated on doing the job the operator is more familiar with—auditing the machine operation.

NC has affected interchangeability. The first time a part is programmed across NC machine tools, manufacturing management is equipped with documentary support in the form of paper tape, magnetic tape, disk or other storage media. All elements involved in applying the NC machining concept provide a basis for easy interchangeability or work between plants, standardized workholding and cutting tools, standarized NC machines, methods, and programming techniques.

In the area of cost estimating, the two major items involved are material cost and the cost of the labor needed to machine the part. When using a numerically controlled machine, the time necessary to machine the part, which is a function of slide feed rates and cutter speeds, is predictable. Therefore, the estimates of cost of parts can be very accurately determined and the cost of the parts more realistically predicted.

Several other costs are involved in machining operations which are not directly connected with the machine. One of these is inspection cost. Because of the increased reliability of numerically controlled machining, and the fact that it is done on one machine, inspection operations are greatly reduced in number. Usually only final inspection on a selected number of pieces is needed to prove out the correctness of the tape and to ensure that the quality that is functionally required is being achieved. Often 100% inspection can be accomplished by the operator during periods between loading and unloading. In this way, NC may contribute to improved part quality by allowing the operator more time to concentrate on inspection functions.

Also, the use of numerically controlled machines reduces material handling costs. Several machining operations can be done in one setup on one machine, and in many cases the only material handling required is bringing raw material to the machine and taking the finished parts from the machine to the stockroom. This is in contrast to the material handling functions required when parts are milled on one machine, drilled on another, and contoured or form-cut on a third. Again, these costs are available immediately and reliably to the cost department so that expenses can be correctly determined.

Another important, but often overlooked, consideration is the depreciation rates which can be established for a numerically controlled machine. On these machines, costs and depreciation rates can be established accurately. This contrasts with several depreciation rates for different machine tools which might be required to produce the same part.

Numerical control also has a positive influence on productivity.

Productivity can be defined as the effective result of the operator in employing machine tools to convert raw materials to useable parts. Numerically controlled machines contribute substantially to increased productivity for a wide variety of reasons. After installation in the reader, the tape or other program storage media is always ready to continue the operation sequence—there is no delay for the operator to implement the operation as is the case in conventional machine operations.

Efficiently programmed operation sequences cause the machine to have a minimum amount of idle time in traversing from one cutting position to another. Where automatic toolchangers are used, toolchanging is performed immediately when desired. The tool is changed in the same amount of time necessary to the operator to change it, yet the automatic changer does not suffer fatigue as would the operator in making manual toolchanges. The tool changes are performed correctly, and correct feeds and speeds are employed on each cut with coolant applied as required. When tool retractions are required to clear chips, as is the case of deep-hole-drilling operations, there are perfomed automatically and consistently from part to part.

Combined, these factors all provide for a higher percentage of time devoted to cutting chips during the operation. Since chip cutting is the prime objective in machine tool operation, it follows that increased productivity is a predictable result.

The preset-tool concept can be applied to conventional machines as well as NC machines. However, the discipline involved in successful installation of NC machining concepts makes preset tooling mandatory. Productivity at the machine spindle is influenced because the preset-tool concept provides both cutting tools and workholding devices to be set up in a remote tool crib away from the machine itself. Cutting tools are set to length and diameter using specific tool drawings with all significant dimensions established. This provides for immediate interchangeability when a tool is dull and virtually removes the need for an operator to reset the tool and interfere with machine operation. When a shuttle-type machine is involved, productivity is increased because there is no unloading and reloading time. The idle station is unloaded and reloaded while the second station is in the cutting position in front of the spindle.

There are advantages to design in using numerical control as well. The use of NC in the shop dramatically impacts the flexibility of part design. For example, more accurate prototypes can be produced when NC machine tools are employed. When the part is put into production, closer tolerances often can be held if NC equipment is used. In addition, workpieces that were impossible to manufacture using conventional machine tools are now routinely produced using NC equipment.

Numerical control machines which have contouring ability can be economically used to eliminate the cost of special form tools. By the elimination of these special tools, the design flexibility to make engineering changes is greatly simplified. If most cases, an engineering change means a change in the part program and the tape which controls the machine.

When a new design is introduced, it is advantageous to the engineering department to see this design in actual hardware as soon as economically possible. Numerical control has made this lead time short because of the elimination of the need for special tools and fixtures. This allows the designer to review his/her design and make any needed engineering and design changes in a short period of time, thus decreasing the time from the drawing board to the finished product.

A major cost in the manufacturing of a prototype is the expense involved in tooling; this includes both capital tooling expenditures and perishable tooling costs. This use of standard tooling, which is common to many parts and universal fixtures, can be used for many parts on numerically controlled machines, has reduced the costs in manufacturing.

When a part is designed, the most important consideration is its function. Many times, functional requirements dictate close tolerances on dimensions. The impact of these tolerances has a direct bearing on manufacturing costs. Numerically controlled machines, being controlled by an electronic system rather than an individual, reduce human errors and move as accurately as the machine slide can position to the commanded dimension. Thus, the tolerance band which the machining system is capable of producing is achieved on all parts. The only time the specified design tolerance enters into the cost of the part is when the tolerance exceeds the manufacturing capabilities of the machining system and the machining operation must be done is steps. This increase in the reliability of the machining system leads to a reduction in scrap and rework costs, which is an important factor in prototype costs.

An additional benefit which is derived from numerically controlled manufacturing is the fact that a prototype part produced on a numerically controlled machine and the production lot produced on the same machine are representative of each other. This is true because the tape or other program storage media, tools, fixtures, and setup that are used for the prototype often are used on the production lot. As a result, the time and expense necessary to introduce new parts into the manufacturing system are reduced.

Parts which are manufactured on numerically controlled machine tools often are more representative of the actual engineering design than those made by conventional means. A major reason for this is that manufacturing of the part and the decisions involved in the manufacturing of the part are removed from the hands of the operator of the machine tool and placed in the hands of the part programmer. The machine operator has little or no control over the sequence of operations or over the tools that are used. The tolerances which are designed into the control tape or other program storage media and into the tooling that is used are repetitive on all the parts. These features lead to manufacturing consistency.

Another important benefit available through the use of NC is that, in the future, parts which are manufactured only for service will be the same as parts made during normal production runs at the time the product was originally manufactured. This allows manufacturing specifications to be established which are consistent throughout the life of the part as well as in any parts that are used for service in the future.

One of the prime advantages of manufacturing with numerically controlled machine tools, from the stand-point of the design engineer, is the ability to design a part which can be quickly and economically manufactured. This is in contrast to the requirements of conventional machine tools which often require form cutters. In many cases, special machines ere formerly required to produce parts which today are produced quickly and economically on machines with contouring ability. It is now possible to a tape to be produced for a machine to manufacture a part economically, whereas in the past, it may have been impossible to justify the necessary expenditures for the manufacturing of the part. New areas are now open to the design engineer who is aware of the manufacturing abilities available on modern machine tools. Any parts which can be defined mathematically can now be manufactured at nominal costs. parts which in the past were impossible to manufacture due to economic considerations have now become everyday, commonplace workpieces.

Current applications include sophisticated parts from models of items which will be molded or cast in the final form to one-of-a-kind models used for demonstrating a design's feasibility. Many CAD/CAM systems can directly output the NC information from a design, including suggesting optimized tool path profiles and depths of each cut to maximize the tool life. Some front-end systems vary the NC program based on the type of machine available, the material to be used, or the priority of parameters considered.

Examples of NC applications are wide and varied.

The use of numerical control with electrical discharge machining is recognized as a method to increase table-positioning efficiency, such as that used in the case of multiple-cavity work, and to cut dies and punches with traveling wire. NC also is used to orbit EDM machine tables, especially for large-mold workpieces. The addition of NC to the vertical ram movement makes it possible to EDM at different angles. Using an automatic electrode charger with the NC machines makes the EDM process completely automatic from roughing to finishing operations.

In milling, programmable milling machines feature pushbutton programming by means of manual data input (MDI). These machines bridge the gap between standard and NC milling machines. Major applications of these machines include tool and die, prototype, and short-run operations.

An NC or CNC system can be designed for as many axes of motion as desired. A major advantage of such machines is that very complex workpieces can frequently be completely machined with one setup. Multiple-spindle machines are also available for increased productivity by simultaniously machining identical parts.

In an article titled *Skin Milling for the Space Shuttle*, in the March 1980 *Manufacturing Engineering*, author Robert L. Vaughn (Lockheed Missiles and Space Company) discussed NC's role in the milling of the tiles for the space shuttles.

Vaughn wrote: "The outer mold line and inner mold line of the array, as well as the boundaries of each tile, are programmed for NC machining via Rockwell's master dimension computer data, which have been converted for use on Lockheed computers. The sizes of the flat blanks from which the tiles are made are also established by these sets of data. NC drafting machine plots are also generated by Lockheed's computer group for each array, depicting the geometry in three-or-five axis modes. These plots are the universal references used by designers, planners, NC programmers, and by inspectors in the final array assembly area."

Authors William Beeby and Phyllis Collier in their book *New Directions Through CAD/CAM* revealed a number of other NC applications.

"Since 1976, Ingersoll Milling Machine Company has used procedures, which they call CADCAM, to develop cutter families. These definitions account for about 60% of the cutters produced by the cutter department. When a new cutter is called for, the appropriate cutter family is retrieved from the computer and the designer applies the parameters of that particular cutter. This system is not interactive, but rather is based on the APT language. Then, the large host computer develops the cutter geometry, which is based on the product design logic, a process that takes about three to five minutes on the CPU. These data can then be transmitted to a graphics terminal for revisions or for plotting.

"The NC programmer retrieves the cutter geometry in the online source file housed in the CPU and generates the required NC tapes for lathe, keyway cutting, and manufacturing. The manufacturing logic is stored in the family-of-parts program with the logic design. Any special, one-of-a-kind cutters that don't fit any of the other programs, are designed on the interactive graphics system.

"Ingersoll, which produces about 90,000 NC tapes a year, expanded the system they are using to the machine division from the cutter department where it has been located for seven years (written in 1985). This expansion was made possible by the fact that the machine division's largely custom production pieces are designed on CAD equipment before being sent to the CAM operation.

"Custom Mold & Machine Company, Kent, Ohio manufactures rubber molds on a contract basis using NC and conventional machine tools. In the beginning days of the company in 1968, NC programming was one using a tape-preparation machine—a typewritter with a tape punch—an antiquated process that required large amounts of time even for simple machining jobs.

"In 1976, Custom Mold began using an integrated NC programming system with manual programming used to define lines, points, circles, and angles. But at least the newer system performed checking functions for errors. Then in 1980, the company brought in additional NC machinery, after investigating the products of four vendors that manufacture and market NC programming systems, eventually settling on a Numeridex LC6000 Numeripower system. The system included a plotter, CRT, microprocessor, printer, and floppy disk storage unit and cost nearly half what comparable equipment on the market at the time cost and contained all the capabilities and features Custom Mold required. And because the system was modular, components could be added or dropped with little trouble.''

The authors also wrote: ''NC and computers have a bright future. Whereas today's NC users have pioneered their use and are leading the way in their continued use, it is becoming clearer that nonusers are moving toward the new production technologies.

''Surveys have shown that the first NC machine installation is the most difficult and that, once over the hurdle, and once the initial NC unit is successfully programmed, maintained, and operated, and management of the system has been mastered, the user is likely to move quickly to the second and subsequent installations of NC equipment.''

Glossary. *absolute system:* NC system in which all positional dimensions, both input and feedback, are measured from a fixed point of origin. Used where point-to-point tolerances are not as important as tolerance on position from origin.

APT: Automatically Programmed Tools, a part programming language.

array: A series of items or elements arranged in a meaningful pattern in one or more dimensions.

BTR: Behind the Tape Reader, a means of inputting data directly into a machine control unit from a source other than a tape reader. This is usually part of a DNC system, in which the instructions are down loaded from a computer.

cathode ray tube: An electronic vacuum tube in which an electron beam can be focused on a small area of luminescent screen and varied in position and intensity to form alphanumeric or graphic representations.

central processing unit: The portion of a computer that is the basic memory or logic. It includes the circuits controlling the interpretation and execution of instructions. Also known as the processor, frame, or mainframe.

circular interpolation: A simplified means of programming circular arcs in one plane, eliminating the calculations required to segment the arc into a series of straight lines.

CL file: Cutter Location file, that defines the path a machine tool will take to produce a part.

CNC: See: Computer Numerical Control.

computer numerical control: A self-contained NC system for a single machine tool utilizing a dedicated computer controlled by stored instructions to perform some or all of the basic NC functions. Punched tape and tape readers are not used except possibly as backup in the event of computer failure. Through a direct link to a central processor, the CNC system can become part of a direct numerical control (DNC) system.

CPU: See: Central Processing Unit.

CRT: See: Cathode Ray Tube.

cutter path: The path of a cutting tool through a part.

direct numerical control: The use of a shared computer to program, service and log a process such as a machine tool cutting operation. Part program data is distributed via data lines to the machine tools.

DNC: See: Direct Numerical Control.

EDM: See: Electrical Discharge Machining.

electrical discharge machining: A nontraditional method of removing metal by a series of rapidly recurring electrical discharges between a tool (electrode) and the workpiece in the presence of a dielectric fluid. Minute particles of metal, generally in the form of hollow spheroids, are removed by melting and vaporization and are continually washed from the gap between tool and workpiece by the dielectric fluid.

fixture: A device to hold and locate a workpiece during inspection or production operations. Many fixtures are used on machine tools to provide a means for cutter setting.

incremental system: A control system in which each successive coordinate or positional dimension, both input and feedback, is taken from the last position rather than from a common datum point, as in an absolute system.

machine tool: A power driven machine, that is used to cut, form, or shape metal.

milling: A machining process that removes material from a workpiece by relative motion between a workpiece and a rotating cutter having multiple cutting edges.

point-to-point control system: An NC system which controls motion only to reach a given end-point, but exercises no path control during the transition from one end-point to the next. Used where point-to-point positional tolerances are critical and position from a datum or path between points are not critical.

post processor: A software program that formats cutter centerline data (CL file) into a form which a specific NC machine controller can interpret.

punched tape: A tape, usually paper or Mylar, on which a pattern of holes or cuts is used to represent data.

ram: The powered, movable portion of the power press brake structure, with die-attachment surface, which imparts the pressing load through male dies onto the piece part and against the stationary portion of the press brake bed. Also the moving part of a forging hammer, forging machine, or press, to which one of the tools is fastened.

setup time: The time required to adjust a machine and attach the proper tooling to prepare it to perform a particular sequence of operations.

tool path: The centerline of the tip of an NC cutting tool as it moves over a workpiece. Tool paths can be created and displayed interactively or automatically by a CAD/CAM system and reformatted into NC tapes.

zero offset: a feature that allows the zero point of an axis to be relocated anywhere within a specified range, thus temporarily redefining the coordinate frame of reference.

Also see: Automatic Tool Change, Automatically Programmed Tools, Broaching, Canned Cycle, Channel, Computer Aided Engineering, Computer Aided Process Planning, Computer Numerical Control, Computer Part Programming, Contouring Control System, Continuous Path Control, Cutter Offset, Diagnostic Routine, Direct Numerical Control, Discrete Jog, Electrical Discharge Machining, Floating Zero, Following Error, Fourth Generation, Full-range Floating Zero, Graphic Input, Group Technology, Hard-wired System, Incremental Dimension, Incremental System, Indexing, Interpolation, Laser, Manual Data Input, Manual Mode, Manual Part Programming, Optimize, Part Program, Plotter, Point-to-point Control System, Positioning/Contouring System, Position Storage, Preparatory Function, Punched Tape, Reference Block, Resolution, Resolver, Retrofit, Rollbending, Sequence Numbers, Setpoint, Stored Program Numerical Control, Subroutine, S Word, Tool Number, Variable Data, Water Jet Cutting, Word Address Format.

Numerical Data. Numerical data is data in which a set of numbers or symbols that assume definite discrete values is used to express information.

O

Object Code. Object code is the resulting output code from a computer language compiler. Object code, sometimes called executable object code, is code that resides in the main memory of the computer during execution. An object is defined as an area in computer memory that serves as a basic structural unit of analysis. Therefore, object code is code which resides in an area of computer memory and performs the analysis which the program was written to accomplish.

Object code may be output from a single-pass compiler directly, or it may be the eventual result of a multipass compiler. Object code cannot be "uncompiled" back into source code, since when source code is compiled, it is both reduced into the basic function logic building blocks of a computer, as well as rendered into binary data which essentially is not human readable. Object code files are linked together to form a complete applications program.

Also see: Compiler, Computers, Computer Languages, Programming.

Object Deck. An object deck is a set of cards containing machine-readable, condensed, computer instructions compiled for handling a specific general processor.

Object Program. An object program is a fully compiled or assembled program, which is the output of an automatic coding system, that is ready for loading into a computer.

Also see: Computers, Programming.

Occupational Safety and Health Administration. The Occupational Safety and Health Administration or OSHA is a government agency which was created by the Occupational Safety and Health Act of 1970. It sets standards and creates regulations to provide for worker safety and health to ensure the preservation of human resources. The federal government is authorized to develop and set mandatory occupational safety and health standards applicable to any business affecting interstate commerce. OSHA conducts tests and keeps statistical records regarding accidents, trends, and other workplace hazards. OSHA has been instrumental in assisting labor unions and corporations to compromise on realistic standards and agreements. OSHA is based in Washington, D.C., with regional offices around the United States. OSHA has regulations for most all industries, with particular attention devoted to understanding exposure to substances that are known or suspected carcinogens. OSHA, along with the labor unions, is responsible for improving the conditions of the workplace. Many times, compensation for workplace hazards are based on information gathered and maintained by OSHA. OSHA tests and certifies safety products and their designs. When new hazards arise, OSHA recommends safety procedures and support products. The OSHA seal of approval is similar to a UL (Underwriter's Laboratories) or Good Housekeeping seal of approval.

OCR. See: Optical Character Recognition.

Octal. An octal is a characteristic of a system in which eight possibilities exist. An octal also pertains to a numbering system that uses a radix or base of 8; only digits 0-7 are used.

Odd Parity. Odd parity is a condition resulting when the sum of 1s in a binary word is odd.

OEM. See: Original Equipment Manufacturer.

Off-delay Timer. An off-delay timer is a program instruction which turns off one or more outputs after a programmed time delay. In relay panel applications, it is a device in which the timing period is initiated upon de-energization of its coil.

Off-line. Off-line is a description of a terminal or computer system which is not maintaining a data communication connection between itself and a host machine. Off-line also refers to the performance of some activity in a time frame which is not controlled by the related activities. An off-line system may be utilized in some cases since the off-line system doesn't have to be fully functional, only being functional to the extent of the activities which are being performed. Doing activities off-line allow a much greater concentration of resources than when restricted to the on-line devices.

Sometimes systems go off-line due to power glitches or other unintended circumstances. When a data collection system ceases to function, it is considered as having been placed off-line. Some critical systems such as life-support apparatus and defense systems cannot be taken off-line without serious consequences resulting.

Also see: Computers, Off-line Programming, On-line, Programming.

Off-line Programming. Off-line programming is the process of creating a program for a robot or numerically controlled machine on a system other than the target system. The off-line system may be identical to the hardware and software of the on-line system, or it may be totally different. The only requirement is that the software must be able to execute on the on-line target system.

Off-line programming has several advantages. One is the possible productivity enhancement of using the ideal and optimized system for the off-line programming, while also being able to use the ideal system for the on-line operation. Some on-line systems don't offer the environment necessary for productive program development. Off-line programming allows the application of multiple resources, even concurrently trying multiple solutions, which is an approach that is impossible when only using the on-line system for programming. Off-line programming is a necessary step to utilizing expert systems and artificial intelligence aids for programming.

Also see: Artificial Intelligence, Expert System, On-line Programming, Programming.

On-delay Timer. An on-delay timer is a program instruction which turns on one or more outputs after a programmed time delay. An on-delay timer may also be used in relay panel applications. In such a reference it is a device in which the timing period is initiated upon energization of its coil.

On-line. On-line describes a condition of a computer or terminal being connected to a host and maintaining an operational data communications line. Since remaining on-line prevents other users from having access to the ports on a computer, most multiuser systems have automatic monitoring and disconnection of nonactive user lines. An on-line data change feature enables the user to change data table values through the program panel while the application is operating normally.

Also see: Computers, Timesharing.

On-line Programming. On-line programming is when a program is created or modified while the computer system is actively executing the program or another program. On-line programming also refers to using the target system as a programming workstation to create the program. Normally this refers to machine controllers such as robot or numerical control machines. On-line programming allows a person to enter a program command and have the system or device respond immediately to the command. On-line programming has some hazards, such as inadvertently entering an improper command which causes the machine to crash into an obstacle or attempt an injurious move. On-line programming has safety hazards to both people and machines. Usually there is no audit trail when performing on-line programming, so a sequence of activities cannot be duplicated. Useful reasons for performing on-line programming are to quickly tweak a program under the dynamic conditions of actual execution. Some programs behave so differently when operating, that off-line programming is prohibitively time-consuming for the final corrections.

Also see: Computers, Industrial Robots, Programming.

Open-loop Control. Open-loop control refers to a control system in which the output of the system under control has no effect upon the input signal, or feedback, to that system. That is, a system for which a control is to be designed will generally have several input variables and several output variables. The control problem generally is to adjust the input variables in such a way that the output variables attain some desired reference values.

Also see: Closed-loop Control.

Open-loop System. An open loop system is a control system that is incapable of comparing output with input for control purposes; that is, no feedback is obtainable.

Open Systems Interconnection. The Open Systems Interconnection (OSI) model was developed by a in 1977 subcommittee (SC 16) of the International Organization for Standardization's (ISO) Technical Committee 97. The model is intended to assist the exchange of information between distributed systems. The idea of OSI is to provide protocols for different vendor's/manufacturer's products to connect with each other—thus allowing an open systems interconnection of user applications.

Currently seven layers have been defined within the OSI model. From the bottom up, the seven layers are: Layer 1, Physical; Layer 2, Data Link; Layer 3, Network; Layer 4, Transport; Layer 5, Session; Layer 6, Presentation, and Layer 7, Application.

The physical layer provides for the transparent bit transmission between the data link entities. Its purpose is to activate, maintain, and deactivate physical connections between the data terminal equipment (DTE) and the data circuit-terminating equipment (DCE). X.21 is a CCITT standard which is an excellent reference for the details of the physical layer.

The data link layer is best described by HDLC (high level data link control). SDLC (synchronous data link control), a specific IBM protocol is very similar to HDLC.

The network layer is implemented by the X.25 packet-switching standard. This specification describes how packet-type data is transferred across the data terminal equipment/data circuit-terminating equipment (DTE/DCE) interface.

The transport layer is not completely defined. General agreement exists that its major function is to relieve the network users of the details of quality and cost-effective service. The major functions provided by this layer are: Mapping transport addresses onto network addresses, multiplexing transport connections onto network layer connections to increase user throughput across the layer, error detection and monitoring of service quality, error recovery, segmentation and blocking, flow control of individual connections of transport layer to network and session layers, and expedited data transfer. In addition the ISO has adopted a five class approach for the transport layer and is based primarily on the amount of error checking and recovery furnished by the lower layers.

The session, presentation, and application layers are not completely defined but essentially provide for higher levels of interaction respectively.

Also see: Local Area Network, Manufacturing Automation Protocol.

Operand. An operand is any quantity entering into or arising from an operation, such as a result, parameter, or an indication of the location of the next instruction.

Operating System. An operating system is a group of programs and/or routines that guide a computer and assist it in accomplishing a task. An operating system controls the basic operations of a computer system such as loading software, executing programs, transferring data between memory and I/O ports, reading and writing data to storage devices, and managing data between the system and the keyboard and monitor. Operating systems allow application software to run.

Some of the more popular operating systems are MS-DOS for microcomputers, UNIX and VMS for minicomputers, and VM and OS for mainframes.

Also see: MS-DOS, UNIX.

Operation Code. An operation code is a recognizable alphanumeric code which is the part of a program instruction designating an operation to be performed.

Optical Character Recognition. Optical character recognition (OCR) is the use of machines to automatically identify human-readable symbols, generally alphanumeric characters and translate their identifities into machine-readable code. OCR is the process by which an electro-optical device is able to translate printed characters into electronic text data. Originally, optical character recognition characters were a special font and character style. The purpose of this special font was to ensure the ability of the reader to distinguish the characteristics of the letters. The technology and ability to optically perceive and speed computing and translation allows optical character readers of today to read many different fonts and type styles. In fact, today almost any text printed page can be read by an OCR machine, even mixed text and graphics can be read and separated into the two respective types of data. The need to use keyboard entry as a means of inputting printed text into a computer has become less and less of a necessity. The price of OCR equipment is continuing to decrease as the technology becomes more readily available and competition increases.

Optical Disk Drive. An optical disk drive is a device which provides a mass memory capability inexpensively. It can provide a storage medium for data which can be accessed optically by a laser beam. The drive utilizes a removable cartridge which consists of a thin metallized-film medium encapsulated in a clear polycarbonate plastic disk. The actual disk is further protected by a high-impact plastic shell with a sliding metal door that allows access for the optical read/write head of the drive. The cartridge size is roughly 5 1/4 inches wide, 6 inches long, and 3/8 inch high. There are three major types of optical disk drives: CD-ROM, erasable optical disk, and WORM drive.

Also see: CD-ROM, Erasable Optical Disk, WORM.

Optical Read-Only Memory. Optical read-only memory or OROM is a CD-ROM (compact disk read-only-memory). A CD-ROM is the only type of optical memory that is a read-only device.

Also see: CD-ROM, ROM.

Optimize. Optimize is the rearrangement of isntructions or data in NC or computer applications to obtain the best set of operating conditions.

Original Equipment Manufacturer. An original equipment manufacturer or OEM is a manufacturer who is the source manufacturer of a product. This distinguishes the OEM from VARs (value added resellers) who purchase the basic equipment from OEMs, add some enhancement which adds monetary value and resell the resulting product.

Output. Output is printed or recorded data resulting from computed source programs. Output may also be data transferred from internal storage to output devices or external storage.

Also see: Computers, Input, I/O, Output Devices, Output Resolution.

Output Devices. Output devices are devices which convey data from a computer to an external device.

Output Resolution. Output resolution is the smallest increment of dimension that can be specified by the interpolation process of a control system and recognized by slide movement.

Oxyfuel Gas Cutting (OFC). Oxyfuel gas cutting is a form of oxygen cutting (OC) in which metals are heated by oxyfuel gas flames, then severed by chemical reaction of the heated metal with a jet of compressed oxygen.

Oxyfuel Gas Welding (OFW). Oxyfuel gas welding is any of several fusion welding processes that join metals by melting them with a flame produced by burning a combustible gas in the presence of oxygen. Pressure may or may not be used, and filler metal may or may not be added. The most common fuel gases are acetylene and hydrogen.

Also see: Welding.

P

Pacer System. A Pacer system is a manual or motorized method of imparting motion to a machine slide through the use of servomechanisms, resolvers, and synchros.

Pad. Pad is a general term used for that part of the die which delivers holding pressure to the metal being worked. In a computer, it is a term used to describe filling a block with what could be termed "dummy data" (usually blanks or zeros). On a printed circuit board, a pad is the conductive area to which a component is soldered.

Also see: Die.

Page. A page is a portion of memory consisting of a fixed number of locations dictated by the direct addressing range of memory reference commands.

Paging. Paging is the division of a program and data into fixed blocks so that data transfer between disc and core can take place in portions rather than as entire programs.

Also see: Programming.

Paint. Paint is a material that when applied as a liquid to a surface, forms a solid film for the purpose of decoration and/or protection. Generally, a paint contains a binder, solvent, and a pigment. Often, other materials are present to give special properties to the paint film. Examples of such additives are rust inhibitors, light stabilizers, and softening agents.

Also see: Industrial Robots, Painting, Spray Painting.

Painting. Painting is a generic term for coating processes that deposit liquid organic films composed largely of a binder (resin), pigment and various additives suspended in a suitable solvent. After drying, the paint is a solid, adherent film that provides a decorative appearance, protects the surface, or both.

PAL. See: Programmable Array Logic.

Paper Tape Reader. A paper tape reader is a device used to translate code perforated on paper tape into electrical signals.

Also see: Numerical Control.

Parabola. A parabola is a U-shaped curve in a plane generated by a point moving so that its distance from a fixed second point is equal to its distance from a fixed line.

Parabolic Interpolation. Parabolic interpolation is a method of controlling contouring which utilizes parabolic arcs to approximate curves by automatic means within the control system. The arcs arc blended automatically.

Parallel Communications. In parallel communications, data is formatted, presented, and transmitted a byte at a time over 8, 16, or sometimes 32 data lines. The essentials of parallel communication are that there are more than one data line in use concurrently when transmitting data, as opposed to serial communication. Serial communication uses a single data transmission channel which forces all data to be transmitted serially (sequentially), no matter how quickly.

Theoretically, parallel communications always has the capability to have a greater data throughput than serial communication, using the same media and transmission methods (given equal speed data lines). On the other hand, parallel communications requires more transmission wires than serial, so the connection cables typically are more expensive.

Parallel communication is used within the bus of CPUs and backplanes. IBM personal computers and compatibles use parallel communication to connect to printers as the default printer port. All modems use serial communication, not parallel.

Also see: Computers.

Parallelism. Parallelism is the condition of a surface, line or axis that is equidistant at all points from a datum surface, line, or axis.

Parallel Processing. Parallel processing is when more than one computer program is being executed at the same time. A multiprocessing environment, which is not the same as a parallel processing environment, is when a CPU switches between multiple tasks, spending a few milliseconds executing a portion of each task. True parallel processing requires multiple processors. Parallel processing is the only method that can reduce the processing time required to perform a computation other than increasing the clock speed. Systems are now being developed which use multiple processors to operate at very high throughputs. Some of the systems use hundreds of processors in a single machine. Also, parallel processing computer systems are designed to provide high availability. High availability means the system is available for use, a high percentage of the time. Parallel processing is one of the latest major techniques to be developed for increasing available computer power.

Also see: Computers, Multiprocessing.

Parallel Transmission. Parallel transmission is the simultaneous availability of two or more bits, channels, or digits.

Parameter. Parameter is a variable that is given a constant value for a specified application. Parameter also is a variable that controls the effect and usage of a command. Parameter has alterable values that control the effect and usage of graphics command.

Parameter Word. A parameter word provides one or several parameters in a direct or indirect manner.

Parity. Parity is a means of testing the accuracy of binary numbers used in transmitted, recorded, or received data. A self-checking code is used on which the total number of ones (1) or zeros (0) is always even or odd.

Also see: Computers, Parity Bit, Parity Check.

Parity Bit. A parity bit is an additional nondata bit appended to an array of bits to make the sum of all ones (1) in a word always even or odd. Parity bits provide a method of error checking for transmitted data. Thus, the number of ones (1) in each array is made always odd or always even. The parity bit is usually associated with a character, word, or block for the purpose of providing a means for checking the accuracy of transmission or storage and retrieval. When using the ASCII, the eighth bit is appended to each character so as to always have an odd or always an even number of 1's. Usually odd parity is used. Thus, the value of the parity bit, 0 or 1, depends on the previous sequence and whether the system operates on odd parity or even parity. The parity bit is also known as the check bit or check digit.

Also see: Check Digit, Data Error Detection, Parity, Parity Check.

Parity Check. Parity check is a check to determine errors in a group of bits. The number of ones (1) or zeros (0) in an array of binary digits should always be even or odd.

Also see: Parity, Parity Bit.

Park. Park is a programmed instruction for moving a tool to a location at which tool and workpiece inspection is safe.

Also see: Programming.

Parting. Parting when used in lathe or screw-machine operations, is the separation of a completed part from chuck-held or collet-fed stock by means of a very narrow, flat-end cutting tool (parting tool). In metalforming, parting is a stamping operation that performs two cutoff operations in the same stroke, producing a small amount of scrap between successive blanks cut from a strip of metal.

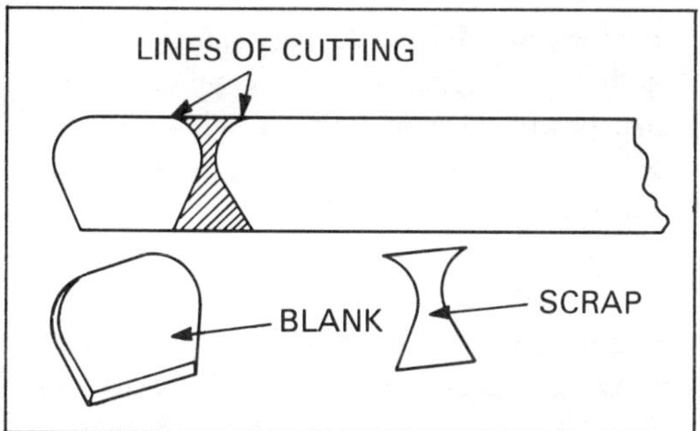

Used in lathe or screw-machine operations, parting involves two cutoff operations which produce blanks from the strip shown in this figure.

Part Program. Part program is a complete set of data and instructions written in source languages for computer processing or written in machine language for manual programming for the manufacturing of parts on an NC machine.

Also see: Numerical Control, Programming.

PASCAL. The computer language PASCAL is a general-purpose, numerically oriented language primarily developed for teaching programming.

Also see: Computer Languages.

Password. A password is a unique string of characters that a program, computer operator, or user must supply to meet security requirements before gaining access to date.

Also see: Computers.

Patch. A patch is a section of temporary coding inserted into a routine to correct or change the routine.

Also see: Programming.

Patenting. Patenting is a heat treatment process for wire and rod prior to drawing or between drafts. It consists of heating to a temperature above the transformation range and then cooling in air or in a bath of molten lead or salt.

Also see: Heat Treatment.

Pattern. A pattern is a form of wood, metal, plastics or other material, around which refractory material is placed to make a mold for casting metals.

Pattern Manipulation. Pattern manipulation is a part programming feature which enables a programmer to reuse a programmed pattern at other positions on the part.

Pattern Recognition. Pattern recognition is a computer vision technique that classifies images into outlines. Pattern recognition is when machine vision systems compare the attributes of an object to stored relationships or pictures as a means of identifying or comparing images. Pattern recognition usually doesn't observe or incorporate color and doesn't relate objects to uses. Pattern recognition uses feature extraction and things such as the circumference, area, maximum and minimum moment axes, and other basic feature attributes to try to match against stored shapes. Pattern recognition is a basic part of Object Recognition Systems (ORS). Optical character readers use pattern recognition to recognize shapes as characters by comparing them to stored images of different fonts.

Also see: Machine Vision, Optical Character Recognition.

PBX. See: Private Branch Exchange.

PC DOS. PC DOS is IBM's proprietary product version of MS-DOS. It is the major operating system or control program for the IBM line of microcomputers. PC DOS was developed by Microsoft in cooperation with IBM and is essentially interchangeable with MS-DOS. Some slight differences exist in the literal code contained within each product version of PC DOS as compared to MS-DOS.

Also see: DOS, MS-DOS.

PCB. See: Printed Circuit Board.

Pecker. A pecker is a pin used in tape or card reading to sense the presence or absence of holes.

Pen Plotter. A pen plotter is a hardcopy output device usually used for computer graphics. A pen plotter uses ink-tipped pens to draw upon the media. Pen plotters may be single pen or multiple pens with four, six, or eight pens typically. The purpose of multiple pens is to provide multiple colors or multiple width lines. Pen plotters may be referred to as flat-bed types in which the media lays flat and stationary on a bed with the pens moved in both the X and Y directions, or moveable media in which the pens only move in one direction, while the media is moved in the other direction. Advantages of moveable media plotters are the smaller footprint of the device and the limited size of the media in one dimension only. Usually, moveable media pen plotters are less expensive than similar capability flat-bed pen plotters. In all types of pen plotters, the pens also have the ability to move up and down (the Z direction) to allow drawing strokes to begin and end.

Also see: Electrostatic Plotter, Plotter.

Percussion Welding. Percussion welding is a resistance welding process in which pressure is applied rapidly (percussively) during or immediately following heating from an electric arc produced by rapid discharge of electricity across the joint.

Also see: Welding.

Perforating. Perforating is the punching of many holes, usually identical and arranged in a regular pattern, in a sheet, workpiece blank, or previously formed part. The holes are usually round, but may be any shape. The operation is also called multiple punching.

Also see: Punching.

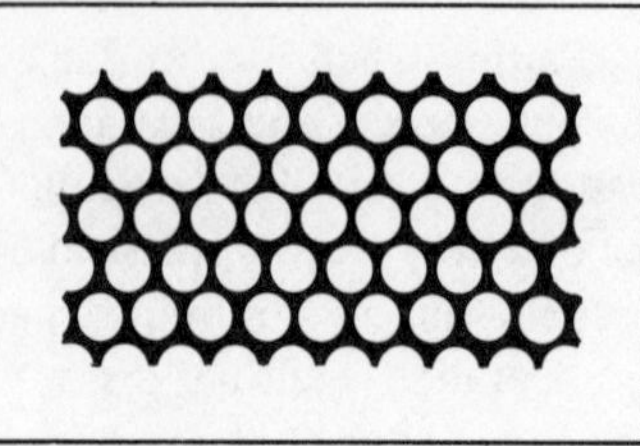

Perforating is used to punch many holes with a specific pattern.

Perforating Press. A perforating press is a high-speed, straight-side crank press usually furnished with a narrow slide, a slide stripper plate, and a special roll feed for progressively feeding sheet, strip or coil stock to be perforated.

Also see: Press.

Peripheral Equipment. Peripheral equipment is auxiliary machines and storage devices which may be placed under control of a central computer and used on-or off-line to provide a system with outside communication; for example, tape readers, high-speed printers, CRTs, magnetic tape feeds, and magnetic drums or disks.

Peripheral Milling. Peripheral milling is a form of milling that produces a finished surface generally in a plane parallel to the rotating axis of a cutter having teeth or inserts on the periphery of the cutter body.

Also see: Milling.

Peripheral Speed. Peripheral speed is the speed at which a point on the face of a wheel is traveling when the wheel is revolving, expressed in surface feet per minute (sfm), or meters per second (m/s).

Permanent-Mold Casting. Permanent-mold casting is a casting process that uses a mold of two or more parts to repeatedly produce many castings. Molds are generally made of graphite, cast iron or steel. If sand cores are used to create recesses or internal cavities in the casting, the process may be termed semipermanent-mold casting.

Personal Computer. Personal computers are completely self-contained computer systems which were originally designed to be used by single users for tasks which usually support activities that are normally performed by an individual.

Personal computers are usually microcomputer-based systems which include all five of the key parts of a computer, which are: the processor, the memory, the input/output (I/O), disk storage, and the programs. Personal computers are usually designed to be a size which can easily fit on top of a desk to be used by a single person although a popular alternate arrangement is to place the system unit on end on the floor, with only the monitor and keyboard on the desktop. Personal computers do not require special environmental conditions to function properly, but can easily operate in most environments which are comfortable to the human operator.

Personal computers are usually physically arranged into three major parts. There is the main box or system unit, which contains the motherboard and any expansion cards, the diskette drives and the hard disk (Winchester disk) drive(s). Other options such as cartridge tape drives or micro-Bernoulli drives may also be installed in the system unit. The keyboard is the second part of the personal computer system, and usually is included as standard with purchase of a system unit. The monitor is the third part of the basic system. Several options exist such as monochrome or color or enhanced graphics displays, with several choices within each category. The monitor is frequently treated as an required option when purchasing a personal computer. Usually a printer is needed to produce hard copy output and a modem may be needed for communication with other computers at remote locations. A vast array of additional optional peripherals such as pen-plotters, mice, digitizing tablets, etc., can also be used with a personal computer.

History. Personal computers first came on the scene through Radio Shack and Apple in the mid-1970s; and in August of 1981, IBM announced the IBM PC. By the spring of 1982, the PC was an astounding success with shipments much greater than predicted. The first "clone" to a PC was the Compaq computer, which was intended to be a portable device, although the difficulty in carrying the heavy machine soon coined the term "luggable." In the spring of 1983, IBM introduced the XT model (XT for extended technology) which contained a high-capacity hard disk in addition to the floppy diskette drive. Compaq matched the XT with another portable called the Compaq Plus.

In 1983, IBM stumbled with the introduction of the PC*jr*, which was not a sales success. The PC*jr*, also known as the "Peanut," which was its internal development name, was discontinued in 1985.

In 1984, Compaq introduced the Compaq DeskPro, which was the first machine to have more computing power than the original IBM PC. Shortly thereafter IBM introduced the AT model (AT for advanced technology) which had much greater computing speed than either the original PCs or the Compaq offerings. Most recently, IBM has introduced a new series of personal computers called the Personal System/2 or PS/2, which uses a new "microchannel" bus. The high-end models incorporate the 80386, which is the latest in the Intel 8086 family series of CPUs.

From about 1985 to the present, many manufacturers, including Asian and European companies have developed "clones" that are personal computers which are sometimes more or less compatible with the IBM products, but usually offer something greater in terms of features and almost always a lesser price. The success of clones has become so great that many institutions including Fortune 500 companies, universities, and the U.S. Government, who need large quantities of personal computers have made decisions to purchase clones instead of machines made by IBM. The competition in the marketplace continues to drive the retail price of personal computers lower.

Another scenario which was occurring during the era of the development of the IBM PC and its clones was the development of the Macintosh personal computer by Apple. The Macintosh or ''Mac'' as it is affectionately called, uses the Motorola 6800 family series of CPUs instead of the Intel 8086 family. The current Mac Plus and ''Fat'' Mac use the 68020 processor chip which is equivalent to the Intel 80386 in terms of general computing capabilities.

Discussion. The functions of the five parts of a basic computer system can be described by the following. The processor is the brains of the machine, which in a personal computer will usually be a variation of the Intel 8086 family. The CPU may sometimes be assisted by a numerical coprocessor chip such as the 8087 or 80287.

The memory is the ''workspace'' of the computer and has several constraints in a PC. The design of the Intel 8086/8088 in the PC and PC XT limits the physical memory to 1024K bytes. An IBM PC AT (or compatible) has a physical addressing limit of 16 megabytes (16,000K) of memory but the DOS operating system can only address 640K of user memory, with 384K reserved for the BIOS and display adapter. Schemes such as expanded and extended memory exist to allow user programs to address more than 640K of memory. Expanded memory and extended memory are discussed in detail in their respective sections. When an operating system is available to operate the 80286 in protected mode, programs will be able to be much larger while running multiple applications simultaneously.

The input/output or I/O section of a personal computer is all the portions which can take in or put out data, such as the keyboard and communications ports for data input and the video display and the same communications ports for data output. Some machines have a parallel printer port, which is usually unidirectional in the output direction only. The parallel port is normally capable of two-way data communication, though, and can be used as a higher-speed data channel than the serial ports. The larger Macintosh computers provide an SCSI (Small Computer Systems Interface) port for daisy chaining devices such as printers and disk drives.

While disk storage devices are also I/O devices, they have such an important role in the functioning of the personal computer that they are described and considered separately. Disk storage is typically the only mass storage for the computer. Disks provide the only permanent storage for when the power is interrupted to the computer and diskettes provide quick and inexpensive means to transfer data between machines. Diskettes are still the primary form of providing software to customers for personal computers. Hard disks also provide access to large quantities of data in a sufficiently fast access mode for many applications such that the 640K memory limit of DOS doesn't cause much problem for the user.

Programs are the last of the five key parts of a computer and that is definitely true with a personal computer. Without programs and the other related application software and the operating systems, personal computers wouldn't perform any useful activities. Some operating environments have been developed for personal computers which run under DOS and are intended to provide a more uniform interface for the development of applications software and a more familiar and consistent user interface. Microsoft Windows is probably the most familiar, with Desqview and GEM desktop being two others. Some of the applications programs in use were first perfected for use on PCs before mainframe and minicomputers. The most notable is spreadsheet software. Lotus 1-2-3 has set the standard for PC-based spreadsheet software and now products with similar capabilities are being developed and are available for minicomputers and mainframes. The reason the spreadsheet was enhanced for the personal computer first is because the concept of a personal computer provides dedicated and available computing resources for a user to constantly perform the interactive calculations required for most normal spreadsheet applications. The former mentality of timesharing and batch computer systems of mainframes is inconsistent with the present thinking of PCs providing available dedicated computing resources.

Applications. Initially, personal computers were used for simpler, basic applications such as creating simple programs in BASIC, word processing, and doing spreadsheet analysis. As with any type of computer, the hardware doesn't really drive the usage of the machine as much as the availability of useful software. With today's vast selection of quality software coupled with the high-speed AT type machines using 80286 and 80386 CPU chips, personal computers are used for applications such as desktop publishing, CAD/CAE entry workstations and design analysis. A PC can easily provide the total computerized support to run a financial package for a small to medium-sized business. PCs have available word processing packages which are more feature-laden than the capabilities of dedicated word processors or mainframe-based wordp rocessors.

Some of the special options available to a PC user are full-page displays to show either a full 8 1/2 by 11-inch page for word processing or a wide 19-inch screen which can display two facing pages for desktop publishing activities. Digitizing cameras and digital scanners can input graphic images while PCs can provide voice and data management for phone message systems. PCs using the 80386 CPU, along with an 80387 numerical coprocessor, can provide capabilities similar some engineering work stations, but at a much reduced cost, plus offering the user all the normal capabilities of a personal computer as a bonus.

Personal computers can be used as standalone machines or they can be networked with other similar or dissimilar PCs, as well as with minicomputers and mainframes. Networking allows the sharing of data and resources such as data storage devices and computing resources or output devices such as printers.

Currently available software for personal computers includes CAD/CAE software to most programming languages and applications software from micro MRPII to complete business accounting packages to sophisticated SQL type database software. PCs can do all the things and more that minicomputers and mainframes did only a few years ago. Currently one of the few areas where micros still can't compete very well with the larger computers is when used as multiuser systems with multiple operator interfaces or terminals being serviced concurrently. With today's low costs of a personal computer, a Turbo XT clone can be purchased for almost the same price as a high-quality terminal and if the PC is used on a network and can boot off the network (determined by the ROM BIOS), which allows it to be diskless, or at least minus a hard disk, the PC can actually be cheaper than a terminal. Once again, a terminal can't provide the functionality available locally with a PC when also using it as a terminal.

Another area where PCs are being used is in the factory on the shop floor. A PC provides an extremely well-appointed foundation upon which to build shop-floor data collection systems. The integration which used to be required for creating shop-floor systems is already provided by the architecture of the PC. In addition, PCs are used in tolerance control, NC programming, cost estimating, machine capabilities studies, regression analyses, and process planning.

One of the attractions of using PCs as standard platforms for products is that the PC will continue to be developed as a generic product at a very fast pace which means the target applications will constantly have a state-of-the-art building block from which the products can be developed. The competition in the marketplace will keep the prices very low for some time which assures PC-based products will be more price/performance competitive than most minicomputer or custom processor-based products.

Glossary. *Bernoulli drive:* A type of storage device which uses a floppy diskette in a rigid, removable cartridge and spins the diskette at much greater speeds than a traditional floppy

unit. The Bernoulli drive provided the first higher-capacity removable storage media.
boot: The initial power-up of a computer during which the system performs self-tests and loads the resident portion of the operating system.
cartridge tape unit: A storage device which uses a removable magnetic tape cartridge. Cartridge tape units can read data in a streaming manner or file-by-file and can restore data in either manner, depending upon the applied software. Cartridge tape units are primarily used to provide back-up for hard disks.
diskette: A thin, flexible disk coated with magnetic recording surfaces top and bottom.
DOS: The disk operating system, usually referring to Microsoft's MS-DOS or PC DOS when talking about an IBM compatible personal computer.
EGA: The Enhanced Graphics Adapter (640X350, 16 of 64 colors), which is one of the higher-resolution graphics standards defined by IBM.
enhanced EGA: A higher resolution EGA standard (640X480, 16 of 64 colors), which was defined by third party manufacturers of video cards. Requires a multiscan monitor.
expanded memory: Expanded memory is the ability of an computer running MS-DOS to address memory above 640K by using bank switching techniques. The expanded memory can physically be RAM or hard disk based.
extended memory: Extended memory is the AT computer's ability to directly address up to 16MB memory. The extended memory address normally starts above 1024K and extends for 15 megabytes.
mouse: A mouse is a 1, 2, or 3-button input device which is moved over a flat surface to input cursor movement. A mouse is named for its mouse-like appearance with the connecting cord like a mouse's tail, and its rapid movement over the surface.
protected mode: Protected mode is when the 80286 and 80386 microprocessors operate in a multitasking mode where each application can have a full 640K of dedicated dynamically managed memory and operate without interfering with the operation of or memory assigned to any other process. Currently no operating system except Xenix can make use of the protected mode of the chips, but OS2, the new forthcoming operating system from Microsoft is supposed to be able to invoke protected mode in the PC.
PS/2: The Personal System 2 which is the new series of personal computers introduced by IBM in 1987.
QIC standard: The QIC or *Quarter Inch Cartridge* standard is a group of defined and accepted standards for cartridge tape units and cassette tape units for personal computer data back-up. Some of the QIC standards are QIC-02 which includes most of the functionality in the drive unit with a smaller interface card, and the QIC-60 standard which has the formatter and other functionality on the interface card, plus the QIC-40 and some others.
real mode: Real mode is when the 80286 and 80386 microprocessor operate by emulating the 8088 and its command set. Protected mode using DOS restricts the CPU to single tasking and a 640K DOS memory limit.
SCSI: The Small Computer Systems Interface is one of the many interfaces defined for use on microprocessor systems. The SCSI interface can be daisy-chained to multiple devices. Currently, disk drives are the primary device which use the SCSI interface. The Macintosh provides an SCSI interface for custom interfacing of peripheral devices.
VGA: The Video Grphics Array IBMs PS/2 display standard (640X480, 16 of 64 colors), which supersedes IBM's EGA display mode. Requires an analog or multiscan monitor.

Also see: Computers, Computer Languages, DOS, Expanded Memory, Extended Memory.

PERT. See: Project Evaluation and Review Technique.

Photochemical Machining. Photochemical machining is a form of chemical milling in which a chemically resistant image of an intricately shaped part is placed on a sheet of metal, which is then exposed to chemical action which dissolves all of the metal except the desired part.

Photo-isolator. A photo-isolator is a solid-state coupling device which permits complete electrical isolation between field wiring and the controller.

Photo-optic Memory. Photo-optic memory is a memory which uses an optical medium for storage; for example, a laser used to record on photographic film.

Physical Properties. Physical properties is a term that pertains to the physics of a material. This may include: melting point, density, electrical and thermal conductivity, specific heat, and coeficient of thermal expansion.

Pick-and-Place Device. A pick-and-place device is usually a simple robot or piece of hard automation which is capable of the simple actions of picking an object from a fixed point and placing the object at another fixed point. The objects must be moved into and away from the two fixed points by other means such as a conveyor or by a human operator. A pick-and-place device usually has little or no intelligence for decision making and no ability to modify its point locations. Pick-and-place devices typically are used to take objects from the end of a feeder and place the object in a fixture or in some other type of repeatable location. Pick-and-place robots were the first types to be developed and utilized limit switches to enable them to make a simple move against two limit stops on each axis. While the sophistication of robotics has advanced greatly since the first products were developed, certain tasks do not require these more sophisticated devices, enabling pick-and-place devices to be widely used where appropriate in today's applications.

Also see: Industrial Robots.

Pick-and-Place Manipulators. Pick and place units are the simplest form of mechanical manipulators. They are either pneumatically operated or cam controlled and are equipped with up to five or six axes of motion; lift and lower, reach and retract, wrist turnover, body rotation, grip and ungrip, and in some cases, up to 10 degrees of vertical pitch.

Significant disagreement exists throughout the robotics industry concerning the definition of characteristics of pick-and-place units versus robots. For purposes of this discussion, pick-and-place units are somewhat arbitrarily defined as being lower cost mechanical manipulators that offer limited shop floor programming flexibility to the user. At the upper end of the manipulator spectrum are industrial robots which are servocontrolled and can be easily reprogrammed by the user to perform different tasks.

The motion of a pneumatic pick-and-place unit is achieved by supplying air to cylinders which in turn control movements such as horizontal stroke, rotation, and vertical lift. The limits of these motions are controlled by adjustable mechanical stops. The motion is then terminated by a sliding or rotating member contacting a mechanical stop, either a pressure-sensitive transducer or a position indicator is activated which signals the controller that the end point of the motion has been reached. When the end point is reached, the air valve remains open to hold position and the controller begins the next motion. This process continues until all the programmed motions of the specific task have been completed. Timer functions are employed when no feedback system is used. On some pneumatic pick-and-place units, hydraulic dampers (shocks) slow motion before impact at end point stops.

Also see: Industrial Robots.

Pickling. Pickling is a form of chemical cleaning that uses an acid solution—usually 10 to 50% concentration—to remove oxides, scale, weld discoloration, smut or corrosion products from metal parts.

PID Control. PID (proportional, integral, derivative) control refers to the process industry where the need exists to provide analog control. The variations of control equations can be proportional (P) only, integral (I) only, derivative (D) only, proportional-integral (PI), proportional-derivative (PD), or proportional-integral-derivative (PID). Any of the following combinations might be utilized, depending upon the nature of the process to be controlled, as well as the sophistication and precision of control required. The basic usage of PID control is to vary the controllable parameters to maintain a setpoint value as closely as possible.

Since PID controllers are analog-controlled, the response and accuracy of the controller is an important factor in selecting a product. Usually the media being controlled are temperature, pressure, or flow, which are all constantly varying properties and therefore typically measured as analog inputs. Temperature can be measured with either a resistive temperature detector (RTD) or a thermocouple. Any device which uses analog circuitry must be calibrated and all PID controllers, when first started up with a control loop, must have the loop tuned to achieve the optimum performance and response. If a PID controller doesn't have appropriate damping, an oscillation may set up about the setpoint as the controller attempts to maintain the setpoint.

Piercing. Piercing is the general term for cutting (shearing or punching) openings, such as holes and slots in sheet material, plate or parts. This operation is similar to blanking; the difference being that the slug or piece produced by piercing is scrap, whereas the blank produced by blanking is the useful part.

PILOT. Developed for computer-aided instruction applications, PILOT stands for Programmed Inquiry, Learning Or Teaching.

Pixel. A pixel (short for picture element) is the display element that can be used to construct a display image, or picture, in the display space on the display surface of a display device. In a bit-mapped display, a pixel is equivalent to the smallest addressable point on the screen. For example, the output of a single optical fiber terminating on the faceplate at the end of a coherent bundle of optical fibers, or the output from a single separate piece of the mosaic that forms the screen of a CRT (cathode ray tube) and whose output can be independently controlled by an electron beam.

In a facsimile transmission system, a pixel is that area of the original document that coincides with the scanning spot at a given instant. In a facsimile transmission system, the area of the finest detail that can be effectively reproduced on the record medium. Both the mark and the space may be considered as separate picture elements.

Also see: Machine Vision.

PL/1. PL/1 is the high-level programming language, designed for use in the wide range of commercial and scientific computer applications, in an attempt to combine the features of both FORTRAN and COBOL.

Also see: Computer Languages.

Planing. Planing is a machining process that produces a flat surfact (horizontal, vertical, or at an angle) by reciprocating the workpiece across a stationary single-point tool.

Plasma-arc Cutting. A method of metal removal, plasma-arc cutting is when plasma is used as the energy source which provides the cutting action for cutting or shaping metals. Plasma-arc cutting uses an extremely high temperature, high-velocity constricted arc between the electrode in the plasma torch and the piece to be cut. When inert gases are used, the cutting process depends upon thermal action alone. When the application is the cutting of such materials as mild steel and cast iron, increased cutting speeds can be achieved by adding oxygen-bearing cutting gases.

This process can be used in general fabrication of all metals in many industrial applications such as shipbuilding, aircraft production, chemical, nuclear and pressure-vessel. It also has found a use in the forging and casting industry. The cutting process can be used in manual or mechanized operations and is nicely adaptable to automation. Plasma-arc cutting usually leaves cut-edge finishes (which are not suitable for finished parts).

Also see: Cutting.

Plasma-arc Machining. Plasma-arc machining is a nontraditional machining process used most often as an alternative to oxyfuel-gas cutting for rough cutting or blanking operations on heavy sheet or plate. In plasma-arc machining, an electric arc between a torch and a workpiece heats and ionizes a gas such as nitrogen, argon-hydrogen or air, which is forced through the torch in a swirling action. Plasma temperatures as high as 50,000 degrees F (27,800 degrees C) melt and vaporize workpiece material. Secondary gases or water flow are often used to clean the kerf of molten metal during cutting. Plasma-arc machining may also be used to replace lathe turning, milling or planing.

Plasma-arc Spraying. Plasma-arc spraying is the most versatile of the thermal spraying processes. In plasma-arc spraying, coating material in wire or powder form is fed into the tip of a plasma-arc torch, where it is melted and propelled against the substrate. Either the torch or the substrate is moved to provide controlled coverage of the target area.

Plasma-arc Welding (PAW). Plasma-arc welding is a fusion welding process that uses a special torch to generate a plasma of ionized gas, which carries welding heat to the joint. PAW is widely used on thinner materials, often without adding filler metal to the joint.

Also see: Welding.

Plasma Spraying. Plasma spraying is a thermal spraying process in which the coating material is melted with heat from a plasma torch that generates a nontransferred arc; molten coating material is propelled against the basis metal by the ionized gas issuing from the torch.

Plastics. Plastics is a generic term for the industry and its products. This term is properly used only as a plural word. Plastics products include polymeric substances, natural or synthetic, and exclude rubber materials.

Plated Wire Memory. Plated wire memory consists of wires coated with a magnetic material that may be magnetized in either of two directions to represent ones (1) and zeros (0).

PLC. See: Programmable Logic Controller.

Plotter. Plotters are used to translate computer generated graphics onto a sheet of paper. A plotter does not generate a bit mapped or character formed representation of information. Instead, a plotter has imbedded intelligence which renders it capable of constructing shapes by vector drawing of geometric primitives of the basic shapes such as circles, lines, fill, and similar functions. A circle may be constructed by a succession of straight segments which approximate a circle. A plotter is much slower than a printer when creating text characters because the plotter must treat every object as a geometric shape and draw the shape segment by segment. By contrast, a printer has the representation of characters stored as internal shapes and reproduces the character shapes when they are called by reference instead of drawn image.

Also see: Electrostatic Plotter, Pen Plotter, Printer.

Point-to-Point Control System. A point-to-point control system is an NC system which controls motion only to move from one point to another without exercising path control during the transition from one end point to the next.

Also see: Numerical Control.

Polar Coordinate System. A polar coordinate system is a mathematical system of coordinates for locating a point in a plane by the length of the plane's radius and the angle the vector makes with a fixed line. In a polar coordinate system points are defined in terms of the distance along a central axis of origin, an angle about the central axis and a distance away from the central axis, along a normal ray. The polar coordinate system is useful when used with devices whose physical configurations lend themselves to fit the polar coordinate system. Some robots have a central vertical axis and an arm which can rotate about the central axis while extending or contracting. The difference between a polar coordinate system and a spherical coordinate system is the existence of an anchor axis instead of an anchor point and the ability to radiate in a plane as opposed to being able to radiate in a spherical pattern. Polar coordinates can be used when describing objects which are turned on a lathe or similarly symmetrically dimensioned.

Also see: Cartesian Coordinate System, Spherical Coordinate System.

Polishing. Simply put, polishing is the act of making a surface smooth. It is any abrading process that removes small to large amounts of stock and generally produces a defined line pattern on the workpiece surface. The term polishing may be interpreted to mean any non-precision procedure providing a glossy surface, but is most commonly used to refer to a flexible abrasive wheel. The wheels may be constructed of felt or rubber with an abrasive band, of multiple coated abrasive disks, of leaves of coated abrasive, of felt or fabric to which loose abrasive is added as needed; or of abrasives in a rubber matrix.

Type and size of abrasives used in polishing is dependent upon the material polished and the type of finish desired. The size of abrasive grain to be used depends on the initial character of the surface to be polished. The first wheel should be coated with an abrasive that will cut down to the bottom of the deepest pits and break through scale. The second wheel should be coated with finer abrasive that will cut out the grain particles made by the coarser wheel. The next wheel should have still finer grain, and so on until the desired surface has been produced. The final wheels can be lubricated to modify the cut and produce a more lustrous surface.

Also see: Finishing.

Polymer. A polymer is a high-molecular-weight organic compound, natural or synthetic, whose structure can be represented by repeated small units (mers). Examples include polyethylene, cellulose, and rubber. Synthetic polymers are formed by addition or condensation polymerization on monomers. If two or more monomers are involved, a copolymer is obtained. Some polymers are elastomers, others are plastics.

Port. A port is an entrance or exit of an electrical network. A port is also a connecting unit between a data link and a device. For example, a port is between an I/O channel, data bus, or interface module and a computer, data terminal or CRT.

Also see: Computers.

Position Analog Unit (PAU). Position analog units feed analog information corresponding to the position of a machine slide back to the servoamplifier for comparison with positional input information.

Positioning/contouring System. A positioning/contouring system is a numerical control system that is able to contour in two axes, without buffer storage, and position in a third axis for operations such as drilling, tapping, and boring.

Also see: Boring, Drilling, Numerical Control.

Position Sensor. Position sensor is a device used in measuring a position and converting the measurement into a form which facilitates transmission.

Position Storage. Position storage is an NC system storage media which contains the coordinate positions read from tape.

Also see: Numerical Control.

Postprocessor. A postprocessor is a computer program that adapts general information received to specific local requirements. Postprocessors are typically used in such applications as electrical CAD systems where the artwork generation is done after the majority of the design work has been completed by other portions of the system. Postprocessors are sometimes the annotation and documentation portion of CAD and CAE systems. As is obvious by the term postprocessor, the assumption is that a major activity which is considered to be the main process has occurred before the activity which is designated as being performed by the postprocessor. A postprocessor can be a portion of code which executes in the same processor as the main process or the postprocessor may be a separate combination of hardware and software. Some vendors market products which are designed to be postprocessors to other vendor's products and provide some enhancement which would not otherwise be available.

Also see: CAD, CAE, Compiler, Computers.

Postprocess Sizing. Postprocess sizing controls sizing by gaging finished parts and making needed corrective changes, commonly by automated functions.

Also see: Gaging.

Power Brushing. Power brushing is any process that uses a power-driven, rotary industrial brush to deburr, clean or finish a metal part. Depending on application, the brush fibers—collectively known as brush fill material—may be metal wires; fiberglass-coated, abrasive-filled plastics; synthetics such as nylon and polypropylene; natural animal hairs such as horsehair; or vegetable fibers such as tampico and bahia.

Power Hacksawing. Power hacksawing is a sawing process that uses the back-and-forth motion of a short, straight toothed blade to cut the workpiece. Hacksawing machines are generally electrically driven, and may or may not provide for application of cutting fluid to the saw blade or workpiece.

Precondition. Precondition is a preparatory condition defined by a PC logic instruction. One or more preconditions may be programmed prior to an output instruction.

Preparatory Function. A preparatory function is an NC command on the input tape that changes the mode of control operation; usually referred to as G function because it is noted at the beginning of a block with the letter character G plus a two-digit number.

Also see: Numerical Control.

Preprocessor. A preprocessor is a computer program effecting preliminary manipulation of data. A preprocessor does some initial processing before another portion of the system continues with the main processing. Preprocessors are used by vision systems to do primary reduction of the data such as thresholding. Software programming language compilers sometimes use precompilers as preprocessors to optimize the source code before the main compiler processes the code. A preprocessor is usually only designated as such when the activity occurs before a major process which would be considered to be more substantial than the activity which is considered to be the preprocessor. A preprocessor can be a portion of code which executes in the same processor as the main process or the preprocesssor may be a separate combination of hardware and software.

Also see: CAD, CAD-D, CAE, Compiler, Computers, Machine Vision, Postprocessor.

Preset. To preset is to establish a value of a variable before it is to be used; to establish an initial condition, such as the control value of a loop.

Press. A press is a machine having a stationary bed and a slide (ram) which has a controlled reciprocating motion toward and away from the bed surface and at a right angle to it, the slide being guided in the frame of the machine to give a definte path of motion.

Also see: Press Brake.

Press Brake. The press brake is a special variation of the gap type press especially developed for making long bends in sheet metal. Reductively complex shapes can be bent from long stock with reasonably simple dies. These type of presses are particularly applicable in the fabrication of long structural parts which must be made from sheet metal such as aircraft, auto and bus parts.

Press brakes are generally constructed in the medium-size range and are available in many

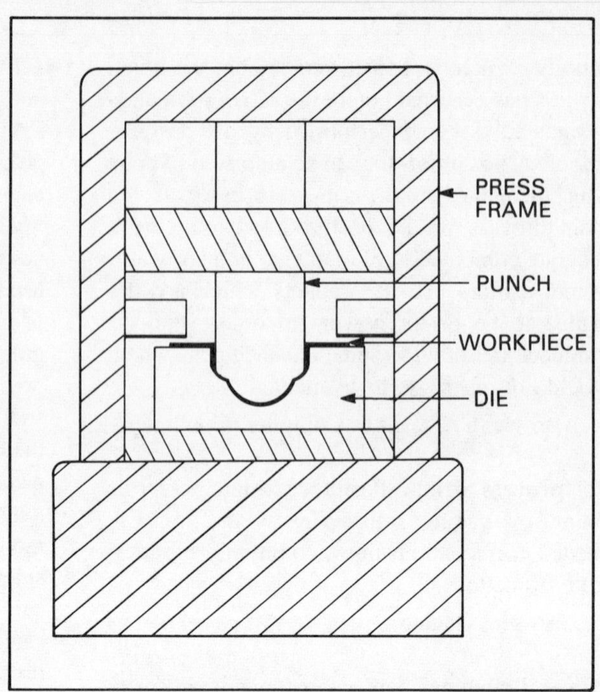

Basic production setup for a stamping operation. Pressure is applied by the press. The sheet metal is formed into the shape determined by the punch and die.

design variations making them adaptable to a very broad range of applications. They may be equipped with automatic indexing dial feeds for assembly work and for performing a number of operations in sequence on simple parts. They may also be obtained with the crankshaft running from front to back and with a solid steel arm or hand extending from the frame for folding or sawing operations. Press brakes are usually available only as a single action press but can be used for every type of bending operation except those which require a long stroke large bed area or exceptional force capacity.

Newer press brakes are equipped with motor-driven variable displacement hydraulic pumps so that the speed and pressure of the operating ram are under instant and automatic control. This is particularly advantageous for deep bending operations. The dies can be brought into initial contact with the work without shock and moved with a uniform, controlled velocity through the bending portion of the cycle.

Press Forming. Press forming is any forming operation performed with tooling by means of a mechanical or hydraulic press.

Also see: Press.

Pressure Bag Molding. Pressure bag molding is a process for molding reinforced plastics in which a tailored, flexible bag is placed over the contact lay-up on the mold, sealed, and clamped in place. Compressed air forces the bag against the part to apply pressure while the part cures.

Pressure Gas Welding (PGW). Pressure gas welding is an oxyfuel gas welding process in which metals are bonded by heating them with flames and applying pressure to the joint. Filler metal is not used.

Also see: Welding.

Printed Circuit Board (PCB). A printed circuit board is an assembled printed wiring board populated with components. It is a board on which a printed pattern of interconnects has been deposited. Some companies refer to printed circuit boards as modules or circuit cards. Printed circuit boards used to always be very thick and rigid, but with today's technology PCBs are sometimes very thin or even flexible and sometimes have multiple layers. Printed circuit boards can be populated with through-hole components; or surface-mounted components, which can be placed on either side of the board. Printed circuit boards can be used by themselves or the PCB may be connected to another by use of a connector and ribbon cable, or edge connectors may be used to connect the PCB to a back-plane or motherboard.

Also see: Printed Wiring Board.

Printed Wiring Board (PWB). A printed wiring board describes the unpopulated printed circuit board. While some companies use various terms such as circuit cards and modules to describe the finished product, most companies use the same terminology of printed wiring board when referring to the finished bare board. A printed wiring board is composed of multiple layers of copper sheets and fiberglass (for electrical insulation and strength). The PWB may contain multiple layers, sometimes as many as 12 layers, although three to six are probably the most common. Usually in a multilayer board, the outer layers on both sides are ground planes to assist in the noise immunity of the circuitry and decrease the susceptibility to radiated interference. Printed wiring boards are designed using CAD systems. The process of manufacturing them uses photo-etching and chemically removing the unwanted conductor on the various layers. A more recent form of PWB uses copper and mylar which makes the result very flexible. These flex circuits are used in such items as folding pocket calculators and circuits which must fit into odd-shaped spaces. Most PWBs use etched sheets of copper. Some PWBs use individual insulated wires which are run from each point to the connecting location and glued to the board. This type of PWB is called a multiwire board.

Also see: CAD, Postprocessing, Printed Circuit Board.

Printer. A printer is a device which generates hard-copy output from a computer. Printers exist in various types including thermal, impact (such as dot matrix, daisy wheel, or ball element), and laser printers. A common reference regarding the quality of text generated by a printer is draft, NLQ (near-letter quality) and LQ (letter-quality). Usually slower speeds and sometimes multiple passes are utilized to generate the better quality output. Some printers can only generate ASCII text characters, while others are capable of receiving and printing bit-mapped graphics. By using multicolor ribbons, dot-matrix printers can produce color output, something a laser printer cannot do.

A printer usually retains the description of the characters internally, and upon receiving the ASCII code for characters and control characters, generates the characters, spaces, line feeds, carriage returns, etc. When in graphic mode, a printer is sent bytes of information which describe a binary pattern for firing the pins of its printhead. This explains the difference between a text character and a graphic character, which even though the output may appear, or in fact be identical, the scenario utilized in conveying the information to the output device is different.

Also see: Bit Map, Computers, Ink-jet Printer, Laser Printer, Line Printer, Personal Computer, Plotter, Thermal Printer.

Priority Processing. Priority processing is the processing of a sequence of jobs on the basis of assigned priorities.

Private Branch Exchange (PBX). A private branch exchange is a manually operated telephone exchange that is internal to and serves a single organization and usually has connections to another telephone exchange or switching center. The outside connection may be made via a dial mode or dial service assistance switchboard.

A private automatic branch exchange or PABX is a private branch exchange (PBX) in which the connections are made automatically and there is no operator.

A computerized branch exchange or CBX is a private branch exchange (PBX) or private automatic branch exchange (PABX) in which a computer provides the switching functions as well as any other services such as call forwarding, call waiting, message handling, data manipulation, and the branch exchange functions digitally as opposed to switching analog circuits.

Problem Orientation Language. A problem orientation language is a programming language which allows a program to be specified as problem components rather than procedures for solving. Such languages are limited to a particular class of problem (e.g., jet engine simulation) but are easier to use, especially by personnel knowledgeable in the problem area, but not in computing.

Procedure Oriented Language. A procedure oriented language is a programming language (e.g., BASIC, FORTRAN, COBOL) whose programs specify the procedure for solving a problem, using computing techniques. These are lower level than the problem-oriented languages.

Process. A process is continuous and regular actions executed in a predetermined, uninterrupted manner. A process also is a PC application involving assembling, compiling, generating, interpreting, and computing information.

Process Control. Process control is the activity of controlling a process (the combination or alteration of ingredients to alter their state). Process control relates to formal systems implented to provide automation of continuous production operations. Process control may be the creation of paper from pulp or the refining of petroleum into gasoline, kerosine, and other byproducts. Process control normally doesn't refer to discrete parts manufacturing or machining operations, where the major activity isn't to change the chemical state of the product. Process control typically requires control of a continuous process, with no specifically defined starting or stopping point, whereas discrete parts manufacturing has natural units of definition for starting and stopping depending upon the incremental quantity desired. A second type of process control is batch, where the entire process is carried out on the entire batch of material as a single operation.

Also see: Controller, Manual Backup, Processor.

Processor. A processor monitors/scans all inputs and outputs and then acts depending on the program instructions. It is a device which is capable of performing calculations or otherwise "processing" a predefined program. A processor can be thought of as the CPU when focusing at a more micro level, or a processor may consist of the entire controller when considering the macro view. If the focus is on a system which consists of various functional devices and modules, the processor is the portion which performs the logic execution operations. In an industrial system the processor may be the programmable logic controller (PLC).

A processor receives input and, after acting upon the input, provides output. The manner of the output depends upon the type of input and the capabilities and instructions given to the processor. A processor is a required part of a system when the state or form of something is altered such as when data is reformatted or a chemical substance is processed.

Also see: Central Processing Unit, Computers, Programmable Logic Controller.

Production Control. Production control is the control and direction of activities in the movement and scheduling of material through a manufacturing cycle. Production control includes activities from the requisitioning of raw materials to the completion of product manufacturing through the shipping cycle. Production control is a part of the overall manufacturing and planning and control systems. This term commonly describes manufacturing organizations charted to support the production planning and shop floor control functions. Both of these functions are part of the coordination and control of the material and labor in manufacturing. Production planning involves the scheduling of manufacturing departments or work centers taking into consideration demand requirements and capacity limitations and capabilities. Shop floor control activities are directed toward material dispatching and priority setting of the manufacturing schedule. The production control process can include the measurement of product output to customer demand or to a predetermined production rates.

Also see: Manufacturing Resource Planning.

Program Panel. A program panel is a device used to insert data into a program, monitor a program, or edit it.

Program Scan Time. Program Scan Time is the time necessary for a processor to execute all program instructions once. The scan repeats continuously. Inputs are monitored and outputs controlled through the input/output table.

Also see: Programming.

Program Stop. A program stop is a miscellaneous function command to stop the spindle, coolant, and feed after completion of the dimensional move in the block. To continue the program, the operator must push a button.

Programmable Array Logic (PAL). Programmable array logic is an integrated circuit chip which has generic array logic which can be configured or programmed to provide certain logical functions. PALs are programmed in much the same manner as programmable read-only memory (PROM), which is by ''blowing'' fused links inside the chips to create a certain matrix of connected gates or a gate array. The difference between a PAL and a PROM is that a PROM is a memory device which contains data that can be read while a PAL is a logic device capable of performing computational operations or otherwise solving logic problems. PALs are used for interface logic between CPUs and peripheral equipment or when initially testing the functionality of a new design before committing to developing the full-custom chip.

Also see: Application-specific Integrated Circuit, Programmable Read-only Memory.

Programmable Controller. A programmable controller or PC (now more often referred to as a programmable logic controller or PLC, to distinguish it from a personal computer) is an event-driven controller intended for industrial usage which can be programmed and interfaced to a variety of devices and other computers.

The original programmable controllers were designed to be replacements for hard-wired relay logic which required a large amount of physical labor to wire the control logic and also required a substantial amount of space to contain the relays and terminal strips. These first programmable controllers were general purpose to the extent they represented ''soft'' automation instead of ''hard'' automation, but otherwise, once programmed, they simply provided the same simple logical control sequences which were also obtainable using hardwired logic.

A programmable controller is a computer with special features and capabilities as compared to other types of general-purpose or special-purpose computers. A programmable controller has a central processor with the program memory and an I/O section which can have a large quantity of I/O points which can be analog, digital, AC, DC, high voltage, or millivolt. Another major part of a programmable controller is the ''man-machine interface,'' which is the operator input and output device. Many modern programmable controllers can use personal computers as a programming device and output display device. Another attribute of programmable controllers is that they are ''factory hardened,'' which means they can survive the factory floor environment which may have elevated temperatures and humidity, EMI and RFI noise, and other contaminants such as dust and caustic vapors. A programmable controller is usually considered to be more ''bullet proof'' in its operating capability and reliability than a personal computer or other nonfactory-hardened devices. Since the production equipment that could be controlled by a programmable controller could cost the factory substantially for any downtime, MTBF (mean time between failure) and MTTR (mean time to repair) are important features considered by the user when selecting a controller.

While the early programmable controllers were intended to be programmed by the electricians who had been connecting the hardwired logic, today's devices are being designed with more sophisticated programming capabilities, such as their ability to receive programs downloaded from a host computer or cell controller.

Programmable controllers are distinctly different from other devices in that they typically use a form of programming known as ladder logic, which graphically appears similar to the physical wiring logic which they replaced. This was done originally with the thought that electricians would have an easier time of becoming familiar and proficient with the new devices. In this age of computer literacy, manufacturers and users are questioning the need to utilize such a restricted method of programming, especially with the capability of programming devices such as personal computers to compile or automatically generate the programs based on higher level input by the user.

Bob Eisenbrown of Allen-Bradley Company reports in the January 1988 *Manufacturing Engineering* article titled, *Programmable Controllers Move to Systems Solutions*, "Programmable controllers are the heart of today's automated factory. Enhancements in programmable controller technology have transformed them from relay replacers to integrated machining and process controllers."

"In the past, application decisions hinged on individual machine automation and hardware costs. Today, system integration and total system life-cycle costs have a greater effect on the selection of automation control equipment."

History. The origin of programmable controllers is considered to begin with the Modicon "relay replacer" which was demonstrated to General Motors. The Hydra-Matic Division of GM placed an order for several machines to use as replacements for hard-wired relay logic control systems. This product launched the era of programmable controllers.

The first programmable controllers used small 8-bit CPU chips and had a small amount of memory for program and operating systems. Quantities of memory such as 2K and 4K were considered sizeable. With these restrictions, the operating system was custom-designed and written in assembly language to obtain the greatest possible performance from the machine. The architecture of the machine was designed to utilize binary numbers in registers as the means of input and output, as well as internal processing. ASCII characters and data were not accommodated by the early programmable controllers.

During their evolution, programmable controllers not only used more general-purpose computer components, but the design architecture started to become more standard like a general-purpose computer. The main reasons for the merging similarities were that the increased speed, performance, and ruggedness of standard computer components had become acceptable for the needs of the shop floor. Standard operating systems were now capable of handling ladder-diagram compilers and other routines running as tasks under them without the resulting performance being undesirable.

As the environment in factories improved (with regards to the harshness for electronic equipment), personal computer-type products have eroded the application of programmable controllers. Many users simply do not see the need to spend more money for a product which isn't as capable as the personal computers of today, with other penalties due to economies of scale in the demand for each type of product.

With the continuing advances in the integration of information and control systems in the factory, concepts such as networks, including MAP (Manufacturing Automation Protocol), are making the traditional programmable controller less capable and desirable than general purpose or personal computers for certain applications. The programmable controller device of today must be capable of upwards communication with cell and supervisory controllers and downward communication with peers and intelligent I/O. Modules such as PID control modules and stepper and servocontroller modules sometimes have as substantial processing and memory capabilities as the central processor. Some intelligent modules can communicate directly with a programming device, which allows them to operate as stand-alone devices. An intelligent module is one which has local processing and decision making capability instead of being a passive extension of the main controller.

Manufacturers of programmable controllers are starting to focus more on the I/O and the devices at the point of control. This portion of the system is required, whether the main controller is a programmable controller or a personal computer. The smaller capacity (relative to points of I/O) and slower (scan time) programmable controller market is now being dominated by Japanese companies such as Omron, Toshiba, Sharp, Meiden, Matsushita, and Mitsubishi among others. Many U.S. programmable controller companies import Japanese products and brand label them for sale as their own low-end products.

Author Eisenbrown noted: "Before looking at the future trends in programmable controllers, a review of the original requirements that drove their development is appropriate. Those early requirements were (1) a control-oriented programming language (relay ladder logic); (2) high speed, consistent user program execution cycles (a scan rate on the order of milliseconds); (3) dedicated processing for consistent input/output (I/O) updating (part of the processor's continuous scanning of I/O); (4) continuous error checking and fault monitoring; (5) immediate, orderly shutdown with last state retention in the event of workstation error or failure; and (6) an industrially robust package.

"While the programmable controller of the future will retain these characteristics, it also will be called upon to address and answer new demands. From an engineering point of view, the programmable controller no longer has to contend just with automating a machine. It must now be equally adept at providing hooks for system integration. For example, we now are beginning to see programmable controllers used not only for control, but also for data acquisition.

"The programmable controller-based control system can no longer sell itself as low-cost hardware alone. Rather, its use in a system must provide low cost for the manufacturing system throughout its life cycle, including design, installation, operation (where reliability is a particular requirement), diagnostics, and maintenance."

Discussion. The traditional programmable controller consists of the following basic segments: power supply, CPU, various types of I/O, and a programming panel.

While the power needs of the programmable controller's main processor are essentially those of a typical computer, enough current must be supplied in the appropriate voltages to power the I/O rack. Since power to a factory typically isn't as stable as the power into a home or office environment, the power supply must be capable of smoothing significant fluctuations in supply voltage. In some factory situations, electric welders and other large electrical equipment cause large dips in the voltage supply of a plant when starting up. This can cause the PLC to "crash." Many PLC power supplies have enough capacitive storage capability (hold up) to be able to conduct an orderly shut down of the system when power is interrupted. This shut down procedure might include putting all I/O points into a prescribed fail-safe state, and saving any information currently in volatile memory to nonvolatile memory, including all register values. This process thereby allows a restart from the state of interruption, if desired, without having to clear out the work-in-process before restarting, which might result in considerable waste of partially finished product. This hold-up capability also allows the PLC to withstand brief power stutters and transient power drops without interrupting operation.

The CPU or controller of the PLC unit is responsible for maintaining and solving the logic program. The controller usually integrates all the other components together logically and functionally, with most of the communication between other devices, except for field wiring, going through the controller. Field wiring, which is simply the connection between an I/O point and a device such as a limit switch or sensor, is direct between the I/O module and the device.

Some programmable controllers that provide redundant or fault-tolerant computing have multiple identical controller cards which reside in the same machine. A multiple controller has a second controller which is installed in the system but it is only idling (powered up with program loaded, but not processing), and can take over when the primary controller goes off-line. A redundant system usually uses three identical controllers, all running the same program with all output from the controllers being compared by a ''voter'' (usually in hardware) which passes the majority message from the three controllers to the I/O of the personal computer. If one of the three deviates from the other two, its performance is investigated by the system due to the possibility it might not be functioning properly.

The I/O modules are the connection points for the field wiring and provide the proper electrical and logic connections for the control devices and input devices and sensors. Some I/O modules have special capabilities such as high-speed signal input capabilities. I/O modules usually provide for some multiple of duplicate functions, referred to as points. A two-wire input or output module might offer 4, 8, 12, 16, 24, or 32 points. The types of I/O might include 120 volt AC, 240 volt AC, millivolt, 5 volt DC, 24 volt DC, and other function modules such as servo and stepper controllers. Some manufacturers have tried to provide modules which can either be bank switched or individually configured for either input or output function on the same I/O module card. The purpose of this is to accommodate the situations when the application requirements call for a quantity of I/O points of a certain type which doesn't happen to be an even multiple of the standard offering.

Another feature unique to programmable controllers is the ability to physically place I/O racks in a remote location from the main processor unit. This allows a single communication cable to connect the remote I/O rack and the main rack instead of having to run field wiring from a single central location to all field devices.

A programming panel is required for the configuration of the programmable controller, including its I/O and communications ports as well as loading software into the system. A programming panel is usually the means for inputting the applications program into the processor. However, in the last few years, downloading programs from another computer has become desirable and more expedient. The early programmable controllers had specially designed programming panels. Most of the controllers which are sold today can use a personal computer as a programming panel. Some companies now only offer a personal computer as the means to program their product. The vast array of existing software for a personal computer make it very attractive for such a task.

The operation of the programmable controller consists of repeatedly processing a looping program which consists of: a self-test program, writing to the I/O, the logic solving program, reading the I/O, and communications handling.

Upon power-up, the programmable controller usually does extensive self-tests including testing the communications to each I/O point in the rack. One of the needs of an unmanned controller on the factory floor is to monitor the system to ensure all parts continue to be functional, and to appropriately conduct a safe shut-down if necessary. If the controller fails, the desire is that the failure be in a safe manner with respect to whatever the controller is handling. To ensure the system, its memory, and the I/O continue to function properly after the initial power-up tests, a ''watch-dog timer'' is used. This ''watch-dog timer'' is a portion of the operating system which periodically monitors a signal response from the CPU that, by virtue of the fact it exists, indicates the system is still alive. The monitoring portion has been programmed regarding how to react if the watch-dog signal is not detected after a certain elapsed time (time-out).

The second major part of the control loop is writing to the I/O. During this phase, the controller transfers all I/O state data residing in the output buffer from the central processor to the I/O modules, where the information is processed and the state of each I/O point is made to correspond. Programmable controllers usually write to the I/O before they read input to ensure the control system begins at a proper and known state. Each time the program is sequentially solved, the information is written to the I/O.

Logic solving is the third portion of a typical programmable controller "scan" loop. A scan is a single time of looping through the self-test, I/O write, logic solve, I/O read, and communications handling. A ladder-logic program is solved, starting at the top left corner and proceeding downward solving the first element in each line, then returning to the top and solving the second element from top to bottom. This is done even though the program is written starting at the top-left corner and writing statements across to the right, then going to the second line.

After the logic is solved, the central processor reads all the status buffers from the I/O which contains any input modules which may be installed. Considerable filtering may be done on the information coming from and going to the I/O modules to ensure correct interpretation.

Usually the last task to be performed during a scan is to service the external communication ports such as to the programming panel or a cell controller or other type of higher-level supervisory computer. Sometimes this communication is inhibited to allow a faster scan.

This programmable controller logic solve loop is repeated over and over. It is easy to see that increasing the number of I/O points can extend the overall scan time since more time must then be spent to handle the increased quantity of information. Some PLCs can be programmed to use a *segmented scan* which means some I/O points are not scanned on every loop. Only the critical items are scanned every loop. In some applications, the programmable controller must be able to receive information within a certain time frame which would either require a buffer of sorts or a programmable controller with a scan time fast enough to be acceptable. With the use of parallel and multiprocessors, having to perform all tasks in one large loop becomes unnecessary.

Applications. Programmable controllers are used in everything for controlling factory machines in all industries. Any industry that uses machines which must be controlled by a programmable, intelligent device is a candidate for a programmable controller.

Other applications for programmable controllers are as traffic light controllers, oil field pumping control remote terminal units, irrigation controllers, and elevator controls.

As people realize programmable controllers are simply specialized computers, and as the price-performance of personal computers continues to improve, PLCs are being replaced with PCs. The addition of an interface to discrete I/O points, a ladder diagram compiler and display driver, and control software convert a personal computer into most of the things provided by a programmable controller. The trends indicate that in the future, programmable controller manufacturers may need to start to use standard platforms such as the many varieties of personal computers.

In the SME Manufacturing Insights Videotape *Programmable Controllers*, Richard Morley, Director of Advanced Technologies at Gould Electronics' Instrumentation and Automation Systems, and the inventor of the programmable controller in 1969, explained: "A programmable controller does many things, but its primary function is to examine the world that the industrial process looks at. Take those inputs into its system, make a logical interconnection between all those sensing devices.

"Typically the ratio of the inputs to outputs in 2 to 1. You have 1,000 inputs and 500 outputs and the programmable controller itself interconnects all of those inputs such that it makes an output.

"For example, it essentially makes the statement, 'If the temperature in department A exceeds 140 degrees F, ring the fire alarm;' (it) says, 'If my limit switch on my transfer motion has been activated, shut down servo power.' It's a very simple relationship.

"Regular computers have to treat all of the relationships together. Programmable controller programming language, and its concept, says each statement can be made all by itself.

"Programmable controllers initially were conceived as stand-alone devices. Although we use computer technology to build programmable controllers, they were not computers, they were appliances.

"So I wanted to build an appliance. This appliance did logic things that my friends the electricians, the machinists, the process engineers and chemical engineers could use without having to think about the technology itself. For example, the modern automobile has between one to 10 computers in it. Am I driving a computer or driving an automobile? I'm driving an automobile.

"The modern microwave oven has one to two computers in it. A modern TV set has three computers in it. They are not computers; they are appliances designed for a specialized vertical functionality. A programmable controller is one of the first such things.

In his *Manufacturing Engineering* article, Eisenbrown wrote: "Today, the programmable controller must meet the growing needs for greater levels of control automation. This involves distributed processing, which helps simplify control system implementation and design and results in greater reliability and system availability. Another developing need in programmable controller control systems is the requirement for central access for system monitoring and control, fault diagnostics for fault conditions, and program development. Although control systems will be more distributed over time, a single point or location for getting into or accessing the system will be necessary to allow the user to interact with the system.

"A third requirement of future control systems is the integration of multiple control functions. While the control system architecture becomes more distributed, each part of the system will have to be capable of a variety of control types, such as sequential and motion control, and intelligent sensing.

"The database that develops as a programmable controller works will take on additional functionality because it can be processed into an information database for supervisory and management use.

"A final trend in programmable controller systems will be enhanced programming methods. A way of developing programmable controller programs to deliver increased functionality quickly and inexpensively will greatly expand the utilization of these systems on the factory floor.

"Each of the developments just mentioned has to manifest itself in one of the three basic elements of a programmable controller: the processor, the I/O structure, and the programming system. Embedded in all three of these is the concept of communication and integration. The ability to easily combine components of automation into an integrated system is a growing requirement.

"One way of making programmable controllers perform in a wide range of applications more easily and inexpensively is commonality and reusability. Processor trends include the use of similar hardware/firmware, program compatibility and transportability, a common user interface, and shared peripheral devices.

"The current trend is to more product line standardization because similar hardware in a programmable controller family makes system integration easier. If each processor platform within the family can use the same software program, a library of reusable functions can evolve. Those functions can be called upon to make application programming easier. In addition to a common software base, user interfaces as well as peripheral devices can be common and shared among the programmable controllers in the factory floor system.

"The size of the processors is also changing. A single large programmable controller with thousands of I/O points in some cases might be less expedient than utilizing a distributed system. Today, for large control applications, we see a movement to distribution of control through multiple small and medium-sized programmable controller processors.

"Using multiple processors in a distributed architecture allows the user to segment and partition the system in an organized fashion. A distributed control architecture can require the programmable controller to have enhanced instruction sets, integral peer-to-peer communication, and a flexible control scheme. This architecture allows faster program execution, which in turn speeds response time and makes the system more deterministic.''

Programming systems include both software and hardware operating platforms. Those platforms are becoming more generic. The programmable controller industry is moving to MS-DOS-based systems, and future systems will probably run under OS-2 to take advantage of that new operating system's features.

Eisenbrown wrote: ''More effective use of distributed programmable controller systems involves both central monitoring and a central point of programming. Instead of plugging into processors dispersed all over the plant floor, a peer-to-peer link will allow a programmer to plug in, at one point and address and program each device. Allen-Bradley's Data Highway Plus network allows programming from one terminal. It also permits access to any program for modifications and changes. Finally, it further permits multiple programming terminals.

''Segmentation of software parallels the segmentation of hardware used in distributed control. This partitioning of software will involve modular program development in an organized structure, allowing the programmer to use a building-block approach to developing control software.

''This has been implemented in some PLC-5 family processors and will be employed in future products. Called sequential function chart programming, it permits the use of a group of subroutines or segments in a program to be developed into a sequential function chart. That programming method encourages an organized approach to integrating the segments into a complete program.

''Sequential function chart programming has transition points based on certain commands or actions in the system. This permits the programmable controller to work on only the part of the program that is active. This in turn improves programmable controller performance compared to a system that requires the processor to scan the entire program each time. A library function on a PC AT personal computer, combined with sequential function chart programming, will help to improve engineering productivity by cutting program development time.

''The implementation of these programmable controller product trends will provide simpler and easier paths to control system integration and will allow modular implementation of automation systems. It will also help reduce integration time and costs and improve engineering productivity. The ultimate benefit will be an increased availability of production information, allowing people more time to think about running the plant rather than worrying about the nuts and bolts of system implementation.''

Glossary. *I/O:* The input/output modules on a PLC. These include digital, analog, AC, DC, and other specialized types of modules with various numbers of connection points.
ladder diagram: A program which uses logic elements whose logical connections are similar to the rungs on a ladder.
safety shut-down: Safety shut-down is when a process must be brought to a stop, but in a required sequence to ensure safety.
Also see: Ladder Diagram, Manufacturing Automation Protocol, Programmable Logic Controllers, Programming.

Programmable Gaging. Programmable gaging is back-gaging in which the gage bar is driven by an encoder signal or servomotor that can be programmed by the operator.

Programmable Logic Controller. Programmable logic controllers or PLCs are industrial controllers. They were originally developed as a less expensive programmable replacement for limit switch control logic. PLCs are now solid-state, user programmable units with memory used to implement predetermined actions. The first programmable controller was made by Modicon (now the Programmable Control division of Gould Electronics) and sold to General Motor's Hydra-Matic Division. Since the beginning, companies such as Allen-Bradley, General

Electric, Westinghouse and many other domestic and foreign companies have developed and marketed programmable logic controllers. PLC is a term used to generically describe the category of device known as a programmable logic controller or simply as a programmable controller. Allen-Bradley began to use the term PLC since their competitor Modicon was already using the acronym PC. But since the advent of the personal computer or PC, the programmable controllers are now mostly referred to as PLCs to distinguish them from the personal computers. PLCs execute programs which have the task of solving some logical activity and therefore are named as such.

Also see: Ladder Diagram, Programmable Controller, Programming.

Programmable Read-only Memory (PROM). Programmable read-only memory is read-only memory which can be reprogrammed after the device is completely manufactured, as opposed to ROMs which must have their data put into the device during the manufacturing process. Programmable read-only memory or PROM is programmed by ''blowing'' miniature fuse-links inside the chip. The presence or absence of these fuse-link connections indicates ones (1) or zeros (0) which constitute the binary data in the chip's memory. PROMs are ''blown'' or programmed by special equipment capable of properly altering the chips. PROM cannot be erased or reprogrammed since the blowing of the fuse links is an irreversible action. PROM is a less expensive form of nonvolatile memory chip than EPROM, and in small quantities is more cost effective than ROM. At a certain quantity, the unit cost of programming each PROM adds up to a greater cost than the one-time masking cost of making a ROM. Unlike EPROM (Erasable Programmable Read-only Memory), PROM cannot be altered once it has been programmed which prevents the unauthorized altering of the data contents.

Also see: Read-only Memory.

Programmed Acceleration. Programmed acceleration is a controlled velocity increase to the programmed feed rate of an NC machine.

Programmed Dwell. Programmed dwell is a delay in program execution for a programmable length of time.

Programmer. A programmer is a workpiece device which writes the instructions for the computer to act upon to develop a program tape. It is also, a computer device which develops the routines that give the computer the basic intelligence to act upon instructions when they are prepared by the workpiece programmer.

Also see: Programming.

Programming. Programming is the function of developing and writing a sequence of commands which can be executed automatically by a computer or other device with similar capabilities to follow instructions. Programming begins with the translation of the desired concepts and ideas into functional modules which can be interrelated and converted into code. Programming is all of the activities and tasks which must be performed to eventually have properly executing code. The testing, debugging, and documentation of the program source code are integral and necessary aspects of the programming task. Some portions of programming can be automated. Syntax checkers and debuggers can review structured code and indicate typing mistakes and obvious violations of the requirements of the programming language.

Also see: Ada, APL Application Program, Call, Computer, Computer Languages, Computer Part Programming, Debug, End of Program, Macro Instruction, Manual Part Programming, Mechanical Drum Programmer, Multiprogramming, Nonlinear Programming, Off-line Programming, On-line, Part Program, Personal Computer, Programmer, Program Panel, Program Scan Time, Segmented Program, Source Code, Source Program, Syntax, Teach.

Program Stop. Program stop refers to a miscellaneous function command to stop the spindle, coolant, and feed after completion of the dimensional move in the block. To continue the program, the operator must push a button.

Also see: Program.

```
1 REM PROGRAM NAME REG1
2 REM WRITTEN BY J.E.NICKS
3 REM CONPUTER APPLICATIONS FOR THE MANUFACTURING ENGINEER
4 REM COPYRIGHT 1981 ALL RIGHTS RESERVED
10 CLS
20 PRINT"THIS PROGRAM CALCULATES REGRESSION LINES"
30 PRINT"FOR ONE UNKNOWN"
40 PRINT"THE PROLGRAM IS DEVELOPED IN 2 PARTS"
50 PRINT"THE FIRST PART THE USER INPUTS X AND Y DATA"
60 PRINT"THEN THE COMPUTER CALCULATES A CORRECTED Y"
65 PRINT
70 PRINT"THE USER THEN HAS THE OPTION OF TYPING IN A"
80 PRINT"SINGLE ENTRY FOR A FORECAST OF Y"
85 PRINT
90 PRINT"X VALUES ARE INDEPENDENT DATA"
91 PRINT"Y VALUES ARE DEPENDENT DATA"
92 FORI=1TO3:PRINT:NEXT
93 INPUT"ENTER THE NAME OF THE X VALUES";X$
94 INPUT"ENTER THE NAME OF THE Y VALUES";Y$
95 DIMX(25),Y(25),Z(25)
96 CLS
100 INPUT"ENTER THE TOTAL NO. OF X & Y DATA POINTS";N
110 FORI=1TON
120 PRINT"TYPE IN X & Y VALUES SEPARATED BY A COMMA"
130 INPUTX(I),Y(I)
135 REMX1=SUM X,X2=SUM X SQ
140 X1=X1+X(I):X2=X2+X(I)[2
150 Y1=Y1+Y(I):Y2=Y2+Y(I)[2
155 REM X3=SUM X* SUM Y
160 X3=X3+(X(I)*Y(I))
170 NEXTI
175 REMX4=X BAR
180 X4=X1/N:Y4=Y1/N
185 REM X5=SUM A-X BAR SQ
190 X5=X2-((X1[2)/N)
200 Y5=Y2-((Y1[2)/N)
205 REM X6=SUM X-X BAR * SUM Y-Y BAR
210 X6=X3-((X1)*(Y1))/N
220 B1=X6/X5
230 B2=Y4-(B1*X4)
240 REM Z(I)=CORRECTED Y
260 LPRINTCHR$(27);CHR$(14);"REGRESSION ANALYSIS"
261 LPRINT"  "
265 LPRINT"SAMPLE";TAB(15);"X";TAB(29);"Y";TAB(43);"CORRECTED Y"
266 LPRINT TAB(15);X$;TAB(29);Y$
267 LPRINT"  "
270 FORI=1TON
275 Z(I)=B2+(B1*X(I))
280 LPRINTI,X(I),Y(I),Z(I)
285 NEXT
289 LPRINT"  "
290 LPRINT"B1 =";B1,"B2 =";B2
300 R2=(B1*(X3-((X1*Y1)/N)))/(Y2-((Y1)[2)/N)
310 R1=(X3-((X1*Y1)/N))/SQR((Y2-((Y1)[2)/N)*(X2-((X1)[2)/N))
320 LPRINT"THE CORRELATION COEFFICIENT IS";R1
330 LPRINT"THE CLOSER TO 1 THE BETTER THE DATA FITS Y=A+BX"
340 LPRINT"  "
350 LPRINT"THE PREDICTION Y EXPLAINS";R2;"% OF THE"
360 LPRINT"VARIATION OF THE DATA INPUTTED AS Y"
370 CLS
380 PRINT"TO INPUT A SINGLE VALUE TYPE 1 OR TYPE 2 TO EXIT"
400 INPUTP
410 ONPGOTO420,500
420 INPUT"THE VALUE OF X IS";X
430 Y=B2+(B1*X)
440 LPRINT"FOR A VALUE OF";X;"Y EQUALS";Y
450 CLS:PRINT"FOR A VALUE OF";X;"Y EQUALS";Y
460 PRINT"FOR ANOTHER VALUE TYPE 1 OR 2 TO EXIT"
470 INPUTP
480 ONPGOTO420,500
500 END
```

Programming is the developing and writing of a sequence of commands which can be executed automatically by a computer or other device with similar capabilities to follow instructions. A sample program is shown here. This program, written in BASIC, calculates regression lines.

Project Evaluation and Review Technique (PERT). Project evaluation and review technique is a project management technique. PERT is a procedure which was developed in 1958 by collaboration between the United States Navy Special Projects Office, consultants from Booz-Allen-Hamilton, and the Lockheed Missile and Space Division. The PERT approach is "event-oriented" which means events are represented as milestone points, and the events are defined and connected to show relationship. Each connecting arrow shows precedence and duration of a task. In a time-scaled diagram, the lengths of the arrows represents the relative time for each activity. The durations are usually expressed as three time estimates for each arrow. They are the pessimistic duration (10% likelihood of missing), the most likely, and the optimistic (100% likelihood of occurring). The calculated duration is then the result of adding the pessimistic duration time to four times the most likely duration time, plus the addition of the optimistic duration time and dividing by six. Statistical evaluation of these durations allows for a determination of the probability of meeting any scheduled completion date for the project.

Also see: Project Management.

Project Management. Project management involves the coordination of group activity wherein the project manager plans, organizes, staffs, directs, and controls to achieve an objective within constraints on time, cost, and performance of the end product. *Project Planning* is the process of preparing for the commitment of resources in the most effective fashion. *Controlling* is the process of making events conform to schedules by coordinating the action of all parts of the organization according to the plan established for attaining the objective. Project management can also be considered to be a blend of art and science: the art of getting things done through and with people in formally organized groups; and the science of handling large amounts of data to plan and control so that project duration and cost are balanced, and excessive and disruptive demands on scarce resources are avoided.

A project is a set of tasks or activities related to the achievement of some planned objective, normally where the objective is unique or nonrepetitive. A project can be distinguished from continuous production by the characteristic of uniqueness, which might be interpreted to mean that the first time an activity is attempted, it is a *project*, whereas successive times of doing the same activity makes it become a *process*, not a project.

The two major aspects of project management, planning and analysis, and control and reporting, are significantly different. People and tools that are excellent at one aspect of project management may not be as good with the other aspect. Project management was originally done using pencil and paper, with charts being drafted to show network diagrams and timescaled PERTs. Each reporting period that the diagrams were updated required either marking the existing diagrams or redrawing the entire network. On large projects, several people were employed full-time in handling the periodic revisions. When computer-based tools were introduced which utilized CAD capabilities to draw the diagrams and edit and show updates to the schedule, the time required to perform detailed reporting was dramatically reduced.

History. Project management in a formal sense has been carried out for some time, but the contemporary methods of project management including PERT and CPM were developed in the late 195zeros (0). CPM is the process of managing an entire schedule by only focusing on those activities which would affect the finish data if they are delayed.

Vast projects such as the space shuttle program and the SDI (strategic defense initiative) cannot be managed without using project management techniques and tools to handle and coordinate the vast numbers of activities.

Today project management is carried out by both private industry and the government sector, with military standards existing with specifications and requirements regarding the reporting and tracking requirements for government contractors.

Discussion. The Gantt chart is a bar chart which lists the activities in a column along the left side, starting at the top with the items listed in ascending order based on start times. A timeline is usually indicated along the horizontal axis. To determine what the status should be on a certain day, a vertical line is drawn from top to bottom on the Gantt chart on the day in question. To then show the actual status, a vertical line is drawn with horizontal adjustments on each bar to show the actual status of each task. The Gantt chart is the simplest form of representation of project management which is limited by not being able to easily depict some of the information. Unless the entire chart is reviewed, either manually or by computer, activities may be overlooked by being physically shown far from their related tasks.

The project network is a method of drawing circles or nodes to represent events and connecting the numbered or lettered nodes with arrows to represent activities and the connectivity. This type of chart is referred to as a CPM (critical path method) chart because the "critical" path or sequence of activities, which is pacing the project, can be highlighted easily.

When there are multiple dependencies, each is shown with a solid arrow. If a precedence relationship requires no time to complete, it is shown with an arrow labelled "dummy," which means it represents precedence only, with zero completion time required.

Activity-on-arrow representation uses arrows to denote the activities with the arrows being labelled by their starting and ending nodes. These nodes signify events which are a unique pair of nodes for each activity. This is referred to as the i-j notation with the i node representing the starting node and the j node representing the ending node. Obviously, the j node for one activity (ending) is the i node for another activity (beginning). Convention for activity-on-arrow or AOA diagramming is to place the name of the task over the arrow and the time estimated to complete the task under the arrow.

Activity-on-node diagramming uses the reverse of the AOA method with the node representing the task to be accomplished and the arrows simply showing connectivity. Activity-on-node diagrams tend to be more cluttered than activity-on-arrow diagrams since the same information must be concentrated about the node instead of spread along an arrow.

Time-scaled network diagrams are charts which use a time line along the horizontal axis of the diagram. Arrows are used to show the activity's length of time, which is represented by the length of the arrow. The arrow is located in the diagram with the end of the tail at the beginning in calendar time of the activity and the tip of the arrow head at the end of the allotted time for the activity. Time-scaled networks use vertical sections of arrows to connect to successive tasks. The time required for a task is shown with a solid line, but when a line must be longer than the representative time to connect horizontally with its successor, due to a dependency, the additional time is represented with a dashed line. This dashed portion indicates "slack" time which is the time from completion of one activity and until the next activity can begin, due to other restrictions.

Hierarchical methods and structure are used when generating large project schedules, such as using subprojects and grouping or "hammocking" related projects together for reporting purposes.

One of the influences of the development of project management techniques and methodologies has been systems approaches to project management. By applying some of the basic control systems methodologies, a more scientific and structured method can be developed for everything from the time estimates to optimizing the updating and reporting frequency.

Some organizations use a project or program administrator to perform the project management tasks while others simply have each individual or group who executes a project to manage the planning and tracking of their own project. As more productive computer-based tools are developed, the need for a highly trained individual devoted solely to managing the projects for others becomes less of a necessity. By having the doers of a project also manage the project, project information and analysis becomes available more directly to the people performing the project.

Also see: Critical Path Method, Project Evaluation and Review Technique.

PROLOG. PROLOG is a programming language used in the area of artificial intelligence. It is based on predicate calculus. PROLOG is one of the most used languages for research in artificial intelligence.

PROM. See: Programmable Read-only Memory.

Protocol. In communication systems, protocol is the rules of format and sequence of electronic signals from computer-to-computer which allows data exchange. Protocol must be followed if communication is to be effected. Protocols may govern portions of a network, types of service, or administrative procedures. For example, a data link protocol is the specification of methods whereby data communications over a data link are performed in terms of the particular transmission mode, control procedures, and recovery procedures. Protocols are designed to control the layers of a communication network or to control the exchange of data between computers in an application network.

There are high-level protocols (HLP), layered protocols, link protocols, and low-level protocols.

Some of the functions handled by communications protocols are error detection, error correction, data integrity, congestion control, routing, traffic flow control, network access services, security control, interprocess communication, data transfer, terminal support, and the applications programs themselves.

Products and standards such as IBM's BSC & SDLC and MAP/TOP are protocols.

Also see: Binary Synchronous Communications Protocol, High-level Data Link Control, Manufacturing Automation Protocol, Technical and Office Protocol.

Proximity Sensor. A proximity sensor is a device that detects the presence of an object at a distance or that measures the distance the object is away from the sensor.

Pseudocode. Pseudocode is the name used to describe an abbreviated form of writing which a programmer uses to represent the programming statements. Pseudocode is not completely correct code with regards to syntax and structure or spelling, but simply allows the programmer to record and remember the intended functionality of each program statement. The pseudocode is then later translated into the proper programming statements for the particular language being used, whether high-level or low-level. Pseudocode is not any fixed or rigid type of defined dictionary of terms, but rather can be created by any programmer as that person sees fit. Some functions usually are indicated with similar abbreviations by most programmers since the functions logically suggest such abbreviations, such as ''read'' and ''write'' statements or basic arithmetic functions. Pseudocode is often used when flow-charting or when doing the higher-level activities in structured programming.

Also see: Programming.

Pulse-Width Modulation. Pulse-width modulation, also known as pulse-length modulation is modulation in which the pulse length (time duration or spatial width) of a carrier pulse is varied in accordance with an attribute of a modulating signal. The modulating signal may vary the time of occurrence of the leading edge or trailing edge of the carrier pulse, or both. 0's or 1's may be represented by a carrier wave that is on or off for longer or shorter periods of time. In pulse-length modulation, however, a pulse is transmitted for each sampling of the signal. The width (duration) of the pulse is proportional to the amplitude of the signal. Pulse-length modulation yields the same signal-to-noise improvement as pulse-position modulation of the same peak power. However, the average power that is required is greater in pulse-length modulation because of the longer average duration of the pulses. Pulse-length modulation is rarely used for direct modulation of a radio carrier, but it has been applied in the intermediate processes of some pulse-position modulation receiving devices. The transformation is usually made by having the position of the short pulse control the starting time of the two longer pulses with a fixed terminating time. Pulse-width modulation (PWM) is synonymous with pulse-duration modulation.

Also see: Modulation.

Pulsed Power Welding. Pulsed power welding is an arc welding process variation in which the power is cyclically programmed to pulse so that effective but short duration values of a parameter can be utilized. Such short duration values are significantly different from the average value of the parameter. Equivalent terms are pulsed voltage or pulsed current welding.

Also see: Pulsed Spray Welding.

Pulsed Spray Welding. Pulsed spray welding is an arc welding process variation in which the current is pulsed to utilize the advantages of the spray mode of metal transfer at average currents equal to or less than the globular-to-spray transition current.

Punched Card. A punched card is a card of constant size and shape on which information is represented by holes in specific positions.

Punched Tape. Punched tape is a form of data storage media which was originally used mostly by teletypewriter data terminals and numerically controlled machines. Punched tape is usually a one inch wide paper or mylar tape in which a pattern of holes represents bits and bytes of information. Punched tapes were used before magnetic media was perfected and cost-reduced to the point of making it a viable media to be used in a harsh factory environment. Punched tape, while slower to be read, is much cheaper and is not susceptible to improper reads of data. Punched tape is not used as much any more since both magnetic media such as data diskettes, cassettes, and cartridges, as well as direct line connections for downloading information provide a more efficient means to transfer data, both with respect to cost and productivity.

Also see: Read, Read-only Memory.

Punching. Punching involves the cutting of holes and results in scrap slugs. It can be performed with punching presses specifically designed to hold the tooling or with stamping presses and unitized tooling. Operations related to punching include nibbling, notching, piercing, perforating, slotting, pointing, and marking. Both ferrous and nonferrous metals are punched, as well as nonmetallic materials.

Punching on a punch press is fast and economical. A variety of shapes and sizes can be punched with standard tooling. Many presses are capable of nibbling large cutouts and contours in workpieces that would generally be produced with other, more costly metalcutting techniques. Plasma and laser cutting attachments permit small internal angles, scrolls, and spirals to be cut on the punch press.

Discussion. The size of the workpiece is generally determined by the throat opening in the punch press (distance from punch center to rear of press), weight of workpiece, and area in which the carriage moves. Several NC punch presses are capable of repositioning the workpiece, which in effect enables any length of material to be punched as long as the workpiece does not exceed the weight limit capacity of the machine and auxiliary tables are used to support the overhanging material. Manually rotating workpieces on presses with rear address clamping enables the width of the workpiece to be doubled. The largest workpiece capable of being punched prior to repositioning or rotation is 82 inches (2083 mm) wide by 100 inches (2540 mm) long.

Incorporating numerical control (NC) on the punch press increases the accuracy of the hole locations. Punch presses are made with different force capabilities, frame configurations, tool-mounting capabilities, and controls. However, the mode of operation for all punch presses is the same.

The workpiece is generally positioned and firmly held down on the worktable prior to punching. When the controls are actuated, the ram descends and the punch knocks out the material as determined by the size of the punch and die. The stripper holds the material firmly in place until the punch has fully withdrawn. If the press is numerically controlled, the workpiece is then automatically moved to the next punching location. On certain presses, lifting devices in the die block prevent burrs created during punching from holding the workpiece on the die.

Turret presses have the capability of holding tools of a variety of shapes and sizes. The variety of tools enables the operator to punch many different hole shapes and sizes without having to change the tools in the turret. However, tools of the proper size and shape must be mounted in the proper location in the turret. The number of tools contained in the turret ranges from 12 to a maximum of 72 depending on the manufacturer. Since the tools can be stored in the turret when they are not used, they are less likely to be damaged due to improper storage.

Three methods are used to control the operation of the press: manual, semiautomatic (manual data input), and automatic control.

On a manually controlled press, the operator is in complete control and is involved with each aspect of part production.

The semiautomatic, sometimes referred to as manual data input (MDI), punch press offers productive capabilities between a manual punch press and an automatic punch press.

The operator initially selects the reference position for the workpiece and then loads the part on the worktable. The press is then set to operate automatically, and the operator inputs the data from the drawing into the press using programmable codes. When the data input is completed, the cycle is started and the part is positioned and/or punched automatically. If more than one part is to be punched, the press can be programmed to repeat the cycle over again.

Two of the main advantages of the semiautomatic punch press are the cost of the machine and simplicity of operation. The cost of a semiautomatic punch press is approximately one-half that of a CNC punch press. Programming and editing are performed right at the machine using the programming codes instead of at a computer terminal. This enables an operator to learn how to program and operate the press in about two hours. The codes used indicate the work positioning speed, subroutines used, type of data input (inch, metric, or absolute), stops, returns to initial positions, positional coordinates, and the tool that is used. Although most CNC presses have MDI capabilities, their inputting procedure is more difficult and they require more time than the semiautomatic punch press. The MDI press is more accurate than the manually controlled press and can achieve the same degree of accuracy as the CNC press.

Generally, the maximum work positioning speed of an MDI press is limited to about 1000 ipm (25 400 mm/min) whereas the CNC press attains speeds as high as 4400 ipm (111 760 mm/min). Since the press uses fewer programmable codes than the CNC press, the operations that can be performed with the semiautomatic punch press are less than those that can be programmed on a CNC press.

Automatic punch press control is achieved using either NC or CNC. The punch press generally consists of a punching mechanism, a workpiece positioner, and a structure upon which these mechanisms are mounted. Automatic control is used for both single-station and turret punch presses.

Presses with C-type frames have the workpiece positioned in both the X-axis and the Y-axis directions while the turret or tool adapter remains in a fixed position. The X axis is generally used in conjunction with the workpiece length; and the Y axis, with the workpiece width. The clamps that grip the workpiece can be located on the front or back of the press table; this feature is referred to as front address or rear address clamping. Presses with rear address clamping are capable of punching parts that are twice the specified press Y axis or throat depth by manually rotating the workpiece. Presses with bridgetype frames have the workpiece positioned in the X-axis and Y-axis directions also, with the exception of some presses that position the workpiece in the X axis and the turret in the Y axis.

The turrets are rotated automatically to properly position the selected punch and die. Tool changes on single-station presses are also achieved automatically with specially designed removal/insertion mechanisms.

The first step in processing a part is to generate a program containing the dimensional

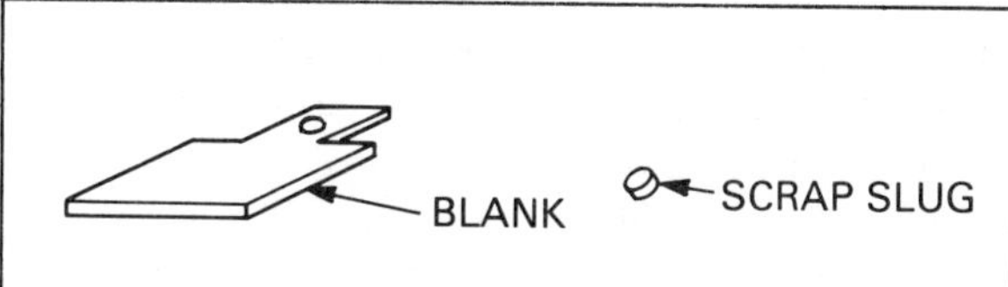

Punching involves the cutting of clean holes with resulting scrap slugs.

Power to the ram is derived manually, mechanically, or hydraulically. Manual punch presses are equipped with a long lever that converts the pull of a human hand to the forces necessary to punch the workpiece. Punching force for these presses is usually limited to about 4 tons (36 kN). Mechanical punch presses utilize flywheel energy which is transferred to the ram by gears, cranks, or J-frame design which eliminates the need for a special machine foundation normally associated with C-frame presses. The most popular design among manufacturers is the C-frame which is designed with either front address or rear address clamping. However, the bridge-frame design minimizes the deflection that occurs as a result of the uneven opposing forces acting upon the upper and lower frame members.

The two types of punch presses built are the single-station press and the multiple-station press. Multiple-station punch presses are generally referred to as turret presses. However, some multiple-station presses are built that contain two or three punching stations. The tooling is mounted individually in the toolholder. These presses fall into a classification between single-station presses and turret presses and are generally used when heavy plate, angles, or beams are being punched for structural steel fabrication.

Another type of multiple-station press contains two punching stations, but incorporates removable cartridges to mount the tooling. The cartridges are capable of holding up to 12 different styles of punches and are positioned under the punching head with servodrive motors. The advantage of this type over the turret press is that the tooling can be mounted in the cartridges for another workpiece while the press is in operation. This type of press is particularly useful when two identical workpieces are being punched simultaneously.

Single-station punch presses are equipped with a single, rigid tool adapter in which a punch of the required size and configuration is mounted. The corresponding die is mounted in the machine pedestal. Toolchanging can be performed manually, semiautomatically, or automatically.

The single tool adapter allows the tooling to be changed quickly during programmed pauses on NC presses. It also provides good punch guidance and allows standard punches to be used in heavy, off-center-loaded situations without excessive tool wear. On presses with manual toolchanging, it is important to store the tools in a convenient place to provide ready access to them during the punching process.

The *turret punch press* derives its name from the manner in which the tools are mounted in the press. An upper turret holds the punches, and a lower turret holds the respective dies.

Turret movement is achieved manually on small, hand-operated punch presses and semiautomatically or automatically on NC or CNC punch presses. Automatic turret movement is obtained by connecting a closed-loop, direct or alternating-current drive to the upper and lower turret assemblies. This movement can be unidirectional or bidirectional, and the direction of the turret rotation is generally determined by the programmed punching sequence. Certain CNC presses offer an ''optimization'' feature which automatically determines the most efficient punching sequence for a particular workpiece. When a hole is to be punched, the turret is rotated until the proper tool is positioned under the ram.

and geometric data of the part from the print. This is usually performed by a programmer rather than a press operator. The program is stored in binary code on a narrow strip of paper or mylar tape. Cassette floppy disks are also used for storing programs. The tape and a list of the required tools are given to the press operator who inserts the tape into the tape reader on the press. After the tools and workpiece are loaded, the press is activated and the part is punched out according to the tape. The program can be stored in the press control to facilitate the production of multiple parts.

Numerically controlled punch presses ensure greater productivity capabilities in most short to medium runs and greater workpiece accuracy. Prototype parts can be made quickly, and the need for models and templates is eliminated. Cost estimating is also simplified because of the precise time keeping.

With the addition of CNC, programming is easier and requires 1/5 to 1/4 the time. Canned cycles and subroutines enable the programmer to program the punching operation in fewer steps. An ''optimization'' feature enables the programmer to program the punching operations in any order. The computer then determines the most efficient punching sequence. The control unit is also used in conjunction with the various press accessories that can be incorporated with the press to permit the press to be more versatile. The control unit has built-in diagnostic features to inform the operator of press malfunctions.

The cost of the press is generally four times the cost of a manual press. The time required to learn how to program and operate the press is greater than that necessary for manual and semiautomatic presses.

Many different punch presses are available, ranging from a manually controlled, single-station press to a computer-controlled turret press. To properly select a punch press, it is therefore beneficial to consider press construction, capacity and capabilities, and controls as they relate to the type of work currently being performed and the production requirements. It is also important to select a press based on production growth that is anticipated in the future.

The two styles of press construction used are the C-frame design and the bridge-frame design. Although the C-frame is a more popular design among manufacturers, some of the larger presses require a special foundation to maintain accuracy between press frame and worktable. However, presses with this style of frame and rear address clamping enable the press to punch workpieces that are twice the rated throat depth by manually rotating the workpiece. Bridge-frame presses and C-frame presses with front address clamping do not permit punching workpieces that are wider than the rated throat depth.

The force rating of a press directly affects the thickness of material that can be punched and also the maximum size of hole that can be punched in a workpiece. Whether the press is single-station or multistation determines the amount of time required for tool changes. Hydraulic presses have a more even punching force than mechanical presses which is useful when punching thick material. However, mechanical presses permit faster speed, strokes per minute (spm), than hydraulic presses. A higher stroke speed is advantageous when large cutouts and contours are being nibbled. Other capabilities to consider are workpiece weight limits, workpiece positioning speed, workpiece repositioning, and the ability to do minor forming. Plasma arc and laser cutting attachments permit the punch press to be more versatile.

How the press is controlled determines the accuracy and the speed at which parts can be produced. Automatic and semiautomatic controls provide the greatest accuracy in hole positioning and part repeatability. Automatic controls on presses provide the greatest flexibility in the contours that can be produced. However, they require more time to program than semiautomatic presses. Since semiautomatic presses are simple to program, they can be used when producing noncomplex and simple prototype parts and production parts on a limited basis.

The versatility of a punch press to produce holes in sheet metal is dependent on the number and shape of the punches available. A laser cutting attachment further enhances the punch press such that virtually any part shape is achievable through programmed motion of the workpiece with respect to the cutting beam.

The most popular type of laser for cutting sheet metal is the CO_2 laser. Its popularity derives from the fact that relatively high-powered beams can be generated at a laser frequency which is suitable for cutting a wide range of materials. Generally, CO_2 industrial lasers used in combination with a punch press range in output power from 400-1500 watts. Elevating the punch press on air pads during punching and nibbling operations reduces vibrations and prevents mirror misalignment.

In laser operation, a vacuum is applied to a series of tubes located in the laser cabinet and small amounts of carbon dioxide, helium, and nitrogen are then introduced into these tubes. A high voltage potential is applied to the gas mixture causing a light to be given off in random directions. This light is reflected by mirrors and strikes gas atoms within the tube. The striking of the gas atoms provides more energy to the light beam. When the energy of the light beam is great enough, the light beam is transmitted through the mirror and conducted toward the cutting head of the laser attachment.

The unfocused beam, generally about 0.4 inch (10 mm) diameter, is concentrated to a beam of about 0.005-0.020 inch (0.13-0.51 mm) diameter by a focusing lens. When the focused beam contacts the material surface, the surface temperatures increase to over 18,000 degrees F (10 000 degrees C), causing the material to melt and vaporize. An assist gas injected into the nozzle aids in the vaporizing of the material being cut. For ferrous materials, oxygen is used as the assist gas and helps to keep the nozzle clean, create an exothermic reaction, and blow away molten material. Compressed air is generally used as an assist gas when cutting nonferrous materials. Bearings, mounted in the cutting head, ride on the material surface to maintain the focusing lens at the proper height with respect to the material.

A laser cutting attachment provides the fabricator with a greater flexibility in producing holes and contours in the workpiece. This is because the laser attachment eliminates the need for the special tools that are required when large holes, cut-outs, or contours are being produced. Changing the assist gas, gas pressure, or cutting speed enables the laser attachment to cut a wide range of material types and thicknesses. Since the laser is not in contact with the workpiece, tool wear is eliminated and sharp corners can be produced. The quality of the cut edge is good and generally eliminates the need for secondary operations. The narrow kerf of the focused beam minimizes the heat-affected zone of the workpiece which in turn reduces workpiece distortion.

However, materials that reflect light rather than absorb it, such as aluminum, copper, and brass, are difficult to cut. Since the maximum power attainable with a laser attachment is around 1500 watts, the thickest mild steel that can be cut is 3/8 inch (9.5 mm). Cutting speeds are generally lower than those using the plasma cutting attachment. Initial alignment through the cutting head is difficult due to the invisible laser beam.

Improving the methods by which the material is handled increases the production rate of the NC punch press. Incorporating a workpiece load/unload system and a workpiece repositioning system in a punch press enables the press to operate automatically. Some presses can also be equipped with an automatic part sorter to separate and sort different parts produced from the same workpiece.

A load/unload system enables the workpiece to be automatically loaded, positioned, and unloaded from the punch press. Depending on the design, the various components can be positioned on the same side of the press or on opposite sides. The load/unload components are interfaced with the CNC unit for the press to provide a completely interactive system.

The loading component consists of a carrier mounted on an overhead rail. Suction cups, located in the retractable head of the carrier, transfer the workpiece from the storage rack to the worktable of the punching machine. When the CNC unit calls for a new workpiece, the retractable head lowers the workpiece to its initial position on the worktable. Generally, the workpiece is loaded and held in a ready position during the punching cycle. Pneumatically operated slides position the workpiece for punching, and sensors in the work clamps and end-stop ensure that the workpiece is positioned correctly and ready for punching.

When the punching is completed, the table moves to the load/unload position and the unload arm grips the finished workpiece. The work clamps are then released, and the workpiece is pulled off onto the scissor table. Cycle time for the loading and unloading sequence varies from 15-30 seconds. Part-discharging systems utilize a trap door and a conveyor belt controlled by the CNC unit to remove cut-out parts or scrap created by nibbling, laser cutting, or plasma arc cutting.

Automatic repositioning makes it possible for the work clamps to move along the edge of the sheet, regrip the sheet, and bring into the working area a portion of the workpiece which was not within punching range previously. There is no theoretical limit to the number of repositioning moves that can be made within a part nor to the length of the piece that can be handled. In practice, however, repositioning approximately doubles the length of the sheet that can be punched. When punching long sheets, it is necessary to use auxiliary tables to fully support the sheets.

The NC punch press is one of the most productive machine tools in the modern sheet metal fabricating industry. The machines are universally accepted for their ability to rapidly and precisely punch parts in runs of 1-100,000 and more. In addition to simple hole punching, they can be used to perform forming operations, including forming the interior of a workpiece with the proper tooling.

Forming done on an NC punch press must always be performed upward so that the forms generated are on the top face of the workpiece. The rapid movement of the workpiece in the machine does not allow for any projections to exist on the underside of the sheet. The underside of the sheet slides across the dies in the lower turret, and any form projecting downward can be caught in the openings of these dies. Any design requiring downward forming should be carefully evaluated; if downward forming is still required, careful programming and, perhaps, special workpiece handling are necessary. However, on a manually controlled press, downward forming is permissible because the workpiece is moved manually.

Applications. Punch presses are used to punch holes of different shapes and sizes in various types of materials. Some of the applications for punch presses are in the production of electronic metal work, electrical boxes, appliances, construction equipment, farm machinery, trucks, office furniture, and vending machines.

Other operations that can be performed on the punch press include notching, forming, tapping, nibbling, and louvering.

Semiautomatic punch presses are especially suited for small job shops in which cost and training prohibit the use of CNC punch presses. They can also be used by large firms that perform prototype work and produce parts in limited quantities.

Automatic presses are used in a wide range of fabricating shops where many different contours and cutouts are required. The ability of the CNC press to store the punching program for different workpieces enables this type of press to be used in limited production runs as well as long production runs. Plasma and laser cutting sometimes permit the press to produce intricate cutouts and contours that are not normally permissible with standard tooling.

Materials which absorb the laser energy to create heat can be cut using the laser attachment.

The cut materials are not confined to metalsonly. Plastics, wood, leather, rubber, and mica can be successfully cut with the laser. Titanium, spring steel, high-carbon steel, and high-nickel alloys are also easily cut by laser.

The cutting speed of the laser depends on available power, material characteristics, material thickness, and desired quality of cut. Speeds range from about 20-400 ipm (500-10 000 mm/min).

The laser cutting attachment is employed to make large contours and cutouts in material that normally requires nibbling. Small contours and special shapes are possible without the use of special punches. The laser can also be used for etching material by employing a faster cutting speed or by regulating the pulse rate and width of the laser if so equipped. However, it is difficult to regulate the depth of the cut.

Glossary. *CNC (Computer Numerical Control):* A self-contained NC system for a single machine tool utilizing a dedicated computer controlled by sorted instructions to perform some or all of the basic NC functions. Punched tape and tape readers are not used except possibly as a backup in the event of a computer failure.

floppy disk: A flexible, magnetic-based disk used to store data input to NC machine machine control units.

MDI (Manual Data Input): A means of manually inserting commands and other data into an NC control.

NC (Numerical Control): A technique for controlling actions of machine tools and similar equipment by the direct insertion of numerical data at a given point. Data is automatically interpreted.

press: A machine having a stationary bed and a slide (ram) which has controlled reciprocating motion toward and away from the bed surface and at a right angle to it, the slide being guided in the frame of the machine to give a definite path of motion.

ram: The powered, movable portion of the power press break structure, with die-attachment surface, which imparts the pressing lead through male dies onto the piece part and against the stationary portion of the press brake bed.

tapping: A process for producing internal threads using a tool (tap) that has teeth on its periphery to cut threads in a predrilled hole. Threads are formed by combined rotary and axial relative motion between tap and workpiece.

tooling: A set of required standard or special tools needed to produce a particular part. Tooling includes jigs, fixtures, gages, cutting tools, but excludes machine tools.

Also see: Computer Numerical Control, Numerical Control.

Purchase Specifications. A purchase specification is a formal written detailed definition of the complete requirements for a part, product or component. Purchase specifications supply adequate information to acquire the part from a supplier. This commonly entails a listing of basic requirements for materials which sometimes includes the product composition, dimension requirements and test conditions. Commonly supplied as part of the purchase specification is a published set of standard specifications from the requisitioning company or from industry group standards. A quantitative description of required characteristics or preferred processes to be used may be included. Purchase specifications are widely used in sourcing new suppliers and evaluating supplier performance to quality standards and cost estimates.

Purchase specifications can be used as a legal record of the definition of the requested commodity. Purchase specifications sometimes allow ''equivalent or better'' substitutions. Government purchases usually involve detailed purchase specifications whose format is, in turn, defined in other government specifications.

PWB. See: Printed Wiring Board.

Q

QBE. Query-by-example, a language developed for use in organizing and query applications in database information handling.

Qbus. The Qbus is a type of backplane bus popularized by Digital Equipment Corporation with their low-end PDP-11, LSI-11, Micro PDP-11, and MicroVax lines of computers.

Also see: Bus, Computers.

QIC Standard. The quarter inch cartridge (QIC) standard is a group of standards for cartridge tape units and cassette tape units used for personal computer data backup. Some of the QIC standards are: QIC-02, which includes most of the functionality in the drive unit with a smaller interface card; and QIC-60, which has the formatter and other functionality on the interface card.

Quality. Vital to the understanding of the implications of quality in a manufacturing enterprise is a perception of the meaning of quality. According to the American Society for Quality Control (ASQC) and their publication *Quality Systems Terminology, ANSI/ASQC A3*, quality is the totality of features and characteristics of a product or service that bear on its ability to satisfy given needs.

The definition implies that the needs of the customer must be identified first because satisfaction of those needs is the ''bottom line'' of achieving quality. Customer needs should then be transformed into product features and characteristics so that a design and the controlled product specifications can be prepared. It should be carefully noted that it is really a composite of product features and characteristics taken as a whole that must satisfy the needs of a broad range of customers.

The product design and specifications are a translation of the product features and characteristics into performance and manufacturing terms. Assessments prior to and during the development and manufacture of the product determine if it conforms to the requirements of the design and specifications. If the needs of the customer have been properly translated, these assessments will predict whether or not the needs will be met.

To satisfy the given needs of the customer, the product must allow the customer to achieve the purpose for which it was bought. The life of the product and its dependability also are of concern.

Other considerations include safe use of the product, ease of performing maintenance, and the availability of spare parts and service facilities.

Another concern is the availability of clear instructions so that the product is used correctly.

Product designs should address an appropriate market segment of customers. For example, customers participating in a lower price market segment for appliances may not need the additional features found in appliances in the higher price market segment. For some, it may be a balance between the financially affordable and the operationally adequate. Therefore, the desires of that type of customer translate into a product with few features. Some features are desirable to both the lower and the higher price market segments; an example of this is the expected life of a product.

Another very important consideration in analyzing customer needs is what is offered by the competition in the same price market segment to better satisfy customer expectations.

Many factors must be taken into consideration when developing a quality system. Some of the most important are: prevention of defects, customer focus, standards, legal liability, total involvement, as well as selection and training.

Central to the development of quality products at low cost is the focus of attention on the product design function (the optimization of design for function) and its relationship to the manufacturing process (product and process design for improved manufacturability). The ability to use scientifically sound concepts and methods for experimentation and the use of statistical thinking and methods to study product and process performance are important factors in establishing a well-founded approach to quality planning design, and control. In particular, emphasis on quality upstream at the product design stage has been a key to the success of the Japanese in establishing the trend of increasing share over a broad range of consumer markets.

Currently, considerable attention is being placed on process design and improvement using statistical thinking and methods. The use of control charts is well-recognized as an effective means of studying process variation and, in particular, of isolating major sources of variation so that quality and productivity improvement opportunities may be revealed. Often the stability of a process (that is, a process in good statistical control) signals but the beginning of a major improvement effort aimed more fundamentally at determining appropriate system changes to effect overall performance improvements. In this regard, the role of statistical design of experiments in both product and process design is now becoming better understood as a tool to reveal significant opportunities for improvement of performance.

The presence of statistical control provides the condition of stability or predictability of the system under study. It provides a measure of constancy with respect to average performance and consistency in terms of variation in performance.

Statistical control suggests that the process performance is free of special or sporadic variations and is governed only by a constant system of common or chronic causes of variation. Design of experiments, concepts, and methods provides a powerful approach to the discovery of improvement opportunities for already stabilized processes. The manner and extent to which a set of factors may singly or in concert influence the process mean level (location effects) and/or process variability level (dispersion effects) can be efficiently revealed using statistically designed experiments. Conversely, design of experimental techniques can be of great help in identifying the root cause(s) or special causes of variation in unstable processes so that actions may be taken to bring such processes into statistical control.

For many problems, the concept of statistical control or stability may be premature. In engineering design and prototype testing, a process may not really exist; but again, planned experimentation can provide valuable insights into the ways in which changes in important product and process design parameters can improve overall quality and cost performance. The wide-ranging applicability of design of experimental techniques for dealing with both off-line issues of quality design and on-line issues of process control and continuing improvements are only just beginning to be recognized broadly.

Quality Assurance. The American Society for Quality Control in its publication *Quality Systems Terminology, ANSI/ASQC A3* explains that quality assurance includes all the planned or systematic actions necessary to provide adequate confidence that a product or service will satisfy given needs.

These actions are aimed at providing confidence that the quality system is working properly and includes evaluating the adequacy of the designs and specifications or auditing the production operations for capability. Internal quality assurance aims at providing confidence to the management of a company, while external quality assurance provides assurance of product quality to those who buy from a company.

Quality Assurance is a term frequently identifying the department or function in a company that is responsible for assuring that the product or service meets the fitness-for-use requirements for the intended market. To accomplish this responsibility, Quality Assurance must be organized to support the full product life cycle, from development to production and then on to actual use in the field.

During development, Quality Assurance professionals typically participate in the design review process. This can include providing input on qualification test requirements, quality standards that the product will be subject to, production quality goals, participation in vendor selection, and quality reporting requirements. Quality Assurance is usually given the responsibility for determining when the product is ready to go into production.

When the product reaches the production phase, Quality Assurance's role typically becomes training, auditing, and reporting. Quality Assurance professionals train people from Quality Control, Production, Engineering, and other areas in skills that can be used to prevent problems. Training areas include statistical process control, process capability analysis, control chart techniques, problem solving, and quality reporting. Quality Assurance will audit both manufacturing processes and completed products for conformance to requirements and specifications. If problems are identified during the audits, Quality Assurance will organize and manage the required corrective action.

One of the major activities for Quality Assurance is status reporting to management, particularly of the customer's perception of actual fitness for use. Quality Assurance will develop and manage a quality reporting system covering all phases of the product life cycle. The quality reporting system must be clear and concise, and provide the right information for decision making. Quality Assurance typically works very closely with Field Service and directly with customers to identify any fitness-for-use problems. Appropriate corrective action is then managed to eliminate the problem from current products and prevent it from occurring in new products.

Quality Control. *Quality Systems Terminology, ANSI/ASQC A3* from the American Society for Quality Control notes that quality control comprises the operational techniques and activities that sustain a quality of product or service so that the product will satisfy given needs. Quality control is also is the use of such techniques or activities.

The quality control function is closest to the product in that various techniques and activities monitor the process and pursue the elimination of unsatisfactory sources of quality performance.

Quality control is also a term identifying the department or function in a company that is responsible for assuring that the product conforms to defined specifications. Traditionally, quality control has been confined to manufacturing operations. Conformance to specifications is measured by specially trained individuals called Quality Control Inspectors. Specific instructions for measuring conformance are usually provided by a Quality Control Engineering Department.

For Quality Control to assure that a product conforms to its specifications, the entire process, all individual piece parts, and all tools and equipment used in the process must be controlled. The quality of piece parts is often measured in a special area called *incoming inspection*. The conformance to specification is determined by measuring characteristics of the part and recording the results. Parts that conform to their specification are sent to the next level of production. Parts that do not conform should not be used, because they could cause the end product to fail to meet its specifications.

In the manufacturing process, conformance to specification is usually measured by a series of in-process inspections and a final inspection of the finished product. In-process inspections measure that the proper parts have been put together in the proper sequence, that any adjustments were properly made, and that no cosmetic imperfections exist. The final inspection checks that the completed product performs the intended function and that all documentation is properly completed.

Also see: Inspection, Machine Vision.

Query. A query is a request for information which may be given by a person or a system. A query usually specifies the conditions and details of the requested information, which may or may not be capable of being satisfied by the recipient. Queries are normally associated with database management software, where large databases are searched for various information as needed. SQL or *structured query language* (sometimes called standard query language) is an accepted set of standard commands used by database software to locate information in a particular database. SQL allows a programmer to develop the query software independently of being familiar with the particular database software, or for the same front-end user software to be used with any SQL-compatible database.

In the communications field, queries may be the ''request to send'' or INQ sent from one station to another. The ''ringing'' of a station by another in telephone-type communications is a query to establish a communications link between the two stations.

Also see: Databases.

Queue. A queue is a line of data elements that are waiting to be brought into service. It is a temporary storage location where various quantities of a certain item are stored. Queues can be used when asynchronous systems interface with each other, or when the desire is to keep an operation supplied with data if the upstream portion of the system slows in its productivity. The use of queues allows the operation of expensive equipment at its maximum output, despite any fluctuations in operations before or after the particular operation. Queues may be FIFO (first in, first out) or LIFO (last in, first out). When a queue is used as an isolation mechanism between two entities, the queue is specifically a buffer. Electrical devices and computers sometimes use buffers between devices.

In some computers which have high-speed processors, memory caches are used. A cache is typically a small amount (compared to the total main memory) of higher-speed memory, which acts as a buffer between the high-speed CPU and the lower-speed main memory.

Also see: Buffer Memory, FIFO, LIFO.

R

Radiation Curing. Radiation curing is a process in which specifically formulated inks, coatings and adhesives are dried and converted to a usable coating by means of high-energy electrons or short wavelengths of light.

Radiographic Testing (RT). Radiographic testing is a nondestructive testing method for detecting internal defects in test objects such as castings, forgings or welds. In radiographic testing, the test object is exposed to ionizing radiation such as X rays or gamma rays. The absorption or scattering of radiation is detected on photographic film or a fluorescent screen for visual interpretation.

RAM. See: Random-access Memory.

RAM Disk. A RAM disk is when RAM memory is used by a computer and accessed as if the memory were a disk drive. The memory is given a logical drive name and files are written to and read from it as if the RAM disk were a floppy disk or hard disk. The advantage in using a RAM disk is that the physical access time is much less than when accessing a hard disk. When running applications which cannot fit into the maximum main system RAM, as with the 640K limit of MS DOS, the user has no choice but to use expanded memory or extended memory, or to use a RAM disk. Many software programs can utilize a RAM disk, or a person can simply use a RAM disk as if it were a regular disk, as the application program has no knowledge that the RAM disk isn't a physical disk. The disadvantage of using a RAM disk is that the data stored in the RAM disk may be lost if power is interrupted to the system. Also, before ending each session, and turning the machine off, the entire contents of the RAM disk must be saved on a physical disk to maintain permanent retention of the data. A casual mistake could lose large quantities of important data.

Also see: Computers, Random-access Memory.

Random-access Memory (RAM). Random-access memory is memory whose stored data can be accessed in a random manner, as opposed to sequentially or in a structured defined manner. Random-access memory usually refers to integrated circuit chips that provide memory capabilities. RAM chips typically are either static or dynamic, which indicates whether the chip requires refreshing or not. Electrically Alterable RAM (EARAM) is a type of memory which does not require any power to be applied to the chip to maintain the data. EARAM is similar to EPROM or ROM, but can be erased and rewritten to by the system which uses the EAROM memory chip. Other types of RAM are bubble memory, CDROMs, floppy diskettes, hard disks, and magnetic tape. The type of media isn't what really makes memory random-access, but the control software and the methods by which the data contained by the memory is "catalogued" by the file handling software either allow or disallow files and other information to be accessed in a random manner.

Also see: Computers, Dynamic Random-access Memory, RAM Disk, Static Random-access Memory.

Rated Voltage. Rated voltage is the maximum voltage at which an electrical component can function over an extended period without degradation.

Rated Weight. Rated weight is the weight limit a robot can handle utilizing any or all of its axes while still maintaining all characteristics (including repeatable placement accuracy established at specified speeds and reaches claimed in specifications).

Also see: Industrial Robots.

Raw Data. Raw data is data that has not been processed.

Also see: Data.

RCC. See: Remote Center Compliance.

Reactance. Reactance is the opposition offered to the flow of alternating current by inductance or capacitance of a component or circuit. Also known as AC resistance.

Reaction Injection Molding (RIM). Reaction injection molding is a molding process which allows the rapid molding of liquid materials. The injection molding process consists of heating and homogenizing plastic granules in a cylinder until they are sufficiently fluid to allow for pressure injection into a relatively cold mold where they solidify and take the shape of the mold cavity. For thermoplastics, no chemical changes occur within the plastic, and consequently the process is repeatable. The major advantages of the injection molding process are the speed of production, minimal requirements for postmolding operations, and simultaneous multipart molding.

In the RIM processes, cold or warm, two highly reactive, low-molecular-weight, low-viscosity resin systems are first injected into a mixing head and from there into a heated mold, where the reaction to a solid is completed. Polymerization and cross-linking occur in the mold and this process has been proven particularly effective for high-speed molding of such materials as polyurethanes, epoxies, polyesters, and nylons.

Read. The term read describes the function of a computer inputting into its main memory, information which is stored on a peripheral device such as a disk drive or tape drive. When data is read, a "read" head scans the media and interprets the stored representations while the associated circuitry reconstructs the information and presents the data to the computer system. Unlike a *write* activity, if a *read* is done with error, the data isn't disturbed and usually another read can accurately interpret the stored data, provided the same error conditions don't still exist.

When a read is performed on a random-access memory device, the first task is to look in the directory to determine the physical tracks and sectors which hold the desired data. Then the device must position the read head over the appropriate locations in sequence (the access time) and read the data. In a sequentially accessed memory device, each read operation must start with the beginning of the storage device and begin reading the entire media sequentially, until the desired data is read.

Also see: Computers, Write.

Read-only Memory (ROM). Read-only memory describes memory which can be written to once, and then read, but not erased or rewritten to. Examples of read-only memory are integrated circuit chip ROMs which must have their data written to them at time of manufacture, but cannot be written to without using special equipment before the chip's manufacturing process is completed. Other types of read-only memory are CDROMs (compact-disk read-only memory) and PROMs (programmable read-only memory). CDROMs can be written to by the user's drive, but the media cannot be erased. The portion of the CDROM which has already been written to can be read while portions of the CDROM are still blank. CDROMs are excellent for data archiving where the data will not change once it is written to the media, but new data continues to be created, stored, and archived. The records of financial institutions are an example of data which must be permanently archived. EPROMs are devices which can be erased and reprogrammed by the user, but special circuitry must be used to reprogram the device and the chip can only be erased by removing the chip from the system and exposing the chip to ultraviolet light for at least 30 minutes. EPROM programmers can be either separate devices, or they can be additional circuitry in the machine. EPROMs, unlike CDROMs, must be completely erased and then the entire chip must be reprogrammed at the same time.

Also see: CDROM, PROM.

Read-out. Read-out is the presentation of output data by means of visual displays, punched tape, etc.

Also see: Data, Computers.

Real-time Processing. Real-time processing is the ability of a computer to execute a program fast enough that neither the operator nor other operations perceive having to wait for any task to complete within the CPU. This response requirement will necessarily differ between different machines and applications. Real-time processing also is dependent upon the needs and requirements of the inputs and input devices. For example, analog process control requires much slower response times than digital data sampling or circuit timing analysis. Graphic workstations and computer-aided engineering (CAE) functions require substantially greater CPU response times to support real-time processing.

In general, if a queue is required to hold incoming data before it can be processed, the system is not doing real-time processing. With the continuing advance in computer hardware, real-time processing is becoming a reality for more and more applications. Some of the physical limits regarding the speed of CPUs and therefore real-time processing, are the internal circuit-path distances which the electrical signals must travel within the integrated circuit chip. Since the speed of light is the fastest these signals can possibly travel, unless the calculations can be simplified, or parallel processors utilized, certain processes are simply too computationally intensive to be real-time at the present time.

Also see: Computers, Hardware, Queue.

Reaming. Reaming is a machining process that uses a multi-edge fluted cutting tool to smooth, enlarge, or accurately size an existing hole. Reaming is performed using the same types of machines as drilling.

Also see: Drilling.

Recording Density. Recording density is the concentration of information which is placed on magnetic media. When referring to nine-track magnetic tape, recording density is indicated in bits-per-inch (bpi), such as 900 bpi, 1200, 1600 bpi or more.

When referring to floppy diskettes, especially 5 1/4 inch personal computer diskettes, single density, double density, and quad density are the terms which are used.

Other factors than just recording density contribute to determining the total amount of data which can be stored on a piece of recording media. These other factors include, the track width (track density), whether one or both (single or double) sides are used, and the minimum size required by the media to reliably maintain the information regarding one bit. With laser compact disks (CDs) as well as larger laser disks, the resolution and recording density achievable on media has dramatically increased. Furthermore, the durability of laser disks is extremely better than magnetic media, as well as by it being nonmagnetic, it is impervious to magnetic erasure problems.

Also see: Bits-per-inch, Computers, Personal Computers.

Rectangular Coordinates. Rectangular coordinates, which is another name for cartesian coordinates in two planes, are coordinates which define a rectangle. (A square is a rectangle with equal-length sides.) Latitude and longitude are rectangular coordinates. Architectural floor plans (plan views) are dimensioned in rectangular coordinates. An X-Y table used by an assembly system or a numerically controlled machining center is another example of a device with rectangular coordinates. Most vision systems can only discern objects in two dimensions. Depth or height of an object isn't readily resolvable by machine vision cameras and the associated computer circuitry. Automatic insertion equipment used to assemble electrical components into and onto printed circuit boards uses rectangular coordinates to move the board to the proper location, with a fixed motion in the third dimension for insertion/onsertion of the devices. Rectangular coordinates are the first step in working with multiple dimensions, such as length and width.

Also see: Cartesian Coordinates.

Redundancy. Redundancy is a backup control that provides a second method of control for certain critical functions.

Reed Relay. A reed relay is an assembly combining a reed switch with an electromagnetic operating coil.

Reference Block. A reference block is a block within an NC program which has enough data to permit resumption of the program following an interruption.

Also see: Numerical Control.

Refresh Tube Display. A refresh tube display is a CRT in which the display is regenerated many times a second. An auxiliary processor is required for the refresh operation. This type of display allows the viewing of dynamic images.

Register. Register is a portion of memory which temporarily stores data (usually one word). Registers are used as ''scratch-pad'' type storage locations as the computer is performing calculations and operations. Registers are usually included in the CPU chip as part of the circuitry. Registers in different CPUs vary in size depending upon the size of the CPU in bits, such as 8-bit, 16-bit, or 32-bit, etc.

Also see: Computers.

Relative Accuracy. Relative accuracy is accuracy that is measured relative to some other parameter. Usually the relative parameters are also changing, since typically if the relative parameter is static, the defined accuracy could be considered to be absolute accuracy instead of relative accuracy. Sometimes a particular parameter cannot be isolated to occur by itself, and therefore the only method of describing the accuracy is to describe the accuracy parameter of interest in combination with, or relative to, another parameter.

Sometimes a combination of parameters and accuracies contribute to the total value of something. Since each accuracy variable has a range and typically has both a plus and a minus value (a tolerance), a combination of plus errors and minus errors could partially or totally cancel each other. This alteration affects the relative accuracy as different boundaries are considered. An inclusion of an additional parameter could sometimes improve the relative accuracy, especially if some of the tolerances have an inverse relationship to each other.

Relative Coordinates. Relative coordinates describe a point location which is defined relative to another object or point which in turn is defined relative to another coordinate system. In a sense, even absolute coordinates are relative coordinates, they are just defined relative to their own coordinate system. The value of using relative coordinates can be realized when doing part programming for a numerically controlled machine or robot, whereby the programming can be done off-line and relative to the part and then translated into the coordinate system of the machine. Relative coordinates are also useful when moving programs from one machine to another. One usage of relative coordinates is to calibrate a machine to an imaginary set of ''zero'' coordinates and then to relate the program to this calibrated coordinated system. When programming complex shapes and surfaces, various program segments can be created, each in a different relative coordinate system, which then can all be related to each other through their relationship to the machine's coordinate system.

Also see: Absolute Coordinates.

Relay. A relay is an electronic device operated by a variation in the conditions of one electric circuit to effect the operation of other devices in the same circuit or another circuit.

Relay Logic. Relay logic is a representation of the program or other logic in a form normally used for relays.

Relay Symbology. Relay symbology is the use of the symbols developed for the hardwired relay.

Remote Access. Remote access is the access to a data processing facility enabling communication by one or more stations that are distant from the facility.

Remote Center Compliance (RCC). Remote center compliance describes a design for a device which can be used to insert an object into a close-tolerance hole. This hole consists of a geometrical arrangement which provides a theoretical pivot point which is some distance from the plane of insertion allowing the inserted device to be pivoted and translated sideways about this "remote" point to assist in the alignment of the two objects. Remote center compliance devices are especially helpful for machine insertion activities such as robotic assembly and electrical component assembly. Since machines do not have the hand-eye coordination or tactile sensing of a human, the ability to passively comply with misalignment is a necessity for effectively completing the operation. While various designs have been experimented with including using polymers and wire-frame designs, RCC devices require a minimum geometrical superstructure to provide the necessary offset distance for the needed remote location of the compliance point.

Also see: Gripper, Industrial Robots.

Remote Computing. Remote computing is the utilization of a computer through a terminal, remotely located from the computer, via a teleprocessing link.

Also see: Computers.

Remote Input/Output. Remote input/output is the capability to place the preset for a timer or counter line into a register and refer to the register in the C (coil) element of the logic line. The preset is not fixed since the register contents (and therefore the preset) can be altered at any time.

Remote Job Entry. Remote job entry is the submission of jobs through an input unit that has access to a computer through a data link. It also is the submission of job control statements and data from a remote terminal, causing the jobs described to be scheduled and executed as though encountered in the input stream at the main computer.

Also see: Computers.

Repeatability. Repeatability is the ability to return to a previous location or orientation. For a robot or machine tool, the repeatability is how well the machine can return to a previously taught point location. Repeatability is different from accuracy, since accuracy is a measure of how well a device can move to a previously unvisited programmed location based upon the calibration of the device. Repeatability is the ability to repeat a move regardless of the accuracy of the move. If a device can be taught or led through the motions for programming, adequate repeatability is the minimum requirement. If a device is to be programmed by off-line programming, accuracy is a must as well as repeatability. For machining operations, the repeatability has a direct impact on the surface finish of the part. With today's convention of making mass-produced interchangeable-parts products, repeatability in processes is a requirement. Repeatability is usually either described in absolute measurements or as a percentage of the overall motion.

Also see: Accuracy.

Report. A report is an application data display or printout containing information in a user-designed format. Initially entered as messages, data for the report is stored in a memory area separate from the user's program.

Also see: Report Generation.

Report Generation. Report generation is the printing or displaying of user-formatted application data. Initiation is by means of a user's program or a data terminal keyboard.

Also see: Report.

Reprogrammable Robot. A reprogrammable robot is a robot that possesses the ability to change performance duties without physically modifying the inherent design of the control, physical structure, and/or its power source.

Also see: Industrial Robots.

Resistance Projection Welding (RPW). Resistance projection welding is a resistance welding process that uses a projection, embossment or intersection in or on both parts to localize and direct the flow of electric current; and control nugget size and contour, instead of depending on electrode size and shape, as in spot welding.

Also see: Welding.

Resistance Seam Welding (RSEW). Resistance seam welding is a resistance welding process that produces a series of overlapping spot welds along a joint using rotating electrode wheels instead of cylindrical or shaped electrodes.

Also see: Welding.

Resistance Spot Welding (RSW). Resistance spot welding is a resistance welding process which produces bonding at one spot. The size and shape of the weld nugget are influenced largely by size and contour of the electrodes that introduce low-voltage, high-amperage electric current into the assembly.

Industrial robots have been used in spot welding on automobile assembly lines. In this application, an auto body is already fixtured on a "body truck" which is moved by a floor conveyor. A rail system is used to guide the body truck and maintain proper alignment to the robot. The welding operation is done along the seam edges while the body is moving or stationary. The body truck position is measured relative to the robot base. The ability to program the body when it is stationary, execute the program when the body is in motion, and a minimum of modification to the line has made this application economical.

Also see: Industrial Robots, Welding.

Resistance Welding (RW). Resistance welding is any of several fusion welding processes in which melting is produced in the joint using heat obtained from resistance of the work to electric current flowing in a circuit that the work is a part of, and then applying pressure. No flux, filler metal, or shielding is used, but the process is sometimes performed in vacuum or inert gas in welding reactive metals.

Also see: Welding.

Resistive Unbalance. Resistive unbalance is the difference in resistances of two or more conductors in a cable. This unbalance is expressed as a percentage of the resistance of a single conductor.

Resistor. A resistor is a device having electrical resistance and used in an electric circuit for protection, operation, or current control.

Resistor-Capacitor-Transistor Logic (RCTL). Resistor-capacitor-transistor logic is logic performed by several resistors, a transistor, and a diode. Transistors are used to produce an inverted output; capacitors are used to enhance switching speed.

Also see: Resistor.

Resistor-transistor Logic (RTL). Resistor-transistor logic is logic performed by resistors, with transistors used to produce an inverted output.

Also see: Resistor.

Resolution. The resolution of monitors varies from lows of commercial TV or home computers in the 320 X 240 pixels range to 640 X 350 which is the IBM EGA (Enhanced Graphics Adapter) type resolution, to 1024 X 1024 and higher which is common to most CAD systems. Resolutions of other commonly used video monitors are 720 X 348 for Hercules monochrome graphics and 640 X 480 for enhanced EGA and VGA modes. The visual perception of resolution is affected by the viewing distance, so the home television is not objectionable due to the viewing relief, while a home computer monitor of the same resolution is undesirable.

One aspect regarding monitors is *interlacing*. Interlacing is the function of skipping every other line as the image is retraced or refreshed and then retracing the skipped lines during the next scan. This is possible because the human eye does not have the ability to discern the decay of the light at a rate fast enough to notice the fading picture before it is refreshed. The on-off rate at which flicker is noticeable can vary from 20 Hz to 100 Hz depending on the light source. Sixty Hz is typical of today's monitors.

Resolution also is the minimum position motion that can be specified by a numerical control system.

Resolver. The resolver, also called a *synchro*, is a feedback component of the drive system which is a part of a group of elements comprising the servomechanism that converts numerical control tape information into precision mechanical displacements of machine tools. These elements include motors (hydraulic or electric), gear trains, and transducers (velocity or position). In NC machine tools, the most common position sensing device is the rotary transducer. Its popularity can be attributed to compact size and the flexibility with which it can be used. The two primary types of rotary transducers used are resolvers and encoders. The resolver is an analog device whose output is converted to digital form, whereas the encoder is a numerical device that outputs digital data directly. The transducer is connected to the lead screw so that the measure of shaft rotation in combination with the specified lead of the screen yields the position of the machine element.

The resolver consists of an assembly which resembles a small electric motor with a stator-rotor configuration. The stator contains two coils excited by voltages of equal amplitude but 90 degrees out of phase. As the rotor turns, the phase relationship between the stator and the rotor voltages corresponds to the shaft angle, so that one electrical degree of phase shift corresponds to one mechanical degree of lead screw rotation. Thus, the phase shift between the rotor and stator coils is an exact measure of the input shaft rotation. Resolvers with accuracies of two to three seconds of arc are common.

Resource. A resource is any facility of the computing system or operating system required by a job or task. It includes main storage, input/output devices, the CPU, data sets, and control or processing programs.

Response Time. The response time is considered to be the elapsed time from the end of a stimulus or action and the beginning of the response or reaction.

In a data system, it is the elapsed time between the end of transmission of an enquiry message and the beginning of the receipt of a response message, measured at the enquiry originating station.

In a disk drive, the response time is the elapsed time from the end of the request for data and the beginning of the respondent data stream reaching the data bus.

In an analog process control system, the response time is the time required for the control system to vary the outputs based on a change in the inputs to the control system. Remember, the outputs of the process are the inputs to the control system and respectively, the outputs of the control system are the inputs to the process.

Also see: Computers.

Restore. To restore is to reset a storage device to a prescribed initial state.

Retentive Output. A retentive output is an output that remains in its last state (on or off) depending upon which of its two program rungs was last to be true. It remains in its last state while both rungs are false and also if power is removed from, then restored to, a PLC.

Retrofit. Retrofit is the modification of a machine originally constructed for one purpose, or to operate in a certain manner, to a machine for a different purpose or method of operation. An example is to change a machine originally operated by manual or tracer control to one that operates by NC controls.

Also see: Numerical Control.

Reverse Flanging. Reverse flanging is a bending operation that produces a narrow flange having at least one shrink flange and one stretch flange connected by a joggle. The radii of curvature through the joggle generally are not sharp.

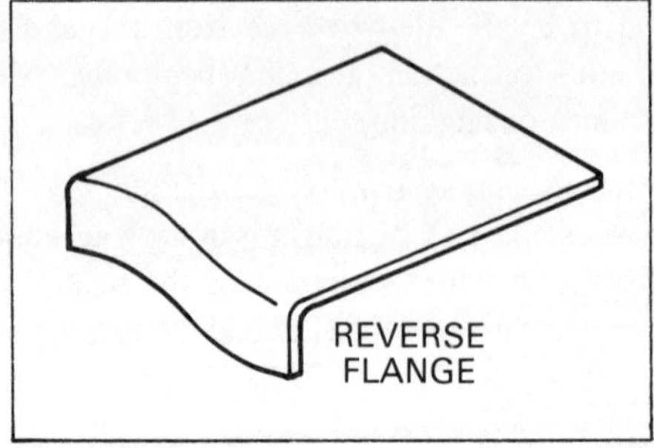

In this case of reverse flanging, the product has at least one shrink flange and one strength flange.

Revision. Revision is a program configuration control identifier. The revision within a given program version is incremented whenever software modifications are made that do not change any program functions. The program version is changed when a significant change is made in program functionality.

RGB. RGB (which stands for red, green, blue) describes the three basic colors used in combination to create the colors and shades within a color monitor. RGB monitors differ from composite video monitors because in RGB monitors each of the three color guns is controlled by a separate video signal. This provides a sharper image with more distinct color than using a single composite video signal.

In an RGB color monitor, the phosphor coating of the CRT (cathode ray tube) contains triplets of phosphor dots that glow in each of the three colors (red, green, and blue) when hit by an electron beam. A shadow mask (a perforated sheet of metal just behind the phosphor screen) allows the three electron beams to impact the appropriate dots.

The three electron guns and the phosphor dots can be arranged in one of two ways. The first, when both are in a triangular (delta) arrangement, is called a *delta gun*. The other position is when both the guns and the phosphor are in a horizontal arrangement, which is termed *in-line*. Regardless of the physical arrangement of the guns and the phosphor dots, all three beams must pass through the screen at exactly the same point in order to turn on the appropriate colors at the coincidental location.

Also see: Resolution.

RIA. See: Robotic Industries Association.

RIM. See: Reaction Injection Molding.

Ring Network. A ring network is a physical communications topography which consists of all the nodes of the communications network arranged in a ring with a physical cable connecting each node to two other nodes. A node is a physical device on a network (i.e. PC, printer, plotter, minicomputer, modem). To add nodes to the ring, the physical cable must be separated and the additional node(s) added. The advantage of a ring network over a bus is that if a single break occurs in the physical network, the logical network is still intact and can function unimpaired. A ring network is a peer type network when there is no physical master. The types of logical networks that can be operated on a physical ring network are both logical bus and logical ring, and even a logical star. A physical ring network can have a master/slave relationship. One of the disadvantages of a ring network is the additional cable requirements to provide the completion of the ring. Depending upon the cabling arrangement, a ring network may use twice the length of cable as a bus arrangement.

Also see: Network.

RI/SME. See: Robotics International of SME.

Riveting. A rivet is a one-piece, unthreaded, permanent fastener consisting of a head and a body. A rivet holds two parts together by passing the body through matching holes in two parts, then forming a second head on the body end.

Manual or automatic riveting machines, bench or floor-mounted (see figure on this page), use the same principle of operation as portable riveting tools. The rivet is fed from the hopper to a track that deposits it, shank down, in the center of the upper jaws. As the cycle proceeds, a driver descends and contacts the rivet head, forcing it downward past the point where the jaws come to a stop. This action pushes the rivet through the jaws and onto a spring-mounted plunger on the lower arm of the machine. With continued downward motion, the plunger pin retracts and guides the rivet through the work until it bottoms against the lower die and is clinched. When the driver and the jaws retract, the plunger pushes the rivet and the work off the die.

The high cost, low production rates, and inconsistent results associated with manual riveting, especially for joining aluminum components in the aircraft industry, has led to the extensive use of automation in fastening. Machines are available that automatically position and clamp the assembly, drill and/or countersink the required holes, feed and insert the proper rivets, head the rivets by upsetting, and unclamp the assembly. Industrial robots, some with computer vision as a sensory feedback, also are being used for automated fastening.

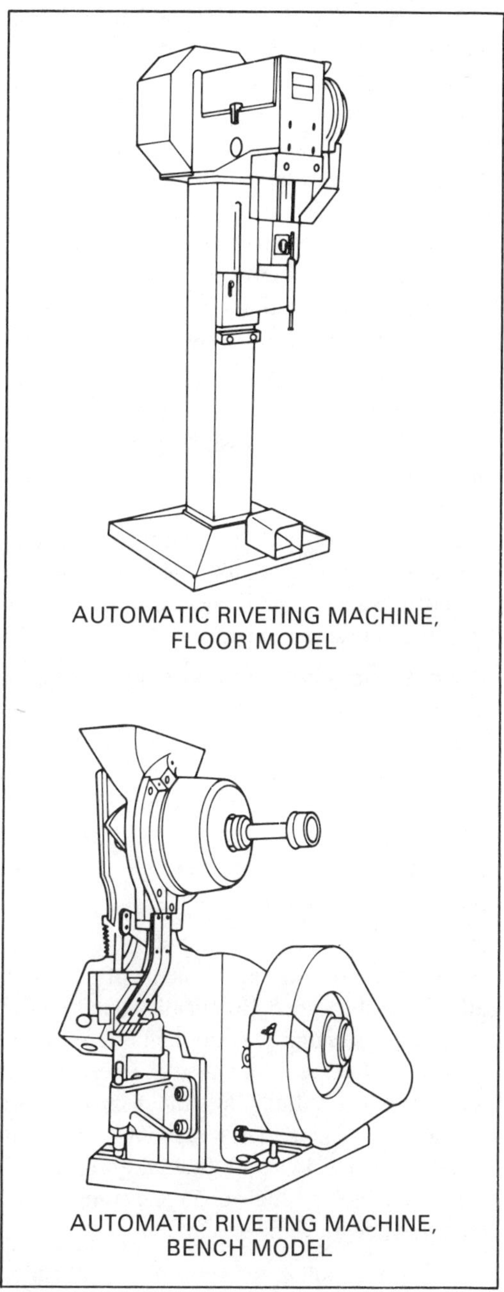

Floor and bench models of automatic riveting machines.

The automatic drilling and riveting process permits faster production at lower cost. In many applications, improvements in production rates of up to 10 times have been realized. Another important advantage is the improved quality of the joints produced. Hole diameters can be held to 0.001 inch (0.03 mm) and countersunk depths maintained to a tolerance of 0.0005 inch (0.013 mm). Close control of the countersunk depths often eliminates the need for shaving the rivet heads, and rigid clamping during drilling makes deburring unnecessary. These machines, with proper upper and lower tooling, will handle all standard type rivets and all materials capable of being squeezed. Controlled parameters of machine operation contribute to high-order repeatability, with a rivet reject rate of less than 1%.

In addition to extensive use by the commercial aircraft industry, automatic fastening is being applied in a growing number of other industries. Materials joined include aluminum, titanium, and other metal alloys, as well as composite materials. The automatic machines are also being used for the application of threaded fasteners, nut plates, and blind fasteners, in addition to rivets.

Applications include the assembly of helicopter tail sections, cargo-bay doors for the space shuttle, dish-shaped antennas, and components for military vehicles. The technology is also being used in the primary assembly line for the all-aluminum Hummer vehicle, which is now being produced to replace the U.S. Army jeep.

Automatic drilling and riveting machines are also being used for the application of slug rivets—low-cost, headless slugs of aluminum alloy. The slugs are hydraulically squeezed from both ends, forming leakproof interference fits. This technique has eliminated the need for fuel tanks previously built into the wings of large aircraft. By making fuel storage integral with the wings (called wet wings), fuel-carrying capacity has been increased and weight and complexity of the wings decreased.

The basic tooling for a machine is illustrated in the figure on the this page. With the workpieces to be joined firmly clamped in the machine, a drilling unit moves into place to produce the hole. The drilling unit is then moved aside and, with the clamping force still applied, a riveting unit is moved into position. A rivet is automatically fed from one or more hoppers, down a track, and into feed fingers that insert it into the hole. In this position, the rivet is hydraulically squeezed against the upper anvil to form the required head. Clamping pressure is then released, and the assembly is removed or repositioned for additional fastening.

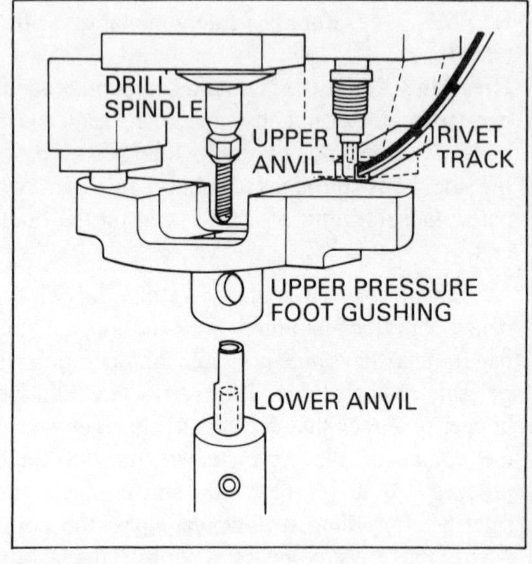

Tooling components for automatic drilling and riveting.

The machines exert as much as 6000 lb (27 kN) of clamping force and up to 50,000 lb (222 kN) of force for heading the rivets. Average cycle time is 3 1/2 seconds, and production rates of more than 24 rivets per minute are attained. When required, the formed rivet heads are shaved flush on the same automatic machine. For hot dimpling, a three-position tooling head is used. Tooling at the first position has a built-in cartridge heater and produces a male dimple profile. Drilling is performed at the second position and upsetting at the third.

Methods of moving assemblies for automatic drilling and riveting vary from simple manual means to sophisticated electronic controls. Manual positioning is used extensively for smaller and lighter assemblies, as well as those with access problems. These systems are generally equipped with bearings and/or rollers to minimize friction, and chains or straps and a sling arrangement are often used for vertical movements.

Semiautomatic positioning systems are generally equipped with hydraulic and/or pneumatic cylinders, and electric motors to jog the assemblies into position. In positioning with computer numerical control (CNC), a preprogrammed microprocessor automatically energizes servomotors to accurately control positioning. The CNC units can also be used to control cycling of the machine.

Rivets must also be removed for rework and repair.

The Robotic Deriveter System (RDS) was built for the U.S. Navy by Southwest Research Institute. The RDS uses a mobile robot as well as machine vision and other perception methods which allow adaptation to individual conditions on aircraft.

Author Ernest A. Franke of Southwest Research Institute described the system in his SME Technical Paper, *Update on the Robotic Deriveter*.

Writes Franke: "Although the RDS was originally intended for work on large, flat panels such as F-4 stabilators, since its arrival at North Island (San Diego), it has been successfully applied to deriveting of curved parts such as the F-14 forward fixed cowling on a fixture and the engine nacelles on F-14 aircraft. The fasteners (B-100 rivets) on the engine nacelles cannot be removed with a single drilling operation. The flexibility of the RDS became apparent when the deriveting cycle was modified to first drill to shallow depth and punch the central stem of the fastener and then drill deeper and punch out the shank.

"Use of the RDS at North Island has demonstrated the value of a mobile system. Many operations are required on complete airframes that cannot be moved to and from a fixed robot work cell. In addition, removal of rivets is only a small part of the total rework process that is carried out in different shops depending on the particular part. The ability to bring a robotic system to the work eliminates the time delays

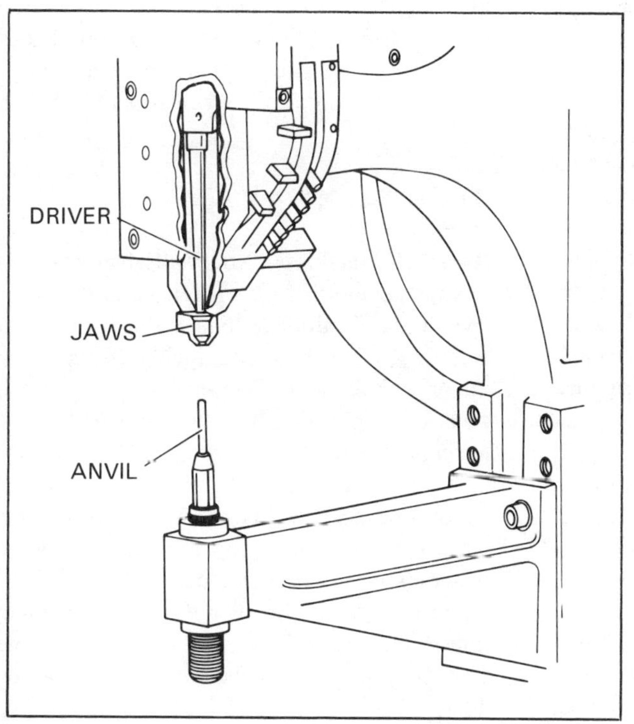

Drive, jaws, and anvil mounted on a riveting machine.

and logistics requirements of moving parts from one hangar to another and allows operation on parts that cannot be moved.

"Finally, it must be noted that the Robotic Deriveter System is the first of its kind ever built. Since there was not a base of previous experience to draw on, the system was designed to be as flexible as possible to allow its use in unanticipated applications. The value of the approach can be seen by the ease with which the system was modified for operation on B-100 fasteners. As is often the case, this increased flexibility is only achieved by reducing the efficiency of specific tasks. This is most apparent in the size of the RDS end effector. It includes modules for bolt installation, inspection, and marking that are not needed to simply perform deriveting. This added capability increases the size of the end effector, restricting the regions it can access and requires greater turret motions between modules, increasing the time required for fastener removal.

"As experience is gained in use of the RDS and specific applications are identified, additional end effectors can be designed to perform only the needed functions. This will allow the system to be optimized for access, speed, and capability for a variety of tasks."

Robotic Industries Association (RIA). Robotic Industries Association is a trade association in North America organized specifically to promote the use and development of robotic technology. Founded in 1974, RIA represents member companies including leading U.S. robot manufacturers, distributors, accessory equipment and systems suppliers, users, consultants, service companies, and research organizations.

In 1984, RIA established two new trade groups: the National Personal Robot Association and the Automated Vision Association. Both groups reflect RIA's goal of serving all industries that are active in developing robotic technology.

Robotics International of SME (RI/SME). Robotics International of the Society of Manufacturing Engineers or RI/SME is a member association founded in 1980 dedicated to the advancement of industrial robotics technology. In addition, RI/SME serves as a liaison between industry, government, and education to determine areas where technological development is needed. RI/SME is the vital link in the chain that connects the individual to the many aspects of robotic-oriented manufacturing.

Members of RI/SME are involved in these three categories:

- Planning, selecting, and applying robots.
- Research and development leading to the creation or improvement of robots.
- Other creative robotic activities in administration, education, or government.

The members of RI/SME update their knowledge and skills through educational programs, chapter meetings, and conferences.

The largest American scientific and educational association exploring the use of robots, RI/SME is an official association of the Society of Manufacturing Engineers and is headquartered in Dearborn, Michigan. RI/SME addresses the informational needs of the engineer involved in the application of robots and provides a forum for discussion among concerned individuals. RI/SME is a partner to industry and a major force in an engineer's professional advancement.

Robots. See: Industrial Robots.

Rockwell Hardness. Rockwell Hardness is an arbitrary number, which is related to the difference in the depth of penetration on a surface of a test specimen by a penetrator subjected to a minor (initial) and a major (final) load under specified conditions. The two classifications of testing in the Rockwell System are the Rockwell Test and the Rockwell Superficial Hardness Test. The Superficial Hardness Test is designed particularly for measuring the hardness of thin materials or of case-hardened metals. The difference between the two tests is that lighter minor and major loads and a tester of higher sensitivity is used for the Superficial Hardness Test.

The minor load in a Rockwell Hardness Test is always 10 kg (22 lb). The major load varies and may be 60 (132.28), 100 (220.46) or 150 kg (330.7 lb). The penetrators most commonly used are a steel ball 0.0625 inch (15.875 mm) in diameter and a spheroconical diamond with a spherical apex of 0.2mm (0.00788 inch) radius and an included angle of 120 degrees. The diamond penetrator is generally called a "brale" penetrator. Steel ball penetrators of 0.125 inch (0.175 mm), 0.250 inch (6.35 mm) and 0.500 inch (12.7mm) are used in tests with varying loads on softer materials such as very soft metals, plastics, or wood.

There are numerous scales associated with the Rockwell Hardness Test. The scale that is appropriate for the particular test being performed is selected according to the material being tested and the General Range of Hardness Test results expected. Each scale specifies a particular combination of penetrator and major load to be used in the test. The test workpiece is positioned on a staging surface under the tester penetrator spindle and the penetrator is placed on the workpiece surface with the minor load applied. An impulse force is applied to the surface of the workpiece by the sudden addition of the specified major load. The depth of penetration is measured by the difference in spindle location after application of the major load. The scale divisions in the various Rockwell Hardness scales correspond to a depth of penetration of 0.000080 inch (0.002032 mm) for each scale division. The total depth of penetration of the test will then yield a number which corresponds to the relative hardness of materials tested using the appropriate scale combination of load and penetrator.

Roll Bending. Roll bending is a process for forming cylindrical or semicylindrical shapes from sheet or plate by feeding the work edgewise into the gap between a cluster of at least three relatively small diameter straight rolls. The size of the gap between rolls, and its relation to sheet thickness, determines the bend radius in the finished cylinder. Roll bending also is used to form solid bars, rods, heavy-wall tubes and special shapes into right-shape parts.

Roll Forming. Roll forming is a continuous process for forming metal from sheet, strip, or coiled stock into shapes of essentially uniform cross section. The material is fed between successive pairs of rolls, which progressively shape it until the desired cross section is produced. During the process, only bending takes place; the material thickness is not changed except for a slight thinning at bend radii.

The two methods used when shaped parts are roll formed are the precut or cut-to-length method and the post-cut method. Method selection is based on the complexity of the cross section and the production length specification.

In precut operations, the material is cut to length prior to entering the roll forming machine. This process usually incorporates a stacking and feeding system to move the blanks into the roll forming machine, a roll forming machine running at a fixed speed of about 50–250 fpm (15–76 m/min), and an exit conveyor and stacking system. The cut-to-length process is used primarily for lower volume parts and whenever notching cannot be easily accomplished in a post-cut line; for example, miter cuts in vertical legs. Many times, the material is run from coil into a shear or blanking press and then mechanically fed into the roll former.

Tooling cost is inexpensive with this method because cutting requires only a flat shear die or end notch die. However, end flare is more pronounced and side roll tooling is required to obtain a good finished shape.

Even though some configurations require the cut-to-length method, the most efficient, most productive, most consistent, and least troublesome is the post-cut method. This method requires an uncoiler, a roll forming machine, a cutoff machine, and runout table. In most segments of the industry, this is the most widely used method. It can be augmented by various auxiliary operations, including prenotching, punching, embossing, marking, trimming, welding, curving, coiling, and die forming. Any or all of these procedures can be combined to eliminate the need for secondary operations, resulting in a complete or net shape product. However, the cost of tooling and the tooling changeover time for this method are greater than the tooling cost and changeover time for the precut method.

Computers are becoming an important aid in the design of roll forming tooling. Consistency, accuracy, and speed enable the designer to determine the optimum design for each roll pass in less time than is required when the calculations are performed by hand. The capability to display the profile of the part enables the designer to see how the material flows through each pass. This profile enables the designer to determine whether too much work is being performed at a particular pass.

The numerical information compiled by the computer can be employed to produce punched tapes used by numerically controlled machines. These tapes ensure that the rolls are accurately produced. The computer also aids in the setup of the rolls on the machine by specifying the size and locations of the required shims and spacers. All this information and data can be stored for future use and reproduced whenever necessary.

To design the rolls, information about the roll forming machine, the cross section of the final shape, and the initial forming sequence are entered in the computer program. The computer defines the coordinates of each corner numerically and displays the profile of the part at each pass and various perspectives of the flower diagram on the computer terminal. Input changes can be made to vary the material flow through the roll forming machine so that the optimum flow is achieved. Computer output includes flower diagrams, drawings of the cross-sectional shape (see the figure to the right), drawings of the rolls, and the tabular data defining the material and rolls. The computer can also produce the tapes used in the manufacturing of the rolls on numerically controlled machines.

Also see: Numerical Control.

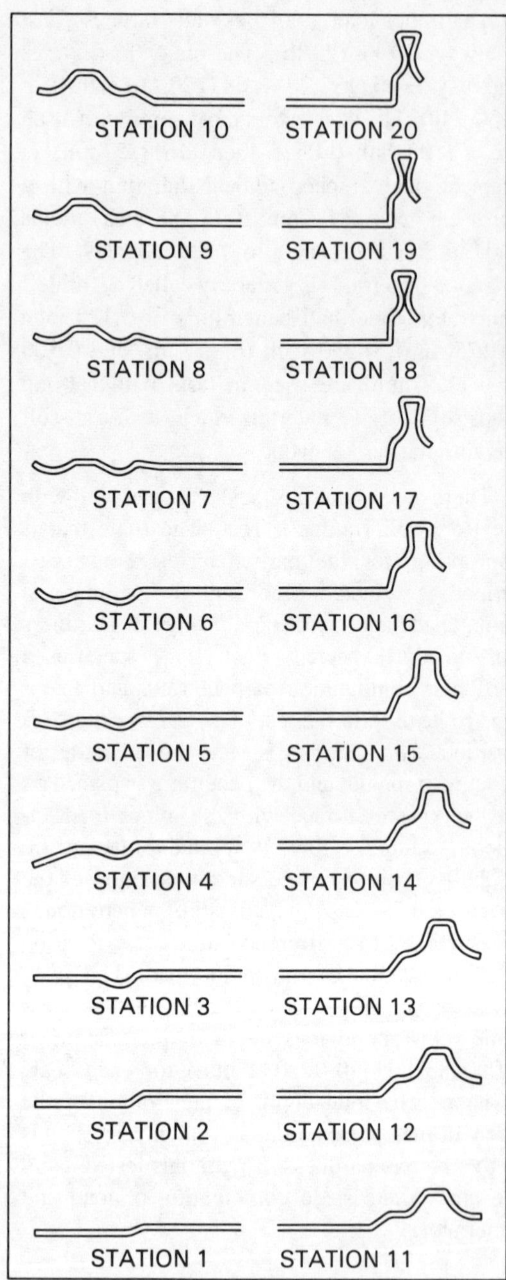

Using the computer for roll design provides the designer with drawings of the cross-sectional shape of the part at each station. This is the output for a part requiring 20 passes. (Only half of the cross section is shown since the part is symmetrical.)

S

Safety. Safety is the ensuring of an nonhazardous environment or system. Safety regarding manually operated equipment consists of providing for ease of use and a lack of hazardous emissions or circumstances. Safety of automated equipment is often different than manually operated machines since people aren't always assumed to be in contact with the machine, but since the equipment may start suddenly or move unexpectedly, the potential safety hazards are more complex than simpler manually operated equipment. Safety hazards such as carcinogens (products which can cause cancer) continue to become more of a concern. Some of the safety practices have been propagated by labor unions who have sought to establish better working conditions for their members. The use of automation and robotics has traditionally allowed some of the less safe jobs to be done by mechanical means instead of manual labor. As technology and manufacturing practices continue to advance, the safety of the average job will continue to improve as humans become removed from the physical aspects of doing the work.

Also see: NIOSH, OSHA.

Safety Shut Down. Safety shut down is a term which describes a situation when control systems must shut down a process or operation in a prescribed safe manner, such as closing down a nuclear reactor without allowing portions to overheat, or stopping a continuous flow process in a coordinated manner. Some operations may have human operators involved in a manner which requires a certain sequence of stopping such as chemical processes. Safety shut down is something which isn't a part of the normal operation of a system, but is being required more by users to ensure this capability. The safety shut down capabilities usually incorporate the ability to sense the occurrence of the need for an emergency enactment of the safety shut down procedure, as well as being able to start when commanded to by an operator. Since safety shut down is sometimes required when control systems fail, fault-tolerant computers or redundant systems are often installed to perform safety shut down.

Also see: Fault-tolerant Computer, Safety.

Sampled Data. Sampled data are data in which the information content is ascertainable only at discrete intervals of time.

Also see: Data.

Sampling Interval. In systematic sampling, the fixed interval of time, output, running hours, etc. between samples.

Sampling Rate. Sampling rate, in a communication system, is the rate at which signals in a given channel are measured or sampled for subsequent modulation, quantization, coding, or conversion (analog-digital). The sampling rate is usually specified as the number of samples that are taken per unit of time. Sampling rate is the same as the sampling frequency.

The reciprocal of the sampling rate is the sampling time. It is the time interval between corresponding points on two successive sampling pulses of the sampling signal.

The sampling theorem states that for a waveform to be sampled and then effectively or satisfactorily reconstituted from the sampled data, it is necessary that the sampling rate be equal to or greater than twice the highest significant frequency component of the waveform.

In analog process control systems, the sampling rate is the rate of accepting updated inputs to the control system, usually monitoring the setpoint(s).

Also see: Analog to Digital, Digital Communications.

Sand Casting. Sand casting is the most widely used of all methods of producing cast metal products. The process consists of making a mold by ramming sand around a three-dimensional pattern of the desired cast shape, removing the pattern, cutting in a gating and riser system to provide passages for the molten metal to reach the mold cavity, and lastly, filling the mold with molten metal and allowing it to solidify. If the sand contains residual moisture at the time of pouring, it is termed green-sand casting; if the sand is intentionally dried before pouring, it is termed dry-sand casting.

Scale. Scale is a ratio of present drawing size to real dimensions.

Scaling. Scaling is the reduction or enlargement of an entity by using a constant value. A range of values may be scaled to a larger range to allow easier resolution. A range of value may be scaled to a smaller range to enable fitting within certain boundary limits. Another reason to scale values is to translate the range into a known or relevant scale to allow easier interpretation.

If extremely large or small values are being considered, scaling the values can provide ease of handling and interpretation. A typical example of scaling is the omission of zeros in financial statements. Another example of scaling is the conversion from one type of units to another, such as from Metric to English.

In an analog process control system which contains an analog display, the input to the display is sometimes scaled to allow the use of the entire display device to represent the signal. In this type of application, a determination is made to decide the maximum and minimum values, whereupon the input signal is scaled to utilize the full-scale of the display device.

Scanner. A scanner is equipment used to digitize coordinate information from a master and convert it to tape for subsequent recreation of the master shape on an NC machine. A scanner also is an instrument that automatically samples or interrogates the state of various processes, files, conditions, or physical states and then initiates action based on the information obtained.

Also see: Numerical Control.

Schematic. A schematic is a graphical pattern or representation of a logical or physical entity where symbols are used to represent the actual components of the system. Schematics are typically considered to describe electronic product designs. Schematics may be electrical or mechanical, or even architectural, such as a building plans, drawings or blueprints. A schematic normally uses standard symbols in a particular trade or profession. The use of standardized-symbol schematics provide for a medium of communication within the discipline. Schematics verify, design, test and depict the design before actually building the product. A schematic is usually a two-dimensional diagram or drawing. With the availability of color and enhanced graphics in computers, schematic symbols are becoming more refined and complex instead of only being able to use a small number of crude symbols. The continuing popularity of the Macintosh computer with its icon graphics type interface makes schematic depiction a more widely used convention.

Also see: CAD, CAD-D, Graphics.

SCR. See: Silicon Controlled Rectifier.

SCSI. See: Small Computer Systems Interface.

Search Function. Search function is a PLC, CNC, or computer programming equipment feature allowing a user to quickly display and/or edit any instruction in the program.

Second Generation. In NC, the second generation is the period of technology associated with transistors.

Also see: Numerical Control.

Second Generation Computers. Second generation computers are the first solid-state component computers. Second generation computers used transistor technology and higher-level languages. They processed only one program at a time, but were more reliable, smaller and required less power than first generation computers. Circa mid 1950s to mid 1960s.

Also see: Computers.

SECS. See: Semiconductor Equipment Communications Standards.

Security. Computer security refers to the technological safeguards and managerial procedures which can be applied to computer hardware, programs and data to assure that organizational assets and individual privacy are protected.

Segmented Program. A segmented program is divided so that each section is self-contained. Interchange of information between segments is by means of data tables in known memory locations; each segment contains instructions that cause the transfer to the next segment.

Selective Compliance Assembly Robot Arm (SCARA). The selective compliance assembly robot arm robot was designed primarily for assembly applications and is gaining worldwide acceptance for use in such applications. It has an asymmetrical, horizontal work plane and offers vertical axis insertion at the end of the wrist from any location within the work plane. Many models are available with symmetrical work planes. Joint configuration allows a multiaxis movement that is often faster than other robot kinematic designs.

Initially, the design intent of the SCARA robot was to position itself over the workpiece and then the drives would disengage so that the arm would be free to act as a compliant device. Some manufacturers now offer SCARA-type horizontally articulated robots. The high positional repeatability for some models and fast motion speeds make this manipulator configuration very desirable for high-volume assembly applications.

Also see: Industrial Robots.

Self-diagnostic. Self-diagnostic is the hardware and firmware within a controller enabling it to monitor its own status and indicate any fault that occurs within it. Monitoring may be continuous or only upon command from the operator.

SEMI. See: Semiconductor Equipment and Materials Institute, Inc.

Semiconductor. See: Diode, Metal-oxide Semiconductor, Negative Metal-oxide Semiconductor, Semiconductor Equipment and Materials Institute, Inc., Semiconductor Equipment Communications Standards.

Semiconductor Equipment and Materials Institute, Inc. (SEMI). Semiconductor Equipment and Materials Institute is an international trade organization serving suppliers of semiconductor equipment, materials and services. SEMI has over 1,100 member companies who have some relationship or interest in the semiconductor industry in its global membership network. SEMI's wide range of member services include six annual SEMICON trade shows, an industry-wide standards program, government relations and technical education program. The standards are available in five volumes which include: Chemicals, Equipment, Materials, Packaging, and Photomasks. The various newsletters, annual publications, and technical reports provide the largest assortment of information relevant to the semiconductor industry from a single source.

The SEMI organization has driven the development of such things as the communication standards for the interconnection of semiconductor manufacturing equipment. A subset of Volume 2 (Equipment), available in both English and Japanese translation, defines the Semiconductor Equipment Communications Standards (SECS I & II) which provides detailed information of the communications standards.

Also see: SECS, Standards.

Semiconductor Equipment Communcations Standards (SECS). The Semiconductor Equipment Communications Standards is a two part specification which defines a communication protocol and implementation for data communication between different types of computer-controlled equipment used in the manufacture, handling, assembly, and testing of semiconductors. SECS is a part of the Semiconductor Equipment and Materials Institute (SEMI). The standards allow manufacturers to purchase their equipment from a variety of manufacturers and ensure the different equipment can be interfaced to each other. The SECS protocol defines commands for most of the functions needed by any of the types of equipment used in the semiconductor manufacturing process. A standing committee with membership consisting of interested parties continually defines and enhances the SECS standards. The SECS protocol is one of the major components of automation standards which, along with the wafer carrier specifications, constitute the major items of compatibility between manufacturer's equipment. The SEMI standards committees meet three times per year to approve changes to the standards.

Also see: SEMI.

Sensors. The term sensor is defined as any device that detects or senses the value or change in the value of any variable, parameter or physical characteristic that is being measured. The sensor is the first and possibly the most important link in a measurement-system.

A measurement-system may require a sensor with an output that is not directly usable. In these cases, the output of the sensor may have to be converted (or transduced) from one form of energy to another. In even more complex systems the transduced signal may then have to be amplified or conditioned in some way to provide a useful output.

Among the things being measured by various types of sensors in use today are time, speed, temperature, light, pressure, sound, direction, mass, force, power, distance, chemical reactions, electrical or magnetic intensity, and various body functions.

History. The first sensors used by mankind were purely biological and the first use was for learning. This use of sensor in learning is so important that it has been given a special classification: *sensory learning*. This topic is described as learning in which a living creature is trained to respond to changes in, or distinguish between differences in, some aspect of physical stimulus presented to one of the sense organs. In humans, of course, the sense organs are the eyes, ears, nose, mouth, and skin.

Early man used these powerful sense organs to perceive everything that was important in his world; the location, direction, distance, amount and type of food or danger. All of these terms are quantitative in nature and have one characteristic in common; they all relate in some way to measurement.

Measurement ranks as one of the human species oldest skills. Much of the information people seek to obtain every day is orientated toward measurement. Questions such as "How many?", "How fast?", "What does it weigh?", "Exactly where is that?", all require a measurement of some type for an accurate answer. The need for information that can only be provided by a measurement drove the development of the first simple sensors.

The first sensors were developed by early man to provide consistency in the measurement of length (or distance), weight and time. To detect or sense the value of the variable under consideration, a comparison was made with something common. Several ancient societies developed measurement systems and, hence, sensors based on units that were the length of a part of the body. Ancient Egyptians based the "cubit" on the length of a man's forearm from the elbow to the tip of the middle finger. Examples of cubit sensors have been found out on wooden rods or stone slabs dating back as early as 3000 B.C. The Roman "uncia," which was the width of a man's thumb, became the English inch. Someone discovered that 12 uncia were roughly the length of a foot and that it was about three feet from the nose to the tip of the middle finger of an outstretched arm and the yard was invented.

Person-based sensing systems quickly failed because all people were not exactly the same, so measurements varied from person to person. This drove the development of sensors based on fixed and agreed upon standards. Today, both the English and Metric Systems use the meter as a standard. A meter is defined as the distance light travels in a vacuum in 1/299,792,458 of a second. Ancient man depended on the sensing of weight to trade commodities such as meat, fruits or grain. Ancient Egyptians developed the balance to sense equal weights of items. Records of weights appear in Egyptian graves dating back to about 4000 B.C.

Measuring and sensors stayed relatively simple for hundreds of years. Advances in sensing technology kept pace with advances in the understanding and use of simple machines. Adaptations of the lever, screw, inclined plane and pulley resulted in much improved sensors for the detection of weight, motion, speed, and direction. Sensing the direction and intensity of wind has always been important for predicting the weather, but the invention of the airplane added significant importance to that topic, and created a need for new sensors. Weathervanes have been used since colonial times to sense wind direction and provide decoration, but an incredibly simple sensor, the windsock, is much more important to a pilot. That one sensor can not only detect the direction of the wind, but provide a relative indication of its intensity.

Moving from these simple early sensors to today's complex detecting equipment has been made possible by two things: the advances in the use of electricity and the development of sophisticated control systems. Virtually every person in the modern world comes in contact with sensors and sensor-actuator control systems every day. Advances in the use of electricity have lead to computers, which have lead to a host of developments. One is the "pulsed laser time-of-flight atom probe," which is the other end of the complexity spectrum from the windsock. This device was used to measure one of the fastest bond breaking chemical reactions ever observed, dissociation of charged molecules of helium and rhodium in eight ten-trillionths of a second. Development of control sensors has made possible refrigerators, space ships and biofeedback systems such as pacemakers.

Discussion. Advances in electrical technology lead to advances in sensor technology. One such advance was the development of devices known as transducers. A transducer is a device which converts a measured quantity of energy or information in one form into a corresponding value of quantity in another form. A simple example is a liquid-in-glass thermometer which senses the temperature of the surroundings and converts that value into a measurable length of liquid in a calibrated capillary tube. This is the same example used to explain a sensor. Therefore, transducers are conveniently divided into two categories, sensors and actuators. An actuator responds to a command signal (in essence, a measurement) and controls the value of a quantity. Together, sensors and actuators form the basis of control systems.

Many transducer applications in both industrial and commercial setting were much more complex than the simple on-off system. In a modern manufacturing plant, sensors are required to measure a wide variety of variables such as speed, temperature, position, location, chemical concentrations, etc. In an automated factory, the output of these sensors would be sent into a central computer system which would be programmed to interpret the signals. The results of this activity takes the form of command signals sent to actuators throughout the factory. These actuators could then move material, adjust speeds, and chemicals, or in the case of quality control applications, stop the process if problems are encountered.

To make the sophisticated control systems available today, many different types of sensors had to be developed. Transducers for measurement and control of physical, chemical and biological processes are available in a wide variety of complexity, sensitivity and accuracy. A few of the principle types of transducers are presented in the following discussion.

In the area of *motion transducers*, many types of transducers for the sensing of linear or angular motion have been developed. They can be converted easily by means of a rack and pinion or lead screw. The automobile odometer may be the linear motion sensor that is familiar to most people. That device converts the angular motion of the driveshaft into a readout that provides information on linear distance traveled. For years the odometer's output device was the simple step wheel counter. Today many auto manufacturers are offering digital readouts utilizing liquid crystal or fluorescent displays.

Many other methods of detecting linear or angular movement exist. Linear potentiometers provide moderate accuracy while interferometric methods provide much greater accuracy but are harder to use. Ultrasonic waves provide accurate thickness measurements in certain materials. X-rays are used in several types of detectors, including thickness measurement in steel mills and airport security systems. Capacitance gages, circular potentiometers, and rotating capacitors are being employed in very accurate sensors.

Flow, pressure, and force transducers measure by the same basic means; mechanical displacement of part of the sensor by means of an object, liquid or gas incident upon a surface. The most familiar of these transducers is probably the bathroom scale. As in many types of force and pressure sensors, the scale depends upon known deformation characteristics of springs or other mechanisms.

Strain gages are a special type of force transducer that are extensively used in aeronautical and mechanical engineering. They measure the deformation of materials in response to applied forces. Without these devices, many advances in the design of buildings, airplanes, and space ships would not have been possible.

Virtually everyone should be familiar with the liquid-in-glass and bimetallic strip thermometers in use today. Most materials have some characteristic that changes with temperature. Changes in size (expansion and contraction) and electrical properties (resistance) of many different materials have become widely used to sense and make use of temperature changes. Some of the more widely known involve our personal comfort, heating, air conditioning and cooking. These types of sensors are *temperature transducers*.

Acoustic transducers are not used strictly for measurement. Speakers convert electrical signals to acoustical waves while microphones do the opposite. Ultrasonic sensor are frequently used for measurement in sonar, nondestructive testing and medicine.

Lastly, consider *chemical and biological sensors*. Chemical field effect transistors, electrochemical and piezoelectric devices sense different types of chemicals. Common applications include checking emission levels of automobiles and various factory outputs. Biotrodes are a new class of specialized sensors being developed to detect important chemicals.

Applications. Noncontact proximity sensors that can measure the range, orientation, and shape of a surface are of potential usefulness in robotic applications such as assembly, inspection, seam tracking, and pick-and-place tasks. Noncontact sensing can be performed by either optical or ultrasonic devices. High resolution requirements are met only by optical means.

Robots installed in flexible assembly systems are instrumented with sensors to ensure that parts are being assembled properly. The sensing systems are often custom-designed for each assembly task and discarded when a different product is assembled. The cost of flexible assembly can be reduced by eliminating the fabrication of a new sensing system for each new assembly job.

One application of a sensor was illustrated by Dennis Locker and John T. McCabe of Mechanical Technology, Incorporated in their SME Technical Paper *Real Time Control of Milling Through Edge Sensing*.

The authors illustrate the application of an optical sensor to control the quality of parts produced by a production machine.

The authors explained: "The edge detection system consists of a focused light source and the optical sensor system (arranged as shown in the figure on the next page). Because of space and operational considerations, the light source and the sensor had to be located outside of the orifice tube. Therefore the edge was detected by reflected light. To minimize the effect of reflectivity differences from tube-to-tube, a high-intensity light source was focused on the newly machined side of the edge.

"The accuracy of the edge sensor for determining the width of the orifice opening was strongly dependent on the condition of the edge left by the cutter and upon the effectiveness of the light sources in delineating the edge. The condition of the edge was dependent on the type of cutter and the state of its cutting edges. Once the cutter was completely specified and the speeds and feeds fixed, then the useful life of the cutting edge with regard to its influence on the accuracy of the measurement system was determined statistically. In selecting the edge illuminator system and its arrangement, a number of trial-and-error tests were made before the configuration was fixed.

"The in-process, real-time control of the path of a milling cutter, actuated by a microcomputer and based on data measured by an electro-optical edge sensor, was shown to be technically and economically feasible. When used in combination with a high-speed spindle with a carbide cutter and an accurate, rigid machine, the sensor controlled system produced a high-quality part in a short period of time, as compared to the existing production process. Finally, using the sensor/computer/printer as the initial inspection process showed great promise as a quick, low-cost method for assuring continuing part quality before the part leaves the machining center."

Another application of sensor technology was provided by Dave Olender (ComTel, St. Paul, MN) in his *Robotics Today* (June 1987) article *Force/Torque Sensor Aids Adhesive/Solder Mask Dispensing*. In the article he reported: "The use of robots to apply adhesive and solder masking on printed circuit (PC) boards is not new. But a robotic dispensing system recently developed by ComTel, Inc. (St. Paul) is unique in its use of a force/torque sensor, rather than machine vision, to locate components on each board. The big advantage over vision is that the sensor approach is significantly more economical. And it's anticipated that, compared to today's vision technology, the sensor will be more reliable over the long term.

"The system is designed to offer maximum flexibility; the robot can be quickly reprogrammed to handle a wide variety of PC board configurations. Some boards carry as many as 100 components. With the force/torque sensor attached to the end of the robot arm, the robot is able to feel for components rather than locate them visually. This approach is particularly useful for applications in which components are densely populated and may be difficult to locate visually. Components that may be tipped or skewed following the flow solder operation can still be found quickly. A three-dimensional vision system would have been required to handle the same task.

"The robot used in this system is an Adept One direct-drive unit built by Adept Technology, Inc. (Mountain View, CA). A fifth axis was added to the robot to provide a yaw capability. This permits the 0.030 inch (0.76 mm) diameter dispensing needle to be tipped at essentially any angle when applying the two-part epoxy adhesive.

"The force/torque sensor is a Model 15/50 unit supplied by Lord Industrial Automation Division, Lord Corporation (Cary, NC). Mounted between the end effector and the mounting plate on the robot arm, the sensor

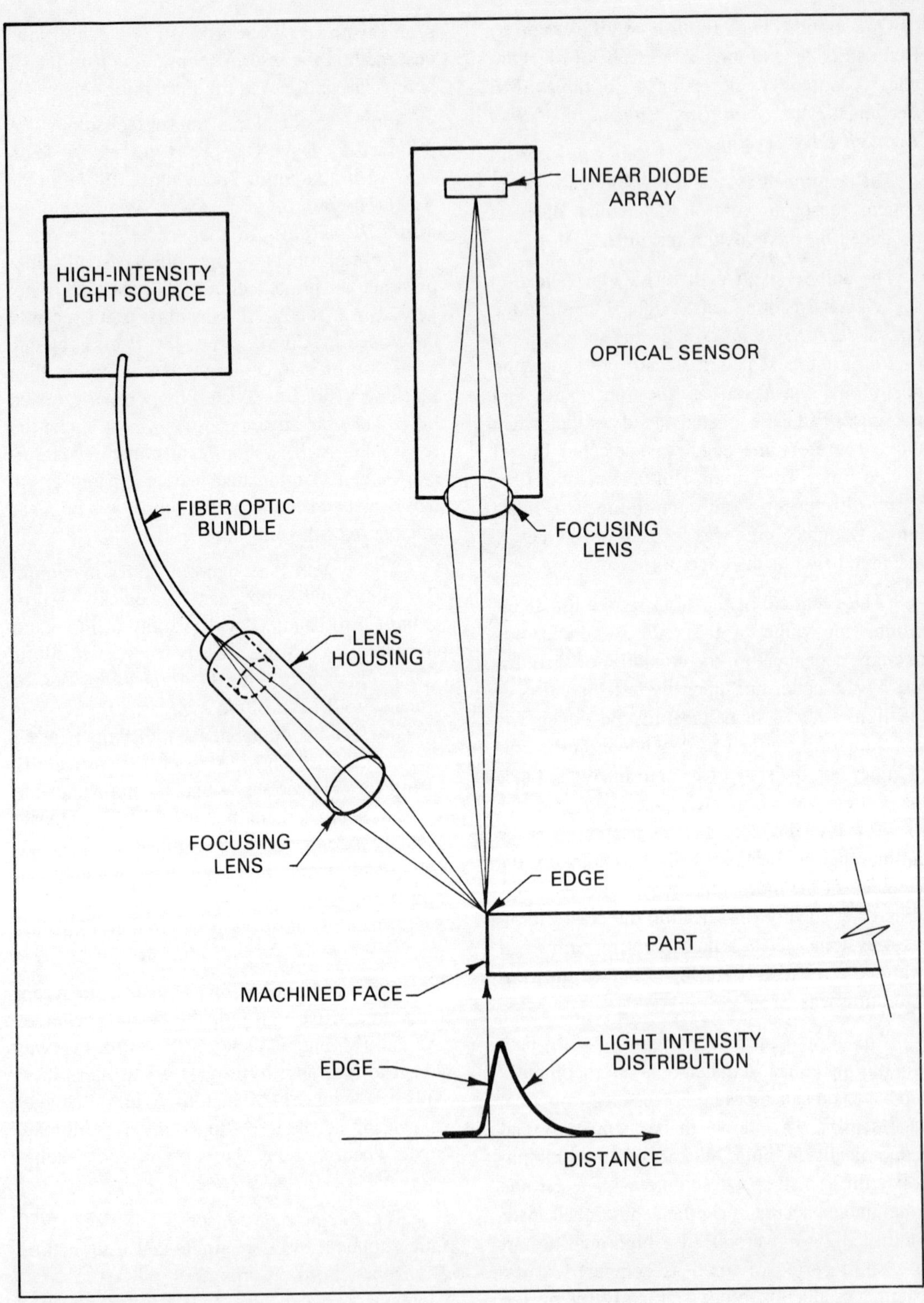

Edge detection system.

provides six-axis force/torque sensing. In a typical sequence, the robot moves the dispensing needle into contact with a component. This produces a force at the sensor, indicating the edge of a component. The location is recorded, and a second edge is then found in the same manner. If the component is detected as having a skewed condition, the robot's path is skewed accordingly as it proceeds with the dispensing operation. This touch routine also is used by the robot to determine whether or not a board is located in each fixture when it arrives at the dispensing station.

"Boards are carried through the work cell on a six-station index table, which is manually loaded and unloaded. A bar code system can be used to identify each board as it arrives at the robot station. Or, the operator can handle the same function by entering the board number on a CRT terminal keyboard. Programs for all of the different boards are created in a manual teach mode and reside in the robot controller. The operator simply holds down a pushbutton on the end effector and moves the robot around the board to each component. There is no need to define any datum or zero point.

"Toolchanging also is automatic. A Quick Change Adaptor system from EOA Systems, Incorporated (Dallas) facilitates changeover between the adhesive dispensing and solder mask tools. When not in use, the tool is stored in an end effector nest near the robot.

"This versatile dispensing system was justified primarily on the basis of improved PC board quality. Experience in the end user's plant has confirmed the sounded of the decision."

In the same *Robotics Today* issue, Robert N. Stauffer also reported low sensor control for robots. The article, *Sensor Controls Robot Speed* reported: "A technique for sensing forces created by the unsynchronized velocities of a rotary seam welder and a robot provides adaptive control of the process for improved weld quality. Installed in a Midwest automotive plant, the seam welder/robot combination is one of four in a line for assembling fuel tanks. The robotic seam welding systems were developed and built by Kuka Welding Systems and Robot Corporation (Sterling Heights, MI), working in cooperation with the customer's process engineers.

"The installation includes six Kuka IR 662/100 Series robots—four that are used to move fuel tank components through the seam welders and two for loading and unloading operations. The complete system, which also includes washing, soldering, and leak testing stations, produces about 200 fuel tanks per hour. The welders are supplied by Resistance Welding Corporation (Bay City, MI).

"In a typical sequence, the top and bottom halves of the fuel tank are loaded into the system and routed through a washer in preparation for assembly. A robot located at a turnover station picks up the bottom half of a tank as it leaves the washer and deposits it upside down on an overhead monorail carrier. The carrier then moves to another station where a plastic insert is placed in the part. From there, the bottom half is moved to a mating station. The top half of the tank is delivered to this station by a shuttle, and the two halves are clinched together in preparation for welding. The mated halves are then moved via the monorail carrier to one of the welders.

"Using a specially designed end effector, the robot grips the tank securely on the bottom and moves it through the welder while holding the flange in a horizontal plane. Each welder uses an AC transformer with wheel-type electrodes driven by a constant-speed peripheral drive. An automatic wheel-height adjustment compensates for wheel wear.

"When the fuel tank contacts the welding wheels, a torsional strain gage located under the end effector senses the forces created by the unsynchronized velocities of the welder and robot. An output signal from the strain gage is directed to the robot controller, causing the robot to instantly speed up or slow down as required to equalize the welder and robot speeds.

"The efficiency of this adaptive feedback system allows for speed variations of up to 15% of the average weld velocity, which is approximately 215 ipm (5461 mm/min). An estimated 15 lb (67 N) of pushing or pulling force can cause weld problems. Forces of this magnitude can be generated easily with just a 2% change in speed on either side, making control of the velocity differential all the more critical.

After the tank has been welded, the carrier transports the assembled product to another Kuka handling robot equipped with a dual end effector. The robot picks a fuel tank from the carrier and swings around to an indexing table at a station where the fuel filler and vent tubes are automatically fed into openings in the tank and soldered. During this move, the robot wrist rotates to bring the empty side of the end effector into position and ready to pick a soldered tank from the index table. The part that is lifted from the index table is then taken back and deposited on the monorail carrier.

"One of the key features of the system is the software/hardware package that accepts signals from the torsional strain gage built into the end effector. This enables the robot to respond quickly and accurately to instantaneous velocity changes while carrying a 200 lb (90 kg) end effector and the part itself.

"The robots are programmed off line, using a combined drawing/calculation procedure that is then converted for direct alphanumerical input into the robot controller. Alternately, a computer program can be used to input data on tank size. The software then generates the proper world coordinates to define the precise robot path.

"The system has proved successful in improving weld quality by reducing extraneous forces at the weld. Fuel tank assembly productivity is higher and equipment cost lower than other systems that were considered. The system is expected to have a two-year payback."

In the same publication (February 1987 issue) Stauffer authored *Force Sensing Keeps Robot on Track in Sanding Operation*. This article reported: Industrial robots are now new on the scene at Cascade Engineering (Grand Rapids, MI), a progressive manufacturer of engineered plastic products primarily for the automotive and furniture industries. The company uses robots for tending a number of its injection molding machines and also has one in a foam dispensing application. its newest installation is a cut above the others, however, in terms of work cell integration and overall sophistication. It also reflects the company's strong commitment to on-time delivery of quality products. Properly engineered and implemented robotic systems help to support such efforts.

"Initially, flash was removed by hand, using a scraper and a file. The file was required to work on the gate area of each molded part. Hand scraping and filling presented two problems, however. One was inconsistent finishes; the other was operator fatigue. Studies showed the manual operation could be improved with an orbital sander, but this tool could not be used to work in two narrow slots in the chair. Hand scraping would still be required in those areas. The idea of robotic system with special tooling to do the complete chair—the periphery and slots—looked increasingly attractive.

"The sanding tool is a two-sided, dual-action sanding pad supplied by National-Detroit, Inc. (Rockford IL). It operates with a reciprocating motion under heavy loads and an orbital motion under lighter loads. The two-sided feature contributes significantly to reduced cycle time, because the sanding pad can do both sides of each slot without having to be withdrawn and reoriented.

"Chair delivered to the system on an indexing conveyor are picked up by the operator and placed on a shuttle transfer for transport into the workstation. with a chair clamped in an inverted position, the Allen-Bradley PLC-2 programmable controller signals both robots to start their programs. By the time the trim (sanding) robot has completed its work on the lower portion of the chair, the Charley robot has finished driving the four inserts and has moved out of the way.

"Force sensing not only provides consistent loading during operation, but makes it possible to change the loading at numerous points around the chair. This feature is used when programming the robot to accommodate variations in slash. Importantly, the sensing system permits fine-tuning of the flash removal process. for example, the operator can make small changes in speed, loading, or the angle of the sanding head to continue to improve the product.

"During programming of the sanding robot, part of the software package allows "flags" to be set at different points around the chair. These are used to indicate preprogrammed escape routes for the robot to follow when the system is shut down, whether intentionally in normal system operation or for an emergency or inadvertent situation. This involves a simple set of robot moves to extract the sanding tool from the chair, get out of the jig, and return to home position. Between 20 and 30 escape routes have been taught. The software quickly determines and updates which of these points would be used for the next escape route. A large amount of data can be tied to various points of the chair without exceeding the memory capability."

Sequence Controllers. Sequence controllers are a class of electromechanical and electronic devices used to control the operation of a machine tool or other equipment in a predetermined step-by-step manner. Characteristic of these devices is the method of establishing the desired control sequence and the manner in which the controller functions. The more common types of sequence controllers available today are electromechanical stepping-drum programmers, perforated wide-paper-tape programmers, and diode-matrix pinboard programmers.

In the drum programmer, the desired control sequence is commonly established by inserting pins into appropriate rows in the surface of a cylinder. Mounted over one row of the cylinder surface are momentary contact switches so that, as each row moves into position under the switches, the pins in that row activate the switches corresponding to the position of the pins present. As the cylinder rotates or steps to the next row, the pins in that row cause the connection of certain input devices, such as pushbuttons, limit switches, and timer contacts, to the logic section of the controller. The logic section, as a result of the inputs, causes the closure of circuits to output devices such as solenoids and motor starters. when the logic section senses that selected inputs in that row are in the proper condition, the controller then advances, or "steps," the cylinder by rotating it to the next row. The pins in the next row then present the next set of input conditions to the controller and cause the closure of the corresponding desired output circuits.

In a perforated wide-paper-tape programmer, the desired control sequence is established by the pattern of holes which are punched into the tape. The operation of this type of device is similar to the operation of the old player piano. In the diode-matrix pinboard programmer, the desired control sequence is established by inserting small plastic pins (each containing a diode) into a plugboard. Alteration of the desired control sequence is accomplished by changing the positions of the plastic plugs, the pattern of holes, or the position of the diode pins. All types of sequence controllers are typically used for applications having a fixed sequence of operation for a large number of repetitions.

Sequence Number. A sequence number is a multidigit "n" number identifying the block or group of blocks on the NC tape. This sequence number is displayed on the operator's console.

Also see: Numerical Control.

Sequencer. A sequencer is a controller that operates an application through a fixed sequence of events or states. In contrast, a programmable controller (PLC) functions according to varying I/O patterns.

Sequential-access Memory. Sequential-access memory is any type of memory which can physically or logically be read from or written to only by starting at the beginning address location and sequentially proceeding forwards. If the situation is a logical restriction only, then the memory or file must be started at the beginning. Typically, this is due to the lack of headers and identifiers which allow random-access of the data. If the situation is a physical restriction, such as magnetic tape, the reason also is due to a lack of a routing directory which can reliably position the read head at a desired spot. Instead, the tape is read from the beginning, discarding all unwanted information, until the desired data is detected. Once the desired data is completely obtained, the remainder of the tape doesn't have to be read. Other types of sequential-access memory are paper tape (also mylar tape), bar codes, and magnetic cards. Sequential-access memory isn't practical or desirable for large memory applications since the inefficiencies of starting at the beginning and sequentially searching for information are unacceptable.

Also see: Random-access Memory.

Serial Communication. Serial communication is the transmission at successive time intervals of the individual signal elements that constitute the same data or telegraph signal. The characters are transmitted in a time sequence over a single line. The serial signal elements may be transmitted with or without interruption, provided that they are not transmitted simultaneously. Thus the transmission of a group of bits that constitute a single character, or other entity of data, over a data circuit, one element at a time, constitutes serial transmission.

Serial communication used by computers is a standard by the RS232-C communications protocol. In serial communication, there is a single transmission line instead of multiple lines as in parallel communications. Serial communications also requires more flow control than parallel. Since the data is in a parallel format within a computer, the data with be structured into a continuous serial form. Most memory devices, however, store data in a serial form.

Also see: Computers, Parallel Communications, RS232-C.

Service Bureau. A service bureau is a company which offers software support service to owners of NC equipment. This support can range from simply supplying program tapes to a complete counseling and computer part programming service.

Servo. A servo is a system used to convert electrical signals into mechanical motion.

Servo Amplifier. A servo amplifier is the part of a servo system that increases the magnitude of the control signal into the servo and provides the output power to drive the actuator, machine slides, or servo valve controlling a hydraulic drive in an NC system.

Servocontrol. A servocontrol system is a method of controlling a powered, moving device either rotary in motion or translating, in which the ability exists to vary the speed and reverse directions at will when desired, subject to some basic physical restraints. Some of the restrictions are simply inertial properties which sometimes require a large mass motor to be slowed to stop using mechanical brakes before the driving fields can be reversed, to avoid damage to the motor. The servosystem has no problem with instantaneous reversal, but the windings of the motor might suffer from overheating.

Servocontrol systems are required for adaptive controls and while stepper motors or limit switches can be used, they do not offer the sophistication and resolution of a servocontrol system. Limit switches typically only provide two end positions of motion and require physical adjustment. Stepper controls cannot provide the same level of resolution as a servosystem. An analogy is that a stepper control system is a digital system (variable in discrete uniform steps) whereas a servo system is an analog system (continuously variable).

In a servocontrol system, lag and lead are two important parameters of sophisticated control applications, which can be adjusted to compensate for dynamic overshoot and undershoot.

Also see: Lag, Lead, Servomotor.

Servomechanism. A power device for effecting machine motion. It embodies a closed-loop system in which the controlled variable is mechanical position and velocity.

A servomechanism (often termed a "servo") is a group of elements which convert the NC input into precision mechanical displacements. These elements include motors (hydraulic or electric), gear trains, and transducers (velocity or position).

The drive to spindles and slides in NC tools is usually provided by either hydraulic or electric motors.

Servomechanisms may be either open or closed loop as shown in the figure below.

In the case of the open-loop servo, there is no feedback signal to assure that the machine axis actually moved the distance programmed. For instance, if the servo is designed to move 0.0001 inch (0.003mm) for each input pulse, and 100 pulses are programmed, the servo will move a table 0.010 inch (0.25mm). The only assurance that the table actually moved 0.010 inch in this type of system is the reliability of the system.

The closed-loop servo, on the other hand, compares information feedback from the machine slide with programmed information to assure that the motion has actually been performed. The signal to the drive motor is modified by the feedback signal.

The basic or main elements of a numerical control system are shown in the figure on the next page. The principles here essentially are the same for positioning and contouring, although the principles of contouring are somewhat more complex.

From the machine control unit, the signals proceed via the *command-signal circuit* to the servomechanism and drive unit. Each machine slide or movement that is to be controlled by the system has its own servomechanism and drive.

A servomechanism amplifies the incoming signal and provides power to move the desired machine slide or carry out a mechanical movement as required. Commonly, it can be electric or hydraulic. The servomechanism may be an electric motor which drives a machine table through a leadscrew; or the system may involve hydraulic motors, hydraulic rams, or other devices for moving the controlled machine elements or slides as required. Motors may drive

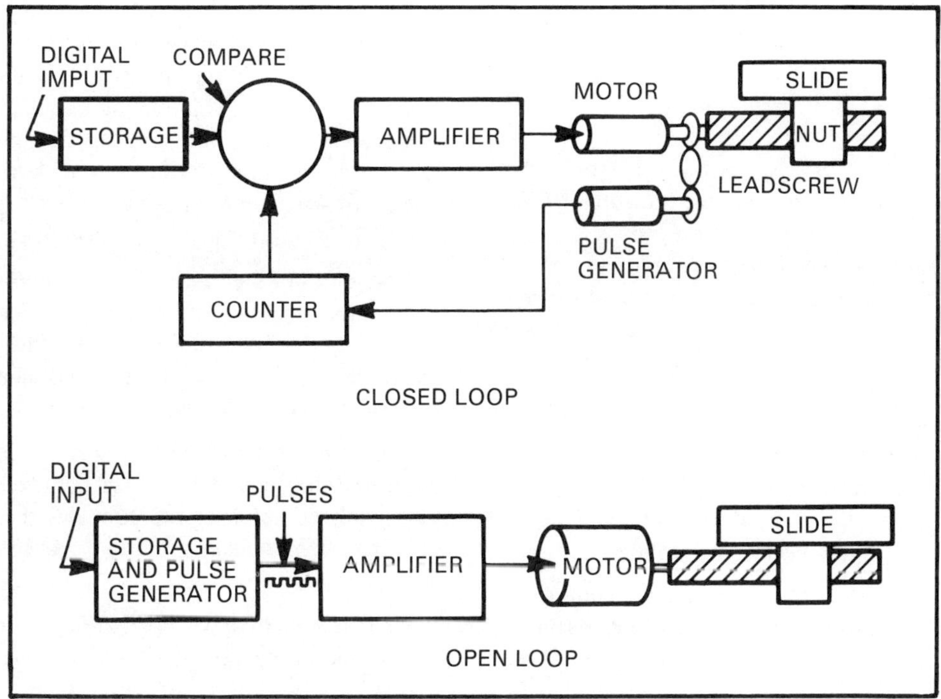

Open and closed-loop servomechanisms.

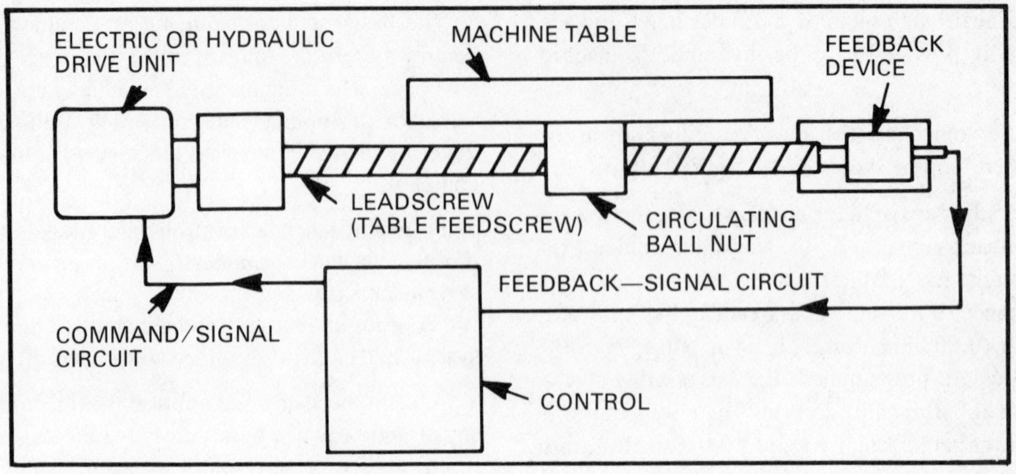

Mechanical elements of a computer numerical control system.

the slides through low-friction leadscrews employing circulating ball nuts or through rack-and-pinion arrangements, or still other devices can be employed.

The controlled machine slide is any controlled part of the machine, or the controlled machine itself. It can be an addition to a machine tool, such as a retrofit positioning table or a conversion, but the controlled unit must be designed for numerical control in any case.

In the electric or hydraulic drive unit, the rotary hydraulic motor uses the pressure of fluid flowing through gears or against pistons to effect a shaft rotation. These motors are usually of the positive-displacement variety. The motor itself is very small for the horsepower it is capable of developing. Oil flow to the motor is controlled by a servovalve. A servovalve is typically an electrohydraulic unit using an electrical solenoid to actuate a small flapper which controls hydraulic pressure on a spool valve. The spool valve directs oil to the hydraulic motor, which produces the desired mechanical motion.

Both ac and dc elecric motors are commonly used for powering NC tools. Recently, printed-circuit motors have also become popular. These motors are high-torque devices which can be directly coupled to a leadscrew to drive a machine-tool slide. A stepping motor is a particular type of electric motor which is actuated by pulses and moves a fixed angular unit for each electrical pulse. The motor is usually clamped magnetically at fixed angular positions.

Precision slides are usually powered by a motor which drives the slide through a gearbox and *leadscrew*. The leadscrew is almost universally used to convert rotary motion from electric or hydraulic motors into linear movement of a machine slide. The recirculating-ball leadscrew is a precision screw with very low friction. The leadscrew is precisely made to provide a given linear displacement to the slide for each revolution. This distance can be very accurate.

Two kinds of transducers are commonly used in machine tools for *feedback devices*. Velocity transducers are used to measure spindle speed and slide velocity, and position transducers are used to measure slide displacement.

The most common velocity transducer is an electric tachometer. Electric tachometers provide a voltage which is proportional to the speed of a shaft. The voltage produced may be either ac or dc.

There are many types of position transducers in use. These include synchros and resolvers, digitizers, and linear induction types. Since measurement of position is one of the key elements of NC, this area has received a great deal of attention.

Synchros and resolvers are electric transformers which provide a voltage output that varies with the angular position of a shaft. This voltage is used as a measure of displacement. Coupled to a leadscrew through a gear train, the resolver or synchro provides position information on a slide.

A linear resolver is a precision-made scale which utilizes the resolver principle of induction. Signals are induced on the linear scale by an adacent exciter scale.

A rotary digitizer is a device which provides a pulse for a given angular displacement at its shaft. For example, a digitizer might generate a pulse for each degree of angular rotation of its shaft. These pulses can be counted and the relationship of the linear displacement to the digitizer angle can be used to determine the slide displacement.

The machine motion, as provided through the servomechanism, is recorded or monitored by a feedback (measuring) unit, *feedback signal circuit* which may be electrical, mechanical, or optical; corrections are automatically made by the machine control unit. Systems with feedback are generally classed as closed-loop types, whereas open-loop systems do not incorporate feedback.

When a machine slide and servo are coupled through a leadscrew; the leadscrew, the ways of the machine, and the accuracy of the servo (including the accuracy of the measuring system) all contribute to the machine error. A closed-loop servo can achieve a high degree of precision when a feedback element such as a linear system is attached to the machine slide. Such a system eliminates gearing, leadscrew, and backlash errors. In many cases, however, the drive and feedback units are attached to the end of a leadscrew, in which case the open-loop and closed loop systems may have equal accuracy at the machine slide. The accuracy of the system must be designed for the jobs that the machine performs. Very high accuracy can be achieved—limited primarily by the ability to measure. The cost of the machine, however, is related to the machine accuracy and increases as the accuracy is improved.

Specialization in planning the details of an operation, in setting up the cutting and workholding tools, and in operating the NC machine tool all contribute to greater operator safety. Under direction of the tape or other program storage media, the machine establishes the cutter-workpiece relationship. The operator is not required to interfere with the operation to adjust measuring devices in hazardous locations. These measuring devices are generally located near the cutter or workpiece on conventional machines for greatest accuracy, but at the same time in the most hazardous location for operator safety.

Transfer of the detailed operation-planning function from the shop environment to an office environment has direct influence on operator safety, since it removes the need for concentration on work for which the operator has no special training, in an environment that is highly distracting. When this detailed planning function is eliminated from the machine operator's scope of responsibility, full attention can be concentrated on doing the job the operator is most familiar with—auditing the machine operation.

Also see: AC Servomotor, DC Servomotor, Industrial Robots, Numerical Control, Servocontrol, Stepper Motor.

Servomotor. A servomotor is a motor, usually electrically powered, which is controlled by a servocontrol system. A servocontrolled electric motor can be either AC or DC type. A motor without any speed control system simply runs at the maximum (slew) speed which is allowed by its power source, while being constrained by its load. A servomotor can be operated at varying speeds or revolutions per minute (rpm) while maintaining high available torque. A problem of certain control systems is to maximize torque at all speeds, especially the lower speeds. For AC type motors, servocontrol systems can vary the frequency of the alternating current, while keeping a constant voltage. With the availability of microprocessor technology to control motors, sophisticated controls based on reference time base oscillators are easily incorporated into the servocontrols.

Also see: AC Servomotor, DC Servomotor, Servocontrol, Stepper Motor.

Setpoint. Setpoint is the final or target value of a controlled variable usually preset in the computer by the operator. Setpoint also is the starting point for a program on an NC machine, as established by the operator.

Settling Time. In robotics, settling time is the amount of time it takes to reach a steady-state position of the end effector once it has reached a programmed step or the time lag required for oscillation to dampen.

Also see: End Effector, Industrial Robots.

Shaping. Shaping is the process of using a machine tool (a shaper) primarily for the production of flat surfaces in horizontal, vertical, or angular planes. It can also include the machining of curved surfaces, helixes, serrations, and special work involving odd and irregular shapes.

The metalworking shaper, like the planer, develops cutting action from straight-line reciprocating motion between the tool and the work. On the shaper, the tool is reciprocated and the work is fed sidewise. The horizontal shaper with the ram movement in the horizontal plane is the most common form of shaper. Vertical shapers, with ram movements in a vertical plane, also come in the metalworking shaper group. Vertical shapers are also known as slotters.

Types of shaper work often may be done on a milling machine or a planer. Shaping however is particularly appropriate for the handling of toolroom work, die-shop work, and small-lot manufacturing, due to the great versatility and speed of setup associated with a shaper. Planing, contouring, internal shaping and high-speed shaping are accomplished rapidly and economically with little need for special fixtures and with simple cutting tools. The metalworking shaper's unusual flexibility and its power to perform work which might be difficult or impossible on other machines make it a basic machine tool. The shaper is particularly advantageous in machining tough die steels of great hardness and alloyed high-strength cast iron.

The most common shaper is the horizontal crank-operated push-cut type (see figure on the next page) with ram movement in the horizontal plane. These shapers range in maximum cutting stroke from seven or eight inches (178 or 203 mm) in bench models to 36 inches (914 mm) in heavy-duty models. They are built with either mechanical or hydraulic drive and with a plain box table or universal table permitting angular tilt in addition to horizontal and vertical adjustments.

A cutting tool is mounted on a shaper head that is attached to the ram, which reciprocates the tool. Workpieces are held in a vise on the shaper table or directly on the table. Power or hand feeding of the table is provided parallel to the table top and perpendicular to the stroke of the ram. The tool cuts on the forward stroke of the ram, and the table feeds the workpiece in a direction perpendicular to the ram motion, for the next cut, on the return stroke of the ram. Many shapers have a rapid traverse for moving the table to various positions along a crossrail fitted to the front of the machine column.

Vertical shapers operate somewhat like horizontal shapers except that the ram reciprocates vertically rather than horizontally. Most machines of this type have provisions for adjustable inclination of their rams, and rotary tables are practically standard equipment.

Shaving. Shaving involves cutting off metal in a chip fashion to obtain accurate dimensions and also to remove the rough fractured edge of the sheet metal. Shaving is performed using dies with a very small clearance, as shown in the figure on the next page. It is considered to be a secondary shearing operation.

Shear Forming. Shear forming, also known as flow turning, is similar to spinning, but during flow turning the metal is intentionally thinned by shear forces. This is also known as power spinning. It is sometimes done at hotworking

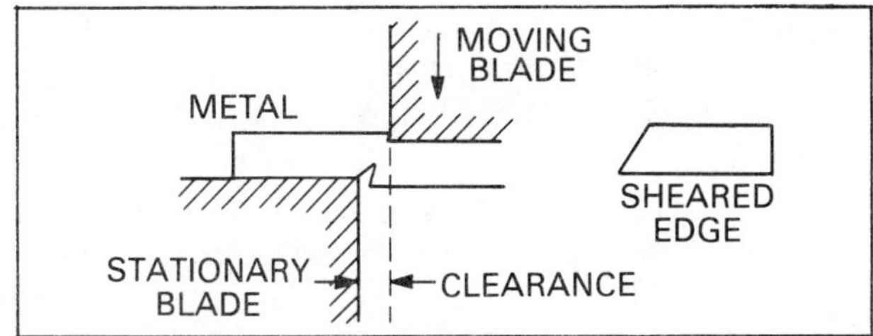

Shearing is the cutting action along a straight line to separate metal by two moving blades.

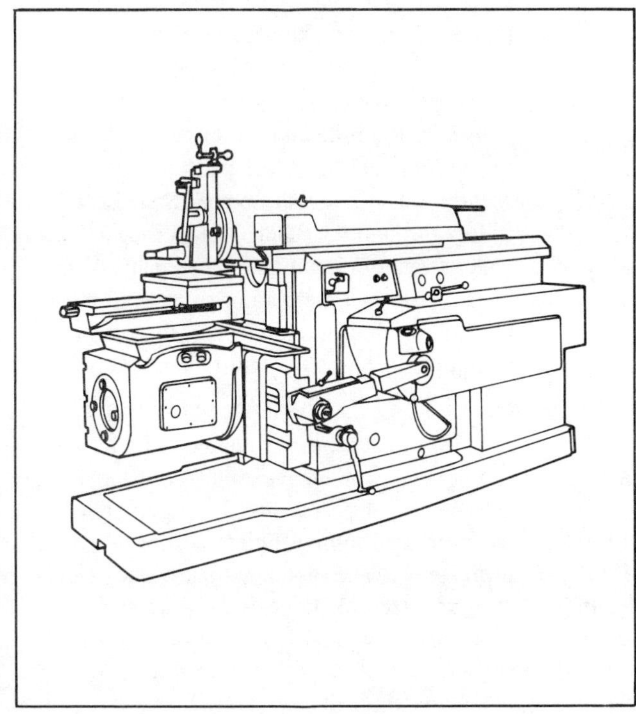

Typical horizontal crank-operated punch-cut shaper.

condition because the metal undergoes sever shear deformation which demands high ductility. This process is done in a lathe.

Shearing. Shearing is a process by which large sheets of material are cut into pieces of smaller length and width. These pieces are often used in subsequent operations such as punching and forming. Shearing also produces blanks or slugs used in subsequent forming and machining processes. Because shearing is often the initial step in a series of processes, it is essential that the operating procedures result in an accurate workpiece.

Computer numerical control (CNC) technology applied to shearing is generally limited to a single-axis positioning table attached to a conventional hydraulic or mechanically driven shear. A sheet to be sheared is placed in the workholders of a direct-current servo driven carriage and positioned in the shear for each cut. The CNC unit can be programmed so that a variety of lengths can be sheared from a single sheet, as well as a variety of initial trimming cuts.

All the advantages of CNC machining apply to CNC shearing, including accuracy, repeatability, increased production, hands-off safe operation, and predictable production rates. When prepunched parts are sheared, errors are not cumulative, as could be the case when backgaging is used with manually operated shears. Accuracy and repeatability are generally good.

Occasionally, two single-axis shears are operated in tandem or at 90 degree angles, enabling the second shear to cut the material into smaller parts as it comes from the first shear.

Sheet Metal. Sheet metal can be defined as any material or piece of uniform thickness and of considerable length and breadth as compared to its thickness is called a sheet or plate. In reference to metal, such pieces under 1/4 in. thick are called sheets and those 1/4 in. thick and over are called plates. Occasionally, the limiting thickness for steel to be designated as ''sheet steel'' is number ten Manufacturer's Standard Gage for sheet steel, which is 0.1345 in. (3.42 mm) thick.

In his SME Technical Paper titled *Sheet Metal FMS*, authors James N. Brecker, Vincent P. Valeri and Dwayne F. Rife wrote: ''Most sheet metal manufacturing operations are job shops that require sizeable quantities of each part to obtain some degree of efficiency. The usual sheet metal part cycle time is thus 1-2 weeks. Lot size is frequently dictated by the number of pieces that can be cut (figure on this page) from a prime sheet, typically a 48 by 96 inch or 60 by 120 inch sheet. The parts are then punched individually on an NC turret punch press.

''Although shear and punch process times are on the order of a minute each, two or three days or more may be required to obtain a sheared/punched part because of move and queue times. It may take another week for a sheet metal part to be bent, welded, and/or painted. A rush part can be 'walked' through the plant in a day with considerable disruption due to breaking setups on the punch and press brakes and any other special operations.

''The strong emphasis on minimizing the labor applied to a given part has led to large inventory and long cycle times since parts are processed in large batches to obtain lower unit time. There is also typically 35% scrap metal due to part size—prime sheet size mismatches. It is usually too time consuming and or too costly to optimize the layout of parts on sheets to be sheared. In addition, custom parts are difficult to follow through the plant. This is especially true when a random input automated paint system is used. There is a plethora or paperwork attached to parts as they move through the plants from operation to operation. Still, it is likely that a custom part will be misplaced or mislabeled.

''There have been a number of advances in recent years which have improved the efficiency of sheet metal operations. The right angle shear has permitted more efficient nesting of parts on prime sheet. The 'notching' shear allows different part sizes to be nested on the same prime sheet with a typical reduction in scrap to 25%. Because the 'shear' is NC in contrast to the usual manual straight shears, prime sheets can also be processed in significantly less time. In the press shop, NC press brakes and panel benders are enhancing the productivity in bending by reducing setup effort and decreasing process time.

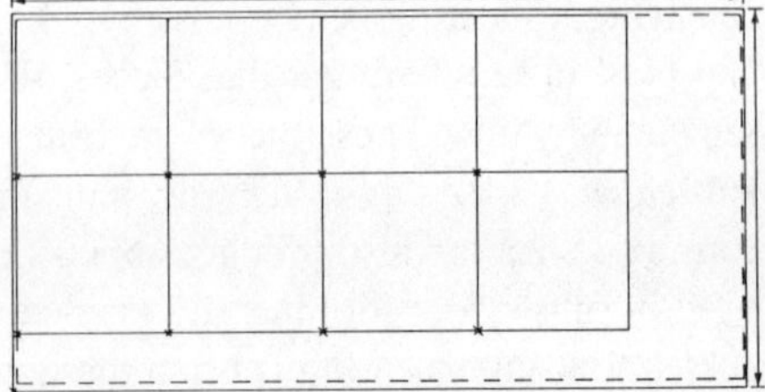

Manually nested blanks for a straight shear.

"Several sheet metal systems have been implemented in Japan in which conveyors have been used to move sheets from shear to punch. In addition, the concept of punching several identical parts as a whole and then shearing them was extended to punching several different parts as a whole then separating them on a right angle shear (see figure). Since considerable NC programming was required, only moderate volume, standard products are processed this way. There has been much standardization of materials, tools, and tolerances to enable these systems to be utilized effectively. A similar system with the additional capability of nesting of random parts has been installed in Canada.

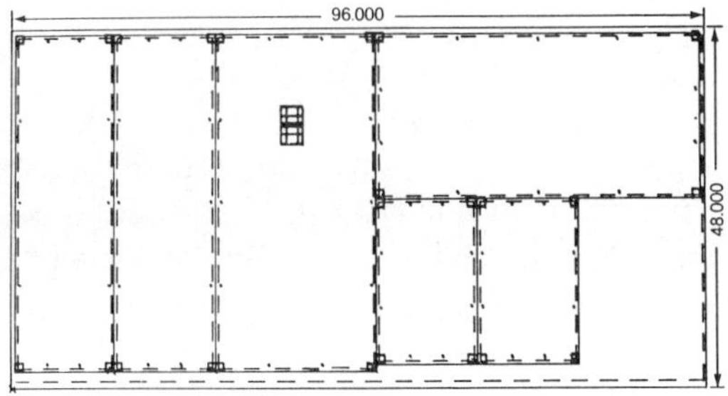

Nested parts for a single assembly.

"An extensive evaluation of existing FMS hardware and software led to the conclusion that greater flexibility was required for efficient sheet metal shop operation than was being offered by equipment vendors. There were definite needs to handle random parts in quantities of one and to mark these parts for traceability in the shop. A higher quality DNC system was needed as well as software to reduce the manpower needed to operate the system. Consequently, an in-house effort was begun to develop a sheet metal FMS with these hardware and software capabilities."

Shielded Metal Arc Welding (SMAW). Shielded metal arc welding is an arc welding process that uses a rod of filler metal coated with fluxing agents as the welding electrode. The arc heat causes chemicals in the coating to vaporize or decompose to provide arc shielding.

Also see: Welding.

Shielding. Shielding is the confining of the electrical field around a conductor to the primary insulation of the cable by putting a conducting layer over and/or under the cable insulation.

Shift Register. A shift register is a register in which characters may be moved serially one or more positions to the right or left, resulting in the loss of characters on the end toward which characters are moved.

Short-circuiting Transfer. A short-circuiting transfer is an arc welding (metal transfer) in which molten metal from a consumable electrode is deposited during repeated short circuits.

Shrink Flanging. During shrink flanging, the line of bending is a convex curve and the metal is under severe compressive stress, so wrinkles are likely to occur. Therefore, the flange width must be limited. An alternative is the design of preplanned offsets in the flange to take up excess metal.

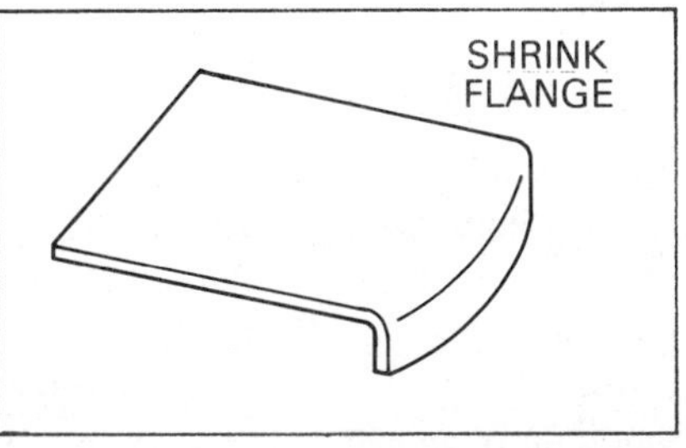

In shrink flanging, the line of bending is a convex curve.

Significant Digit. A significant digit contributes to the precision of a number. The number of significant digits is counted beginning with the digit having the most value, called the most significant digit, and ending with the one having the least value, called the least significant digit.

Silicon Controlled Rectifier (SCR). A silicon controlled rectifier is an electronic device generally used in control systems for high-power loads. It is an electrical value that can be turned on by a signal and will turn off when the power is removed or reverses direction.

Silicone. Silicone is one of a family of polymeric materials in which the recurring chemical group contains silicon and oxygen atoms as links in the main chain. Silicons are derived from silica (sand) and methyl chloride.

Simulation. Simulation is an experimental technique that uses mathematical models of real systems to develop inferences about these real systems. Simulation models can be used for design, procedural analysis, and performance assessment. Generally, simulation models are built for digital computers. However, analog computers, physical models and other representations of real systems are sometimes employed.

A wide variety of real systems are studied using simulation. These include air traffic control, electronic circuit design, warehouse location, manufacturing process design, financial forecasting, police response system design and political voting prediction. A number of computer programming languages have been designed specifically for simulation. Some of the better known simulation languages are SIMSCRIPT, GPSS, GASP, SLAM, Q-GERT, DYNAMO, and CSSL.

History. Military planning and the development of military systems provided the major impetus for the initial growth of simulation. Aircraft and submarine simulators were first developed in the 1940s, and Monte Carlo analysis came into prominence during the late 1940s in connection with nuclear shielding problems. Game theory and operations research also grew in importance during that period and in the 1950s.

In the late 1950s and early 1960s, general purpose computational languages such as ALGOL and FORTRAN were probably the most widely used simulation languages. During this period, numerous special purpose simulation languages were developed to deal with specific areas.

As the benefits of digital computer simulation became more widely recognized, the need for better languages increased. The existing languages were too specialized, and they were not easily learned. At the same time, it had become evident that many techniques and logical elements were common to a wide variety of simulation problems. The early 1960s saw the advent of several new general purpose simulation languages. Most were adaptations of FORTRAN (e.g., SIMSCRIPT, DYNAMO, GASP) or ALGOL (e.g., SIMULA). Some were entirely new. For example, GPS soon evolved into GPSS.

The main advantage of these general purpose simulation languages over general purpose programming languages is that they incorporate means for controlling the sequence in which events occur. This, and other aspects of simulation models, introduce many complexities when an ordinary programming language is being used.

During the 1970s and into the 1980s more powerful simulation languages were developed, improving the ability to employ various world views for simulation modeling. These include the three discrete modeling world views: ''event oriented,'' ''activity scanning,'' and ''process oriented;'' plus the continuous model. The culmination of these efforts is represented in the SLAM language which provides a unified modeling framework that allows mixing any, and all, of these views in the same model.

In the 1980s, simulation systems have emerged that incorporate graphical output approximating the activity of the real system. This helps visualize the effects of various experimental conditions and is an aid in communicating the results of a simulation to nontechnical managers who might make decisions based on results of the simulation.

Discussion. The simulation process begins with a simple model which is embellished in an evolutionary fashion to meet problem solving requirements. Within this process the following stages of development can be identified.

Step one is *problem formulation*, which is the definition of the problem to be studied, including a statement of the problem solving objective.

Step two is *model building*, where the abstraction of the system into mathematical-logical relationships in accordance with the problem formulation is done.

Data acquisition, the third step, is the identification, specification, and collection of data.

Step four, *modes translation*, is the preparation of the model for computer processing.

Then step five, *verification*, is the process of verifying that the computer program processes as intended.

Validation, step six, is the process of establishing that the desired accuracy or correspondence exists between the simulation and the real system.

Step seven involves *strategic and tactical planning*. This is the process of establishing the experimental conditions for using the model.

Experimentation is step eight. Experimentation involves the execution of the simulation model to obtain output values.

Step nine is *analysis of results*—the process of analyzing the simulation outputs to draw inferences and make recommendations for problem resolution.

Lastly, step ten is *implementation and documentation*, the process of implementing decisions resulting from the simulation and documenting the model and its use.

Systems to be simulated are generally composed of one or more elements that have uncertainty associated with them. Such systems evolve through time in a manner that is not completely predictable and are referred to as stochastic systems. The simulation of stochastic systems requires that the variability of the elements in the system be characterized using probability concepts. The outputs from a simulation model are also probabilistic, and therefore statistical interpretations about them are usually required.

Applications. One example of simulation was presented by Nathan A. Yoffa of Deneb Robotics in his SME Technical Paper *Workcell Simulation Case Study: A Spot Welding Application*. In it Yoffa wrote: "Six degree of freedom GMF robots (S-480R) programmed for a spot welding application work simultaneously in closely confined areas. When programmed entirely on-line, approximately 300 man hours are required to have the robots working to meet required cycle times. Using the Interactive Graphics Robot Instruction Program (GRIP), the same application was simulated, and actual programs generated from the simulation were downloaded to all six robots in only 60 man hours. A dynamic computer graphics system such as IGRIP pertaining to Offline Robot Programming and Workcell Layout Design are beneficial. IGRIP allows the user to use data from the actual spot welding workcell to calibrate the simulation system for workcell setup error. This calibration allowed the user to download data to the robot controller taking into account potential errors in workcell setup."

The body framing station includes six robots each with a weldgun mounted to a clutch which is attached to the robots' mounting plate. The robots are set up two on either side of the car body, one above the front motor compartment, and one above the rear trunk section. The front and rear robots are actually hanging from the structure which also supports all of the fixtures used to lock the car body in place for the welding operations. During the body framing spot welding process, the car body, sitting on an automated guided vehicle (AGV), moves into the body framing structure. The fixtures lock the body into place, the robots run through their programs, the fixtures release, and the AGV moves out of the structure.

In relation to automated guided vehicles, author Richard Miller in his book *Automated Guided Vehicles and Automated Manufacturing* discussed simulation.

Miller wrote: "A simulation is a model of a system. This model represents the importance of the performance of a system in a way which permits it to be easily manipulated. The main value of simulations is that they can be conveniently manipulated until the desired results are achieved."

"It is possible to perform simulations by hand. Most of the simulations that are interesting, however, are of the size that manipulating the markers and keeping "score" becomes cumbersome. Thus, computers are used for virtually all manufacturing simulations. This chapter describes the components of the computer simulation system and how they are used."

"With simulation, a model of the system under study is constructed using a simulation language. This language gives structure to the model building process by providing special modeling constructs that relate to the system under study.

"Many different types of simulation languages are available. They range from general-purpose languages, such as SLAM II (Pritsker & Associates), SIMSCRIPT II.5 (C.A.C.I.), GPSS (IBM's General Purpose System Simulation), and SIMAN (Systems Modeling Corporation) to more specialized, manufacturing-oriented languages such as MAP/I (Pritsker & Associates), SPEED (Horizon Software), and MAST (CMS Research)."

"For a simulation to be portable, meaning that it is able to run on different computer models, it must be written in a popular high-level language such as FORTRAN, or a simulation language that is written in FORTRAN."

"Some of the early simulation languages had the disadvantage of being difficult to learn. The emphasis on newer systems is to be user-friendly through interactive graphics. The AutoSimulations, Inc. system uses a natural language front-end so that the user can program in near-English."

"The General Electric Company is one of the biggest users of simulation for robotics and manufacturing in the United States. Based on the experience of performing simulation for plants around the country, General Electric has developed the following eleven steps that have proven to result in accurate models of flexible manufacturing systems:

1. Define the goal of the project so the simulation can be tailored to meet the goal.
2. Make assumptions on the equipment to be installed.
3. Build the software model of the actual facility and equipment.
4. Gather data on the various known parameters of the physical system being simulated.
5. Develop the program using the most appropriate computer language.
6. Verify the program by running it under simple assumptions for which the results are known or easily computed.
7. Validate the program using goodness-of-fit tests to check that the computer program's output resembles output data taken from a real system.
8. Test the model under varying conditions and assumptions usually at extremes, to check on the sensitivity at both ends of the model.
9. Analyze the output data, all event times and changes in the simulation project by means of statistical and graphical techniques.
10. Record the results of the simulation to aid decision making and the successful implementation of the system.
11. Train the user of the system so that the software used for simulation purposes can also be hooked up for actual scheduling systems used for factory production.

"A flexible manufacturing system (FMS) involves the integration of materials handling, storage, inspection, and maintenance with manufacturing operations to produce a variety of products. Because the systems are difficult to design, involving both hardware components and control, million-dollar mistakes are easy.

"Computer simulation provides an effective means to study all components within an FMS and their interactions with control algorithms. The statistical output from simulation can be used to measure overall capacity of the FMS. The use of animation in the form of computer graphics allows the system designer the opportunity to observe the operation of a proposed system to ensure design feasibility.

"Of all the analytical techniques that may be used to help design material handling systems, simulation is probably the most versatile and widely applicable. It enables the user to build a system on the computer—the same principle as a pilot plant—and to test the model by actually running it. Simulations can provide the answers to questions such as: What will happen if production volumes double next year? Will the plant still operate if the main conveyor goes down?

"Models for simulating material handling systems may be classified as either continuous or discrete. A continuous model contains closed form activity rate functions and these rates are ultimately used to calculate traffic through every resource entity in the system. A continuous model is much like water flowing through a pipe.

"A discrete model is as complicated as a continuous model because it includes traffic generation and routing definitions. In addition, it also contains rules for the matching of the system demands to system resources through time as those demands occur. Watching a discrete model is much like watching pieces move around a checkerboard. One reason for using a discrete-based approach is that it provides an analytical framework to capture the dynamic behavior of systems. It also designs a system more intelligently by making as-needed provisions for batch-like patterns in system resources (machines, space, etc.).

"One of the fundamental questions in the design of an automated material handling system is the determination of the number of automated guided vehicles that will be required. The first step in sizing the number of vehicles required in an AGVS is typically the development of a design horizon (e.g.,per hour) traffic requirement in the form of a from/to chart. The chart should show the number of loaded trips from every pickup station to every deposit station.

"The definition of any material handling process involves four primary elements:

- Scaled and dimensional drawings or layouts of the process.
- Brief narratives of data, material flow, control algorithms, and management strategies.
- Statistical descriptions of system parameters, such as desired inputs, outputs, component performance, and timing of events in the system.
- A list of specific questions and information desired to be provided by the simulation.

"This information should be developed prior to the development of a simulation model. In most cases, all information can be provided on a single drawing accompanied by appropriate notes and comments."

Simultaneous Engineering. Simultaneous engineering is a relatively new buzz term in manufacturing, which really boils down to a matter of improved, timely communications between product designers, manufacturing engineers, and quality engineers. It can be, and probably should be, extended even further upstream in the total manufacturing cycle to include suppliers of equipment that will do the actual processing. And in some cases, the concept is being enhanced by the participation of marketing and finance personnel. This cooperative approach provides the best assurance of resolving problems as the product, equipment, and process are designed and developed.

Simultaneous engineering gets down to the basics and gives individuals a key role in the scheme of things. It tends to discourage empire building and emphasizes the benefits of a cooperative relationship between disciplines. The result is an environment where individuals are given a greater opportunity to participate and have their ideas for improvement and solutions to problems given appropriate attention.

Closing the loop between all key personnel and departments is more essential than ever in the cost-effective manufacture of quality products for today's highly competitive marketplace. In addition to higher quality, simultaneous engineering can lead to increased product flexibility, reduced response time to customer demands, and the savings that result from doing things right the first time. Teamwork, stressing communication and commitment, is the key. For maximum effectiveness, this teamwork should extend beyond the company's own team to include members from outside supplier and support organizations.

Single-inline Package (SIP). A single-inline package is an electrical component package which has connection pins in a single row, (hence the single-inline reference). As with dual-inline packages, SIPs were designed as a way of using a standard size package which can be easily inserted (with through-hole devices) or onserted (with surface-mount devices). The adoption of standard sizes and pin counts for SIPs furthers these original goals and intents. SIPs typically are devices which have few leads such as resistor packs and generally do not need to have large numbers of devices and pins in the package. Some designs of memory boards use SIP style configuration for the memory packs which populate the board to allow greater memory density on the circuit board. While through-hole SIPs don't pose much of an assembly problem, surface-mounted SIPs are a dilemma as far as keeping them positioned upright while they are soldered, when the base area isn't sufficient to keep the component in the proper orientation. As a result, the "popcorn" shape of surface-mounted devices is used, where the profile and aspect ratio is much lower.

Also see: Dual-inline Package.

SIP. See: Single-inline Package.

Slew Rate. The slew rate when applied to the movement of a mechanical device or robot refers to the maximum velocity at which a manipulator joint can move. This rate is imposed by saturation somewhere in the servo loop controlling that joint (e.g., by a value reaching its maximum open setting). When describing a machine tool, the slew rate is the maximum speed at which the tool tip can move.

The phrase "slew across" means to move through some distance at the maximum capable speed, ignoring all other properties of controlled motion during that move. The object is to get from point A to point B as quickly as possible.

When applied to electrical signals, communication, and transmitted sine-wave signals, the slew is typically the change between two similar properties, such as the change in relative phase between a given clock signal and a phase-stable reference signal, which is known as clock phase slew.

Sliding Window. A sliding window data link control is one which allows multiple messages to be in the process of transmission at the same time, which are usually numbered so a specific message can be retransmitted if an error is detected. The sliding window concept has much great efficiency than the stop-and-wait alternating sequence of data link control.

Slitting. Slitting is cutting along single lines. A gang of circular blades cuts strips from a sheet. This operation is also used to cut along lines of given length of contour in a sheet or in products.

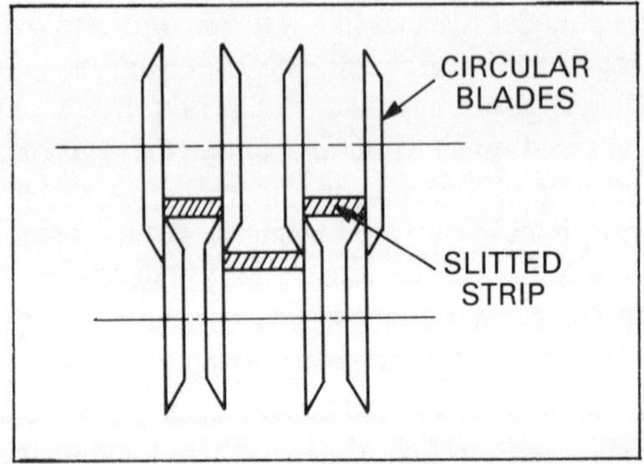

Slitting is cutting along single lines.

Slotting. Slotting is a punching operation in which elongated and rectangular holes are uct.

Small Computer Systems Interface (SCSI). The Small Computer Systems Interface is a complete bus structure which is subordinate to the rest of the system architecture. The SCSI is an outgrowth of the Shugart Associates System Interface (SASI) which was originally created in 1979 as a solution to the interface problems with eight inch Winchester disks. When the SASI was presented to ANSI, the American National

Standards Committee (ANSC) X3T9.2 was assigned the task of developing a refined document for its parent committee, X3T9. Since the use of a manufacturer's name in a standard (Shugart) was objectionable, the name Small Computer Systems Interface was adopted. The SCSI document is being adopted by the ISO committee TC-97 to develop the SCSI as an international standard.

The major disk drive interfaces include SA450, ST506, ST412, ESDI, SMD, SASI, SCSI, and IPI. The SA (Shugart Associates) 450 interface was developed by Shugart, as a continuation of their proprietary floppy and hard disk drive interfaces. The ST506 and ST412 was developed by Shugart Technology (now Seagate Technology) and has become a standard for low-performance Winchester disk drives. The ESDI (Enhanced Small Device Interface) is for 5 1/4 inch hard drives and is an outgrowth of the SA450 interface. The SMD (Storage Module Drive) interface was developed by Control Data Corporation, and is an enhancement of the ST506/412 interface. SMD is the only interface that is appropriate for disk drives from 5 1/4 thru 14 inches and is the de facto standard on high-performance Winchester disk drives. The SCSI controller is an intelligent controller which relieves the main system CPU of having to control the I/O to peripheral devices, as well as being capable of such tasks as servicing multiple hosts, servicing multiple devices, and full arbitration and message passing. The IPI (Intelligent Peripheral Interface) extends the data rates of the SMD type interface. RLL (run length limited) encoding allows more data to be stored or the media.

Small Scale Integration (SSI). Small scale integration is any integrated circuit having fewer than 12 equivalent gates.

SMAW. See: Shielded Metal Arc Welding.

SME. See: Society of Manufacturing Engineers.

Snap Gage. A snap gage is a type of fixed functional gage that has the gaging members arranged for measuring diameter, length, thickness or width. An external snap gage consists of a C-frame with an anvil at one side of the opening and gage contacts at the other. the outer contact is set to the Go dimension and the inner contact to the No Go dimension, both measured from the anvil.

Also see: Gaging.

Society of Manufacturing Engineers. The Society of Manufacturing Engineers (SME) is an international technical society dedicated to advancing scientific knowledge in the field of manufacturing. Found in 1932, SME has more than 77,000 members in 70 countries and sponsors more than 300 chapters and 186 student chapters and units worldwide.

Soft Automation. Sometimes called programmable automation, soft automation is a process automated by software programming instead of physically designing and building a fixed purpose machine in which the automation is hard coded by limit switches or a drum sequence controller or similar types of devices. While a software program may be put into EPROM or ROM to prevent inadvertent changes, the principle is still soft automation. Soft automation has the advantage of being able to be changed and modified more quickly than hard automation. Soft automation is typically easier and quicker to design and implement. With the availability of CAD systems and computer simulation, soft automation can sometimes be directly derived from the design information. The replication of soft automation is easier and the capital investment, especially in single-use, special-purpose machines is often substantially less than hard automation. Both soft automation and hard automation can be fixed automation or flexible automation, which describes the ability of the automation to do only one, or more than one, type of activity on a randomly defined schedule.

Also see: APAS, Flexible Manufacturing System, Hard Automation.

Soft Copy. Soft copy refers to copy which is not human readable without the aid of some device such as a microfiche reader if the soft copy is microfilm, or a computer monitor or printer if the soft copy is a computer data file. Soft copy can be manipulated and transmitted electronically, which is one of the major reasons for its use. Soft copy is usually quicker to generate and much less expensive to create and distribute. Soft copy doesn't require the large physical space which hard copy does. The ability to easily edit and modify soft copy is both an advantage and a disadvantage. The advantage is that reams of old versions of hard copy are eliminated. The disadvantage is that an accurate audit trail and record of the exact scenario of the sequence of changes and existence of copy is much more difficult to maintain. Furthermore, application of copyright laws and regulations are more difficult to interpret and apply to soft copy than to hard copy.

Also see: Hard Copy.

Soft Error. A soft error is when an error occurs in the operation of a device or computer which doesn't stop the operation of the device. Recovery from a soft error does not involve a system reset and starting the program from the beginning. Soft errors in devices such as printers could be running out of paper or having to replace a ribbon. The operation can be resumed as if no interruption had occurred. Soft errors can be parity check errors in data transmission, which do not reoccur upon retransmission of data. Soft errors which occur during data transmission can significantly impact the effective performance of the data communication, but may be entirely unknown to the user since the faults are automatically corrected by the system. While no errors are desirable, soft errors are preferably to hard errors since the effect is usually less severe.

Also see: Computers.

Software. Software is an often mysterious and misunderstood entity. Software cannot be touched or felt and software cannot perform any useful activity by itself. Software must be executed on some computer hardware and the useful function occurs when the software is used to carry out some application. Only when software is combined with hardware can the software be used to affect properties of matter and sources of energy. The term software describes something which has a"soft'' or nonspecific physical form. Webster's dictionary defines software as the entire set of programs, procedures, and related documentation associated with a system and especially a computer system.

Software programs are a nonarbitrary arrangement of electronic information or electronic computer data which can be stored on a computer and transferred between computers or executed as application programs. Software programs are computer information which provide some useful function. As opposed to hardware, which can be touched and felt because it has many physical properties, software cannot be touched or felt. Software doesn't really have physical properties, but an extreme case might describe software as the logical scheme of orientation of the memory devices which contain the program data.

Software engineering is the application of science and mathematics by which the potential and capabilities of computer based equipment can provide useful functions for humans by using computer programs, procedures, and associated documentation. The science of software engineering includes a software life-cycle which usually includes the following steps: Planning, Specification, Design, Coding, Integration, Test, Implementation, and Maintenance.

Software is classified as copyrighted works, the same as literary or art material. Software is considered to be the creation of a meaningful and useful product using the generic statements of a programming language. The value and usefulness comes from the arrangement of the program statements which when executed, perform a valuable and useful function.

Software is usually licensed, not sold, to the customer which means the buyer is purchasing a license of limited or unlimited duration to use the software in any manner he or she desires, provided the use isn't prohibited by the terms of the license. Sometimes making copies of the software for any reason, including archival backups, is strictly prohibited. When the right to use something is licensed, the ownership of the items, in this case the data on the distribution media, the media itself, and any accompanying documentation, remain with the seller. The seller can sometimes demand the return of all items if certain conditions arise or misuse occurs under the terms of the license agreement.

History. Software didn't exist until computers themselves were created. Early software was embedded inside computers or had to be manually entered step by step. Software represented a minor portion of the cost of the entire system. The first software programs were written in assembly language since the computers which ran them had limited memory space and low performance speed capabilities. As higher-level languages were developed and the price/performance relationship of computers became more attractive to the consumer, software became much more sophisticated with programs being developed which required more man-hours to develop than the computer hardware which executed the software.

Early software development was carried out as if the final product were an art form. Programs were developed by programmers as they saw fit, with no particular structure or guidelines as to how the development of the software was carried out. The final code tended to be a collection of sequential functions with "GOTO" statements being used extensively as the functionality became more complex. Early software was difficult to debug or modify by anyone other than the creator, and sometimes even the original programmer had difficulty deciphering the intent of the code, especially if some time had elapsed since the time of creation. When new applications were developed, as well as when a different product was designed which included some functionality of previous products, the entire software program was created from scratch.

As structured programming languages such as Pascal and C were developed, so was the methodology of structured design software development. A field known as Computer Aided Software Engineering or (CASE) includes computer based tools used to enhance the productivity and maintain the structured design process.

The next major evolution in the area of software development will be when automatic code-generating tools are developed to reliably generate the code based on an input of a structured, logical description of the desired functionality to be included in the software program. Most of the products available today which are in that area output COBOL and are intended for business software programs.

Another area which will continue to become a major part of many software programs involves artificial intelligence. Artificial intelligence will allow very sophisticated functions and situational decision processes to be performed without having explicitly programmed the specific situation and parameters into the system. Artificial intelligence capabilities will give a software program greatly enhanced capabilities over a typical software program which has no ability to adapt and respond to novel stimulus and input.

Discussion. Software is the critical ingredient which makes a computer hardware system capable of functioning and performing useful and desired output. Neither hardware or software can function and be useful without the other.

Software development doesn't involve creation of the software so much as it involves "growing" the final product. Software programs, especially large and complex applications, are now developed using a structured approach which consists of outlining the basic functionality of the entire package with respect to the desired interaction by the user. The concept of growing software describes the practice of developing a shell of the basic functionality and then continuing to add modules of definable functionality to the basic shell. If a module of functionality has been previously developed, the existing module is adapted and reused, instead of recreating a module of software. With this methodology, the functionality and value of a particular piece of software increases. Software is then portrayed as a living product, which is never finished, but is modified with enhancements to its functionality and corrections of any "bugs" or malfunctions discovered in the original code.

The structured approach to software development which is used more and more today allows the desired product to be segmented into multiple logical modules which can be developed separately and then integrated into a final product. Modular design and development of software makes modification and enhancement of certain functionality portions easier since only a portion of the code needs to be disturbed. This is especially desirable since the proper method of testing software code requires all portions of the program source (which were affected by changes in lines of code) be retested for proper functioning of the new code and ensure no errors have been introduced into previously properly functioning code.

The cost of software is sometimes difficult to define. The actual unit cost of the distributed copy is very low compared to the development cost of the software. The unit distribution cost may only be the media (a floppy disk), the user's manual and the shipping and handling. Sometimes software is value priced, which means the selling or licensing cost is chosen based on the perceived value to the customer of using the software. Another aspect of determining a price for selling software is based on the size and type of machine the end product will be used on. For example, a software package which will be used on a mainframe computer by multiple users will cost much more than the same software package when used on a microcomputer.

When one looks at the relationship between the cost of the software compared to the cost of the hardware, the relationship between the two in the early days of software development shared the software representing a small fraction of the cost of the entire system. Today's software, which is very sophisticated compared to earlier software, represents a much larger cost and portion of the development time. Some software efforts require much greater resources than the development of the hardware which will run the software.

Application. Several of the rules, laws, and regulations for governing the use, sale, and distribution of software do not seem to be sufficient for the contemporary software industry. One of the greatest problems with software use is the unauthorized copying and distribution of software. Major products in the marketplace have been illegally copied and used in domestic and foreign corporations with a significant loss of income by the software manufacturer. Since software is invisible unless the media which holds the code is allowed to be read by a computer, there is great difficulty in locating illegal or "bootleg" copies of software. Since licensed use of the software isn't illegal, just observing someone using the product doesn't automatically mean a violation is occurring.

Several schemes for copy protection have been developed and are used by many manufacturers. Some of the major ones use hidden files on the distribution disk which have a "counter" of the allowable number of times the software can be installed. One of the problems of this scheme is that when backups of the hard disk are performed, the "installed" copy-protected software cannot be restored to the hard disk using the DOS or other RESTORE commands. This problem has caused many users to insist on the removal of copy protection by the manufacturer. Many of the most popular programs have had their copy protection removed. Another scheme for copy protection uses a hidden routine which makes a "fingerprint" of the original hard disk. This method allows the software to be copied to other directories, backed up, or otherwise manipulated and as long as the original hard disk is still the device which the program is attempted to be executed from, the program will still work. Only reformatting the hard disk will destroy the fingerprint and require reinstallation.

Another practice used by some manufacturers is to limit the timeframe that a demo software package is used by putting a routine at the startup of the program which looks at the system date, and after a prescribed length of time has elapsed, the program is scuttled at the next attempt to execute.

The applications of software are a part of any device which contains a computer. From imbedded systems to general-purpose computers, whether mainframes or micros, all require software to operate. The emphasis today when selecting and buying a computer system is to focus more on the available software packages and to choose hardware based on the available applications software. This also has been driven by the fact that in some situations, the software portion of a system may cost more than the hardware, especially if the software is licensed using an annual license fee.

An application of software was noted by Michael J. McQuire of the Octagon Corporation in his SME Technical Paper titled *Computer Software in the Metals Industry*. McQuire said: "Production programs are the newest wave in the software for metals processors. They could open a whole new approach to the customer supplier relationship. Production scheduling or production optimization programs are a mixed bag. Some systems are written for the industry as a whole and you must fit the mold. Some use codes known only within the system and a company must change all of the records and the way of procuring documents to reflect the code. Still others do optimize but it is a long and tedious operation because you can only do one coil at a time or one group of orders."

Glossary. *bug:* A bug is an error is the execution of a software program.
imbedded system: A system with computer hardware and software inside which controls functions of the system.

Also see: Computers, Computer Languages, Firmware, Hardware, Programming.

Soldering. Soldering is any of several joining processes similar to welding in which a filler metal melting below 840 degrees F (450 degrees C) is used to create a metallurgical bond. In soldering, the molten filler metal is distributed within the joint by capillary action or wetting or both. Flux is almost always required to ensure wetting by the molten solder, and filler metal (solder) may be preplaced in the joint or fed from an external source.

Soldering is one of the oldest methods of joining metals and still finds varied and extensive use in industry. Most soldering operations produce a metallurgical intermetallic-type bond between the filler metal and the base material. Joints can be made to surfaces without this bond as in glass-to-metal joining, where surface activity and adhesion are the main mechanisms of joining.

Selection of soldering as a joining method over mechanical fastening, welding, brazing, or adhesive bonding depends primarily on end-use requirements with respect to joint strength, application, and operating temperature and environment, as well as production costs.

Application areas of soldering vary widely. They include copper plumbing systems, automotive copper and brass radiators, aluminum refrigeration components, and electrical and electronic connections. The most sophisticated computers contain many thousands of soldered joints.

The methods of producing the soldered joints and the availability of a wide range of soldering alloys facilitates these uses. The technology is still changing, and new joining materials in the soldering group continue to appear and find application.

Electrical and electronic applications, including printed circuit boards, represent the major use of soldering. Common products produced by soldering include television and radio sets, car radiators, light bulbs, telephones, typewriters, and automotive fuel and ignition systems.

Most automated systems for soldering are custom built for specific applications. With most systems, an operator loads and unloads components, and the application of solder, heat, and cooling is done automatically.

On rotary index machines, an index table conveys fixtured parts through a timed sequence of operations, resulting in production rates to 1000 assemblies per hour. Fixtures at each station on the table facilitate loading and unloading and hold the components in proper alignment for soldering. The number of stations on a table depends on production requirements and the operator's ability to unload and reload each fixture during the index cycle. The loading and unloading can also be automated with vibratory feeders, pick-and-place units, robots, or other means. Unloading of soldered assemblies can also be done with ejection devices.

The application of solder and flux or paste is generally done at the first station adjacent to the loading and unloading position. Automatic applicators are frequently used for paste solders, with one or more guns mounted on a slide or slides. Heating is usually performed at several stations to progressively bring the parts to soldering temperature. Open-flame burners supplied with a natural gas and air mixture are a common means of heating, but radiant, induction, resistance, and oven heating are also used. Cooling is often done with compressed air jets or a water wash.

In-line systems are generally desirable for high production rates and for soldering certain assembly configurations, such as long and narrow parts. They are generally designed for continuous operation with chain-driven fixtures that return to the operator or a mesh belt to carry self-locating parts. Methods of heating include gas-fired burners, heating chambers (gas or electric), or controlled atmospheres. Vacuum furnaces can be used for low vapor pressure metals.

Fixed-station machines for soldering one assembly at a time, are often used for large assemblies and when production requirements are moderate.

Shuttle machines are generally designed to solder two assemblies simultaneously, with the parts being moved from the loading and unloading station to the heating station. They are suitable for low to moderate production requirements, with maximum production rates of about 200 assemblies per hour.

Solid Modeling. CIMdata's Robert H. Johnson presented a description of solid modeling in his paper presented at AUTOFACT 87. The paper is titled *The Acceptance of Solid Modeling in Industry*. It noted: ''A solid modeler is used to define complete 3D shapes of mechanical parts and assemblies. The resulting solid model:

- Is a computer representation of a fully-enclosed 3D (i.e., solid) shape.
- Is complete and unambiguous. Because it is complete, it is a single representation useful for all CAD/CAM tasks. Because it is unambiguous, it is suitable for the automation of many (if not all) CAD/CAM tasks. Fully realistic image projections may be automatically created for visualization by the designer for communication to others.
- Assures that only physically-realizable parts are modeled. A solid model cannot have a face missing, a dangling edge, or an edge going inside the part. Because of this feature, errors or design 'impossibilities' can be avoided.''

He noted: ''A solid-modeling system is an interactive computer-graphic system used to create, manipulate and apply solid models. Since the solid model is complete and unambiguous, it can form the basis of a highly automated application program—which requires little human intervention, is fast and productive, and can make consistent, correct, and complete judgements. Solid modeling systems have the potential to further automate disparate engineering and manufacturing tasks.

''With Boolean and other high-level operators, designs of complex parts and assemblies may be created, and modified more quickly and with fewer errors. The faces and edges on the resulting model do not have to be individually developed or checked; if the solid modeler is working, they must be correct.

''Overall, with solid modeling, errors frequent in wireframe modeling are avoided, and significant savings in time are realized.

''Solid modeling is important to industry because of its potential role in companies who design and manufacture mechanical products:

- A solid model is created by the designer, as part of the layout or preliminary design process. This model is stored in the CAD/CAM system's database.

- This single model's data is then used for many application tasks, including: documentation, generation of detailed engineering drawings and illustrations; manufacturing engineering, to create process plans and numerical control (NC) programs; and for analysis to evaluate the design (or the model) for articulation, structural adequacy or other aspects of performance.''

The SME Blue Book *Solid Modeling Round Table* noted: ''Solid modeling will play a key role in the process of computer integrated manufacturing (CIM). This key role is attributed to the fact that solid modeling has the potential to provide a complete, unambiguous description of a part. Unambiguous product descriptions are necessary to automate these processes by removing human interpretation. These descriptions will then be utilized for several key elements of CIM, such as NC tool path generation/verification; detailed design/analysis using the same database; process planning and all other elements of CIM. Solid modeling systems are currently providing the key elements required for CIM.''

Solid State. Solid state pertains to an electrical circuit having no moving parts, relays, vacuum tubes, or gaseous tube components.

Sonic Sensor. A sonic sensor is a device or system that converts sound into an electrical signal.

Also see: Industrial Robots, Sensor.

Sort. To sort is to segregate items into groups according to specified criteria. Sorting involves ordering, but need not involve sequencing, for the groups may be arranged in an arbitrary order. Sort also is to arrange a set of items according to keys, which are used as a basis for determining the sequence of the items.

Source Code. Source code, sometimes called a source program, is the original code which is written as an input to a compiler or translator. The source code is usually written in a high-level language, which may then be compiled into machine language. To revise the finished or compiled program, a person must return to the original source code. After modification, the new source code must be recompiled, then executed and debugged. Some programming aids examine the syntax of source code and provide some preliminary, high-level debugging. Since source code is critical to understand how the code is written or to change the resulting compiled code, protection of the source code is crucial to eliminating unauthorized use or change of all or part of a computer program. The resulting operation of a computer executing some object code (compiled source code) might have more than one possible sequence of source code.

Also see: Compiler, High-level Language, Syntax.

Source Language. A source language is a computer input language comprised of statements and formulas used to specify computer processing. It is translated into object language by an assembler, compiler, or an interpreter.

Source Program. A source program is a program written in a symbolic language designed for ease of expression by humans. A source program also is the input program to be processed.

SPC. See: Statistical Process Control.

Special Purpose Logic. Special purpose logic are those proprietary features of a controller which allow it to do things not normally found in relay ladder logic.

Spherical Coordinates. Spherical coordinates are a coordinate system consisting of a central point of origin with the coordinates defined in terms of three angles from the three perpendicular axes and a length along the resulting ray. Spherical coordinates describe the motion of a device having either fixed or variable length.

When the anchor reference point can translate along one a single axis instead of being fixed, the coordinate system is polar, not spherical. Spherical coordinates can define any point within the volume of a sphere.

Maneuverability within a spherical area can be achieved through robot end effectors by combining a rotational and a pivot joint in combination with the rotating joint closer to the base than the pivoting joint. Applications such as spray painting require dexterous and fluid motion and utilize spherical coordinate motion patterns by the spray gun.

In robotics, a spherical coordinate is a type of system whose arm motions to achieve X axis and Z axis movement create an angular relationship to their center point of origin. This creates two plane dimensions necessary to produce a spherical shape.

Also see: Cartesian Coordinates, Polar Coordinate System, Rectangular Coordinates.

Spindle Finishing. Spindle finishing is a mass finishing process in which workpieces are individually mounted on spindles, then lowered into a rotating tub containing the finishing media. In most applications, the spindles rotate at 10 to 3000 rpm, but in some cases the spindles oscillate up and down instead of rotating. The process is sometimes automated for robotic loading and unloading.

Also see: Finishing, Industrial Robots, Mass Finishing.

Spinning. Spinning is a chipless production method of forming axially symmetrical metal shapes. It is a point deformation process by which a metal disc, cylindrical workpiece, or preform is plastically deformed by being brought into contact with a rotating chuck (mandrel) by axial or axial-radial motions or a tool or rollers. Shapes produced include cones, hemispheres, tubes, cylinders, and other radially symmetrical, hollow parts in a wide variety of sizes and contours. Elliptical shapes are also spun; however, forming these shapes is not as easy as producing symmetrical cylindrical parts.

Spinning is an economical, efficient, versatile method of producing parts, especially if the cost of stamping or deep-drawing dies would be substantial. The method is used for requirements ranging from prototypes to high production parts. Parts ranging from 1/4 inch (6.4 mm) to 26 feet (7.9 m) diameter have been spun from metals up to 3 inch (76 mm) or more in thickness.

Semiautomatic or fully automatic spinning is being used to attain higher production rates and to maintain closer tolerances. For semiautomatic operation, automatic positioning of the saddles, cross slides, and tailstocks during the cycle can be controlled by limit switches. Fully automatic spinning is attained by electrical sequential programming, NC, or CNC.

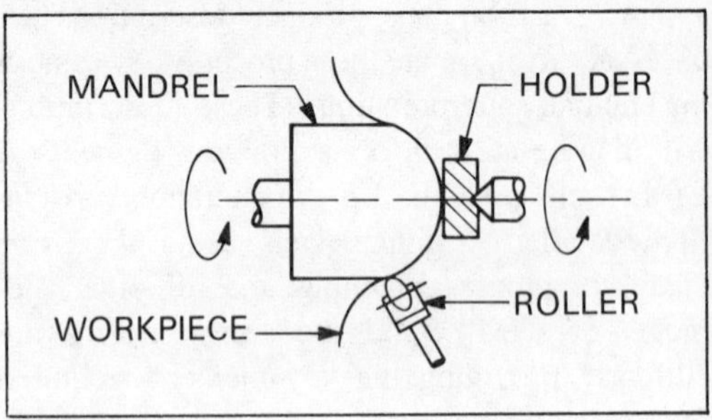

Spinning is a method of forming sheet metal or tubing.

On some machines equipped for sequential programming, plug-in diode pins are inserted in pegboard-type matrix control panels. Once the machine has been programmed, it is only necessary to load blanks or preforms and unload spun parts. Program charts indicating the proper locations for inserting the diode pins are retained for future setups.

Some machines are equipped with digital playback systems for automatic control. The machine is manually operated to produce the first part. A control unit automatically records and stores all movements (changed into electrical signals) on a cassette. The program data is stored on a floppy disc with some systems. For immediate or future production requirements, the control is switched to automatic replay.

Computer numerical control (CNC) is being applied extensively to spinning for many of the same reasons that have made it such a popular control technology for metal removal. Primary advantages include improved accuracy, consistent repeatability, and elimination of possible human errors. The additional cost of such controls is offset by reduced labor costs, quick setup, and increased productivity.

With CNC, there is a transition of technical expertise from the shop floor to an engineering database. In an engineering environment, the process can be studied, programmed, and reviewed in the form of graphic plots before execution on the machine tool. Once the program is at the machine, CNC editing capabilities permit a fast change to an acceptable sequence of forming geometries. Furthermore, more complex forming geometries can be accommodated than are possible with multiple or swiveling template controls.

Once programmed, CNC spinning machines can be used economically for limited production runs because their setup time is short. Minimum labor requirements, especially when automated loading/unloading means are provided, also allow CNC spinning to be used for large-volume production, competing successfully with deep-draw operations in many instances. Spinning machines with CNC have been built for simultaneous contouring control of two, three, and four rollers, requiring as many as eight axes.

Programming a CNC for spinning is more complex than for metalcutting. Because of the plastic deformation of material and deflections of machine components in spinning, the resulting spun geometry is not identical to the programmed contour. Programming is facilitated by integral interactive-graphics hardware and software.

Software modules for spinning are available from machine and control manufacturers. Standard features available on some CNC units include contouring selection, spindle speed control, and axis speed orientation and control. Most controls can be interfaced with and react to other on-line auxiliary equipment, such as optical pyrometers, pressure transducers, and other force or electrical current measuring devices, often used in spinning superalloys. With the addition of a computerized data-acquisition system, it is possible to provide process traceability of the parts produced.

Spray Coating. Spray coating is the process of applying a liquid coating by causing the liquid coating material to be broken up or atomized into a fine mist or spray that is then deposited onto the part surface. The individual droplets of material flow together on the surface to form the coating film. The three basic methods of breaking up or atomizing liquid coatings are: (1) conventional or air atomization, (2) airless or hydraulic atomization, and (3) rotary (centrifugal) atomization. There are also two adaptations of these basic methods that are gaining noted use today. In the newest adaptation, the primary atomization is with hydraulic pressure, while air pressure is used to control the spray pattern. This method is normally referred to as air-assisted airless. The application of an electrostatic charge to the basic methods of atomization can aid in the efficiency of the droplet transfer to the part being painted. This method is usually referred to as electrostatic spraying.

Before a coating material is sprayed, it is necessary to make sure that the coating is well mixed and of the proper viscosity. Generally, small containers, less than 5 gal (18 L), can be mixed by hand using paddles. Large containers, holding more than 5 gal (18L), should always be mixed using a mechanical mixer. Some materials require constant mixing to prevent settling.

Viscosity is resistance to the flow of the material and refers to the thickness or thinness of the material. The viscosity of the material is expressed in a term unit called poise (1 dyne sec/cm2); however, a unit called centipoise is more commonly used. Materials of very low viscosities can cause runs and sags on the painted surfaces. The viscosity of the coating material is dependent on the amount of solvent present and the temperature of the material. The optimum application viscosity of the coating material is generally supplied by the material manufacturer.

In painting operations, there are two viscosities that affect the resulting finish of the material on the part: the viscosity of the material when it is atomized and the viscosity of the material when it reaches the part. The viscosity of the material when it is atomized is critical because it determines the fineness of the material particle. The viscosity of the material when it reaches the part can cause sags and runs, but it can be controlled by the quantity of material applied. The difference between these two viscosities is the result of solvent evaporation as the particle goes to the part.

The viscosity of a coating material is typically measured by a method that measures the time for a given volume of material to flow through a given orifice. In the paint industry, the Ford or Zahn cup method of viscosity measurement is widely used. The value obtained from these methods is then converted to a centipoise number for general reference. To make these measurements, a cup-like device with a hole in it is dipped into the coating material until it is completely full. It is then withdrawn, and the time from the moment the cup clears the surface of material until the time the solid stream of material breaks up or stops is recorded. Depending upon the type of cup used, this time can be converted into a different scale if required.

If the viscosity must be adjusted, the operator can either add a proper solvent to lower the viscosity or add paint solids to raise the viscosity. Generally, a material is either thinned or used as is, as adding paint solids is not normally recommended. Another method of reducing the viscosity is to heat the material. The addition of heat causes the viscosity to lower, but it is important to check and adjust the viscosity for the temperature at which the material will be used.

In conventional air spray, the material is usually supplied from a container in one of two ways. In the first way, the container may use pressure, up to l00 psi (690 k Pa), to force the material to the spraying equipment via a pressure vessel or a pumping device. The other method of material supply uses a vacuum created by the spraying device to pull the material from the container to the atomizing area.

In manual spraying, the operator selects the system variables such as fluid flow rate, atomizing pressure, fan shape, paint temperature, and the sweep of the gun. In unmanned spraying processes, these parameters are preset to coat a preselected number or type of parts, with the operator checking that the parameters stay within the specified tollerances.

Two methods of unmanned spraying are in current use, automated booths and robotic systems. A part identification system is necessary for both. These systems size the part with mechanical fingers, limit switches, photocells, magnetic strips, bar codes, or by visual observation. A color-change system is also required for unmanned spraying processes; in it, a manual or automatic signal interacts with the color-change mechanism, which ejects the paint, cleans the paint line with solvent and air, and then refills the line with the newly selected paint. The time required for the color change varies with the paint viscosity and length of supply lines. In addition, unmanned systems require the parts to be hung uniformly since missing or improperly hung parts cannot be detected. Conveyors must also operate smoothly because these systems cannot compensate for swaying parts or other unforeseen motion.

Automated spraying methods ensure greater uniformity of finish from part to part as well as from day to day, while reducing the amount of manpower required. In addition, ventilation can be reduced; unmanned booths may require as little as one-third the amount of air as manned booths.

Automated booths are used when a limited number of different parts have to be painted and when model changes are infrequent. They use spray painting guns that are mounted in either fixed positions or on reciprocating arms. In some instances, the arm tilts the gun to follow the contour of the part, using curved rails, cam-operated valves, or timers, as well as some programmable controls. Air, airless, or electrostatic spray guns can be used in these booths.

Theoretically, it is possible to set the equipment so that all the parts are coated acceptably. Practically, however, in order to reach certain areas of the part, other areas may receive an unnecessarily thick coat. It is therefore preferable to use a manual touch-up. In some applications, touch-up is performed before machine spraying.

The safety of automated booths is relatively high since all the operations are carried out within an enclosure. To guard against entering the booth during spraying, interlocks on doors and equipment are provided, as well as pressure-sensitive mats and photocells.

Robotic spraying is performed using a six-axis servocontrolled robot that is capable of reproducing the wrist, arm, and waist movements of a human painter. Since the robot operates in the presence of flammable solvents, it is hydraulically activated. The volume of space that the robot can reach is called the working envelope.

Training a robot can be performed using continuous programming or point-to-point programming. In continuous programming, the robot's memory is activated, and then the robot arm is moved through the required motion pattern. On most robots, minor changes in the program can be made during each operating cycle until the optimum program has been achieved.

Point-to-point programming is much slower than continuous programming and is generally only used when straight passes of the gun are required. In either case, the program can be speeded up or slowed down to meet the variations in the conveyor speed. However, care must be taken when increasing or decreasing the robot's speed so that the other parameters—such as proper spraying technique and, especially, the stroking speed (too fast-dry spray, too slow=runs and sags) and the curing cycle are not adversely affected.

Robots are currently being used to apply topcoats to truck beds and steel office furniture, primers to truck cabs, hoods, and fenders, stains on wooden furniture, sound deadener on appliances, and porcelain enamel on bathtubs. Robots are most practical in coating applications that would be hazardous to a human painter, that are monotonous and repetitive, and that require the coating to be applied completely and uniformly. Air, airless, and electrostatic spray guns can be used with most robots.

Spray Painting. Spray painting is a generic term for coating processes that deposit liquid organic films composed largely of a binder (resin), pigment and various additives suspended in a suitable solvent. After drying, the paint is a solid, adherent film that provides a decorative appearance, protects the surface or both.

Spraying paint was one of the early, successful applications of industrial robots.

In an article titled *Expanding the Capabilities of Spray Painting Robot* (*Robotics Today*, April 1982), Hadi Akeel of General Motors reported: "The past ten years have seen the development and marketing of several spray painting robots. In automotive production plants, these robots have helped automate the painting of small parts such as fascias and body moldings and noncritical areas such as engines, pickup boxes, truck compartments, and underbody coating. Occasionally they have been used to do finish painting of Class A surfaces."

Akeel reported on a paint spray system which was developed with direct numerical control technology to provide a means for automating the paint process. Because of the similarity of the paint and sealer application requirements and commonality of the workpiece (car body), the system was also suitable for automating the sealant application process and thus goes a long way toward fully automating the paint shop.

The GM robot system consists of the following elements: a seven-axis NC paint spray machine; two-axis door opener/closer; model recognition camera; teach booth equipped with NC painter, door operator, and teach devices;

machine control computers and input panels; supervisory computers with central high speed data storage disks; and conventional automatic roof and side spray machines equipped with automatic color changers. A typical robot installation is shown in the figure below. NC painters and door operators are shown to augment the operation of the automatic machines. Note that the grates can be removed to reduce booth cleaning costs. A typical installation for 60 jobs per hour production requires eight robots (four pairs) on-line; one more pair is needed for backup.

When bodies approach the paint booth, the model recognition camera identifies the style and communicates the information to the central supervisor computer. The supervisor checks the style against the plant schedule.

If disagreement occurs, the supervisor alerts the human operator and requests manual input of correct information. Otherwise, the supervisor tracks the body along the line and communicates necessary path programs to each control computer as needed for appropriate paint coats at individual stations. Should one machine or station fail, its operation is switched to other stations down the line and the backup station is activated.

Job assignment to the stations is done automatically by the supervisor in accordance with sequences preplanned for each failure. Optimum process sequence can thus be maintained. Paint path programs are developed in an off-line teach booth, where a machine identical to those on-line can be taught all programs. Programs can be verified through actual painting. Once accepted, they are transmitted to the supervisory computer for operation by the on-line machines.

The June 1982 *Robotics Today* reported on another application of a robotic painting system at Fairchild Republic Co., Farmingdale, N.Y. The system is unique in that it's used for painting large, complex structures rather than small details. Also unique is the fact that an air bearing-supported carriage is used to transport the robot along a rail to a predetermined work

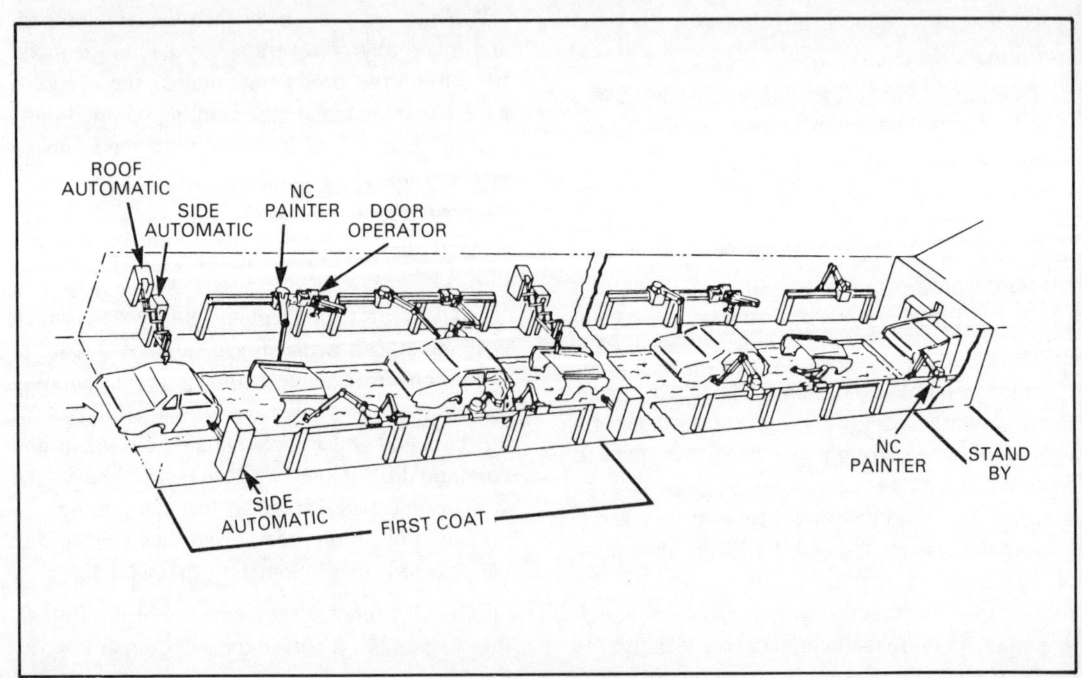

Typical layout of the DNC paint system employs four pairs of robots.

position. Other features include a programmable logic system and a triple security sensing system to deactivate the power systems in the event of human intrusion into the paint booth, or robot arm contact with the structure being painted.

The structures painted include the nose section, armor tub, and assorted assemblies for the A-10 attack aircraft, and flaps and ailerons for the Boeing 747 aircraft. The A-10 sections include numerous cavities—large and small, internal and external. Prime and finish coats are applied in these areas. Some of the problems encountered in previous manual painting operations are limited access to confined areas, operator discomfort in an oppressive environment, the need for extensive rework, and touchup, and excessive time and labor costs.

The paint shop layout consists of individual booths for painting large assemblies. Thus, the robot painting system must be moved from booth to booth depending on work flow. The equipment must also traverse within the booth to perform its functions over the entire length of the assembly. Finally, every piece of equipment must be explosion-proof.

A decision was made to introduce the robot paint spray system immediately into production, bypassing the laboratory phase typically used for familiarization and system debugging. Much useful information was gleaned in discussion with the foreman and paint shop personnel. Based on these discussions, a list was prepared suggesting the type of automation to be selected and outlining design, safety, and human factors considerations. Because they were involved in the planning stage, paint shop personnel exhibited a highly positive attitude toward the automated system.

The robot required for the project had to be extremely flexible, easily programmed, and readily adaptable to the harsh environment and abusive treatment of traversing from one end of the paint booth to the other.

Also see: Industrial Robots.

Spray Transfer. Spray transfer, in arc welding, is metal transfer in which molten metal from a consumable electrode is propelled axially across the arc in small droplets.

Also see: Arc Welding.

SRAM. See: Static Random Access Memory.

Stand-alone System. A complete operational system that does not require support from other devices or systems.

Standards. A standard is a formally accepted or decreed specification regarding something which can be described by words, diagrams, pictures, etc. A standard can be established by authority, custom, or general consent.

Many agencies and groups exist for the purpose of establishing standards. The CCITT (Comite Consultiaf Internationale de Telegraphique et Telephonique) is a standards body under the International Telecommunications Union, an agency of the United Nations. The CCITT is the primary organization for developing standards on telephone and data communications systems. The ISO (International Standards Organization) is a voluntary organization consisting of national standards committees of each member country. ANSI (American National Standards Institute) is a voluntary standards body in the United States and a member of the ISO. SEMI (Semiconductor Equipment and Materials Institute) is a body which establishes standards for the semiconductor industry. The EIA (Electronic Industries Association) is a national trade association. EIA is very influential in developing standards for North American countries. EIA focuses primarily on electrical standards. The NBS (National Bureau of Standards) is a government agency responsible for developing, testing, and validating standards.

Also see: ANSI, CCITT, EIA, IEEE, ISO, NBS, NEMA, SEMI.

Star Network. A star network refers to the physical configuration of a network topography. In a star network, there is only one node from which there is only one path to each endpoint node on the network. A star network is the same as a centralized network.

A star network requires that the central node provide a channel connection to each of the endpoint nodes. This is similar to how mainframe computers were connected to user terminals. The disadvantages of a star network is the requirement of a one-to-one relationship between the number of end nodes and the interface ports of the central node. Another disadvantage is the requirement of the central node to handle the communications overhead for all nodes, whether a node wants to communicate with the central node or with another end node. One advantage of a star network over a bus is that a break in the physical connection only disables one end node. Another plus: the physical media only has to carry the bandwidth of the node's traffic, not multiple multiplexed information.

Also see: Star Network.

Start Bit. A start bit is utilized in asynchronous communication to precede the data character. The purpose of the start bit is to notify the receiving site that data is on the path (start bit detection). The path or line is said to be idle prior to the arrival of the start bit and remains in the idle state until a start bit is transmitted. During the idle state, the signal voltage is held at the high level. The low signal state of start (bit) initiates mechanisms in the receiving device for sampling, counting, and receiving the bits of the data stream (bit counter). A start bit precedes each message transmission, and a stop bit ends each message transmission.

This type of communication is called *asynchronous communication* due to the absence of continuous synchronization between the sender and receiver. This allows data to be sent at any time without regards to any previous timing signal.

Also see: Asynchronous Communication, Stop Bit.

State. State is the logic condition, 0 or 1, in a PLC memory or at a circuit's input or output. State also is the logic condition of any device which may be a number greater than 0 or 1. A sequencer may have many states.

Statement. A statement is a meaningful expression or generalized instruction in a source language.

Static Random Access Memory (SRAM). Static random access memory is a type of random access memory which doesn't require a refresh signal to maintain its contents. One of the limitations of dynamic RAM is that time must be allowed for the refresh to occur in-between accessing the memory contents. This allowance restricts the maximum speed at which data can be read from the dynamic memory. Usually a ''wait-state'' is inserted in the operation of the CPU to accommodate the refresh cycle. Static RAM doesn't require the refresh cycle and therefore can be continuously accessed at the maximum frequency and rate of data transfer which the memory chip is capable of handling. The maximum speed of operations is defined in milliseconds such as 150ms, 120ms, or 100ms. Static RAM is more expensive than dynamic RAM due to the additional circuitry in the chip required to maintain the memory contents, therefore most computer manufacturers utilize dynamic RAM instead of static.

Also see: Computers, Dynamic Random Access Memory, Personal Computers.

Statistical Process Control (SPC). Statistical process control is a system allowing for economically sound decision making about actions affecting a process. The process may be described as the entire combination of people, equipment, input materials, methods, and environment that work together to produce output. The economic decision is made by balancing the risks of taking action when action is not necessary (overcontrol) versus failing to take action when action is necessary (under control). The decisions regarding these risks are made in the context of the concept of variation and specifically with respect to the two sources of process variation called common (system) and special (assignable).

Common causes refer to the many sources of variation within a process that is in statistical control. A process is said to be in a state of statistical control when all variation present in the process originates from common causes. Common causes will cause the process to behave like a constant system of random chance causes. While individual measured values are all different, as a group they tend to form a pattern that can be described as a distribution. This distribution is characterized by its location on a measurement scale, its spread (amount by which smaller values differ from larger units), and its shape (pattern of variation whether it is symmetrical, peaked, etc.,).

Special causes refer to any factors causing variation that cannot be adequately described by any single distribution of the process output, as would be the case if the process were in statistical control. Unless all the special causes of variation are identified and corrected, they will affect the process output in unpredictable ways and the process may be described as being out of statistical control.

A statistical process control system provides a statistical signal when special causes of variation are present and avoid giving false signals when they are not present. Dr. Walter Shewhart of Bell Laboratories, first made the distinction between controlled and uncontrolled variation, due to common and special causes. He developed tools called control charts to statistically separate and identify the two sources of variation in a process. There are several types of control charts to analyze both variables (continuous distributions) and attributes (discrete distributions). All of the various types of control charts have the same two basic uses. The first is a judgement tool to provide evidence as to whether a process has been operating in a state of statistical control and to signal the presence of special causes of variation so that corrective action may be taken. The second is as an operation to maintain the state of statistical control by extending control limits as a basis for real-time decisions.

For many years the primary role of the quality control function was to ensure that the manufactured product conformed to engineering specifications. The primary instrument of this function was inspection, using either ad hoc methods or plans for sampling based on statistical methods and probability theory. In either case, the principal result was the assurance of a satisfactory outgoing quality level for the goods being inspected.

For years many in U.S. industry embraced the quality control function through these methods of product control. As a result, from a productivity standpoint, inefficiency and waste were widespread and perhaps even increased over the two decades prior to 1980. Competitive balance was maintained because everyone was doing an equally poor job. This situation slowly changed through the 1970s as worldwide competition in the marketplace intensified. A competitive edge grew slowly but steadily until the problem gained alarming proportions. It soon became obvious that many world competitors of the U.S. were using a different management philosophy and a total systems approach to quality in both engineering design and manufacturing.

Since 1980, important progress has been made in the U.S. to regain a leadership position in many markets. As techniques such as statistical process control (SPC) become better understood conceptually and then implemented across a broader and broader base, real differences are being made.

The techniques of SPC, were put forth more than 60 years ago by Dr. Walter Shewhart. SPC methods use basic probability laws as well as statistical methods. These two combine to develop models related to the behavior of the variations found in the quality of manufactured goods as these goods are produced. Through the statistical study of these patterns of variation, it is possible to attribute certain types of variations to certain types of fault sources and develop a knowledge and understanding of what types of corrections may be required. In this regard, the most fundamental principle in the application of SPC methods is the partitioning of the total variation pattern into two major sources: the system sources (chronic problems or common causes) and more localized sources (sporadic problems or special causes). Because system variation comes from many ever-present sources, it appears as a stable, random, and well-behaved pattern of variation. On the other hand, sporadic problems being of external disturbance orgin produce unusual patterns of variation that are visible when statistical methods are used to study the total variation pattern.

It is important to recognize the proper assignment of responsibility for corrective action of common and special causes. Because common causes are system faults, they require the attention of management. It is likely that some major breakthrough is required to affect a significant improvement in quality and productivity when common causes are involved. Only management can institute the changes required to remove major sources of common cause variability.

The roles and responsibilities of management in precipitating breakthrough are substantial because the vast majority of all the problems that must be dealt with on a continuing basis are chronic problems.

Understanding the need for the pursuit of statistical control or process stability when the process is already producing parts within the specifications is essential to the continual improvement of quality and productivity. In this regard, central to the acceptance of SPC as an essential concept and manufacturing management tool is the recognition of the quality/productivity relationship. Statistical process control helps identify the fundamental product and process faults that lead to the production of defective material and helps formulate corrective actions to prevent defective materials from being made in the first place.

Every fault or source of variation not only erodes quality but is a source of waste and inefficiency in one form or another. As a result, the elimination of each and every source of variation leads to improvements in productivity as well as in quality. It is therefore advantageous to adopt the philosophy of never-ending improvement using SPC rather than the notion of attaining and then maintaining an acceptable level of improvement. The quality/productivity relationship is a clear motivation for the continual pursuit of process stability.

A stable process is a predictable process, and hence the management of scheduling, inventory control, and maintenance management are all directly influenced by the presence (or absence) of process control. Improvements in the efficiency of production management can obviously be a tremendous force in the attainment of a positive competitive position. The lack of process control/stability will always have a deleterious effect on the ability to efficiently manage and control the total resources of a manufacturing operation.

Statistical control also is essential to the proper evaluation of process capability. Process capability is a measure of the consistent ability of the process to produce at a certain level. Consistency means stability, the ability to extrapolate performance into the future with a strong degree of belief that predictions of future performance levels will be realized. If the process is behaving in an erratic and unstable fashion, it is impossible to assess the process capability. Process data collected at different times may give totally different pictures of the process capability.

Attaining process stability is by no means sufficient to ensure high levels of quality and productivity. A process may be stable or predictable, but not at all capable in terms of its ability to produce a high percentage of acceptable product. A process in a state of good statistical control simply means that the best is being done under the present system. Because a controlled process means that the process is subject only to common causes of variation, the system will have to be changed to make any additional improvement.

The continual pursuit of variation reduction in a manufacturing process provides for still another important benefit—the existence of a quiet, low noise level system. Systems with low noise levels (high signal-to-noise ratios) are more sensitive to change and therefore provide stronger signals when either purposeful change through experimentation is introduced or when sporadic change occurs in the form of special causes of variation. In other words, the ability to efficiently and reliably reveal improvement opportunities through the techniques of statistical process control and experimental design depends strongly on the ability to continually affect noise reduction in the system.

It is not uncommon today to hear of efforts to reduce the number of defective items produced to levels that seem on the surface to defy all reason and practical judgment. If variation reduction was only done for the sake of reducing scrap from one defect per 10,000 items to one defect per 1 million items, the economic basis for continual improvement may be considered weak; clearly it is the other broad-based benefits of variation reduction that overwhelmingly support its continuance on a never-ending basis.

Stepper Motor. A stepper motor is a rotary electric motor (usually) which moves in a sequence of uniform increments called *steps*. Each step is some fraction of a degree of rotation. The steps are resolved by the physical construction of the motor which has poles opposite the windings to coincide with the location of each step. A desirable advantage of using stepper motors, which are open-loop control devices, is the lack of the need for a feedback mechanism to perform the motion. This absence simplifies the control system, speeds the control function by eliminating the latent time required for a feedback loop, and eliminates other aspects such as lead/lag variations and servoing or "hunting" for a position when stationary.

One stepper motor drawback: The motor can only stop at a step, which defines the minimum increment of motion, as well as the coincidence of where the motor can physically stop. Servo motors on the other hand can stop at any point in their rotation and can move as small of an increment as the control system is capable of deciphering and controlling. An enhancement to typical stepper motor control is called microstepping, in which the motor is stopped at locations equivalent to precisely at, slightly before, or slightly after, a primary step. This allows the stepper motor to exhibit more of the finer resolution of a servomotor, without the servocontrols.

Also see: Servocontrol, Servomotor.

Stop Bit. A stop bit is actually used as multiples, never as a single bit, and is a sequence of one or more consecutive mark signals. Stop bits provide for a time lapse in which the mechanism in the receiving site (of older equipment) can readjust for the next character. Following the stop bits, the signal returns to the high or idle level, thus guaranteeing that the next character will begin with a 1 to 0 transition. If the preceding character had been all 0s the start bit detection would become confused if a stop bit were not present to return the voltage to a high or idle level. Stop bits are essentially the time during which the transmission line is maintained in a high state immediately following data transmission. Depending upon the baud (reciprocal of the bits/second) rate, a stop bit will vary in the actual length of time required to be interpreted as a single bit or multiple bits.

Also see: Asynchronous Communication, Computers, Start Bit.

Storage Medium. A storage medium is any device or recording medium on which data can be stored for subsequent retrieval.

Also see: Computers.

Storage Tube Display. A storage tube display is a CRT where the data is stored on the tube face. To change the display, a special signal must be applied to the tube face to erase it. The display is then reprinted with new data included.

Stored-field, Read-only Alterable Memory. Stored-field, read-only alterable memory, generally read-only in nature, which may be reprogrammed in a limited fashion.

Stored Program Numerical Control. Stored program numerical control is the same as CNC except that it features an internal memory which can be altered by receiving new instructions.

Also see: Numerical Control.

Straight Flanging. Straight flanging is a bending operation in which the line of bending is straight. It is the easiest operation to perform, and there are few restrictions on the flange width.

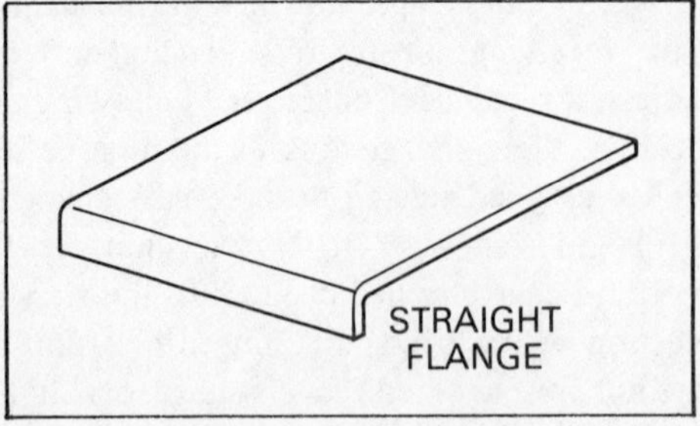

Straight flanging is a bending operation in which the line of bending is straight.

Stretch Flanging. Stretch flanging is also a bending operation, but the line of bending is not straight. It is a concave curvature, as shown is the figure. The flange is known as a stretch flange and the process is called stretch flanging because the material undergoes tension when the flange is being formed. Tearing or breaking of the edge is common. To reduce the possibility of tension tears, the width of stretch flanges must be limited.

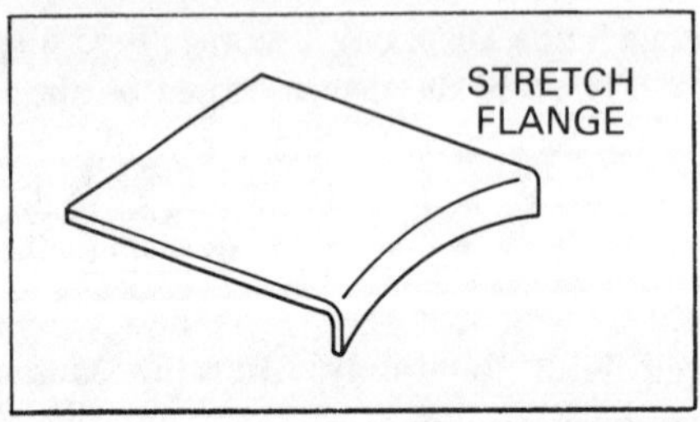

Stretch flanging is a bending operation in which the line of bending is not straight.

Stretch Forming. Stretch forming is the process of forming by the application of primarily tensile forces to stretch the sheet metal over a tool or form block as shown in the figure on the next page. The process is an outgrowth of the stretcher leveling of rolled sheet. This operation is used most extensively in the aircraft industry to produce parts of large radius of curvature, frequently with double curvature.

Stud Arc Welding (SW). Stud arc welding is a fusion welding process that uses a stud or other fastener as the electrode. In stud welding, an arc is drawn briefly between the stud and the work using a high-current DC source or capacitor discharge. Then, before the weld puddle can solidify the stud in plunged into the puddle to complete the joint.

Also see: Welding.

Subroutine. A subroutine is a portion of a software program which usually performs a particular isolated and definable function or utility. A subroutine is usually "called" by the main program or a higher level subroutine. When the execution of the subroutine is completed, the program flow returns to the next sequential program step in the calling program, immediately after the statement which caused the branch to the subroutine. Since subroutines usually perform certain useful functions, they are used as building blocks in successive programs. Using subroutines is part of the methodology of structured programming, which advocates using GOSUB statements instead of GOTO statements. GOTOs can cause the program flow to become so confused as to be indecipherable, whereas GOSUBs maintain a top-down flow of the general program. Using subroutines allows for easy substitution of different or enhanced segments of code. Subroutines provide the ability to easily allow multiple resources to contribute to the development of the code with easily manageable chunks of code. A subroutine is a portion of a numerical control program, stored in memory and capable of being called up to accomplish a particular operation. It reverts to the master routine upon completion.

Also see: Computer Languages, Computers, Macro Instruction, Programming, Software, Utility Program.

Susceptibility Testing. Susceptibility testing is when products or devices are tested to observe if they will function properly when subjected to emitted radiation such as RFI (Radio Frequency Interference) and EMI (electromagnetic interference). Susceptibility testing requires the creation

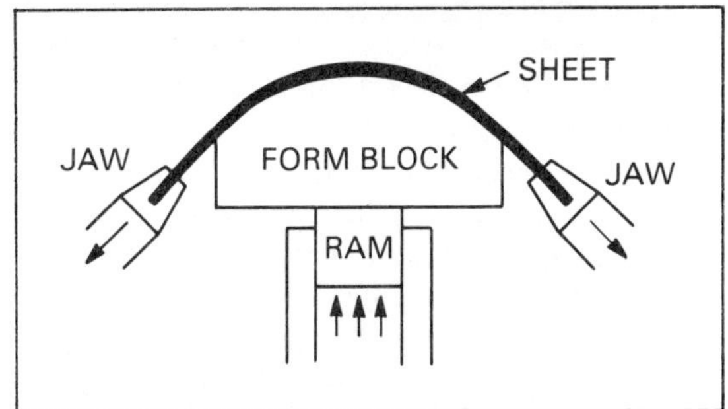

Stretch forming applies primarily tensile forces to stretch the sheet metal over a tool or formblock.

of simulated interferences to observe and measure the response of the product. Static electricity is a major problem with electronic products that come into contact with human operators. The voltage levels on a dry winter day can be as great as 20,000 volts. While static electricity has very low amperage (current), and therefore is not harmful to individuals, integrated circuit chips, LCD displays and similar components can be harmed by the voltage spikes.

Equipment that is certified to be used as life-support or life-sustaining equipment must not be affected by the typical disturbances and interferences existing in its working environment. Since this is a critical usage of equipment, most devices (such as computer controllers) are sold with a specific disclaimer referring to their use as part of life-support equipment to avoid potential lawsuits if the product is used in a life-sustaining application and malfunctions.

Also see: Standards.

SW. See: Stud Arc Welding.

Switch Sequencing. Switch sequencing is a programming technique for robots. Switch sequencing is a method of establishing logic sequences by setting a limit switch sequencing device and physical hardstops to control axes travel. It is considered reprogrammable if switch sequencing and stop settings can be adapted to effect new motions without modifying the basic design.

Also see: Industrial Robots.

S Word. S Word is a NC program code that determines spindle speed.

Also see: Numerical Control.

Symbolic Coding. Symbolic Coding are instructions written in nonmachine language.

Symbolic Control. Symbolic control is programmer-to-computer communication by means of an abstract language using symbols rather than numerical codes.

Symbolic Instruction. A symbolic instruction is an instruction in an assembly language directly translatable into a machine code.

Symbolic Language. Symbolic language is computer language which is representative of something such as an icon type graphic language. Fifth-generation programming languages such as PROLOG, and LISP attempt to provide symbolic representation of information. The field of artificial intelligence uses symbolic languages extensively and several computer companies have developed computers with the design intent of taking advantage of symbolic languages. One company, Symbolics, Incorporated in Cambridge, MA even has named itself the same as the type of programming they propagate.

Machines such as the Macintosh computer by Apple utilize symbolic representation of data to make operation and programming easier for the less knowledgeable computer user, as well as the user who prefers to not be forced to interact with a computer in "computerese." As the capabilities of computers continues to increase, with the price decreasing, symbolic languages and symbolic programming will evolve. The current languages used are much more symbolic and natural than assembly language programming or interacting with a computer console to set bits and enter binary words as was required with the first computers.

Also see: Computer Languages, PROLOG.

Synchronizer. A synchronizer is a storage device used to compensate for a difference in the flow rate of information when it is being transmitted.

Synchronous Communication. Synchronous communication is a method of communication which requires continuous synchronization between the sender and receiver. Synchronous communication is visibly distinguishable from asynchronous communication by the existence of a separate clocking signal at the sending and receiving stations. Also the entire data field is surrounded by control bits (called synchronization or SYN bytes; where a byte is a character). The SYN bytes provide a similar function to the start bit in asynchronous transmission by notifying the receiver that a message is arriving. As in the asynchronous mode, the receiving device searches for the unique bit pattern of the SYN character but uses a locally generated clocking signal to determine when to sample the incoming signal. The SYN patterns provide the means to synchronize the clocking devices (oscillators) at the sending and receiving devices. Once the oscillators are synchronized, they are usually very stable. Most communications devices resynchronize against drift every one or two seconds by periodically inserting SYN bytes into the message.

The SYN character or bytes must be distinguished from user data to allow the receiver to search for the SYN bit patterns. Vendors may use different bit streams to represent the control character. The American National Standards Institute's code (ASCII) uses a 100010110 byte. Many vendors now use the ASCII code standard.

Also see: Asynchronous Communication.

Syntax. Syntax (or linkage) is one of the five interrelated components that make up human language. The other four are meaning (or semantics), medium, expression, and participants. Languages contain precise (and typically complex) sets of "rules" that dictate which linguistic units can be linked and their order. The study of such linkage is called syntax (or sometimes grammar). Syntax determines the order of words in a sentence. It modifies meanings, changing singulars to plurals or present to future tense. Syntax also allows variety in constructing different sentences with the same meaning.

Computer languages differ from natural languages in the rules used by their syntax. Instead of rules for pluralization or word order, computer languages have rules to instruct computers to add numbers, sort lists, etc. Another difference is the number of rules involved. Natural languages have many thousands of rules (and exceptions). However, even the most sophisticated computer languages have only a fraction of that quantity. Computer languages also have very little leeway in how a sentence can be constructed. At this time, computers continue to be simply capable of generating one structured output in response to specific input.

Also see: Computer Languages, Programming.

System Custodian. The system custodian is the organization which is the focal point for the problems which the system provides solutions to. Custodianship implies strong responsibility to provide communication, training and change management for the system.

System Development. System development is a formal, phased approach to producing a significant new system or major changes to an existing system. It stresses teamwork among users and technical personnel, a series of major milestones, and thorough documentation to assure compliance with performance and schedule goals.

Systems Engineering. The systems engineering method recognizes each system as an integrated whole even though it is composed of diverse, specialized objectives and that a balance must be achieved that optimizes the overall system, functions according to the weighted objectives to achieve maximum compatibility

The overall activity of systems engineering is comprised of two parts, the system engineering associated with the way that the operating system itself works and, the systematic process of performing the engineering and the associated work in producing the operating system. It is apparent that the two parts may strongly influence each other.

History. In the past, systems engineering techniques and applications have often been focused on the activities surrounding the operation of a system. Earlier emphasis has been on the necessary engineering, mathematics and judgement principles involved in performing the detailed operation of a system.

As systems have become more complex with respect to their nature and operation, as well as far more extensive in their scope, the nature and scope of systems engineering activities has grown proportionally. Additionally, such changes have created the need for greater activity in the development of systems. Increased emphasis has been placed on the organization and performance of tasks that are a part of designing and building the system. Expanding breadth and complexity dictate that some objective means of evaluating which of the alternative methods for producing the system is most appropriate in a given case. It is essential to understand the many judgement factors involved, such as the value of the system being planned, its cost, the time required to produce it and its reliability.

As systems engineering has evolved, increased levels of resources have been devoted to the preimplementation phase. The focus is on the higher-level decision-making process of systems engineering which helps to establish the goals and the trade-offs between conflicting objectives. Systems judgements must contend with the reliance on all varieties of systems. With the rising costs associated with the development of more complicated systems, it is essential that many issues be addressed. Long before actual development is underway, fundamental questions must be answered. What is the value or worth of the system being contemplated? Should it be built at all? How should it be structured and organized to operate most effectively? How is one to ensure that the system, when complete, will operate to provide the desired results and associated benefits? Are the necessary technologies available to support the systems development? If not, what level of effort is necessary to develop them? Finally, what are the reasonable values of costs and time for developing, implementing, and operating (on an ongoing basis) the system in comparison to the benefits that one expects to derive from it?

Discussion. In a world in which the training and the functions of individuals and groups are growing more specialized, the number of ways to accomplish any particular result is increasing. Different designs, different facilities, different equipments, different patterns and various organizational means meet particular needs. It is very desirable to have trained individuals to review the various alternatives, to compare their effectiveness and to provide a logical path to sound engineering decisions. The objective of systems engineering is to provide a broad, structured approach to support the decision making process.

Each system is unique. However, by capitalizing on similarities between them, time, effort and therefore cost of many of them can be reduced through the use of structural design methods.

Included in the problems of systems engineering are those of complexity and of choice. It becomes important to review all the facts concerning a system and all the requirements of the system to determine those that will have the greatest impact on the design and operation of the system. To do this, a structured approach is required. This structured method can be grouped into six major categories. They are: establishment of the system need and value; estimation of the system cost; estimating the time and resources required to develop the system; designing and structuring the system; organization and implementation planning; and implementation to ensure a reliable system.

Each area is addressed in turn. First attention is given to the establishment of system need and value.

Systems typically involve a significant amount of resources or cost to the organization which is obtaining them. These resource requirements may include money, manpower, equipment, material or a combination of several of these. Therefore it is essential to obtain management or corporate approval to secure and/or use such resources. Without it, the ability to develop and to implement a system of any size or significance is extremely limited.

To secure management approval for the allocation of required resources, the proposed system must be justified to management. That is, the worth or value of the system relative to its cost must be demonstrated to warrant the expenditure. It also is critical that the technical feasibility of the undertaking be established—even if only on a broad basis initially—before large quantities of resources have been expended.

Approval of systems of increasing size or cost involves the higher levels of corporate management. To obtain this higher-level corporate or organizational management interest and support, the values of the system must be expressed in terms of objectives that are meaningful to people in such positions of responsibility. Return on investment (ROI) profit ratios, product leadership, improved technology or functionality, and improved quality are typical of these values.

Initial go/no go decisions with respect to a proposed large-scale system is generally made by a relatively small number of people on basis which are essentially nontechnical. Since a significant amount of time may elapse between the initiation and completion of a system, precise technical data may be impossible to obtain. Therefore it is important for the systems engineer to ensure the technical soundness of his or her proposed system, and to provide an understandable and attractive sets of values and/or value/cost considerations to satisfy decision makers.

Another important factor in establishing the overall worth of a system is the magnitude of its cost (estimation of the system cost). Once the value of the system has been established, it is necessary to determine the system's cost as a basis for comparison.

System cost can be considered in a variety of ways. For example, the cost of developing a system and the cost to operate it on an ongoing basis may be combined to arrive at an estimate of total cost. Total cost may be derived via another method that sums the fixed costs, variable costs, overhead and past investment changes. It is important to consider that for many large systems projects, several time phases may be involved, each with significantly different cost factors.

Cost evaluation can have a technical and engineering impact as well as a financial one. Costs can help establish a basis by which value is packaged and to which the performance and other characteristics obtained from a system are linked. System costs include the necessary engineering, technical studies, and investigation to determine performance, manufacturing, and reliability. Inclusion of these costs provides a greater likelihood of systems success than if such costs have been omitted. Cost constraints can have a significant bearing on the attainability of technical accomplishments within a system.

Cost has a very important influence on the judgement given a system during the decision-making process as well as on the technical achievements what that system can ultimately realize. It is an important consideration in systems engineering.

In estimating the time and resources required to develop the system, the value of the system and the system costs are important factors. So also are the time and resources required to develop the system. In reality, the time and resources required and the value established for the system are somewhat dependent on the available time and resources to perform the development. The system's cost is based on the assumption of some ''reasonable'' availability of time and resources.

It is important to be specific and detailed when estimating time and resources required to develop the system. Since systems costs may involve large sums of money and/or enormous amounts of resources, additional time and resources consumed by one part of a system, even at no direct expense to that part, may result in large changes being incurred by the overall project because of constraint expenditure rates in these other areas. The expression ''time is money'' is very appropriate here and proper attention in a systems sense must be given to the time and resources required to develop and implement most projects.

It is important to recognize that alternatives are available to achieve many systems objectives. For example, if additional manpower is allocated to a job, it will in general be possible to shorten the development time. Conversely, if less manpower is available, the development time may be increased. Similarly varying shifts in manpower and associated skill levels impact the overall gravity of the job. Therefore, the optimum mix of various resources and time (not just absolute amounts) must be considered when evaluating a systems project.

In the designing and structuring of the system, the job of developing and implementing a system is a very expensive activity. Only when a system has definitely been authorized, or perhaps when a specific contract has been sent for a detailed design study, can the specific formulation and structuring of the system in detail be performed. This stage includes the determination of inputs and outputs, requirements objectives and constraints. Structuring the system establishes one or more method of organizing the solution to the method of operation, the selection of parts/components and the nature of their performance requirements.

The process includes gathering of information about the problem, and about the system. More specifically, the design and structuring process consists of a detailed, technical appraisal of the specific needs of the system. The technical approval is much more strongly oriented in terms of what is needed for the system to accomplish its objective than in terms of the more customer-oriented considerations of value, cost and time that have previously been discussed.

From the design and structuring process, detailed specifications provide information about the inputs and outputs and subsystem requirements. The structure provides a means of having data about one part of the system available to provide inputs to parts subsequently related within the system. This part of the technical process is a highly creative one and is strongly dependent on the state of the art. It also must include considerations of what will be necessary to support, that is, train ,ervice and maintain the system.

Once a definite idea has been established as to the system's makeup, and the system's specifications, it is possible to formulate an organization and pinpoint the efforts required to implement it. This is the phase of organization and implementation planning.

The implementation effort is comprised of many particular tasks all with the objective of making the system work. One area supplies the necessary functional analysis and systems thinking which involves supplying the support systems which back up the operational system. Another aspect requires the outlining of the effort to develop schedules, review procedures and establishing methods of identifying system components in terms of drawings, critical specifications data and other means of specifying where they fit in the overall system.

Associated with identification of components is the need for a record-keeping procedure to maintain up-to-date information about the system as it becomes available. Further preparation is required to specify how the system is to be tested and to provide for potential growth and change via future alternatives.

All these requirements for organization and planning should be established initially. A systematic and clearly stated implementation plan provides the basis for communicating to people and organizations exactly what is expected of them in terms of development installation and eventual operation of the system.

Lastly, the discussion of implementation to ensure a reliable system. Reliability refers to the probability that a device or system to perform adequately for the period of time intended under the operating conditions encountered.

With the value of systems growing larger and larger, it is apparent that to justify the heavy expense of systems of the sizes that are currently being designed and built, such systems must continue to operate a high percentage of the time. That is to say, a system which does not operate satisfactorily may incur losses of significant amounts and tie up large amounts of money which might otherwise be gainfully employed. It is extremely important from cost and operational standpoints for commercial systems to be reliable. Without question, reliability is an essential consideration both in developing and implementing.

Reliability considerations must be included in a system during its conception as well as during its design phases. Likewise, reliability must be considered during the ongoing phases of development, installation, testing, operation and maintenance. Therefore, a reliable system must be conceptualized, understood and controlled throughout its overall period of existence from conception to operation. Each step must be performed in a systematic and careful fashion.

Through the use of recording field data on prior and existing systems, it is possible to develop information as to what performance has been obtained from them. Thus, facts can be provided to ensure that the current and proposed designs will be adequate and reliable. From a management perspective it is essential that reliability be considered a prime objective of design and implementation similar to performance, cost, time, and maintainability.

The reliability aspect of systems engineering is a significant factor in the way the overall system is implemented and in the results that can be obtained from the system's operation.

T

Tab. A tab is a nonprinitng spacing action on a tape preparation device whose code separates groups of characters in a tab sequential format.

Also see: Computers.

Table. A table is an array of values which can be utilized by a computer program to relate values to certain basic conditions. A table might be used to provide correction values to a servosystem based upon the range of weight of the load. Tables are used to convert data from one situation or environment to another such as when using a program on more than one size or type of part. Sometimes a generic template is created in a program with specific values being added from a table. Tables are a convenient way to present a large quantity of orderly data in a minimal amount of space.

A table also refers to the moveable portion of a machine tool upon which the workpiece is mounted in the case of a milling machine or a drill press or similar type of machine which typically has a horizontally oriented workholding platform.

Also see: Array.

Tablet Digitizer. An input device common to drafting organizations is the large tablet digitizer or board on which full-size drawings may be traced or created. Drawings can be created without the use of a telewriter by designating squares on a menu, digitizing directly onto the board, or a combination of both.

Tab Sequential Format. A tab sequential format is a means of identifying a word by the number of tab characters that precede it in a block. The first character of all words is a tab character. Words are presented in a specific order, but all characters in a word, except the tab character, may be omitted when the command represented by the word is not desired.

Tachogenerator. A tachogenerator is a resolving device which deciphers dynamic speed (usually rotary motion) and generates a signal. The signal is either an analog signal proportional to the rate of speed or a digital signal which uses binary representation to indicate the speed relative to a predefined range and scale. Tachogenerators are used in motors and on robot axes and conveyors to determine the speed for coordination with other devices. A robot arm may need to continuously determine the speed of its joints to perform programmed-speed movement of its tool tip. Tachogenerators monitor motors to protect against ''run-away,'' which is the operation at an undesirably, and possibly damaging, high rate of speed especially when no load resistance is being applied to the motor.

Also see: Encoder, Resolver.

Tag. A tag is one or more characters, attached to a set of data, that contains information about the set, including its identification. A tag also is an alphanumeric name attached to an entity.

Taguchi Method. Developing high-quality products in a way that reduces costs is a major tenet behind the Taguchi Method. This approach to quality was instituted by Dr. Genichi Taguchi, executive director of the American Supplier Institute, Inc. (ASI) of Dearborn, MI. Taguchi's approach to quality is called quality engineering. It involves combining engineering and statistical methods to achieve improvements in cost and quality by optimizing product design and manufacturing processes. These statistical methods—known as the Taguchi methods—are currently being implemented by a growing number of companies.

Nancy E. Ryan explained more about Taguchi methods in an article in the May 1987 *Manufacturing Engineering*. The article titled *Tapping into Taguchi* explained: "Taguchi defines quality as cost or loss to society. Inherent within the cost/loss category is customer dissatisfaction, which leads to a supplier's lack of credibility and an eventual loss of market share. Taguchi's motive in the pursuit of quality is not monetary; it's based on responsibility to society. When he proposed his methods to AT&T Bell Laboratories (Holmdel, NJ) in 1980, financial compensation was not an issue.

"Bell Laboratories and Xerox Corp. (Stamford, CT) experimented with the Taguchi methods in the early 1980s. Lawrence P. Sullivan, chair of ASI—home to the Center for Taguchi Methods—learned of the Taguchi methods while conducting a study mission in Japan for the Ford Motor Co. (Dearborn, MI). In October 1982, Ford introduced the technology to its suppliers. The 10 case studies presented at ASI's First Symposium on Taguchi Methods in 1983 were automotive related. At the Fourth Symposium on Taguchi Methods, held last October in Dearborn, MI, 11 of the 43 presentations were from nonautomotive companies."

In an article in the same publication titled *Taguchi Methods Quantify the Relationship Between Design and Manufacturing*, John Vergoz of the Budd Company explained: "A definite benefit to the Taguchi methods is that design engineers and process engineers learn how to talk to each other in a common language. Not only can the two groups quantify their requirements from each other, but at the same time, they can quantify the relationships between the manufacturing process and design requirements."

Vergoz adds that design and process engineers can pinpoint which variables have the strongest functional relationship to the products' requirements. The Taguchi methods isolate the effects on the product of adjusting manufacturing variables that can be controlled. The methods also determine what effect uncontrollable variation in the manufacturing process has on quality.

Vergoz points out three strengths of using the system. First, the methods help determine the functional relationship between those things that can be controlled and the outcome of the process. Second, the methods can be used to move the mean of the process results to the desired position by changing controllable variables. Third, the Taguchi methods determine the relationship of noise—data and variables that cannot be controlled, including the stackup of normal processing tolerances—to the variation in the product as manufactured.

For example, the Taguchi methods were used to optimize the machining of a disk brake rotor. The part is machined to tolerances beyond anything that can be developed from machining tables or database lookups. With the methodology, engineers determined correct feeds and speeds, coolants, and cutting tool materials.

Vergoz: "We're pushing day-in and day-out operations beyond what the standard tables will permit." Total brake rotor runout—measured on parallel surfaces perpendicular to the axis of the conical section of the bearing seat—is held to very tight tolerances. The functional specifications to be met include predictable and quantifiable braking action along with absence of brake pedal vibration. The braking action and pedal smoothness are both dependent on predictable surface finish, parallelism, and perpendicularity.

The Budd Co. has developed the following summary of the benefits of the Taguchi methods:

- The methods enhance communication between functional groups by providing a uniform format of communication.
- They help the company determine the best methods of process control by rapidly determining the optimum setting for a number of control variables.
- They permit a more efficient method for fine-tuning processes, permitting fewer adjustments with more predictable effect from each adjustment.
- They reduce process costs with essentially no capital expenditures.
- They can be applied to a wide variety of processes.

At the Budd Co., many manufacturing processes, including metalforming, machining, foundry work, and plastic molding, are being optimized with the Taguchi methods as a means of reducing product variability with optimized process control.

Tape. See: Computer Grade Tape, Loop Tape, Magnetic Tape, Numerical Control, Punched Tape.

Tapping. Tapping is a process for producing internal threads using a tool (tap) that has teeth on its periphery to cut threads in a predrilled hole. Threads are formed by a combined rotary and axial-relative motion between tap and workpiece.

Teach. To teach is to show or cause something or someone to learn an activity or some information. The term teach in the industrial world is sometimes used when describing the process of creating a program in a robot or numerically controlled machine which must execute a set of programmed moves to perform the task. Most robots have a "teach" mode which allows an operator to move the robot through the desired path of motions while the robot's controller records the orientations of the arm and other necessary attributes such as speed of movement along the motion segments, dwell time at any points of interruption in the motion, and any other actions the robot must perform when executing the programmed moves.

An expert system or a system with artificial intelligence may be capable of being taught information to allow it to make decisions. A vision system used for recognition of parts must first be taught the attributes of the objects before the system can perform the desired tasks.

Also see: Artificial Intelligence, Expert Systems, Machine Vision, Industrial Robots.

Technical and Office Protocol (TOP). Technical and Office Protocol or TOP is a companion to the Manufacturing Automation Protocol (MAP) specification. While MAP is intended to be used on the factory floor, TOP is intended to be used in the office environment. While the MAP portion of the MAP/TOP specification is spearheaded by General Motors, the TOP portion is being lead by Boeing. The idea behind TOP is to develop a protocol specification for data communication in the office environment, which can be integrated with the data flow to and from the factory floor which uses MAP.

TOP is based on the International Standards Organization (ISO) seven-layer model for Open Systems Interconnection (OSI). TOP specifies an approved or emerging international standard for each layer of the ISO model. While layers two through six of the ISO standard are essentially identical for MAP and TOP, layers one and seven have some significant differences between the TOP and the MAP protocol specification. While both protocols have seven layers based on the ISO standard, TOP uses IEEE 802.3 CSMA/CD (Carrier Sense, Multiple Access/Collision Detect) on a baseband media, while MAP uses IEEE 802.4 token access on a broadband media. The IEEE 802.3 standard is similar, but not identical to, Ethernet communication local area networks (LANs). Broadband is being studied for use by the CSMA/CD (IEEE 802.3). Layers two through six are essentially identical with the MAP standards, however, layers one and seven are different.

At layer one (Physical Access) TOP specifies IEEE 802.3—CSMA/CD (Carrier Sense Multiple Access with Collision Detection)—using a baseband (single-channel) transmission medium. MAP specifies IEEE 802.4—token access on a broadband (multichannel) medium. Broadband is being studied for use in IEEE 802.3 and TOP. In some instances, broadband is currently used for Ethernet.

Layer seven (Application) with FTAM (file transfer access and management) which uses the same ISO standard in both MAP and TOP. A user can access and modify files in either MAP or TOP nodes, from either type of access node. While the same basic ISO FTAM standard is used by MAP and TOP at layer seven TOP has limited file management and can handle ASCII and binary data only.

TOP is expected to accomplish several major architectural and operational objectives. TOP will allow the interconnection of multiple office LANs, which can be integrated with wide-area networks (WANs) and private branch exchanges (PBXs). TOP will make possible an office communications network in which equipment from multiple suppliers could be utilized by facilitating free and easy data access and the interchange of data, by equipment from different suppliers. The cost of office systems is expected to be less by reducing the need for multiple cables and customized networking software. The flexibility and adaptability of production systems will be improved to meet changing demands and shorten lead time for designing and implementing integrated office systems.

TOP is expected to be developed in three phases, with all three drawing from work done in MAP. Phase one includes basic file transfer and limited file management. Phase two covers file access, enhanced file management, and message handling. Phase three covers document revision and exchange, directory services, graphics, and database management.

Also see: Manufacturing Automation Protocol.

Technology Transfer. Technology transfer is when technological information is communicated from one entity to another such as when a research and development group transfers the knowledge of a new process to the manufacturing groups. Technology transfer may occur between the academic world and the industrial world. Technology transfer is sometimes difficult since usually it involves interaction between significantly different environments and individuals with varied and sometimes conflicting interests. Technology transfer sometimes involves training people in new methods and processes, while other times, it may require that individuals with different skills be utilized. Since the primary objective of manufacturing entities is to produce products using existing methods and established procedures, technology transfer will always be a necessary activity in the industrial world. Some of the aids to technology transfer are professional trade shows and conferences which present speakers with case histories and tutorials. Industry conferences and their written proceedings are one of the major tools used for technology transfer.

Telecommunications. Telecommunications is communications over relatively large distances by any transmission, emission, or reception of signals, signs, writing, images, and sounds, or intelligence of any nature by wire, radio, visual, or other electrical, electromagnetic, optical, acoustic, or mechanical means. The process enables one or more users to pass to one or more other users information of any nature delivered in any usable form, such as written or printed matter, fixed or moving pictures, words, music, visible or audible signals, or signals that can control the functioning of equipment or mechanisms.

A telecommunication carrier is an organization that provides telecommunication services. The carrier may be a private organization, such as the Teletypewriter Exchange Service (TWX) or the International Teleprinter Network (TELEX), or a government agency. It may serve the general public, such as a common carrier, or it may serve only a select set of users (customers, subscribers). Compuserve and The Source are examples of private organizations which provide telecommunications service to its subscribers through the use of their bulletin board services.

Also see: Network.

Teleoperated Robotics. The role of teleoperated robotics has been to control hazardous materials as well as explore the reaches of space and the depths of the seas. These devices allow direct human control over manipulation activities that occur at a remote location.

H. Lee Martin and W. R. Hamel noted in their SME Technical Paper *Joining Teleoperation with Robotics for Remote Manipulation in Hostile Environments*: "A teleoperator system is not a robot. Most teleoperator systems consist of a manipulator that is similar to a robot in many aspects. What sets teleoperators apart is the man-machine interface which allows real-time interaction between the human operator and the mechanical manipulator. The most common form of teleoperator man-machine interface is the replica master arm. This interface allows the user to operate the slave throughout the workspace simply by moving the master manipulator arm, which is normally a kinematic replica of the slave. Other forms of control include rate controlling joysticks and joint switch controls.

The advent of nuclear energy brought the initial development of mechanical master/slave manipulators as a means of reducing human exposure to radioactive materials. As larger nuclear processing facilities were built, the need for large-volume manipulation led to the development of electromechanical manipulators in the 1950s. Teleoperated manipulations handle radioactive materials as well as work in radioactive environments."

Several example applications illustrate the type of work these manipulators are performing.

For the past two decades, the United States Navy has used remote controlled submarine-type vehicles to inspect and repair underwater cable. Work has also been done in the area of robotic cleaning of underwater surfaces (usually the surfaces of ships).

Another teleoperated application is the Surface Sampler Subsystem aboard the Viking landers which explored the surface of Mars in the mid-1970s. The "robot arm" of the Space Shuttle is probably the best-known example of a teleoperated manipulator.

In a more down to earth example, manipulators were added to foundry applications to remove sprues and runners. They are removed from a table in a shakeout area—an area where workers are subject to intense heat, air-borne dust, and noise.

Most recently, teleoperated manipulators on submersible vehicles have been used to retrieve artifacts from the Titanic wreckage on the ocean floor. This application extends man's grasp to environments that are too highly pressurized for human occupation.

Finally, one company has remote automation manufacturing technology for material handling and process control in the area of remote demilitarization of chemical munitions.

Also see: Forging, Industrial Robots

Teletypewriter (TTY). A teletypewriter is a printing telegraph instrument that has a signal-actuated mechanism for automatically printing received strings of characters in the form of messages and that usually has a keyboard similar to that of a typewriter for sending messages by manually depressing the keys or by using a perforated tape transmitter. A teleprinter is a receive-only unit that has no keyboard. Teletype is a trademark of The Teletype Corporation.

Template. A template is an accurately shaped profiled plate used for either making or checking the product of machining or other manufacturing operations.

The uses of templates range from large lathe applications to miniature precision contour gage applications. Steel rolling mills use templates to control roll-turning tracer lathes and to check the finished parts for size and accuracy.

Building aircraft engines requries many airfoil-shaped templates to control tracer machines, construct prototype models, and check final shapes. Without electrical discharge wire cutting (EDWC), these templates would be laid out by a skilled layout person, then sawed and filed to shape. A considerable number of attempts may be required before the final shape is approved. This hand process is both tedious and marginally repeatable from an accuracy standpoint.

Electrical discharge wire cutting can cut templates accurately, quickly, and without the need for highly skilled labor. A further benefit is that changes can be effected quickly by simple NC tape alterations and, possibly more important, any number of templates can be produced identically when NC is employed.

One of the best applications of a true CAD-/CAM link in many companies is the manufacture of sheetmetal components. Traditionally, such parts have been routed manually on broken-arm routers using individual routing templates as masters.

For typical parts, such an approach takes 14 to 21 days from engineering release to finished part. Using automated template systems, manufacturing engineering accesses the design database, adds changes and information required for manufacturing, and then automatically generates control data for the NC routers. Flow time using such a system for any part can be reduced to as little as one hour.

Templates are typically prepared by digitizing undimensioned drawings or photocontact masters released by engineering. In addition to the digitizing process, manufacturing engineering must add hold-down tabs and other manufacturing details.

With CAD/CAM, manufacturing engineering can bring up the engineering release on a graphics system—the same kind of system used by engineering to produce the initial design. It can add the manufacturing information graphically and nest the pattern, also in a graphics mode. When the nesting has been completed, the data is transmitted to a floppy disk that runs the special NC router.

Terminal. A terminal is usually an operator interface which is physically located at an endpoint of a communication line. A terminal or DTE (Data Terminal Equipment) device is a data input/output device. Some terminals are "receive only" such as monitors without keyboards or printers. Some terminals are "send/receive" such as the typical CRT terminal with keyboard or a printer with keyboard. Most terminals are capable of several basic functions such as being able to translate the data which is received and convert it into viewable form if the terminal has a monitor, or drive a printing mechanism if the terminal is a printer. Other functions such as answer-back identification are standard functions necessary to communicate over modem lines. Terminals are referred to as *dumb* and *smart*, dumb terminals providing the basics of input and output with direct control by the host. Smart terminals have capabilities to do such things as maintain more than one screen's amount of data, provide horizontal and vertical scrolling without interacting with the host, generate graphics locally, perform editing, drive a printer with an integral printer port, display in split screen mode, and provide dual session connectivity.

Also see: Cathode Ray Tube, Computer Graphics, Computers, Data-circuit Terminating Equipment, Data Terminal Equipment, Intelligent Terminal, Liquid Crystal Display, Monitor, Monochrome, Multimedia Terminal, Storage Tube Display.

Testing. See: Acceptance Sampling, Acceptance Test, Computer Aided Testing, Eddy-current Testing, Holographic Nondestructive Testing, Machine Vision, Radiographic Testing, Suspectibility Testing, Underwriter's Laboratories.

Thermal Printer. A thermal printer is a non-impact printer that uses heated printing elements to cause specially coated paper to change colors and display the desired characters and graphic images. Most thermal printers use a print head which has a dot-matrix of heater wires which are in a similar configuration to typical dot-matrix printers which fire solenoid pins into the back of an inked ribbon. The difference with thermal printers is there are no solenoids. Some thermal printers use a line matrix of elements, to print a line at a time instead of a character at a time.

Thermal printers are very quiet compared to impact printers. Thermal printers can be manufactured at very low cost and when the quality of the output is not extremely critical or does not require color, thermal printers are a low-cost alternative. One of the major detractions of thermal printers is that the printout, which is done on specially coated paper, tends to yellow significantly when exposed to light, since the same attributes which allow the paper to respond to the heated elements also are sensitive to sunlight.

Thermit Welding (TW). Thermit welding is a fusion welding process that produces the joint by an exothermic reaction between finely divided iron oxide (or copper oxide) and aluminum. The reaction produces superheated molten metal, which is allowed to flow into a mold surrounding the joint where it solidifies to metallurgically bond the parts together.

Third Generation Computers. Third generation computers use microminiature integrated circuitry, process multiple programs concurrently and use high-level, fairly standardized languages to allow the use of a given program on different computers. High-speed disks are used for magnetic media storage. Computers are smaller, faster, more reliable, use less power, and are less expensive per computing task. These attributes have broadly expanded the areas of computing applications.

Threading. Threading is a process of both external and internal (tapping) cutting, turning, and rolling of threads from a particular material. Standardized specifications are available to determine the desired results of the threading process. Numerous thread series designations are indexed for specific applications. A common threading process is with the use of a lathe. Specifications such as thread height are critical in determining the strength of the threads. The material used is taken into consideration in determining the expected results of any particular application for that threaded piece. In external threading a calculated depth is required as well as a particular angle to the cut. To perform internal threading the exact diameter to bore the hole is critical before threading. The threads are distinguished from each other by the amount of tolerance and/or allowance that is specified. Threading machines are designed to take into consideration a range of specifications such as the maximum diameter, threading pitch, number of threading cuts and speed. Threading in context with pipe threading takes into consideration pressure ratings using published standards.

CNC lathes and turning machines are employed for faster and higher volume threading.

Also see: Turning.

Thresholding. The simplest (and least effective) way of generating a binary image is known as thresholding. In this method, any pixel with an intensity above a certain value is made white, and below that value, black. Windowing is a derivative of thresholding where the pixel is made white if it is above a certain value (bottom threshold) and below a second value (top threshold).

Also see: Machine Vision.

Throughput. Throughput is the number of bits, bytes, characters, words, blocks, messages, calls, or other units of data that pass through all or part of a communication system. The maximum throughput occurs when the system is working at optimum capacity, which may be at saturation. Throughput is usually expressed in data units per unit of time, such as blocks per second or messages per day.

Timesharing. Timesharing is when a large, usually mainframe computer, is used to provide computing services and the user must share the computing resource with others. Timesharing originated when the cost of early computers was prohibitive for any single user to own a system, and therefore the only economical method was for a user to pay for computing time on an as-needed basis to a service bureau who owned and maintained the system.

Timesharing systems used to run in batch mode where the users must submit their desired tasks which are then placed in a queue to be run on the computer when the task's respective turn occurs. Sometimes priorities are included as a defined attribute of the submitted job, which allow a job entered in the queue at a later time to be executed before an earlier entry. The results of batch processing are then returned to the user at a later time. When timesharing was provided in an on-line mode, the cost was extremely high as the user was charged for both CPU processing time and connect time. Since terminal servers and local editors didn't exist, every activity required CPU time.

The development of personal computers and other lower cost computer systems has put computing power within the reach of individual users, which has mostly obsoleted the need and viability of timesharing.

Also see: Computers, Distributed Data Processing, Numerical Control.

Token. A token is a time slot or packet of data that is passed from one station to the next station on a ring network. The packet may contain the address of the station or may simply be an ''empty slot'' available to any station that has traffic to place in the packet. Upon using the packet, the station sets a flag and places a destination address in the packet header to indicate the slot is full. The token moves around the physical ring, is checked at each intermediate station for a relevant address, and is eventually passed to the destination node. The receiving station checks for errors and relays an ACK or NAK back to the originator, which then sets the token to ''empty.'' As with any protocol, some form of window control is needed with the token passing scheme. For example, multiple tokens may not be allowed; a station is not allowed to use the same token twice; and stations may be limited in the number of sessions that can exist between end users at any one time. Duplicate tokens are usually not allowed due to the complexity of the ACK/NAK logic.

Token Bus. A token bus is a physical bus topography network which logically functions with the use of a token, which is passed from station to station. Using a bus path, the stations pass the token by placing the address of the next logical recipient in the header of the token packet. The signal passes along the bus and is physically monitored by all stations, but the token is made available to a receiving station based on the sending station's placement of a station address in the destination header. In the event a token is passed to a failed node, the originator will time out, retransmit a given number of times, and eventually pass over the defective station. The IEEE 802 Standard supports both CSMA/CD and token passing. Bus and ring topologies are included in the recommended standard. The 802 Standard also includes specifications for token bus and token ring access.

Also see: Local Area Network, Manufacturing Automation Protocol, Network, Technical and Office Protocol, Token.

Tolerance. Tolerance is the measurement of the allowable error in a measurement, location, occurrence, weight, distance or other measurable quantity. Tolerance implies that less than perfection is acceptable or usable. Tolerances are typically specified to allow interchangeability of like parts and to allow a product or system to be designed and built in pieces and portions, while allowing the merged assemblies to fit and function adequately. Before tolerances were a normal part of specifications, a product had to be individually crafted to allow hand-fitting of each unique assembly. Tolerances are also used when describing performance characteristics of processes and testing results. To avoid focusing on less than optimum use of resources, tolerances are specified to allow certain activities with minimal impact to be less than perfect while focusing efforts on improving the activities with a greater impact on the critical factors of interest. In certain industries, generally accepted standards have been created for understood tolerances in the absence of some specified values. Tolerances are sometimes indicated by absolute values and other times defined as a percentage of the total dimension to which the tolerance applies.

Also see: Acceptable Quality Level, Tolerance Chart, Tolerance Control.

Tolerance Chart. A tolerance chart is a graphic tool used to ensure the orderly and accurate development of mean sizes and working tolerances required by a new machining process or to analyze a set of existing mean dimensions and their tolerances to determine if a part can be made to print.

The latter task is by far the easier of the two applications and is not discussed further in this section since the technique for building a tolerance chart for a new process also permits one to analyze existing processes.

The tolerance chart assumes the character of an accountant's worksheet except that tolerances are manipulated rather than dollars. Just as the entries in the accountant's worksheets are entered in strict conformance to rules and procedures to arrive at an ironclad picture of the results and the manner in which they were obtained, every numerical entry in the tolerance chart is also based on rules and procedures.

Also see: Tolerance, Tolerance Control.

Tolerance Control. Today, more than ever before in industry, there is a growing need for a systematic approach to tolerance control in machined parts manufacturing. This need is based on the self-evident importance of ''doing it right the first time''—which implies that structured, methodical, analytic techniques are to be properly used at the right point in time to locate, define, and resolve tolerance problems.

It is no longer practical or economically feasible to use hit-or-miss tolerance calculations, to pluck tolerances from the air because accurate machines are available, or to wait for tool tryout on new processes to reveal the preseence of a tolerance control problem. By that time the situation is cast in concrete, the tooling and gaging have been built, and the best of frantic corrective actions will seldom yield an ideal, cost-effective solution to the tolerance problem. These ''debugging'' actions for tolerance control problems are characterized by maximized start-up times, demands that product design open up tolerances to suit the process deficiencies and thus legitimize them, and an increase in manufacturing costs over those that could have been possible if proper tolerance control techniques had been used at the ''paper stage'' of planning.

Until this point the term *tolerance* has been used in an all-inclusive sense. However, tolerances can be defined as being concerned either with physical sizes of features on a part or with the geometric characteristics of those features. Complete coverage of standard practices for dimensioning of sizes and geometric characteristics is given in the ANSI Standard Y14. 5-1973.

This standard applies to the dimensioning and tolerancing of machined part drawings and is universally accepted by American industry and by the U.S. Department of Defense. Agreement between this standard and the British Standard BS 308, the Canadian Standard CSA B78.2, and the International Standard ISO-R1101 is very close or equal in most respects. Evolved over the years, these standards make it possible to unambiguously define the design intent in a standardized, uniform manner and thereby communicate precise understandings related to workpiece drawings.

Many companies find it desirable, however, to develop and assign their own in-house specification numbers to the same body of methodology. Comparison of most in-house specifications on dimensioning and tolerancing with specifications in the ANSI Standard Y14.5 will show almost complete correspondence except in those special situations unique to the company's product line.

Whether a part is made completely to print in one operation or routed over a series of machines, some of which may be NC, the process engineer must be capable of recognizing that a tolerance stackup situation has been created which will affect the tolerances assigned to the machining cuts. When tolerance stackup problems must be handled, the easiest, quickest, and most foolproof way is by use of the tolerance chart.

Whether the tolerance chart is built manually or by a computer program, it is only built after all the initial engineering decisions have been made concerning the process. These decision include:

1. The sequence of operations to be performed.
2. The machine selection for each operation, based on its capacities and known accuracies. (Note: If a transfer line is to be designed, the same considerations apply except that the stations of the line will be custom designed to accuracy levels based on the tolerances developed by construction of the tolerance chart.)
3. The dimensioning patterns for the cuts to be made in each operation. (These patterns do not always correspond to the blueprint.)
4. The selection of the locating surface to be used in each operation. (These datum surfaces do not always conform to those shown on the blueprint.)
5. The kind and type of tooling to be used in each operation to control geometric characteristics such as squareness, parrallelism, concentricity, and symmetry.

Once these decisions have been made, and possibly subjected to critiquing by tool engineers, the master mechanic, and others. A tolerance chart can be constructed to generate the dimensions and tolerances required by each process cut. Properly constructed, the tolerance chart will verify that the following criteria for economical production have been satisfied:

1. Within the framework of the process/tooling decisions, as much as possible of the blueprint maximum tolerance has been allocated among the in-process cuts, which results in the maximum possible tolerance being assigned to each cut in the process.
2. The minimum and maximum stock removals on secondary cuts are practical and acceptable to the shop.
3. Every tolerance assigned is equal to and preferably larger than the estimated process capability for the cut in question. Since the relationship between the working tolerance and the process capability has direct bearing on the frequency of tool changes or adjustments, many companies have in-house rules that call for the working tolerance to be 1.5-2.0 times the process capability value.
4. During the course of building the tolerance chart, it may become obvious that one or more of the initial process/tooling decisions results in assigning an impossibly tight tolerance to an in-process dimension. When this happens, it is necessary to change these decisions to satisfy the criteria for economic production.
5. Since all these decisions are still in the paper stage, that is, no tooling has yet been designed, no great time or dollar loss will occur if a process change is required. However, failure to respond to the clear signals from the tolerance chart will result in the problems on paper being transferred into iron on the shop floor.
6. The widespread and growing use of NC machining, when it can be applied, has reduced the extent of the tolerance stackup control problem by allowing cuts to be machined as shown on the blueprint dimensioning schemes, by eliminating manual control of machine decisions affecting the cuts, and by reducing the number of location surface changes and the attendant fixturing required by non-NC machining. In general, it has also improved size control and control of geometric characteristics of part features. However, not all tolerance stackups are eliminated by using NC machines.
7. Numerical control machines are often combined with conventional machines to form a grouping of machines to make a part. If the volume justifies it, special-purpose transfer lines are built which incorporate all milling, drilling, boring, and broaching, so that the transfer line is equivalent to a selection from a large inventory of machine tools. Designing such a transfer line without basing the design on a tolerance chart is far more risky than processing a part on the basis of stand-

alone machines selected from a large inventory since each station is custom built for that part alone. Imagine the excess cost if, when the last stations of the machine are built and tested, it is found that failure to meet blueprint specifications is due to a tolerance stackup problem that reaches back to the early station. During testing is no time to find out that the sequence of stations is wrong or that the selection of locating surfaces is responsible for the tolerance stackup problem.

8. Process planning for machining is undergoing major changes as computer programs are being developed to do the complete processing job on either the basis of a family design of parts or the basis of a true generative approach.
9. These programs, to be complete, should include subroutines that will handle the tolerance charting function for development of required mean sizes and tolerances for machining cuts, and that will also analyze the process and tooling capabilities relative to control of squareness, parallelism, and roundness.
10. If predictions are correct, it will come about in time—near, not far—that the design engineer's job will fuse with the production engineer's job. A common database will be in computer memory, along with design and production methodologies. From this amalgam will flow not only the product design but all production engineering outputs of operations sheets, tool designes, tolerance charts, and NC tapes.

Also see: Tolerance, Tolerance Chart.

Tool. A tool is something which can be used to assist in performing some activity. A tool is usually some item used directly to perform the work by the person or object doing the task. A machine tool is the cutting element which is mounted in the tool holder which contacts the workpiece and cuts into the surface to remove the material. Tooling on a robot could consist of a gripper or a painting nozzle, or possibly a welding tip.

A tool for CAD/CAE systems may be a software routine which performs an operation which otherwise would have to be done manually or which may be too complicated to achieve without the assistance of the tool, such as simulation of several logic circuit configurations over an accelerated period of time. Due to the inherent limitations of a person or machine, a tool may simply be a necessity to perform a task.

A *tool assembly* is a complete assembly usually consisting of the toolholder with collet; where necessary, the cutter, and if applicable, the tool insert. The toolholder fits directly into the spindle nose of the machine.

A *tool function* is a command identifying a tool and calling for its selection.

Tooling resins are resins that have applications as tooling aids, coreboxes, prototypes, hammer forms, stretch forms, or foundry patterns. Epoxy and silicone are common examples.

In his book *Fundamentals of Tool Design*, author Edward G. Hoffman describes tool design as "a specialized area of manufacturing engineering which comprises the analysis, planning, design, construction, and application of tools, methods, and procedures necessary to increase manufacturing productivity. To carry out these responsibilities, today's tool designer must have a working knowledge of machine shop practices, toolmaking procedures, machine tool design, manufacturing procedures and methods, as well as the more conventional engineering disciplines of planning, designing, engineering graphics and drawing, and cost analysis."

Since the advent of numerical control, a new generation of tool holders ranging from general to highly specialized use, has been developed. Varieties and combinations are almost endless.

Most NC machine builders use their own type of spindle-to-tool locking and mounting devices. These are categorized as either straight shank or tapered shank tooling, and many variations of each exist.

Most toolholders have a special device in the back which allows for some type of locking drawbar to pull the tool firmly into the spindle and to release the tool automatically upon tape command to allow automatic tool changing.

NC machining centers, probably the most versatile machine on the market today, can perform many tasks. As a result, the toolholders must be capable of holding a variety of tools.

Almost all NC machines are more productive if all the tooling is set at approximately the same length. This allows the distance between the tool and workpiece to be set at a minimum, always using the longest tool.

While most NC machines select tools from a numbered pocket in the tool magazine, some magazines select tools by a set of code rings on the tool itself. NC lathes use a special tool holder for turning applications called a qualified tool holder.

The four main objectives in designing a workholding fixture for NC are: accuracy, rigidity, accessibility, and speed and ease of clamping and changing the workpiece.

The major impact of computer aided design (CAD) on tool design will be in the areas of tools and techniques used in accomplishing the design process. Such design tools as pencils, scales, sketch pads, and pocket calculators will eventually become obsolete as they are replaced by computer tools such as keyboards, system displays, digitizers, and plotters. In addition, the techniques used in designing a tool and documenting the design will also undergo significant change.

In designing a tool, the designer goes through several stages. While in reality these stages overlap, for the purpose of instruction, it is helpful to separate them into five distinct stages (see figure):

1. Identify the problem.
2. Develop tentative solutions or designs.
3. Accept the best design.
4. Make models or prototypes for testing.
5. Produce working drawings.

CAD will affect all five stages of the design process. In the initial phase—problem identification—the computer will give designers easy access to data concerning similar problems which have already been confronted, designed for, and solved. The computer gives designers immediate access to previous problems and the tools which were designed to solve the problems. This will allow designers to avoid the problem of redundant effort or ''reinventing the wheel'' each time a design problem occurs.

In stage two—tentative solutions—designers will use the keyboard, system display, digitizer, and a function menu to begin roughing out design ideas and performing calculations.

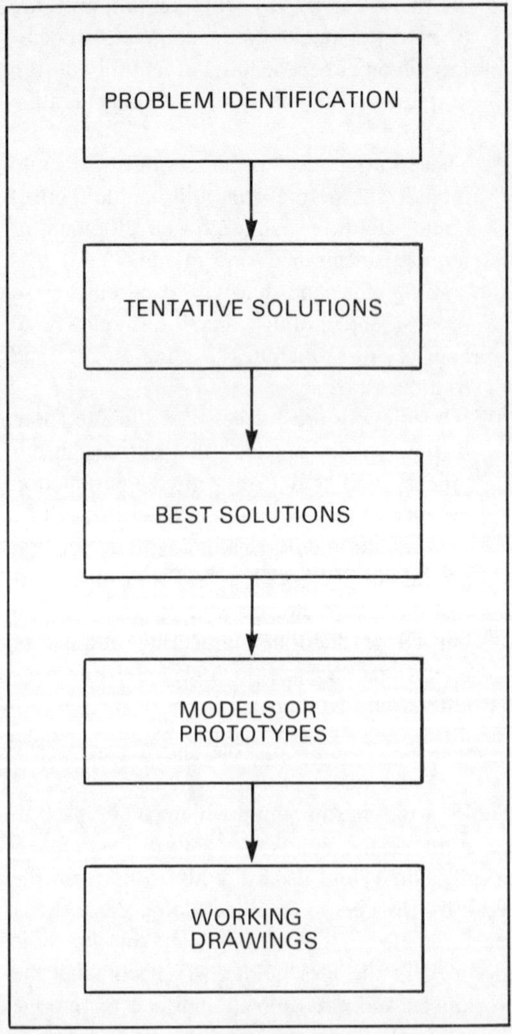

The five stages of tool design.

In stage three—selection of the best design—one of the preliminary designs will be accepted or a compromise which includes characteristics of several preliminary designs will be adopted. Tentative drawings will be called up from storage and displayed. Corrections and revisions will be made as needed copies of the finished preliminary drawings will be plotted and circulated for final input.

The most drastic changes resulting from the computer will take place in stage four—models or prototypes. Rather than making and testing actual models or prototypes of a tool, designers create a three-dimensional model of the tool in the computer. The model is displayed on the CRT and tested through the use of computer simulation.

In stage five—working drawings—the files containing all of the preliminary drawings and the three-dimensional model of the tool will be given to drafting personnel so that they can prepare completely dimensioned and annotated working drawings of the tool. If the tool involves several parts, a parts list will be produced along with the working drawings.

Hoffman also noted the future trends in CAD tooling, "CAD is a rapidly developing and changing technology. The future of CAD in tool design will be one of continuous change and improvement. Hardware development has produced equipment that is vastly superior to its predecessors, and hardware technology can be expected to continue improving.

"Improvements to processors will provide more memory allowing many more workstations to be interfaced with even smaller more powerful central processors. Along with the additional memory will come faster processing time, decreasing even further the amount of time required to design closely toleranced tools. Color displays with larger, thinner screens, but higher resolution can also be expected. Faster, more accurate plotters will become available as research and development in the area of graphic output begins to catch up with input technology. And, finally, voice-activated function menus will become a standard feature of CAD systems.

"Software developments will probably overshadow hardware developments, however. Improved three-dimensional modeling programs will become more readily available and less expensive as will programs for motion and testing through simulation. However, the most significant trend with regard to the computer in tool design will be the continuously improving partnership between CAD and its manufacturing counterpart CAM.

"Automation in the manufacturing of tools is a concept that is even older than automation in tool design. Numerical control (NC) machining processes have been used since the early 1950s. These processes were improved to include computer numerical control (CNC) and direct numerical control (DNC). The current phase is computer-aided manufacturing or CAM.

"CAM involves using the database created when designing a tool on the computer to prepare the instructions for operating automated machines which will make the tool.

"As improvements continue to emanate from the research and development departments of companies which produce CAD/CAM technology, expect to see a situation in which a designer designs a tool, produces the necessary documentation, commands robots to set up the manufacturing equipment, and directs the equipment in the machining of the tool—all from one computer terminal."

Electronic tool management is an important part of modern FMS technology. Shorter cycle times, smaller batch sizes, and increasing part varieties are all problems for tool handling in an FMS. Advanced schedule planning is essential for a responsive tool delivery system. A reasoned application of data management, tool identification, scheduled tool changeover delivery, tool migration, and tool maintenance technologies can provide payback to system implementors.

Also see: Die, Gripper, Machine Tool, Numerical Control, Tool Crib, Tool Migration, Tool Maintenance, Tool Number.

Tool Crib. A tool crib is the place or places in the factory where tools are stored, inventoried and dispensed to shop personnel. In recent years, the lines waiting at the tool crib have grown because the crib's job has increased in complexity faster than manual systems can control. Behind that rise in complexity are tooling decisions that make for faster, more precise, more predictable, and therefore higher quality metal removal. Modular approaches to tooling, along with modular approaches to machine, controller, and software design, have contributed to cutting metal faster. Indexable inserts in a variety of shapes, specialized toolholders, and a variety of tooling, turret, and tool handling systems have made it possible to cut some tough geometries with greater precision and better surface finish.

David J. Gayman reported in the September 1986 *Manufacturing Engineering* in an article titled *Computers in the Tool Crib* that: "Rapid matching of number to number, precise and rapid storage and retrieval of data, consistent memory of storage locations, and lightning search capabilities are what is required. Those attributes are special strengths of the computer. So it is no wonder that software has come from a number of vendors, running on computers big and small and designed for manufacturers big and small, whose intent is to make the tool crib another node in computer-integrated manufacturing (CIM).

"Fisher Control's (Clayton, MO) Sherman, TX, facility is one of 23 manufacturing facilities in 13 countries. A division of Monsanto, Fisher makes automatic control valves and actuators, gas regulators, field and control room instrumentation, and process control systems. The Sherman plant cuts metal. The facility employs computer numerical control (CNC) and conventional turning, milling, and other metalworking machines. Production is in batches; lots run from 1–800, averaging about 100 pieces. The plant operates three shifts, five days a week, and machines are expected to be cutting metal 108 hours per week. The plant's on-time delivery rate is better than 98%.

"Tooling inventory consists of 7200 tool items, of which about 2000 are fixtures. Fixturing is specialized, with some fixtures weighing up to 1700 lb (765 kg). Small tools are stored in 50 modular storage cabinets. The tool crib has three attendants and two issue windows.

"Ross Long, Fisher's supervisor of machining methods, headed a task force charged with bringing computer tool management to the shop floor. He held meetings first with production supervisors, then with machine operators, and finally with crib attendants. Monthly meeting were held with the entire shop force, on company time, to gather suggestions. The recommendations from these meetings were reviewed by the task force and implemented where appropriate. Every recommendation received consideration and a response.

"Installation of the software began in October 1983 and was completed gradually, in modules. The modules Fisher selected were tool inventory control, bill of tooling, and purchase information. Tooling inventory began with Fisher's fixturing because of its low volume and slow turnover. Thus, the shop force was introduced slowly to the operations of the software."

Another *Manufacturing Engineering* article titled *Bar Coding Labeled Efficient and Flexible* described a bar code system at use in a tool crib. The article appeared in the June 1987 issue of the publication and author Nancy Ryan wrote: "General Dynamics Land Systems Division (Troy, MI) turned to bar coding to improve availability of materials needed in production, to provide better control of materials, to standardize tool crib inventory control operations, and to reduce related production costs. The Land Systems Division, which has an estimated $9.1 million tool inventory, has plants in Warren and Sterling Heights, MI; Lima, OH; and Eynon, PA. Bar coding is currently in use at the Detroit Army Tank Plant (DATP) using ATICTS (Automated Tool Inventory Control and Tracking System) application software from Data Enterprise (Bellevue, WA).

"The DATP tool crib is serviced by two terminals located at the crib's front window, each of which supports 20 satellite bar code wands. Production workers at DATP wear special bar-coded badges that feature their six-digit employee numbers. To check out tools, a worker brings a completed requisition form to the tool crib window and displays his or her badge to an attendant stationed at one of the two terminals. The tool crib attendant wands the worker's badge; the worker then proceeds to the high-density tool cabinets that house the tools. Bar code labels representing tool identification numbers are adhered to the cabinets' interiors; attaching labels directly to tools is not effective. Labels representing quantities are located atop the cabinets, as are wands with extension cords. The worker scans the appropriate labels with a wand, indicating the tools being checked out, and returns to the front window to complete the transaction. The worker can check in tools at the front window, too; the tools are designated as OK, lost, or broken, which is then recorded on the ATICTS system.''

Also see: Bar Code, Tool.

Tool ID. A tool ID is a unique physical tool identification, usually machine-readable, which is directly or indirectly referenced by the *tool number*.

Tool Migration. Tool migration is a function of tool management and tool delivery which can collect and transport unneeded tools from some stations in order to fill processing needs at other stations.

Also see: Tool.

Tool Maintenance. Tool maintenance is a function of tool management and tool delivery which can collect and transport worn and broken tools from stations for return to the tool staging area for inspection and/or repair, and then return to service.

Also see: Tool.

Tool Number. A tool number is a logical tool identification which is referenced by NC programs and tool management system.

Also see: Numerical Control, Tool.

TOP. See: Technical and Office Protocol.

Top Stop. A top stop is a machine-generated signal for stopping a press at the top of a stroke (after each cycle).

Torch Brazing. Torch brazing is a process for producing a brazed joint in which the parts being brazed are heated by burning gas combinations such as oxygen-acetylene, air-natural gas or air-propane in a torch assembly.

Torch Soldering. Torch soldering is a soldering process that uses heat from a torch flame to raise the components to soldering temperature.

Total Quality Commitment (TQC). Total quality commitment is a term which has been used to describe the more complete commitment to quality in all aspects of a manufacturing environment or a business. TQC includes all the traditional aspects of quality assurance, the necessary portions of quality control (quality assurance eliminates the need for 100% inspection and other stringent quality control procedures) and other crucial aspects such as defining and supporting the critical success factors of the business, longer range planning, and continuing cost reductions and quality improvements. JIT (Just In Time) methodologies are an integral part of TQC. A TQC program involves an environment larger than just the plant or business itself, but also the employees, suppliers, vendors, and customers. TQC is a way of doing business which can be applied equally well to ''white-collar'' business services as well as manufacturing and engineering environments. Sophisticated statistical analysis and mathematical prediction methods are employed to carry out the practices of TQC.

Also see: Quality Assurance, Quality Control.

Touch Screen. Touch screen technology allows the user to input data (usually a response) by touching a terminal screen. Touch screen technologies available are: transparent switch matrix, analog resistive, capacitive, surface acoustic wave, and infrared.

In the SME Technical Paper *A Graphical User Interface with Touch Screen and Icon Driven Software*, Jeanne Temesan Merchant (General Motors) wrote: "A transparent switch matrix consists of two transparent sheets of parallel conductors (facing each other at right angles) which are slightly separated by air. Finger pressure shorts the top to the bottom sheet, closing the switch. This is low-cost implementation, with only a simple decoder circuit required to detect touch location. Unfortunately, this material (e.g. Mylar) is not abrasion resistant, reduces image clarity, and does not adapt well to curved screens.

"A related technology, analog resistive, uses a resistive sheet overlaid by a conductive sheet, with a slight separation. Voltage is imposed on the back resistive sheet, first along the X axis, then the Y axis. Finger pressure deforms the top conductor onto the back resistor, forming a voltage divider. The detected voltage in each axis is digitized to represent the position of the touch. This analog technique makes high resolution possible, with software selectable switch locations. Unfortunately, the electronics required to implement it are expensive. It also suffers from the same disadvantages as the switch matrix.

"Low-cost capacitive overlays have a limited number of metalized touch areas on the screen. The touch of a finger causes a signal to be shunted to ground through the human body capacitance. The cost of the electronics to interpret the change in capacitance is low, but there are several disadvantages to this technique that make it unacceptable for the factory floor. A conductive stylus is required—a gloved hand, for example, will not work. Detection is affected by variables which can not be controlled, such as humidity and the user's body capacitance.

"The least common (and most exotic) technology is surface acoustic wave. Piezoelectric transducers induce a surface acoustic wave in a glass overlay. The touch of a finger produces and echo wave. the time difference between transmission and reception is measured on each axis, and screen position is calculated based on the speed of sound in glass. The clear glass overlay does not produce visual distortion, although it is thick and bulky. Switch locations are software selectable, and there are no parallax problems. The complexity and high cost of this technique are disadvantages, but one other characteristic rules it out for factory floor applications—dirt or other foreign substances on the glass may cause false touches to register.

"Infrared screens use a static matrix of sequentially pulsed infrared light-emitting diodes (LEDs) and photodetectors, or a single infrared LED light source and detector which sweeps across a ninety degree arc of the screen. These sensors are packaged in a bezel which is often purchased as an add-on kit. When the beams are interrupted by a finger or other stylus, its screen location may be mapped to (X, Y) coordinates.

"Compared to other technologies, an infrared touch screen appears to be the best choice for the factory floor. Since nothing covers the monitor screen, there is no possibility of punctures and no image distortion. A finger, pen, or pencil can be used to select options. We installed a flip-down glass cover to provide additional protection when the screen was not in use. Ambient light compensation may be required, but this is usually provided by the manufacturer in the hardware power-up sequence. Infrared touch screens have finite resolution, based largely upon the space available to install emitter/detector pairs, but it has proven to be more than adequate."

Also see: Touch Sensitive Display.

Touch Sensitive Display. A touch-sensitive display (sometimes referred to as a touch-screen) is an input device which allows a person to use their finger as the device which interacts with the touch-sensitive display. There are essentially two types of touch-screen technology; in one type, a person's finger breaks infrared light beams, in the other type the pressure from the finger closes a circuit on the screen's electrically conductive surface. Both types of touch-screens then convert the position of the person's finger into x-y coordinates and presents the coordinate location information to the computer software.

In the type which uses infrared light beam, the touch-screen bezel has rows of LEDs along two adjacent sides of the screen with LED receivers along the two opposite sides. These rows of LEDs create a matrix of infrared light beams, which when the beam is broken by a finger, the location can be determined. Infrared touch-screens don't actually have to be touched to activate the input, just the light beam has to be broken. Infrared touch-screens can be activated by insulated, inanimate objects such as pencils.

In the second type, a transparent coating is fused to the CRT screen that acts as one plate of a capacitor. When a finger touches the coating, it causes a current drain at the contact spot. Electrodes placed at the four corners of the screen detect changes in the current caused by the touch and determine the position of the finger on the screen. Capacitance type screens have to have pressure applied at a location before the screen can register a location. Capacitance screens must also be touched by a conductive device capable of draining current.

Also see: Touch Screen.

TQA (Total Quality Assurance). See: Quality Assurance.

TQC. See: Total Quality Commitment.

Transceiver. A transceiver is the combination of transmitting and receiving equipment in a common housing, usually for portable or mobile use, employing common circuit components for both transmitting and receiving. It is a device that is designed for simplex (one direction as a time) operation. A transceiver may be stationary, portable, or mobile. Early transceivers were analog devices, but modern transceivers utilize digital communications and digitize all information before transmitting.

An example of a stationary transceiver is a facsimile machine, which is used to transmit as well as receive documents over standard telephone lines, however not concurrently.

An example of a portable transceiver is a walkie-talkie, which can be easily hand-carried. The push-to-talk feature is due to its simplex operating design.

A mobile transceiver would be a device such as the portable printing terminals in modern police cars which allow an officer to request and receive a hard-copy printout of information.

Also see: Network.

Transducer. A transducer is a device for converting energy from one form to another. In NC, a transducer is a device for measuring output and converting it into a signal acceptable to an error detector.

Also see: Numerical Control.

Transfer Line. A transfer line is a manufacturing system in which individual stations are used for dedicated purposes. In more complex, integrated manufacturing systems, a series of robotic work cells are linked to form process continuity. This continuity, which could be called a flexible transfer line, involves a number of considerations as listed in the following discussion. Not all of the factors listed are necessarily applicable, but a well-conceived system would give them due consideration. Proper selection of the robot(s) in terms of size, payload, and dexterity is assumed.

Typically, more than three discrete machining operations are to be performed. Ordinarily a single robot can service at least three machine tools in a work cell. Equipment size and machining cycle times are major determining factors.

Production requirements (rate and volume) may require several machine tools performing the same operation. Load balancing is essential for achieving the required throughput. Operations requiring long time cycles may require more than one machine tool. Conventional or NC machine tools can be mixed. This consideration is a matter of flexibility, economy, and availability.

Reorganization of the plant (layout) may be required. Grouping of the machinery to accommodate the robot's sphere of influence as well as to optimize material flow usually requires equipment relocation.

Batch processing and similar parts grouping (group technology) must be considered, and system flexibility must be inherent.

Applied CAD/CAM may be involved. The application of computer-aided design and manufacturing techniques facilitates much of the system design in terms of tooling and end effector development and graphic tryout.

Product design for manufacturability should be considered together with compatible tooling. Product design for manufacturability is an essential ingredient for cost reduction and system reliability regardless of the type of automation applied. CAD/CAM is a highly useful tool to accomplish this.

Parts feeders, orienting devices, and buffer storage devices must be provided. System flexibility not only facilitates batch processing of a variety of parts but also (unlike conventional transfer lines) provides the ability within the system to maintain throughput even when portions of the system are down for repair, maintenance, or tool dressing. Parts feeders, buffer storage towers, and pallets are essential to achieving this productivity-building benefit.

In-process inspection and gaging must be considered. In-process inspection and gaging provisions also contribute to maintaining high levels of throughput by anticipating performance deterioration. Product quality is enhanced, and scrap is minimized.

Compatible robot end effector designs must be developed. The end effectors must reliably grip and maintain orientation of the workpiece, and they must be designed to work compatibly with the machine tool chucks and tooling fixtures.

Machine reliability and repair, preventive maintenance, and tool refurbishing and changing procedures must be devised. The system design must take into account the inherent reliability and serviceability of the equipment involved so that the desired throughput is achieved. The mean-time-between-failure (MTBF) and mean-time-to-repair (MTTR) significantly influence the design criteria for parts feeders, and orientation devices. Well-conceived and scheduled preventive maintenance and tool refurbishing procedures significantly influence the MTBF and MTTR.

System operation must be integrated through a central supervisory control. Perhaps the most critical single element of the system is the central supervisory control. Its job is to coordinate the sequence of events and flow of material throughout the process. For example, if part of the system work flow is interrupted, the central control interprets the resulting input signals and provides appropriate instructions for alternate actions to be taken upstream and downstream from the point of interruption. Similarly, if in-process inspection results indicate performance deterioration, corrective action and/or alarms are instituted through the control. The importance of a well-designed central control cannot be overstressed.

Factory data (performance feedback) must be provided. An extension of the central control functions is to provide factory data to the management team. Such information as the number and type of failure incidents, changes in parts quality, cycle times, load imbalance, and other trend data is vital to maintaining and improving system performance.

The example illustrated on the next page consists of three robots, a transfer conveyor, two CNC center-driven, double-ended lathes, and brazing, grinding, broaching, and turning machines.

The three robots (10 ft, 3 m reach) are used to transfer the parts between the conveyor and the machine tools. Nine basic programs accommodate the transfer of an entire family of parts as well as alternate bypass transfer programs to cope with machine tool downtime. A library of programs is maintained on cassettes to permit rapid reloading of programs as required by product changes.

Automatic program selection, interlocking of robot movements with other line operations, cycle initiation, fault monitoring, and emergency shutdown procedures are handled by a supervisory programmable controller.

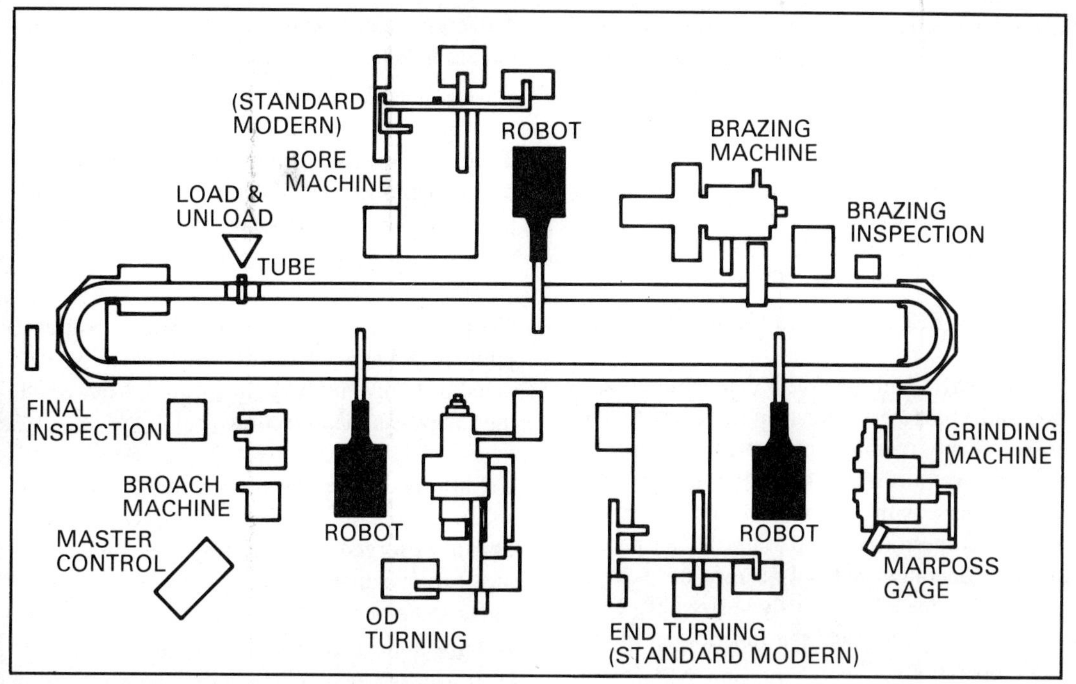

Flexible transfer line—three robots, conveyor, and six machine tools.

Each robot has two hands which independently clamp and unclamp the parts. This allows rapid handling of finished and unfinished parts at each operation, minimizing the load and unload cycle time.

Also see: Industrial Robots.

Transformation. A transformation is when a set of point locations which are originally referenced to one coordinate system are referenced to a different coordinate system, which may require changes to the values of the cartesian (or polar or cylindrical) coordinates. If the relationship between the two coordinate systems is known, the transformation can be done mathematically.

Robots and numerically controlled machines use coordinate transformations to allow a program to be written with moves relative to the workpiece, instead of possibly more complex motion definitions when viewed relative to the machine's base. A common transformation is to define motions in tool coordinates, which enable the tool tip to be moved along simple rectangular coordinates relative to either the tool or the workpiece.

Transformations also allow a program to be developed for a specific application and then replicated for additional similar applications such as mirror-image conditions, larger or smaller scale sized parts or for different machines capable of the same program tasks.

Also see: Coordinates, Industrial Robots, Numerical Control.

Transistor-to-transistor Logic (TTL). Transistor-to-transistor logic is one of the first forms of digital switching logic which allowed the development of integrated circuits. The logic consists of connecting transistors (switches) together in different sequences and arrays which allow certain logical functions to be performed. TTL logic is also characterized by utilizing 0 to 5 volts as the required voltage levels. Other types of ICs such as CMOS utilize up to 12 volts.

TTL logic has been improved upon and several other types of logic connection schemes exist today to serve particular purposes. Emitter-coupled logic (ECL) is a very high-speed logic connection scheme used in high-speed, high-performance computers. At one extreme, gate arrays and other standard cells are used to design a particular functional of anIC, while at the other extreme, ASICs and other types of semicustom and full-custom chips are designed to optimize the creation of a particular IC.

Also see: ECL, Integrated Circuit.

Transparent. With regard to data communications and computers, transparent refers to the absence of any required interaction by a user or operator. A transparent operation is also one which occurs with no outward indication of when it is performed. At times, transparent computer operation not only shows no explicit signs, it also doesn't obviously utilize CPU time or in other ways degrade the functioning of the visible operations.

When transparent is used relative to software and computer programs, the term refers to a change which shows no effect on the output of the program or one which doesn't affect the functioning of the program in its normal fashion within the computer.

In data communications, transparent functions might be the contention or error detection and correction functions which occur with no visible indication of occurrence to the user.

Also see: Computers.

Transportability. Transportability is the ability of a program to be executed on different computers, without major changes and without error.

Also see: Computers, Programming.

Trap. A trap is an unprogrammed conditional jump to a known location, automatically activated by hardware, with the location from which the jump was made recorded.

Trimming. Trimming is a secondary cutting or shearing operation on previously formed, drawn, or forged parts. In trimming, the surplus metal or irregular outline or edge is sheared off to form the desired shape and size.

Excess material beyond or outside the finished part is termed *trimming allowance*. This is necessary because of variations or irregularities in material and processing. This excess is removed in a secondary or finishing operation.

R. F. Humphreys, R. J. Leifeld, and C.A. Ross of the McDonnell Douglas Corporation describe a robot trimming operation in their SME Technical Paper titled *Robotic Sheet Metal Routing*. The authors wrote: ''McDonnell Air-

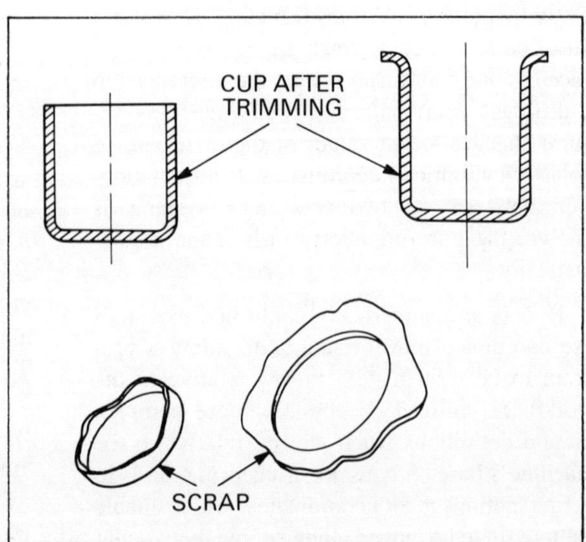

Trimming is the operation of cutting scrap off a fully or partially formed product to establish a trim line. The edge of a cup is sometimes trimmed by pinching or pushing the flange or lip of the cup over the edge of a stationary punch.

craft is currently using a six-axis gantry robot to perform routing or trimming of contoured sheet metal aircraft skins. The working envelope of the robot is divided into four identical subenvelopes. The operator may perform setup in one zone while the robot simultaneously trims a part in an adjacent zone, resulting in free routing time. Part programs are taught by the operator on the production site. The cell is controlled by a stand-alone microcomputer. Special features of the cell include a constant force compliant end effector and automatic coordinate transformation, allowing the teaching and running of any part in any of the four zones.

"The robotic routing cell was installed in the sheet metal fabrication facility in the Fall of 1986 to provide quality and productivity improvements over the conventional manual operations. Contoured aluminum panels for three aircraft programs are being trimmed by the robotic cell. The cell uses a gantry type robot and conventional trim fixtures to perform template guided routing. Cost reductions are realized through labor reductions achieved by simultaneous setup and routing operations. Union hand router operators have been retrained and now are responsible for complete operation of the cell, including both programming and production modes.

"The routing cell is based on a six-axis, electrically operated gantry type robot. The robot is equipped with three linear axes and three rotational axes. The linear axes provide an overall working envelope of 178 inches (X axis) by 170 inches (Y axis) by 36 inches (Z axis). The three rotational axes provide 330 degrees, 210 degrees and 800 degrees of freedom, respectively.

"The working envelope is subdivided into four indentical workstations or quadrants. This is accomplished by bisecting both the X and Y axes, with programmed soft limits as well as mechanical hard stops. This allows the simultaneous operations of routing and tool/part setup, in adjacent workstations.

"A two-axis, hydraulically compliant, end effector is used to interface the cutting tool and the robot. This custom designed unit provides one inch of tool motion (in a linear fashion) in the planes parallel and normal to the robot faceplate. The compliant feature of the end effector is used to provide an adjustable force which causes the router guide bushing to be held against the trim fixture, emulating human operation. The force is adjustable from 5 to 35 pounds. Additionally, mild contours in part geometry can be followed between programmed points, which reduces the quantity of taught points by as much as 50%. Automatic changing of tools (router motors) is also provided by the end effector.

"Utilizing conventional manual trim fixtures, four workstations and automatic tool changing, the cell provides a high degree of efficiency by allowing simultaneous robotic trimming and manual setup."

Turning Centers. Turning centers are modern lathes or turning machines equipped with computer numerical control (CNC), multiple tooling, and flexible automation equipment. With rotary tooling now available, milling, drilling, tapping, and other operations can be performed in the same setup. This can sometimes eliminate the need for a machining center or other machine tools to perform secondary operations.

Many turning centers are of modular design that permits expansion to higher automation levels when required. Turning centers are most frequently used as flexible stand-alone machines. However, they are also being integrated as components of flexible manufacturing cells (FMCs) consisting of two or more machines, a material handling system, and a cell controller.

Advantages of modern turning centers include improved quality workpieces, increased productivity, greater flexibility, and cost savings. Improved quality results from the improved accuracy and reliability of the machines and controls and the integration of in-process or post-process gaging. Increased productivity is attained from faster setups, reduced manual toolchanging requirements, more rapid metal removal rates, and shorter cycles. The capability of producing a wide variety of parts increases their flexibility. Optional equipment is available for maximizing automation with minimum changeover time.

Means of automatically changing worn tools or replacing certain tools with others are not as common on turning centers as they are on machining centers. This is because most turning operations can be completed with fewer tools. However, the increasing demand for smaller lot sizes of a variety of different parts and the availability of power-driven rotary tools for secondary operations are making increased tooling capacity more essential. To meet this need, automatic toolchangers are now available for most turning centers.

Automatic toolchangers for turning centers are of the same basic design concept as for machining centers. They generally consist of a swing arm, double arm, or other automatic changing device equipped with grippers for transferring tools between a storage magazine, rack, drum, or carousel and the machine turret(s).

The cutting units of quick-change tooling used on turning centers are generally preset outside the machines. Then, any deviations from required settings are often automatically adjusted by means of the tool offsets, as directed by the CNC unit, after the tools are placed on the machine. In this way, no measuring cuts or other machine settings are necessary, thus reducing downtime.

TTL. See: Transistor-to-transistor Logic.

TTY. See: Teletypewriter.

Turning. The turning process is a method of metal removal where a work piece is gripped in a suitable holding device and rotated under power against a single point cutting medium which is fed radically or longitudinally to the axis of the work piece. The process is essentially one of surface generation and thus true cylindrical, conical, spherical or planar forms with coincident axes are reproduced in the work piece, depending on the direction of cutter travel.

Single-point shark-type tools are used for turning, boring, facing, forming, cutting off and some types of threading. Cutting tools are commonly used for drilling, reaming, tapping, and treading. Cutting tools used in the turning process may be made from various materials, the choice of which depends on such production factors as power available and condition of the machine tool, work finish required, work piece material, speed of machining as related to production requirements and type of cutting operation. In the turning process, these cutter materials include hardened carbon and high-speed tool steel, the cast nonferrous alloys, cemented carbides and to a limited extent, diamonds.

Power chucks operated by a pneumatically or hydraulically powered drawbar or tube, or having a self-contained power actuating device, are better suited for medium-to-long, repetitive production runs. Many NC lathes are provided with chucks such as these.

Many different designs of power chucks are available from various manufacturers. The higher speed capability of modern NC lathes has necessitated the development of improved power chucks to provide better retention of gripping force under increased centrifugal forces. Most power chucks are either wedge or lever type.

Many power chucks are actuated by an air or hydraulic cylinder mounted on the rear of the lathe spindle, with a drawbar or tube in the spindle connecting the cylinder to the chuck. Many modern NC lathes have spindle-ported hydraulic systems, and power chucks are available that contain a hydraulic cylinder within the chuck body. This eliminates the need for a rear-mounted cylinder and reduces space requirements. The spindle bore is also left open to accept maximum-size bar or tube stock.

Turning Machines. The main types of turning machines (lathes) are engine, turret, single-spindle, automatic screw machine, multispindle automatic, multistation machines, boring, vertical, and tracer. The level of automation can range from semiautomatic to tape-controlled machining centers. The engine lathe is considered to be the basic turning machine. Its main components are the head stock, bed, carriage, and quick-change gear box. The turret lathe has a turret mounted in place of the tailstock. Tools are mounted on the turret and can be indexed so that the appropriate tool is brought to the work when required. The automatic screw machine has five radically mounted tools that are cam-controlled. The stock can be made to feed as it is being cut, thus any desired cylindrical shape may be generated. The tracer lathes are basically the same as engine or turret lathes, but a two or three-dimensional template is used to control the path of the cutting tool. The numerically controlled (NC) machines are capable of producing complicated design with a great degree of accuracy and repetition. The control provides automatic functioning of speed, feed, depth of cut, tool path turret entwining, oil application and other necessary functions.

A turning center is a lathe-type NC machine tool capable of automatically boring, turning outer and inner diameters, threading, and facing parts. It is often equipped with a system for automatically changing or indexing cutting tools.

Equipped with NC or CNC, lathes and turning machines are being increasingly applied because of their capabilities for greater productivity, improved quality parts, cost savings, and increased versatility.

Also see: Turning Centers.

Turnkey System. A turnkey system describes a system whose ownership is transferred to the recipient after installation and test and only requires production setup and startup. A turnkey NC or computer system is installed by a supplier who has total responsibility for building, installing, and testing the system. In some cases, the installer may have to run the machine for weeks, producing trial parts, before proper installation and adjustment is obtained. Upon acceptance by the buyer, the system is ready for normal production usage.

The idea of a turnkey system has been popularized by promotional material depicting the simple step of turning a key. A term being used synonymously with turnkey systems is turnkey software. This term is used to promote/describe software which does not require configuration, linkage of modules, or other installation activities. Typically, turnkey software may have a batch file for installation which the user can invoke with a single command. Some turnkey software requires the user to answer a few questions to indicate the type of hardware and options present. The more sophisticated turnkey software packages query and test the system to determine the hardware present. Ease of installation with no custom programming permits the immediate use of the software after installation.

TW. See: Thermit Welding.

U

U (Count Per Unit). U is the average count, or average number of events of a given classification, per unit (unit area of opportunity) occurring within a sample. More than one event may occur in a unit (unit area or opportunity), and each such event is counted.

Ultrasonically Assisted Machining. Ultrasonically assisted machining is machining in which an outside source of vibratory energy is coupled to the tool or tool holder in a traditional machining process such as turning, drilling, or gundrilling to increase cutting quality and speed in machining hard, tough materials.

Ultrasonic Cleaning. Ultrasonic cleaning is a method of cleaning metal or plastic parts by immersing them in an aqueous or solvent-based cleaning solution and imposing ultrasound energy on the bath to enhance cleaning by creating cavitation conditions at the part surface, which imparts a strong scouring action to remove tenacious soils.

Ultrasonic Machining. Ultrasonic machining is a mechanical machining process that creates an exact shape by the cutting action of an abrasive slurry driven by a tool vibrating at high frequency in line with its longitudinal axis. Also known as ultrasonic abrasive machining.

Also see: Rotary Ultrasonic Machining.

Ultrasonic Welding (USW). Ultrasonic welding is a solid-state welding process which uses ultrasonic frequency shear stresses parallel to a joint that is clamped under moderate normal force to disperse surface films and produce intimate contact and bonding. USW can be used not only on metals (especially aluminum, copper, precious metals, and other soft alloys) but also on thermoplastics.

Undercut. Undercut is a cut shorter than the programmed cut resulting after a command change in direction. A condition in generated gear teeth when any part of the fillet curve lies inside of a line drawn tangent to the working profile at its point of juncture with the fillet. Undercut may be deliberately introduced to facilitate finishing operations, as in preshaving. A groove melted into the base metal adjacent to the toe or root of a weld and left unfilled by weld metal. Undercut is actually a washout effect usually caused by motion of the weld puddle when improper welding techniques are used.

Underwriter's Laboratories. The Underwriter's Laboratories (UL) is an independent testing laboratory that tests electrical products for safety and adequacy for the intended purpose. Standards on the construction and assembly of many types of electrical equipment, materials, and appliances are set forth in literature issued by the Underwriter's Laboratories, Inc. Each year the UL publishes three volumes listing commercially available electrical products which have been found acceptable with reference to fire and accident hazards and which conform with the application and installation requirements of the code. The three volumes are titled: "Electrical

Construction Materials Directory,'' ''Electrical Appliance and Utilization Equipment Directory,'' and ''Hazardous Location Equipment Directory.'' The UL publishes other literature such as ''Gas and Oil Equipment Directory'' and "Fire Protection Equipment Directory,'' dealing with special equipment involving hazard to life or property.

When products are submitted for testing, the UL may respond with requests and suggestions for changes to the product before endorsing the product as having passed the UL testing and certification standards. Some companies and localities have adopted a position of requiring UL approval on products before they will purchase the products. Both consumer products and industrial products are tested and reviewed by the UL for their certification. Part of the reputation of the UL is built upon the track record of instilling safety and quality in products which meet all the applicable UL standards. Since the UL is independent, they are not influenced by any manufacturers to allow certain products to be approved without meeting the uniformly stringent standards.

Also see: Standards.

Unibus. Unibus is a type of backplane bus popularized by Digital Equipment Corporation with their high end PDP-11 and VAX lines of computers. The latest VAX computers utilize a more advanced bus known as the VAXBI (VAX Bus Interconnect) bus.

Also see: Computers.

Universal Fixturing. Universal fixturing is a set of components resembling building blocks from which a fixture or fixture setup may be constructed.

Also see: Fixture.

Universal Product Code. The Universal Product Code (UPC) is a coded combination of bar code and human readable numerals which define the type of product. The UPC is a defined and accepted standard which puts products into different classed and subclasses. By using the UPC, there is no longer a need to describe the product using its longer word description for a computer system to determine the product and the container size. The UPC code bar code is a unique code format, different from the others such as Code 39. The UPC is a standard which is utilized in countries other than the United States, which allows a generic and comprehensible description of the products to be used. The laser bar code scanners, used in many sales check-out lines today, can read the UPC codes and with the particular store's translation table of prices, feed the point-of-sale data terminal (cash register) with the proper description and the current pricing.

Also see: Bar Coding.

Universal Synchronous Asynchronous Receive Transmit. Universal Synchronous Asynchronous Receive Transmit or USART is usually an integrated-circuit chip which provides the control for duplex serial data transmission. A USART chip can be used in a synchronous or asynchronous application, while a USART can only be used in an asynchronous application. A USART can be utilized in a simplex or duplex circuit.

Prior to the development of USART chips, the functional circuitry required to perform the functions of receiving and transmitting serial communication had to be implemented with discrete components. Especially in communication software, external noise ground planes are of critical importance. With the utilization of an integrated circuit chip, not only are most of these concerns alleviated, but now greater consistency in design implementation is the rule due to accepted standard chips from manufacturers such as Intel, Motorola, NEC, and Texas Instruments. The generally accepted standards now allow a much more brief definition of the functional and electrical interface when defining a new product.

UNIX. UNIX is an operating system which has some unique attributes as compared to other operating systems. One of the characteristics of the UNIX operating system is that it can be fairly easily ported to many different computers, from mainframes to microcomputers. One of the reasons the software is so easily portable is that it is written primarily in the C programming language, which is a portable language—that is, programs written in the C programming language can be easily run on many different machines with a minimum of customization and machine-dependent code being written.

While the UNIX operation system was developed on a particular computer, the DEC PDP-7, the operation system was not optimized for any particular machine and a version which could be easily transferred (or "ported") to different machines was developed. This was accomplished by not making assumptions regarding the specific architecture of the target computer hardware and by writing most of the operating system in a high-level language, C. Today, the UNIX operating system exists in many versions and flavors and is running on many different machines from microcomputers to mainframes. While the operating system itself requires very little modifications when porting from one machine to another, application programs require no modification when being installed and run on different computers which all run the same version of UNIX.

Some of the different versions of the UNIX operating system are UNIX System V, by AT&T Bell Labs, which is an evolution of the original UNIX operating system. The most recent flavor of System V is System V Release 2. Berkeley University created an enhanced version of the original UNIX which has become known as UNIX 4.2 BSD.

UNIX is considered to be currently the most contemporary, general-purpose operating system. While DOS reigns supreme in the microcomputer world and Digital and IBM have their preferred proprietary operating systems, most companies are offering UNIX operating system options for their products so as to maximize hardware sales. In the microcomputer world, XENIX is a version of UNIX which runs on personal computers, and in the minicomputer area, ULTRIX is the VAX version of UNIX. In many cases, applications written to run under a UNIX operating system can be run unaltered on machines running XENIX or ULTRIX.

History. The UNIX operating system was developed by two software engineers at Bell Laboratories in the latter 1960s. Ken Thompson and Dennis Ritchie designed the UNIX operating system to provide an environment which promoted efficient application program development. Other important attributes were that the code be small in size and memory efficient with easy maintenance. Dennis Ritchie, along with Brian W. Kernighan, created the C programming language at Bell Labs to have the use of general-purpose language which features economy of expression, modern control flow and data structures, and a rich set of operators. While both the C programming language and the UNIX operating system were developed during the same general time frame, the C language was created first, and UNIX is in turn written in C.

While the first original version of the UNIX operating system was developed with a specific computer's hardware in mind, shortly after a version was developed which was easily portable to different computers. Since UNIX is written in C, and UNIX was developed with the functions of C in mind, running programs written in C on machines running UNIX is a natural combination.

The UNIX operating system consists of three major parts: the scheduler, the file system, and the shell. There are three shells which dominate the majority of UNIX users, however one can easily write their own shell, or modify and enhance an existing shell. The three most popular shells are the "Bourne shell" (sh), the "C shell" (csh), and the "Korn shell" (ksh). The first shell was invented by Stephen Bourne and the last one by David Korn. The C shell was developed by the University of California at Berkeley and its name comes from its programming language which resembles the C language in syntax. The Korn shell is compatible with the Bourne shell, while the C shell isn't. The Bourne shell is currently being distributed with standard AT&T UNIX System V. The C shell is characterized by, among other things, a percent sign (%) as the command prompt.

Discussion. An operating system is a collection of programs that coordinate the operating functions of a computer's hardware and software. Operating systems provide basic functions such as communication, input/output handling, command interpretation, data management, security, time-sharing, accounting, and provide program development tools.

Communication refers to the ability of one computer to transfer data with other computers or data terminal devices. One of the functions included with the basic UNIX operating system is the UNIX mail facility, which allows mail to be sent between different users of the same machine, or different machines if both are running UNIX, via the operating system. The mail facility is actually one of the number of commands which execute under the UUCP (UNIX-to-UNIX copy).

Input/output handling is the storing and retrieving of information between the main portion of the computer and peripheral devices such as printers, terminals, disks, and tapes. UNIX has what is referred to as standard input and standard output, which allows a function known as pipelining, which is the redirection of I/O such as having the standard output of one program become the standard input of another program. Once I/O is redirected, the tasks can be executed with no concern as to whether or not the data is being conveyed to the proper places.

Command interpretation is when the computer reads the characters which are entered from the keyboard by the user and executes the appropriate tasks. In the UNIX operating system, the shell is the command interpreter, which translates operator keyboard input into machine language that can be executed by the system. In a multiuser UNIX environment, each user has his own copy of the shell, which can be modified and customized and not be restricted by the desires of other users.

Data management in a computer is the organizing of data into manageable and logical groupings called files. In the UNIX operating system, a hierarchical, logical arrangement of directories and subdirectories is used to manage files. There are three types of UNIX files: directory files, ordinary files, and special files. A directory file contains the directory and subdirectory information. An ordinary file is any file which contains data, text, program instructions, or most any other type of information.

Security protects the operating system from the users as well as protecting users from each other. Security allows multiple users to actively use the same computer or files at the same time without corrupting each other's data and processes. UNIX has password protection and security levels built-in. File locking and record locking which provides for the integrity of the data when multiple users are reading and writing the same files is an integral part of the data security.

Time-sharing is the basic ability to allow more than one person to use the computer from different terminal devices at the same time. UNIX can accommodate multiple users with no modification. Since each user has his own copy of the shell, or any shell which can run under the UNIX language, the user interface can be optimized for the individual user and a particular application.

Accounting by an operating system is the keeping track of what each person does, and the amount of time a person is logged on. In some operating systems, and an audit trail is maintained of the activities of a user. The UNIX operating system automatically keeps track of the modification of and access to, all files. Logins, login attempts, and the performance of executing programs are some of the data which is maintained.

Program development tools are things such as compilers, assemblers, debuggers, and code management tools such as editors and the UNIX "make" facility. The "make" command or facility creates an executable object code module from many different component software modules, with the ability to select the most current or released version is each case, or to select the desired module if a specific revision is indicated.

The UNIX operating system can be divided into three basic components: the shell, the file system, and the scheduler.

The UNIX shell is a command interpreter which reads the lines of code which are typed by the user, translates them into machine language, and directs the response to standard I/O. The shell handles program execution by translating the lines of code which are typed in at the terminal each time the RETURN key is pressed. The general format of all UNIX commands is to state the command, followed by the arguments (if any). The shell handles file name substitution, I/O redirection, pipelining hookup, environmental control, and interpretive programming language functions in addition to program execution. The interpretive language is a fairly powerful programming language whose commands can be executed line by line when typed in at the command prompt, or entered into a file first, and then the file executed. Environmental control is the ability to allow the user flexibility in customizing the operating environment, such as specifying the location of the home directory.

Pipelining and I/O redirection are similar in that the end result is that standard output from a process or file becomes standard input for another file or process. Pipelining uses an operator (the vertical bar) which is typed between two programs, whereupon the shell automatically connects the standard output of the first to the standard input of the second.

The UNIX file system is a simple and fundamental part of the UNIX operating system. There are three types of files, directory files, general files, and special files. The command "ls" (list) command can be used to list the files in a directory, with a modifier, -1, providing more detailed information such as the file type and size.

The UNIX scheduler is a program which handles time-sharing functions of the operating system. The scheduler handles data and memory swapping when the system doesn't have sufficient memory as multiple tasks are being run makes it more desirable to decrease the number of tasks to be able to run all concurrently in main memory and be able to use the computer time efficiently executing the programs instead of swapping memory.

Applications. UNIX applications range from general-purpose uses such as word processing and mail type communication to code development to being used as the operating system for embedded computer systems in products. Since the file and data handling are very good compared to many other operating systems and due to the portability of the UNIX operating system, many manufacturers who develop software products to market to users who may have different computers, develop the applications to run under UNIX to minimize the code changes required for each port to another machine.

The imbedded networking and multiuser capabilities of the operating system allow software which may have been written as a single-user application, to be accessed by multiple users on the UNIX system, with some file locking and record locking being handled by the operating system. Other operating systems can be run as tasks under the UNIX operating system, which allows the use of software written specifically for certain other operating systems. This feature can be especially useful when using a UNIX environment to develop code for systems which may be non-UNIX.

Some of the outputting functions such as "nroff" and "troff" allow data output to be formatted for a standard type of printer or to be sent to a phototypesetter. While originally troff was designed with the Wang Labs C/A/T phototypesetter in mind, troff is now device independent. Another function called "pic" is a program for drawing figures, which are described in a special language and then run through the "pic" program. The output from "pic" is "troff" input. The WRITER'S WORKBENCH is a group of programs which assist in analyzing and creating documents, including editors, spelling checkers, and writing style analyzers.

Also see: Operating Systems.

Unlatch Instruction. An unlatch instruction is a PC instruction causing a latched output to change state. A latched output is one that is state-retentive on power failure. If a latched output is ON when power fails, it stays ON. Similarly, if it is OFF when power fails, it stays OFF. In addition, the latch instruction itself does not cause the output to stay in the new state. A latched output can change state during the normal course of program execution, staying in a particular state until a situation causing it to change state occurs. It only latches in a state when power fails.

Unloading Arm. An unloading arm, on some machines, is an arm that unloads the workpart.

Unloading Chute. An unloading chute is a discharge chute for finished workparts.

UPC. See: Universal Product Code.

Updating. Updating is a process of changing a master file by adding, modifying, or deleting information.

Upset Welding (UW). Upset welding is a resistance welding process for making butt or seam welds in which pressure is applied to the joint before application of electric current and is maintained throughout the heating cycle.

USART. See: Universal Synchronous Asynchronous Receive Transmit.

Utility Program. A utility program is a computer program whose purpose is to perform some utilitarian task such as copying files or disk housekeeping tasks. Other typical utilities for computer systems are diagnostic programs to check a disk for errors, formatting and unformatting routines, and programs which gather all fragments of files on disks and rearrange all data files in contiguous physical files. Database software is another candidate for different utilities since the database is usually used to store precious data, the loss of which could be catastrophic.

A well known group of utility programs are the Norton Utilities, which include such things as an unerase program to recover accidentally erased files (provided the file has not been overwritten) and an emerging defacto standard, the Norton System Index which measures the major performance characteristics of a personal computer using defined representative tests.

Another well known utility program is Sidekick by Borland International which provides memory resident utilities such as a calculator, notepad, and editor.

V

Vacuum Bag Molding. A process for molding reinforced plastics in which a sheet of flexible, transparent material is placed over the lay-up on the mold and sealed. A vacuum is created between the sheet and the lay-up. The entrapped air is next mechanically worked out of the lay-up and removed by the vacuum; finally, the part is cured.

Vacuum Forming. A thermoforming method of sheet forming in which the plastics sheet is clamped in a stationary frame, heated, and drawn down by vacuum into a mold.

Vacuum Metalizing. A process in which surfaces are thinly coated with metal by exposing them to the vapor of metal that has been evaporated under vacuum.

Vacuum Refining. Melting and/or casting in vacuum to remove gaseous contaminants from a metal.

Value. A number representing a computed or assigned quantity.

Variable Block Format. A tape format allowing the number and order of words in a block of tape to vary from one block to the next.

Variable Data. Numerical information that can be changed during application operation.

Variables, Method of. Measurement of quality by the method of variables consists of measuring and recording the numerical magnitude of a quality characteristic for each of the units in the group under consideration. This involves reference to a continuous scale of some kind.

Variable Word Length. A data storage method in which each character or digit is stored is one storage location. The number of the location of the storage varies with the field size.

Variance. A measure of variability (dispersion) of observations based upon the mean of the squared deviations from the arithmetic mean.

Variation Allowance. The permissible effective variation.

Vector. A quantity that has magnitude and direction and is usually represented by a directed line segment whose length represents the magnitude and whose orientation in space represents the direction.

Verify. To check, usually by automatic means, one typing or recording of data against another to minimize the number of errors in the data transcription.

Version. A program configuration control identifier. The version identifier is changed whenever software modifications alter the set of functions performed by the program.

Very Large-Scale Integration. High density integrated circuits characterized by relatively large size and high complexity (over 10,000 to 100,000 gates). A review of the size of the memory cells associated with VLSI is shown in the figure below.

Also see: Computers, Large-Scale Integration.

Vibratory Feeder. A vibratory feeder is a type of conveyor that utilizes a standing wave of vibration to cause objects to translate along its surface. The speed of movement of the standing wave and its direction control the speed and direction of the objects that are being conveyed. Vibratory feeders are often used when the need is for a storage conveyor where the parts may sit on the conveyor for some length of time, but must be available at the outlet end for whenever the next process requires parts. In this type of application, vibratory feeders provide an easy means to cue parts. A disadvantage of vibratory feeders is their tendency to damage fragile parts due to the constant vibration and chafing of the parts. In some instances, when the process needs to separate fine particles from parts or from coarser particles, a vibratory feeder provides both the conveying needs as well as the separation needs.

Virtual Device Interface. The VDI or virtual device interface is a device-level graphics interface that has become the de facto standard in North America today. When designing electronic and computer hardware, the ability to utilize the virtual device interface speeds the process by allowing the designer to concentrate on the other portions of the design, while being able to utilize previously developed interface standards. A virtual device interface also allows new and different devices to be designed and introduced easily into the user community by simply complying with an existing interface.

A tape back-up unit, which ''looks'' like a floppy disk drive and can be interfaced to a floppy controller, is an example of a physical device (the tape back-up unit) utilizing the definitions of a VDI to interface as a virtual device (the floppy disk drive). Other examples of utilizing VDI are with regards to RAM disks, mice, laser printers, keyboards, numerical coprocessors, and similar devices.

Virtual Memory. A combination of core memory and secondary memory that can be treated as a single memory, thereby giving the ''virtual''

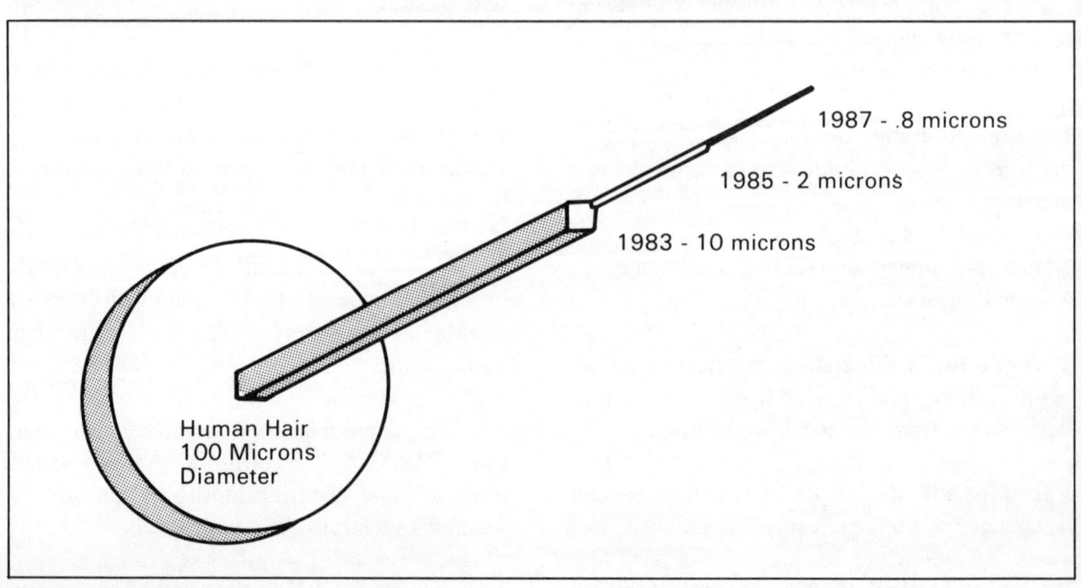

VLSI memory cells.

appearance of a larger core memory to the programmer.

Virtual Storage. The notion of space on storage devices that may be regarded as main storage by the user of a computing system, in which virtual addresses are mapped into real addresses. The size of the virtual storage is limited only by the addressing scheme of the computing system and by the amount of auxiliary storage available, rather than by the actual number of main storage locations.

Vision. See: Machine Vision.

VLSI. See: Very Large-Scale Integration.

Voice Data Entry. The ability for a person to speak aurally into a microphone, which is connected to a computer and capable of translating the verbal input into digital information which can control a program. Usually voice data entry is faster and easier than keyboard or touch-screen data entry. If an operator has no available hands to operate a keyboard, or is working with substances that are harmful to a keyboard, voice data entry may be not only a desirable, but an only solution.

Voice data entry is especially desirable for workers whose hands are occupied with their task and therefore are unable to utilize a keyboard. If the person is an inspector, voice data entry allows them to easily and quickly enter brief information regarding the items as they inspect them.

Currently voice data entry is reliable when using a small vocabulary and when the input device has been tuned to a single operator. Voice data entry is different from voice response, which is the construction of a voice output. Voice response is farther advanced in development than voice data entry.

Also see: Voice Input Device.

Voice Input. A computer input device allowing vocal input to be interpreted by a computer.

Voice Input Device. A voice input device is the hardware apparatus that must be connected to a computer to enable voice data entry. Typically this apparatus includes a microphone and digitizing circuitry. Operation software and translation dictionaries are usually installed within the host computer system.

Because the amplitude (volume) of the spoken word, as well as the speech itself, is an analog quantity, the first stage of the voice input device changes the analog signal to digital data. A voice input device will usually try to filter out ambient and background noises and will digitize the input signal for comparison with taught words and phonemes. Phonemes are the basic sound elements in a language or dialect. Due to people's accents, voice inflection, slang, and other variations, voice input devices sometimes have difficulty discerning the words that are being spoken. Several voice data entry systems are available for use with personal computers as well as data terminals.

Also see: Voice Data Entry.

Voice Output. Voice output is a device that converts data in storage to a vocal response.

Volatile Memory. Volatile memory is memory that loses its contents when power being applied to it is removed. Volatile memory is typically RAM memory. While battery-backed CMOS RAM doesn't lose its contents when the main power removed, it still is volatile memory because if the batteries are removed, the memory would be lost. Currently, volatile memory components are the fastest and least expensive memory components. As other types of memory such as bubble memory and laser memory become more price/performance attractive, certain uses of volatile memory may cease to exist in as prevalent a fashion.

Also see: Computers.

Voltage. A term used in place of electromotive force, potential, potential difference, or voltage drop to describe the electric pressure existing between two points. It is capable of producing a flow or current when a closed circuit is connected between the two points.

Voltage Drop. Voltage drop occurring between two points because of the flow of current through a resistance connected between the points. Also known as IR drop.

Voltage Rating. The maximum amount of voltage that can be safely applied to a given device during continuous use in a normal manner.

von Neumann Bottleneck. Dr. John von Neumann, a mathematician, was one of the early pioneers in computer science. He is credited with preparing the first documentation in written form of the stored program. He was a pioneer in the development of the program flowchart system.

In his 1987 SME Blue Book paper *Fifth Generation Management for Fifth Generation Technology*, Dr. Charles Savage wrote: "Essentially, the first four generation of the computer must pass all information through a single Central Processing Unit (CPU). This single CPU has been described as the "von Neumann bottleneck." It is named after John von Neumann, computer pioneer and mathematician, who essentially designed the architecture which became embodied in the first through fourth generation computers.

"The key to the Fifth Generation Computer, is in the networking of multiple processing units (PU). This linking provides a new task, that of dividing the problem in such a way that these PUs can work on portions of the same problem concurrently and in parallel, then piece together the whole solution."

VS. See: Virtual Storage.

W

Wafer. A wafer is the large uncut piece of substrate that consists of multiple integrated circuit chips before they are tested and trimmed. A wafer is usually flat and round. The wafer is the part that is handled and placed in the forge to create the integrated circuit. Because certain operations are done on the wafer as a whole regardless of the number of integrated circuits on the wafer, the larger the wafer, the cheaper the unit price of an individual integrated circuit chip from the wafer. Also, as technology is able to create smaller integrated circuit chips, larger numbers of integrated circuits can be contained within the same size wafer. By utilizing wafers, the material handling process for manufacturing integrated circuit chips can remain on a macro scale until the cutting of the wafer into individual integrated circuit (IC) chips causes a need for the handling of micro scale devices.

Also see: Circuit, Computers, Integrated Circuits.

Walk Through. A walk through is when a person or group of people are figuratively led through a sequence of something such as the plan to develop a product or the schedule of events of a presentation. The walk through doesn't actually go through the details of every aspect, but a walk through does pause at each and all steps in the sequence. The purpose of a walk through is to perform an abbreviated simulation of the actual sequence, this simulation ensuring that no events are obviously missing or improper. Missing or improper events may only be visible when physically checking the details of each step. Walk throughs are typically done with regards to complex software modules, to ensure the complex functionality and interrelations exist. A walk through may be a prescribed step of the inspection and quality control process. Statistical quality assurance teachings tend to eliminate the walk through because statistical methodologies do not perpetuate the need to test or inspect 100%, but use probabilistic calculations to predict when and where testing and inspection should take place.

Washer. A simple fastener component, generally consisting of a cylindrical piece of metal with a hole in its center. Washers are used to provide seats under bolts or nuts to distribute loads and stresses or to control tightening torque. Washers may also be used to compensate for oversized bolt holes, provide spring tension, protect a surfaced from physical damage, seal the assembly, provide electrical connection, insulate the assembly from the fastener, or serve as a corrosion barrier.

Water Jet Cutting. Water jet cutting uses a fine, high-pressure, high-velocity jet of water directed by a small nozzle to cut material. The pressure of the water jet is usually thousands of psi and the velocity of the stream can exceed twice the speed of sound. The small nozzle opening (from between 0.004 inch to 0.016 inch, or 0.10mm to 0.41mm produces a very narrow kerf.

Three units make up the water jet cutting system: (1) the pump, to generate high pressure; (2) the cutting unit, to actually cut the material with the jet nozzle; and (3) the filtration unit, to remove particles and impurities from the cutting fluid.

Many variables affect the performance of water jet cutting, nozzle orifice diameter, water pressure, flow rate and narrow stand-off distance. Tests should be therefore run to determine the most productive levels and combinations of the various parameters.

In WJC, there are two entirely different ways to generate high pressure: the hydraulic pump and the plunger pump.

In a hydraulic pump, a hydraulic intensifier pressurizes the cutting water. The water then passes through an accumulator to prevent pulsation in the cutting stream.

A plunger pump uses water for cutting directly pressurized by the crank chain. It is also called the crank reciprocating pump or mechanical driven pump.

The jet nozzle assembly in WJC consists of a nozzle holder and jewel nozzle. They are made of stainless steel and industrial jewel respectively. Among the jewels, the diamond nozzle is the most expensive. Its working life, however, is longer than other jewel nozzles such as ruby or sapphire.

Several types of water jet cutting units are available. Control of the unit may include manual operation, optical tracing systems, or a numerical control system. These cutting units can be used with both the hydraulic pump and the plunger pump. The cutting unit also can be used for robotic applications.

The selection of cutting fluid, depending on the operational requirements, quality of finish, cutting speed, and overall cost, will be either plain water or a polymer solution (water with an additive). Polymer solutions tend to work better when a sharp edge is required, because polymer solutions provide better jet coherency. Glycerine, polyethylene oxide and long chain polymers are commonly used additives. Penetration also is influenced by fluid choice. Cutting fluids must be filtered to maintain the quality of the jet stream. Nozzle life is also affected by fluid purity.

Optical tracing systems employ an optical scanner that traces a line drawing and produces electronix signals that control the X-Y axis. With an optical tracing system, the line drawing may be done in pencil. Changes can easily be made in part shape by simply erasing a line and redrawing it. The figure below is a schematic of an optical control system.

NC systems are well-suited for mass production using WJC since they can cut any shape

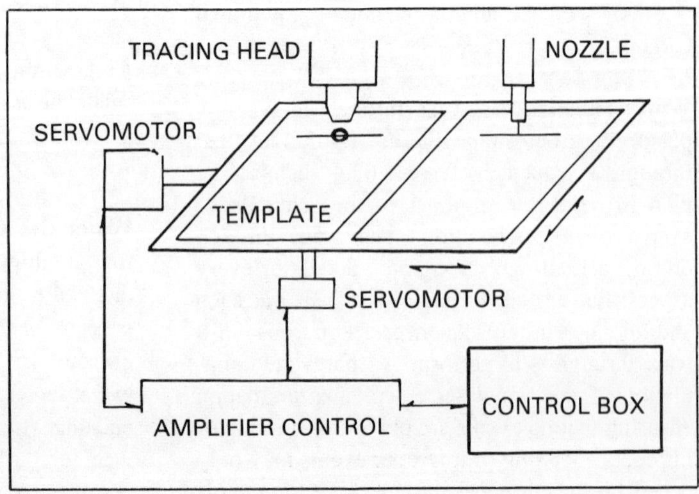

A schematic of an optical control system for water jet cutting.

continuously, repetitively, and precisely. Three-dimensional (three-axis) controls are possible with NC. Such capability provides the advantage of being able to work on a nonflat surface. The nozzle is maintained at a constant distance from the workpiece surface, resulting in a uniform cut.

Automated machines use patterns to accurately cut large quantities of materials such as fabrics, foams, and plastics. The use of water doesn't contaminate or stain the material as might occur with a metal cutting tool, which might contain an oil-based lubricant.

WJC is used to cut many nonmetallic materials: kevlar, glass epoxy, graphite, boron, fiber-reinforced plastic, corrugated board, leather, and many others. Brittle materials, such as glass, are unsuitable in most cases for WJC because they crack or break during processing. Soft and friable materials can be easily cut using WJC and will yield good edge quality.

Weighted Value. The numerical value assigned a bit as a function of its code-word position.

Welding. Welding is a metal-joining process wherein heat is used to melt and join metals with or without the application of pressure and the use of filler metal. Most industrial welding involves fusion. The edges of the surfaces which are to be welded are brought to the molten state. The molten metal bridges the gap between the parts. After the source of heat from the welding head has been taken away, the liquid solidifies, thereby joining or welding the parts together. There are principally five fusion welding process: arc welding, resistance welding, gas welding, laser welding, and electron-beam welding. Other fusion welding processes include high-frequency, thermal, and chemical reaction welding such as used on plastics.

Weldability is the capacity of similar or dissimilar materials to be welded under the fabrication conditions imposed and to perform satisfactorily in the intended service.

In arc welding, electrical current, is applied across an air gap by first contacting the workpiece, and then separating by a small distance. This creates a gap across which the current arcs, such as in a spark plug. The arcing current produces heat which melts the metal at the point of electrical contact. Gas welding such as MIG or TIG uses inert gases such as CO_2, argon or helium to replace air as the medium across which the arc occurs. This eliminates some of the impurities which may come from the air and are deposited in the weld. Some material such as aluminum cannot be successfully welded unless they are done in an inert gas environment.

Automated welding machines can perform arc welding including seam tracking by sensing the distance of the tip from the metal on both sides. Many of the arc welding equipment senses touching the workpiece and then draws back the proper distance to commence welding. This capability allows the system to weld assemblies with considerable differences in uniformity from piece to piece, as well as errors in location within the welding fixture.

The first major usage of automated resistance spot welding equipment was General Motor's Lordstown, Ohio assembly plant which uses industrial robots with resistance spot welding guns to respot automotive bodies. After the initial relationship between the panels is established by using sidegate fixtures and a few welds and clamps, the robots then perform the remainder of the majority of welds. While the robots are reprogrammable, and therefore capable of having their programs, tweaked to improve or alter the welding pattern, the major benefit of using robots is their consistency and uniformity of output at a tireless, continuing pace.

The laser has also become a useful tool in the welding process. Laser beam welding is a fusion welding process for making structural, assembly, sealing, and conduction welds in material by directing a high-power density beams of laser light onto the joint. Inert gas is generally used to reduce oxidation, but filler metal is rarely used.

The industrial robot has also been used in the welding process.

Vernon L. Melton of Melton Machine & Control Company, noted in his RI/SME Tech-

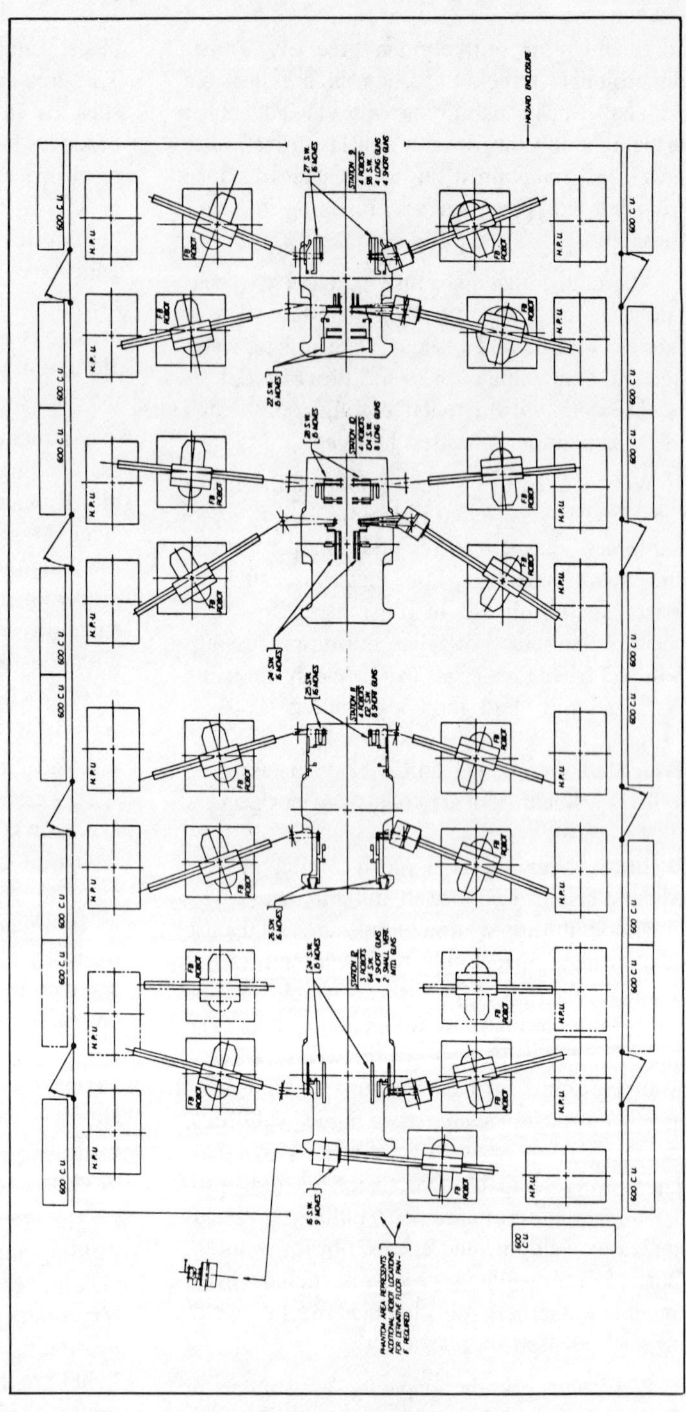

A portion of an assembly line using robots for welding.

nical Paper *Robots Versus Dedicated Automated Welding* that: ''robots are frequently associated with high-volume production welding, such as automotive manufacturing. While the giant manufacturing firms may have been very instrumental in the early development and implementation of robots, it is increasingly apparent that the smaller firms have good justification. Robots, with their extreme versatility and dexterity, are well-suited to handle the variety type welding found in small to medium-size companies. Robots are frequently easier to justify, with less chance for obsolescence, than dedicated machines in a job shop environment.

"Dedicated machines usually need substantially higher production requirements than robots for justification. The ability to operate two or more torches simultaneously enhances the capability of the dedicated machine. When production requirements are high and multiple torches can be used, the advantages of a dedicated system are hard to beat."

Melton also pointed out in his paper: "An often overlooked cost reduction opportunity exists in the materials handling and workstation arrangement. Obscured material flow problems begin to stand out when applying robots or other automated welding. Careful attention to the overall station arrangement will assure the-maximum return on investment in welding automation."

Robots and other automated machines are capable of weld patterns which are unattainable by manual means, such as arc welding in perfect circles and feeding and welding wire and moving the tip at constant speeds. Automated equipment can weld seams much faster than a human, due to their greater maximum inches-per-second speed of performance.

In relation to resistance spot welding, heavy and large-size spot welding guns have, in the past, proved cumbersome for human workers. The hard automation of robotics offered some solutions but restraining factors included: limited flexibility because different body styles could not be run down the same line; expensive model changeover, and backup in the event of failure.

Development of the industrial robot in the past few years has minimized these problems. In his SME paper, *High Capacity Robots in Demanding Resistance Welding Applications*, Joseph Messer of Prab Robots, Incorporated, provided a diagram of a portion of an assembly line using robots for welding (see the figure on the previous page). In this layout, Station 9 is the first station to accomplish robotic welding. It applies approximately 98 welds.

Like station 9, Stations 10 and 11 are made up of four robots equipped with dual mounted weld guns. Station 10 applies 104 welds, while Station 11 applies 102 welds. The final station, Station 12, has three primary robots with a fourth robot serving as a back-up.

Messer also presented some robot and process considerations (see figure on the next page).

Also see: Arc Welding, Backhand Welding, Braze Welding, Carbon Arc Welding, Coextrusion Welding, Cold Welding, Continuous Weld, Cross-wire Welding, Diffusion Welding, Dilution, Electrogas Welding, Electron Beam Welding, Explosion Welding, Fillet Weld, Flash Welding, Flux-cored Arc Welding, Forge Welding, Friction Welding, Fusion Welding, Gas Tungsten Arc Welding, High-frequency Induction Welding, High-frequency Resistance Welding, Hot Pressure Welding, Hot Wire Welding, Industrial Robots, Lasers, Oxyfuel Gas Welding, Percussion Welding, Plasma Arc Welding, Pulsed Gas Welding, Pulsed Power Welding, Resistance Projection Welding, Resistance Seam Welding, Resistance Spot Welding, Resistance Welding, Shielded Metal Arc Welding, Short-circuiting Transfer, Spray Transfer, Stud Arc Welding, Submerged Arc Welding, Thermit Welding.

Winchester Disk. A Winchester disk is a hard disk drive. The reference of Winchester disk originated from the fact the first hard disks were designed as 30MB fixed and 30MB removable, hence the reference to the 30/30 Winchester rifle, or simple Winchester disk. Even though the configuration of most hard disks isn't necessarily 30MB and most disks do not include a removable companion aspect, the title of Winchester disks has remained as an understood reference to the type of disk drive.

A hard disk uses multiple platters that rotate on a single spindle in a sealed environment. Because the read/write heads of a hard disk do not contact the platter, but float aerodynamically

ROBOT CONSIDERATIONS:

1. Style
   - — Cylindrical
   - — Spherical
   - — Rectalinear
   - — Jointed Arm
   - — Hydraulic
   - — Electric
2. Performance
   - — Demonstration/Mock-Ups
   - — Field Installation
   - — Reliability
   - — Payload Capacity
   - — Control Capabilities
   - — Speed
   - — Axis Requirements
3. Personnel Support
   - — Training
   - — Field Service
   - — Experience
   - — Engineering Support

PROCESSING CONSIDERATIONS:

1. Floor Space
   - — Number Robots Required
   - — Line Layout
   - — Number Weld Stations Required
2. Welding Parameters
   - — 36″ Throat Weld Guns
   - — 225 Pounds per Weld Gun
   - — Floor Pan 0.032″ Thickness
   - — 2″ to 3″ Spot Spacing
   - — Approximately 460 Spots
   - — 0.6 to 0.8 Seconds Weld Time
   - — Double Galvanized Material
   - — Planes of required Welding
   - — Welding Transformer Location
3. Interfacing
   - — Weld Controllers
   - — Robot Controllers
   - — Master Control Unit
4. Back-up Capabilities
   - — Number Robots Required
   - — Maintainable Line Rates
   - — Cost

Industrial robot and processing considerations for welding applications.

a few millimeters from the surface, even a speck of dust could wedge between the head and the platter and cause catastrophic consequences (a head crash). Some drives provide a dedicated area which contains no data to be used for the head to rest on when the disk stops spinning. This is referred to as a *dedicated landing zone*.

Several of the major manufacturer of hard disks are Maxtor, Miniscribe, Micropolis, Seagate, and Control Data Corp. Several of the larger computer companies such as Digital and IBM design and build many of their own disks drives.

Also see: Floppy Disk, Mass Memory.

Wire Feed Machine. A numerically controlled EDM machine that uses copper or brass wire for the electrode. Cutting much like a band saw, the wire cuts any electrical conductive material. This machine is primarily used in manufacturing tools, dies, and extrusion dies and works in an accuracy of ±0.005 in. (0.001mm).

Word. A string of characters, bytes, bits, or other symbols that is considered, processed, or handled as a unit, that is, as a single entity. Usually meaning can be assigned to a word to represent information. In telegraph communications, six character intervals are considered as a word when computing traffic capacity in words per minute (wpm). The traffic capacity in wpm is computed by multiplying the modulation rate in baud by 10 and dividing the resulting product by the number of unit intervals per character. For example, if the modulation rate is 50 baud, that is, 50 unit intervals per second, and there are 5 unit intervals per character, the traffic capacity or traffic rate is (50)(10)/5 = 100 wpm at an implied 6 characters per word. The 50 unit intervals per second and the 5 unit intervals per character is equivalent to 10 characters per second, or 600 characters per minute (cpm), which is equivalent to 100 wpm at 6 characters per word as obtained before.

In computer data, a word depends on the design of the central processing unit. For example, some computers that have 8 bit bytes consider 16 bits as a word, while other computers consider 32 bits as a word, or even 64 bits (sometimes called a longword).

Word Address Format. An NC tape format in which each word in a block is identified by one or more preceding characters.

Word Length. The number of bits or characters in a word.

Work Cell. A work cell is a cell consisting of a grouping of one or more workstations in which work is performed. Other types of cells might be storage cells or testing cells. A work cell usually has equipment that is intended to perform some subfunction of the total manufacturing process.

A work cell might be a group of machines dedicated to the assembly process, or painting and coating, or testing of a product or subassembly. Work cells are sometimes normal groupings of machines based on processes or they may be a grouping of machines used to do a certain portion of work relative to a manufacturing process. In the latter case, the machines in the work cell might be very dissimilar to each other with regards to purpose and function, but are related to each other in that together they provide a particular function or subfunction.

Also see: Cell, Flexible Manufacturing System, Industrial Robots, Transfer Line, Workstation.

Workholder. The chuck collet or arbor attached to the work spindle that holds and may also rotate the workpart.

Also see: Jig, Fixture.

Work In Process. Work In Process (WIP) is material in diverse stages of completion throughout a manufacturing process. This may include raw material released to production for initial processing, subassemblies in the production process where manufacturing has added value to the material, subcontracted parts out-sourced at vendors, and completely processed material in final inspection waiting for final acceptance as finished goods or shipment to customers.

Work-in-process may also include semifinished stocked material where production has added value or components. Work in-process is considered part of the total manufacturing inventory. The principle areas of inventory investment is raw material in stock, work-in-process, finished goods, and consignment inventory.

Processing methods, transportation, setup and queue time, schedules, order sizing, release decisions, and buffer stock policies all impact the WIP inventory investment. The Work In Process inventory investment is part of the overall policy and strategies involving the manufacturing inventory control goals.

Also see: Just-in-Time, Manufacturing Resource Planning.

Workpiece Program. A program that provides instruction for machining a specific workpiece.

Workstation. A workstation is a single-user terminal that is usually designed for a narrowly defined activity such as CAD design or typesetting. A workstation usually contains optimized hardware and software, which sometimes has greater capability than the typical general purpose terminal or personal computer. Workstation technology and product offerings of today usually consist of 32-bit computers with enhanced video drivers and high-resolution monitors. Workstations can usually be networked to other workstations or to host computers and file servers and network servers. Workstation prices continue to dramatically decrease, while the performance of workstations continues to increase at a rapid pace, especially because the capabilities of 80386 and 68020 based personal computers (which are 32-bit machines) can rival the speed and functionality of the major workstation vendors such as Apollo, Digital, and Sun. The trend has moved towards using standard platforms from one of the major vendors and developing software to run on the platform as the value-added aspect.

World Coordinate System. The world coordinate system is the reference system of coordinates with respect to the earth. The world coordinate system is used as a common reference between multiple relative coordinate systems. While the world coordinate system itself is relative to the earth as we know it, which is spinning and rotating with respect to the solar system, we generally consider the earth as a basic reference frame. The world coordinate system is typically used when referencing robot coordinate systems and when zeroing relative coordinate systems on numerically controlled machines. Without the reference system, programs written for one machine might not work on another machine. The world coordinate system is also used for surveying highways and large buildings where their orientation relative to the earth is a necessarily determined quantity. The world coordinate system is also used for alignment and coordination of communication equipment such as line-of-sight microwave equipment and scientific telescopes.

WORM. See: Write Once, Read Mostly.

Write. The term *write* is used to describe the function of transferring data onto a memory storage device such as RAM memory of a hard disk or floppy disk. Writing is the initial transfer of data to the media, while reading is the transfer of that information from the data's memory to some other memory. Writing the magnetic media consists of using electrical signals to alter the orientation of the magnetic particles of the memory device's media in a manner that can be read to indicate the data originally written to the media. Magnetic media should not be subjected to a stray magnetic field in close proximity, because the magnetic field may inadvertently alter or erase the data. Writing the RAM or ROM utilizes electrical signals that open and close switches in the memory circuits to indicate ones and zeros, which are the binary representations of the data.

Also see: Read.

Write Once, Read Mostly. Write once, read mostly (sometimes referred to as write once, read many) is a term applied to optical disk drives that are written to, and read by, a laser beam. This type of optical disk drive allows the user to write information onto a segment of the disk once, but to read the information from the disk many times. Information can continually be written to the disk until the disk is full. This type of storage media cannot be erased.

Write once, read mostly or WORM drives are designed for application requiring storage of large quantities of data that are never modified, but are later retrieved for reference, preferably available in an on-line manner

WORM drives can provide storage of up to 400 megabytes, which is the equivalent of nearly 10,000 pages, into cartridges the size of a stack of four floppy disks. The typical WORM drive unit is similar in size to a 5 1/4 floppy disk drive.

Also see: CD-ROM, Erasable Optical Disk, Optical Disk Drive, Read, Write.

X Axis. X axis is the axis of motion that is horizontal and parallel to the workholding surface.

XTABL. The APT vocabulary table containing the code numbers which are used to represent the vocabulary words of the APT language, as used internally by APT, and passed along to the postprocessor.

X.25 Protocol. X.25 is a packet-switching standard that refers to the network layer within the ISO OSI model. X.25 describes how packet-type data is transferred across the data terminal equipment (DTE)/data circuit-terminating equipment (DCE) interface. X.25 establishes the packet format, packet control identifiers, call setup, data flow management, packet windows, call termination, and many other features. The X.25 standard is rapidly gaining acceptance in the industry and will see implementation in many networks and vendor products. Essentially, X.25 provides the logical channels for communication when using packet-switching networks.

An end user-to-end user session is established under X.25 protocol by the following events. The source site transmits a control packet to request a session or connection. The Call Request packet is transmitted to the receiving site where it is either accepted or rejected. X.25 defines limits to the number of end user sessions allowed at one time. Assuming the request is accepted, a Call Accepted packet is returned to the requester. Both of these control packets contain identifiers to provide for a session or call binding. After call establishment, the packets containing user data are exchanged, with X.25 defining the packet flow control and window rules. After all data have been transmitted, a Clear Request control packet is sent to the receiving site and the session can be terminated by a Clear Confirmation packet.

Y Axis. The Y axis is the axis of motion that is perpendicular to the X and Z axes.

Also see: Axis.

Yaw. In robotics, yaw refers to a degrees of freedom. It is the movement of the robot's wrist about a vertical centerline. Yaw is an angular displacement (either to the left or to the right) as seen along the principal axis of an object (such as a robot) which has a top side.

Also see: Industrial Robots.

Yield. Yield is the ratio of output from a manufacturing process compared to the materials of input to the process. Usually expressed as a percentage, yield may be in terms of the total input or in terms of a specific raw material. Yield may also be referred to as a *yield factor*.

Z

Z Axis. The axis of motion that is parallel to the principal spindle of the machine.

Zero Offset. A characteristic of a numerical control machine which permits the zero point on an axis to be shifted readily over a specified range.

Also see: Numerical Control.

Zoom. A capability of computer aided design (CAD) systems that proportionately enlarges or reduces a figure displayed on a CRT Screen. This change of viewing resolution may be accomplished by the simpler task of displaying a smaller section of the drawing as a full screen image, or it may in some situations be done by enlarging the representation of the drawing.

Normally a drawing image can be zoomed (enlarged, or reduced) if it is constructed by mathematically describing the outline of the images and the type of fill. The mathematical equations can be proportionately enlarged or reduced. If an image is a bit-map image, its size cannot be sealed in the same manner. A display device can however project a smaller section of the same relative bit-mapped image over the full screen. Some systems are able to interpolate the scaling of a bit-mapped image, but this isn't usually done.

The major difference between a drawing program and a painting program, which is utilized by personal computers, is that a paint program cannot scale or enlarge and reduce the drawing, but can only zoom the image on the screen. A drawing or drafting program can enlarge and reduce the drawing and sections of the drawing, relative to the drawing workspace, as well as reduce and enlarge the viewable screen image.

Also see: Computer Aided Design.

Figure Sources

A

Automated Assembly.
Some possible advantages of automating assembly operations. C. Wick, R. F. Veilleux, *Tool and Manufacturing Engineers Handbook*, Fourth Ed., Vol. 4, "Quality Control and Assembly", (Dearborn, MI: Society of Manufacturing Engineers, 1987), p. 12-1.

Two part assembly is replaced by a single stamping. C. Wick, R. F. Veilleux, *Tool and Manufacturing Engineers Handbook*, Fourth Ed., Vol. 4, "Quality Control and Assembly", (Dearborn, MI: Society of Manufacturing Engineers, 1987), p. 12-10.

Robotic assembly system for small electric motors. C. Wick, R. F. Veilleux, *Tool and Manufacturing Engineers Handbook*, Fourth Ed., Vol. 4, "Quality Control and Assembly", (Dearborn, MI: Society of Manufacturing Engineers, 1987), p. 12-24.

An automated assembly line schematic. T.W. Leverett. "The Architecture of a Small Automated Assembly Line," Technical Paper MS87-0198, Society of Manufacturing Engineers, 1987.

A sample automated assembly workstation. T.W. Leverett. "The Architecture of a Small Automated Assembly Line," Technical Paper MS87-0198, Society of Manufacturing Engineers, 1987.

Automated Guided Vehicle.
Sample automated guided vehicles. Top. Adapted from: S. M. Vaccaro, D. M. Cox. "FMC's Venture into FMS: A Case Study," Technical Paper MS85-0152, Society of Manufacturing Engineers, 1985.

Sample automated guided vehicles. Bottom, left. Adapted from: R. K. Miller, *Automated Guided Vehicles and Automated Manufacturing*, First Ed.,(Dearborn: Society of Manufacturing Engineers, 1987) p. 86.

Sample automated guided vehicles. Bottom, right. Adapted from: R. Olker. "Managing Tooling in an FMS," Technical Paper TE87-0849, Society of Manufacturing Engineers, 1987.

Axis.
A robot performing plasma arc contouring with the X, Y, and Z axis illustrated. W. Wu, S. Nebergall, *15th North American Manufacturing Research Conference Proceedings* from the paper *Improvement of Accuracy in Robotic Contouring Using a Pre-gain Control Factor*. Dearborn, MI, Society of Manufacturing Engineers, p. 682.

B

Bar Coding.
Bar coding at work in a warehouse. Adapted from "Bar Coding, Focus on New Products," *Manufacturing Engineering*, Vol. 98, No. 6; June 1987, p. 41.

A bar code and scanner system is used at the transfer of tools from a delivery system to a machine station storage. The system verifies correct delivery and establishes physical presence in this storage system. Adapted from R. Olker. "Managing Tooling in an FMS," Technical Paper TE87-0849, Society of Manufacturing Engineers, 1987.

C

Coining.
Coining is used for sheet metal working as well as for bulk forming. C. Wick, J. T. Benedict, R. F. Veilleux, *Tool and Manufacturing Engineers Handbook*. Fourth Ed., Vol. 2, "Forming", (Dearborn, MI: Society of Manufacturing Engineers, 1984), p. 4-8.

Computer Integrated Manufacturing.
The CIM Enterprise Wheel. D. Appleton, *Introducing the New CIM Wheel*, Second Ed., (Dearborn, MI: Computer and Automated Systems Association of the Society of Manufacturing Engineers, 1985), p. 5.

The general business management family of process portion of the CIM Enterprise Wheel. D. Appleton, *Introducing the New CIM Wheel*, Second Ed., (Dearborn, MI: Computer and Automated Systems Association of the Society of Manufacturing Engineers, 1985), p. 7.

The product and process definition portion of the CIM Enterprise Wheel. D. Appleton, *Introducing the New CIM Wheel*, Second Ed., (Dearborn, MI: Computer and Automated Systems Association of the Society of Manufacturing Engineers, 1985), p. 7.

The manufacturing planning and control portion of the CIM Enterprise Wheel. D. Appleton, *Introducing the New CIM Wheel*, Second Ed., (Dearborn, MI: Computer and Automated Systems Association of the Society of Manufacturing Engineers, 1985), p. 8.

The factory automation portion of the CIM Enterprise Wheel. D. Appleton, *Introducing the New CIM Wheel*, Second Ed., (Dearborn, MI: Computer and Automated Systems Association of the Society of Manufacturing Engineers, 1985), p. 8.

The Information Resource Management portion of the CIM Enterprise Wheel. D. Appleton, *Introducing the New CIM Wheel*, Second Ed., (Dearborn, MI: Computer and Automated Systems Association of the Society of Manufacturing Engineers, 1985), p. 10.

The benefits of CIM. L. Bertain, H. L. Hales, *A Program Guide for CIM Implementation*, Second Ed., (Dearborn, MI: Society of Manufacturing Engineers, 1987), p. iii.

Computer Numerical Control.

Elements of a computer numerical control system. T. J. Drozda, C. Wick, *Tool and Manufacturing Engineers Handbook*, Fourth Ed., Vol. 1, "Machining", (Dearborn, MI: Society of Manufacturing Engineers, 1983), p. 5-25.

Five major functional units of a computer numerical control system. T. J. Drozda, C. Wick, *Tool and Manufacturing Engineers Handbook*, Fourth Ed., Vol. 1, "Machining", (Dearborn, MI: Society of Manufacturing Engineers, 1983), p. 5-25.

Operator interface devices used in a computer numerical control system. T. J. Drozda, C. Wick, *Tool and Manufacturing Engineers Handbook*, Fourth Ed., Vol. 1, "Machining", (Dearborn, MI: Society of Manufacturing Engineers, 1983), p. 5-26.

Open and closed loop servomechanisms. T. J. Drozda, C. Wick, *Tool and Manufacturing Engineers Handbook*, Fourth Ed., Vol. 1, "Machining", (Dearborn, MI: Society of Manufacturing Engineers, 1983), p. 5-28.

Mechanical elements of a computer numerical control system. T. J. Drozda, C. Wick, *Tool and Manufacturing Engineers Handbook*, Fourth Ed., Vol. 1, "Machining", (Dearborn, MI: Society of Manufacturing Engineers, 1983), p. 5-28.

Cost Estimating.

A cost estimate of producing a part. This estimate, displayed several different ways, is produced by using a personal computer. J. E. Nicks, *Manufacturing Cost Estimating Software Manual*, First Ed., (Dearborn, MI: Society of Manufacturing Engineers, 1986), p. 13.

A cost estimating program. J. E. Nicks, *Manufacturing Cost Estimating Software Manual*, First Ed., (Dearborn, MI: Society of Manufacturing Engineers, 1986), p. 83.

D

Deburring.

A robotic arm with live spindle for deburring gear box cases. L. K. Gillespie, *Robotic Deburring Handbook*, Second Ed., (Dearborn, MI: Society of Manufacturing Engineers, 1987), p. 4.

Deburring tool mounted in a robot gripper. F. M. Puls, M. M. Barash. "An Adaptive Control Algorithm for Robotic Deburring," *Journal of Manufacturing Systems*, Vol 4, No. 2, 1985, p. 174.

Depreciation.

A computer-generated depreciation schedule. J. E. Nicks, *BASIC Programming Solutions for Manufacturing*, First Ed., (Dearborn, MI: Society of Manufacturing Engineers, 1982), p. 123.

Dies.

Return-blank blanking dies. C. Wick, J. T. Benedict, R. F. Veilleux, *Tool and Manufacturing Engineers Handbook*, Fourth Ed., Vol. 2, "Forming", (Dearborn, MI: Society of Manufacturing Engineers, 1984), p. 6-24.

Drilling.

Knowledge-based order-specific NC drilling system. L. F. Pau, D. Paus. "Knowledge-based Order-specific NC Drilling System," Technical Paper MS87-0781, Society of Manufacturing Engineers, 1987.

E

Electrical Discharge Machining.

Components of an electrical discharge machine. T. J. Drozda, C. Wick, *Tool and Manufacturing Engineers Handbook*, Fourth Ed., Vol. 1, "Machining", (Dearborn, MI: Society of Manufacturing Engineers, 1983), p. 14-42.

Points to consider for evaluating numerical control and electrical discharge machining. T. Bryce. "The Basics of NC EDM," Technical Paper MR87-0644, Society of Manufacturing Engineers, 1987.

Electron Beam Machining.

Elements of an electron beam machine. T. J. Drozda, C. Wick, *Tool and Manufacturing Engineers Handbook*, Fourth Ed., Vol. 1, "Machining", (Dearborn, MI: Society of Manufacturing Engineers, 1983), p. 14-37.

End Effector.
Sample end effectors. Top, left. W. Hessler. "Application of Commercially Available End of Arm Tooling," Technical Paper MS86-0413, Society of Manufacturing Engineers, 1986.

Sample end effectors. Top, right. W. Hessler. "Application of Commercially Available End of Arm Tooling," Technical Paper MS86-0413, Society of Manufacturing Engineers, 1986.

Sample end effectors. Middle, left. W. Hessler. "Application of Commercially Available End of Arm Tooling," Technical Paper MS86-0413, Society of Manufacturing Engineers, 1986.

Sample end effectors. Middle, right. A. J. Wright. "Robotic End Effectors: Mechanisms for Cassette Load/Unload and Disk Handling," Technical Paper MS84-0333, Society of Manufacturing Engineers, 1984.

Sample end effectors. Bottom, left. Y. S. Liu. "Robot Assembly; Motion Time," Technical Paper MS86-0385, Society of Manufacturing Engineers, 1986.

Engraving.
The four zones of a sample laser engraving system. C. Anderson. "Laser Engraving Using Vision Processing and Robotic Manipulation," Technical Paper MS87-0333, Society of Manufacturing Engineers, 1987.

F

Finishing.
Advantages, applications and limitations of finishing processes. C. Wick, R. F. Veilleux, *Tool and Manufacturing Engineers Handbook*, Fourth Ed., Vol. 3, "Materials Finishing and Coating", (Dearborn, MI: Society of Manufacturing Engineers, 1985), p. 15-24.

Cross section of finishing barrel showing flow of parts and media. C. Wick, R. F. Veilleux, *Tool and Manufacturing Engineers Handbook*, Fourth Ed., Vol. 3, "Materials Finishing and Coating", (Dearborn, MI: Society of Manufacturing Engineers, 1985), p. 16-65.

Fixture.
Pads on a casting are accessible. W. E. Boyes, *Low Cost Jigs, Fixtures and Gages for Limited Production*, First Ed. (Dearborn, MI: Society of Manufacturing Engineers, 1986), p. 41.

Tool chucked as short as possible. W. E. Boyes, *Low Cost Jigs, Fixtures and Gages for Limited Production*, First Ed. (Dearborn, MI: Society of Manufacturing Engineers, 1986), p. 41.

Minimize tool travel time over clamp. W. E. Boyes, *Low Cost Jigs, Fixtures and Gages for Limited Production*, First Ed. (Dearborn, MI: Society of Manufacturing Engineers, 1986), p. 41.

Blanking a fixture against a stop. W. E. Boyes, *Low Cost Jigs, Fixtures and Gages for Limited Production*, First Ed. (Dearborn, MI: Society of Manufacturing Engineers, 1986), p. 46.

Keying to T slots. W. E. Boyes, *Low Cost Jigs, Fixtures and Gages for Limited Production*, First Ed. (Dearborn, MI: Society of Manufacturing Engineers, 1986), p. 46.

Pinning a worktable. W. E. Boyes, *Low Cost Jigs, Fixtures and Gages for Limited Production*, First Ed. (Dearborn, MI: Society of Manufacturing Engineers, 1986), p. 47.

Indicating edge. W. E. Boyes, *Low Cost Jigs, Fixtures and Gages for Limited Production*, First Ed. (Dearborn, MI: Society of Manufacturing Engineers, 1986), p. 47.

Multistation fixturing. W. E. Boyes, *Low Cost Jigs, Fixtures and Gages for Limited Production*, First Ed. (Dearborn, MI: Society of Manufacturing Engineers, 1986), p. 48.

Flexible Manufacturing System.
Flexible manufacturing system controller scheme. O. Z. Maimon. "Real-time Operational Control of Flexible Manufacturing System," *Journal of Manufacturing Systems*, Vol. 6, No. 2, 1987, p. 126.

A sample FMS layout. V. E. Lott, D. L. Manack. "Expert System Schedules Automated Cell," Technical Paper MS87-0844, Society of Manufacturing Engineers, 1987.

An FMS layout. K. Chui. "Case Report on Integrating FMS and Traditional Machine Tools," Technical Paper MS88-0106, Society of Manufacturing Engineers, 1988.

Flowchart.
A sample flow chart. Original drawing.

Forging.
Material handling and inspection system used in a forging operation. J. T. Barczak, J. T. Merchant. "Using Machine Vision to Inspect Automobile Forgings," Technical Paper IQ86-0603, Society of Manufacturing Engineers, 1986.

G

Gaging.
Typical plug gages. C. Wick, R. F. Veilleux, *Tool and Manufacturing Engineers Handbook*, Fourth Ed., Vol. 4, "Quality Control and Assembly", (Dearborn, MI: Society of Manufacturing Engineers, 1987), p. 3-20.

Ring gage set used to inspect the diameter of shafts. C. Wick, R. F. Veilleux, *Tool and Manufacturing Engineers Handbook*, Fourth Ed., Vol. 4, "Quality Control and Assembly", (Dearborn, MI: Society of Manufacturing Engineers, 1987), p. 3-21.

Gantry Robot.
A gantry robot at work. G. K. Sweet, L. A. Hutto. "Robots Refurbish Space Shuttle Hardware," Technical Paper MS87-0284, Society of Manufacturing Engineers, 1987.

Gripper.
A sample gripper. D. M. Fullmer. "Signature Analysis of Robotic Assembly Forces—A Flexible Sensor System," Technical Paper MS87-0174, Society of Manufacturing Engineers, 1987.

Gundrilling.
Various types of holes produced by gundrilling. T. J. Drozda, C. Wick, *Tool and Manufacturing Engineers Handbook*, Fourth Ed., Vol. 1, "Machining", (Dearborn, MI: Society of Manufacturing Engineers, 1983), p. 9-57.

H

Hexadecimal Numbering System.
The hexadecimal numbering system. Original drawing.

I

Industrial Robots.
Industrial robot arm configurations. T. J. Drozda, C. Wick, *Tool and Manufacturing Engineers Handbook*, Fourth Ed., Vol. 1, "Machining", (Dearborn, MI: Society of Manufacturing Engineers, 1983), p. 16-14.

Wrist articulations on an industrial robot. T. J. Drozda, C. Wick, *Tool and Manufacturing Engineers Handbook*, Fourth Ed., Vol. 1, "Machining", (Dearborn, MI: Society of Manufacturing Engineers, 1983), p. 16-15.

Coordinate systems. T. J. Drozda, C. Wick, *Tool and Manufacturing Engineers Handbook*, Fourth Ed., Vol. 1, "Machining", (Dearborn, MI: Society of Manufacturing Engineers, 1983), p. 16-15.

Robot-driven work cell—vertical milling machine, broach and drill loaded and unloaded with a robot. T. J. Drozda, C. Wick, *Tool and Manufacturing Engineers Handbook*, Fourth Ed., Vol. 1, "Machining", (Dearborn, MI: Society of Manufacturing Engineers, 1983), p. 16-17.

Robot door assembly system. C. Wick, R. F. Veilleux, *Tool and Manufacturing Engineers Handbook*, Fourth Ed., Vol. 1, "Quality Control and Assembly", (Dearborn, MI: Society of Manufacturing Engineers, 1987), p. 11-24.

Spot welding guns are positioned accurately and automatically by means of robots in assembly of automobile bodies. C. Wick, R. F. Veilleux, *Tool and Manufacturing Engineers Handbook*, Fourth Ed., Vol. 1, "Quality Control and Assembly", (Dearborn, MI: Society of Manufacturing Engineers, 1987), p. 12-23.

Underbody bolt securing using a vision-guided robot. C. Wick, R. F. Veilleux, *Tool and Manufacturing Engineers Handbook*, Fourth Ed., Vol. 4, "Quality Control and Assembly", (Dearborn, MI: Society of Manufacturing Engineers, 1987), p. 12-23.

A single-disk, internal-diameter gripper that will pick and place disks in clean room applications. E. Morz, R. B. Stevens. "Robots Handling Magnetic Disks in Clean Rooms", Technical Paper MS86-1017, Society of Manufacturing Engineers, 1986.

Inspection.
Comparison of robot position repeatability for two different setting times. C. Wick, R. F. Veilleux, *Tool and Manufacturing Engineers Handbook*, Fourth Ed., Vol. 4, "Quality Control and Assembly", (Dearborn, MI: Society of Manufacturing Engineers, 1987), p. 3-60.

J

Jig Boring.
Jig boring machine of open-sided construction. T. J. Drozda, C. Wick, *Tool and Manufacturing Engineers Handbook*, Fourth Ed., Vol. 1, "Machining", (Dearborn, MI: Society of Manufacturing Engineers, 1983), p. 8-102.

Adjustable-rail or planer-type jig boring machine equipped with graduated-scale measuring system. T. J. Drozda, C. Wick, *Tool and Manufacturing Engineers Handbook*, Fourth Ed., Vol. 1, "Machining", (Dearborn, MI: Society of Manufacturing Engineers, 1983), p. 8-103.

Fixed-bridge design of a jig boring machine on which the worktable traverses in the longitudinal direction. T. J. Drozda, C. Wick, *Tool and Manufacturing Engineers Handbook*, Fourth Ed., Vol. 1, "Machining", (Dearborn, MI: Society of Manufacturing Engineers, 1983), p. 8-103.

Numerically controlled jig boring machine of the open-side type. T. J. Drozda, C. Wick, *Tool and Manufacturing Engineers Handbook*, Fourth Ed., Vol. 1, "Machining", (Dearborn, MI: Society of Manufacturing Engineers, 1983), p. 8-103.

Just In Time.
An overview of Just In Time. B. J. Nicholls. "Computer and Materials Scheduling at Nissan Australia," Technical Paper MS86-0473, Society of Manufacturing Engineers, 1986.

L

Lancing.
Lancing is cutting along a line in the product without freeing the scrap from the product. C. Wick, J. T. Benedict, R. F. Veilleux, *Tool and Manufacturing Engineers Handbook*, Fourth Ed., Vol. 2, "Forming", (Dearborn, MI: Society of Manufacturing Engineers, 1984), p. 4-5.

Lasers.
The basics of laser material processing. W. E. Lawson. "Laser Processing of Materials," Technical Paper MS87-0461, Society of Manufacturing Engineers, 1987.

Top and end view of a typical Nd:YAG laser. E. J. Weller, *Nontraditional Machining Processes*, Second Ed., (Dearborn, MI: Society of Manufacturing Engineers, 1984), p. 141.

Eliptical focusing structure using linear lamp. E. J. Weller, *Nontraditional Machining Processes*, Second Ed., (Dearborn, MI: Society of Manufacturing Engineers, 1984), p. 142.

A sample laser/robotic system. G. H. Parsons. "CO_2 Lasers in the Automobile Press Shop," Technical Paper MS87-0332, Society of Manufacturing Engineers, 1987.

Automated laser cutting and welding work cell. P. A. Lovio. "Automated Laser Robotic Workcell for Jet Engine Rework," Technical Paper MS87-0331, Society of Manufacturing Engineers, 1987.

M

Machine Vision.
A sample machine vision/inspection system. I. N. Tansel, B. Clark. "Detection of the Machining Problems by a Machine Vision System," Technical Paper MS87-0687, Society of Manufacturing Engineers, 1987.

Five brake shoes at random locations and orientations. R. J. Schlichtig. "A Hardware Based Machine Vision System," Technical Paper MS86-0597, Society of Manufacturing Engineers, 1986.

A machine vision system using a laser, galvometer scanners, detector and computer control. P. A. Lovio. "Automated Laser Robotic Workcell for Jet Engine Component Rework," Technical Paper MS 87-0331, Society of Manufacturing Engineers, 1987.

Machining Cell.
Machining cell layout. "Robot Loading/Unloading Pays Off in Machining Cell," *Robotics Today*, Vol. 4, No. 2, February 1982, p. 32.

Maintenance Management.
A computer-assisted maintenance management schedule. K. Bagadia. "Microcomputer Aided Maintenance Management and Inventory Control," Technical Paper MS86-0479, Society of Manufacturing Engineers, 1986.

Manufacturing Automation Protocol.
MAP specifications summary by layer (2.2 Reference Spec). World Federation of MAP Users Group, *MAP Reference Specification*, Based on GM MAP 2.2 Specification. (Dearborn, MI: Society of Manufacturing Engineers, 1988), p. 3-2.

The Deere Harvester MAP Network. D. C. Scott. "Making MAP a Reality, A User's View," Technical Paper TM86-3311, Society of Manufacturing Engineers, 1986.

Milling.
Milling machine having five axes of motion that are tape controlled. T. J. Drozda, C. Wick, *Tool and Manufacturing Engineers Handbook*, Fourth Ed., Vol. 1, "Machining", (Dearborn, MI: Society of Manufacturing Engineers, 1983), p. 10-16.

N

Necking.
Necking and swaging will reduce the diameter on the tube or drawn-cup. Necking is accomplished between dies in a manner that is similar to other drawing operations. C. Wick, J. T. Benedict, R. F. Veilleux, *Tool and Manufacturing Engineers Handbook*, Fourth Ed., Vol. 2, "Forming", (Society of Manufacturing Engineers, 1984), p. 4-8.

Numerical Control.
Diagram of a vertical spindle machine tool showing he axes—X, longitudinal; y, transverse; and z, vertical. T. J. Drozda, C. Wick, *Tool and Manufacturing Engineers Handbook*, Fourth Ed., Vol. 1, "Machining", (Dearborn, MI: Society of Manufacturing Engineers, 1983), p. 5-34.

Two-axis, tape-controlled drilling machine. T. J. Drozda, C. Wick, *Tool and Manufacturing Engineers Handbook*, Fourth Ed., Vol. 1, "Machining", (Dearborn, MI: Society of Manufacturing Engineers, 1983), p. 5-35.

Three-axis, tape-controlled drilling machine with turret. T. J. Drozda, C. Wick, *Tool and Manufacturing Engineers Handbook*, Fourth Ed., Vol. 1, "Machining", (Dearborn, MI: Society of Manufacturing Engineers, 1983), p. 5-35.

Sample consumable tool costs—typical as experienced by one major manufacturer in the United States. T. J. Drozda, C. Wick, *Tool and Manufacturing Engineers Handbook*, Fourth Ed., Vol. 1, "Machining", (Dearborn, MI: Society of Manufacturing Engineers, 1983), p. 5-17.

Standard eight-channel tape specifications. T. J. Drozda, C. Wick, *Tool and Manufacturing Engineers Handbook*, Fourth Ed., Vol. 1, "Machining", (Dearborn, MI: Society of Manufacturing Engineers, 1983), p. 5-41.

P

Parting.
Used in lathe or screw-machine operations, parting involves two cutoff operations which produce blanks from the strip shown in this figure. C. Wick, J. T. Benedict, R. F. Veilleux, *Tool and Manufacturing Engineers Handbook*, Fourth Ed.,

Vol. 2, "Forming", (Dearborn, MI: Society of Manufacturing Engineers, 1984), p. 4-4.

Perforating.
Perforating is used to punch many holes with a specific pattern. C. Wick, J. T. Benedict, R. F. Veilleux, *Tool and Manufacturing Engineers Handbook*, Fourth Ed., Vol. 2, "Forming", (Dearborn, MI: Society of Manufacturing Engineers, 1984), p. 4-4.

Press.
Basic production setup for a stamping operation. Pressure is applied by the press. The sheet metal is formed into the shape determined by the punch and die. C. Wick, J. T. Benedict, R. F. Veilleux, *Tool and Manufacturing Engineers Handbook*, Fourth Ed., Vol. 2, "Forming", (Dearborn, MI: Society of Manufacturing Engineers, 1984), p. 4-3.

Programming.
Programming is the development and writing of a sequence of commands which can be executed automatically by a computer or other device with similar capabilities to follow instructions. A sample program is shown here. This program, written in BASIC, calculates regression lines. J. E. Nicks, *BASIC Programming Solutions for Manufacturing*, First Ed. (Dearborn, MI: Society of Manufacturing Engineers, 1982), pp. 149, 150.

Punching.
Punching involves the cutting of clean holes with resulting scrap slugs. C. Wick, J. T. Benedict, R. F. Veilleux, *Tool and Manufacturing Engineers Handbook*, Fourth Ed., Vol. 2, "Forming", (Dearborn, MI: Society of Manufacturing Engineers, 1984), p. 4-4.

R

Reverse Flanging.
In this case of reverse flanging, the product has at least one shrink flange and one strength flange. C. Wick, J. T. Benedict, R. F. Veilleux, *Tool and Manufacturing Engineers Handbook*, Fourth Ed., Vol. 2, "Forming", (Dearborn, MI: Society of Manufacturing Engineers, 1984), p. 4-6.

Riveting.
Floor and bench models of automatic riveting machines. C. Wick, R. F. Veilleux, *Tool and Manufacturing Engineers Handbook*, Fourth Ed., Vol. 4, "Quality Control and Assembly", (Dearborn, MI: Society of Manufacturing Engineers, 1987), p. 8-52.

Tooling components for automatic drilling and riveting. C. Wick, R. F. Veilleux, *Tool and Manufacturing Engineers Handbook*, Fourth Ed., Vol. 4, "Quality Control and Assembly", (Dearborn, MI: Society of Manufacturing Engineers, 1987), p. 8-53.

Drive, jaws, and anvil mounted on a riveting machine. C. Wick, R. F. Veilleux, *Tool and Manufacturing Engineers Handbook*, Fourth Ed., Vol. 4, "Quality Control and Assembly", (Dearborn, MI: Society of Manufacturing Engineers, 1987), p. 8-52.

Roll Forming.
Using the computer for roll design provides the designer with drawings of the cross-sectional shape of the part at each station. This is the output for a part requiring 20 passes. (Only half of the cross section is shown since the part is symmetrical.) C. Wick, J. T. Benedict, R. F. Veilleux, *Tool and Manufacturing Engineers Handbook*, Fourth Ed., Vol. 2, "Forming", (Dearborn, MI: Society of Manufacturing Engineers, 1984), p. 8-21.

S

Sensor.
Edge detection system. D. Locke, J. T. McCabe. "Real Time Control of Milling through Edge Sensing," Technical Paper MS85-1002, Society of Manufacturing Engineers, 1985.

Servomechanisms.
Open and closed-loop servomechanisms. T. J. Drozda, C. Wick, *Tool and Manufacturing Engineers Handbook*, Fourth Ed., Vol. 1, "Machining", (Dearborn, MI: Society of Manufacturing Engineers, 1983), p. 5-28.

Mechanical elements of a computer numerical control system. T. J. Drozda, C Wick, *Tool and Manufacturing Engineers Handbook*, Fourth Ed., Vol. 1, "Machining", (Dearborn, MI: Society of Manufacturing Engineers, 1983), p. 5-28.

Shaping.
Typical horizontal crank-operated push-cut shaper. T. J. Drozda, C. Wick, *Tool and Manufacturing Engineers Handbook*, Fourth Ed., Vol. 1, "Machining", (Dearborn, MI: Society of Manufacturing Engineers, 1983), p. 7-49.

Shearing.
Shearing is the cutting action along a straight line to separate metal by two moving blades. C. Wick, J. T. Benedict, R. F. Veilleux, *Tool and Manufacturing Engineers Handbook*, Fourth Ed., Vol. 2, "Forming", (Dearborn, MI: Society of Manufacturing Engineers, 1984), p. 4-3.

Sheet Metal.
Manually nested blanks for a straight shear. J. N. Brecker, D. F. Rife, V. P. Valeri. "Sheet Metal FMS," Technical Paper MS84-0790, Society of Manufacturing Engineers, 1984.

Nested parts for a single assembly. J. N. Brecker, D. F. Rife, V. P. Valeri. "Sheet Metal FMS," Technical Paper MS84-0790, Society of Manufacturing Engineers, 1984.

Shrink Flanging.
In shrink flanging, the line of bending is a convex curve. C. Wick, J. T. Benedict, R. F. Veilleux, *Tool and Manufacturing Engineers Handbook*, Fourth Ed., Vol. 2, "Forming", (Dearborn, MI: Society of Manufacturing Engineers, 1984), p. 4-6.

Slitting.
Slitting is cutting along single lines. C. Wick, J. T. Benedict, R. F. Veilleux, *Tool and Manufacturing Engineers Handbook*, Fourth Ed., Vol. 2, "Forming", (Dearborn, MI: Society of Manufacturing Engineers, 1984), p. 4-5.

Spinning.
Spinning is a method of forming sheet metal or tubing. C. Wick, J. T. Benedict, R. F. Veilleux, *Tool and Manufacturing Engineers Handbook*, Fourth Ed., Vol. 2, "Forming", (Dearborn, MI: Society of Manufacturing Engineers, 1984), p. 4-8.

Spray Painting.
Typical layout of the DNC paint system employs four pairs of robots. H. Akeel. "Expanding the Capabilities of Spray Painting Robots," *Robotics Today*, Vol 4, No. 2, April 1982, p. 78.

Straight Flanging.
*Straight flanging is a bending operation in which the line of bending is straight.*C. Wick, J. T. Benedict, R. F. Veilleux, *Tool and Manufacturing Engineers Handbook*, Fourth Ed., Vol. 2, "Forming", (Dearborn, MI: Society of Manufacturing Engineers, 1984), p. 4-6.

Stretch Flanging.
Stretch flanging is a bending operation in which the line of bending is not straight. C. Wick, J. T. Benedict, R. F. Veilleux, *Tool and Manufacturing Engineers Handbook*, Fourth Ed., Vol. 2, "Forming", (Dearborn, MI: Society of Manufacturing Engineers, 1984), p. 4-6.

Stretch Forming.
Stretch forming applies primarily tensile forces to stretch the sheet metal over a tool or formblock. C. Wick, J. T. Benedict, R. F. Veilleux, *Tool and Manufacturing Engineers Handbook*, Fourth Ed., Vol. 2, "Forming", (Dearborn, MI: Society of Manufacturing Engineers, 1984), p. 4-7.

T

Tooling.
Five stages of tool design. E. G. Hoffman, *Fundamentals of Tool Design*, Second Ed., (Dearborn, MI: Society of Manufacturing Engineers, 1984), p. 705.

Transfer Line.
Flexible transfer line—three robots, conveyor, and six machine tools. T. J. Drozda, C. Wick, *Tool and Manufacturing Engineers Handbook*, Fourth Ed., Vol. 1, "Machining", (Dearborn, MI: Society of Manufacturing Engineers, 1983), p. 16-18.

Trimming.
Trimming is the operation of cutting scrap off a fully or partially formed product to establish a trim line. The edge of a cup is sometimes trimmed by pinching or pushing the flange or lip of the cup over the edge of a stationary punch. C. Wick, J. T. Benedict, R. F. Veilleux, *Tool and Manufacturing Engineers Handbook*, Fourth Ed., Vol. 2, "Forming", (Dearborn, MI: Society of Manufacturing Engineers, 1984), p. 4-5.

V

Very Large-scale Integration.
VLSI memory cells. G. C. Dunlap. "CIM—Communication Integrated Manufacturing," Technical Paper MM87-0697, Society of Manufacturing Engineers, 1987.

W

Water Jet Cutting.
A schematic of an optical control system for water jet cutting. E. J. Weller, *Nontraditional Machining Processes*, Second Ed., (Dearborn, MI: Society of Manufacturing Engineers, 1984), p. 141.

Welding.
A portion of an assembly line using robots for welding. J. S. Messer. "High Capacity Robots in Demanding Resistance Welding Applications," Technical Paper MS83-0822, Society of Manufacturing Engineers, 1983.

Industrial robots and processing considerations for welding applications. J. S. Messer. "High Capacity Robots in Demanding Resistance Welding Applications," Technical Paper MS83-0822, Society of Manufacturing Engineers, 1983.